THE REGIONAL GEOGRAPHY OF CANADA

THE REGIONAL GEOGRAPHY OF CANADA

SIXTH EDITION

ROBERT M. BONE

OXFORD
UNIVERSITY PRESS

Oxford University Press is a department of the University of Oxford.
It furthers the University's objective of excellence in research, scholarship,
and education by publishing worldwide. Oxford is a registered trade mark of
Oxford University Press in the UK and in certain other countries.

Published in Canada by
Oxford University Press
8 Sampson Mews, Suite 204,
Don Mills, Ontario M3C 0H5 Canada

www.oupcanada.com

Copyright © Oxford University Press Canada 2014

The moral rights of the author have been asserted

Database right Oxford University Press (maker)

First Edition published in 2000
Second Edition published in 2002
Third Edition published in 2005
Fourth Edition published in 2008
Fifth Edition published in 2011

All rights reserved. No part of this publication may be reproduced, stored in
a retrieval system, or transmitted, in any form or by any means, without the
prior permission in writing of Oxford University Press, or as expressly permitted
by law, by licence, or under terms agreed with the appropriate reprographics
rights organization. Enquiries concerning reproduction outside the scope of the
above should be sent to the Permissions Department at the address above
or through the following url: www.oupcanada.com/permission/permission_request.php

Every effort has been made to determine and contact copyright holders.
In the case of any omissions, the publisher will be pleased to make
suitable acknowledgement in future editions.

Library and Archives Canada Cataloguing in Publication
Bone, Robert M.
The regional geography of Canada / Robert M. Bone. — 6th ed.

Includes bibliographical references and index.
ISBN 978–0–19–900242–9

1. Canada—Geography—Textbooks. I. Title.

FC76.B66 2013   917.1   C2013-901503-5

Cover image: View of Signal Hill, the easternmost part of Canada in St John's,
Newfoundland by Rex Montalban Photography/Getty Images

Oxford University Press is committed to our environment.
This book is printed on Forest Stewardship Council® certified paper
and comes from responsible sources.

Printed and bound in the United States of America

1 2 3 4 — 17 16 15 14

BRIEF CONTENTS

FIGURES x
TABLES xii
BOXES xiv
PREFACE xvi
IMPORTANT FEATURES OF THIS EDITION 1

1 **Regions of Canada** 2

2 **Canada's Physical Base** 28

3 **Canada's Historical Geography** 66

4 **Canada's Human Face** 120

5 **Ontario** 164

6 **Québec** 214

7 **British Columbia** 264

8 **Western Canada** 310

9 **Atlantic Canada** 356

10 **The Territorial North** 400

11 **Canada: A Country of Regions within a Global Economy** 438

GLOSSARY 448
WEBSITES 460
NOTES 463
BIBLIOGRAPHY 472
INDEX 496

CONTENTS

FIGURES x | TABLES xii | BOXES xiv | PREFACE xvi | IMPORTANT FEATURES OF THIS EDITION 1

1 Regions of Canada
INTRODUCTION 3
CHAPTER OVERVIEW 3
Geography as a Discipline 4
Regional Geography 4
Canada's Geographic Regions 5
The Dynamic Nature of Regions 8
Sense of Place 8
Faultlines within Canada 10
The Core/Periphery Theory 17
The Global Economy 17
Canada in the Global World 19
Canada–US Trade Relations 23
Understanding Canada's Regions 24
SUMMARY 26
CHALLENGE QUESTIONS 26
FURTHER READING 27

2 Canada's Physical Base
INTRODUCTION 29
CHAPTER OVERVIEW 29
Physical Variations within Canada 30
The Nature of Landforms 31
Physiographic Regions 32
Geographic Location 44
Climate 46
Major Drainage Basins 56
Environmental Challenges 58
Canada and Global Warming 60
SUMMARY 64
CHALLENGE QUESTIONS 65
FURTHER READING 65

3 Canada's Historical Geography
INTRODUCTION 67
CHAPTER OVERVIEW 67
The First People 68
The Second People 74
The Third People 78
The Territorial Evolution of Canada 79
Faultlines in Canada's Early Years 85
The Centralist/Decentralist Faultline 89
The Aboriginal/Non-Aboriginal Faultline 92
The Immigration Faultline 103
The French/English Faultline 110
SUMMARY 118
CHALLENGE QUESTIONS 118
FURTHER READING 119

4 Canada's Human Face

INTRODUCTION 121
CHAPTER OVERVIEW 121
Canada's Population 122
Urban Population 129
Population Change 132
Immigration 137
Canada's Aging Population 140
Sense of Place: Now and Then 147
Canada's Economic Face 157
SUMMARY 162
CHALLENGE QUESTIONS 163
FURTHER READING 163

5 Ontario

INTRODUCTION 165
CHAPTER OVERVIEW 165
Ontario within Canada 166
Ontario's Physical Geography 168
Climate and Agriculture 171
Environmental Challenges 175

Ontario's Historical Geography 178
Ontario Today 185
Key Topic: The Automobile Industry 192
Northern Ontario 199
Ontario's Urban Geography 204
SUMMARY 212
CHALLENGE QUESTIONS 213
FURTHER READING 213

6 Québec

INTRODUCTION 215
CHAPTER OVERVIEW 215
Québec's Culture, Identity, and Language 216
Québec's Place within Canada 219
Québec's Physical Geography 225
Environmental Challenges 228
Québec's Historical Geography 228
Québec Today 236
Québec's Economy 237
Key Topic: Hydro-Québec 240
Tourism 248
Southern Québec 250
Québec's Urban Geography 257
The French/English Faultline in Québec 262
SUMMARY 262
CHALLENGE QUESTIONS 263
FURTHER READING 263

7 British Columbia

INTRODUCTION 265
CHAPTER OVERVIEW 265
British Columbia within Canada 266
British Columbia's Physical Geography 268
Environmental Challenges 273
British Columbia's Historical Geography 277
British Columbia Today 283
British Columbia's Wealth 285
Key Topic: Forestry 297
British Columbia's Urban Geography 303
Faultlines in BC 307
SUMMARY 308
CHALLENGE QUESTIONS 308
FURTHER READING 309

9 Atlantic Canada

INTRODUCTION 357
CHAPTER OVERVIEW 357
Atlantic Canada within Canada 358
Atlantic Canada's Physical Geography 361
Environmental Challenges 366
Atlantic Canada's Historical Geography 367
Atlantic Canada Today 373
Key Topic: The Fishing Industry 381
Atlantic Canada's Resource Wealth 387
Atlantic Canada's Population 395
Atlantic Canada's Urban Geography 396
SUMMARY 399
CHALLENGE QUESTIONS 399
FURTHER READING 399

8 Western Canada

INTRODUCTION 311
CHAPTER OVERVIEW 311
Western Canada within Canada 312
Western Canada's Physical Geography 314
Environmental Challenges 318
Western Canada's Historical Geography 321
Key Topic: Agriculture 330
Western Canada's Resource Base 340
Western Canada's Population 351
Western Canada's Urban Geography 351
Faultline: Aboriginal/Non-Aboriginal Populations 353
SUMMARY 354
CHALLENGE QUESTIONS 354
FURTHER READING 355

10 The Territorial North

INTRODUCTION 401
CHAPTER OVERVIEW 401
The Territorial North within Canada 402
Physical Geography of the Territorial North 405
Environmental Challenge: Global Warming 407
Historical Geography of the Territorial North 408
Arctic Sovereignty: The Arctic Ocean and the Northwest Passage 414
The Territorial North Today 418
Resource Development in the Territorial North 427

Key Topic: Megaprojects 431
SUMMARY 436
CHALLENGE QUESTIONS 436
FURTHER READING 437

Canada and the Global Economy 443
The Future 444
CHALLENGE QUESTIONS 447
FURTHER READING 447

GLOSSARY 448

WEBSITES 460

NOTES 463

BIBLIOGRAPHY 472

INDEX 496

11 Canada: A Country of Regions within a Global Economy

INTRODUCTION 439
Regional Character and Structure 440
Canada's Faultlines 442

FIGURES

1.1 Gable's regions of Canada 5
1.2 The six geographic regions of Canada 6
1.3 Regional populations by percentage, 1871 and 2011 8
1.4 The burden of Confederation 12
1.5 CN: Symbol of the integrated North American economy 21
1.6 The Kyoto Protocol and the Northern Gateway Pipeline 22
1.7 Share of exports to the United States, 2001–2010 23
1.8 An eagle with clipped talons 24
2.1 Physiographic regions and continental shelves in Canada 33
2.2 Maximum extent of ice, 18,000 BP 34
2.3 Time zones 45
2.4 Climatic zones of Canada 47
2.5 Seasonal temperatures in Celsius, January 49
2.6 Seasonal temperatures in Celsius, July 50
2.7 Annual precipitation in millimetres 51
2.8 Natural vegetation zones 53
2.9 Soil zones 54
2.10 Permafrost zones 55
2.11 Drainage basins of Canada 56
2.12 Worldwide change in volume (km^3) of glaciers, 1960–2004 63
3.1 Migration routes into North America 68
3.2 Culture regions of Aboriginal peoples 73
3.3 Aboriginal language families 74
3.4 Canada, 1867 80
3.5 Canada, 1873 81
3.6 Canada, 1905 85
3.7 Canada, 1927 86
3.8 Canada, 1999 87
3.9 Sharing the wealth—Alberta's wealth? 90
3.10 Historic treaties 98
3.11 Modern treaties 99
3.12 Western Canada and the Northwest Rebellion of 1885 107
4.1 Population of Canada, 1851–2011 122
4.2 Canada's population zones and highway system 124
4.3 Capital cities 128
4.4 Percentage of Canadian population in urban regions, 1901–2011 130
4.5 Immigration: An increasingly important component of Canadian population growth 133
4.6 Annual number of immigrants admitted to Canada, 1901–2010 138
4.7 New Canadians: Permanent residents by top 10 source countries, 2008–2010 139
4.8 The unrelenting advance of seniors 140
4.9 Number and share of the foreign-born population in Canada, 1901–2031 149
4.10 Aboriginal population by ancestry, 1901–2011 151
4.11 Reset the relationship 152
4.12 Percentage of Aboriginal people in the population, Canada, provinces, and territories, 2006 154
4.13 Manufacturing's declining share of employment 157
4.14 The cyclical nature of Canada's economy 158
4.15 Global competition and hollowing-out 160
4.16 Economic bunker 161
5.1 Ontario's economy to 2020: Which way? 166
5.2 Ontario basic statistics, 2011 167
5.3 The province of Ontario 169
5.4 Central Canada 170
5.5 The Great Lakes Basin 179
5.6 The Haldimand Tract 184
5.7 Manufacturing sales: Ontario 190
5.8 Employment in Canada's automobile industry 193
5.9 Automobile assembly centres in Ontario 197
5.10 Northern Ontario's Ring of Fire 203
5.11 Major urban centres in Central Canada 206
6.1 Pastagate: Language inspector rejects "pasta" on Italian restaurant menu 217
6.2 Québec 219
6.3 The St Lawrence River and Lake Ontario 220
6.4 Québec basic statistics, 2011 222
6.5 Québec monthly real GDP 224
6.6 Champlain's map of La Nouvelle France, 1632 230
6.7 Percentage change in real GDP by province, 2012 237
6.8 Overview of the Romaine complex 242
6.9 Hydroelectric power in Central Canada 243
6.10 Overview of the Eastmain-1-A/Sarcelle/Rupert project 246
6.11 Cree communities of Québec 248
7.1 British Columbia 266
7.2 British Columbia basic statistics, 2011 268
7.3 The Second Narrows Bridge: Getting supertankers to port 269
7.4 Physiography of the Cordillera 270
7.5 Westridge oil terminal expansion, Burnaby, BC 274
7.6 Railways in British Columbia 280
7.7 BC capture fishing by value, 2010 289
7.8 Mines in British Columbia 291
7.9 Thermal coal prices, 2001–2012 293
7.10 BC coal production and value, 1980–2011 293
7.11 BC's shrinking forestry exports, 1991–2011 298
7.12 A downward spiral for paper products 299
7.13 Coast and Interior timber harvests 299
7.14 Forest regions in British Columbia 301
7.15 British Columbia's Lower Mainland 304
8.1 Western Canada 312
8.2 Western Canada basic statistics, 2011 313
8.3 Chernozemic soils in Western Canada 318
8.4 Agricultural regions in Western Canada 319
8.5 World Food Price Index: Higher prices for short term or long term? 337
8.6 Saskatchewan summerfallow area, 1992–2012 338

8.7   Alberta's hydrocarbon resources: Oil sands and oil fields 341
8.8   World's largest oil reserves by country (billion barrels), 2010 342
8.9   Crude oil by province, 2011 343
8.10 Cyclic steam stimulation (CSS) 344
8.11 Oil sands deposits and proposed pipelines 344
8.12 Crude oil prices, 1947–March 2013 345
8.13 Thomas Mulcair comes to Alberta 347
9.1   Atlantic Canada 359
9.2   Atlantic Canada basic statistics, 2011 361
9.3   The Labrador Current and the Gulf Stream 364
9.4   Atlantic Canada in 1750 368
9.5   The three Maritime provinces: First to enter Confederation 371
9.6   Newfoundland and Labrador 374
9.7   Newfoundland jackpot: Offshore oil royalties as a percentage of provincial revenues 375
9.8   Lower Churchill hydroelectric projects 378
9.9   Major fishing banks in Atlantic Canada 382
9.10 Natural resources in Atlantic Canada 388
9.11 Voisey's Bay mine site 392
10.1 The Territorial North 402
10.2 The Territorial North basic statistics, 2011 403
10.3 Pilotless drones for the Canadian Arctic? 415
10.4 The Arctic Basin and national boundaries 416
10.5 Major urban centres in the Territorial North 419
10.6 Inuit Nunangat 423
10.7 Resource development in the Territorial North 430
10.8 Oil production at Norman Wells and Cameron Hills 433
11.1 Fighting over oil revenues 443
11.2 China's foreign trade, 2001–10 (2010 $ trillions) 444
11.3 The proposed Northern Gateway Pipeline route 444
11.4 Top 20 destinations for Canadian merchandise exports, 2009 and 2040 445
11.5 Selling Canada's energy to the world 446

TABLES

1.1 General Characteristics of the Six Canadian Regions, 2011 7
1.2 Social Characteristics of the Six Canadian Regions, 2011 13
1.3 National Version of the Core/Periphery Model 18
1.4 Economic Structure of Canada and Its Six Geographic Regions, 2011 18
1.5 Changing Trade Direction: From Continental to Global Trade 19
2.1 Geological Time Chart 33
2.2 Latitude and Longitude of Selected Centres 45
2.3 Climatic Types 48
2.4 Air Masses Affecting Canada 48
2.5 Canada's Drainage Basins 57
2.6 Industrial Emissions, Provinces and Territories, 2006 and 2010 61
3.1 Timeline: Old World Hunters to Contact with Europeans 70
3.2 Population of the Red River Settlement, 1869 77
3.3 Canada's Population by Provinces and Territories, 1901 and 1921 79
3.4 Timeline: Territorial Evolution of Canada 82
3.5 Timeline: Evolution of Canada's Internal Boundaries 87
3.6 Members of the House of Commons by Geographic Region, 1911 and 2011 88
3.7 Total Equalization Payments ($ millions), 2000–1 to 2012–13 92
3.8 Modern Land Claim Agreements, 1975–2011 100
3.9 Population in Western Canada by Province, 1871–1911 108
3.10 Population of Western Canada by Ethnic Group, 1916 109
3.11 Population by Colony or Province, 1841–1871 (%) 113
4.1 Population Size, Percentage, and Density, Canada and Regions, 2011 124
4.2 Population Zones, 2011 127
4.3 Population of Census Metropolitan Areas, 2006 and 2011 131
4.4 Percentage of Urban Population by Region, 1901–2011 133
4.5 Population Increase, 1951–2011 135
4.6 Canada's Rate of Natural Increase, 1851–2011 136
4.7 Population Increase, Selected Years 137
4.8 Phases in the Demographic Transition Theory 139
4.9 Ethnic Origins of Canadians, 1996 and 2006 142
4.10 Major Phases for the Aboriginal Population in Canada 153
4.11 Aboriginal Population by Identity, Canada and Regions, 2001 and 2006 153
4.12 Population with French Mother Tongue, 1951–2011 156
4.13 Changes in Regional Populations, 2001–2011 157
4.14 Traditional View of Economic Structure 159
4.15 Sectoral Changes in Canada's Labour Force, 1881–2011 160
5.1 Before and After the Crash: Unemployment Rates by Province, 2007, 2009, and 2012 166
5.2 Ontario Employment by Industrial Sector, Number of Workers, 2005 and 2011 191
5.3 Ontario Employment by Industrial Sector, 2005 and 2011 (%) 191
5.4 Canadian Motor Vehicle Production, 1999–2011 194
5.5 Ontario Automobile Assembly Plants, 2012 196
5.6 Mineral Production by Province and Territory, 2011 ($ millions) 202
5.7 Census Metropolitan Areas in Ontario, 2001–2011 205
5.8 Top 10 Countries by Birth of Immigrants to Toronto, 2006 209
5.9 Urban Centres in Northern Ontario, 2001–2011 210
6.1 Québec Election Results, 4 September 2012 218
6.2 Timeline: Historical Milestones in New France 229
6.3 Timeline: Historical Milestones in the British Colony of Lower Canada 232
6.4 Timeline: Historical Milestones for Québec in Confederation 235
6.5 Employment by Industrial Sector in Québec, 2011 240
6.6 Shift of Ontario and Québec Industrial Structures, 2005–2011 (%) 240
6.7 Technical Potential Hydro Power by Province and Territory 244
6.8 Overview of the Romaine complex 258
6.9 Population Change: Montréal and Toronto, 1951–2011 (000s) 259
6.10 Major Cities in the Western Appalachian Uplands, 2001–2011 260
6.11 Population of Cities of Northern Québec, the North Shore, and Gaspé Peninsula, 2001–2011 260
7.1 Employment by Economic Sector in British Columbia, 2005 and 2011 285
7.2 Exports through British Columbia, 2001, 2006, 2009, and 2011 286
7.3 Canada's Leading Minerals (Including Coal) by Value of Production, 2011 294
7.4 The Declining Value of Softwood Lumber Exports to the US 302
7.5 Census Metropolitan Areas in British Columbia, 2001–2011 304
7.6 Other Urban Centres in British Columbia, 2001–2011 305
8.1 Basic Statistics for Western Canada, 2011 328
8.2 Employment by Industrial Sector, Western Canada, 2005 and 2011 329
8.3 Number of Farms in Western Canada, 1971–2011 330
8.4 Average Acreage of Farms in Western Canada, 1971–2011 331

- 8.5 Seeded Area and Production for Western Canadian Canola 338
- 8.6 Oil Production in Western Canada to 2030 (millions bbl/d) 343
- 8.7 Foreign Direct Investment in Oil Sands, 2009–2012 346
- 8.8 Upgraders Located in Alberta 347
- 8.9 Proposed Oil Pipelines to the West Coast 348
- 8.10 Uranium Mines and Mills in Northern Saskatchewan 350
- 8.11 Forested Areas by Province, Western Canada 350
- 8.12 Major Urban Centres in Western Canada, 2001–2011 352
- 8.13 Aboriginal Population by Province, Western Canada 353
- 9.1 Basic Statistics for Atlantic Canada by Province, 2011 361
- 9.2 Equalization Payments to Atlantic Canada Provinces ($ millions) 377
- 9.3 Employment by Economic Sectors in Atlantic Canada, 2005, 2008, and 2011 380
- 9.4 Unemployment Rates in Alberta and Atlantic Provinces, 1996, 2006, and 2011 380
- 9.5 Value of Atlantic Coast Commercial Landings, by Region, 2011 ($000s) 384
- 9.6 Cod Landings for Newfoundland/Labrador and Atlantic Canada, 1990–2011 (metric tonnes live weight) 385
- 9.7 Agriculture, Fishing, Forestry, Mineral, and Petroleum Production in Atlantic Canada, 2011 ($ millions) 389
- 9.8 Population Change in Atlantic Canada, 1996–2011 396
- 9.9 Urban Centres in Atlantic Canada, 2001 and 2011 398
- 10.1 Petroleum Resources in the Territorial North 417
- 10.2 Capital Cities Dominate the Demographics of the Territorial North 419
- 10.3 Population and Aboriginal Population, Territorial North 420
- 10.4 Components of Population Growth for the Territories, 2011–2012 420
- 10.5 Estimated Employment by Economic Sector, Territorial North, 2011 421
- 10.6 Comprehensive Land Claim Agreements in the Territorial North 425
- 10.7 Mineral and Petroleum Production in the Territorial North, 2011 ($ millions) 435

BOXES

Vignette 1.1	Curiosity: The Starting Point for Geography 4		Vignette 4.3	Dual Loyalties and Dual Passports 143
Vignette 1.2	Is Central Canada Losing Its Dominant Position in Confederation? 12		Vignette 4.4	Charles Taylor on Multiculturalism 145
Vignette 1.3	The Challenge of Cultural Adjustment 16		Vignette 4.5	Canadian Identity 146
			Contested Terrain 4.2	Immigration and Multiculturalism 147
Vignette 1.4	The Staples Thesis 17		Vignette 4.6	Russell Peters, a Premier South Asian-Canadian Comic 150
Vignette 1.5	The New World Order and the Super Cycle Theory 19		Vignette 4.7	The Assembly of First Nations 152
Contested Terrain 1.1	McGuinty–Redford War of Words 20		Vignette 4.8	Early Estimates of Aboriginal Population 154
Contested Terrain 1.2	Ambassador Bridge and the Proposed New International Trade Crossing 25		Vignette 4.9	Culture, Stability, and Aboriginal Families 155
Vignette 2.1	Two Different Geographies 30		Vignette 4.10	Canada's Growing Age/Wealth Divide 159
Vignette 2.2	The Earth's Crust and Major Types of Rocks 31		Vignette 4.11	Public Support for Scientific Research and Knowledge-based Nodes 161
Contested Terrain 2.1	The Northern Gateway Pipeline 32		Vignette 5.1	The Burden of Confederation 170
Vignette 2.3	Alpine Glaciation, Glaciers, and Water for the Prairies 37		Vignette 5.2	Contrasting Sub-Regions: Northern and Southern Ontario 171
Vignette 2.4	Cypress Hills 40		Vignette 5.3	Ontario's Snowbelts 172
Vignette 2.5	Isostatic Rebound 41		Contested Terrain 5.1	Urban Sprawl and Farmland 173
Vignette 2.6	Prince Edward Island 42		Vignette 5.4	The Niagara Fruit Belt 174
Vignette 2.7	Champlain Sea 43		Vignette 5.5	The Tragedy of Urban Sprawl 175
Vignette 2.8	Climate Trend Mapper 46		Contested Terrain 5.2	Cleaner Air versus Higher Electrical Costs 177
Vignette 2.9	Types of Precipitation 52		Vignette 5.6	The Welland Canal 180
Vignette 2.10	Global Warming and Climate Change: What Is the Difference? 59		Vignette 5.7	Historical Timeline of the Caledonia Dispute 186
Vignette 2.11	Natural Factors Affecting Global Warming 60		Vignette 5.8	High-Tech Stars and the Creative Class 188
Contested Terrain 2.2	Beyond the 1997 Kyoto Protocol 61		Vignette 5.9	Importance of Central Location to Ontario 189
Vignette 2.12	Fluctuations in World Temperatures 63		Vignette 5.10	The Wave of the Future: Clusters and High Technology 192
Vignette 3.1	Contact between Cartier and Donnacona 72		Vignette 5.11	Automobile Production: Industrial Core or Dead End? 194
Vignette 3.2	The Land Survey System and the Settling of the Prairies 78		Vignette 5.12	The Bailout of Chrysler and GM: Sound Public Policy? 195
Vignette 3.3	America's Manifest Destiny 80		Vignette 5.13	The Threat of Hollowing-Out 199
Vignette 3.4	The Transfer of the Arctic Archipelago to Canada 82		Vignette 5.14	Hamilton: Steel City or Rust Town? 207
Vignette 3.5	Times Change for Canada's Regions 88		Vignette 5.15	Population Distribution in Northwestern Ontario 211
Contested Terrain 3.1	Equalization Payments 91		Contested Terrain 6.1	Fortress Québec? 218
Vignette 3.6	What Is Equalization? 91		Vignette 6.1	Sovereignty: Now or Tomorrow? 218
Vignette 3.7	Indian Residential Schools 94		Vignette 6.2	The St Lawrence River 220
Contested Terrain 3.2	The Supreme Court and the Métis 97		Vignette 6.3	Québec Students March 221
Vignette 3.8	The Origin of the Métis Nation 101		Vignette 6.4	Variations in Agriculture Resources in Québec's Physiographic Regions 226
Vignette 3.9	From a Colonial Straitjacket to Aboriginal Power 102		Vignette 6.5	The Future of Beluga Whales in the St Lawrence 229
Vignette 3.10	The Results of the 30 October 1995 Referendum 117		Vignette 6.6	The Rebellions of 1837–8 233
Contested Terrain 4.1	Social Engineering Often Backfires 129			
Vignette 4.1	The Baby Boom 135			
Vignette 4.2	The Concept of Replacement Fertility and Canada's Sex Ratio 136			

Vignette 6.7	SNC-Lavalin: Building a New City and Transport System 239	Vignette 8.9	Potash: Saskatchewan's Underground Wealth 349
Vignette 6.8	Natural Advantages for Hydroelectric Developments in Québec 244	Vignette 8.10	Winnipeg 352
Vignette 6.9	Mapping the Road to Nunavik 254	Vignette 9.1	Consequences of the Geographic Fragmentation of Atlantic Canada 362
Vignette 6.10	Nunavimmiut Benefit from Resource Profit-Sharing Agreement 255	Vignette 9.2	Weather in St John's 365
Vignette 7.1	BC's Precipitation: Too Much or Too Little? 269	Vignette 9.3	The Annapolis Valley 366
Contested Terrain 7.1	Cartographic simplification or purposely misleading? 275	Vignette 9.4	Mi'kmaq Become Lobster Fishers 370
Vignette 7.2	Who Owns BC? 278	Vignette 9.5	Steel, Iron, and Coal—The Rise and Fall of Nova Scotia's Industrial Base 372
Vignette 7.3	Facts about the Nisga'a Agreement 279	Contested Terrain 9.1	Whose Energy? Whose Revenue? 379
Vignette 7.4	Is the Manufacture of BC Natural Resources Important? 288	Vignette 9.6	The End of Great Harbour Deep 381
Vignette 7.5	Aboriginal Fisheries Strategy: A Temporary Arrangement? 290	Vignette 9.7	Georges Bank 383
Vignette 7.6	Northeast Coal Project: A Failed Megaproject 292	Vignette 9.8	Ecological Crises and the Resource Cycle 386
Vignette 7.7	Granville Island 307	Vignette 9.9	The Hibernia Platform 390
Vignette 8.1	Water Deficit and Evapotranspiration 313	Vignette 9.10	Land Claims and the Labrador Inuit and Innu 392
Vignette 8.2	Palliser's Triangle 317	Vignette 9.11	The Big Commute 396
Vignette 8.3	"Next Year Country" 322	Vignette 9.12	Halifax 398
Vignette 8.4	The Hudson Bay Railway 326	Vignette 10.1	Sedimentary Basins 406
Vignette 8.5	Reaching for the Sky 327	Contested Terrain 10.1	Less Ice, More Whales 407
Vignette 8.6	Technological Gamble: Carbon Capture and Storage 329	Vignette 10.2	Toxic Time Bombs: The Hidden Cost of Mining 408
Vignette 8.7	The Great Sand Hills 334	Vignette 10.3	The Northwest Passage and the Franklin Search 409
Contested Terrain 8.1	Living with Pigs? 340	Vignette 10.4	European Diseases 410
Vignette 8.8	Technological Breakthrough: Horizontal Drilling in Oil Shale 341	Vignette 10.5	The Arctic Council and the Circumpolar World 417
Contested Terrain 8.2	Who Pays for and Who Profits from the Oil Sands? 347	Vignette 10.6	Sea Transportation on the Arctic Ocean 430
		Vignette 11.1	Canadian Identities 443

PREFACE

The purpose of this book is to introduce university students to Canada's regional geography. In studying the regional geography of Canada, the student gains an appreciation of the country's amazing diversity; learns how its regions interact with one another; and grasps how regions change over time. By developing the central theme that Canada is a country of regions, this text presents a number of images of Canada, revealing its physical, cultural, and economic diversity as well as its regional complexity. A number of features such as photos, maps, vignettes, tables, graphs, further readings, chapter bibliographies, and glossary are designed to facilitate and enrich student learning.

The Regional Geography of Canada divides Canada into six geographic regions: Ontario, Québec, British Columbia, Western Canada, Atlantic Canada, and the Territorial North. Each region has a particular regional geography, history, population, and a unique location. These factors have determined each region's character, set the direction for its development, and created a sense of place. In examining these themes, this book underscores the dynamic nature of Canada's regional geography, which is marked by a shift in power relations among Canada's six regions. World trade opened Canada to global influences, which, in turn, transformed each region and the relationships between the regions. This text employs a core/periphery framework. Such an approach allows the reader to comprehend more easily the economic relations between regions as well as modifications in these relations that occur over time. A simplified version of the core/peripheral framework takes the form of "have" and "have-not" provinces. At the same time, social cracks within Canadian society provide a different insight into the nature of Canada and its regions. Each faultline has deep historic roots in Canadian society. While they may rest dormant for some time, these raw tensions can erupt into national crises. Four such stress points exist between Aboriginal and non-Aboriginal Canadians; French and English Canadians; centralist (Ottawa and/or Central Canada) and decentralist (the other, less powerful regions) forces; and recent immigrants (newcomers) and those born in Canada (old-timers). This book explores the nature of these faultlines, the need to reach compromises, and the fact that reaching compromises provides the country with its greatest strength—diversity. While more progress in resolving differences is required, these faultlines are shown to be not divisive forces but forces of change that ensure Canada's existence as an open society within the context of a country of regions.

Organization of the Text

This book consists of 11 chapters. Chapters 1 through 4 deal with general topics related to Canada's national and regional geographies—Canada's physical, historical, and human geography—thereby setting the stage for a discussion of the six main geographic regions of Canada. Chapters 5 through 10 focus on these six geographic regions. The core/periphery model provides a guide for the ordering of these six regions. The regional discussion begins with Ontario and Québec, which represent the traditional demographic, economic, and political core of Canada. The core regions are followed by British Columbia, Western Canada, Atlantic Canada, and the Territorial North. Chapter 11 provides a conclusion.

Chapter 1 discusses the nature of regions and regional geography, including the core/periphery model and its applications. Chapter 2 introduces the major physiographic regions of Canada and other elements of physical geography that affect Canada and its regions. Chapter 3 is devoted to Canada's historical geography, such as its territorial evolution and the emergence of regional tensions and regionalism. This discussion is followed, in Chapter 4, by an examination of the basic demographic, economic, and social factors that influence both Canada and its regions. This chapter explores the national and global forces that have shaped Canada's regions as well as the features of its population (size, urbanization, etc.). To sharpen our awareness of how economic forces affect local and regional developments, four major themes running throughout this text are introduced in these first four chapters. The primary theme is that Canada is a country of regions. Two secondary themes—the integration of the North American economy and the changing world economy—reflect the recent shift in economic circumstances and its effects on regional geography. These two economic forces, described as continentalism and globalization, exert both positive and negative impacts on Canada and its regions, and are explored through the core/periphery model—a model introduced in the first edition back in 2000 that has had its basic premises shaken within a decade by the uneven effects of global trade on Canada's regions.

In Chapters 5 to 10, the text moves from a broad, national overview to a more regional focus. Each of these six chapters profiles one of Canada's large geographic regions: Ontario, Québec, British Columbia, Western Canada, Atlantic Canada, and the Territorial North. These regional chapters explore the physical and human characteristics that distinguish each region from the others and that give each region its special sense of place. To emphasize the economic specialty of each region, a predominant economic activity is identified and explored through a "Key Topic." They are: the automobile industry in Ontario; Hydro-Québec in Québec; forestry in British Columbia; agricultural transition in Western Canada; the fishing industry in Atlantic Canada; and megaprojects in the Territorial North. From this presentation, the unique character of each region emerges. The book concludes with Chapter 11, which discusses the future of Canada and its regions within the rapidly changing global economy.

Sixth Edition

For Canada's regions, the consequences of recent global economic developments—notably the remarkable industrialization of Asian countries, especially China—are twofold. First, Western Canada (led by Alberta and Saskatchewan) and British Columbia are enjoying high prices for their resources. The resulting boom has meant that the economies of Alberta, Saskatchewan, and British Columbia are attracting record numbers of newcomers: migrants from other parts of Canada and immigrants from abroad. With the spillover of Alberta's economic boom into Saskatchewan, this traditionally "have-not" province is undergoing its own mini-boom. High oil prices have also benefited Newfoundland and Labrador, though its population continues to decline.

Second, while Ontario and Québec remain the economic and population pillars of Canada, a shift of regional power seems to be in the wind. Over the last decade, Ontario and Québec have suffered a decline in their manufacturing sectors. This decline began with the relocation of many firms offshore, where labour costs are significantly lower. Canadian manufactured goods also are troubled by the so-called "Dutch disease"—a combination of high energy prices and revenues and a rising Canadian dollar—which has made their production and export more difficult, thus magnifying the problems facing the industrial heartland of Canada. All of these troubling trends came to a head in late 2008 when a global economic crisis caused havoc in Canada and its regions. Ontario's automobile industry was badly wounded and its recovery, so crucial to Ontario, was still underway in 2013.

Reviews play an important role in crafting an expansive, inclusive text such as this, and reviewers of the fifth edition have provided me with new insights and fresh challenges. How to recast Canada's regions within the global economic crisis was one challenge. The call for a more sophisticated text was made, but I am conscious of the broad undergraduate audience who seek a better understanding of Canada's geography while pursuing their own academic goals. My aim, in the sixth edition, is to focus on who we are, where we have been, and where we are headed—individually, collectively, and as a country of regions—all from a regional perspective.

Consequently, besides many new photos, tables, figures, and examples throughout the text, this new edition has experienced a major overhaul in content to account for changes both globally and within Canada. As well as features from the previous edition that helped students make connections and understand historical and contemporary processes, "Contested Terrain" boxes are a new feature to the sixth edition. These boxes highlight specific issues that make the regional geography of Canada dynamic and at times difficult for the major political actors and for citizens of Canada.

Acknowledgements

With each edition, I have benefited from the constructive comments of anonymous reviewers selected by Oxford University Press. I especially owe a debt of thanks to one of those reviewers who spiced his critical comments with words of encouragement that kept me going. As Canada has changed, so has each edition of this book. When I look back at the first edition, I see a much different Canada from today. This transformation process is often captured in the constructive comments of reviewers.

I have called on the resources of The National Atlas of Canada and Statistics Canada to provide maps and statistics. As well, both organizations have created important websites for geography students. These websites provide access to a wide range of geographic data and maps that, because they are constantly updated, allows the student to access the most recently available information on Canada and its regions.

The staff at Oxford University Press, but particularly Phyllis Wilson, made the preparation of the sixth edition a pleasant and rewarding task. Lisa Peterson, the developmental editor who worked with me in the process of revising the text and selecting new photographs, deserves special thanks. Richard Tallman, who diligently and skilfully has edited my manuscripts into polished finished products for each of the last five editions, deserves special mention. As copy editor, Richard has become an old friend who often pushes me to clarify my ideas.

Finally, a special note of appreciation to my wife, Karen, is in order, as well as to our three wonderful grandchildren, Casey, Davis, and Austyn.

IMPORTANT FEATURES OF THIS EDITION

The sixth edition of *The Regional Geography of Canada* has been fully revised to incorporate the newest census data and reflect Canada's ever-changing role in the global economy. Building on the strengths of previous editions, the text takes into account key factors in human geography such as the slow but continuing recovery from the global economic crisis and the significant industrialization of Asian countries, which are contributing factors in shifting and reshaping the balance of power across the nation.

As in previous editions, the sixth edition incorporates a wide range of resources for students that complement and enhance the text. Features appearing throughout the text include

- **new "Contested Terrain" boxes** that draw attention to specific issues in the regional geography of Canada;
- **new and updated vignettes** that focus on issues specific to each chapter;
- **new and updated "Think About It" questions** that prompt students to analyze the material both in and out of the classroom;
- **new and updated cross-chapter references** that highlight the interconnectedness of content across chapters to ensure a comprehensive study of the material;
- **new and updated maps** that highlight the characteristics of various regions across Canada; and
- **updated colour photographs** that engage the reader and provide strong visual references tied to the material.

The result is a new edition that retains the strengths that have made *The Regional Geography of Canada* a best-selling text while introducing new concepts and exploring topics of interest to today's student.

REGIONS OF CANADA

Introduction

Geography helps us understand our world. Since Canada is such a huge and diverse country, its geography is best understood from a regional perspective. In fact, the image of Canada as "a country of regions" runs deep in Canadian thought and literature, and even in the national psyche. This image is, in fact, political reality as geography and history have forged Canada into a complex and diverse set of regions. Each has its own political agenda and economic objectives and, as a consequence, these differing agendas and objectives sometimes collide, causing tensions between regions.

Canada consists of six regions. Each differs by location, physical geography, resources, and historical development. From these differences, a strong sense of regional identity exists in each region. These identities, shaped over time as people came face to face with challenges presented by their economic, physical, and social environments, produced a unique sense of place in each region. At the same time, each part of Canada contains centripetal forces that, from time to time, erupt into fractious disagreements between the federal government and/or provinces, and these tensions pose challenges—some serious and other less so—to Canadian unity.

Canada's place in the world economy remains dependent on exports. Most exports flow to the United States but Canada's trade is tilting slowly but surely towards China and other fast-growing Asian economies. What is behind this shift in trade? The answer lies in two recent developments. First, the global realignment of world power that is relentlessly gravitating towards Asia offers trade opportunities for Canada. Second, the slow recovery of the US economy from the 2008–9 global meltdown has reduced Canadian exports to the United States. Canada, so dependent on US trade, had no choice but to turn to other countries. The consequences are twofold: exports to other countries filled the gap created by the loss of sales to the United States; and new foreign markets reduced Canadian dependency on the US market and, at the same time, appeared to position Canada's world trade for decades to come. In turn, this global trade realignment has affected Canada's regions differently. The resource-rich regions of Western Canada and British Columbia have gained the most from this realignment while the manufacturing belt in Central Canada has lost the most.

CHAPTER OVERVIEW

Geographic analysis and synthesis of our world require an intuitive grasp of its regional nature and spatial relationships, especially as Canada's regions relate to contiguous US regions. The following topics are examined in Chapter 1:

- Geography as a discipline.
- Regional geography.
- Canada's geographic regions.
- The dynamic nature of regions.
- Sense of place.
- Faultlines within Canada.
- Core/periphery theory.
- The global economy.
- Canada in the global world.
- Canada–US trade relations.
- Understanding Canada's regions.

The Crowsnest Highway, British Columbia and Alberta. The Crowsnest Highway is 1,161 km long and runs in an east-west direction from Hope, BC to Medicine Hat, Alberta. It takes its name from Crowsnest Pass located at the BC/Alberta border. Photo: Barrett & MacKay/All Canada Photos.

Geography as a Discipline

THINK ABOUT IT
Is Canadian identity based on place? Does this concept still apply if a person moves from the region of his/her birth and resettles in another part of Canada? Does it have any meaning for a recent immigrant to Canada? Within the concept of place, do you see a relationship between identity, **regionalism**, and nationalism?

Geography provides a description and explanation of lands, places, and peoples beyond our personal experience. De Blij and Murphy (2006: 3) go so far as to state that "Geography is destiny," meaning that for most people, place is the most powerful determinant of their life chances, experiences, and opportunities. In that sense, geography sets the parameters for a person's life opportunities and this concept transfers easily to regions and nations. The concept of "place" is much more than an area; rather, "place" refers to the community/region where one was born and raised, and emphasizes that this geographical fact, this "placeness," has played a role in shaping the attitudes and values of the inhabitants of that community/region.

Not surprisingly, then, people have always been curious about distant places and foreign lands. Back in the days of the ancient Greeks, information from travellers was gathered to provide a better understanding of distant places and kingdoms in their world (Vignette 1.1). The concept of place is based on living, working, and sharing together in a common space, and that experience inevitably leads to the formation of a **regional identity** and **consciousness**. Both are products of a region's physical geography, historical events, and economic situation. As such, they form one of the cornerstones of regional geography.

Regional Geography

The geographic study of a particular part of the world is called **regional geography**. In such studies, people, interacting with their economic, physical, and social environments, are perceived as placing their imprint on the landscape just as the landscape helps to determine their lives and activities. In layperson's terms, the goal of regional geography is to find out what makes a region "tick." By achieving such an understanding, we gain a fuller appreciation of the complexity, diversity, and interconnectivity of our world.

Regional geography has evolved over time.[1] Originally, geographers focused their attention on the physical aspects of a **region** that affected and shaped the people and their institutions. Today, geographers place more emphasis on the human side because the physical environment is largely mediated through culture, economy, and technology (Agnew, 2002; Paasi, 2003). Over time, a multitude of profound and often repeated extreme experiences, whether they are economic, natural, or political events, mark the people, forcing them to respond. In turn, these responses help create a common sense of regional belonging and regional consciousness. In Canada, the vast size of the country means that each region has its own version of belonging and consciousness. From a theoretical perspective, expressions of regional belonging and consciousness are contained in the geographic concept of **sense of place** where local control over regional and community affairs is the order of the day.

Regional self-interest, a logical outcome of regional identity and consciousness, often results in conflicts between the provincial and federal governments and quite different views of the "nation" in different regions of the country, as suggested in Figure 1.1. The struggle for economic

VIGNETTE 1.1

Curiosity: The Starting Point for Geography

Curiosity about distant places is not a new phenomenon. The ancient Greeks were curious about the world around them. From reports of travellers, they recognized that the earth varied from place to place and that different peoples inhabited each place. Stimulated by the travels, writings, and map-making of scholars such as Herodotus (484–c. 425 BCE), Aristotle (384–332 BCE), Thales (c. 625–c. 547 BCE), Ptolemy (90–168 CE), and Eratosthenes (c. 276–c. 192 BCE), the ancient Greeks coined the word "geography" and mapped their known world. By considering both human and physical aspects of a region, geographers have developed an integrative approach to the study of our world. This approach, which is the essence of geography, separates geography from other disciplines. The richness and excitement of geography are revealed in Canada's six regions—each region is the product of its physical setting, past events, and contemporary issues that combine to produce a set of unique regional identities.

REGIONS OF CANADA

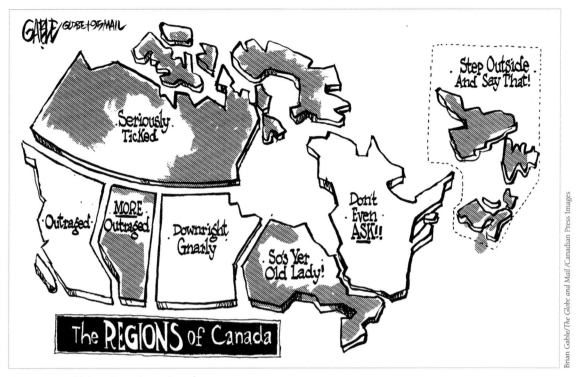

Figure 1.1 Gable's regions of Canada
Political cartoonist Brian Gable aptly captured the occasionally fractured relationships between provinces and territories with his map of Canada. In 1985, regional tensions reached the boiling point over the threat of Québec separating from Canada. The results of the 1995 referendum were very close, but afterwards the heated political scene cooled somewhat. Another fractious issue arose in the twenty-first century, when the economic success of the resource-rich western provinces greatly outpaced that of Central Canada's manufacturing provinces. Hot words were exchanged between provincial leaders with the Premier of Ontario charging that the oil sands caused the high Canadian dollar, which in turn resulted in a decline in the manufacturing sector of his province. Adding fuel to the fire, the federal NDP leader entered the fray by declaring that the oil sands created a Canadian version of the Dutch disease, which linked a high Canadian dollar with the oil sands and led to a declining manufacturing industry. Such would not be the case, the NDP leader said, if the oil sands companies paid the full cost of the environmental damage.

well-being was at the heart of a dispute that took place nearly a decade ago, where economic gains in one region and losses in another region were attributed to the **Dutch disease**. In this case, no-holds-barred public brawling between the federal government and the provincial government of Newfoundland and Labrador took place over the proposed federal clawback of Newfoundland and Labrador's equalization payments. Ottawa argued that, because of much larger provincial energy royalties, the province's revenue no longer justified the level of past annual equalization payments. Premier Danny Williams, however, claimed that Ottawa's policy would deny the province an opportunity to climb out of its dire economic condition. In typical Canadian fashion, a political compromise was achieved in an agreement that reduced the federal clawback for a limited period of time, thus allowing the provincial government to enjoy its oil royalties and equalization payments.

See Vignette 3.5 for a fuller discussion of the Dutch disease and Chapter 3, "Major Federal-Provincial Programs," page 91, for more on equalization and transfer payments.

Canada's Geographic Regions

The geographer's challenge is to divide a large spatial unit like Canada into a series of "like places." To do so, a regional geographer selects the critical physical and human characteristics that logically divide a large spatial unit into a series of regions and that distinguish each region from adjacent ones. Towards the margins of a region, its core characteristics become less distinct

THINK ABOUT IT
Each region has had its struggles with Ottawa. What event(s) affected relations between your region and Ottawa?

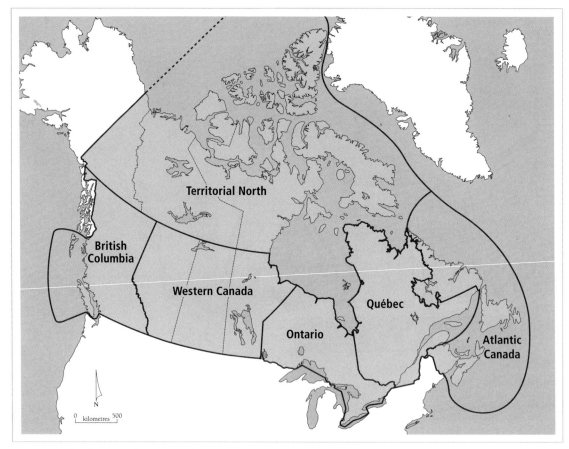

Figure 1.2 **The six geographic regions of Canada**
The coastal boundaries of Canada are recognized by other nations except for the "sector" boundary in the Arctic Ocean, which is shown as a dashed boundary. In the near future, the Territorial North may extend well into the Arctic Ocean and its seabed. In 2013, Canada will submit its claim to the "international" portion of the continental shelf of the Arctic Ocean. If successful, Canada may gain a portion of the Arctic Ocean's seabed as large as the Maritimes. The vast petroleum resources beneath the Arctic seabed may, by the end of this century, become as important to Canada as the Alberta oil sands are today. See Chapter 10 for more on this topic.

THINK ABOUT IT
There are many ways to create regions. Can you think of another way to divide Canada into a set of regions? Consider physiographic regions or climatic zones, shown in Figures 2.1 and 2.6, respectively.

and merge with those characteristics of a neighbouring region. In that sense, boundaries separating regions are best considered transition zones rather than finite limits.

In this book, we examine Canada as composed of six geographic regions (Figure 1.2):

- Atlantic Canada
- Québec
- Ontario
- Western Canada
- British Columbia
- Territorial North

The six regions were selected for several reasons. First, a huge Canada needs to be divided into a set of manageable segments. Too many regions would distract the reader from the goal of easily grasping the basic nature of Canada's regional geography. Six regions allow us to readily comprehend Canada's regional geography and to place these regions within a conceptual framework based on the core/periphery model, discussed later in this chapter. This is not to say that there are not internal regions or sub-regions. In Chapter 5, Ontario provides such an example. Ontario is subdivided into southern Ontario (the industrial **core** of Canada) and northern Ontario (a resource **hinterland**). Southern Ontario is Canada's most densely populated area and contains the bulk of the nation's manufacturing industries. Northern Ontario, on the other hand, is sparsely populated and is losing population because of the decline of its mining and forestry activities.

Second, an effort has been made to balance these regions by their geographic size, economic

Table 1.1 General Characteristics of the Six Canadian Regions, 2011

Geographic Region	Area* (000 km²)	Area (%)	Population	Population (%)	GDP (%)
Ontario	1,076.4	10.8	12,851,821	38.4	39.6
Québec	1,542.1	15.4	7,903,001	23.6	20.5
British Columbia	944.7	9.5	4,400,057	13.1	12.2
Western Canada	1,960.7	19.6	5,886,906	17.6	21.5
Atlantic Canada	539.1	5.4	2,327,638	7.0	5.7
Territorial North	3,909.8	39.3	107,265	0.3	0.5
Canada	9,972.8	100.0	33,476,688	100.0	100.0

*Includes freshwater bodies such as the Canadian portion of the Great Lakes.
Source: Statistics Canada (2006a, 2012a, 2012b).

importance, and population size, thus allowing for comparisons (see Table 1.1). For this reason, Alberta is combined with Saskatchewan and Manitoba to form Western Canada, while Newfoundland and Labrador along with Prince Edward Island, New Brunswick, and Nova Scotia comprise Atlantic Canada. The Territorial North, consisting of three territories, makes up a single region. Three provinces, Ontario, Québec, and British Columbia, have the geographic size, economic importance, and population size to form separate geographic regions.

This set of regions is readily understood by Canadians because of their frequent use in the media. These regions:

- are associated with distinctive physical features, natural resources, and economic activities;
- reflect the political structure of Canada;
- facilitate the use of statistical data;
- are linked to regional identity;
- are associated with reoccurring regional complaints and disputes;
- reveal regional economic strengths and cultural presence.

The six geographic regions form the heart of this book (Chapters 5–10). The critical question is: What distinguishes each region? Certainly geographic location and historical development play a key role. Equally important are contemporary elements such as variations in area, population, and economic strength (Table 1.1), while the proportions of French-speaking and Aboriginal peoples in each region form another essential part of the puzzle (Table 1.2). These basic geographic elements provide a start to understanding the nature of the six regions. Further understanding is provided by analysis of an important economic activity found in each region. Under the heading of "Key Topic," fuller analysis and detailed insights into the nature and strength of each regional economy are presented, as are the challenges faced by these economies. These key economic activities are:

- Ontario: automobile manufacturing
- Québec: hydroelectric power
- British Columbia: forest industry
- Western Canada: agriculture
- Atlantic Canada: fisheries
- The Territorial North: megaprojects

The task of interpreting Canada and its six regions poses a challenge, especially because of shifting global power, the changing demographic and economic strengths of the six regions, and the implications of shifting centres of power for political relationships within Canada. A spatial conceptual framework based on the core/periphery model helps us to understand the nature of this regional diversity within the national and global economies. At the same time, the social dimensions of Canada are captured in the concept of faultlines that identify and address deep-rooted tensions in Canadian society that sometimes stir negative feelings towards Ottawa and even other provinces. At the same time, such tensions represent regional consciousness and this sense of place moulds each region's unique identity.

The Dynamic Nature of Regions

Canada and its regions are not static entities. Since Confederation, Canada's population has increased approximately 10 times. In 1871 Canada's population was 3.7 million and by 2011 it had reached 34 million (Statistics Canada, 2012a). Yet, this population increase was not distributed evenly across the country (Figure 1.3). In 1871, for instance, Ontario and Québec housed three-quarters of Canada's 3.7 million people (Statistics Canada, 2012a). By 2011, while the combined population of these two regions had increased to over 20 million, their percentage of Canada's population had dropped to 62 per cent (Statistics Canada, 2012a). Over the same span of time, the western half of the country saw its population jump from less than 100,000 to just over 10 million, forming 31 per cent of Canada's population. This dramatic demographic shift mirrors the major realignment of Canada's economy caused by global forces. Such changes pull at the ropes holding the political balance of power, and in 2011 Ottawa announced that Western Canada and British Columbia, along with Ontario and Québec, would receive more seats by the time of the next election (see Chapter 4 for a more complete examination of electoral redistribution).

Without a doubt global forces are gathering steam, with economic power increasing in Asia. The central question for Canada and its regions is how best to respond to this rapidly evolving world. Responses to the question vary from one region to another. This critical topic is broached in each regional chapter and discussed further in the concluding chapter.

THINK ABOUT IT

In the 1980s, economists were convinced that globalization of the world's economy and the resulting global trade would benefit all countries and their peoples. Fast forward to February 2012 in London, Ontario, where the workers at the International Caterpillar plant were faced with a 50 per cent reduction in their wages; they refused; and the company closed the plant less than a month later. Do you think that the **Hobson's choice** given the workers (accept sharply lower wages or face plant closure) was an inevitable result of globalization and global trade?

Sense of Place

In spite of our globalized world with its homogenized urban landscapes, the historic character of cities still matters. The term "sense of place" embodies this perspective. Sense of place has deep roots in cultural and regional geography. Leading scholars in this area include Agnew, Cresswell, Paasi, Relph, and Tuan. While "sense of place" has been defined and used in different ways, in this text the term reflects a deeply felt attachment to a region or area by local residents who have, over time, bonded to their environment and resulting institutions. As such, sense of place provides some protection from the landscapes produced by economic and cultural **globalization**. These landscapes are associated with a sense of **placelessness** (Relph). Sense of place, on the other hand, involves a powerful psychological bond between people and locale; globalized or generic landscapes do not have local roots. Local roots stem from the physical nature, human activities, and institutional bodies found in a region. Yellowknife, for example, located on the rocky shores of Great Slave Lake, exhibits a uniquely northern character (see photo 1.1).

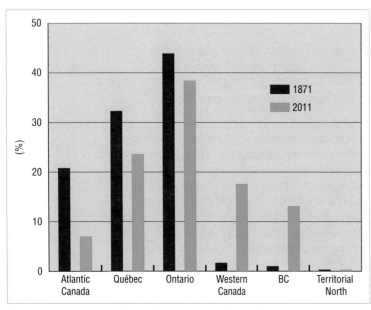

Figure 1.3 Regional populations by percentage, 1871 and 2011
At the time of Confederation, virtually all Canadians lived in Central and Atlantic Canada. By 2011, nearly one-third resided west of Ontario. In just over 100 years, the country's centre of population gravity shifted westward, a trend that is likely to continue for the foreseeable future. *Sources:* Table 3.3 and Statistics Canada (1997, 2012a).

Photo 1.1
Yellowknife, Northwest Territories, is no longer a rough-and-tumble mining town. This capital city also serves as a regional service centre, providing goods and services to surrounding villages and towns as well as to the mines and tourist camps. The public sector dominates the economy, with most workers employed by the territorial and federal governments. Expensive housing exists along its waterfront where, in the warm summer months, pleasure craft, sailboats, and float planes are moored along its sheltered coves on the north shore of Great Slave Lake. On the other hand, Yellowknife, like other Canadian cities, has its share of public housing and homeless people.

The concept of a sense of place, then, recognizes that people living in a region have undergone collective experiences that have led to shared aspirations, concerns, goals, and values. Many experiences are a product of the geography found in each region. Over time, such experiences develop into a social cohesiveness among those people living within a spatial unit.

Atlantic Canada and Western Canada provide two examples where nature has made its mark on the people and their sense of place. The sea plays a fundamental role in the sense of place in Atlantic Canada. With its rich cod fishery, early settlements dotted the coast and provided the economic basis for many coastal communities right up to the end of the twentieth century. The tragic demise of the northern cod fishery by overfishing brought great economic hardship to the inhabitants of the many small fishing communities in Atlantic Canada, but especially to the outports in Newfoundland (for more on this subject, see Chapter 9). Across the country, a similar natural event coloured Western Canada's sense of place. The continental climate of the Prairies is characterized by dry spells, making farming difficult and, at times, impossible. One such time was the Dust Bowl of the 1930s that forced many to abandon their homesteads. While the drought conditions of that earlier era have faded into past collective memory, this semi-arid environment continues to influence life on the Prairies and shape its institutions (for more on this topic, see Chapter 8).

An equally powerful expression of place and alienation exists in Québec, where history and geography have had four centuries to nurture a strong sense of place and to give birth to a nationalist movement that has sought to separate Québec from the rest of Canada. New France was born and grew within the confines of the St Lawrence Lowland. Since Québec's early days, the St Lawrence River has played a key role in its settlement and economic development. Prior to the British Conquest of New France, formalized in 1763 by the Treaty of Paris, this mighty river promoted the interests of the French colony by providing a supply route to France, allowing the fur traders (coureurs du bois) to penetrate far inland to trade with distant Indian tribes, and giving French explorers access into the heart of North America.

A sense of place, however, is much more than bad memories, resentment, and alienation. Indeed, in its truest sense, it is about home and fond memories, about security and the strength to endure the vagaries of the natural world in a particular locale. Peggy's Cove in Nova Scotia provides an example of a community with close ties to the sea. Originally a tiny fishing village, Peggy's Cove, with its sense of history and its location on the sea, has become an important tourist destination (see photo 1.2).

A region, then, is a synthesis of physical and human characteristics that, combined with its distinctiveness from surrounding regions, produces a unique character, including a sense of place and power. People living and working in a region

THINK ABOUT IT
In which geographic region do you live? Is this an important part of your self-identity? Is it as important as your national identity?

Photo 1.2

Peggy's Cove, Nova Scotia. Along its deep but sheltered waters, small fishing boats are moored while lobster traps line the dock. Brightly painted houses and buildings provide a picturesque background.

are conscious of belonging to that place and frequently demonstrate an attachment and commitment to their "home" region. A sense of place, falling somewhere between regional tribalism and commitment to Canada, unites people on common issues and challenges, and compels them to seek solutions to these issues and challenges. For the Inuit, a sense of place is embodied in their natural homeland of the Arctic and in their search for a place within Canadian society; for new Canadians, a sense of place may seem elusive at first, but, with time, they too will develop such feelings as they (or more likely their children) sink their roots into Canadian society. Indeed, the theme of this book is that Canada is a country of regions, each of which has a strong sense of regional pride, but also a commitment to Canadian federalism.

Faultlines within Canada

Tensions exist within Canada, especially between its regions. In 1993, *Globe and Mail* columnist Jeffrey Simpson applied the term "**faultlines**"—the geological phenomenon of cracks in the earth's crust caused by tectonic forces—to the economic, social, and political cracks that divide regions and people in Canada and threaten to destabilize Canada's integrity as a **nation**. In this text, "faultlines" refers to four fractious tensions that lie deep inside Canada's psyche. For long periods of time, these faultlines can remain dormant, but they can shift at any time, dividing the country into wrangling factions.

While many tensions within Canadian society emanate from the plight of the disabled, the homeless, the rural/urban divide, and the seemingly relentless growing gap between the very rich and the rest of Canada's people, our discussion is confined to four principal faultlines that have had profound regional consequences and that have, from time to time, challenged our national unity. These four faultlines represent struggles between: centralist and decentralist visions of Canada; English and French; old and new Canadians; and Aboriginal and non-Aboriginal Canadians. Commencing in Chapter 3, specific examples of faultlines are explored within their regional setting.

Each weak spot has played a fundamental role in Canada's historic evolution, and they remain critical elements of Canada's character in the twenty-first century. In extreme cases, these cracks have threatened the cohesiveness of Canada and, by doing so, have shaken the very pillars of federalism. Under these circumstances, compromise between these differing positions was essential to Canada's survival. From these traumatic experiences, Canada, over time, has become what John Ralston Saul (1997: 8–9) describes as a "soft" country, meaning a society where conflicts are, more often than not, resolved through discussion and negotiations.[2]

Disagreement over the nature of Canada—is it a partnership between the two so-called founding societies of the French and the English or is it composed of 10 equal provinces?—has troubled the country since Confederation in 1867 and in recent years has come to a head twice with the sovereignty-association and independence referendums in Québec in 1980 and 1995, the latter of which was won by the federalist side by the narrowest margin—a mere percentage point. At the height of the referendum campaigns, uneasy relationships tore at the very fabric of Canada. But the ensuing dialogue and goodwill between Québec and the rest of the country have led to an unspoken and uneasy compromise that cements the country together. Still, tensions do arise, such as the appointment by the federal government in 2011 of two non-French speaking Supreme Court judges. The two perspectives on these appointments are: (1) qualifications—the most qualified candidates were selected, and (2) balance between French and English Canada—how can the Supreme Court properly evaluate cases that involve documents written in French if judges are not bilingual? A broad and historical-based perspective on this language issue and the notion of a Canadian partnership is found in Chapter 3 under the heading, "One Country, Two Visions."

Centralist/Decentralist Faultline

Of all the faultlines, the centralist/decentralist one leans the most heavily on Canada's geography and its political system. Canada's size and its varied physical geography and resource base provide the stage for regional differences that can—and have—led to federal–provincial feuds. Adding another dimension to such feuds is provincial control over natural resources—at the moment, oil-rich provinces have an advantage over other provinces. These disputes often flare up between particular provinces and Ottawa, or are reflected in the occasional potshots directed at other provinces by some provincial political leaders. A closer examination of the underlying forces driving the centralist/decentralist faultline follows.

First, quarrels with Ottawa often revolve around federal transfer payments. As pointed out earlier, provincially administered post-secondary, health, and social programs far outstrip their financial capacity. Here we have contested ground with two different objectives. On the one hand, provinces seek an increase in transfer payments, while on the other hand, Ottawa seeks ways to reduce its annual expenditures in order to balance its budget. Health care provides a perfect example. On 20 December 2011 the federal government announced its next 10-year program for transfer payments. Most provinces, including Ontario and Québec, were shocked because the 6 per cent per annum increase for health-care transfers that was so highly prized by provinces in the first 10-year accord (2004–5 to 2014–15) and that was, in their opinion sacrosanct, could end in the second year of the next 10-year plan (Bailey and Curry, 2011:A1). From the federal government's position, this reduction was an important plank in reducing the federal deficit.

Second, another bone of contention exists between Central Canada, where the majority of national population (and voters) resides, and the rest of Canada over the extensive public support/subsidies for Central Canadian industries. From the perspective of Ottawa and the provinces of Québec and Ontario, this public support is based on the long-held premise that economic success in Central Canada will benefit the nation as a whole. Not surprisingly, the premise is less well received in the rest of the country. Whether true or not, from this hinterland perspective, Central Canada often benefits from federal policies that support industries representing the "national interest" while other provinces are left out in the cold. Federal support for Ontario's automotive industry and for Québec's aerospace/rail industries are two examples. In the 2008–9 global downturn, both General Motors and Chrysler plants in Ontario were saved from collapse by huge loans from Ottawa and Ontario totalling nearly $15 billion.

The financial loans to Chrysler and General Motors are described in Chapter 5, Vignette 5.12, "The Bailout of Chrysler and GM."

Third, maybe Bob Dylan's 1964 song, "The Times Are A-Changin'," applies to the centralist/decentralist faultline? As the centre of population gravity edges westward, the population advantage of Central Canada is slowly eroding and, coupled with the rapidly growing economies in Alberta, British Columbia, and Saskatchewan, Ottawa sees the "national interest" taking the form of a "super energy power" on the international stage. Central Canada, but especially Ontario, sees the rise of western power as a threat to its traditional position within Confederation and, hence, is lukewarm to the federal government's support for the Alberta oil sands and the construction of gas and oil pipelines across the Rocky Mountains to the Pacific coast. One sign of this power shift revolves around equalization payments to Ontario. Given Ontario's long-standing position as the "powerhouse" of Canada, who would have thought that Ontario would have this "have-not" collar hung around its neck? (Vignette 1.2)

Fourth, faultlines do exist between provinces. The long-standing dispute between Québec and Newfoundland and Labrador over the 1969 Churchill Falls Agreement illustrates this point. Within this contested terrain lurks the contentious boundary settlement made in 1927 when Newfoundland was still a British colony. Québec has never accepted this boundary and it acts like a burr in the psyche of the Québec nation. Could it be a hidden factor in preventing Hydro-Québec from reopening the 1969 agreement to seek a "fair" pricing arrangement, as well as a factor in Québec's denying space for the transmission of future hydroelectricity from Labrador along its transmission lines to Canadian and US markets? From the perspective of Newfoundlanders and Labradorians, the unfairness of the agreement sees billions flowing to Québec and only a pittance to them. Why is the agreement so one-sided? The 1969 agreement gave a set amount of electric power—some 31 billion kilowatt hours per year—generated by Churchill Falls to Hydro-Québec for 65 years at a very low fixed price, which, instead of increasing over time, actually decreases. Shortly after the signing of this agreement, world oil prices more than doubled, and the jump in prices cascaded into much higher hydro prices. Hydro-Québec gained a huge windfall and this windfall will continue to 2041. To Newfoundland and Labrador, the agreement is a perfect example of exploitation by a bigger, more powerful province. Yet, as the Supreme Court determined in 1984, "a contract is a

VIGNETTE 1.2

Is Central Canada Losing Its Dominant Position in Confederation?

The world is now a global economy. Canada's place in that global world has turned its regional economies on their heads. On the one hand, Central Canada has witnessed the erosion of its manufacturing base while resource-rich provinces have bounded ahead based on high global prices for their resources. Much of this change in the economic landscape is due to the incredible rise in China's industrial might and its export of manufactured goods to Canada and other developed countries. On the other hand, resource-rich provinces have benefited from the seemingly insatiable appetite for energy and raw material from China, Japan, and South Korea, which has pushed both the demand and price for these goods upward. While Western Canada and British Columbia have been swept along in this economic tsunami, even the former "have-not" province of Newfoundland and Labrador has seen its oil-driven economy make great gains. If oil prices continue to climb, the vast offshore oil reserves of Newfoundland and Labrador will turn into a huge revenue bonanza for the government of Newfoundland and Labrador, thus sustaining the province's newly won status as a "have" province.

Figure 1.4 **The burden of Confederation**
For the first time, Ontario became a "have-not" province in 2009.

contract," and the dispute was legally settled—though Hydro-Québec held all the cards in negotiations and virtually dictated the terms. Fast forward to 2013. Will this bitter pill drive the smaller province to build an expensive—perhaps costing as much as $3.3 billion—underwater transmission system from Labrador to Newfoundland and then on to Nova Scotia and eventually to the huge market in New England, or will Hydro-Québec strike a deal?

For more on this ongoing dispute, see Chapter 3, "Internal Boundaries," pages 83–85, and Chapter 9, page 377, "Can The Lower Churchill Hydro Project Mend Atlantic Canada's Fractured Geography?"

English-speaking/French-speaking Canadians

Canada is a bilingual country. Yet, English is spoken in most parts of the country. History accounts for the two languages, though it was not until 1969 that the Official Languages Act recognized English and French as having equal status in the government of Canada. A few years later, in 1974, the Québec government passed its Official Language Act, making French the sole official language in the province. The rationale for this action was the desire to ensure and foster the French language and therefore the Québécois culture. In fact, only one province, New Brunswick, officially recognizes both official languages. The explanation for this particular "quilt-work" of languages across Canada is found in the geographic fact that relatively few French-speaking Canadians live outside of New Brunswick and Québec. In the Québec case, this led to the political decision in 1974 to proclaim French as the official language of the province. As seen in Table 1.2, the French fact is highly concentrated in Québec, where the francophone culture and French language thrive, thus allowing the province to form a distinct cultural region within Canada.

Language remains a sensitive issue and, because the proportion of French-speaking Canadians has declined over time, it forms a faultline. Back in 1867, the population of Canada consisted of two main groups: British, comprising 61 per cent, and French, making up 31 per cent, formed nearly 92 per cent of the population (*Atlas of Canada*, 2009). Today, though the total number of French-speaking Canadians has increased, reaching just over 7 million in 2011, their proportion of the total population has declined to 21.3 per cent. This drop represents a serious dilemma for that community and signals

Table 1.2 **Social Characteristics of the Six Canadian Regions, 2011**

Geographic Region	French (000s)	French (% of regional population)	Aboriginal Peoples (000s)*	Aboriginal Peoples (% of regional population)
Ontario	493,300	3.9	301,425	2.4
Québec	6,102,210	78.1	141,915	1.8
British Columbia	57,280	1.3	232,290	5.3
Western Canada	126,915	2.2	574,335	9.8
Atlantic Canada	272,315	11.9	94,490	2.6
Territorial North	2,970	2.8	56,225	52.4
Canada	7,054,975	21.3	1,400,685	4.3

*These population statistics from Statistics Canada are based on the **Aboriginal identity** question rather than the somewhat higher figures generated from the ethnic-based census question known as **Aboriginal ancestry**. In 2011, Aboriginal identity population was based on a census question that defined identity by three groups of Aboriginal peoples: First Nations, Métis, and Inuit while Aboriginal ancestry is based on a self-declaration of ethnicity. In previous censuses, Aboriginal population data were collected from the obligatory long form census, but in 2011, census questions that determine Aboriginal populations were modified and assigned to a voluntary survey known as the *National Household Survey*.

Sources: Statistics Canada (2012d, 2013).

a greater and greater erosion of their political position within Canada.

What about the future? While French is one of the official languages of Canada, the prospect for the future does not augur well and the proportion of French-speaking Canadians might well fall below 20 per cent by the 2016 census. If so, how would this demographic development affect the balance between the two language groups and the unity of Canada?

Not surprisingly, then, tensions between English- and French-speaking Canadians erupt over language issues. Within Québec, a faultline exists between the two language groups. The political and cultural desire to maintain French as a viable language in a principally English-speaking continent resulted in the controversial Bill 22 in 1974 followed by Bill 101 in 1977, which proclaimed French as the official language in Québec for just about every facet of life: government, the judicial system, education, advertising, and business. In 1982, the Supreme Court of Canada struck down Bill 101. However, the Liberal government of Robert Bourassa overturned this ruling by employing the "notwithstanding clause" (section 33 of the 1982 Charter of Rights and Freedoms). While language is less of a hot-button issue in Québec today, fears that the traditional Québécois way of life is slipping away came to a head in 2007 with the town of Hérouxville's "code of standards," the subsequent Bouchard-Taylor Commission on reasonable accommodation, and Gatineau's recent code of conduct for immigrants.

For a discussion of language and culture, see the following sections: Chapter 1, Vignette 1.3, "The Challenge of Cultural Adjustment in Québec"; in Chapter 4, "Language," pages 142–143, Vignette 4.4, "Charles Taylor on Multiculturalism," and "French/English Language Imbalance," page 156.

Aboriginal Peoples and the Non-Aboriginal Majority

As the first people to occupy the territory now called Canada, many Aboriginal peoples—now legally referred to as "the Indian, Inuit and Métis peoples of Canada"—find themselves stuck on the margins of Canadian society. While Canada prides itself on its open society, upward social mobility, and economic opportunities, few Aboriginal people have followed that path. That raises the question, "Why have so few Aboriginal individuals, such as the Minister of Health Canada, Leona Aglukkaq, and their communities, such as the Whitecap Dakota First Nation near Saskatoon, found that path while most remain lost in the economic wilderness?" The majority, both individuals and communities, remain dependent on the Canadian state, which leaves them in a virtual state of poverty and underdevelopment. While there is no simple answer to this question, several factors have played a major role.

First and foremost, only in the late twentieth century did the doors to the economic and social opportunities available to other Canadians begin to open to Aboriginal people. Until then, the Indian Act served as the federal government's means to control, dominate, and manage First Nations peoples and their lands as well as to keep them restricted to reserves. The treaty land selection process in earlier centuries and the relocation of northern Aboriginal peoples to settlements in the 1950s reinforced the segregation of Aboriginal people from the rest of Canadian society. The relocation of those still on the land to trading posts in the 1950s produced mixed blessings—access to public services and secure food supplies was a plus but these villages had no economic foundation and therefore left the adults in no man's land—they were neither hunters and trappers with a self-sustaining culture and way of life nor could they become employed workers. Consequently, dependency on social assistance became the foundation of their new economic system.

Second, the first crack in the door took place in 1960 when First Nation men and women were allowed to vote in federal elections. Since then, civil and human rights common to Canadians were extended to Aboriginal peoples, though the Indian Act still leaves a role for the federal government in the affairs of Aboriginal peoples. While the Métis and Inuit were not included in the Indian Act, they were, for the most part, treated equally badly and they too suffered from this quasi-apartheid policy. For instance, all three groups suffered through the long period of the residential schools from the late nineteenth century to the late twentieth century. These schools were designed to equip young

Aboriginal students to find and accept a place on the bottom rungs of the larger society but resulted in the loss of their language and culture as well as the connection to the land and their parents, many of whom continued a traditional life of hunting and trapping. In fact, the residential schools created "lost generations" who fitted into neither world.

Third, Aboriginal peoples began to gain control of their traditional lands through modern land claims agreements and through impact benefit agreements. The first breakthrough took place in 1975 with the James Bay and Northern Québec Agreement, followed in 1989 by the Inuvialuit Final Agreement in the western Arctic and then, in 1999, the establishment of Nunavut in the central and eastern Arctic. But not all agreements are equal. The Attawapiskat First Nation, for example, does not have a profit-sharing plan with De Beers Canada, which owns and operates the Victor diamond mine located on their traditional lands, while the Inuit of Nunavik in Arctic Québec do receive annual profit-sharing payments from Xstrata's nickel/copper mine located on their traditional lands.

Social change is not easy and may take generations. Complicating matters, Aboriginal peoples and their organizations are a diverse group with many different needs, rights, and wants. This diversity includes 615 First Nations scattered across Canada, many with populations under 1,000 and some near major urban centres, while many others are in remote locations far from such centres of economic opportunity. Clearly, a single path to successful development does not exist. Perhaps the first step is to revamp the Indian Act. Yet, while some Aboriginal leaders at the 2011 Aboriginal summit with the federal government called for striking down the Indian Act, most did not. Until the Indian Act is somehow reconfigured to provide an opportunity for First Nations to grow, the underdeveloped state of First Nation reserves will remain, as it will, by extension, for the Inuit and Métis.

Newcomers and Old-Timers

While Canada is a land of immigrants, the first European immigrants, the British and, to a lesser degree, the French, established the economic, political, and social structure of Canada. Old-timers have also set the rules of the game—a secular state and, for the past three decades, a Charter of Rights and Freedoms governing the relationship between the state and its citizens. Newcomers must adjust to these political and structural facts. For newcomers, finding a job and a place to live provided a basic comfort level that would allow them to be themselves, support themselves in families, and gain a sense of belonging. This challenge was not as easy for the many non-anglophone or francophone immigrants from the developing world. As well, a sense of belonging, so important to feeling part of the larger Canadian society, was easier for English- and French-speaking immigrants, while the others often left this search for belonging to the next generation, who would grow up in Canada speaking either English or French.

All cultures change with time. What is different in Canada are the continuous waves of newcomers, each bringing their own set of cultures, languages, and religions. The interaction between newcomers and old-timers represents a faultline. This social faultline involves a lively interaction: the cultural rubbing and bumping between those whose cultural roots are in distant overseas homelands and those whose roots developed in Canada. Of course, this interaction necessarily requires adjustments and compromises from both groups. For the most part, old-timers are not always prepared to give ground and sometimes, as in the case of Québec, efforts have been made to establish a code of standards and values (Vignette 1.3). While the Charter of Rights and Freedoms protects minority groups from the tyranny of the majority, there are limits—one clear limit is honour killings; another is **shariah law**, which would replace the state justice system with that of a particular ethno-religious group. The wearing of the burqa and the niqab—both cover the face—falls into the grey area of acceptance, though the federal government may require Muslim women to show their faces when voting and at citizenship ceremonies when they declare their Oath of Allegiance. Within the Canadian Muslim community, support for the burqa and niqab is mixed: the Muslim Canadian Congress has stated that these garments have no basis in Islam while more conservative Muslims accept them and even insist that women be so covered.

THINK ABOUT IT
Since very few Muslim women in Canada wear the burqa or niqab, is this a tempest in a teapot or are these garments a form of female oppression/male control?

The sometimes unpleasant interactions between newcomers and old-timers are a fact of Canadian life and signal that a dialogue—and eventually an accommodation—is taking place. Generally, the second and subsequent generations of immigrant groups, by being born and raised in Canada, have had a much easier time feeling connected to Canada. Yet, as with those nicknamed CBCs (Chinese born in Canada), a cultural gap emerges within the diasporic community, with those of later generations more acclimated to Canada but less attuned to the culture and language of their forebears. The same challenge faces conservative Muslim immigrants today.

VIGNETTE 1.3

The Challenge of Cultural Adjustment

All regions of Canada face the challenge of cultural adjustment. Those born and raised in Canada watch such adjustments with both pleasure and trepidation. Pleasure comes from the enrichment of Canadian society; trepidations come from fear of losing control of "their world." Cultural adjustment is a particularly sensitive matter in Québec. In 2007, matters came to a boil in Québec when the Hérouxville municipal council announced its "code of standards." The purpose of the code is "to inform the new arrivals that the lifestyle that they left behind in their birth country cannot be brought here with them and they would have to adapt to their new social identity" (Hamilton, 2007: A7). In fact, the "code" clearly targeted Middle Eastern, North African, and South Asian immigrants with its banning of the veil and of carrying weapons to school (Sikh boys and men wear ceremonial daggers) and its censure of such practices as stoning women and genital mutilation. The resulting firestorm within Québec society and beyond caused the provincial government to launch the Bouchard-Taylor Commission[3] to address the issue of "accommodation practices related to cultural differences in response to public discontent concerning reasonable accommodation." The Commission's report received mixed reviews and some felt that its recommendations were unreasonably accommodating. Yet, the Québec public remains uneasy, and in December 2011 the City of Gatineau announced its "Statement of Values" for immigrants coming to the city.

Photo 1.3

Emotion ran high during a protest in Toronto against shariah law on 8 September 2005. In the photo, a female protestor points out that the book Imin Mubin Shaikh is using to support the application of shariah law to Muslim women was written by a man. This protest is a culmination of a series of events. In the early 1990s, the Ontario government allowed Catholic and Jewish faith-based arbitrations to settle family disputes, such as divorce, custody, and inheritances, outside of the Ontario court system—if the parties involved agreed. In 2004, the Islamic Institute of Civil Justice applied to establish its own faith-based arbitration panel under the Ontario Arbitration Act. Reaction from the public was swift, and strong opposition was voiced by officials from both the Muslim community and women's organizations: the Muslim Canadian Congress, the Canadian Council of Muslim Women, and the National Association of Women and the Law vigorously opposed such a move because, in their judgement, women are not treated equally to men. In September 2005, Ontario Premier Dalton McGuinty rejected the request from the Islamic Institute of Civil Justice and, at the same time, rescinded the operation of Catholic and Jewish tribunals.

The Core/Periphery Theory

Canada is a complex country. In 1930, the acclaimed political economist, Harold Innis, wrote that "The economic history of Canada has been dominated by the discrepancy between the centre and the margins of western civilizations" (Innis, 1930: 385). Innis also devised the **staples thesis** to explain the process of development in the early stages of Canada's industrialization (see Vignette 1.4). His centre/margins concept and his view of Canadian regional development remain central to an understanding of our economic life.

To understand our complex country and its regional nature, a modified **core/periphery model** helps us grasp the broad economic relationships between regions.[4] This model is based on John Friedmann's 1960 adaptation of the core/periphery model to Venezuela, where he expanded the number of **periphery** regions from one to three (Table 1.3).

These four regions—a core and three peripheries—are easily adapted to Canada:

- core region centred on manufacturing (Ontario and Québec);
- rapidly growing region based on an expanding resource base (British Columbia and Western Canada);
- slow growing region based on a declining resource base (Atlantic Canada);
- resource frontier region where many resources exist but few are viable (the Territorial North).

The Global Economy

The new global order has turned the traditional ratio of prices of manufactured goods to resource products on its head. In fact, the core/periphery theory—a product of Wallerstein's **capitalist world-systems theory**—is based on the assumption that manufacturing cores have an inherent advantage over resource-based peripheries because, over time, prices for manufactured goods increase more rapidly than those for resources. In this way, Wallerstein argued, industrial centres become richer and richer over time compared to their peripheries because of this price advantage. However, the global economy has reversed this by creating an unprecedented demand for energy and resources from rapidly industrializing countries, thus pushing prices for those products to record levels, and, at the same time, by producing manufactured goods with low-cost labour, thus reducing the prices of those products.

Let us examine global prices over time more closely. Since the Industrial Revolution, world

THINK ABOUT IT
Do you favour a strict separation of state and church? If so, where do you stand on public support for Roman Catholic schools?

VIGNETTE 1.4

The Staples Thesis

Early in the twentieth century, Harold Innis (1930) presented his theory of Canada's economic development to explain why Canada developed distinct regional economies and political institutions. Known as the staples thesis, Innis based his theory on resource development in the hinterland (Canada) and trade with the **heartland** (Great Britain). Over time, the exploitation of these staples led to regional economic diversification and the formation of institutions that defined the political culture of each region. The staples thesis accomplished two important objectives. First, Innis articulated an explanation for Canada's economic development, which was initially based on the exploitation of resources (staples). Over time, the development of primary extraction industries led to a more broadly based economy. Second, Innis argued that each resource had a different impact on its region, its institutions, and its political culture.

The staples thesis, then, provided Innis with a Canadian version of regional development within the context of a core/periphery framework (Watkins, 1963, 1977; Barnes, 1993; Hayter and Barnes, 2001). In his 1963 article, Watkins introduced three concepts to the staples thesis: **backward**, **forward**, and **final demand linkages**; in his 1977 article, Watkins made a pessimistic interpretation of regional development in a frontier area where he felt a **staples trap**, that is, a collapse of an economy based on non-renewable resources, was inevitable.

Table 1.3 **National Version of the Core/Periphery Model**

Regions	Characteristics
Core Region	The core region is the focus of economic, political, and social activity. Most people live in the core, which is highly urbanized and industrialized. The core has a high capacity for innovation and economic change. Innovations and economic advances are disseminated downward through the national urban hierarchical system to the periphery. Ontario and Québec form Canada's core regions.
Rapidly Growing Region	The rapidly growing region's economy and population expand as both capital and labour flow into this area. While initial development occurred in the primary sector, there is now a greater emphasis on manufacturing and service activities. British Columbia and Western Canada are examples of rapidly growing regions.
Slow Growing Region	In a slow growing region, the economy is declining, unemployment is rising, and out-migration is occurring. Often this is an "old" region dependent on resource development for its economic growth. Now that these resources have passed their prime or have been exhausted, the regional economy has stalled. Atlantic Canada is an example of a slow growing region.
Resource Frontier	Located far from the core region, few people live in this frontier and little development has taken place. Resource companies are just beginning to penetrate into this remote area. As energy and mineral deposits are discovered, the prospects for economic growth are enhanced. The Territorial North represents Canada's last resource frontier.

prices have favoured manufactured goods over resources, that is, prices for manufactured goods have risen more than prices for resources. But the world economy reached a tipping point with the emergence especially of China as a new superpower, and the economy entered a super cycle, thus reversing this price relationship. One argument supporting the **super cycle theory** is that prices for resources, while declining during the 2008–9 global crisis, remained at relatively high levels because of sustained demand from rapidly industrializing countries such as China, India, and Brazil (Parkinson, 2009).

Without a doubt, globalization has transformed Canada's economy and, in doing so, created an economic divide within Canada. The manufacturing-oriented provinces of Ontario and Québec have fallen well behind the resource-rich western provinces in terms of economic performance. While the manufacturing sector may find its feet, the old sector could not compete with manufacturers in China and other low-wage countries. Canada recognizes that the balance of economic power is shifting swiftly to China and other Asian countries and is taking measures to reposition itself within the global economy, in the hope of being on the right side of history as the twenty-first century progresses. But just how will Canada reposition itself to take advantage of the new economic order? The following section addresses this complex question.

Table 1.4 **Economic Structure of Canada and Its Six Geographic Regions, 2011**

Economic Sector	Ontario	Québec	British Columbia	Western Canada	Atlantic Canada	Territorial North	Canada (000s)	Canada (%)
Primary	1.9	2.3	2.9	9.3	5.1	13.0	642.8	3.7
Secondary	19.2	19.1	16.8	17.2	15.8	2.0	3,162.2	18.3
Tertiary	78.9	78.6	80.3	73.5	79.1	85.0	13,501.3	78.0
Total	100.0	100.0	100.0	100.0	100.0	100.0	17,306.2	100.0

Sources: Statistics Canada (2012c); for Territorial North, author's estimates based on Yukon (2011) and Northwest Territories (2011).

Table 1.5 Changing Trade Direction: From Continental to Global Trade

Date	Continental Orientation
1965	The Auto Pact between Canada and the United States creates a continental market for automobiles. An important provision for Canada is that its share of automobile production is to be maintained. The three large American car manufacturers (General Motors, Ford, and Chrysler) could now expand their production to a continental scale. In this process, Canadian branch plants cease to produce a wide variety of automobiles for the Canadian market and begin to produce only a few models (but more of them) for the continental market.
1989	The Canada–US **Free Trade Agreement** (FTA) is approved. The aim of the FTA is to integrate the two economies. Unlike the Auto Pact, however, there are no assurances that Canada's share of economic activities will be maintained. Instead, the North American marketplace determines the type and amount of economic activity in Canada.
1994	Superseding the FTA, the **North American Free Trade Agreement** (NAFTA) broadens the geographic area of the FTA to include Mexico. The North American economy now has a member with much lower wage rates, causing many labour-intensive industries to close their operations in Canada and reopen in Mexico.
2001	The end of the Auto Pact results from a WTO ruling that the Auto Pact discriminates against foreign companies.
2011	Beyond Borders Agreement outlines joint Canada–US efforts to strengthen shared security and improve the flow of people, goods, and services across the shared border.

Date	Global Orientation
1997	Israel and Chile free trade agreements reached.
2002	Costa Rica Free Trade Agreement is achieved; Canada signs the Kyoto Protocol committing 38 industrialized countries to cut their emissions of greenhouse gases between 2008 and 2012 to levels 5.2 per cent below 1990 levels.
2008	Colombia Free Trade Agreement reached.
2009	Jordan and Peru free trade agreements are signed.
2010	Panama Free Trade Agreement is made.
2011	Canada withdraws from the Kyoto Protocol.
Pending	Free trade agreements with European Union, Pacific Rim countries (Trans-Pacific Partnership), China, India, and South Korea are in the works, as are more free trade agreements with Central American countries.

Canada in the Global World

Canada's place in the world has changed. With global economic power shifting to Asia, Canada is rapidly realigning itself to take advantage of new trade opportunities. One mark of this shift in federal policy is the number of trade missions to China; another is the number of major and minor trade agreements with other countries in recent years (Table 1.5). Two major trade agreements are

VIGNETTE 1.5

The New World Order and the Super Cycle Theory

The explosion of economic growth in Asia, but particularly in China and India, has driven commodity and energy prices to new highs. Have we entered a new super cycle? In the past, two super cycles took place, one between 1870 and 1913 and the other from 1946 to 1973. What are super cycles? Economists consider super cycles as extended periods of high global growth driven by the emergence of large, new economies. The current super cycle is driven by the rapid industrialization of formerly Third World countries such as China and India. In the process of industrialization, the demand for energy and raw materials soars, thus increasing prices for commodities.

> **CONTESTED TERRAIN 1.1**
>
> **McGuinty–Redford War of Words**
>
> In late February 2012, a skirmish broke out between the premiers of Alberta and Ontario over the oil sands. Premier Redford asked Ontario to support the development of Alberta's oil sands, which is under constant attack from environmentalists. Rather than offering Redford support, Premier McGuinty responded by claiming that the "petro-dollar" was killing Ontario's manufacturing base. While both premiers were defending the interests of their respective provinces, this skirmish underscores the widening economic gap between Alberta and Ontario.
>
> *Source:* McParland (2012).

pending. One with the European Union may be finalized in 2013. Another major one, **Trans-Pacific Partnership**, is on the drawing board, and, if successful, would draw Canada more deeply into trade with Pacific Rim countries.

Diversification of trade is now a top priority of the federal government, and private companies, responding to Asian demand, are engaged in redirecting their products to the Asian market. Prior to the US economic decline, Canada's access to the world's largest market, the United States, was seen as an advantage. Now, with the US economy struggling to regain its footing in the world economy, exports to the United States have declined, causing some to see this trade dependency as a weakness. To reverse the heavy reliance on trade with one country, diversification of trade is now a key component of Canada's foreign policy. Already many free trade agreements have been concluded and more are on the way. By promoting world trade, international agreements form a key element in Ottawa's strategy to ensure Canada's place in the world and, at the same time, reduce its dependency on the United States. These efforts range from trade agreements with Europe, India, and a number of smaller countries, plus entering into negotiations with Pacific Rim countries. Still, the United States remains Canada's principal market and, in December 2011, Canada signed a border pact with the United States known as "Beyond Borders" that will facilitate cross-border trade. Also, companies in the transportation and financial sectors logically have sought to expand in the US market (see Figure 1.5).

In recognition of this trade shift, Canada seeks to take advantage of the high demand (and prices) for agricultural products, energy, and other raw materials in short supply in China and other Asian countries. The proposed natural gas and oil (bitumen) pipelines to the Pacific coast reflect this new world order. The completion of the natural gas pipeline seems assured while the Northern Gateway Pipeline faces strong opposition from First Nations groups and environmental organizations. With the Prime Minister declaring that the Northern Gateway Pipeline is in the national interest, the stakes are extremely high.

Another priority for the Canadian government is to expand the infrastructure needed to protect Canada's Arctic sovereignty. Canada is securing its Arctic sovereignty in two ways. First, like the other four Arctic countries with coastlines on the Arctic Ocean, Canada has been preparing its claim for a large part of the Arctic Ocean seabed, to be submitted in 2013 under the United Nations Convention on the Law of the Sea (UNCLOS). This statement is based on extensive scientific investigation in Canada's Arctic waters for the past five years. Second, the government has sought to increase its naval capacity in the Arctic by committing $33 billion over 30 years to build an Arctic navy, including a polar icebreaker. These ships will be constructed in Canadian shipyards—one in Halifax, the other in Vancouver—thus giving a huge long-term boost to the economies of these two regions.

Beyond trade and the protection of Arctic sovereignty, Canada is playing a more active role on the world stage. Hosting the 2010 Winter Olympics in Vancouver turned into a national celebration, while the G20 meeting in Toronto in 2010 brought with it the world spotlight but also protests and controversy. In fact, a few described

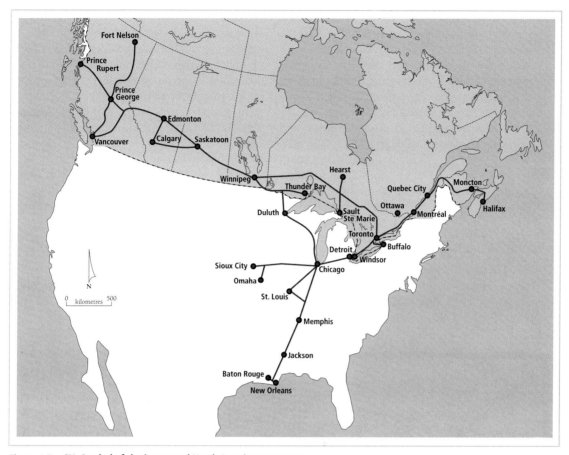

Figure 1.5 **CN: Symbol of the integrated North American economy**
With the Free Trade Agreement, CN switched its emphasis from a predominantly Canadian east–west railway system to a North American continental system, and by doing so it gained economies of scale. CN now operates in eight Canadian provinces and 16 US states. Not surprisingly, Canadian banks have followed suit, purchasing regional banks whose value had dropped sharply following the recent financial crisis in the United States, and now have substantial holdings in the US. The high Canadian dollar made these purchases even more attractive. The Bank of Montreal, for instance, added to its US financial holdings, which consisted of the BMO Harris Bank, with the purchase of Marshall & Ilsey Corporation in 2011. Now its US operations are equal in size and value to those in Canada.

the G20 meeting as a black eye for Canada. Also, some Canadians have had mixed feelings about recent military involvements in Afghanistan and Libya. Military spending—the butter/guns argument—is never popular in Canada with the general public. No doubt relief was felt with the Prime Minister's announcement in May 2012 that Canada will withdraw its remaining troops—those training Afghanistan soldiers—as planned in 2014. However, the war burden will continue after 2014 as Canada agreed to contribute to the cost of maintaining the Afghanistan army.

While Canadians support the concept of a well-equipped armed force, expensive purchases raise questions. One example has been Ottawa's plan to purchase F-35 fighter jets to replace the aging CF-18 fleet, but the soaring price, as well as questions about the military role of these fighter aircraft, led the government at the end of 2012 to back away from its previous commitment to purchase 65 of the jets: the lifetime cost had reached $45 billion, a 500 per cent increase in the span of only a few years.

On the domestic front, the simmering Aboriginal file remains a challenge. For instance, what is the role of the Indian Act in the coming decades? Stung by the dreadful Third World conditions found in many Aboriginal communities, Ottawa and the Aboriginal leaders are still struggling to find solutions and to define a new relationship that will move Aboriginal people from the margins of Canadian society, ensure that Aboriginal youth help solve Canada's labour shortage, and, in doing so, break the cycle of dependency on

Ottawa. A recent study by the Ottawa-based Centre for the Study of Living Standards, which measures "quality of life," provides more evidence of the socio-economic gap between Aboriginal people and the rest of Canadians. The study found that Nunavut—the only political unit with a majority Aboriginal population, at about 80 per cent—falls well below all other territories and provinces (Hazell et al., 2012).

The extent to which Aboriginal affairs could no longer remain Ottawa's unwanted stepchild became ever more apparent as the year 2012 turned to 2013. Chief Theresa Spence of northern Ontario's benighted Attawapiskat First Nation, from her perch on a Native-owned island in the Ottawa River just opposite the capital, was into the fourth week of a hunger strike demanding an executive-level meeting between Aboriginal leaders and the Prime Minister and Governor General (the Queen's representative) to address, yet again, the many needs of Aboriginal communities. The Idle No More movement, formed in November 2012 as a coalition of grassroots Aboriginal activists and non-Aboriginal sympathizers with a broad agenda focused on government recognition of treaty rights and environment-related issues, supported Chief Spence's singular effort, and at the same time it had blockaded rail lines and highways in parts of the country and had created round-dancing flash mobs in malls full of Christmas shoppers. When Prime Minister Harper finally, on 4 January 2013, announced a meeting with Aboriginal leaders for a week later, the Idle No More activists reiterated their tactics would continue, and Chief Spence issued a statement that she wouldn't attend the meeting since the Queen's representative would not be there. The Idle No More movement found supporters and traction in the United States and in some other countries around the world.

Figure 1.6 The Kyoto Protocol and the Northern Gateway Pipeline
Canada is looking for new markets for its oil and gas reserves. From Ottawa's perspective, oil and gas pipelines to the Pacific coast would allow these products to reach China and other Asian markets. The tipping point for the federal government backing of west coast pipelines shifting from lukewarm to urgent was the US delay in approving the Keystone XL Pipeline from Alberta's oil sands to the heavy oil refineries on the Texas Gulf coast. The Northern Gateway Pipeline would ship bitumen from the oil sands to Kitimat on the British Columbia coast while the Pacific Trail Pipeline would transport natural gas from northern BC to the coast. As Prime Minister Harper put it, "Canada will not become a 'captive supplier' to US energy needs" (Blanchfield, 2011).

On a different international front but still a controversial one—and one that also impinges on Aboriginal rights and issues—the 2011 decision by Ottawa to withdraw from the Kyoto Protocol, the government's strong support for oil sands developments, regardless of the nationality of the major oil sands players, and its insistence that an oil pipeline to the west coast is in "the national interest" were blows to many Canadians and to environmental organizations.

 See Chapter 2, pages 60 and 61, for a fuller discussion of Canada's withdrawal from the Kyoto Protocol.

Canada–US Trade Relations

Living in the northern half of North America, Canadians have long recognized two facts: first, that the "grain of the land"—the topography, from the Appalachians to the Rockies—has a north–south orientation, thus facilitating north–south trade routes; and second, that exports to the huge US market are essential for Canadian industries to achieve **economies of scale** and therefore low per unit prices. For those reasons alone, **continentalism** has always lurked just below the surface of serious political thought. Consequently, trade between Canada and the United States remains a fundamental force in Canada's economy, so much so, in fact, that the value of Canadian–US trade remains the largest among any two countries in the world. More than that, Canada and the United States have been close allies for a long time. As President Kennedy remarked in his speech to Canada's Parliament in 1961: "Geography has made us neighbors. History has made us friends. Economics has made us partners. And necessity has made us allies."

When its major trade partner's economy started to sputter beginning in 2008–9, Canada was forced to look to other world markets. The drop in exports to the United States was dramatic. The value of Canada's total exports fell from 2006 to 2010 by nearly 10 per cent while its US exports dropped by 17 per cent (Statistics Canada, 2011). Not surprisingly, the federal government was scrambling for trade agreements with other countries to make up the loss of exports south of the border. As shown in Figure 1.7, Canada's exports

THINK ABOUT IT
Once the American economy regains its strength, would it be a wise or foolish policy to continue efforts to expand trade with other countries?

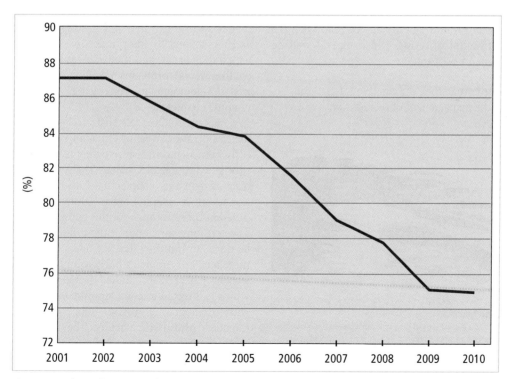

Figure 1.7 **Share of exports to the United States, 2001–2010**
Source: Statistics Canada (2011).

to the US hit a high-water mark of 87 per cent in 2001 but then declined to just under 75 per cent by 2010. Until the US economy recovers, exports to the US will remain in the doldrums.

Why is this trade so strong? The primary reason is that North America provides a natural economic trade zone. A second factor is that the economies of the two countries complement each other. For example, the United States requires large quantities of Canadian energy and commodities, plus the integration of the automobile industry generates much cross-border trade. Canada is anxious to regain former levels of trade with the United States, which, of course, is dependent on a US economic recovery. Trade between the two countries grew after the signing of the FTA in 1989, but slowed sharply following the 2008–9 economic collapse (Figure 1.8). A return to those heady days when 80 per cent or more of Canada's exports moved south are long gone. Another barrier to that level of trade remains trade disagreements—in the last decade, such disputes have involved lumber, grain, and beef. Trade disputes sometimes turned nasty.[5] Still, Derek Burney, former Canadian Ambassador to the United States, and policy expert Fen Hampson (2012:A13) express the significance of the long-standing Canada–US relationship in the strongest of terms, which extend well beyond matters of trade:

> The cornerstone of Canada's foreign policy is the management of relations with the United States, not for reasons of sentiment but because that's how we preserve our most vital economic and security interests and our capacity for global influence.

More detail on the New International Trade Crossing is found in Chapter 5 in the section "Automobile Parts Firms," page 197.

Understanding Canada's Regions

As a prelude to understanding the many interrelated physical and human dimensions of Canada's geographic regions, the next three chapters focus on Canada's physical geography, history, and human geography. Chapter 2 ("Canada's Physical Base") explores the wide range of natural and geomorphic processes that account for wide variations in landforms across Canada. Seven physiographic regions are described with brief accounts of their geology, soils, natural vegetation, and climate. A major theme is the physical environment's influence on human occupancy of the land and the effects of human occupancy on the physical environment. Chapter 2, therefore, is more than a mere backdrop for the remainder of this text; rather, it illuminates how human activities and human decision-making both shape and are shaped by the physical environment in which they occur, and it introduces the extent of environmental problems resulting from industrial development, as described in each of the regional chapters.

Chapter 3 ("Canada's Historical Geography") focuses on the arrival of Canada's First Peoples, early French and British settlements, and the territorial evolution of Canada. The four faultlines discussed in this chapter have affected Canada and its six geographic regions differently and at different times throughout this country's history. The tensions surrounding these faultlines often lie dormant but then flare up unexpectedly, and these tensions provide a background for the subject of Chapter 4 ("Canada's Human Face"), which is directed to contemporary issues dealing with demography (baby boom and bust), Canada's **pluralistic society** (the impact of a more open immigration policy), a continental economy (the effect of the Canada–US Free Trade Agreement), and the global economy (the effect of free trade agreements with other countries). This chapter also looks at the population trends

Figure 1.8 An eagle with clipped talons
A "hurting" American bald eagle symbolizes the sad state of the American economy following the 2008–9 global crisis. By 2012, the American economy, hobbled by high unemployment rates and congressional failures to act, continued to struggle. Imports from Canada, such as lumber and a variety of manufactured goods, declined sharply, forcing Ottawa to reassess its international trading policies. The continental approach based on exports to the United States is under review while efforts to diversify Canadian exports to Asian countries have taken centre stage in Ottawa. *Source:* Simpson (2011).

CONTESTED TERRAIN 1.2

Ambassador Bridge and the Proposed New International Trade Crossing

Built in 1930, the Ambassador Bridge over the Detroit River connecting Windsor, Ontario, and Detroit, Michigan, is a vital transportation link between Canada and the United States. An estimated one-quarter of the merchandise trade between the two countries travels over this bridge. The Ambassador Bridge is privately owned by a Detroit-based billionaire and his family. They strongly opposed the construction of the New International Trade Crossing, a proposed six-lane span financed by public funds with the Canadian government playing a leading role. From their perspective, another bridge may cut their profits or even result in losses. On the other hand, Ottawa sees the new bridge as Canada's most important infrastructure project in Ontario and one that will facilitate trade, including auto parts and vehicles, with the United States (Potter, 2010). The new bridge, south of the Ambassador Bridge, will tie in directly with Highway 401, thus bypassing the congested downtown streets of Windsor.

Photo 1.4

The Ambassador Bridge over the Detroit River is the most important trade artery between Canada and the United States. Yet, the volume of traffic, including that associated with the automobile industry, is simply too much for a single bridge. Plans to build a second span, known as the New International Trade Crossing, should relieve the traffic congestion at this critical border crossing as well as handle the expected increase in vehicle crossings in the years to come. With truck crossings alone expected to more than double by 2035—to 6 million trucks a year—a second bridge, funded largely by Ottawa, is welcomed (Chase and Keenan, 2012). On 12 April 2013 US President Barack Obama gave the official green light to build the bridge—the final political approval required for broadening this vital trade conduit between the US and Canada.

associated with the tensions that arise from centralist and decentralist visions for Canada, French and English views of "nation," immigration and Canada's pluralistic society, and claims to the land and their place on it from Aboriginal populations.

SUMMARY

Canada's six regions provide a vehicle to explore the geographic essence of Canada. To simplify the complexities of space, regional geographers divide the world and countries into regions, which vary by scale but often are interrelated in a hierarchical order. A regional geographer selects critical physical, historic, and human characteristics that logically divide a large spatial unit into a series of regions. Towards the margins of a region, its main characteristics become less distinct and merge with those of a neighbouring region. For that reason, boundaries are best considered as transition zones. Canada is a country of regions. Shaped by its history and physical geography, Canada is distinguished by six geographic regions: Ontario, Québec, British Columbia, Western Canada, Atlantic Canada, and the Territorial North. Within Canada and its regions, four key tensions exist that, through their interaction, demonstrate the very essence of Canada as a "soft" nation, where conflicts are usually resolved or ameliorated through compromise rather than by political or military power.

The essential foundation for studying regional geography is to conceptualize places and regions as components of a constantly changing global system. The core/periphery model provides an abstract spatial framework for understanding the general workings of the modern capitalist system. It consists of an interlocking set of industrial cores and resource peripheries. This model can function at different geographic scales and serves as an economic framework for interpreting Canada's regional nature. In addition to the core region, three types of regions devised by Friedmann extend our appreciation of the diversity of the Canadian periphery. They are: (1) a rapidly growing region, (2) a slow growing region, and (3) a resource frontier.

Time affects geographic regions. Perhaps the most dramatic events to affect Canada and its regions have been (1) changes to immigration regulations, (2) the shift over time from a closed economy to an open one, and (3) the dominating impact of China on the global economy, and, in turn, Canada and its regions.

Trade clearly dominates the regional economies in Canada, but its impact is far from even. On the one hand, resource-rich provinces are benefiting from high prices due to growing world demand for energy and **primary products**. On the other hand, Québec and Ontario manufacturers are facing stiff competition from other countries where costs of labour are extremely low.

CHALLENGE QUESTIONS

1. What does Saul mean by referring to Canada as a "soft" country?

2. Given that culture is an essential element of human societies, does the concept of "placelessness" reflect favourably or unfavourably on the future well-being of the diversity of global cultures?

3. Does the core/periphery model measure up to Nobel laureate Paul Samuelson's observation that "Every theory, whether in the physical or biological or social sciences, distorts reality in that it oversimplifies. But if it is a good theory, what is omitted is outweighed by the beam of illumination and understanding thrown over the diverse empirical data."

4. De Blij and Murphy state that "Geography is destiny." To what extent and how does this imply that if you were born and raised on a First Nation reserve in a remote part of northern Ontario, you would have a different set of life opportunities compared to those who were raised in a large Canadian city like Toronto?

5. Trade defines Canada. Should Ottawa continue its efforts to find new places for its exports or should it focus on regaining its past levels of exports to the United States?

FURTHER READING

Heap, Alan. 2005. "China—The Engine of a Commodities Super Cycle," Citigroup, 31 Mar. At: <www.fallstreet.com/Commodities_China_Engine0331.pdf>.

When the Peoples' Republic of China was admitted to the World Trade Organization in December 2001, global trade was radically changed as Chinese manufactured goods became able to penetrate markets around the globe. Canada and its main manufacturing provinces of Ontario and Québec felt the brunt of low-cost Chinese goods displacing local manufacturers. In turn, China's industrial capacity, fuelled in part by its ever increasing volume of exports, created a commodities super cycle.

The first economist to recognize this fundamental shift in global economic power was Alan Heap. In 2005, Heap predicted a prolonged rise in commodity prices driven by the unprecedented economic growth in China. As Heap pointed out in his presentation to the Mineral Economics and Management Society annual conference in Washington, DC, in April 2005, prior to the current super cycle, "There have been two super cycles in the last 150 years: late 1800s–early 1900s, driving economic growth in the USA; 1945–1975, prompted by post-war construction in Europe and by Japan's later, massive economic expansion." In all three cycles, rapid economic growth created a high demand for energy, minerals, and other commodities, thus pushing their prices higher. In the case of Canada, resource-rich provinces led by Alberta benefited from the super cycle. But like all economic cycles, what goes up, eventually comes down. This question was raised by Heap and his answer was based on past events, namely, the time span of the previous super cycles. Accordingly, Heap roughly estimated that the present super cycle, like the previous two, will last for several decades, coming to an end when China's economic expansion slows to a crawl. While Heap indicated that China's economic surge received a huge boost from its membership in the World Trade Organization in 2001, China's industrialization began at least a decade earlier, which means China has entered its third decade of rapid economic expansion.

2

CANADA'S PHYSICAL BASE

Introduction

The earth provides a wide variety of natural settings for human beings. For that reason, physical geography helps us understand the regional nature of our world. The basic question posed in this chapter is: *Why is Canada's physical geography so essential to an understanding of its regional geography?* Much of the answer lies in the spatial character of Canada as exemplified by its seven physiographic regions.[1] Physical geography, but especially our northern location, climate, and physiography, poses an underlying framework that moulds Canada's national and regional character. For instance, physical geography provides a fundamental explanation for the location of Canada's **ecumene**, illustrating that Canada's population hugs a narrow zone just north of the border with the United States, leaving the rest of the country sparsely populated. Extending this argument, one can see that population differences, in both size and distribution, between Canada and the United States can be attributed, in part, to their very different physical geographies (Vignette 2.1).

In this text, physical geography provides the raison d'être for the basis of the core/periphery model. The argument is a simple one: regions with a more favourable physical base are more likely to develop into core regions that contain large populations. Regions with less favourable physical conditions have fewer opportunities to encourage settlement and economic development. As pointed out in Chapter 1, circumstances defining a favourable physical base can change over time and such changes have the power to alter the prospects for regional expansion and contraction.

CHAPTER OVERVIEW

This chapter provides a basic introduction to Canada's physical geography, emphasizing how it has shaped the regional nature of Canada. At the same time, Chapter 2 lays the foundation for our discussion of the six regional chapters. In Chapter 2 we will examine the following topics:

- The geological structure, origins, and characteristics of Canada's physical base and its seven physiographic regions.
- The nature of Canada's climate and its seven climatic zones.
- The concept of extreme weather events and how these shape regional consciousness.
- Canada's cold environment as illustrated by the presence of permafrost in two-thirds of Canada's land mass.
- The five drainage basins that empty vast quantities of fresh water into Canada's three oceans—the Arctic, Atlantic, and Pacific.
- Environmental challenges take the form of human impact on Canada's natural environment but the most pressing one for the twenty-first century is global warming.

Cap-aux-Meules, Les îles de la Madeleine, Québec. This small archipelago in the Gulf of the Saint Lawrence is composed of 12 islands. Six of the islands are connected by long, thin sand dunes. The cliffs featured in this image are susceptible to erosion from wind and rain, resulting in highly unstable though stunning formations. Photo: Allen McEachern/All Canada Photos.

Physical Variations within Canada

The physical geography varies across Canada. Climate provides one example. While Canada has a cold, northern climate, the types of climate vary from place to place with temperate climates in southern Canada and polar ones in northern Canada. The Maritimes, for instance, has a mild, wet climate, while the Arctic has a cold, dry climate. Climate also affects the shape of landforms (mountains, plateaus, and lowlands) through a variety of weathering and erosional processes. Millions of years of weathering and erosion have produced the Appalachian Uplands, which in distant geological times was a young, rugged mountain range similar in many ways to the Rocky Mountains. Major landforms are the basis of seven physiographic regions in Canada and they illustrate the regional distinctiveness of Canada's physical geography. For instance, the flat to gently rolling topography of the Prairies is totally different from that found in the Canadian Shield, which consists of rugged, rocky, hilly terrain.

Geographers perceive an interaction between people and the physical world. This interactive two-way relationship is a fundamental component of regional geography. Favourable physical conditions can make a region more attractive for human settlement. The combination of a mild climate and fertile soils in the Great Lakes–St Lawrence Lowlands encourages agricultural settlement, while the St Lawrence River and the Great Lakes provide low-cost water transportation to local, American, and world markets. The favourable physical features of this region have allowed it to become Canada's industrial heartland.

As scientists who study the spatial aspects of nature and the processes that shape nature, physical geographers are concerned with all aspects of the physical world: **physiography** (landforms), bodies of water, climate, soils, and natural vegetation. Regional geographers, however, are more interested in how physical geography varies and subsequently influences human settlement of the land. The Rocky Mountains, for instance, offer few opportunities for agricultural settlement, but the spectacular scenery has led to the emergence of an economy based on tourism. Nature often works slowly and change may take centuries, but nature can work quickly. Floods and storms have had sudden and dramatic impacts on human occupation of the land while climate warming is an example of relatively rapid change to our environment that is affecting the human landscape.

Regional geographers also are concerned about the effect of human activities on the natural environment. In most cases, humans have a negative impact on the environment. For example, within the Bow Valley of the Rocky Mountains, extensive

VIGNETTE 2.1

Two Different Geographies

Canada and the United States occupy the northern and central parts of North America, yet the two countries have strikingly different geographies. Canada, while larger in geographic area, has a much smaller area suitable for agriculture and settlement. A significant portion of Canada lies in high latitudes where polar climates and permafrost place these lands far beyond the limits of commercial agriculture and settlement. Consequently, most Canadians live in a narrow zone close to the border with the United States (see Figure 4.1). Here, more temperate climates prevail. Geography, therefore, has been kinder to the United States, if this is to be judged by carrying capacity—how many people the land can sustain—and economic potential. The US simply has more suitable physical space for settlement, which has allowed its population to reach 312 million by 2011, compared to 35 million for Canada. Thus, Canada has a population density of just under 4 people per square kilometre compared to 30 in the United States. This physical reality, best described as Canada's northern handicap, limits the areas suitable for settlement. Recent immigrants to Canada have recognized this geographic fact and most take up residence in one of Canada's three largest cities (Toronto, Montréal, and Vancouver), though the economic boom in Western Canada has turned the tables with many immigrants now settling in Calgary, Edmonton, Winnipeg, Saskatoon, and Regina.

land developments have reduced the size of the natural habitat of wild animals such as bears and elk. Ironically, if more land is converted into golf courses, resort facilities, and housing developments, the animals that make this wilderness region so unique and attractive to tourists may no longer be able to survive. Another example is urban sprawl, which has gobbled up some of Canada's best farmland in the Niagara Peninsula, the Fraser Valley, and the Okanagan Valley. In our contemporary world, therefore, humans are the most active and, some would say, the most dangerous agents of environmental change.

The discussion of physical geography in this chapter and in the six regional chapters is designed to provide basic information about the natural environment and its essential role in the regional geography of Canada. To that end, the following points are emphasized:

- Physical geography has distinct and unique regional patterns across Canada.
- Physiographic regions represent one aspect of this natural diversity.
- Climate, soils, and natural vegetation provide other natural components and spatial patterns and, in doing so, provide the basis for a wide range of biodiversity across Canada.
- Human activity is changing the natural environment into an urban industrial landscape as well as causing air, soil, and water pollution for which there are long-term negative implications for all life forms; this global pollution represents a major environmental challenge for the twenty-first century.
- Lastly, our natural world varies widely. The implication for Canadians is clear—certain areas are more conducive for agriculture and manufacturing, and this natural spatial variation accounts for Canada's ecumene.

We begin our discussion of physical geography by examining the nature of landforms.

The Nature of Landforms

The earth's surface features a variety of landforms. A simple classification of landforms results in three principal types: mountains, plateaus, and lowlands. These landforms are subject to change by various physical processes. Some processes create new landforms while others reduce them. From a geological perspective, the earth, then, is a dynamic planet, and its surface is actively shaped and reshaped over thousands and millions of years (see Vignette 2.2 for more on the earth's origin and types of rocks). From a human perspective, however, the earth is relatively stable with few changes

VIGNETTE 2.2

The Earth's Crust and Major Types of Rocks

The earth's crust, which forms less than 0.01 per cent of the earth and is its thin solidified shell, consists of three types of rocks: igneous, sedimentary, and metamorphic. When the earth's crust cooled about 4.5 billion years ago, **igneous rocks** were formed from molten rock known as magma. Some 3 billion years later, **sedimentary rocks** were formed from particles derived from previously existing rock. Through denudation (weathering and erosion), rocks are broken down and transported by water, wind, or ice and then deposited in a lake or sea. At the bottom of a water body, these sediments form a soft substance or mud. In geological time, they harden into rocks. Hardening occurs because of the pressure exerted by the weight of additional layers of sediments and because of chemical action that cements the particles together. Since only sedimentary rocks are formed in layers (called **strata**), this feature is unique to this type of rock. **Metamorphic rocks** are distinguished from the other two types of rock by their origin: they are igneous or sedimentary rocks that have been transformed into metamorphic rocks by the tremendous pressures and high temperatures beneath the earth's surface. Metamorphic rocks are often produced when the earth's crust is subjected to folding and faulting. Lava from volcanoes constitutes a metamorphic rock and basalt is a product of the cooling of a thick lava flow.

CONTESTED TERRAIN 2.1

The Northern Gateway Pipeline

Changes to the earth's surface take place every day. Natural changes, such as earthquakes and volcanoes, are beyond the control of human beings. On the other hand, industrial projects can be approved, modified, or rejected. The economic attraction of industrial projects provides the basis for their approval while potential environmental damage may outweigh those economic gains, causing the project to be rejected. Moving from the general to the specific, the proposed Northern Gateway Pipeline project was declared by the Prime Minister as "in the national interest" because it would provide access to Asian markets for Alberta's oil sands. Yet, First Nations and environmental organizations are strongly opposed, fearing that a bitumen spill could cause irreparable damage to the land and sea. On this contested terrain, which "camp" do you favour and why?

THINK ABOUT IT
In Chapter 1, the term "faultline" is used. How is this geographic term similar to yet different from the geologic term "fault line"?

observable over a person's the lifetime. For instance, the Appalachian Uplands in Atlantic Canada and Québec are undergoing the very slow process known as **denudation**, which is the gradual wearing down of mountains by erosion and weathering over millions of years. How did this happen? First, over millions of years, **weathering** broke down the solid rock of these ancient mountains into smaller particles. Second, **erosion** transported these smaller particles by means of air, ice, and water to lower locations where they were deposited. The result was a much subdued mountain chain from what once resembled the Rocky Mountains. Denudation and **deposition**, then, are constantly at work and, over long periods of time, dramatically reshape the earth's surface.

Physiographic Regions

The earth's surface can be classified into a series of physiographic regions. **A physiographic region** is a large area of the earth's crust that has three key characteristics:

- It extends over a large, contiguous area with similar relief features.
- Its landform has been shaped by a common set of geomorphic processes.
- It possesses a common geological structure and history.

Canada has seven physiographic regions (Figure 2.1). The Canadian Shield is by far the largest region, while the Great Lakes—St Lawrence Lowlands is the smallest. Perhaps the most spectacular and varied **topography** occurs in the Cordillera, while the Hudson Bay Lowlands has the most uniform **relief**. The remaining three regions are the Interior Plains, Arctic Lands, and Appalachian Uplands. Most significantly, three of these physiographic regions—the Cordillera, Interior Plains, and Appalachian Uplands—display a strong north–south orientation to the topography of North America.

Each physiographic region has a different geological structure. These structural differences have produced a particular set of mineral resources in each physiographic region. For example, formed from the solidification of the earth's crust about 4.5 billion years ago, the Precambrian crystalline rock that makes up the Canadian Shield contains deposits of copper, diamonds, gold, nickel, iron, and uranium. Other physiographic regions were formed much later, as shown in the geological time chart (Table 2.1). The formation of the Interior Plains began about 500 million years ago when ancient rivers deposited sediment in a shallow sea that existed in this area. Over a period of about 300 million years, more and more material was deposited into this inland sea, including massive amounts of vegetation and the remains of dinosaurs and other creatures. Eventually, these deposits were solidified into layers of sedimentary rock 1 to 3 km thick. As a result, the Interior Plains have a sedimentary structure that contains vast oil and gas deposits plus the Alberta oil sands, which have propelled Canada into the leading ranks of global oil producers. Furthermore, as these regions developed their energy and mineral resources, differences in regional economies began to take shape and these differences were magnified by the global

CANADA'S PHYSICAL BASE

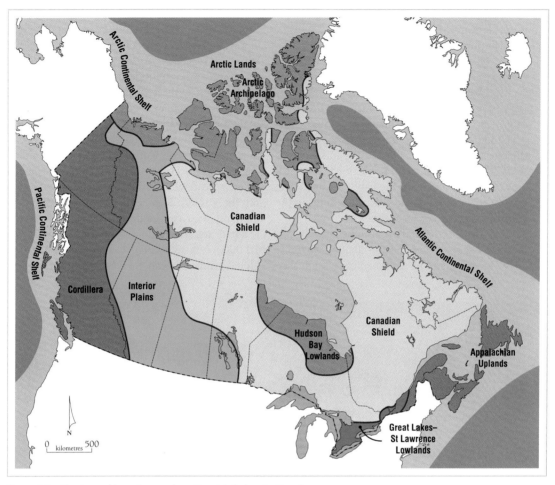

Figure 2.1 **Physiographic regions and continental shelves in Canada**
The seven physiographic regions are all different. The Arctic Lands is the most complex, consisting of a dozen large islands and numerous small islands that have been subjected to various geological events resulting in a mix of lowlands, uplands, and mountains. Together, these islands are known as the Arctic Archipelago. The Canadian Shield is the largest physiographic region and it extends beneath the Interior Plains, the Hudson Bay Lowlands, and the Great Lakes–St Lawrence Lowlands. The Cordillera and the Appalachian Uplands are products of plate tectonic activities—in the former case, less than 200 million years ago. The mountains of the Appalachian Uplands, on the other hand, formed nearly 500 million years ago.

Table 2.1 **Geological Time Chart**

Geological Era	Geological Time (millions of years ago)	Physiographic Region(s) Formed
Precambrian	600 to 4,500	Canadian Shield
Paleozoic	250 to 600	Appalachian Uplands, Arctic Lands
Mesozoic	100 to 250	Interior Plains
Cenozoic	0 to 100	Cordillera
Quaternary	0 to 2.5	The Great Lakes–St Lawrence Lowlands
Pleistocene	0.01 to 2.5	Hudson Bay Lowlands
Holocene	0.01 to present	

economy, which greatly increased demand (and prices) for these subterranean resources.

Each physiographic region has its own topography. The most dramatic difference is between the mountainous Cordillera and the relatively flat Hudson Bay Lowlands and, to a less degree, the Interior Plains. The surface material found in each region varies in hardness and thus resistance to erosional forces. Another factor affecting the rate of erosion is that erosion agents, such as wind, water, and ice, are more active in some regions than others. The Arctic Lands, for instance, are frozen for most of the year, thus limiting the activities of all erosional agents.

One common natural event—the last advance of the Wisconsin ice sheets—shaped virtually all of the topography of Canada. This advance marked the peak of the last ice age. It began some 30,000 years ago and represents the end of the **Pleistocene epoch** (Table 2.1). The late Wisconsin ice advance consisted of two major ice sheets, the Laurentide and the Cordillera. The Laurentide Ice Sheet was centred in the Hudson Bay area. As its mass increased, the sheer weight of the ice sheet caused it to move, eventually covering much of Canada east of the Rocky Mountains. In the Cordillera, a series of alpine glaciers coalesced into the Cordillera Ice Sheet, which spread westward into the continental shelf off the Pacific coast and eastward, eventually merging with the Laurentide Ice Sheet, which reached its maximum southern extent about 18,000 years ago (Figure 2.2). Gradually, the global climate began to warm and the grip of

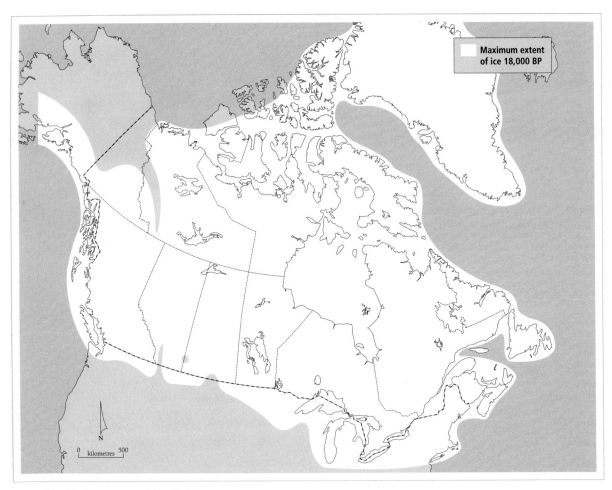

Figure 2.2 **Maximum extent of ice, 18,000 BP**

The last advance of the late Wisconsin ice age (the combined Laurentide and Cordillera ice sheets) covered almost all of Canada and extended into the northern part of the United States around 18,000 BP. Geologists have believed that the present "warm" climate is an interlude before the next ice age, but climate scientists see the current warming as a product of industrialization. (For more on global warming, see the discussion on this topic later in this chapter under the subheading, "Canada and Global Warming.")

colossal ice sheets weakened. Ice sheets retreated first in the Interior Plains and much later in Ontario and Québec. The major remnants of these massive ice sheets are found as glaciers in the Cordillera and Arctic Lands.

Glaciation from these two huge ice sheets radically altered the geography of Canada. Glacial scouring and deposition took place everywhere (see photo 2.1). The Laurentide Ice Sheet slowly pushed southward across the Canadian Shield, stripping away its surface material and depositing it much further south. When the ice sheet began to melt, it deposited material in situ, and meltwaters formed glacial lakes, including Lake Agassiz. As the world's greatest glacial lake, it created the nearly flat topography of the Manitoba Lowland in Western Canada. The Great Lakes provide another example of the impact of glaciation. The bottoms of these lakes were formed by glacial scouring and then filled with meltwater from the receding ice sheet. Only a few remnants remain of these huge ice sheets. Today, the largest glaciers and ice fields are in the mountains of Ellesmere Island.

The Canadian Shield

As noted previously, the Canadian Shield is the largest physiographic region in Canada. It extends over nearly half of the country's land mass. The Canadian Shield also forms the ancient geological core of North America. More than 4.5 billion years ago, molten rock solidified into the Canadian Shield (Table 2.1). Today, these ancient Precambrian rocks are not only exposed at the surface of the Shield but also underlie many of Canada's other physiographic regions. Beneath this core rock is the molten heart of our planet.

The rock-like surface of the Canadian Shield consists mainly of a rugged, rolling upland. Shaped like an inverted saucer, the region's lowest elevations are along the shoreline of Hudson Bay, while its highest elevations occur in Labrador and on Baffin Island, where the most rugged and scenic landforms of the Canadian Shield are found. The Torngat Mountains in northern Labrador, for instance, provide spectacular scenery with a coastline of fjords (photo 2.2). These mountains reach elevations of 1,600 m, making them the highest land east of the Rocky Mountains. The water divide of the Torngat Mountains represents the political boundary between northern Québec and Labrador.

Photo 2.1
Beyond the treeline, the rugged nature of the Canadian Shield, stripped of most overlying material, exposes bare bedrock on Melville Peninsula, Nunavut. As observed in the photograph, the Laurentide Ice Sheet dramatically altered the surface by scouring, scratching, and polishing its surface. Only a few rocks and boulders were deposited when the ice sheet melted. These boulders are called erratics.

As shown in Chapter 9 (photo 9.3, page 362), the Torngat Mountains form an impressive mountain chain extending in a north–south direction in Labrador and the adjacent area of Québec. While this boundary, defined by the watershed whereby some waters drain to Hudson Bay and others to the Atlantic Ocean, was established by the British Privy Council in 1927, Québec still has not recognized this decision (see Chapter 6, "Confederation, 1867–Present," page 234).

During the last ice advance, the surfaces of the Canadian Shield and those of other physiographic

Photo 2.2
Arctic landscape with tundra vegetation in the foreground, Saglek Fjord, Torngat National Park Reserve, Labrador.

regions were subjected to **glacial erosion** and deposition (Vignette 2.2). Glacial erosion and deposition are caused by giant ice sheets slowly grinding over the earth's surface (see photo 2.1). As the ice sheet moved over the surface of the Canadian Shield, the ice scraped, scoured, and scratched its massive rock surface. During the movement of an ice sheet, huge quantities of various loose materials such as sand, gravel, and boulders were trapped within the ice sheet. As the ice sheet reached its maximum extent, the edge of the ice sheet melted, depositing rocks, soil, and other debris. This debris is called **till**. Towards the end of the **ice age**, these ice sheets melted in situ, depositing whatever debris they contained. Sometimes the water from the melting ice was blocked from reaching the sea by the retreating ice sheet. These waters then formed temporary lakes. Once this ice was removed, these waters surged towards the sea.

Evidence of the impact of these processes on the surface of the Canadian Shield is widespread. Drumlins and eskers, both depositional landforms, are common to this region. **Drumlins** are low, elliptical hills (also called whalebacks or hogbacks) composed of till (material deposited and shaped by the movement of an ice sheet and subglacial megafloods), while **eskers** are long, narrow mounds of sand and gravel deposited by meltwater streams found under a glacier. There are also **glacial striations**, which are scratches in the rock surface caused by large rocks embedded in the slowly moving ice sheet.

The wealth of the Canadian Shield is in its vast and varied mineral resources. Along its southern fringe, huge deposits were sufficiently close to markets to permit exploitation. As an example, the rich nickel deposit near Sudbury has sustained that city for over 100 years and counting. In more remote areas, single-industry mining towns, such as the iron-mining town of Labrador City in Newfoundland and Labrador, were connected to global markets by rail and then sea transportation. While these towns flourished, those deposits beyond modern transportation networks employ a fly-in and fly-out labour force. For mines with a highly valued per unit product, such as diamonds, air transportation delivers the mineral to the marketplace. In the case of low-value per unit product, such as iron ore, the mine must have access to ocean transportation. The Raglan nickel mine in Arctic Québec falls into that category while the "Ring of Fire" mineral zone in Pre-Cambrian rocks in remote northern Ontario west of James Bay may well exceed the size and value of the Sudbury area, but transport issues have held up large-scale production except at the Victor diamond mine.

Like most of the Canadian Shield, the Laurentides, located just north of Montréal, contain many lakes and hills within a forested environment. Because of its close proximity to major cities in Québec, Ontario, and New England, the landscape, like that of the Muskoka region in Ontario, is principally exploited for recreation and tourism, with local residents and tourists enjoying these surroundings in both summer and winter.

For more on the Laurentides recreational tourist industry, see Chapter 6, "Tourism," page 248.

The Cordillera

The Cordillera, a complex region of mountains, plateaus, and valleys, occupies over 16 per cent of Canada's territory. With its north–south alignment, the Cordillera extends from southern British Columbia to Yukon; its western border is the Pacific Ocean (photo 2.3). The rugged nature of the Cordillera is illustrated in Figure 7.4, showing north–south aligned mountain ranges from the Vancouver Island Mountains on the Pacific coast to the Alberta border with the Rocky Mountains.

Plate tectonics played a critical role in the formation of the Cordillera. Beginning some 175 million years ago and ending around 85 million years ago, the Cordillera was created out of the horizontal sedimentary rocks of the North American Plate with final tectonic push crumbling these rocks into the Alberta foothills. This extremely slow movement of the two plates was caused by the almost imperceptible motion of the molten mass below the earth's crust. Severe **folding** and **faulting** of the North American Plate ensued and its flat-lying sedimentary rocks were twisted and broken into the series of mountains found in the Cordillera.

Along the **fault line** separating the Pacific and North American plates, tectonic movement continues, making the coast of British Columbia vulnerable to both earthquakes and volcanic activity. With the vast majority of the population and human-built environment of this region clustered along the coast in the cities of Vancouver, Victoria, New Westminster, and Nanaimo, the

THINK ABOUT IT
If you were Canada's first Prime Minister, John A. Macdonald, would you have seen the Canadian Shield as a bridge or a barrier to the Red River Settlement (now part of Manitoba)?

CANADA'S PHYSICAL BASE

Photo 2.3
Located along the Continental Divide between British Columbia and Alberta, the Athabasca Glacier forms part of the massive Columbia Icefield. Known as the "mother of rivers," the meltwaters from the Columbia Icefield nourish the Saskatchewan, Columbia, Athabasca, and Fraser river systems, the waters of which empty into three oceans—the Atlantic, Arctic, and Pacific oceans.

damage and loss of life from a major earthquake (measuring 7.0 or greater on the Richter scale) would be the worst natural disaster to strike Canada. The strongest earthquake ever recorded in Canada shook the sparsely populated Haida Gwaii (formerly called the Queen Charlotte Islands) in August 1949. This earthquake measured 8.1 on the Richter scale. Located along the Pacific Ring of Fire, a volatile region of active volcanoes, fault lines, and shifting tectonic plates, the densely populated Lower Mainland faces the threat of a powerful earthquake and possible tsunami sometime in the twenty-first century. The unknown question facing British Columbians is, when will the "Big One" strike?

In more recent geological times, the Cordillera Ice Sheet altered the landforms of the region. Over the last 20,000 years, alpine glaciation has sharpened the features of the mountain ranges in the Cordillera and broadened its many river valleys (Vignette 2.3). The Rocky Mountains are the best known of these mountain ranges. Most have

VIGNETTE 2.3

Alpine Glaciation, Glaciers, and Water for the Prairies

While glaciers still exist in the Rocky Mountains, they are slowly melting and retreating. During the late Wisconsin ice advance about 18,000 years ago, these glaciers grew in size and eventually covered the entire Cordillera. At that time, alpine glaciers advanced down slopes, carving out hollows called **cirques**. As the glaciers increased in size, they spread downward into the main valleys, creating **arêtes**, steep-sided ridges formed between two cirques. As these glaciers advanced, they eroded the sides of the river valleys, creating distinctive U-shaped glacial valleys known as **glacial troughs**. The Bow Valley is one of Canada's most famous glacial troughs. Cutting through the Rocky Mountains, the Bow Valley now serves as a major transportation corridor. It has also developed into an international tourist area. The centre of this tourist trade is the world-famous ski resort of Banff. While the many rivers but especially the South Saskatchewan River rely on these melting glaciers for fresh water that eventually flows to the cities and towns of the Canadian Prairies, as well as to irrigation works and industrial/mining operations, the concern is that, at some future time—perhaps in the last decades of the twenty-first century—these glaciers will disappear, thus greatly reducing the water supply for the Interior Plains.

Photo 2.4

In Kluane National Park Reserve in southwest Yukon, the St Elias Mountains are Canada's highest mountain range. Abundant snowfall from Pacific air masses, extremely high elevation, and latitudes above 60°N account for the extensive glaciers.

elevations between 3,000 and 4,000 metres. Their sharp, jagged peaks create some of the most striking landscape in North America. The highest mountain in Canada—at nearly 6,000 metres—is Mount Logan, part of the St Elias Mountain Range in southwest Yukon (photo 2.4).

The Interior Plains

The Interior Plains region is a vast and geologically stable sedimentary plain that covers nearly 20 per cent of Canada's land mass. This physiographic region is wedged between the Canadian Shield and the Cordillera, extending from the Canada–US border to the Arctic Ocean. Within the Interior Plains, most of the population lives in the southern area where a longer growing season permits grain farming and cattle ranching but low rainfall can affect yields and pastures. To those who are not native to this region, its topography often seems featureless and bright sunshine does not offset the frigid winter months.

Millions of years ago, a huge shallow inland sea occupied the Interior Plains. Over the course of time, sediments were deposited into this sea. Eventually, the sheer weight of these deposits produced sufficient heat and pressure to transform these sediments into sedimentary rocks. The oldest sedimentary rocks were formed during the Paleozoic era, about 500 million years ago (Table 2.1). Since then, other sedimentary deposits have settled on top of them, including those associated with the Mesozoic and Jurassic eras when dinosaurs roamed the earth. Unlike the Cordillera, the Interior Plains occupies a stable zone of the North American Plate where tectonic forces are not at play.

Tectonic forces associated with the collision of huge plates have had little effect on the geology of this region. For that reason, the Interior Plains is described as a stable geological region. For example, sedimentary rocks formed millions of years ago remain as a series of flat rock layers within the earth's crust. Geologists have used such sedimentary structures as geological time charts. In Alberta and Saskatchewan, rivers have cut deeply into these soft rocks, exposing Cretaceous rock strata. The Alberta Badlands provide an example of this rough and arid terrain (photo 2.5). Archaeologists have discovered many dinosaur fossils within these Mesozoic rocks in southern Alberta and Saskatchewan.

Beneath the surface of the Interior Plains, valuable deposits of oil and gas are in sedimentary structures called **basins**. Known as fossil fuels, oil and gas deposits are the result of the capture of the sun's energy by plants and animals in earlier geologic time. The storage of this energy in the form of hydrocarbon compounds takes place in sedimentary basins. The Western Sedimentary Basin is the largest such basin. Most oil and gas production in Alberta comes from this basin, with the oil sands playing an increasingly important role. Fossil fuels are non-renewable resources, meaning that they cannot regenerate themselves. Renewable resources, such as trees, can reproduce themselves.

As the Laurentide Ice Sheet melted and began to retreat from the Interior Plains about 12,000 years ago, the surface of the region was covered

CANADA'S PHYSICAL BASE

Photo 2.5
The sedimentary strata dating back to the late Cretaceous Period remain virtually undisturbed in the Alberta Badlands, but these horizontal strata have been exposed by stream erosion when vast meltwaters associated with the melting of the two ice sheets flowed through the Red Deer River and its tributaries. These quick-moving waters easily cut through soft sedimentary rocks of the Interior Plains to reach rock layers that date back to the days of dinosaurs, some 70 million years ago. The Dinosaur Trail that explores these badlands and the Royal Tyrrell Museum of Paleontology are located near Drumheller, Alberta.

with as much as 300 m of debris deposited by the ice sheet. Huge glacial lakes were formed in a few places. Later, the meltwater from these lakes drained to the sea, leaving behind an exposed lakebed. **Lake Agassiz**, for example, was once the largest glacial lake in North America and covered much of Manitoba, northwestern Ontario, and eastern Saskatchewan—its lakebed is now flat and fertile land that provides some of the best farmland in Manitoba but, as part of the Red River flood plain, is subject to frequent spring floods. When glacial waters escaped into the existing drainage system, they cut deeply into the glacial till and sedimentary rocks, creating huge river valleys known as **glacial spillways**.

Just north of Edmonton, the Interior Plains slopes towards the Arctic Ocean while east of Edmonton, the land tilts towards Hudson Bay and the Atlantic Ocean. Across this west–east cross-section of the Interior Plains, elevations decline from 1,600 m at Kicking Horse Pass of the Rocky Mountains to about 200 m near Lake Winnipeg and then to sea level at the mouth of the Nelson River. A south–north cross-section has a much smaller elevation drop—from 1,100 m at Yellowhead Pass to sea level at the mouth of the Mackenzie River. The principal rivers draining the Interior Plains are the northward-flowing Athabasca and Peace rivers, whose waters eventually enter the Mackenzie River and proceed to the Arctic Ocean, and the eastward-flowing North and South Saskatchewan rivers, which rise in the Rocky Mountains, join in central Saskatchewan, and empty into Lake Winnipeg.

Then, by the Nelson River, these waters drain into Hudson Bay.

Three sub-regions based on sharp changes in elevations take place within the Canadian Prairies: the Manitoba Lowland, the Saskatchewan Plain, and the Alberta Plateau. Typical elevations are 250 m in the Manitoba Lowland, 550 m on the Saskatchewan Plain, and 900 m on the Alberta Plateau. The Cypress Hills, which reach elevations of nearly 1,500 m, provide a sharp contrast to the surrounding flat to rolling terrain of the Alberta Plateau (Vignette 2.4).

The Hudson Bay Lowlands

The Hudson Bay Lowlands comprises about 3.5 per cent of the area of Canada. Underlain by the Canadian Shield, these lowlands consist of a thin cover of marine sediments deposited by the Atlantic Ocean some 10,000 to 12,000 years ago. The Hudson Bay Lowlands lies mainly in northern Ontario, though small portions stretch into Manitoba and Québec. This region extends from James Bay along the west coast of Hudson Bay to just north of the Churchill River. Permafrost is widespread and the northern half lies beyond the treeline.

Surface water is everywhere in the short summer but a frozen landscape exists in the long winter months. **Muskeg**, a type of peat, is the dominate ground cover, beneath which lies permafrost (photo 2.6). Low ridges of sand and gravel—remnants of former beaches of the **Tyrrell Sea**—separate these extensive areas of

Photo 2.6

The Hudson Bay Lowlands is a vast wetland where the lack of slope and the presence of permafrost restrict the development of a drainage system. Consequently, this lowland is dotted with myriad ponds and lakes. Muskeg prevails while black spruce occupies the higher, better-drained land made up of terraces (old sea beaches) and drumlins. The northern half of the Hudson Bay Lowlands lies beyond the treeline.

Reproduced with the permission of Natural Resources Canada 2013, courtesy of the Geological Survey of Canada (Photo Ice-wedge polygons and thermokarst ponds in peatlands, Hudson Bay Lowlands, Manitoba, "2001-124" by Lynda Dredge)

muskeg. Because of its almost level surface, the presence of permafrost, and its immature drainage system, the Hudson Bay Lowlands is poorly drained. Underneath the muskeg are recently deposited marine sediments mixed with glacial till.

The Hudson Bay Lowlands, by far the youngest physiographic region, was formed around 10,000 years ago by two events. First, a warmer climate appeared some 15,000 years ago, causing the Laurentide Ice Sheet to melt and recede. By 12,000 years ago, this area was free of ice but submerged beneath the Atlantic Ocean. This inland extension of the Atlantic Ocean over what is now the Hudson Bay Lowlands is called the Tyrrell Sea. With the tremendous weight of the ice gone from this submerged land, the second event began as the Hudson Bay Lowlands slowly rose above sea level.

This inland saltwater sea reached its maximum extent about 7,000 years ago, extending over much of the lowlands surrounding Hudson and James bays. As the earth's crust began to rise, the Tyrrell Sea retreated. This process is called **isostatic rebound** (Vignette 2.5). Slowly the isostatic rebound caused the seabed of the Tyrrell Sea to rise above sea level, thus exposing a low, poorly drained coastal plain (most of which is called the Hudson Bay Lowlands). This process of isostatic rebound has slowed over time from 600 cm per century to 100 cm per century. This relatively recent geomorphic process makes the Hudson Bay Lowlands the youngest of the physiographic regions in Canada (Table 2.1).

With few resources to support human activities, the region has only a handful of tiny settlements. From this perspective, the Hudson Bay Lowlands is one of the least favourably endowed physiographic regions of Canada. Moosonee (at the mouth of the Moose River in northern Ontario) and Churchill (at the mouth of the Churchill River in northern Manitoba) are the largest settlements in the region, each with a population of just over 1,000 people. These two settlements, formerly fur-trading posts, have an

VIGNETTE 2.4

Cypress Hills

The Cypress Hills, a sub-region of the Interior Plains, consist of a rolling plateau-like upland that is deeply incised by fast-flowing streams. Situated in southern Alberta and Saskatchewan, this area is the highest point in Canada between the Rocky Mountains and Labrador. The hills are an erosion-produced remnant of an ancient higher-level plain formed in the Cenozoic era (see Table 2.1). With a maximum elevation close to 1,500 m, these hills rise 600 m above the surrounding plain. During the maximum extent of the Laurentide Ice Sheet about 18,000 years ago, the Cypress Hills were enclosed by the ice sheet but the higher parts remained above the ice sheet. Known as nunataks, these areas served as refuge for animals and plants. As the alpine glacier melted, streams flowing from the Rocky Mountains deposited a layer of gravel up to 100 m thick on these hills.

Today, the Cypress Hills area is a humid "island" surrounded by a semi-arid environment and has entirely different natural vegetation compared to the area surrounding it. Unlike the grasslands, the Cypress Hills have a mixed forest of lodgepole pine, white spruce, balsam poplar, and aspen. The Cypress Hills also contain many varieties of plants and animals found in the Rocky Mountains. For the Plains Aboriginal peoples, these hills were and are a sacred place.

economic function as the termini of two northern railways (the Ontario Northland Railway and the Hudson Bay Railway, respectively). Regrettably, First Nation reserves have no similar economic functions and are a sad product of Ottawa's relocation policy of the 1960s. The plight of the First Nations of Attawapiskat and Kashechewan is, unfortunately, only the tip of the iceberg in this region of northern Ontario.

 For more on the subject of the Kashechewan reserve, see Chapter 5, page 212.

Arctic Lands

The Arctic Lands region stretches over nearly 10 per cent of the area of Canada. Centred in the Canadian Arctic Archipelago, this region lies north of the **Arctic Circle**. It is a complex composite of coastal plains, plateaus, and mountains. The Arctic Platform, the Arctic Coastal Plain, and the Innuitian Mountain Complex are the three principal physiographic sub-regions. The Arctic Platform consists of a series of plateaus composed of sedimentary rocks. This sub-region is in the western half of the Arctic Archipelago around Victoria Island. The Arctic Coastal Plain extends from the Yukon coast and the adjacent area of the Northwest Territories into the islands located in the western part of the Beaufort Sea. The third sub-region, the Innuitian Mountain Complex, is located in the eastern half of the Arctic Archipelago. It is composed of ancient sedimentary rocks. Like the Rocky Mountains, its sedimentary rocks were folded and faulted. However, unlike the Rocky Mountains, the plateaus and mountains in the Innuitian sub-region were formed in the early Paleozoic era (Table 2.1). During this geological time, volcanic activity took place as the world island broke into North America, Eurasia, and Africa, leaving behind vast areas of basaltic rocks exposed at the surface (photo 2.7). At 2,616 m, Mount Barbeau on Ellesmere Island is the highest point in the Arctic Lands region.

Photo 2.7
Basalt on Axel Heiberg Island, Nunavut. Basalt is a hard, black volcanic rock that, when cooled, can form various shapes, including tabular columns. Because they are resistant to erosion, basalt columns often form prominent cliffs. These weathered basalt columns date to the Paleozoic era (Table 2.1). At that geological time, North America, Greenland, and Eurasia broke into separate landmasses.

Across these lands, the ground is permanently frozen to great depths, never thawing, except at the surface during the short summer season. This cold thermal condition is called **permafrost**. Physical weathering, consisting mainly of differential heating and frost action, shatters bedrock and produces various forms of **patterned ground**. Patterned ground consists of rocks arranged in

---

VIGNETTE 2.5

Isostatic Rebound

At its maximum extent about 18,000 years ago, the weight of the huge Laurentide Ice Sheet caused a depression in the earth's crust. When the ice sheet covering northern Canada melted, this enormous weight was removed, and the elastic nature of the earth's crust allowed it to return to its original shape. This process, known as isostatic rebound or uplift, follows a specific cycle. As the ice mass slowly diminishes, the isostatic recovery begins. This phase is called a **restrained rebound**. Once the ice mass is gone, the rate of uplift reaches a maximum. This phase is called a **postglacial uplift**. It is followed by a period of final adjustment called the **residual uplift**. Eventually the earth's crust reaches an equilibrium point and the isostatic process ceases. In the Canadian North, this process began about 12,000 years ago and has not yet completed its cycle.

polygonal forms by minute movements of the ground caused by repeated freezing and thawing. Patterned ground and **pingos** (ice-cored mounds or hills) give the Arctic Lands a unique landscape.

The climate in this region is cold and dry. In the mountainous zone of Ellesmere Island, glaciers are still active. That is, as these alpine glaciers advance from the land into the sea, the ice is "calved" or broken from the glacier, forming icebergs. On the plains and plateaus, it is a polar desert environment. The term "polar desert" describes barren areas of bare rock, shattered bedrock, and sterile gravel. Except for primitive plants known as lichens, no vegetation grows. Aside from frost action, there are no other geomorphic processes, such as water erosion, to disturb the patterned ground.

Most people live in the coastal plain in the western part of this physiographic region. The three largest settlements are situated at the mouth of the Mackenzie River. Inuvik has a population of almost 3,000, while Aklavik and Tuktoyaktuk are smaller communities.

The Appalachian Uplands

The Appalachian Uplands region represents only about 2 per cent of Canada's land mass. Sometimes known as Appalachia, this physiographic region consists of the northern section of the Appalachian Mountains, though few mountains are found in the Canadian section. The Appalachians extend south in the eastern United States to northern Georgia and Alabama. With the exception of Prince Edward Island (Vignette 2.6), its terrain in Canada is a mosaic of rounded uplands and narrow river valleys. Typical Appalachian Uplands terrain is found in Cape Breton, Nova Scotia (photo 2.8). These weathered uplands are either rounded or flat-topped. They are the remnants of ancient mountains that underwent a variety of weathering and erosional processes over a period of almost 500 million years. Together, weathering and erosion (including transportation of loose material by water, wind, and ice to lower elevations) have worn down these mountains, creating a much subdued mountain landscape with **peneplain** features. The highest elevations are on the Gaspé Peninsula in Québec, where Mount Jacques Cartier rises to an elevation of 1,268 m. The coastal area has been slightly submerged; consequently, ocean waters have invaded the lower valleys, creating bays or estuaries, with a number of excellent small harbours and a few large ones, such as Halifax harbour. The island of Newfoundland consists of a rocky upland with only pockets of soil found in valleys. Like the Maritimes, it has an indented coastline where small harbours abound. The nature of this physiographic region favoured early European settlement along the heavily indented coastline where there was easy access to the vast cod stocks. With the demise of the cod stocks, these tiny settlements are declining or, like Great Harbour Deep, have been abandoned (see Vignette 9.6).

The Great Lakes–St Lawrence Lowlands

The Great Lakes–St Lawrence Lowlands physiographic region is small but important. Extending from the St Lawrence River near Québec City to Windsor, this narrow strip of land is wedged between the Appalachians, the Canadian Shield, and the Great Lakes.[2] Near the eastern end of Lake Ontario, the Canadian Shield extends across this region into the United States, where it forms the Adirondack Mountains in New York State. Known as the Frontenac Axis, this part of the Canadian Shield divides the Great Lakes–St Lawrence Lowlands into two distinct sub-regions.

VIGNETTE 2.6

Prince Edward Island

Unlike other areas of Appalachian Uplands, Prince Edward Island has a flat to rolling landscape. While the island is underlain by sedimentary strata, these rocks consist of relatively soft, red-coloured sandstone that is quickly broken down by weathering and erosional processes. Occasionally, outcrops of this sandstone are exposed, but for the most part the surface is covered by reddish soil that contains a large amount of sand and clay. The heavy concentrations of iron oxides in the rock and soil give the island its distinctive reddish-brown hue. Prince Edward Island, unlike the other provinces in this region, has an abundance of arable land.

CANADA'S PHYSICAL BASE 43

Photo 2.8
The Appalachian Uplands have sustained much erosion and the resulting landscape represents "worn-down" mountains. In rugged Cape Breton Highlands National Park, a table-like surface or peneplain lies between steep valleys carved by streams.

As the smallest physiographic region in Canada, the Great Lakes–St Lawrence Lowlands comprises less than 2 per cent of the area of Canada. As its name suggests, the landscape is flat to rolling. This topography reflects the underlying sedimentary strata and its thin cover of glacial deposits. In the Great Lakes sub-region, flat sedimentary rocks are found just below the surface. This slightly tilted sedimentary rock, which consists of limestone, is exposed at the surface in southern Ontario, forming the Niagara Escarpment. A thin layer of glacial and lacustrine (i.e., lake) material, deposited after the melting of the Laurentide Ice Sheet in this area about 12,000 years ago, forms the surface, covering the sedimentary rocks.

In the St Lawrence sub-region, the landscape was shaped by the Champlain Sea, which occupied this area for about 2,000 years. It retreated about 10,000 years ago and left broad **terraces** that slope gently towards the St Lawrence River (Vignette 2.7). The sandy to clay surface materials are a mixture of recently deposited sea, river, or glacial

VIGNETTE 2.7

Champlain Sea

About 12,000 years ago, vast quantities of glacial water from the melting ice sheets around the world drained into the world's oceans. Sea levels rose, causing the Atlantic Ocean to surge into the St Lawrence and Ottawa valleys, perhaps as far west as the edge of Lake Ontario. Known as the Champlain Sea, this body of water occupied the depressed land between Québec City and Cornwall and extended up the Ottawa River Valley to Pembroke. These lands had been depressed earlier by the weight of the Laurentide Ice Sheet. About 10,000 years ago, as the earth's crust rebounded, the Champlain Sea retreated. However, the sea left behind marine deposits, which today form the basis of the fertile soils in the St Lawrence Lowlands.

Photo 2.9

In the lower Great Lakes section of the Great Lakes–St Lawrence Lowlands, the last glaciation formed a hummocky moraine landscape with widespread till deposits. Stranded ice blocks from the last glaciation left behind depressions that became filled with water. The gently rolling landscape is underlain by limestone.

Joseph Gareri/iStock

materials. For the most part, this sub-region's soils are fertile, which, when combined with a long growing season, allows agricultural activities to flourish.

The physiographic region lies well south of the forty-ninth parallel, which forms the US–Canada border from west of Ontario to Vancouver Island. The Great Lakes sub-region extends from 42° N to 45° N, while the St Lawrence sub-region lies somewhat further north, reaching towards 47° N. As a result of its southerly location, its proximity to the industrial heartland of the United States, and its favourable physical setting, the Great Lakes–St Lawrence region is home to Canada's main ecumene and manufacturing core.

Geographic Location

Canada occupies the northern portion of North America. As the two different geographies of Canada and the United States reveal (Vignette 2.1), this geographic fact has had profound implications for the course of development in the two countries. Simply stated, Canada's cooler climate translates into less area suitable for agriculture and settlement than is the case in the US.

A measure of geographic location on the earth's surface is provided by latitude and longitude. Because of the size of Canada, latitude and longitude vary enormously. Taking longitude as an example, the longitudinal distance between the westernmost area of Canada (as represented by Whitehorse, Yukon, at 135°08′ west of the prime meridian in Greenwich, England) and the easternmost point (as represented by St John's at 52°43′ W) is nearly 83 degrees. In kilometres, the distance between the two cities is about 5,200 km. From a standard time perspective, six and a half time zones stretch across the country, from the Newfoundland Time Zone in the east to the Pacific Time Zone in the west (Figure 2.3).

How did the concept of standard time emerge? The need for time zones stems from the need to organize time to accommodate the scheduling of rapid transportation, such as for buses, railways, and air flights. Time zones provide a solution. In

THINK ABOUT IT
Which physiographic region do you live in?

fact, we have Sir Sanford Fleming, a Scottish-born Canadian civil engineer, to thank for inventing a system of standard time (as against individual time for each place as calculated from the sun). Fleming, in 1878, proposed the system of worldwide time zones based on the premise that since the earth rotates once every 24 hours and there are 360 degrees of longitude in a sphere, each hour the earth rotates one-twenty-fourth of a circle or 15 degrees of longitude.

Because the earth is a spherical body, this measure is given in degrees (°) and minutes (′) for both latitude and longitude. By **longitude**, we mean the distance east or west of the prime meridian. As the equator represents zero latitude, the prime meridian represents zero longitude. It is an imaginary line that runs from the North Pole to the South Pole and passes through the Royal Observatory at Greenwich, England. Canada lies entirely in the area of west longitude. Ottawa, for example, is 75°28′ west of the prime meridian. The distance between longitudes varies, being greatest at the equator and reaching zero at the North Pole.[3] Latitudes and longitudes for some other Canadian cities are found in Table 2.2.

By **latitude**, we mean the measure of distance north and south of the equator. For example, Ottawa is 45 degrees 24 minutes north of the equator. How do we translate these latitudes into an understanding of Canada's northern location and its cold environment? One way is to examine the southernmost latitudes near Windsor, Ontario, which is close to 42°N. Another way is to realize that over half of Canada lies north of

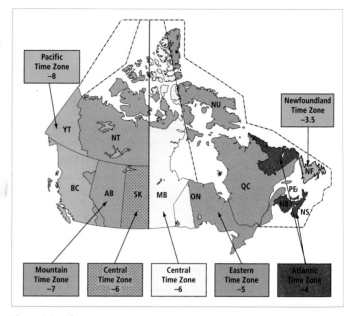

Figure 2.3 Time zones
Most of Saskatchewan observes Central Standard Time year-round. Lloydminster, which uses Mountain Standard Time and observes daylight saving time, is an exception. Some communities in Canada may choose not to observe official time zones and this map does not reflect all such variances. *Source:* TimeTemperature.com, at: <www.timetemperature.com/canada/canada_time_zone.shtml>.

the 60th parallel. Finally, we can look at the latitude of our capital. Ottawa is nearly 5,000 km north of the equator. Since the distance between each degree of latitude is about 111 km, the middle to high latitudes in Canada have considerable implications for the amount of solar energy received at the surface of the earth, and, hence, Canada's climate and its notorious short summers and long, dark winters.

Table 2.2 Latitude and Longitude of Selected Centres

Centre	Latitude	Longitude
Windsor, Ontario	42°19′ N	83° W
Montréal, Québec	45°32′ N	73°36′ W
St John's, Newfoundland and Labrador	47°35′ N	52°43′ W
Victoria, British Columbia	48°27′ N	123°20′ W
Winnipeg, Manitoba	49°53′ N	97°10′ W
Saskatoon, Saskatchewan	52°10′ N	106°40′ W
Edmonton, Alberta	53°34′ N	113°25′ W
Whitehorse, Yukon	60°42′ N	135°08′ W
Inuvik, Northwest Territories	68°16′ N	133°40′ W
Alert, Nunavut	82°31′ N	62°20′ W

Climate

Our physical world encompasses more than just landforms, physiographic regions, and geographic location. **Climate**, for instance, is a central aspect of the physical world. Climate describes average weather conditions for a specific place or region based on past weather over a very long period of time, perhaps thousands of years. On the other hand, weather refers to the current state of the atmosphere with a focus on weather conditions that affect people living in a particular place for a relatively short period of time. In sum, climate is what we can expect while weather is what we get. Geography students can gain firsthand knowledge of current and historic patterns in climate and weather by using Climate Trend Mapper (Vignette 2.8).

For each part of the world, climate is translated into regional versions known as climatic types and regions. This approach provides a much more detailed presentation of climate. Yet, before we discuss the spatial aspects of climate, a few general remarks are in order. First and foremost, Canadians live in a northern environment where summers are short and winters long. The length of these two seasons varies across the country, with the shortest winters found in the temperate climate of the middle latitudes and the longest winters in the Arctic climate of the high latitudes.

Winters dominate Canada's climate because of their length and temperature severity. "Coldness," wrote French and Slaymaker (1993: i), "is a pervasive Canadian characteristic, part of the nation's culture and history." They note that winter's effects include not only low absolute temperatures but also exposure to wind chill, snow, ice, and permafrost. The spatial extent of Canadian winter was captured on a NASA satellite image on 28 February 2009, which clearly and chillingly illustrates the firm grip of winter. Photo 2.10 shows the extent of snow cover and ice conditions on that day, revealing that only BC's narrow coastal littoral had escaped winter's grip and, thanks to a vigorous Pacific air mass, Vancouverites enjoyed mild temperatures.

As Canadians well know from personal experience, Canada has a cold environment. As seen on the map of climatic zones (Figure 2.4), the bulk of Canada's territory is associated with two northern climatic types, the Arctic and Subarctic zones. Both have extremely long, cold winters. Although the two maritime climatic types—along the coast of BC and in Atlantic Canada—have relatively short periods of cold weather, winter is a fact of life for Canadians. While many Canadians take winter holidays to tropical places, artists, musicians, and novelists have found this northern theme appealing. In "Mon pays," Gilles Vigneault, one of Québec's best-known chansonniers, refers to Québec, his country, in the opening line: *Mon pays ce n'est pas un pays, c'est l'hiver* (My country is not a country, it is winter). Such a sentiment resonates equally well in all regions of Canada. Another appealing theme expresses the physical challenge of a cold and harsh land faced by early explorers and settlers. Stan Rogers's heroic song, "Northwest Passage," captures that spirit. In fact, Rogers's moving lyrics have become one of Canada's unofficial anthems.

Climate Factors

Canada's climate is the product of many factors but three dominate. Most importantly, the energy from the sun sets the parameters. The amount of solar energy absorbed by the earth and its atmosphere is converted into heat. The amount of this energy received at the earth's surface varies by latitude. Low latitudes around the equator have a net surplus of energy (and therefore high temperatures), but in

---

VIGNETTE 2.8

Climate Trend Mapper

Professor Danny Blair and Ryan Smith at the University of Winnipeg have developed a climate program based on weather stations across Canada that allows the user to create temperature and precipitation maps. The data set now extends from 1840 to 2010, though very few weather stations existed until the twentieth century. For more information, visit their website at: <climate.uwinnipeg.ca/>.

CANADA'S PHYSICAL BASE

high latitudes around the North and South poles, more energy is lost through re-radiation than is received, and therefore annual average temperatures are extremely low. Canada, where settlements extend from 42° N (Windsor) to 83° N (Alert), experiences great variation in the amount of solar energy received (and therefore great variation in temperatures, climatic types, and zones).

Second, the **global circulation system** redistributes this energy (i.e., energy transfers) from low latitudes to high latitudes through circulation in the atmosphere (system of winds and air masses) and the oceans (system of ocean currents). For example, the Japan Current warms the Pacific Ocean, bringing milder weather to British Columbia. On Canada's east coast, the opposite process occurs as the Labrador Current

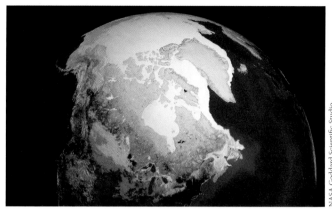

Photo 2.10

This NASA image shows Canada in the grip of winter, 28 February 2009. Canada is a northern nation with winter its dominant season. While most Canadians live close to the Canada–US border, where a more temperate climate prevails, global warming is shortening the annual length of snow coverage and extreme winter cold.

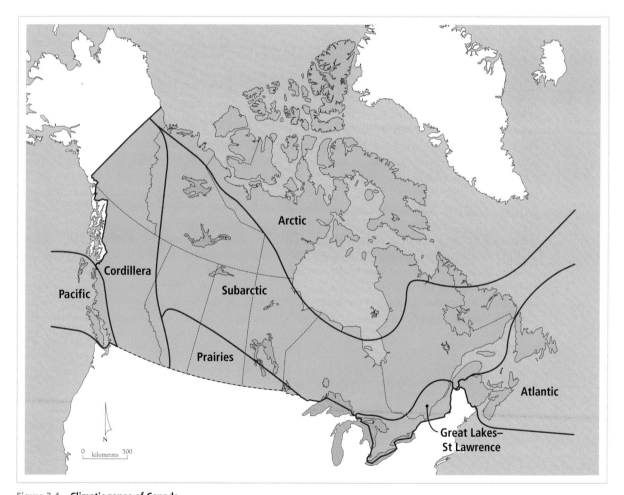

Figure 2.4 **Climatic zones of Canada**

Each climatic zone represents average climatic conditions in that area. Canada's most extensive climatic zone, the Subarctic, is associated with the boreal forest and **podzolic** soils. In spite of warmer annual temperatures over the short-term, boundaries of climatic zones remain unchanged.

Table 2.3 Climatic Types

Köppen Classification	Canadian Climatic Zone	General Characteristics
Marine West Coast	Pacific	Warm to cool summers, mild winters Precipitation throughout the year with a maximum in winter
Highland	Cordillera	Cooler temperature at similar latitudes because of higher elevations
Steppe	Prairies	Hot, dry summers and long cold winters Low annual precipitation
Humid continental	Great Lakes–St Lawrence Lowlands	Hot, humid summers and short, cold winters Moderate annual precipitation with little seasonal variation
Humid continental, cool	Atlantic Canada	Cool to warm, humid summers and short, cool winters
Subarctic	Subarctic	Short, cool summers and long, cold winters Low annual precipitation
Tundra	Arctic	Extremely cool and very short summers; long, cold winters Very low annual precipitation

Source: Adapted from Christopherson (1998); Hare and Thomas (1974).

brings Arctic waters to Atlantic Canada. While Halifax, at 44°39′ N, lies about 500 km closer to the equator than Victoria, at 48°27′ N, Halifax's winter temperatures, on average, are much lower than those experienced in Victoria.

Third, the global circulation system travels in a west-to-east direction in the higher latitudes of the northern hemisphere, causing air masses that develop over large water bodies to bring mild and moist weather to adjacent land masses. For Canada, air masses from the Pacific Ocean cross the Cordillera into the interior of Canada. Such air masses are known as **marine air masses**. In this way, energy transfers ultimately determine regional patterns of global weather and climate (see Tables 2.3 and 2.4). Air masses originating over large land masses are known as **continental air masses**. These air masses are normally very dry and vary in temperature depending on the season. In the winter continental air masses are cold, while in the summer they are associated with hot weather.

Canada experiences warmer and moister weather in its lower latitudes and colder and drier conditions in its higher latitudes. However, coastal areas (particularly the Pacific coast) experience smaller ranges of seasonal temperatures and more annual precipitation than do inland or

Table 2.4 Air Masses Affecting Canada

Air Mass	Type	Characteristics	Season
Pacific	Marine	Mild and wet	All
Atlantic	Marine	Cool and wet	All
Gulf of Mexico	Marine	Hot and wet	Summer
Southwest US	Continental	Hot and dry	Summer
Arctic	Continental	Cold and dry	Winter

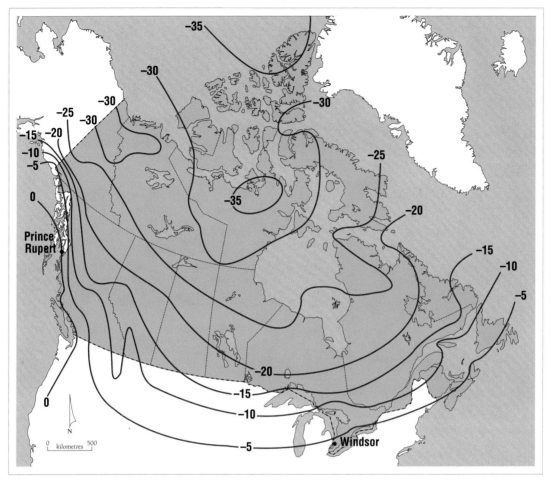

Figure 2.5 **Seasonal temperatures in Celsius, January**
The moderating influence of the Pacific Ocean and its warm air masses is readily apparent in the 0 to –5°C January isotherm. For example, Prince Rupert, located near 55° N, has a warmer January average temperature (0°C) than Windsor (–2°C), which is located near 42° N.

continental areas at the same latitude (Figures 2.5, 2.6, and 2.7). Winnipeg, for example, experiences a much greater daily and annual range in temperature than does Vancouver, even though both lie near 49° N. The principal reason is Vancouver's greater proximity to the ameliorating effects of the Pacific Ocean.

The so-called "continental effect" provides a final factor. Continental effect refers to the fact that land masses heat up and cool more quickly than oceans. In turn, greater distance from an ocean affects temperature and precipitation, that is, as distance from an ocean increases, the daily and seasonal temperature ranges increase and the annual precipitation decreases. For example, the continental position of the Canadian Prairies results in a much greater range of temperatures and much less precipitation than experienced at the same latitude on the Pacific coast of British Columbia.

Climatic Types and Zones

Since climate was relatively stable over thousands of year, the earth developed a series of climatic types and zones. But how many types and zones exist? It all depends on the criteria. In the nineteenth century, well before global climatic stations existed, Wladimir Köppen, a German scientist, developed a climatic classification scheme for the world based on natural vegetation zones. He assumed that these natural vegetation zones required certain temperatures and

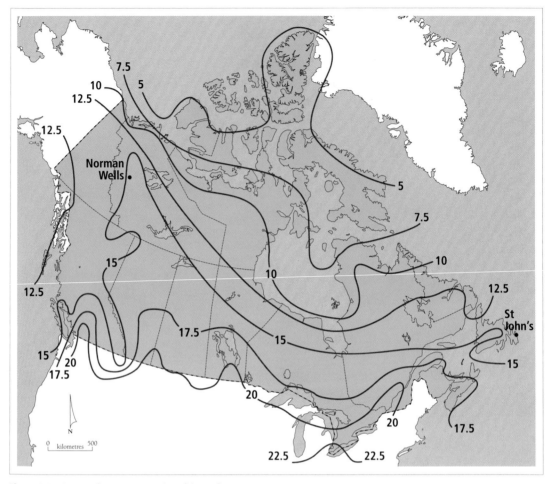

Figure 2.6 Seasonal temperatures in Celsius, July
The continental effect results in very warm summer temperatures that extend into high latitudes, as illustrated by the 15°C July isotherm. For example, Norman Wells, located near the Arctic Circle, has warmer July temperatures than St John's.

precipitation to thrive and therefore these zones were a surrogate for climatic types and zones. Based on world patterns of natural vegetation, Köppen created 25 climate types, each of which was assigned particular temperature regimes and seasonal precipitation patterns. Seven of Köppen's climatic types are found in Canada (Table 2.3).

A **climatic zone** is an area of the earth's surface where similar weather conditions occur.

THINK ABOUT IT
Global warming has begun but so far the boundaries of the Canadian climate zones shown in Table 2.3 have not been altered. Does this mean that global warming has not yet had sufficient impact on our climate or is there a lag by climatologists in updating these boundaries? From your personal experience, can you identify any major changes in the weather patterns in the climatic zone where you live?

Long-term data describing annual, seasonal, and daily temperatures and precipitation are used to define the extent of a climatic zone.

Canada has seven climatic zones (Figure 2.4): the Pacific, Cordillera, Prairies, Great Lakes–St Lawrence, Atlantic, Subarctic, and Arctic. The Subarctic zone is the largest climatic zone. It extends over much of the interior of Canada and is found in each geographic region. Though the Arctic climate exists along the Labrador coast and in the extreme northern reaches of Québec, the Subarctic climate prevails in the northern areas of Atlantic Canada, Québec, Ontario, and Western Canada, and it is present in northeast British Columbia. As well, the Subarctic climate is found in the Territorial North and is the principal climate in the Northwest Territories. The

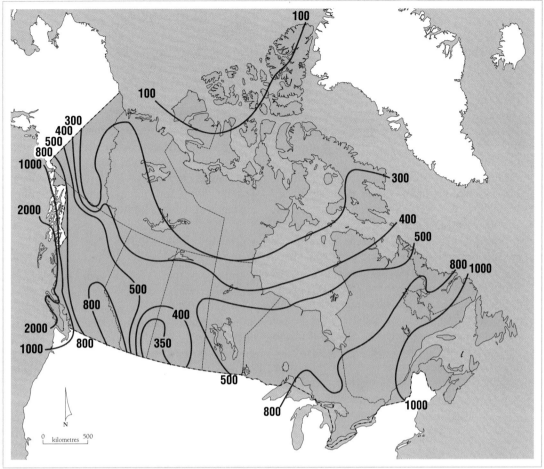

Figure 2.7 Annual precipitation in millimetres
The lowest average annual precipitation occurs in the Territorial North, indicating the dry nature of the Arctic air masses that originate over the ice-covered Arctic Ocean. The highest average annual precipitation occurs along the coast of British Columbia due to the moist marine air masses and the coastal mountains.

Subarctic climatic zone extends into much higher latitudes in northwest Canada than in northeast Canada because of warmer temperatures in the northwest. In northwest Canada, the average July temperature often reaches or exceeds 10°C, thus permitting the growth of trees. In similar latitudes of northeast Canada, summer temperatures are much lower. In the extreme north of Québec, for example, the average July temperature is below 10°C, thus resulting in tundra rather than a tree vegetation cover. The Subarctic climatic zone therefore has a southeast to northwest alignment (Figure 2.4). This alignment is somewhat modified in Québec because of its proximity to the marine influences of the Atlantic Ocean.

Air Masses

Air masses are large sections of the atmosphere with similar temperature and humidity characteristics. They form over large areas with uniform surface features and relatively consistent temperatures where they take on these temperature and humidity characteristics. Such areas are known as source regions. The Pacific Ocean is a marine source region, while the interior of North America is a continental source region. During a period of about a week or so, an air mass may form over a source region. Canada's weather is affected by five air masses associated with the northern hemisphere. For example, Pacific air masses bring mild, wet weather to British

THINK ABOUT IT
Which air mass dominates the summer climate where you live?

Columbia's coast for most of the year. These air masses are much stronger in the winter, so British Columbia normally experiences greater precipitation in winter than in summer. In some years, British Columbia can have a relatively dry summer. The general characteristics of the five major air masses affecting Canada's weather are shown in Table 2.4.

Air masses bring moisture from oceans to land bodies. Across Canada, precipitation is unevenly distributed (Figure 2.7). The lowest average annual precipitation occurs in the Territorial North, indicating the dry nature of the Arctic air masses originating over the ice-covered Arctic Ocean. The highest average annual precipitation is found in coastal British Columbia. Here, much of the precipitation falls as frontal and orographic rainfall due to two factors: (1) the warm Pacific Ocean serves as a source for eastward-moving Pacific air masses that often contain large quantities of water vapour, and (2) along the British Columbia coast, precipitation occurs either as the warm Pacific air mass rises over a colder one or as this same air mass must rise over the coastal mountain ranges. In both cases, the water vapour condenses and falls as rain or, at higher elevations, as snow. The three principal types of precipitation are discussed in Vignette 2.9.

THINK ABOUT IT
Assuming that global warming results in an average world temperature increase of 3 degrees Celsius by the end of the twenty-first century, what factor(s) might delay the altering of the boundaries of the existing global patterns of natural vegetation zones? On the other hand, do you think that boundaries of permafrost might respond more quickly?

Climate, Soils, and Natural Vegetation

As noted earlier, climate affects the development of soils and the growth of natural vegetation. In fact, the interdependency of climate, soils, and natural vegetation is so strong that physical geographers have identified an orderly and interrelated global pattern of climatic, soil, and natural vegetation zones. Recall that Köppen's original classification system for climate was based on world patterns of natural vegetation that had taken thousands of years to develop. This broad geographic relationship is revealed in Figures 2.4, 2.8, and 2.9. Climate determines to a large extent the **soil order** and native vegetation in a given region and hence influences land use, such as crop cultivation, forestry, or grazing. Together with topography, climate determines the land's suitability for human settlement.

Extreme Weather Events

Extreme weather events—such as blizzards, droughts, and ice storms—are also part of climate and often have very powerful impacts on humans. In fact, extreme weather events constitute the most serious of natural hazards, whether they take the form of droughts, floods, ice storms, or tornados. The Intergovernmental Panel on Climate Change, which assesses and synthesizes the research of more than 2,000 climate scientists throughout the world, foresees an increase in extreme weather events because of rising world temperatures: a warmer atmosphere would have the capacity to hold increased moisture, thus supplying the fuel for heavier rainfalls, snowfalls, tornados, and other extreme weather events.

Often, extreme weather events occur with little warning and result in heavy losses of property and sometimes lives. Hurricanes are such extreme

VIGNETTE 2.9

Types of Precipitation

As an air mass rises, its temperature drops. This cooling process triggers condensation of water vapour within the air mass. With sufficient cooling, water droplets are formed. When these droplets reach a sufficient size, precipitation begins. Precipitation refers to rainfall, snow, and hail. There are three types of precipitation. **Convectional precipitation** results when moist air is forced to rise because the ground has become particularly warm. Often this form of precipitation is associated with thunderstorms. **Frontal precipitation** occurs when a warm air mass is forced to rise over a colder (and denser) air mass. **Orographic precipitation** results when an air mass is forced to rise over high mountains. However, as the same air mass descends along the leeward slopes of those mountains (that is, the slopes that lie on the east side of the mountains), the temperature rises and precipitation is less likely to occur. This phenomenon is known as the **rain shadow effect**.

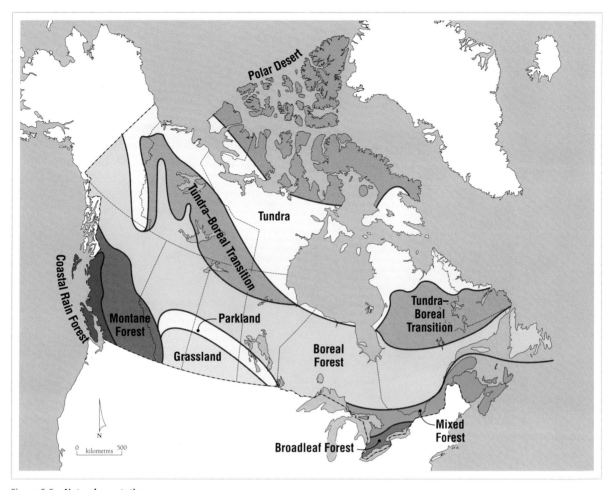

Figure 2.8　**Natural vegetation zones**
These natural vegetation zones have "core" characteristics, which diminish towards their edges. Transitions exist between natural vegetation zones. Two major transition zones shown here are the Tundra–Boreal Transition and the Parkland.

weather events. Atlantic Canada has been the site of many of these destructive tropical cyclones. One recent storm took place on 29 September 2003, when Halifax felt the full fury of Hurricane Juan. As Conrad (2009:163–5) notes, the impact was devastating, with power outages in Nova Scotia and Prince Edward Island lasting for up to two weeks and damage estimates ranging from $100 million to $150 million; eight deaths occurred. In addition, the well-treed Point Pleasant Park in Halifax was particularly hard hit and lost most of its trees. Conrad (2009: 1) places such weather in a broader context:

> Canada's climate is typified by extremes, and thus Canadians are interested in the weather out of necessity and concern. With the inevitable changes in our global climate, scientists as well as the general public are concerned with the impact such change will have on extreme weather events in Canada.

Not surprisingly, extreme weather events often have a cultural impact by providing a common threat and, as people struggle against this threat, creating a common bond. As indicated in Chapter 1, natural disasters have contributed to people's sense of belonging to a region, and often they are recurring phenomena, such as floods in a flood-prone area. As de Loë (2000: 357) explains, "Floods are considered hazards only in cases where human beings occupy floodplains and shoreland." Heavy rainfall combined with rapid snowmelt often triggers catastrophic floods. An excellent example is found in the flat Manitoba Lowland where the normally benign Red River

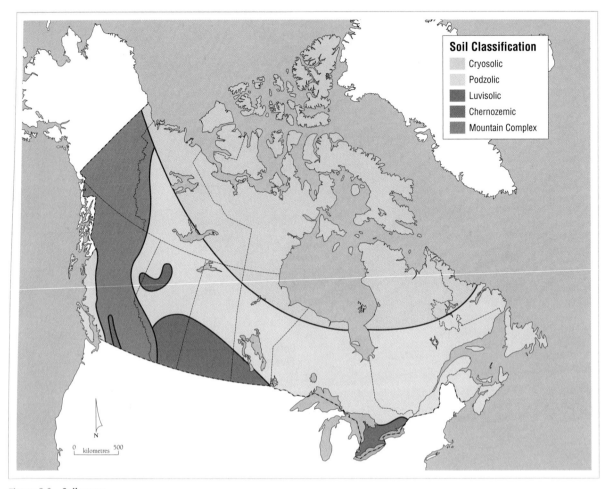

Figure 2.9 **Soil zones**
Most agricultural land is in **luvisolic** and **chernozemic** soil zones, which together comprise about 5 per cent of Canada's land base.

winds its way from North Dakota in the United States northward to Lake Winnipeg. Since 1950, residents of Winnipeg and other communities along the Red River have suffered through eight major spring floods—in 1950, 1979, 1996, 1997, 2001, 2006, 2009, and 2011—and the frequency of these floods appears to be increasing, giving credence to the fears of the Intergovernmental Panel on Climate Change. In 1950, the Red River flood drove over 100,000 people from their homes. Following that disaster, the Red River Floodway, a wide channel nearly 50 km long, was constructed. Its purpose was to divert the flood waters around the city of Winnipeg. However, small communities in the Red River Basin remained vulnerable to flooding. In 1997, the largest flood in the twentieth century occurred (Rasid et al., 2000). While the Red River Floodway saved Winnipeg, the towns of Emerson, Morris, Ste Agathe, and St Adolphe and the surrounding farm buildings and lands were less fortunate. In April 2006, April 2009, and April 2011, the Red River flooded (photo 2.11).

Photo 2.11
The town of St Jean Baptiste, 40 km north of the Canada–US border, is surrounded by a ring dike to protect it from the flooding Red River, 20 April 2011.

Permafrost

Permafrost is a relic from a very cold Pleistocene climate (Table 2.1). This distinctive feature of Canada's physical geography is permanently frozen ground with temperatures at or below zero for at least two years. The vast extent of permafrost in Canada and, in places, its great depth provide a measure of the country's cold environment (Figure 2.10). Permafrost exists in the Arctic and Subarctic climatic zones and occurs at higher elevations in the Cordillera zone. Overall, permafrost is found in just over two-thirds of Canada's land mass.

In more northerly regions, permafrost extends far into the ground. North of the Arctic Circle, the permafrost may be several hundred metres deep. Further south, permafrost is less frequent, and where it occurs it rarely penetrates more than 10 m into the ground. Permafrost is found in all six of Canada's geographic regions and reaches its most southerly position along 50° N in Ontario and Québec. Along the southern edge of permafrost, there is a transition zone where small pockets of frozen ground have a depth of less than 1 m. Further south, these pockets of permafrost disappear.

Permafrost is divided into four types. **Alpine permafrost** is found in mountainous areas and takes on a vertical pattern as elevations of a mountain increase. Over most of Canada, however, permafrost follows a zonal pattern, which does not correspond to latitude but rather to the annual mean temperatures that fall below zero.[4] The zonal pattern has a northwest to southeast alignment, that is, from Yukon to central Québec (see Figure 2.10).

THINK ABOUT IT
Even though the Red River floods regularly, Winnipeg is able to avoid serious flooding. Why?

THINK ABOUT IT
If our climate is warming, what are the consequences for the landscape if ice in the ground known as permafrost melts?

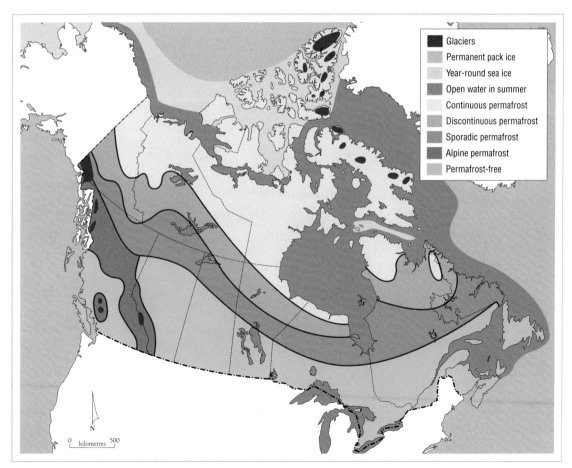

Figure 2.10 **Permafrost zones**
Canada's cold environment is demonstrated by the permanently frozen ground that extends over two-thirds of the country. Sea ice varies in thickness and duration. The most durable and thickest ice is found in the permanent ice pack. Lake ice disappears first in the Great Lakes while sea ice melts in the offshore waters of Atlantic Canada, and last in the Arctic Ocean. In September 2007, satellite imagery indicated that the extent of open water in the Arctic Ocean was greater than in previous decades.

As the mean annual temperature varies, the type of permafrost also changes. **Continuous permafrost** occurs in the higher latitudes of the Arctic climatic zone, where at least 80 per cent of the ground is permanently frozen, although it also extends into northern Québec. Continuous permafrost is associated with very low mean annual air temperatures of –15°C or less. **Discontinuous permafrost** occurs when 30 to 80 per cent of the ground is permanently frozen. It is found in the Subarctic climatic zone where mean annual air temperature ranges from –5°C in the south to –15°C in the north. **Sporadic permafrost** is found mainly in the northern parts of the provinces, where less than 30 per cent of the area is permanently frozen. Sporadic permafrost is associated with mean annual temperatures of zero to –5°C.

Major Drainage Basins

Canada, bounded by the Arctic, Atlantic, and Pacific oceans, is a maritime country (Figure 2.11). It has the longest coastline in the world and has four major drainage basins: the Atlantic Basin, the Hudson Bay Basin, the Arctic Basin, and the Pacific Basin. The Atlantic and Hudson Bay basins both drain into the Atlantic Ocean (Figure 2.11 and Table 2.5). In addition, a small portion of southern Alberta and Saskatchewan drains southward to the Missouri River, which forms part of the Mississippi River system that empties its water into the Gulf of Mexico. A **drainage basin** is land that slopes towards the sea and is separated from other lands by topographic ridges. These ridges

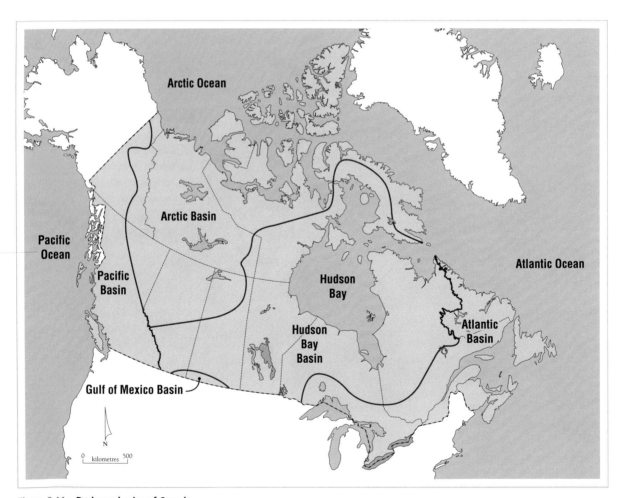

Figure 2.11 **Drainage basins of Canada**
The four divides determine Canada's drainage basins. They are the Continental or Great Divide, the Northern Divide, the Arctic Divide, and the St Lawrence Divide. The Hudson Bay Basin lies between three divides—the Continental Divide, the Arctic Divide, and the Northern Divide—and is by far the largest of the five basins in Canada. It also serves as a boundary between southern Alberta and British Columbia, and between northern Québec and Labrador.

Table 2.5 **Canada's Drainage Basins**

Drainage Basin	Area (million km²)	Stream-flow (m³/second)
Hudson Bay	3.8	30,594
Arctic	3.6	20,491
Atlantic	1.6	29,087
Pacific	1.0	24,951
Gulf of Mexico	<0.1	12
Total	10.0	105,135

Sources: Laycock (1987: 32); Dearden and Mitchell (2012: 136).

form drainage divides. The Continental or Great Divide of the Rocky Mountains, for example, separates those streams flowing to the Pacific Ocean from those flowing to the Arctic and Atlantic oceans. On the east coast, the Northern Divide extends along the Labrador/Québec boundary and separates waters flowing into the Atlantic Ocean and Hudson Bay. Canada's five drainage basins are shown in Figure 2.11.

A few rivers cross the US–Canada border and part of the Great Lakes lies in the United States. The Columbia River leaves British Columbia and continues its journey to the Pacific Ocean through the US states of Idaho, Washington, and Oregon. Two small rivers, the Milk and the Poplar, flow from southern Alberta and Saskatchewan into the Missouri River, which drains much of the northern half of the US Great Plains. The Red River flows from the US states of North Dakota and Minnesota into Manitoba and beyond to Lake Winnipeg and eventually to Hudson Bay by means of the Nelson River. The Richelieu River flows from Lake Champlain, which lies mainly in New York, and drains into the St Lawrence River near Sorel, Québec. The headwaters of the Saint John River partially originate in Maine, and this river empties into the Bay of Fundy at the city of Saint John, New Brunswick.

Historically, the rivers in each basin have played major roles in providing access to the interior of Canada and in the development of the country. For example, Aboriginal peoples and Europeans used the St Lawrence and Mackenzie rivers as transportation routes during the fur trade. Today, the St Lawrence River plays a key role in Canada's internal and foreign shipments of goods to and from Montréal and other cities along the St Lawrence and along the shores of the Great Lakes. River barges bring food, building materials, and other goods to settlements along the Mackenzie River and along the coast of the Beaufort Sea. Oil and gas equipment also is barged to the Norman Wells oil fields and those in the Beaufort Sea. These rivers remain important waterways today.

Water is a scarce commodity, particularly in the dry Southwest of the United States. California, for instance, has had to divert water from the Colorado River to help meet its needs. Under the North American Free Trade Agreement (NAFTA), water was classed as a commodity and therefore Canadian water could be exported to the United States. Since Canada has the world's largest supply of fresh water, large-scale transfer of water from Canada has appeal to water-short American states. Prior to NAFTA, several inter-basin proposals appeared in the media but none were commercially feasible. Two such schemes were the Great Recycling and Northern Development Canal, which called for diversion of water from James Bay to the Great Lakes and then, by pipelines, to the water-short American Southwest. Another continental water diversion scheme, the North American Water and Power Alliance project, proposed diverting the Yukon River and the two major tributaries of the Mackenzie River, the Liard and the Peace, to the US along the Rocky Mountain Trench. Continental diversion of fresh water from one basin to another, however, is not a practical matter because Canadians rivers do not flow south but rather east, west, and north.

The Atlantic Basin

The Atlantic Basin is centred on the Great Lakes and the St Lawrence River and its tributaries, but the basin also includes Labrador. The Atlantic Basin has the third largest drainage area and also the second greatest stream-flow. As seen in Figure 2.11, the Atlantic Basin receives considerable precipitation, making it second only to the Pacific Basin. The largest hydroelectric development in this drainage basin is located at Churchill Falls in Labrador, with the promise of more development at Muskrat Falls on the Lower Churchill River. In January 2009, a five-member federal–provincial Joint Review Panel for the proposed Lower Churchill hydroelectric generation project was established. Earlier hydroelectric developments took place along the St Lawrence River in southern Québec and along its tributary

THINK ABOUT IT
One reason why Canada has the longest coastline in the world is its physical size. But since Russia has an even larger territorial area, why does Russia not have a longer coastline? The second reason why Canada's coasts form 25 per cent of the world's coastline is the number of islands, most of which are situated in the Arctic Archipelago.

rivers that flow out of the Laurentide Upland of the Canadian Shield. Rivers such as the Manicouagan River originate in the higher elevation of the Laurentide Upland. Here, abundant precipitation, natural lakes, and a sharp increase in elevation provide ideal conditions for the generation of hydroelectric power. Because there is a large market for electrical power in the St Lawrence Lowlands, virtually all potential sites in the Laurentides have been developed.

The Hudson Bay Basin

The Hudson Bay Basin is the largest drainage basin in Canada (Table 2.5), covering about 3.8 million km². Precipitation varies greatly across this basin. In the West, precipitation is low while it is greater in the East (Figure 2.7), where the headwaters of its rivers in the uplands of northern Québec flow westward into James Bay. In northern Ontario and Manitoba, rivers drain into James and Hudson bays.

The large rivers and sudden drops in elevation that occur in the Canadian Shield make this part of the basin ideal for developing hydroelectric power stations. In fact, most of Canada's hydroelectric power is generated in the Canadian Shield area of the Hudson Bay Basin—the largest installations are on La Grande Rivière in northern Québec and on the Nelson River in northern Manitoba. La Grande Rivière's hydroelectric developments are the first stage in the James Bay Project. The Great Whale River Project was to follow the completion of the hydroelectric projects on La Grande Rivière, but a variety of circumstances (low energy demand, low prices in New England, and strong opposition from environmental groups and the Cree Indians of northern Québec) stalled its development. Instead, Québec focused its attention on the Eastmain Diversion Project and the Romaine Hydro Complex (see Chapter 6, "Key Topic: Hydro-Québec").

The Arctic Basin

The Arctic Basin is Canada's second-largest drainage basin. The Mackenzie River dominates the drainage system in this basin. Along with its major tributaries (the Athabasca, Liard, and Peace rivers), the Mackenzie River is the second-longest river in North America. However, because of low precipitation in the Arctic, this basin has only the fourth-largest stream-flow. There are few hydroelectric projects in the Arctic Basin because of the long distance to markets, with the exception of the hydroelectric development on the Peace River in British Columbia. Here, power from the Gordon M. Shrum generating facility is transmitted to the population centres in southern British Columbia and to the United States, primarily to the states of Washington, Oregon, and California.

The Pacific Basin

The Pacific Basin is the smallest basin. However, it has the third-highest volume of water draining into the sea. Heavy precipitation along the coastal mountains of British Columbia accounts for this unusually high stream-flow. As a result, the Pacific Basin is the site of one of Canada's largest hydroelectric projects. Located at Kemano, this facility is owned and operated by Rio Tinto, which uses the electrical generating station to supply power to its aluminum smelter at Kitimat. The ice-free, deep-water harbour at Kitimat and low-cost electric power generated at Kemano make Kitimat an ideal location for an aluminum smelter. Because of its deep-water harbour and relatively short shipping distance to Asia, Kitimat may become a terminal port for natural gas and bitumen pipelines.

Environmental Challenges

In our contemporary world, humans are the most active and dangerous agents of environmental change. All human activities affect the environment. Cultivation of the land, building of cities, exploitation of renewable and non-renewable resources, and processing of primary products have forever changed our natural environment into an industrial landscape. Mining, before environmental regulations were in place, left behind toxic wastes. In 2012, Canada's Commissioner of the Environment and Sustainable Development estimated that the cost to the federal government for removing these wastes would exceed the $4 billion already committed by Ottawa and could reach $7.7 billion (Vaughan, 2012).

Yet, without a doubt, global warming offers the greatest anthropogenic change—and threat—to our planet in the twenty-first century. While the process of climate change is not a simple one, annual global temperatures are on the rise. Just how much and how this increase will alter Canadian climates will unfold later in the twenty-first century. Then, too, one of the less understood elements of climate change is precipitation. Warmer

> **THINK ABOUT IT**
> Ottawa, our capital city, lies in which drainage basin?

air masses have a greater capacity to hold moisture and produce precipitation—and violent storms. On top of the interplay of global temperatures and precipitation, different parts of the world are affected differently. In Canada's Arctic, for instance, an indisputable indicator of warmer summers is the summer retreat of **sea ice**, leaving the possibility of an ice-free Arctic Ocean for part of the year. In another part of Canada, the Canadian Prairies, the jury is still out on how these two climatic elements of temperature and precipitation will play out. The semi-arid Prairies have long been subject to a cycle of wet and dry years. Now the question is whether, with longer and warmer summers, this region will turn into an arid environment or a more humid one. In the summer of 2012, for instance, the Grain Belt of the Canadian Prairies received well above average levels of precipitation while the opposite was true for the Corn Belt of the United States, where below-average levels of rainfall created drought conditions. Is this a one-shot anomaly or a sign of the future?

The extent and possible effects of global warming remain controversial. One reason is that tracking the world's average temperature is not an easy task and, as late as the 1960s, the general scientific view was that the world was cooling (Weart, 2008). Ten years later, the scientific community leaned the other way and now sufficient evidence points to a warming trend. Even a leading climate warming skeptic, Richard Muller (2012), concluded that global warming is for real, thus confirming the prediction of Wallace Broecker back in 1975, who asked, "Are we on the brink of a pronounced global warming?" For the twenty-first century, the question focuses on the rate of this warming trend. If, as it appears, this trend is not part of the natural fluctuations in global temperatures but rather a product of the burning of fossil fuels and the release of greenhouse gases into the atmosphere, then the world can expect that global temperatures will be higher by at least 2°C by the end of the twenty-first century. The Intergovernmental Panel on Climate Change argues that world temperatures are increasing at a rapid rate and that, by the end of the century, average world temperatures will be higher within a range of 1.8 to 4°C (Pachauri and Reisinger, 2007). Some few climate change deniers, chiefly those aligned financially or politically with business and resource extraction interests, continue to disagree with this analysis and with the argument that these temperature changes result from the anthropogenic air pollution that largely began with the Industrial Revolution. But even this view is shifting. As Dearden and Mitchell (2012: 206) report:

> The world's largest oil company, Exxon, which up to 10 years ago was questioning the veracity of climate change, in its 2011 annual report . . . concludes that the scientific community is being unrealistically optimistic in its projections [regarding greenhouse gas emissions and the consequences]. Exxon expects carbon emissions to rise a further 25 per cent over the next 20 years . . . , leading to annual increases of 0.9 per cent emissions per year.

VIGNETTE 2.10

Global Warming and Climate Change: What Is the Difference?

The definitions of both "global warming" and "climate change" are centred on the premise that temperatures are increasing, which, in turn, contributes to changes in global climate patterns and an increase in extreme weather events. The principal difference between the two terms is that global warming focuses on temperature change that results from increased emissions of **greenhouse gases** from human activities, notably the burning of fossil fuels. Known as the **greenhouse effect**, this anthropogenic-caused warming trend began with the Industrial Revolution when great quantities of coal first began to be consumed to supply the energy necessary for industry. Originally, European countries were the major consumers of coal but now China leads the world in consumption of coal with India not far behind.

Climate change, on the other hand, is not restricted to human-caused warming but considers natural forces, too. As well, climate change places more emphasis on other elements of climate, such as precipitation and wind patterns.

Canada and Global Warming

Focusing our attention on Canada in regard to global warming, we need to consider two questions in particular:

- How might Canada be affected by global warming in the twenty-first century?
- How well has Canada met its Kyoto target?

Of course, how we will be affected this century will be better known in the years to come, since climate change is a very slow process. Nevertheless, Canada's weather appears to be slowly warming, and if that process continues the country's climatic zones will shift northward by the end of the century, followed very slowly but inevitably by natural vegetation, soil, and wildlife zones (Bouchard, 2001). Most dramatic changes are expected first in northern Canada, where the loss of snow cover and ice accelerates the warming process. Already, the Arctic ice pack has retreated each summer, providing an early warning of changes to come. Eventually, the Arctic zone could shrivel in size, perhaps limited to the Arctic Archipelago, and the Arctic Ocean would be ice-free in the summer. Over this century, permafrost, for example, might only remain in the zone of continuous permafrost (see Figure 2.10). One consequence of the melting of the ice in permafrost would be massive ground **subsidence**, resulting in an irregular relief referred to by physical geographers as "thermokarst topography." During the twenty-first century, such massive melting/subsidence could require major repairs to highways, bridges, and pipeline systems, and play havoc with foundations for buildings, bridges, and other human-made structures (Bone et al., 1997: 265–74). Other potential developments could be a longer growing season for crops; an extended navigation season for the St. Lawrence Seaway and for the **Northwest Passage**; and rising sea levels that could cause flooding of coastal areas.

As for the second question, Canada failed to meet its Kyoto target for reducing its greenhouse gas emissions, and, to avoid stiff penalties, withdrew from the Kyoto Accord. According to the 1997 Kyoto Protocol, Canada was expected to reduce its greenhouse gas emissions by 6 per cent below 1990 levels over the five years from 2008 to 2012. However, Canadian emissions failed to decline at a sufficient rate over this five-year period to meet its target. Curry and McCarthy (2011:A4) estimate that Canada exceeded its Kyoto goal by 30 per cent. Nevertheless, some progress was made as total greenhouse gas emissions from 2006 to 2010 did decline from 273.1 tonnes to 261.8 tonnes CO_2 equivalent (Table 2.6).

Another Environmental Challenge

Canadians are all too familiar with smog and other forms of air pollution. Most air pollution results from industrial emissions and from automobile and truck exhaust. Coal-burning plants and oil sands production account for most industrial pollution. Nanticoke Station in Ontario and TransAlta Plant in Alberta are the two largest coal-burning electric power stations in Canada. As Table 2.6 reveals, the top three provinces by industrial emissions are Alberta, Ontario, and Saskatchewan. Ontario is closing coal-burning plants, which should reduce its industrial emissions significantly. Nanticoke, on Lake Erie, is scheduled to cease burning coal by the end of 2013.

VIGNETTE 2.11

Natural Factors Affecting Global Warming

The physics of global warming in the greenhouse model are elementary, but the actual process of climate change is extremely complex and remains unclear. While the IPCC has declared a global warming trend for the twenty-first century of at least 2°C, what natural factors might alter that predicted trend in the future? One possibility is a reduction in the amount of solar energy reaching the earth's surface, which would follow from massive volcanic eruptions that would release huge amounts of dust into the air, thus reflecting significant amounts of incoming solar energy back into outer space and thereby chilling the world's climate. But we can hardly count on such a natural event to reverse the current warming trend, and so efforts—such as more efficient automobile engines—are necessary to reduce the release of carbon dioxide into the atmosphere. Muller (2012) expects "the rate of warming to proceed at a steady pace, about one and a half degrees over land in the next 50 years."

CONTESTED TERRAIN 2.2

Beyond the 1997 Kyoto Protocol

The present Canadian government has rejected the terms of the 1997 Kyoto Protocol and supports the call for a new protocol that would include all nations. Ottawa claims that the Kyoto agreement was both ineffective and unfair: ineffective because major emitters that account for the bulk of the greenhouse gas emissions, including the US, China, and India, never signed the1997 agreement; unfair because developed nations participant in the Kyoto Protocol must pay "climate reparations" in the form of carbon credits to developing countries, including China and India. In fact, Canada withdrew from the Kyoto Protocol in December 2011 to avoid purchasing international carbon credits—at a cost of around $14 billion—for its emissions beyond its Kyoto target (York, 2011: A1).

The United Nations, sponsor of these climate talks, wanted Canada to remain a member of the Kyoto regime. At the 2011 Durban meeting, Archbishop Desmond Tutu called on Canada to show leadership on climate talks and to reclaim the moral high ground it once held during the anti-apartheid movement. Instead, however, Canada's position on climate change and carbon emissions under the Harper government, according to many observers, has been to delay hard decisions and to obfuscate. On the issue of climate reparations, New Democratic Party environment critic Megan Leslie stated that "the reality is, certain countries have contributed more emissions historically and it is up to them to show leadership. It is the ethical thing to do" (McCarthy, 2011: A4).

The participant countries at the Durban climate change conference determined to move towards a new agreement by 2015 that will include all nations, with binding commitments taking place in 2020. But for now, it remains a pledge and all the details remain to be negotiated. As they say, "the devil is in the details."

Table 2.6 **Industrial Emissions, Provinces and Territories, 2006 and 2010**

Rank	Province/Territory	Million Tonnes CO_2, 2006	Million Tonnes CO_2, 2010	2006 (% of national total)	2010 (% of national total)
1	Alberta	115.9	122.5	42.4	46.8
2	Ontario	71.8	56.2	26.3	21.5
3	Saskatchewan	22.5	22.8	8.2	8.7
4	Québec	22.0	20.7	8.0	7.9
5	British Columbia	11.8	13.7	4.3	5.2
6	Nova Scotia	10.9	10.6	4.0	4.0
7	New Brunswick	10.3	8.2	3.8	3.1
8	Newfoundland and Labrador	5.0	4.6	1.8	1.8
9	Manitoba	2.4	1.9	0.9	0.7
10	Northwest Territories	0.4	0.5	0.2	0.2
11	Prince Edward Island	0.1	0.1	0.1	0.1
12	Yukon	<0.1	<0.1	<0.1	<0.1
13	Nunavut	<0.1	<0.1	<0.1	<0.1
	Total emissions	273.1	261.8	100.0	100.0

Sources: Environment Canada (2012: Table 2). Reproduced with the permission of the Minister of Public Works and Government Services Canada, 2013.

Photo 2.12

Decisions over economic growth or environmental sustainability usually favour economic growth. One example is the oil sands, where vast tailing ponds, such as the one serving the Suncor oil sands operation near Fort McMurray, Alberta, are a consequence of production. While the oil sands produce the bulk of Canada's oil, the environmental cost is high, including huge greenhouse gas emissions.

THINK ABOUT IT

True or false—or somewhere in the middle? According to NDP leader Thomas Mulcair: "If resource companies were required to pay for their pollution, the cost of oil sands bitumen and other natural resource exports would rise and the upward pressure on the dollar would ease" (Bryden, 2012: B6).

While the list of serious environmental problems is long, automobile pollution, coal-burning plants, and oil sands development in Alberta rank at or near the top. The Alberta oil sands supply both Canada and the United States with a large amount of crude oil, and Asian countries, most notably China, have purchased a significant share of the action. However, from an environmental perspective, extraction of oil from these tar sands represents a risky gamble to the local environment and to global warming. Open-pit mining is the worst offender and such mining takes place at several sites, including the Suncor mine (photo 2.12). As well, oil sands extraction is the primary reason why Alberta is the leading province in greenhouse emissions and why Alberta's emissions increased from 2006 to 2010 (see Table 2.6). Environmental organizations, ranging from international ones such as Greenpeace and the Sierra Club to local ones such as the Pembina Institute, have long targeted the oil sands as a major source of greenhouse gas emissions and therefore one of the industrial culprits affecting global warming. Politicians, too, have entered the fray, with US President Barack Obama delaying approval of the Keystone XL Pipeline project and both Ontario Premier Dalton McGuinty and NDP Leader Thomas Mulcair suggesting that exploitation of the oil sands has inflated the Canadian currency and thereby hurt manufacturing in Central Canada. Their argument is based on the so-called Dutch disease, which is discussed more fully in Chapter 4.

Photo 2.13

Glacial retreat of the Athabasca Glacier, 1992 to 2005, can be attributed to warmer annual temperatures.

CANADA'S PHYSICAL BASE

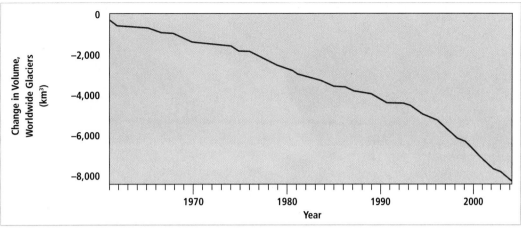

Figure 2.12 **Worldwide change in volume (km³) of glaciers, 1960–2004**
Source: NASA, Earth Today, at: <www.nasa.gov/vision/earth/features/index.html>.

THINK ABOUT IT
Would you advocate the immediate construction of a natural gas pipeline from the Mackenzie Delta and LNG port facilities to allow tankers to transport liquefied natural gas from the Mackenzie Delta through the Beaufort Sea to Asian markets?

Since environmental issues are the result of human actions, solutions are possible (Dearden and Mitchell, 2012: ch. 1). One solution involves establishing more protected areas and parks (Slocombe and Dearden, 2009). Another is for more stringent regulations that will reduce damage to the environment caused by both new and existing projects. But perhaps the most significant solution lies in "going green," which includes recycling waste products, moving towards electric automobiles, and increasing the production of electricity from natural sources such as solar and wind rather than coal.

Vignette 8.6, "Technological Gamble: Carbon Capture and Storage," page 329, discusses the plans of Shell Oil to reduce its greenhouse emissions from its upgrader at Fort Saskatchewan.

VIGNETTE 2.12

Fluctuations in World Temperatures

While it appears that our climate is relatively constant, wide fluctuations in world temperatures are part of our geological history. The last 10,000 years—the **Holocene epoch**—represents a relatively warm period in the earth's history. However, within that epoch, a series of shorter cycles of warm and then colder periods have occurred. Each has lasted around 500 years or so. For instance, a warm period known as the Medieval Warming Period took place between the tenth and fourteenth centuries. During this time, the Vikings established a settlement on the southern tip of Greenland and made voyages to neighbouring parts of Arctic Canada and southward to the northern tip of Newfoundland. At the same time, the Thule occupied much of the Canadian Arctic and were able to hunt the huge bowhead whales in the open summer waters. This warm spell was replaced by a cooler period known as the Little Ice Age, which lasted from about 1450 to around 1850. Since then, however, global warming has become the trend, with occasional interruptions of short periods of cooler weather such as occurred in the 1960s.

The Little Ice Age, perhaps triggered by volcanism and sustained by sea-ice/ocean feedback, chilled the northern hemisphere (Miller et al., 2012). Living on the edge of a cold climate, even a slight drop in annual temperatures had a dramatic impact on human beings. The Vikings were no longer able to sustain themselves in Greenland and the Thule had to make a drastic adjustment to their hunting economy. With a much more extensive and long-lasting ice cover over the Arctic Ocean, the bowhead whales no longer entered these waters for long periods of time. The consequences for the Thule inhabitants were devastating. They were forced to hunt smaller game—seals and caribou. The results were twofold: the new hunting system could not support as many people and it required smaller, more mobile hunting groups. Archaeologists believe that the Inuit, who were established in the Arctic by the mid-sixteenth century, are the descendants of the Thule people.

Photo 2.14

Warm summer temperatures have caused the Arctic sea ice to shrink, thus exposing more open water. In September 2007, for example, Arctic sea ice coverage had reached its lowest extent for the year and the lowest amount recorded since satellite observations began some 30 years ago. Since then, this pattern of open water in the late summer has continued with two routes through the Northwest Passage ice-free, but only for a short time. Differences do exist: the portion of the Arctic Ocean from the Beaufort Sea to Bering Strait is ice-free for a longer time and its ice-free zone is much larger than in the central and eastern Arctic. The possibility of ocean-going vessels plying these western Arctic waters is more likely to occur before commercial cargo ships begin to use the entire Northwest Passage. The most likely scenario would see liquefied natural gas from the Mackenzie Delta gas reserves shipped to Asian markets.

NASA/Goddard Space Flight Center Scientific Visualization Studio; Blue Marble Next Generation data courtesy RetoStockli (NASA/GSFC).

THINK ABOUT IT
Since China and India need more coal-burning thermal electricity plants to keep their economies growing, is it fair to ask those countries to reduce their greenhouse emissions?

Urban Smog

Air pollution is the foremost pollution problem facing urban residents of major cities. The source of air pollution (smog) comes from emissions from automobiles and coal-burning plants. Smog, a mixture of smoke, sulphur dioxide, and other contaminants, takes the form of a brownish haze over cities. Most contaminants are produced by automobiles (carbon dioxide) and coal-fired thermal plants (sulphur dioxide). Vehicles account for over half of urban air pollution. Coal-burning plants are the second principal source of urban air pollution. Densely populated areas like southern Ontario contain millions of automobiles and trucks. Furthermore, southern Ontario is an energy-deficient area. For years, coal thermal plants have produced much of the region's electricity at a relatively low cost. Now that urban air pollution has become a health problem, Ontario has plans to limit sulphur dioxide emissions by closing its coal-burning plants, replacing the electrical production with natural gas, nuclear power, and water power. In addition, Ontario is considering importing electricity from Québec, Michigan, New York, and Manitoba.

SUMMARY

Physical geography varies across Canada. This spatial variation is critical in understanding Canada's regional character. Added to this spatial dimension, a new concern may arise in this century, namely that the land's capacity to supply food and water to Canadians may be affected by global warming.

Physiographic regions represent one measure of Canada's physical diversity where large areas with similar landforms and geological structures provide a broad and simple framework of Canada's physical geography. Climate adds to that geography by creating a zonal arrangement of climate, soils, and natural vegetation types. Climate therefore determines to a large extent the type of soil and native vegetation in a given region and hence influences land use. Together with topography, climate partly determines the land's agricultural ability to support a population. This link between the physical and human worlds identifies those regions having a more favourable mix of physical characteristics for economic development, and translates the abstract core/periphery model into a geographic reality. The Great Lakes–St Lawrence Lowlands region is the most favoured physical region in Canada. Physical barriers, such as the Rocky Mountains, and extreme climatic conditions, such as the very long and cold winters in northern Canada, have also affected the historical settlement of the country and continue to influence contemporary economic activities.

Canada has several physiographic regions and climatic zones. The seven physiographic regions are: the Canadian Shield, the Cordillera, the Interior Plains, the Hudson Bay Lowlands, the Arctic Lands, the Appalachian Uplands, and the Great Lakes–St Lawrence Lowlands. The seven climatic zones are: the Pacific, Cordillera, Prairie, Great Lakes–St Lawrence, Atlantic, Subarctic, and Arctic.

Global warming and the Alberta oil sands represent major environmental challenges. For Ottawa, finding the right balance between huge economic gains generated by the oil sands and the massive environmental damage and pollution caused by that development is not easy. While Canada's greenhouse gas emissions—with the greatest amount coming from Alberta—account for only a relatively small amount of the total greenhouse gases emitted into the atmosphere by all countries, environmental organizations have targeted the oil sands as a major offender and one that should be shut down. For Alberta, the federal government, and the oil sands developers, the connection between global warming and oil sands development cannot be dismissed. While solutions may be both challenging and expensive, the costs to the earth and future generations are too high to ignore.

CHALLENGE QUESTIONS

1. Why do physiographic regions and climatic zones—as expressed in Figures 2.1 and 2.4—have rather different spatial expressions while climatic zones and natural vegetation zones (Figure 2.8) are quite similar?
2. Why have environmentalists focused so much of their attention on the Alberta oil sands?
3. In your opinion, did Canada make the right move in withdrawing from the Kyoto Accord? Explain.
4. What physical evidence indicates that Canada's Arctic is warming?
5. Québec has the second-largest population as well as the second-largest industrial base, but, according to Table 2.6, it ranks fourth in industrial emissions. Why does this region not produce as much industrial emissions as Saskatchewan?

FURTHER READING

Muller, Richard A. 2012. "The Conversion of a Climate-Change Skeptic," *New York Times*, 28 July, at: <www.nytimes.com/2012/07/30/opinion/the-conversion-of-a-climate-change-skeptic.html?pagewanted=all>; and Berkeley Earth, at: <BerkeleyEarth.Org>.

In a startling release, Richard Muller, a professor of physics at the University of California, Berkeley, who for long had been an outspoken skeptic of global warming, announced that he had become a believer! His change of heart was based on the scientific findings of his research team, Berkeley Earth. Their key finding is that the rise in average world land temperature is approximately 1.5°C in the past 250 years, and about 0.9 degrees in the past 50 years.

In his *New York Times* article, Dr Muller muses about the future. As carbon dioxide emissions increase, he states the conventional view that the temperature should continue to rise. Furthermore, Muller expects the rate of warming to proceed at a steady pace, about one and a half degrees over land in the next 50 years. However, if China's development of its vast coal reserves continues at the current high pace, then that same warming could take place in less than 20 years. (In 2010, China accounted for 48 per cent of world coal production, followed by the United States at 15 per cent; Canada produced less than 1 per cent.) Finally, Dr Muller hopes that "the Berkeley Earth analysis will help settle the scientific debate regarding global warming and its human causes." He observes that "the difficult part remains: agreeing across the political and diplomatic spectrum about what can and should be done."

CANADA'S HISTORICAL GEOGRAPHY

Introduction

In one sense, Canada is a young country. Its formal history began in 1867 with the passing of the British North America Act by the British Parliament. In another sense, Canada is an old country with a human history that goes back perhaps as far as 40,000 years. As an old country, its history has followed many twists and turns, but three events stand out because they continue to have a profound impact on the nature of Canadian society. These events are the arrival of the first people in North America and, most importantly, the capacity of their descendants to survive and endure the colonization and assimilation efforts of European settlers; the colonization of North America by France and England and the establishment of British institutions, laws, and values; and, in the early twentieth century, the influx of people from Central Europe and czarist Russia, many of whom settled the prairie lands. In later years, these "outsiders" provided the political push behind government-backed multicultural programs and policies and, at the same time, challenged the previous vision of a French/English Canada.

Given these circumstances, tensions between regions and groups were inevitable. Disagreements arose. Four faultlines stand out and they are based on fundamental differences between English- and French-speakers, centralists and decentralists, "new" and "old" Canadians, and Aboriginal and non-Aboriginal Canadians. With the exception of first-generation "new" Canadians, all have their roots in Canada's history and geography. The search for solutions to these tensions is a dominant feature of modern Canadian society and, to a large degree, this process of seeking a middle ground accounts for Canadians' high degree of acceptance of various religious and ethnic groups as well as their desire to resolve regional disagreements. However, this tolerance did not magically appear; rather, it was "learned" over time—often after reconsidering past intolerant acts towards minority groups, especially towards Aboriginal people and visible minorities. Using John Ralston Saul's metaphor, an underlying theme in this chapter is that Canada evolved from a **"hard" country** to a **"soft" country**.

CHAPTER OVERVIEW

In this chapter we will consider the following topics:

- The arrival of Canada's first people.
- The colonization of Canada by the French and the British.
- The settlement of Canada's West by peoples from Central Europe and czarist Russia.
- The territorial evolution of Canada.
- The four faultlines as they have developed in the context of Canada's geography and history.
- The notion of "One Country, Two Visions."
- Two power struggles: economic and political.
- Possible "solutions" to complex problems.

The Knowledge Totem Pole at the BC Legislature in Victoria, British Columbia. This totem pole was erected in 1994 to welcome visitors to the Commonwealth Games. It was carved by Coast Salish artist Cicero August. Photo: Michael Wheatley/All Canada Photos.

The First People

Around 40,000 years ago, according to some archaeological estimates, the first people to set foot on North American soil were Old World hunters who crossed a land bridge (known as Beringia) into Alaska and Yukon. Thousands of years would pass before they reached the interior of North America.

Beringia was a product of the last ice advance when so much water was contained in the continental glaciers that the sea level dropped, thus exposing the ocean bottom between Siberia and Alaska. Scientists estimate that the sea level dropped by at least 100 metres, thus creating a land bridge between Asia and North America. The Old World hunters were now in Alaska and Yukon, but were blocked from proceeding south by an ice sheet that was perhaps 4 km thick.

Beginning around 15,000 years ago, the Great Melt began. Why the climate warmed remains a puzzle, but the result was clear—the ice sheets began their retreat and eventually the Old World hunters migrated into the heart of North America. But just when did these hunters arrive in the south and then evolve into Paleo-Indian peoples? A second question involves which migration route they took. Solid evidence indicates that, some 13,000 years ago, the Paleo-Indians known as Clovis inhabited New Mexico, and archaeological finds suggest that pre-Clovis peoples may have occupied the Americas. While archaeologists continue to seek the answers to these

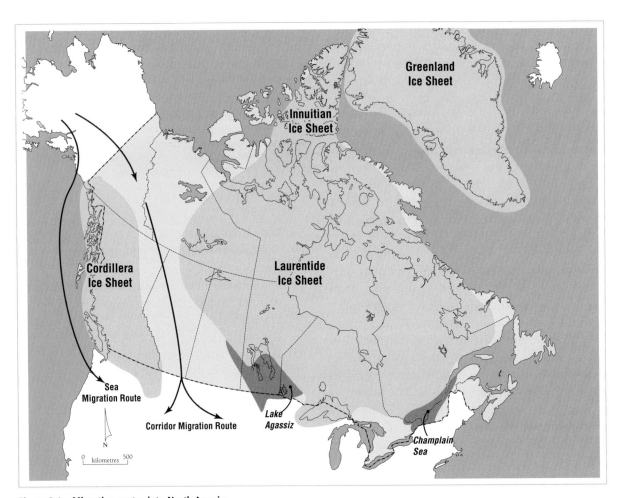

Figure 3.1 **Migration routes into North America**
Archaeologists originally believed that the Corridor Route allowed the descendants of the Old World hunters to reach the heart of North America. More recently, the Sea Route has gained favour because it explains how human beings could have arrived south of the ice sheet before the Great Melt created the ice-free corridor between the two ice sheets. During this Great Melt, huge amounts of fresh water surged to the oceans while some formed huge glacial lakes such as Lake Agassiz. As explained in Vignette 2.6, however, the Champlain Sea was not a glacial lake but an extension of the Atlantic Ocean into the isostatically depressed valley of the St Lawrence River.

questions, two possibilities, the Corridor and Sea routes, dominate this academic discussion.

In the Corridor theory, the Old World hunters who had been blocked from proceeding south by the ice sheet were now able to migrate south (Figure 3.1). Archaeologists have estimated that this ice-free corridor appeared about 13,000 to 14,000 years ago. Shortly thereafter, the first people travelled along this narrow corridor into the heart of North America. While evidence proves conclusively that the ancient hunters reached the interior of North America and evolved into the Paleo-Indians known as Clovis some 13,000 BP (years before the present), no such evidence has yet been discovered in the corridor, leaving the door open to another route.

The Sea Route theory provides one plausible explanation for the Old World hunters reaching the interior of North America before the ice sheet began to melt, namely by island-hopping along the sea edge of the Cordillera Ice Sheet until they reached the unglaciated US Pacific coast. However, archaeological evidence of such a route is lacking because these ancient island campsites, if they existed, are now well below the current sea level.

Although the question of when, precisely, **Paleo-Indians** first reached the unglaciated lands south of the ice sheet is still unsettled, the most recent archaeological evidence suggests that "by about 11,000 BP humans [Paleo-Indians] were inhabiting the length and breadth of the Americas, with the greatest concentrations being along the Pacific coast of the two continents" (Dickason with McNab, 2009: 15).

Paleo-Indians

The Paleo-Indians, the people who devised the fluted spear points characteristic of Clovis culture, were descendants of the Old World hunters. The oldest fluted points found in North America are about 11,500 years old. These spearheads, along with the bones of woolly mammoths, have been discovered in the southern part of the Canadian Prairies. By 9,000 years ago, many of the large species, such as the woolly mammoths and the mastodons, had become extinct, possibly as a result of excessive hunting and/or climatic change. Later Paleo-Indian cultures, which archaeologists refer to collectively as Folsom and Plano, developed a variety of unfluted stone points with stems for attachment to spear shafts. These technological changes made weapons more suitable for hunting buffalo and caribou. About 8,000 years ago, hunters in the grasslands of the interior of Canada pursued the buffalo, while those in the tundra and forest lands of northern and eastern Canada depended on caribou for most of their food. However, these smaller prey species could not support large numbers of people, so the Paleo-Indians had to develop new survival strategies. These strategies involved remaining in one area (and presumably keeping other peoples out of that area), developing effective hunting techniques for the local game, and making extensive use of fish and plants to supplement their principal diet of game. The time frame provided by Thomas (1999: 10) divides Paleo-Indians into three groups:

- Clovis culture from 11,500 to 10,500 BP;
- Folsom culture from 11,000 to 10,200 BP;
- Plano culture from 10,000 to 8,000 BP.

This link between geographic territory and hunting societies marked the development of Paleo-Indian **culture areas** with the following two characteristics: (1) a common set of natural conditions that resulted in similar plants and animals; and (2) inhabitants who used a common set of hunting, fishing, and food-gathering techniques and tools. Under these conditions, Paleo-Indians formed more enduring social units that became the forerunners of the numerous North American Indian tribes at the time of contact with Europeans.

Indians

Most archaeologists support the idea that Algonquians (e.g., Cree, Ojibwa) are direct descendants of Paleo-Indians, but they are less certain about Athapaskans (e.g., Dene, Chipewyan, Gwich'in), whose ancestors may have arrived from Asia some 10,000 years ago. Since the Paleo-Indian culture had emerged some 11,500 years ago, the Athapaskans represent a distinct Indian culture. But how did the early Athapaskans cross the Bering Strait? Archaeologists suggest that the ancient ancestors of the Athapaskans either walked across the frozen Bering Strait or crossed it in small, primitive boats.

Around 10,000 years ago, the Laurentide Ice Sheet had retreated just to the north of the Great Lakes. At that time, climatic zones ranged from a tropical climate in Mexico to an Arctic climate just south of the Laurentide Ice Sheet. These climatic

Table 3.1 Timeline: Old World Hunters to Contact with Europeans

Date	Events: Possible and Actual
(BP = years before present)	
40,000–35,000 BP	Old World hunters from Asia may have crossed the Beringia land bridge into the unglaciated areas of Alaska and southern Yukon but were blocked by the Cordillera Ice Sheet from moving into the rest of North America.
35,000–32,000 BP	Corridor Route became ice-free in this interglacial period, and, in pursuing the woolly mammoth, the Paleo-Indians may have found their way further south.
24,000–18,000 BP	Late Wisconsin Ice Age: Old World hunters from Asia again crossed the Beringia land bridge into the unglaciated areas of Alaska and southern Yukon, but were blocked by the ice sheet from moving into the rest of North America.
18,000 BP	The Wisconsin ice sheets reach a maximum geographic extent, covering virtually all of Canada.
15,000 BP	With climate warming, the ice sheets retreat rapidly in western Canada, exposing a narrow ice-free area along the foothills of the Rocky Mountains known as the Corridor Route. At the same time, the Cordillera Ice Sheet withdrew from the Pacific coast, making island-hopping along the Sea Route more amenable.
13,000–12,000 BP	Carbon dating of stone points from the Clovis culture provides solid evidence of Paleo-Indians presence in New Mexico some 13,000 years ago.
9,000 BP	Mammoths and mastodons become extinct, forcing early inhabitants of North America to adjust their hunting practices and thereby become more mobile and less numerous.
5,000 BP	As the Arctic coast became ice-free, Paleo-Eskimo (known as the Denbigh) hunters were the first people to cross the Bering Strait to the Arctic coast of Alaska. Within 2,000 years, they moved eastward along the Canadian section of the Arctic coast and eventually reached Greenland.
3,000 BP	The Dorset people represented another wave of Arctic immigrants and, with a more advanced technology suited for an Arctic marine environment, they either absorbed or replaced the Denbigh hunters. What are believed by some scholars to be the last Dorset people, known as Sadlermiut, lived in isolation principally on Southampton Island in Hudson Bay and became extinct around 1902.
1,000 BP	A third wave of Arctic hunters, known as the Thule, migrate across the Arctic, eventually reaching the coast of Labrador. Their primary source of food was the bowhead whale. At the same time Vikings reach Greenland and North America, where they establish a settlement on the north coast of Newfoundland (L'Anse aux Meadows).
1450 CE (common era) or thereabouts	Climate cools, marking the onset of the Little Ice Age. Both Thule and Vikings suffered in the colder environment with the Viking settlement disappearing and the Thule culture evolving from hunters of bowhead whales to the Inuit culture of small-game hunters (see Vignette 2.12).
1497 CE	John Cabot lands on the east coast (Newfoundland or Nova Scotia).
1534 CE	Jacques Cartier plants the flag of France near Baie de Chaleur.
1576 CE	Martin Frobisher sails to Baffin Island and makes contact with the Thule.

zones provided different agricultural opportunities for Indian tribes in North America. About 5,000 years ago, Indians living in the tropical climate of Mexico began to domesticate plants and animals. This agricultural system and its people gradually spread northward into areas with more restrictive growing conditions. These climatic differences required Indians to adapt their agricultural system accordingly. About 3,000 years ago, Indians in what is today the eastern United States planted corn, beans, and squash (known as "the three sisters"), which supplemented their diet of game and fish.

Agriculture was not possible north of the Great Lakes–St Lawrence Lowlands because of the shorter growing season for corn and other crops. Algonquian-speaking Indians who lived north of the Great Lakes had to hunt big-game animals, particularly caribou, for sustenance. They also traded with those more sedentary Indians, such as the Huron and Iroquois, who practised a form of slash-and-burn (swidden) agriculture in the Great Lakes–St Lawrence Lowlands and in the Ohio Valley. By the sixteenth century, the Huron controlled the agricultural lands between Lake Simcoe and the southeastern corner of Georgian Bay, where about 7,000 acres were under cultivation and where Indian villages with populations as large as 1,500 persons, and in some cases considerably larger, were commonplace (Dickason with McNab, 2009: 46–7). In western North America, agriculture spread northward along the valleys of the Mississippi River and its major tributary, the Missouri River. Tribes on the Canadian Prairies engaged in trade for agricultural products with tribes along the upper reaches of the Missouri River. In the Northwest, Athapaskan-speaking Indians, whose ancestors probably came from Asia much later (perhaps between 7,000 and 10,000 years ago), continued to practise a nomadic lifestyle of hunting and gathering. They moved about in the forest lands stalking big-game animals and made summer hunting trips to the tundra where the caribou had their calving grounds.

Photo 3.1

A tent ring located near Igloolik, Nunavut. The stone ring indicates the edges of the tent's skin walls, which were weighted down with rocks. Tent rings similar to this one mark sites where Thule families located their tents.

Arctic Migration

Arctic Canada became a human habitat much later than the forested lands of the Subarctic. Before people could occupy the Arctic Lands, two developments were necessary. The first was the melting of the ice sheets that covered this physiographic region. About 8,000 years ago, only small remnants of the great Laurentide Ice Sheet remained in northeastern Canada, leaving the western Arctic free of the ice sheet. The second development was the emergence of a hunting technique that would enable people to live in an Arctic environment. About 5,000 years ago, the Paleo-Eskimos, who were living along the northeast coast of Asia and in Alaska, had developed an Arctic sea-based hunting tradition, and they began to move eastward along the coast of Alaska and then along the Arctic coast of Canada. This Paleo-Eskimoan hunting culture is known as the Denbigh. Unlike previous marine hunting societies, these people invented a harpoon and other tools that enabled them to hunt seals and other marine mammals, though they also relied heavily on terrestrial game such as the caribou. About 3,000 years ago, a second migration from Alaska took place. Known as the Dorset culture, this culture replaced the Denbigh. The third and final Arctic migration took place roughly 1,000 years ago, when the Thule people, who had developed a sophisticated sea-hunting culture, spread eastward from Alaska and gradually succeeded their predecessors. The Thule, the ancestors of the Inuit, hunted the bowhead whale and the walrus. However, the climate began to cool (known as the Little Ice Age) and whales no longer entered the Arctic Ocean in large numbers because the ocean was covered by ice for most of the year (Vignette 2.12). The Inuit became more nomadic and hunted smaller game such as the seal and the caribou.

A fuller discussion of the Little Ice Age and its impact on the Thule is found in Vignette 2.12, "Fluctuations in World Temperatures," page 63.

THINK ABOUT IT
In all societies, food security is a critical element. Which of the two groups—the hunters or the agriculturalists—would have better food security and thus be less likely to face starvation?

Initial Contacts

Long before the time of European contact, the descendants of Old World hunters occupied all of North and South America. Yet, Europeans considered the New World **terra nullius** or empty lands. North American Indian and Inuit tribes met the European explorers searching for a trade route to the Orient. While the total population of these tribes can only be estimated, many scholars now believe there may have been as many as 500,000 Indians and Inuit living in Canada at the time of first contact. The greatest concentrations were found along the Pacific coast, where marine resources provide abundant food, and in the Great Lakes–St Lawrence Lowlands, where agriculture supported relatively large sedentary populations. Following contact, their numbers dropped sharply, perhaps declining to 100,000. By 1871, the Census of Canada reported an Aboriginal population of 122,700 (Romaniuc, 2000: 136). Loss of hunting grounds to European settlers and the spread of new diseases by explorers, fur traders, and missionaries greatly contributed to this depopulation. In the early seventeenth century, the sudden collapse of Huronia, the most powerful Indian group in the region, took place shortly after contact with the French. Their numbers dropped from 21,000 to less than half this number within a decade. This demographic catastrophe was not unique to Huronia, but it does provide one example of the deadly impact of European diseases and the cost of colonial alliances on the Aboriginal peoples. French missionaries brought diseases to the Huron villages, while the Iroquois, who opposed the French–Huron alliance, attacked the Huron villages and eventually destroyed the Huron Confederacy in 1649. In little more than a generation, Huron villages were abandoned, the cornfields of Huronia reverted to forest, and its people were greatly reduced in numbers and scattered across the land, some finding their way to the north shore of the St Lawrence in Québec, others being captured and assimilated by the Iroquois, and another remnant joining related tribes in what are today Michigan and Ohio.

John Cabot, the first European explorer to land in Canada after the brief settlement of the Norse at the tip of Newfoundland around AD 1000, reached the east coast in 1497. Cabot was followed by others, including Jacques Cartier and Martin Frobisher. In 1534 Cartier made contact with two Indian tribes along the Gaspé coast, and in 1576 Frobisher encountered an Inuit encampment along the Arctic coast of southern Baffin Island. Both explorers were searching for a route to Asia, but instead discovered a new continent and peoples. In both instances, contact with North Americans ended badly. Lives were lost on both sides, and the ore that the explorers took back to Paris and London, respectively, proved worthless (Vignette 3.1). Instead of gold, they had found fool's gold (iron pyrites).

VIGNETTE 3.1

Contact between Cartier and Donnacona

In 1534, Jacques Cartier made contact at the Gaspé coast with Donnacona, the leader of the St Lawrence Iroquois. The Iroquois, whose village (Stadacona) was located at the site of Québec City, came to the Gaspé to fish for mackerel. The next year Cartier returned. After reaching the Iroquois village of Hochelaga, which later became the site of Montréal, Cartier realized that the St Lawrence did not provide a sea passage to China. Like the Spanish in Mexico and Peru who discovered gold and silver, Cartier hoped to find wealth in the New World. In 1541, on his third voyage to Stadacona, Cartier hoped to find diamonds and gold. He left France with several hundred colonists. By then, Donnacona and all but one of the Iroquois captives that Cartier had taken with him when he returned to France in 1536 were dead. The French authorities, who wished to conceal this fact, decided not to allow the surviving Indian woman to return to Stadacona with Cartier. The French arrived at Cap Rouge where they established a settlement just west of Stadacona. When the St Lawrence Iroquoians realized that Cartier was not going to return their chief and the other Iroquois captives, relationships quickly soured. Over the winter, the Iroquois killed at least 35 of the French settlers. Cartier and his surviving party left for France as soon as the river ice melted. He took along rocks that he believed contained gold and diamonds, but, like the ore mined on Baffin Island by Frobisher's men several decades later, Cartier brought back fool's gold and quartz. As there seemed to be no prospect for mineral wealth, the King of France lost interest in the New World. Sixty years went by before the French made another attempt at establishing a settlement at Stadacona. In 1608, Samuel de Champlain founded Québec City.

Source: Adapted from Ramsay Cook, *The Voyages of Jacques Cartier* (Toronto: University of Toronto Press, 1993).

CANADA'S HISTORICAL GEOGRAPHY

Culture Regions

At the time of contact with Europeans, Aboriginal peoples occupied specific territories (cultural regions). Within each cultural region, these first human inhabitants developed distinct techniques suitable for the local environment and wildlife. The seven culture regions in present-day Canada are the Eastern Woodlands, Eastern Subarctic, Western Subarctic, Arctic, Plains, Plateau, and Northwest Coast (Figure 3.2). The Inuit occupied the Arctic cultural region. In the Eastern Subarctic, the Cree were the principal Algonquian tribe, and further east the Innu (Naskapi and Montagnais) resided. The Cree in this region had developed a technology—snowshoes—to hunt moose in deep snow. In the Western Subarctic, the Athapaskans, including a number of Dene tribes, hunted caribou and other big-game animals. Northwest Coast Indians harvested the rich marine life found along the Pacific coast. Tribes such as the Haida, Nootka (Nuu-chah-nulth), and Salish comprised the Northwest Coast cultural region. In the southern interior of British Columbia, the Plateau Indians—including the Carrier, Lillooet, Okanagan, and Shuswap—occupied the valleys of the Cordillera, forming the Plateau cultural region. Across the grasslands of the Canadian West, Plains Indians such as the Assiniboine, Blackfoot, Sarcee, and Plains Cree hunted bison. The Iroquois and Huron were among those living in the Eastern Woodlands of southern Ontario and Québec, although the Iroquois were primarily based further south, in present-day New York State. Both groups combined agriculture with hunting. In the Maritimes,

> **THINK ABOUT IT**
> Imagine yourself to be Captain Martin Frobisher. How would you communicate with the Baffin Island Inuit who held several of your sailors?

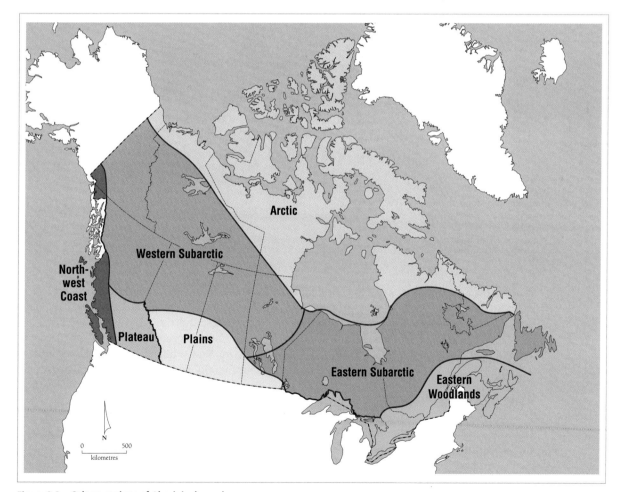

Figure 3.2 **Culture regions of Aboriginal peoples**
Resources and natural conditions found in culture regions provided the foundation for the formation of unique spatial versions of Aboriginal hunting systems and social organizations, and accounted for varying population densities. For example, the reliable and rich sea resources found along the Northwest Coast supported one of the densest Indian populations, while the opposite was true for the Indians in the Subarctic and the Inuit in the Arctic.

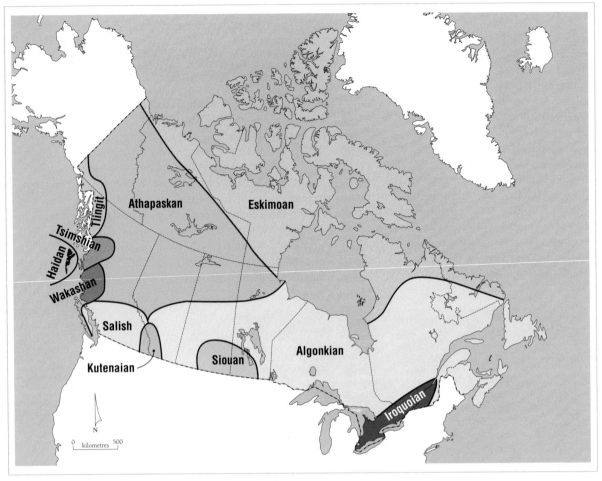

Figure 3.3 **Aboriginal language families**
Aboriginal peoples in Canada form a very diverse population. At the time of contact, there were over 50 distinct Aboriginal languages spoken. These languages formed 11 language families, five of which were in one natural cultural region, the Northwest Coast. Following contact, language loss was swift. By the end of the twentieth century, only three Aboriginal languages, Cree, Inuktitut, and Ojibwa, had over 20,000 speakers.

THINK ABOUT IT
European countries claimed various parts of North America. What was the basis of their claim of sovereignty over lands they "discovered"?

the Mi'kmaq and Maliseet also occupied the Eastern Woodlands, where they hunted and fished. The complexity and diversity of Aboriginal peoples can be gleaned from the spatial arrangement of their languages (Figure 3.3).

The Second People

The colonization of North America by the French and the British was the second major development in Canada's early history. France and England established colonies in North America in the seventeenth century. Québec City, founded in 1608 by Samuel de Champlain, was the first permanent settlement in Canada. By 1663, the French population in New France was about 3,000 compared to a population of about 10,000 Indians (mainly Huron and Iroquois), who lived in the same area of the St Lawrence Valley, the Great Lakes, and the Ohio Basin. By 1750, the French Canadians constituted most of the population in New France, where they numbered about 60,000, while the Indian population continued to drop because of disease and warfare. Following the British Conquest of New France in 1759, the flow of French colonists ceased and British immigrants began to move to what used to be New France. In doing so, they increased the number of British settlers. Meanwhile, the French-Canadian population now had to depend entirely on natural population increase.

The first large wave of British immigrants consisted of refugees from the United States. These Loyalists had supported Britain during the American War of Independence (1775–83). After the defeat of the British army, they sought refuge in other parts of the British Empire, including its North American possessions. In North America, most Loyalists settled in Nova Scotia, while a smaller number moved to the Eastern Townships of Québec and to Montréal. Others settled in what is now southern Ontario. At the end of the American War of Independence, the forty-fifth parallel was established as the border between Lower Canada (Québec) and New York State and Vermont. The St Lawrence River and the Great Lakes became the boundary between Upper Canada and the United States.

The Americans' interest in the remaining territories of North America, including those in British North America, was strong and took the form of Manifest Destiny. In 1803, the United States purchased from France the vast lands west of the Mississippi River known as the Louisiana Purchase. This purchase of 2.14 million km² reinforced the idea that the United States was destined to occupy all of North America. A decade later the War of 1812 was in fact a struggle between competing visions of North America—Americans warring against the British Empire in an attempt to absorb British North America into their young republic; British, Canadians, and their Indian allies fighting to repel these invaders in order to keep British North America intact; and, for the Indians, to maintain their territories free of settlers. The outcome of the War of 1812 was a stalemate, though both the British/Canadians and the Americans claimed victory. As Laxer (2012: 1) points out, however, the Indian forces led by Tecumseh lost by failing to halt the unyielding westward march of settlers, first in America and later in Canada.

The fledgling United States had two objectives in declaring war. The first was to punish Britain while its army and navy were engaged in the Napoleonic Wars. The second and seemingly more attainable objective was to absorb the British colonies to the north. Resistance to the American invasion of the St Lawrence–Great Lakes region of British North America was unexpectedly fierce and American forces were repelled. Historians argue that Canada's national identity was forged during the War of 1812. As well, the battles in Lower Canada (Chateauguay) and Upper Canada (Queenston Heights) tested the limits of America's Manifest Destiny (photo 3.2). Ironically, because of poor communications, what was perhaps the Americans' greatest victory of the war, at the Battle of New Orleans where British forces were routed on 8 January 1815, occurred after the peace was

Photo 3.2

The death of General Brock at Queenston Heights. On 13 October 1812, the first major battle of the War of 1812 took place when American troops crossed the Niagara River with the objective of establishing a military base on Canadian soil before winter set in. These invaders were repelled by British troops, Canadian militia, and Mohawk Indians, forcing the Americans to retreat to the American side of the river. Early in the battle, General Isaac Brock was mortally shot while he led a charge on American forces who had taken a strategic position at the top of Queenston Heights. This C.W. Jefferys painting, c. 1908, romantically and heroically depicts Brock's death.

achieved. At the end of the war, the Treaty of Ghent, signed on 24 December 1814, settled the boundary between the British colonies and the United States. At that time, Upper Canada extended to Lake of the Woods and beyond that the Hudson's Bay Company managed this British territory. Later, with American settlers pouring into the lands beyond the Mississippi, the question of a western boundary became a pressing political question. In 1818, the border from the Lake of the Woods to the Rocky Mountains was set as the forty-ninth parallel. A decision on the boundary to the Pacific coast was left until the Oregon Treaty in 1846, which extended the boundary along the forty-ninth parallel to the Pacific coast—with the exception of Vancouver Island.

The second wave of immigrants came from the British Isles, mainly in the first half of nineteenth century when almost a million people migrated from Britain to British North America. In the 1840s, for example, the potato famines in Ireland added to the problems in the Old World, causing terrible hardships for the Irish people. Thousands fled the countryside and many left for the New World to settle in the towns and cities of British North America and the United States.

These two waves of British immigrants greatly changed Canada, by turning the demographic balance of power from a French-Canadian majority to an English-speaking one. At the time of Confederation in 1867, the population of British North America had reached 3 million, with over 60 per cent of British descent. While the French Canadians were concentrated in Québec and New Brunswick, most English-speakers lived in Upper Canada, the major cities in Québec, and the Maritimes. Across the rest of British North America, Aboriginal people made up most of the population. In the Red River Settlement, a new Aboriginal people, the Métis, who were of Native and European descent, had emerged. By 1869, the Métis, who were split between French- and English-speaking, greatly outnumbered white settlers and fur traders in the settlement, which had a population of nearly 12,000. The Métis formed over 80 per cent of this population (Table 3.2).

But more important, the old balance of demographic power between the French and English had been reversed. British migration to Canada, over the course of 100 years, had changed the demographic balance between French-speaking and English-speaking Canadians. Migration had also changed the ethnic composition of the English-speaking population from almost entirely English to a mixture of English, Irish, Scottish, and Welsh. Moreover, by the 1860s, Canada's ethnic character varied by region. In Atlantic Canada, the Scottish and Irish outnumbered the English. In Québec, the English and Irish formed a sizable minority in the towns and cities, though rural Québec remained solidly French-speaking except in the Eastern Townships. Ontario, like Atlantic Canada, was decidedly British.

Canada began as a collection of four small British colonies—Upper and Lower Canada, New Brunswick, and Nova Scotia. From Great Britain's perspective, union of its colonies was essential to withstand annexation by the United States. Yet, the challenge was how to convince each colony. For Québec, the issue was language, religion, and culture. The Maritimes economy was oriented to the sea, not linked by land to Central Canada. While the Maritimes harboured ill feelings towards Confederation for many decades because its economy suffered, the separation of the British and French in different geographic parts of British

Photo 3.3
Relationships between explorers and Aboriginal peoples were not always peaceful. In this painting by John White, the artist, a member of Frobisher's second expedition in 1577, records a fight between Frobisher's crew and Baffin Island Inuit.

Table 3.2 Population of the Red River Settlement, 1869

Ethnic Group	Population Size	Population Percentage
Whites born in Canada	294	2.5
Whites born in Britain or a foreign country	524	4.4
Indians	558	4.7
Whites born in Red River	747	6.2
English-speaking Métis	4,083	34.1
French-speaking Métis	5,757	48.1
Total Population	11,963	100.0

Source: Adapted from Lower (1983: 96).

North America remains the greatest challenge to national unity and it ensures the continued existence of two visions of Canada. The greatest threat to Canadian unity stems from these two irreconcilable visions.

Further discussion of two visions of Canada is found later in this chapter in the section "The French/English Faultline," page 110.

In 1867, 92 per cent of the population was either British or French. Each linguistic group developed its own vision of Canada: French Canadians saw Canada as two founding peoples, while English-speaking Canadians favoured the notion of equality among provinces. In English Canada, the notion of equal provinces grew out of the following factors:

- The nature of Confederation was such that provincial powers were shared equally.
- The British formed the majority of the population in three of the four provinces,

Photo 3.4
The American Revolution (1775–83) divided the residents of the Thirteen Colonies. With the defeat of British forces, those British subjects who did not support the revolutionary cause were forced to leave, losing their property and sometimes their lives. Considered traitors by Americans, Loyalists were often subjected to mob violence.

- thereby dominating the political affairs of those provinces.
- The British, while a minority within Québec, were the dominant business group.

The Third People

The hegemony of the British and French began to weaken in the early part of the twentieth century when the fertile lands of Western Canada were settled in large part by neither English-Canadian nor French-Canadian farmers. In 1870, Ottawa obtained the vast land of the Hudson's Bay Company and was faced with the question of settling this territory. Little progress was made until after the signing of treaties with various Plains Indian groups and the completion of the Canadian Pacific Railway in November 1885. Now the time was right for massive immigration to settle these lands. For Ottawa, there were two key advantages in encouraging settlement. First, the threat of American settlers moving into the Canadian West and annexing these lands would be diminished. Second, the creation of a grain economy would provide freight for the Canadian Pacific Railway, thereby helping turn it into a viable operation. But where to find the people? Some came from Ontario, Québec, and Atlantic Canada to claim their 160 acres as homesteaders; others came from Britain and the United States (Vignette 3.2).

Still, much of Western Canada was unoccupied. Clifford Sifton, the Minister of the Interior, accepted the challenge to settle the West. By the beginning of the twentieth century, Sifton had launched an aggressive and innovative advertising campaign to lure people from Britain and the United States to "The Last Best West," but this effort failed to bring sufficient immigrants. At that point, he recognized the need to go beyond these two countries. In a break with past immigration policy, Sifton turned his attention to the people of Central Europe, Scandinavia, and czarist Russia. Land-hungry peasants from Ukraine formed the largest single group of immigrants but Doukhobors and German-speaking Mennonites also came from czarist Russia, giving Western Canada a distinct and different mix of ethnic groups and landholdings, with communal rather than individual landholdings the norm among some groups. Sifton recognized that Ukrainian peasants from czarist Russia came from a cold grassland environment and were therefore well suited for settling the Canadian Prairies. He was even willing to accept Russian-speaking Doukhobors, whose religious beliefs and social customs, including communal farming practices, separated them from other homesteaders. From 1901 to 1921, Western Canada's population increased from 400,000 to 2 million, with Saskatchewan becoming the third most populous province by 1921 (Table 3.3). As these ethnic groups and individuals spread across the Prairies, their impact was enormous on a landscape that only recently was populated largely by vast herds of buffalo and semi-nomadic Indian tribes (see Kerr and Holdsworth, 1990: Plate 17).

> **THINK ABOUT IT**
> Homesteaders far from the railway had to use local resources for building materials. What resource was readily available in the Prairie landscape?

VIGNETTE 3.2

The Land Survey System and the Settling of the Prairies

In 1872, the federal government passed the Dominion Lands Act. This legislation established a survey system that divided the land into square townships made up of 36 sections, each measuring 1 mile by 1 mile, with allowances for roads. Each section was further subdivided into four quarters, each quarter section measuring one-half mile by one-half mile and comprising 160 acres. This survey system gave the Canadian Prairies a distinctive "checkerboard" pattern.

Ottawa sought to populate the arable lands of Western Canada by offering "free" land to farmers. The free land took the form of a homestead (a quarter section or 160 acres). A free grant of one quarter section was available to persons 21 years or older with the payment of a $10 fee. Upon fulfilling cultivation and residency requirements within three years of acquiring the property, the homesteader would receive title to the land. Since Ottawa had made substantial land grants to the Canadian Pacific Railway (CPR) and to the Hudson's Bay Company (HBC), not all the land was free. Both the CPR and the HBC sold their land at market prices, thus making a considerable profit.

Table 3.3 **Canada's Population by Provinces and Territories, 1901 and 1921**

Political Unit	Population 1901	%	Population 1921	%
Ontario	2,182,947	40.6	2,933,662	33.4
Québec	1,648,898	30.7	2,360,510	26.9
Nova Scotia	459,574	8.6	523,837	6.0
New Brunswick	331,120	6.2	387,876	4.4
Manitoba	255,211	4.8	610,118	6.9
Northwest Territories*	20,129	0.4	8,143	0.1
Prince Edward Island	103,259	1.9	88,615	1.0
British Columbia	178,657	3.3	524,582	6.0
Yukon	27,219	0.5	4,147	>0.1
Saskatchewan	91,279	1.7	757,510	8.6
Alberta	73,022	1.3	588,454	6.7
Canada	5,371,315	100.0	8,787,949**	100.0

*Saskatchewan and Alberta did not become provinces until 1905 and were officially included in the population of the Northwest Territories.
**Includes 485 members of the armed forces.

Source: Adapted from Statistics Canada (2003).

By opening the door for immigration from European countries without a French or British background, Sifton's immigration policy changed the face of Canada. His goal of settling the West was accomplished, and a new dimension had been added to Canada's social fabric—people with neither a French nor a British background. While the majority of these immigrants were homesteaders, some took jobs in Canada's booming mining and logging industries while others settled in Canada's major cities, especial Montréal and Toronto. Although most of these new settlers soon learned English, this cultural/linguistic difference from the two founding peoples surfaced later in the twentieth century as a powerful force for multiculturalism and pluralism in English-speaking Canada. The immigration to the Canadian Prairies from 1896 to 1914 forms the topic for more detailed discussion later in this chapter under the subheading "The Immigration Faultline."

The Territorial Evolution of Canada

Canada's formal history as a nation began with the proclamation of the British North America Act on 1 July 1867. This Act of the British Parliament united the colonies of New Brunswick, Nova Scotia, and the Province of Canada (formerly Upper Canada and Lower Canada) into the Dominion of Canada.[1] Canada soon acquired more territory. In 1870, the Deed of Surrender transferred Rupert's Land and the North-Western Territory to the federal government, at which time this large expanse that had been under HBC control was renamed the North-West Territories. In 1871, British Columbia joined Confederation, and Prince Edward Island followed two years later. In 1880, Ottawa acquired the Arctic Archipelago from Great Britain. Thus, while Canada began as a small country, consisting of what is now known as southern Ontario, southern Québec, New Brunswick, and Nova Scotia, it quickly became one of the largest in the world (Figures 3.4 and 3.5).

What was behind these real estate deals? For Britain, the **doctrine of Manifest Destiny** first expressed in the early nineteenth century lurked in the political background of British–American diplomacy. More specifically, the union of its North American colonies had three advantages for Britain: (1) a better chance for the political survival of its North American colonies against the growing economic and military strength of the United States; (2) an improved environment for British investment, especially for the proposed trans-Canada

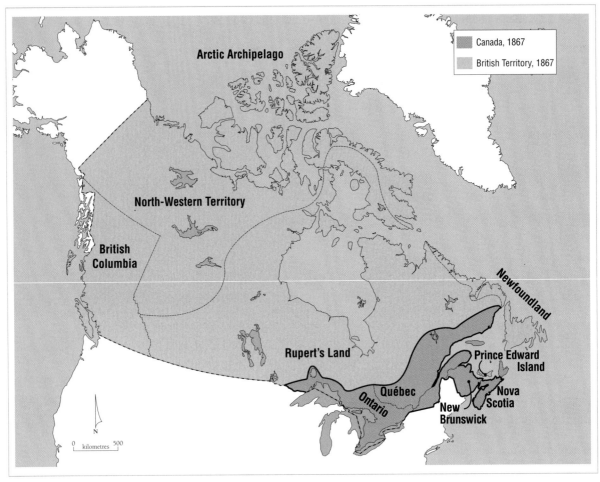

Figure 3.4 **Canada, 1867**
At Confederation, Canada was only a fraction of its current territorial extent. The Hudson's Bay Company controlled most of British North America, including Rupert's Land and the North-Western Territory.

VIGNETTE 3.3

America's Manifest Destiny

The doctrine of Manifest Destiny was based on the belief that the United States would eventually expand to all parts of North America, thus incorporating Canada into the American republic. From its beginnings along the Atlantic seaboard, the United States had greatly increased its territory by a combination of force, negotiation, and purchase. In 1803, the United States purchased the Louisiana Territory (a vast land west of the Mississippi and east of the Rocky Mountains) from France; in 1846, the US gained the Oregon Territory in negotiations with Great Britain; and in 1867 the country bought Alaska from Russia. To Americans, such expansion was an expression of their right to North America. As well, it would rid North America of the much-hated European colonial powers.

Not surprisingly, the Fathers of Confederation were concerned about American designs on British North America. First, in 1866, the Fenians raided Upper Canada, Lower Canada, and New Brunswick with the intention of seizing British North America and holding it for ransom until Ireland was free of British rule. Second, in 1867, the American purchase of Alaska left British Columbia wedged between American territory to its north and south, and the exact boundary along the coastline south of 60° N was uncertain. Canada and the United States settled this final border dispute in 1903.

CANADA'S HISTORICAL GEOGRAPHY

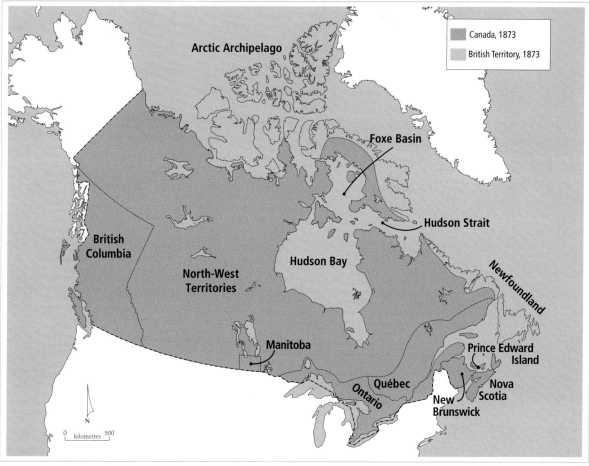

Figure 3.5 **Canada, 1873**
Canada's geographic extent increased between 1867 and 1873. During that short span of time, Canada had obtained the vast Hudson's Bay Company lands (including the Red River Settlement and a small part of the Arctic Archipelago whose streams flow into Hudson Bay and Foxe Basin) as well as two British colonies (British Columbia and Prince Edward Island). For the North-Western Territory and Rupert's Land, the Crown paid the HBC £300,000, granted the Company one-twentieth of the lands in the Canadian Prairies, and allowed it to keep its 120 trading posts and adjoining land. In 1870, these lands were renamed the North-West Territories. In 1880, Britain transferred the Arctic Archipelago to Canada. The details of this transfer are found in Vignette 3.4.

railway; and (3) a reduction in British expenditures for the defence of its North American colonies. The British colonies perceived unification differently. The Province of Canada, led by John A. Macdonald, pushed hard for a united British North America because it would have a larger domestic market for its growing manufacturing industries and a stronger defensive position against a feared American invasion. The Atlantic colonies showed little interest in such a union. As they were part of the British Empire, the attraction of joining the Province of Canada had little appeal. Furthermore, unlike the Canadians, Maritimers continued to base their prosperity on a flourishing transatlantic trading economy with Caribbean countries and Great Britain. From 1840 to 1870, the backbone of the Maritime economy was the construction of wooden sailing-ships. Shipbuilding was so important that this period was known as the "Golden Age of Sail" in the Maritimes. Even diplomatic pressure from Britain to join Confederation had little effect on Maritime politicians. But the Fenian raids into New Brunswick in 1866 and the termination of the Reciprocity Treaty with the United States quickly changed public opinion in the Maritimes.[2] Shortly after the Fenian raids, the legislatures of both New Brunswick and Nova Scotia voted to join Confederation.

As we have seen, within a decade, the territorial extent of Canada expanded from four British colonies to the northern half of North America. The new Dominion grew in size with the addition of other British colonies and territories and the creation of new political jurisdictions (Figure 3.5),

Table 3.4 Timeline: Territorial Evolution of Canada

Date	Event
1867	Ontario, Québec, New Brunswick, and Nova Scotia unite to form the Dominion of Canada.
1870	The Hudson's Bay Company's lands are transferred by Britain to Canada. The Red River colony enters Confederation as the province of Manitoba.
1871	British Columbia joins Canada.
1873	Prince Edward Island becomes the seventh province of Canada.
1880	Great Britain transfers its claim to the Arctic Archipelago to Canada.
1949	Newfoundland joins Canada to become the tenth province.

while the British government transferred its claim to the Arctic Archipelago to Canada in 1880. Negotiations between Ottawa and the British colonies in British Columbia and Prince Edward Island soon brought them into the fold of Confederation, in 1871 and 1873, respectively, and the "numbered treaties" were signed with Indian tribes of the West beginning in 1871. No such negotiations took place with the Red River Métis. The result was the Red River Rebellion of 1869–70, the formation of the Métis Provisional Government, and then negotiations with Ottawa that led to the Manitoba Act of 1870 and entry into Confederation of an initially tiny Manitoba, consisting of the Red River Settlement and a small surrounding area.

THINK ABOUT IT
Is it more accurate historically to refer to the Red River Rebellion as the Red River Resistance?

For further discussion of the first Riel-led resistance, see "The First Clash: Red River Rebellion of 1869–70," page 104.

By 1880, Canada stretched from the Atlantic to the Pacific and north to the Arctic. While still part of the British Empire, Canada had begun the slow journey to independence and nationhood. In 1905 the provinces of Alberta and Saskatchewan were created, and much later, in 1949, Newfoundland joined Canada, completing the union of British North America into a single political entity. (The territorial evolution of Canada is illustrated in Figures 3.4 to 3.8 and Table 3.4.)

Within a decade after Confederation, Canada's territorial extent had burgeoned, making it the second-largest country in the world, which posed a serious problem for the nation's leaders, who had to somehow transform this vast territory into a nation. Sir John A. Macdonald, Canada's first Prime Minister, resolved this in part by authorizing the construction of a transcontinental railway, the Canadian Pacific Railway. The government had to heavily subsidize the building of the railway but, to Macdonald, the high cost was worth it.

National Boundaries

Well before Confederation in 1867, wars and treaties between Britain and the United States shaped

VIGNETTE 3.4

The Transfer of the Arctic Archipelago to Canada

As with the transfer of Hudson's Bay lands in 1870, the territorial size of Canada was greatly increased when Great Britain transferred the Arctic Archipelago to Canada in 1889. Ten years earlier, Canada had acquired the small portion of these lands that drained into Hudson Bay, which included the western half of Baffin Island and several islands located in Hudson Strait, Hudson Bay, and Foxe Basin. Britain's claim to this vast archipelago was based on its naval exploration of the Arctic Ocean. The first venture by Britain into these waters took place in 1576 when Martin Frobisher and his crew sailed to southern Baffin Island. The most intensive exploration took place in the mid-nineteenth century with the search for the lost British naval expedition led by Sir John Franklin (see Vignette 10.3, "The Northwest Passage and the Franklin Search").

many of Canada's boundaries. The southern boundary of New Brunswick, Québec, and Ontario was settled in 1783 when Britain and the United States signed the Treaty of Paris. Under this treaty, the United States gained control of the Indian lands of the Ohio Basin, with Britain controlling Québec lands draining into the St Lawrence River. Earlier, Britain had formally recognized the rights of Indians to the lands of the Ohio Basin, and to the north and west, in the Royal Proclamation of 1763, which provided the constitutional framework for negotiating treaties with Aboriginal peoples. This recognition was the basis of **Aboriginal rights** in Canada (see "Aboriginal Rights" in this chapter). Based on the fur trade route to the western interior, the boundary of 1783 passed through the Great Lakes to the Lake of the Woods. In 1818, Canada's southern boundary was adjusted; it was set at 49° N from the Lake of the Woods to the Rocky Mountains. As for the northwestern boundary, there was some question as to where British territory ended and Russian territory began. British and Russians had come into contact through the fur trade. In the eighteenth century, Russian fur traders established trading posts along the Alaskan coast. Indians travelled along the Yukon River through the interior of Alaska to trade at these Russian forts. By the early nineteenth century, the HBC had reached the tributaries of the Yukon River. In 1825, Britain and Russia set the northern boundary at 141° W (the Treaty of St Petersburg). Russia also agreed to relinquish to Britain its claims to the coastal regions south of 54°40′ N to 42° N. In Atlantic Canada, the boundary between Maine and New Brunswick had not been precisely defined in 1783. Minor adjustments took place in 1842 when Britain and the United States concluded the Webster–Ashburton Treaty.

In the early nineteenth century, the border separating Canada's western territories from the United States was not well defined. In fact, it depended on the natural boundary between Rupert's Land and the Louisiana Territory. Rupert's Land was defined as the lands whose waters flow into Hudson Bay while the lands of the Louisiana Territory were determined by the rivers draining into the Mississippi River system. Rupert's Land included the Red River Valley while the Louisiana Territory had the Milk and Poplar rivers flowing from what is now Alberta and Saskatchewan into the Missouri River. In 1818, Britain and the United States decided on a compromise of the forty-ninth parallel because it was easier to delineate. With the establishment of the North American Boundary Commission in 1872, this boundary was surveyed and marked (see photo 3.5).

The last major territorial dispute between Britain and the United States took place over the **Oregon Territory**. The Oregon Territory stretched along the Pacific coast from Alaska to Mexico—Mexico at that time extended northward to 42° N. Britain's claim to the Oregon Territory hinged on exploration and the fur trade. David Thompson of the North West Company reached this area in 1807, and after the 1821 merger of the North West Company with the Hudson's Bay Company, the HBC established Fort Vancouver at the mouth of the Columbia River in 1825, where it conducted trade with the local Indian tribes. In the 1830s, American settlers crossed the Rocky Mountains into the Columbia Valley, where they proceeded to cultivate the fertile soils of the Willamette Valley. Because both Britain and the United States had a foothold in the Oregon Territory, sovereignty was uncertain. Great Britain's claim was based on two factors: (1) the HBC had established the first European settlement in the region, at Fort Vancouver; and (2) the Company exerted control over the land where the Indians trapped fur-bearing animals. The American claim was centred on the fact that the vast majority of settlers were Americans. In the final outcome, there was no doubt that occupancy was a more powerful claim to disputed lands than claims based on exploration and the presence of a fur-trading economy. Too late, the British urged the HBC to bring settlers from Fort Garry to the Oregon Territory (the **Red River migration**). With the Oregon Boundary Treaty of 1846, the boundary between British and American territory from the Rockies to the Pacific coast was set at 49° N with the exception of Vancouver Island, which extended south of this parallel.

Internal Boundaries

Since Confederation, the internal boundaries of Canada have changed (Figures 3.4–3.8). These changes have created new provinces (Alberta and Saskatchewan) and territories (Yukon and Nunavut). As well, the boundaries of Manitoba, Ontario, and Québec were extended. In all cases, these political changes took land away from the Northwest Territories.

In 1870, the boundary of Manitoba formed a tiny rectangle comprising little more than the Red River Settlement. The province's western

THINK ABOUT IT
Why did Canada and the US set the boundary from the Lake of the Woods to the Pacific coast at the forty-ninth parallel rather than use the natural boundary between the Mississippi/Missouri River Basin and the Hudson Bay drainage basin? As a result, the upper reaches of the Red River went to the United States and the headwaters of the Milk and Poplar rivers were assigned to British North America.

Photo 3.5
A crew from the North American Boundary Commission building a mound marking the border between Canada and the United States, August or September 1873.

THINK ABOUT IT
Examine Figures 3.6 and 3.7. Did the decision of King Charles II in 1670—that the Hudson's Bay Company was given control over lands draining into Hudson Bay—affect the determination by the British Privy Council of the 1927 border between Québec and Labrador?

boundary, while extended in 1881 and 1884, did not reach its present limit until 1912. By this time, Manitoba spread north to the boundary with the Northwest Territories (now Nunavut) and east to the Lake of the Woods.

Québec, too, received northern territories. In 1898, its boundary was extended northward to the Eastmain River and then eastward to Labrador. In 1912, Ottawa assigned Québec more territory that extended its lands to Hudson Strait. Canada also believed that the province of Québec should extend to the narrow coastal strip along the Labrador coast, while the colony of Newfoundland contended that Newfoundland owned all the land draining into the Atlantic Ocean. In 1927, this dispute between two British dominions (Canada and Newfoundland) went to London. The British government ruled in favour of Newfoundland (Figure 3.7). The Québec government has never formally accepted this ruling, though Québec has respected the boundary. However, the long-term contract for Québec to purchase electric power at 1960 prices from the Churchill Falls hydroelectric station located in Labrador remains a sore point for Newfoundland and Labrador. Plans to produce more hydroelectric power in Labrador, sell it to Hydro-Québec, and then transmit the power through Québec to markets in New England remain the most viable, but another option calls for a partially undersea transmission route to Nova Scotia and then to New England.

"The Politics of Hydro Development: The Lower Churchill Project" in Chapter 9, page 395, discusses the hydroelectric relationship between Québec and Newfoundland and Labrador.

In the years following Confederation, Ontario gained two large areas. In 1899, its western boundary was set at the Lake of the Woods (previously this area belonged to Manitoba); at the same time, its northern boundary was extended to the Albany River and James Bay. Then, in 1912, Ontario obtained its vast northern lands, which stretch to Hudson Bay.

In 1905, Canada formed two new provinces, Alberta and Saskatchewan. The final adjustment to Canada's internal boundaries occurred in 1999 with the establishment of the territory of

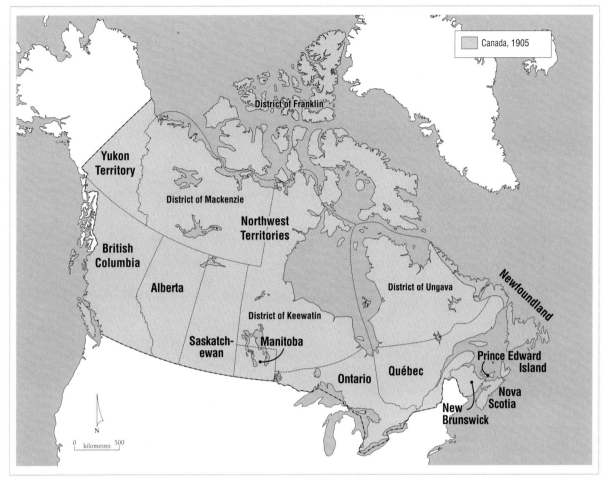

Figure 3.6 **Canada, 1905**
By 1905, two new provinces (Alberta and Saskatchewan) and two territories (Yukon and the Northwest Territories) were created out of the North-West Territories and the Arctic Archipelago, which was ceded to Canada in 1880 and later formed the District of Franklin. As well, the provinces of Ontario, Québec, and Manitoba expanded their boundaries into the former North-West Territories.

Nunavut (Figure 3.8 and Table 3.5), which was hived off from the Northwest Territories in the eastern Arctic.

Faultlines in Canada's Early Years

Canada's regional geography has always been defined by its faultlines, a notion introduced in Chapter 1. For better or worse, this aspect of regionalism is a fact of life in Canada and it may well be the most telling characteristic of Canada's changing national character over the centuries. Four faultlines described in this text have their roots in Canada's historical geography. In all cases, these cracks in Canada's unity pose powerful challenges to the federal government. The federal government, because it is charged with establishing national policies and programs, has the power to settle such issues. The challenge facing Ottawa is to seek a balance between the regional factions, but such efforts rarely satisfy all parts of Canada. Nevertheless, the process of searching for solutions to these tensions informs Canadians of different views and, in doing so, lessens the tension—perhaps not initially but usually over time. Even so, Ottawa is most aware of the political power (i.e., the number of seats in the House of Commons) of Ontario and Québec and, whether real or not, federal policies have seemed to favour these two largest provinces—at least until recently.

Faultlines first appeared in the early days of Canada and have evolved over time. The sheer size

THE REGIONAL GEOGRAPHY OF CANADA

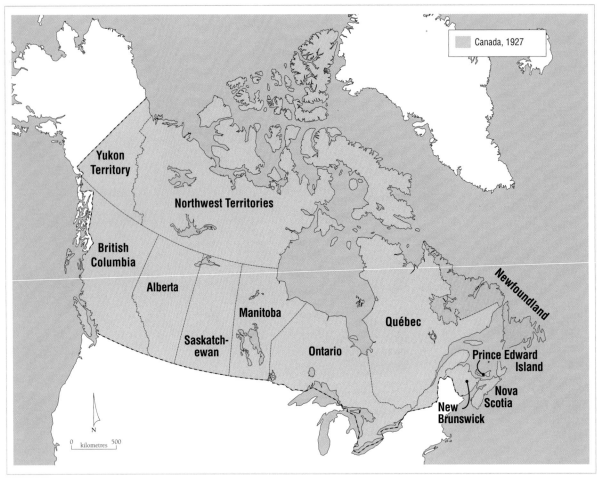

Figure 3.7 Canada, 1927
The complicated history of Lower Canada and Newfoundland provided ample justification for both parties to claim the land between the Northern Divide and the coastal strip associated with the fisheries. In 1927, the Privy Council of the British Parliament ruled in favour of Newfoundland by selecting the watershed boundary, a decision that dated back to 1670 when King Charles II created Rupert's Land. In 1912 Ontario, Québec, and Manitoba gained additional northern lands to reach their current geographic size.

of Canada and its varied physical geography provide the basis for a regional faultline, though the nature of this faultline has changed over time. In the nineteenth century, the federal government launched two national initiatives that changed the course of Canada's history and added fuel to the centralist/decentralist faultline. The first initiative, the CPR, linked the country from the Atlantic to Pacific and thus overcome Canada's vast space. The second challenge was to establish an industrial core in Central Canada through the **National Policy**, which established high tariffs on imported goods and encouraged a home market for the manufactures of the core. These two federal efforts provided the basis of a national core/periphery economic structure that endured for over 100 years.

Rightly or wrongly, Canadians living outside these two provinces, known collectively as Central Canada, believed that Ontario and Québec had an unfair influence over national policies and therefore Ottawa would favour economic development in Central Canada over that in the rest of the country. From a different perspective, given that the majority of voters continue to live in these two provinces, their concerns often translate into national concerns that federal governments must address. In 1911, 62.9 per cent of Canadians lived in Central Canada and they held 68.3 per cent of the seats in the House of Commons. By 2011, these percentages had declined to 62.0 per cent of the population and 58.8 per cent seats in the

CANADA'S HISTORICAL GEOGRAPHY

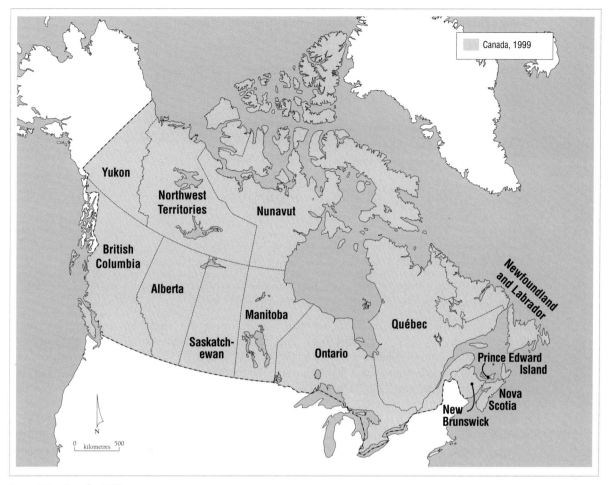

Figure 3.8 **Canada, 1999**
On 1 April 1999, Nunavut became a territory.

Table 3.5 **Timeline: Evolution of Canada's Internal Boundaries**

Date	Event
1881	Ottawa enlarges the boundaries of Manitoba.
1898	Ottawa approves extension of Québec's northern limit to the Eastmain River.
1899	Ottawa decides to set Ontario's western boundary at the Lake of the Woods and extend its northern boundary to the Albany River and James Bay.
1905	Ottawa announces the creation of two new provinces, Alberta and Saskatchewan.
1912	Ottawa redefines the boundaries of Manitoba, Ontario, and Québec, extending them to their present position.
1927	Great Britain sets the boundary between Québec and Labrador as the Northern Divide. Québec has never accepted this decision.
1999	A new territory, Nunavut, is hived off from the Northwest Territories in the eastern Arctic.

Table 3.6 Members of the House of Commons by Geographic Region, 1911 and 2011

Geographic Region	1911		2011	
	Members (no.)	Population	Members (no.)	Population
Territorial North	1	15,019	3	107,265
British Columbia	7	392,480	36	4,000,057
Western Canada	27	1,328,121	56	5,886,906
Atlantic Canada*	35	937,855	32	2,327,638
Québec	65	2,005,776	75	7,903,001
Ontario	86	2,527,292	106	12,851,821
Total	221	7,206,543	308	33,476,688

*In 1911, the province of Newfoundland and Labrador had not yet joined Confederation.

Sources: Parliament of Canada, *Members of the House of Commons*, 2012, at: <www.parl.gc.ca/ParlInfo/Lists/Members.aspx?Language=E&Parliament=66c50d6d-ffa1-419d-b704-a320a9eedadf&New=False&Current=False&First=True&ElectionDate=1911.09.21>; House of Commons, *Party Standings: 41st Parliament*, 2011, at: <www.parl.gc.ca/SenatorsMembers/House/PartyStandings/standings-e.htm>.

Commons (see Table 3.6). Even though their political weight declined somewhat over this span of time, Central Canada and especially Ontario still held sway. For that reason, Canadians should not have been surprised when the federal government reacted to a "national" crisis in the Canadian automobile industry by providing massive financial support in 2009 to General Motors and Chrysler. Equally important, Atlantic Canada, even with the addition of Newfoundland and Labrador, was a net loser of political power over the past century while British Columbia registered the largest gain.

A second prominent faultline emerges from Québec's place within Confederation. While the province of Québec and the Canadian federal system ensure a relatively safe haven for Québécois society and its French language within North America, sharp disagreements with Ottawa and the rest of Canada have surfaced from time to time. In the late nineteenth century the nature of Canada was debated: was the country 10 equal provinces

VIGNETTE 3.5

Times Change for Canada's Regions

By the beginning of the twenty-first century, the twentieth-century model for Canada's economy had been replaced by a global one that affected Canada's regions entirely differently. Resource-rich provinces benefited from the high prices for their commodities while the manufacturing provinces suffered from low-cost imports and a drop in exports. In this sense, globalization rearranged the spatial pattern of wealth in Canada and upset the traditional arrangement of "have" and "have-not" provinces. Not surprisingly, a new round of regional tensions surfaced. The debate over "Dutch disease" provides one insight into these regional tensions and, whether true or false, this doctrine has become a key battleground between federal and provincial leaders, with federal NDP leader Thomas Mulcair leading the charge against the oil sands while Saskatchewan Premier Brad Wall attempted to poke holes in Mulcair's argument. In effect, the Dutch disease argument is that as the exploitation of and revenues from natural resources become central to a national economy, this makes the nation's currency stronger and, consequently, manufactured exports decrease significantly because of the higher value of the currency. The federal government, not wishing to dampen the booming resource economy that has helped keep Canada's head above the dark financial clouds swirling around the global economy, has thrown cold water on the Dutch disease doctrine and remains eager to "make oil while the sun shines."

or a partnership between Québec and the rest of Canada? Without a doubt, this cornerstone of misunderstanding over Confederation has taken different forms of disagreement, including separatism.

The place of Aboriginal peoples within Canada forms another key faultline. As seen in the War of 1812, Indians in Central Canada, and in adjoining sections of the republic to the south, allied with the British to prevent the American invaders from destroying British North America—and the promises inherent in the Royal Proclamation of 1763. But the Indian tribes lost their desired place in North America as their lands east of the Mississippi were opened to a flood of American settlers. Treaties between Ottawa (and the British monarch) and the Indians were made on the Prairies nearly a hundred years later, as they had been decades earlier in Central Canada; but a combination of events including the 1876 Indian Act, a flood of settlers arriving in the Canadian Prairies in the late nineteenth and early twentieth centuries, and the creation of reserves and Indian residential schools represented dark days for Indian peoples.

Lastly, the flood of non-British immigrants to Canada in the late nineteenth and early twentieth centuries added another dimension to what had been primarily the British and French ethnicity of Canada. This newcomer element eroded the traditional balance of power between the two founding European peoples, the English and French, and, as the numbers of newcomers increased, eventually led to a pluralistic Canada.

The Centralist/Decentralist Faultline

The traditional centralist/decentralist faultline has had specific characteristics. On the one hand, centralists advocated a strong central government, national policies that exert a political dominance over provinces, and a strong national economy (which means an expanding industrial core primarily in Ontario and Québec). The underlying premise is that a viable manufacturing sector requires economies of scale, which translates into a large local market and access to distant markets. On the other hand, decentralists have sought to strengthen the powers allocated to provinces. In particular, decentralists over the years have called for a devolution of federal powers to the provincial governments and the expansion and diversification of regional economies (see Figure 3.9).

In the late 1970s, the price of oil rose dramatically, thus propelling Alberta from a "have-not" province to a "have." Although resource development falls under provincial powers, in 1980 the federal Liberal government intervened in the marketplace by imposing the **National Energy Program** on oil-producing provinces. This program had four goals:

- increase national energy security;
- expand Canadian private and public ownership of the oil industry;
- provide Ottawa with a larger share of oil revenues;
- maintain lower oil prices in Central Canada.

The National Energy Program lasted only four years (1980–4), but it altered the political landscape in Canada. Its chief political legacy was to deepen western alienation to the federal government and to cause the Liberal Party to lose its western base for the next 30 years and perhaps longer. This political divide surfaced again in 2012 under the rubric of the Dutch disease debate. Now the focus is on whether the federal government should introduce a carbon tax on the carbon content of energy. Two provinces, Québec in October 2007 and BC in July 2008, have imposed such a tax—and, despite dire warnings from the opponents of carbon taxes, their provincial economic skies did not fall. Ottawa has considered and then rejected another approach known as a cap-and-trade system, whereby Ottawa would set emission quotas and those industries exceeding their quotas would have to buy unused quota from other companies. The explanation for this waffling is that the federal government is waiting for the US to take the lead in the creation of a continental cap-and-trade form of a carbon tax, which would allow a continental system of quotas, reward those industries reducing their carbon footprint, and promote cross-border trading.

The Federal Role

The most critical role of the federal government is to maintain national unity. Yet, because the powers to govern are shared with provinces, this role is complicated, making Canada "a troublesome country to govern." Each Prime Minister, from Sir John A. Macdonald to Stephen Harper, has had to balance national economic interests against regional ones. Federal interventions in the marketplace are attempts to protect national interests, but they sometimes appear to focus on a single region.

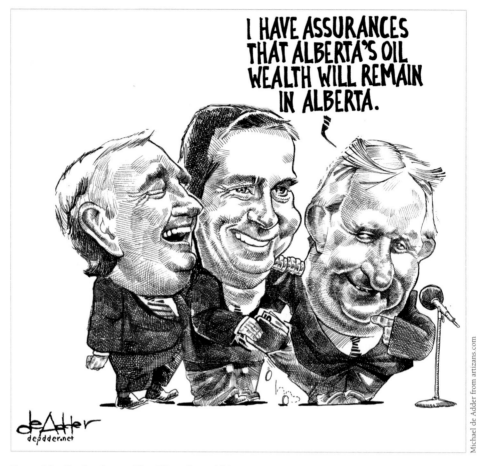

Figure 3.9 Sharing the wealth—Alberta's wealth?
In August 2005, cartoonist Michael de Adder captured the tensions between Alberta and Ontario, with Ottawa caught in the middle. In the last year of the Liberal government under Prime Minister Paul Martin, Ontario's Premier Dalton McGuinty called for a "sharing of Alberta's oil wealth." At the same time, the Prime Minister assured Alberta's Premier Ralph Klein that Alberta's oil revenues would remain in Alberta, but in the back of his mind thoughts of another majority government based on vote-rich Ontario must have lingered. In this case, Ottawa did not repeat a new version of the National Energy Program but that fear remains deeply rooted in Alberta politics and forms one of the elements of western alienation.

A recent example was the 2009 financial bailout of General Motors and Chrysler that "saved" the automobile industry in Ontario. In the larger picture, however, this was a North American bailout that also involved the US government and the US auto industry, and certainly in some instances what is good for one region can be good for the rest of the country. Had the auto industry gone belly up or shrunk drastically, which seemed likely without the bailout, unemployment in Ontario and the loss of tax revenues to that province (and to the federal government) would have meant that other regions, notably Western Canada, would have had to provide much more in equalization payments to Ontario.

Vignette 5.12, "The Bailout of Chrysler and GM: Sound Public Policy?" presents more details on the auto industry bailout.

Under Canada's federal system, the powers of government are divided between the federal government and the 10 provincial governments. All provinces have the same powers over education, health and welfare, highways, civil law (property and civil rights), local government, and natural resources, while Ottawa has a much wider mandate, including authority over defence and foreign affairs, criminal law, money and banking, trade, transportation, citizenship, and Indian affairs. The two levels of government were assigned joint

CANADA'S HISTORICAL GEOGRAPHY

CONTESTED TERRAIN 3.1

Equalization Payments

Why does the Equalization Program cause sparks to fly between Ottawa and the provinces? The calculation of the amount that less prosperous provinces receive is straightforward. It is determined by measuring provinces' ability to raise revenues. Yet, political life in Canada is complicated. In 2004, Ottawa decided to deduct the rapidly growing oil royalties received by Newfoundland and Labrador from its equalization payments. The Premier of Newfoundland and Labrador, Danny Williams, was so enraged by this unilateral decision by the federal government that he ordered the removal of the Canadian flag from provincial buildings. Ultimately, the Atlantic Energy Accord, an agreement that was favourable to the province, was the outcome.

jurisdiction over agriculture, immigration, and taxation. Territorial governments are assigned their powers from Ottawa, while municipal governments obtain theirs from the provincial governments.

Major Federal–Provincial Programs

The federal government offers four programs that transfer funds to provinces. Two programs, the Canada Health Transfer and Canada Social Transfer, are designed to assist provinces in funding their health and social programs, while the Equalization Program delivers money to less prosperous provinces. The three territories benefit from a separate program called Territorial Formula Financing. In 2012–13, the cost of these programs was $59 billion: Canada Health Transfer ($28.6 billion), Equalization Program ($15.4 billion), Canada Social Transfer ($11.9 billion), and Territorial Formula Financing ($3.1 billion) (Canada, Department of Finance, 2011).

Why does Ottawa provide financial support to provincial and territorial governments? First, equalization payments are designed to ensure a reasonable degree of economic equality across the country because some provinces ("have-not" provinces) are less able to meet their fiscal obligations than others ("have" provinces). In 2012–13, these funds went to six provinces (Table 3.7): Quebec ($7.4 billion), Ontario ($3.3 billion), Manitoba ($1.7 billion), New Brunswick ($1.5 billion), Nova Scotia ($1.3 billion), and Prince Edward Island ($337million). The Territorial Formula Financing saw Yukon receive $767 million, Northwest Territories $1.07 billion, and Nunavut $1.27 billion (Canada, Department of Finance, 2011).

VIGNETTE 3.6

What Is Equalization?

Equalization is the government of Canada's transfer program for addressing fiscal disparities among provinces. This program became a formal federal commitment to the provinces in 1957. Its purpose was entrenched in the Constitution Act, 1982 (s. 36[2]):

Parliament and the government of Canada are committed to the principle of making equalization payments to ensure that provincial governments have sufficient revenues to provide reasonably comparable levels of public services at reasonably comparable levels of taxation.

Equalization payments are unconditional—receiving provinces are free to spend the funds according to their own priorities.

Source: Canada, Department of Finance (2011).

Table 3.7 **Total Equalization Payments ($ millions), 2000–1 to 2012–13**

Year	PEI	NB	NL	NS	MB	QC	SK	BC	ON	Total
2000–1	269	1,260	1,112	1,404	1,314	5,380	208	0	0	10,947
2001–2	256	1,202	1,055	1,315	1,362	4,679	200	240	0	10,309
2002–3	235	1,143	875	1,122	1,303	4,004	106	71	0	8,859
2003–4	232	1,142	766	1,130	1,336	3,764	0	320	0	8,690
2004–5	277	1,326	762	1,313	1,607	4,155	652	682	0	10,774
2005–6	277	1,348	861	1,344	1,601	4,798	82	590	0	10,901
2006–7	291	1,386	632	1,451	1,709	5,539	13	260	0	11,281
2007–8	294	1,308	477	1,477	1,826	7,160	226	0	0	12,768
2008–9	322	1,584	197	1,571	2,063	8,028	0	0	0	13,765
2009–10	340	1,689	0	1,571	2,063	8,355	0	0	347	14,365
2010-11	330	1,581	0	1,110	1,826	8,552	0	0	972	14,372
2011-12	329	1,483	0	1,167	1,666	7,815	0	0	2,200	14,659
2012-13	337	1,495	0	1,268	1,671	7,391	0	0	3,261	15,482

Note: Alberta received no equalization payments over this 13-year period.

Source: Canada, Department of Finance (2011). Reproduced with the permission of the Minister of Public Works and Government Services Canada, 2013.

THINK ABOUT IT
Both Manitoba and Québec receive large revenues from their hydro Crown corporations. Should these revenues, like those derived from other natural resources, be included in the calculation of equalization payments?

Second, other transfer payments are aimed at supporting specific public programs delivered by provinces. These provincial programs were assigned to the provinces at the time of Confederation. Since then, Canada's education, health, and social service programs have evolved and now constitute the major expenditure facing each province. During the same time, provincial revenues through taxes and royalties have not kept pace, hence the need for federal transfer payments.

The Aboriginal/Non-Aboriginal Faultline

In many ways, the Aboriginal/non-Aboriginal divide is the most complex one facing Canada. Its complexity results from the tangled historical relations between Aboriginal peoples and European settlers, first through the British Crown and then between Aboriginal peoples and Ottawa; the clash between hunters and settlers for the same land added to the complexity; and the forced assimilation policies of the federal government solidified an Aboriginal distrust of the Canadian state and the Crown. These failed policies, which continued into the late twentieth century, created an enormous divide between Aboriginal peoples and the rest of Canadian society. The net result was a disaster for Aboriginal peoples who, pushed to the margins of Canadian society, caught in a dependency relationship with Ottawa, and facing unrelenting forms of racism, became the invisible and ignored members of Canadian society. The fact that treaty Indians only obtained the vote for federal elections in 1960 is but one example of their isolation from Canadian society.

Canada's Aboriginal citizens face an uphill battle to find a place in Canadian society. This place will vary with each Aboriginal community and individual. Those in remote communities have different needs and goals than those in urban Canada. Key events that offer the prospects of a new path forward for Aboriginal peoples are (1) the recognition by the federal government of past wrongs and (2) the awareness of the general population of the hard times faced by First Nations. One step in that direction has been the Truth and Reconciliation Commission, created as part of the Residential Schools Settlement Agreement. In 2012, this Commission was still holding public meetings across the country with survivors of the residential schools, their children, and other members of the Canadian public. The Commission's goal is to provide closure for these former students and, through this process, better educate the broader Canadian public of past wrongs. Will this Commission help to turn the corner for Aboriginal peoples? While time will tell, a First Nations columnist for the Saskatoon *StarPhoenix*, Doug Cuthand (2012), concluded:

> I realize that you can't bury the past, but we need to move on. It's said that if you can't get over something the best you can do is get through it. We have to think of future generations, and I hope that time is now.

The diversity of Aboriginal peoples, the scattered nature of their reserves and settlements, the erosion of their culture and languages, and what has been their largely passive struggle against assimilation add to their dilemma. However, the rise of the Idle No More movement in late 2012 suggests that passivity—and acceptance of whatever the First Nations leadership determines—may no longer be an option.

Here, our attention is on past relations, on a time when alliances were the order of the day—with each partner having different goals.

The Royal Proclamation and the Haldimand Grant

History sometimes makes strange allies. Shortly after Pontiac, chief of the Odawa, led a successful uprising against the British in 1763, Britain decided to form an alliance with him and other Indian leaders.[3] Pontiac's goal was to keep the Ohio Valley lands free of settlers from New England while Britain could not hold these lands without the support of Pontiac. For strategic reasons, then, George III issued the Royal Proclamation of 1763, which identified a part of British territory west of the Appalachian Mountains as Indian lands. The British also believed that Indians had a limited ownership over the forested lands they inhabited, and that therefore such lands could not be occupied by settlers but must be purchased from the Indian "owners." This somewhat ambiguous concept remains the basis of land claims by Canadian Aboriginal peoples.

With the American victory, the concept of Indian lands in the Ohio Valley quickly disappeared as a flood of land-hungry settlers poured across the Appalachian Mountains. The defeated Indian forces retreated to Canada where they received the first major Indian land grant, the Haldimand Grant of 1784. The purpose was to reward the Iroquois who had served on the British side during the American Revolution. In his proclamation, the Governor of Québec, Lord Haldimand, prohibited the leasing or sale of land to anyone but the government in the tract extending from the source of the Grand River in present-day southwestern Ontario to the point where the river feeds into Lake Erie. However, Joseph Brant, the leader of the Iroquois, insisted that they had the same rights as the colony's Loyalist settlers, that is, freehold land tenure. And so the waters were muddied by the early sale and lease of plots of land in the original Haldimand Grant. This issue has been part of the contemporary conflict between non-Aboriginal residents and Six Nations Iroquois at Caledonia, Ontario.

Canada Takes Over with the Indian Act

In 1867, the British North America Act transferred the responsibility for the Indian tribes from Great Britain to Canada. How was Canada going to manage these tribes? The answer came in 1876, when the government passed the Indian Act. This restrictive legislation had the effect of isolating Indian tribes from the rest of Canada and stripping them of the power to govern themselves. Basing its action on the premise that Indians could not manage their affairs, Ottawa, through the federal Department of Indian Affairs, served as their guardian until First Nations were fully integrated into Canadian society—as defined by Ottawa. As a result, the federal department intervened in band issues, including managing Indian lands, resources, and moneys, with the objective of assimilating Indians into Canadian society. This Act promoted a dependency on Ottawa and left control of band affairs in the hands of local Indian agents, thus stifling Indian initiatives. While Indians were living in Canada, they were isolated from other Canadians and did not have the rights of citizenship, including the right to vote. Perhaps the only positive outcome of the Act was, unlike the Métis, Indian land could not be sold to private individuals unless approved by Ottawa. Oddly enough, the Métis and Inuit did not fall under this Act, but they too fell into this twilight zone of living in Canada but not being fully accepted. Today, the Inuit have a homeland in the territory of Nunavut, as well as in the northern extremes of the Northwest Territories, Québec, and Labrador, and in March 2013 the

THINK ABOUT IT
Why did the Métis sell their scrip instead of converting it to farmland? When did the "responsibility" for the Inuit fall to Ottawa?

Manitoba Métis Federation won a landmark case in the Supreme Court in regard to the government's failure—in 1870 and ever since—to provide the Métis a proper land base.

Residential Schools: An Assimilation Tool

From the beginning, Ottawa's objective was the assimilation of Indian peoples into Canadian society (Milloy, 1999). Education was an important tool in federal efforts to "civilize" Aboriginal people. One such effort, Indian residential schools, stands out. Spread across Canada but concentrated in Western Canada, the residential schools were operated by the major religious groups but especially the Roman Catholic Church. Without a doubt, residential schools were the most painful experience for many Indian children and their parents, and this learning experience has had long-term effects (Vignette 3.7). After a detailed examination of Indian–white relations, J.R. Miller (2000: 269) concludes that:

> While some students of these residential schools were thoroughly converted by the experience, many more absorbed only enough schooling to resist still more effectively. It would be from the ranks of former residential school pupils that most of the leaders of Indian political movements would come in the twentieth century. By any reasonable standard of evaluation, the residential school program from the 1880s to the 1960s failed dismally.

Not only did this assimilation program fail, but many students were abused, some sexually, by their religious teachers. By the 1990s some residential school survivors began to seek reparation for harms through the courts, and the Canadian legal system demanded financial compensation. The churches claimed that they were unable to pay for these claims and Canada offered to pay 70 per cent of compensation in respect of joint government and church liabilities to victims of sexual and physical abuse at Indian residential schools. The churches involved—Roman Catholic, United, Methodist, Presbyterian, and Anglican—negotiated separate financial agreements with Ottawa. The Anglican Church, in 2003, was the first to reach a settlement, for payment of up to $25 million. It had also been the first church to formally apologize, in 1993, for its part in the tragic residential schools history. On 23 November 2005, the Canadian government announced a $1.9 billion compensation package to benefit tens of thousands of survivors of the residential schools. The settlement provides for a lump-sum payment to former students: $10,000 for the first school year plus $3,000 for each additional year. The average payout has been about $28,000. Those who suffered sexual or serious physical

VIGNETTE 3.7

Indian Residential Schools

In 1892, the federal government entered a formal arrangement with several Christian churches—Roman Catholic, United, Anglican, Methodist, and Presbyterian—to provide a boarding school education for young Aboriginal children. The churches ran the schools; Ottawa paid the bills. The plan was to quickly assimilate these young children into society by removing them from their families and home communities and by insisting that they not use their native languages. The effect was to destroy their culture and leave them between two worlds without roots in either one. While some parents wanted their children to attend these schools, many others were forced to send their children. From 1931 to 1996, about 150,000 Indian, Inuit, and Métis children attended boarding schools; at least 3,000 students died at the schools, largely from diseases such as the Spanish flu (CBC News, 2013a).

The federal government (and society in general) believed that Aboriginal children could be successful in modern society if they abandoned their culture and language and adopted Christianity, learned English or French, and had a basic education. Attendance was mandatory and this rule was enforced by Indian agents and other federal officers as well as by missionaries. By the 1980s, the failure of this assimilation program was self-evident. The last school was closed in 1996.

abuses, or other abuses that caused serious psychological effects, could apply for additional compensation or seek redress through the courts. Finally, on 11 June 2008 the Prime Minister made a formal apology in the House of Commons for the harm done to individuals, families, and cultures by the residential schools.

Since the 1970s, Ottawa has adopted a more enlightened policy towards resolving issues related to Canada's First Peoples, stressing three elements: settling of outstanding land claims, recognizing Aboriginal right to self-government, and accepting that the concerns and rights of each Aboriginal people (Indians, Métis, and Inuit) are different and that such concerns and rights require specific solutions.

Aboriginal peoples' struggle to escape from the dependency trap set by the Indian Act and, in doing so, to find an acceptable place within Canada is rooted in two questions:

- Who are the Aboriginal peoples of Canada?
- What are Aboriginal and treaty rights?

Defining "Aboriginal Peoples"

The Constitution Act, 1982 refers to Indians, Métis, and Inuit under the umbrella term **Aboriginal peoples**, that is, those now living in Canada who can trace their ancestry to the original inhabitants who were in North America before the arrival of Europeans in the fifteenth century. From a cultural perspective, the legal terms used to describe First Nations people as status, non-status, and treaty Indians have little meaning in regard to their traditional or current lifeways or their relationship with the land. Indians legally defined as **status (registered) Indians** are recorded by the federal government as Indians, according to the Indian Act as amended in June 1985, and have certain rights acknowledged by the federal government, such as tax exemption for income generated on a reserve. According to data compiled by the Indian Affairs department, the number of status Indians—from 617 First Nations—had grown to 868,206 by 31 December 2011 (AANDC, 2013).

Non-status Indians are those of Indian ancestry who are not registered as Indians and therefore have no rights under the Indian Act. **Treaty Indians** are status or registered Indians who are members of (or can prove descent from) a band that signed a treaty. They have a legal right to live on a reserve and participate in band affairs. Less than half live on reserves. The **Métis** are people of European and North American Indian ancestry. The **Inuit** are Aboriginal people located mainly in the Arctic.

Statistics Canada records Aboriginal people by their identity as declared by those individuals on census day. In 2011, the *National Household Survey* recorded 1,400,685 Aboriginal peoples which was composed of 851,560 First Nations Indians; 451,795 Métis; and 59,445 Inuit (Statistics Canada, 2013). The principal reason for the difference in population size of registered Indians recorded by Aboriginal Affairs and Northern Development Canada (AANDC) and the census figure for First Nations population is due to the two data collection methods. The registry kept by Aboriginal Affairs and Northern Development Canada (AANDC) is based on the list of Indians compiled by each of the 617 bands, while the census records the self-identity of people. Some people are missed in the census survey and a few bands have refused to allow a census enumeration.

The Indians, Inuit, and Métis are a highly diversified population. One indication of their cultural diversity is linguistic classification. As noted earlier, there were approximately 55 distinct Aboriginal languages (of 11 language families) spoken in Canada at the time of original contact (Figure 3.3). The largest language family is Algonkian. There are 15 distinct Algonkian-based languages, the most common of which are Cree and Ojibwa. Inuktitut, the Inuit language, has regional dialects and is spoken across the Canadian Arctic.

Another measure of Aboriginal diversity is self-identification. Many Indians prefer to identify themselves with the name of their tribal group, such as Cree or Iroquois, while others use the name of their First Nation (band) for a more precise identification. For example, the Cree occupy a vast territory that stretches from northern Québec to Alberta. There are many Cree tribes within that territory. A Cree living in northern Saskatchewan might identify himself or herself as a member of a Cree band, such as the Lac La Ronge band.

AANDC (2013) has recorded the population of each band. The largest First Nations are southern Ontario's Six Nations of the Grand River (Iroquois) with a population of 24,384; Qalipa Mi'kmaq of Nova Scotia (21,424); the Mohawk of Akwesasne (11,466) at St Regis on the Ontario–Quebec border near Cornwall; the Blood (11,448) in southern Alberta; Kahnawake (Mohawk) (10,053) near Montréal; the Saddle Lake Cree

THINK ABOUT IT
While the high birth rate of First Nations people allows both the number of Indians living on-reserve and off-reserve to increase, the rate of population increase is higher for the off-reserve population. Soon, perhaps by 2013, more First Nations people will live in cities than on reserves.

reserve (9,574) outside Edmonton; and the Lac La Ronge Cree in northern Saskatchewan (9,408). Approximately half of **First Nations people** live on reserves (51 per cent) while the rest live off-reserve, mainly in cities. The number of First Nations people living on reserves has increased but the percentage has declined. In 1984, for instance, 223,169 Indians or 64 per cent lived on reserves compared to 441,891 Indians or 51 per cent in 2011 (AANDC, 2013). The growing number of "urban" Indians is a significant economic and political factor and the original source of the Idle No More movement.

Aboriginal peoples are reclaiming their identity and place names. Some bands are relinquishing the names given to them by Europeans in favour of their original names, such as Anishinabe (for Ojibwa) and Gwich'in (for Kutchin). The landscape is also being reclaimed. For example, the Arctic community of Frobisher Bay, named after the English explorer, Martin Frobisher, is now Iqaluit ("the place where the fish are"), the capital city of Nunavut ("our land"). On the west coast, the Queen Charlotte Islands have been renamed Haida Gwaii.

Aboriginal and Treaty Rights

Aboriginal rights are group or collective rights that stem from Aboriginal peoples' occupation of the land before contact. Such rights apply most readily to status Indians and Inuit, while Métis are less well protected in regard to rights.

Métis Rights

What is the historic context of the Métis claim of Aboriginal rights? In 1870, Ottawa accepted that the Métis, by virtue of their Indian ancestry, had Aboriginal rights and Ottawa viewed these rights in the narrowest of interpretations by providing individual land grants to the Métis. The agreement had three components. First, land occupied before 1870 became private property. Second, the children of the Métis were eligible for a land grant of 140 acres. Third, each head of a Métis family received 160 acres in scrip, which could either be claimed or sold. The federal government set aside 1.4 million acres for the Métis children, estimated in 1871 at around 10,000. Based on these figures, each Métis child, at adulthood, could claim 140 acres. Before the actual land allocation began, the government ordered a census of the Métis population and this 1872 census identified just over 5,000 eligible Métis children. Accordingly, their individual allocation was increased to 240 acres (Library and Archives of Canada, 2012). Unfortunately for the Métis, few claimed their allocated land and many adults sold their scrip. By 1880, the outcome was clear—the dream of a Métis land base was dead. It remains to be seen how the March 2013 Supreme Court ruling changes this, although the president of the Manitoba Métis Federation stated that at this point recognition and some kind of compensation, not land per se, are of most importance to the Métis (CBC News, 2013b).

While the dispersal of many Métis from the Red River Valley is a historic fact, controversy has surrounded the reasons for this dispersal. The controversy hinges on two interpretations of that historic period in Canadian history. On the one hand, these rights, according to Ottawa, were distinguished by the awarding of **scrip** in accordance with the Manitoba Act of 1870. This position is supported by Flanagan (1991), who argues that the federal government of the day, while often slow in settling the Métis claims, did not act in bad faith. Taking the opposition position, Sprague (1988) claims that the Métis were victims of a deliberate conspiracy by the federal government to prevent a Métis land base in Manitoba. Milne (1995) provides a summary of this controversy. In 2013 the Supreme Court of Canada, as noted above, finally put to rest this controversial page of Canadian history.[4]

For more on the subject of land allotments to the Métis, see pages 104–105, "The First Clash: Red River Rebellion of 1869–70," and page 113, "The Red River Rebellion, 1869–70."

Treaty Rights

Treaty rights are the most generous of Aboriginal rights. Treaties set aside **reserve** land, held collectively by and for the benefit of the band, and define other negotiated rights (benefits).

The reasons for signing treaties varied depending on the historical context. During the late nineteenth century, treaties were signed throughout the Prairies to remove tribes from the land and make way for European settlement, and to provide a place for Indian tribes so that the Indian Wars, which were common south of the border between the US military and various Indian tribes, would not erupt in Canada. For Aboriginal peoples, treaties promised land (reserves) that

CONTESTED TERRAIN 3.2

The Supreme Court and the Métis

The Supreme Court of Canada began hearing the case brought by the Manitoba Métis Federation in December 2011. With the Court ruling in favour of the Métis claim that the federal government of the day did not safeguard the interest of the Métis, Ottawa will likely favour a cash settlement. Although David Chartrand, president of the Federation, stated in the immediate aftermath that land was not at issue after so many years, that sentiment might not be true for many of the Métis, especially those whose forebears were affected by the government's foot-dragging of almost a century and half ago and who, consequently, migrated to the west and north of the original "postage stamp" province of Manitoba.

would not be available to settlers, as well as support to shift from nomadic hunting to sedentary farming. The numbered treaties for the Plains Indians therefore offered protection from the anticipated flood of settlers and some guarantee that the federal government would care for them now that their principal source of food, the buffalo, was gone (Figure 3.10). However, treaty assurances of federal assistance were often not met (see Brownlie, 2003; Carter, 2004).

The terms of each treaty varied, although they generally included cash gratuities and presents at the signing of the treaty, annual payments in perpetuity, the promise of educational and agricultural assistance, and the right to hunt and fish on Crown land until such land was required for other purposes, as well as land reserves to be held by the Crown in trust for the Indians. In Treaty No. 6 of 1876, for example, which covered much of central Saskatchewan and Alberta, the amount of land assigned to each tribe was determined by its population, i.e., each family of five received one square mile. Reserves represent land collectively owned by First Nations bands, though legally the Crown holds the land in trust.

Conflicting ideas as to the significance of treaties between the signing parties largely shaped Aboriginal and non-Aboriginal relations in Canada during the twentieth century. When treaties were signed, Crown authorities viewed them as vehicles for extinguishing Aboriginal rights and titles to land and thus for opening the land to agricultural settlement. First Nations, however, understood them as agreements between "sovereign" powers to share land and resources. With such diverse perceptions, disagreements were inevitable.

Modern or comprehensive treaties came about in the latter part of the twentieth century, the first being the James Bay and North Québec Agreement of 1975, and have continued to be negotiated into the twenty-first century (Figure 3.11). Comprehensive treaties or agreements extend rights to those Aboriginal groups, especially in northern Canada and British Columbia, that had never signed treaties, and generally include large cash settlements, a portion of the group's traditional lands, surrender of the larger portion of traditional lands, self-governance agreements, and environmental/natural resource co-management agreements.

Modern Treaties

See Vignette 5.7, "Historical Timeline of the Caledonia Dispute," page 186.

The legal meaning of Aboriginal title to land has evolved over time. Until the 1970s, Ottawa recognized two forms of land rights. Reserve lands were one type of right or ownership, which the Canadian government held for Indian people. The second type was a usufructuary right to use Crown land for hunting and trapping, in other words, to freely use and enjoy Crown lands without any claim to ownership of these lands. At that time, Crown lands (both provincial and federal) included most of Canada's unsettled areas. Indian, Inuit, and Métis families lived on Crown lands, continuing to hunt, trap, and fish. However, federal and provincial governments could sell such lands to individuals and corporations or grant them a lease to use the land for a specific purpose, such as mineral exploration or logging, without

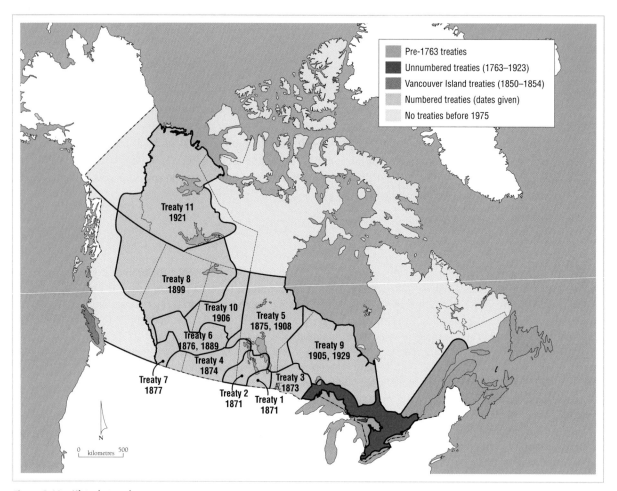

Figure 3.10 **Historic treaties**

The first treaties, made between the British government and Indian tribes, were "friendship" agreements. In Upper Canada the Robinson treaties of 1850 set aside reserve lands in exchange for the title to the remaining lands. With the settlement of lands in the Canadian West, Indians became concerned about their future, so many of the 11 numbered treaties, which spanned a half-century from 1871 to 1921, included provisions for agricultural supplies. When the last numbered treaty was signed, many Aboriginal peoples in Atlantic Canada, Québec, and British Columbia were without treaties.

compensating the Aboriginal users of those lands. By the 1960s, many Aboriginal groups still did not have treaties with the Canadian government. Atlantic Canada, Québec, the Territorial North, and British Columbia contained huge areas where treaties had not been concluded. As a consequence, Aboriginal peoples had no control over developments on these lands.

A combination of events radically changed this situation. One factor was the emergence of Native leaders who understood the political and legal systems. They used the courts to force the federal and provincial governments to address the issue of Aboriginal rights and land claims. The first major event took place in 1969, when Ottawa proposed reforms to the Indian Act in its White Paper on Indian Policy. This galvanized treaty Indians into action. The White Paper proposed to treat all Canadians equally. For Indians, it meant the abolition of their treaty rights and the reserve land system. At about the same time, the Nisga'a in northern British Columbia took their land claim, known as the *Calder* case, to court.

In 1973, the Supreme Court of Canada narrowly ruled (by a vote of four to three) against the Nisga'a argument that the tribe still had a land claim to territory in northern British Columbia. However, in their ruling, six of the seven judges agreed that Aboriginal title to the land had existed in British Columbia at the time of Confederation. Furthermore, three judges said that Aboriginal title still existed in British Columbia because it

CANADA'S HISTORICAL GEOGRAPHY

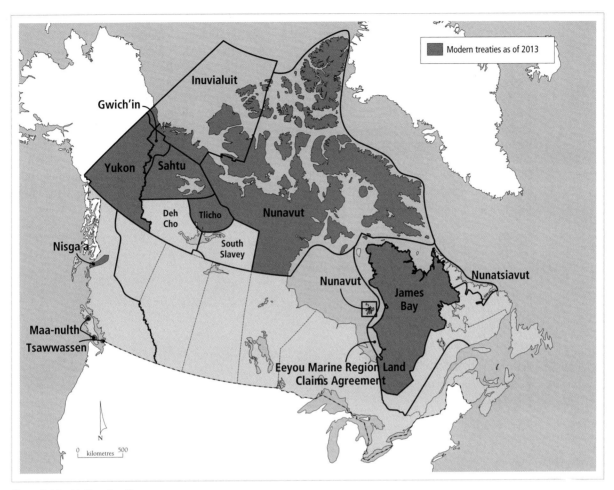

Figure 3.11 **Modern treaties**
The first modern treaty was the James Bay and Northern Québec Agreement, signed in 1975. By 2013, the main areas without treaties were much of BC, part of Labrador, and lands in central and southern Québec.

had not been extinguished by the British Columbia government, while three other judges stated that Aboriginal title had been extinguished by the various laws passed by the British Columbia government since 1871. The seventh judge ruled against the Nisga'a claim on a legal technicality. The Supreme Court's narrow verdict and the legal opinion of three judges that Aboriginal title still existed changed the course of Aboriginal land claims in Canada. In that year, 1973, the federal government agreed that Aboriginal peoples who had not signed a treaty may very well have a legal claim to Crown lands, and in 1974 an Office of Native Claims was first established to deal with both specific and comprehensive land claims.

In the early and mid-1970s the James Bay Project in northern Québec and the proposed Mackenzie Valley Pipeline Project in the Northwest Territories added fuel to the political fire over Aboriginal rights. The possible impact of these industrial projects on Aboriginal peoples was made clear through the Mackenzie Valley Pipeline Inquiry of 1974–7 (the Berger Inquiry) into possible environmental and socio-economic impacts and in the media. Aboriginal organizations obviously were prepared to take action to defend their land claims. Their position in the 1970s was "no development without land claims settlements." All these events changed both the public's views of Aboriginal rights and the government's position. At first grudgingly and then more willingly, governments, corporations, and Canadian society recognized the validity of Aboriginal land claims.

A **comprehensive land claim agreement** is sought when a group of Aboriginal people who

THINK ABOUT IT
From an Indian perspective, what was wrong with the basic concept behind the White Paper?

have not yet signed a treaty can demonstrate a claim to land through past occupancy. Such agreements are considered modern treaties (Table 3.8). The James Bay and Northern Québec Agreement is regarded as the first modern treaty, although negotiations between governments and the James Bay Cree and the Inuit of northern Québec had begun before the government policy was established. As well, the 1978 Northeastern Québec Agreement, signed between the Innu (Naskapi) and governments, can be considered a part of the James Bay and Northern Québec Agreement.

In 1984, the Inuvialuit of the Western Arctic became the first Aboriginal people to settle a comprehensive land claim with the federal government under the comprehensive land claim process. Since then, 20 comprehensive claims have been finalized in Canada, involving over 90 Aboriginal communities with over 70,000 members. Negotiations can be extremely slow and complex, and approximately 60 comprehensive claims are in various stages of negotiation at present. Most of these claims involve First Nations in British Columbia. Virtually the entire province of British Columbia, except for

Table 3.8 **Modern Land Claim Agreements, 1975–2011**

Name of Agreement	Year
James Bay and Northern Québec Agreement	1975
Northeastern Québec Agreement	1978
Inuvialuit Final Agreement	1984
Gwich'in Comprehensive Land Claim Agreement	1992
Nunavut Land Claims Agreement	1993
Yukon First Nations Final Agreements:	
Champagne and Aishihik First Nations	1993
First Nation of Nacho Nyak Dun	1993
Teslin Tlingit Council	1993
Vuntut Gwitchin First Nation	1993
Little Salmon/Carmacks First Nation	1997
Selkirk First Nation	1997
Tr'ondek Hwechin'in First Nation	1998
Ta'an Kwach'an Council	2002
Kluane First Nation	2003
Kwanlin Dun First Nation	2005
Carcross/Tagish First Nation	2005
Sahtu Dene and Métis Comprehensive Land Claim Agreement	1993
Nisga'a Final Agreement	2000
Tlicho Land Claims and Self-Government Agreement	2003
Labrador Inuit Land Claims Agreement	2005
Nunavik Inuit Land Claims Agreement	2007
Tsawwassen First Nation Final Agreement	2009
Eeyou Marine Region Land Claims Agreement	2010
Maa-nulth Final Agreement	2011

Source: Land Claims Agreement Coalition, "Modern Treaties," 2011, at: <www.landclaimscoalition.ca/modern-treaties/>.

Vancouver Island (where 14 treaties—the Douglas treaties—were signed between First Nations and the Hudson's Bay Company in the early 1850s), is claimed by First Nations. In BC, progress has been extremely slow. Until 1992, the provincial government claimed that British occupancy had extinguished Aboriginal title. However, in 1992 the British Columbia government accepted the principle of Aboriginal land claims. The following year, Ottawa and Victoria agreed to a formula for settling outstanding claims. The federal government would pay 90 per cent of the money needed to settle outstanding claims and the province would provide the land. In 2000, the Nisga'a Agreement was finalized, followed by two more—the 600-strong Tsawwassen First Nation signed their final agreement in 2007 and it became law in 2009; the Maa-nulth First Nation signed their final agreement in 2008, which was approved by Parliament in 2011.

For more on BC land claims, see page 278, Vignette 7.2, "Who Owns BC?"

Those Aboriginal groups who have concluded modern treaties are moving forward. They are able to focus on economic and cultural developments rather than expending their energies on land claim negotiations. In 1993, the Nunavut agreement broke new ground by effectively establishing self-government over an entire territory. Since then, modern land claim agreements, such as the Nisga'a Agreement, have included arrangements for self-government. As a result, a gap is emerging within the Aboriginal community between those who have a modern treaty and those who do not, as well as between those on reserve and those who live in urban areas among Canada's increasingly pluralistic majority society. Also, as with countries and with regions, some Aboriginal groups reside on lands that are rich in natural resources, resource developments, and development potential (e.g., oil and gas deposits, oil sands, pipelines, prime timber land) that provide a base for economic growth and considerable wealth, while many other groups live in areas with little resource potential where even subsistence from the land is marginal if not impossible.

Bridging the Aboriginal/Non-Aboriginal Faultline

Aboriginal peoples are taking the control of their affairs away from Ottawa (Vignette 3.9). Some Indian and Inuit peoples have made substantial advances in economic development, while others have gained increasing control over their own affairs through self-government and sovereignty. Unfortunately, some Aboriginal peoples, including the Métis, have not yet begun this self-government process and remain on the political margins of Canadian society. For most, the process of change has started.

In 1996, the Report of the Royal Commission on Aboriginal Peoples identified two major goals: Aboriginal economic development and self-government. The gap between Aboriginal and non-Aboriginal societies will not be bridged until

THINK ABOUT IT
Why have modern treaty negotiations resulted in more benefits and powers than were gained in the numbered treaties?

VIGNETTE 3.8

The Origin of the Métis Nation

The fur trade and the Métis are part of the historical fabric of Western Canada. With their command of English/French and Indian languages, the Métis were logical intermediaries in the fur trade. Over the centuries, the fur trade absorbed the mixed-blood offspring of Cree, Ojibwa, or Saulteaux women with French fur traders from the North West Company or Scottish and English fur traders from the Hudson's Bay Company. In the early nineteenth century, the settlement near the confluence of the Red River and the Assiniboine River consisted mainly of French and Scottish "half-breeds" and Scottish settlers brought from Scotland to fulfill Lord Selkirk's dream of an agricultural community. By 1821, the two fur-trading empires had amalgamated, throwing many English, French, and Scottish half-breeds out of work. Many gathered in the Red River Settlement, which provided the cultural melting pot for the formation of Métis Nation. The Métis culture was neither European nor Indian, but a fusion of the two.

these goals are achieved. The principal factor is transferring power from Ottawa (the political power core) to the various Aboriginal communities (the politically weak periphery). The economic and social well-being of Aboriginal reserves varies widely. Some, such as the Whitecap First Nation near Saskatoon with its casino and world-class golf course, have gained a high level of economic and social stability. The Labrador Inuit provide another example when they gained a share of the royalties from the Voisey's Bay nickel mine within their land claim agreement. Many others remain trapped in poverty and, without an economic base, breaking the dependency on the federal government seems an impossible task. Many individuals and families have made a choice, relocating to cities where a variety of opportunities are available. But for those who remain, developing an economic base on a reserve or in a remote community is not an easy task. Education levels are low with relatively few with high school degrees. Then, too, taking advantage of business and employment opportunities requires experiences and other assets not commonly found in such communities. Consider Attawapiskat First Nation, whose members are

VIGNETTE 3.9

From a Colonial Straitjacket to Aboriginal Power

Until 1969, Canada's Aboriginal peoples were largely invisible to other Canadians. Most were geographically separated, as many status Indians lived on reserves. Métis were in isolated communities, and Inuit still roamed the Arctic. They were outside the political process and thus denied access to political decision-making. In fact, status Indians did not receive the right to vote in federal elections until 1960. Shunned by Canadian society, these marginalized peoples had been subjected to assimilation policies for many years. In 1969, Ottawa made one last attempt to assimilate the Indian people of Canada through its "Statement of the Government of Canada on Indian Policy," more popularly known as the White Paper. The federal government proposed to eliminate the legal distinction between Indians and other Canadians by repealing the Indian Act and amending the British North America Act to remove those parts of the Act that called for separate treatment for Indians, and to abolish the Department of Indian Affairs. Ottawa believed that the separation of Indians from other Canadians was not only divisive but also made the Indians dependent on government and thereby held them back. The remedy was "equality." In the context of the 1960s, when oppressed people in other countries, including blacks in the United States and in South Africa, fiercely sought equality and the American Indian Movement railed against colonial suppression, Prime Minister Trudeau believed that the White Paper was the answer to Canada's Indian problem. And some Indian leaders supported this solution. As it turned out, many others did not. Reaction was swift. In the same year, Harold Cardinal published *The Unjust Society* and the following year, under his leadership, the Alberta chiefs published a formal rebuttal to the White Paper, commonly known as the "Red Paper" and titled *Citizens Plus: A Presentation by the Indian Chiefs of Alberta to the Right Honourable P.E. Trudeau*.

Over the next decade, the debate over the place of Indians in Canadian society took several different directions. First, there was legal support for the Indian position, beginning with the *Calder* case in 1973 when the Supreme Court held that the Nisga'a had Aboriginal rights. Second, the election of the Parti Québécois in 1976 called for "nation-to-nation" discussions between the province and the federal government. Aboriginal leaders seized the opportunity to present their demands in the same constitutional language. Third, recognition of the Aboriginal peoples and their rights in the 1982 Constitution Act dramatically enhanced their status and bargaining power. Fourth, the Constitution Act did not define Aboriginal rights, leaving that task to negotiations or the courts. The courts have been active in defining Aboriginal rights. In 1997, the Supreme Court's landmark decision in the *Delgamuukw* case overturned the earlier decision denying that Indians in British Columbia had Aboriginal title. Furthermore, the Court ruled that Aboriginal title means that Indians have the right to the resources on their lands.

unable to take full advantage of the opportunities provided by the De Beers diamond mine located on their traditional land. Solutions to Aboriginal poverty and dependency, as demonstrated by Attawapiskat, are not necessarily on the horizon.

The Immigration Faultline

The history of non-British immigration to Canada is complex and sometimes controversial. Most importantly, immigration has been a continuous stream of people coming to Canada, with each wave having a distinct impact on the land and society. Before 1867, immigration was often an instrument of colonial power. After the British Conquest of New France, for example, the British government set the immigration policy and the French-speaking majority in Canada did not have a say in the shaping of this policy. This power lay exclusively with the colonial power. The British government's objective was to offset the large French-speaking population by encouraging large-scale immigration from the British Isles and curtailing immigration from France. In the case of the Acadians, the British, beginning in 1755, deported many of these people to England and other English colonies and many others fled to Québec or found their way to the Louisiana Territory. At the same time, the British sought to resettle the area with British subjects. Colonial-style immigration, therefore, not only generated tensions between the existing population and the newcomers, but it also imposed a way of life and a set of institutions on the existing population and often marginalized these people.

After 1867, the Dominion of Canada remained closely tied to the British Empire and its immigration policies continued to reflect the "imperialist" attitude displayed in London, namely that Europeans but especially the British were superior to non-European peoples. This attitude, common to ancient as well as modern empires, flows from the ease with which Britain and other European nations carved out colonies in the rest of the world. Their power and capacity to form colonies stemmed from their industrialized, market-oriented economy and their efficient military system. After the end of slavery, local or imported non-white workers provided the labour force for the plantations and other economic endeavours in European colonies. Rudyard Kipling's concept of the "White Man's Burden" provided a more benevolent interpretation of this racist attitude: that the white man had the burden of extending his superior civilization to the rest of the world. It was an imperialist variant of the conservative, class-based notion of noblesse oblige. In any case, the building of the CPR line across the Cordillera provides a "Canadian" variation of the use of non-white labour. In the sparsely populated western half of Canada, the CPR faced a severe shortage of workers. The solution was to import Chinese labourers, who worked for half the wages of white labourers and often were assigned the more dangerous construction work.

While the existing colonial population asked how the newcomers would benefit them and their society, the colonial power took a rather different view, asking how the existing populations and the colonies would benefit the imperial centre. The economic, military, and social relationship between New France and the Huron Confederacy illustrates this point. In 1609, the Huron chiefs met with Samuel de Champlain to discuss both trade and a military alliance. The Huron had three objectives: (1) to gain access to European goods, including firearms, by supplying the French with beaver pelts; (2) to improve their material well-being with the trade goods and, in turn, trade these goods to more distant Indian tribes for profit; and (3) to strengthen their military position against their traditional enemies, the Iroquois, who were allied with the Dutch and later the English traders based in New York. The French had two objectives: (1) to secure a supply of furs; and (2) to convert the Huron to Christianity. At the height of the fur trade in the seventeenth century, New France greatly prospered and the Huron accounted for around half of the furs shipped to France (Dickason with McNab, 2009: 101). Trade was so important to the Huron tribes that when the French insisted that the Huron allow Jesuit missionaries to live among them as a condition for continued trade, the Huron reluctantly agreed. Unfortunately, the missionaries brought with them smallpox and other diseases that quickly swept through the Huron tribes, causing a sharp decline in their population.

Here, our focus is on the impact of immigration on the settling of Western Canada. The story begins with the purchase of Hudson's Bay lands by Ottawa, the reaction of the Métis in the Red River Settlement, the making of treaties, and then the subsequent settling of the Canadian Prairies

THINK ABOUT IT
Even though the jury found Riel guilty, they recommended mercy. Yet the judge ordered his execution. Prime Minister Macdonald is quoted by Bélanger (2007) as saying: "He shall hang though every dog in Québec bark in his favour."

THINK ABOUT IT
Despite the unfairness of two levels of payment for CPR workers, what do you think motivated these Chinese labourers to leave their homeland to work in the mountainous and dangerous terrain of British Columbia? You might ask yourself another question—why did the CPR not employ Aboriginal workers?

by many people who were not of British ancestry. This historic period stretches from 1870 to 1914. During this time, although the face of British colonialism had changed from London to Ottawa, it had not softened. Immigrants and those being incorporated into the expanding Canada had to conform to the legacy of the British colonial society. The experiences of the original occupants of Western Canada—the Plains Indians and the Métis—and then of the Doukhobors, who were very clearly not British immigrants, ended badly. Even the **Manitoba Act of 1870** did not protect the hard-fought gains of the **Provisional Government** of the Métis led by Louis Riel. In all instances, the pressure to conform to the majority society was both overt and covert; and in each case, the outcome pushed these peoples to the margins of Western Canadian society. The leader of the Métis, Louis Riel, was forced into exile in the United States. Later, Riel led the second Métis uprising in 1885, which was suppressed, and he was convicted of treason and hanged on 16 November 1885. The final irony was that Riel's death came only nine days after the driving of the last spike on the Canadian Pacific rail line at Craigellachie, British Columbia, an event redolent in symbolism as a uniting and tying together, from sea to sea, of the new nation—after the inconvenience of "others" and troublemakers had been dealt with.

The First Clash: Red River Rebellion of 1869–70

With the transfer of the vast lands administered by the Hudson's Bay Company, Canada changed from a small territory to a truly continental country. While the boundary between Western Canada and the United States had been determined earlier, the survey of lands for agricultural settlement took place in the 1880s. The survey system,

Photo 3.6
The confluence of the Red and Assiniboine rivers is known as the Forks. Today, the Forks lies in the heart of Winnipeg. In times past, the strategic location of the Forks provided Plains Indians with ready access by canoe to the lands south of the forty-ninth parallel and to the vast western interior. In 1738, the French explorer La Vérendrye established Fort Rouge at the Forks. With the founding of the Red River Settlement in 1812, the Forks became its focal point.

Ron Garnett/AirScapes.ca

based on a township and range model, stamped a rectangular-shaped grid on the cultural landscape, thus determining the shape and placement of farms, roads, and towns (Vignette 3.2). As Moffat (2002: 204) points out, this survey system "enabled the division of western lands among the HBC, the Canadian Pacific Railway (CPR) and homesteaders, and set aside two sections in each township for the future of local education." However, when the federal surveyors set foot in the Red River Colony in 1869, Ottawa had failed to acknowledge the presence and rights of the Métis. As well, the federal government had not yet set in process negotiations with the Prairie Indians. In fact, Ottawa did not inform the residents of the Red River Colony of its plans for the Hudson's Bay lands nor did the government signal that it recognized local landholdings. With the clash between the surveyors and the Métis, events quickly spun out of control, resulting in the Métis Provincial Government and the Red River Rebellion.

The Red River Rebellion pitted the existing population of the Red River Colony against Ottawa, whose land survey and agricultural plans posed a potentially fatal threat to the existing Métis settlement and its hunting economy. Even before the arrival of settlers, surveyors sent by Ottawa ignored the long-lot holdings of the Métis along the Red and Assiniboine rivers. In 1869, the Red River Colony was the only settled area of any size in the North-Western Territory, with a population of nearly 12,000 evenly divided between French- and English-speaking residents (Table 3.2). Most consisted of mixed-blood people, born of French and British fur traders and Indian mothers, who had settled in long lots along the banks of the two major rivers, and whose economy was based on the buffalo hunt and subsistence farming. By early 1869, news of the pending transfer of Hudson's Bay Company lands to Ottawa had reached the colony, and the arrival of land surveyors resulted in open hostility. When Canadian surveyors began to survey Métis occupied lands, the Métis feared for their rights to those lands and even for their place in the new society. Matters came to a boil when, in October 1869, Louis Riel put his foot on a surveyor's chain and told them to leave. Thus, the Red River Rebellion began, during which the Métis took control of Upper Fort Garry, the HBC headquarters, and William McDougall, who had been appointed lieutenant-governor of the HBC lands soon to be passed over to Canada, was turned back at the border in his attempt to claim Canadian sovereignty over the territory.

Two months later, the Métis under Riel formed a Provisional Government and soon began to negotiate with Canada over the terms of entry into Confederation. The three-man delegation sent to Ottawa by Riel's Provisional Government gained much of what they sought, including agreement to the establishment of a new province, but Orange Order elements from Ontario who had come to the Red River area were not pleased that the French-speaking Métis "half-breeds" were in charge, and one man, Thomas Scott, who had been arrested by the Métis but persisted in being belligerent and unruly, was summarily executed after a brief trial. This inevitably led to further difficulties.

One advantage Riel had had in his negotiations with the Canadian government was "remoteness." Without rail connection to the settlement, Ottawa could not rush troops to quell the resistance, which, with the execution of Scott, seemed on the verge of full-scale warfare. Although a rail line reached St Paul in Minnesota, the US government refused to allow Canadian troops to cross the border. In April 1870, Macdonald authorized a military force of 1,000 troops—the Wolseley Expedition—to advance on Red River and assert Canada's sovereignty over the colony. The Canadian troops followed an old fur trade route and took four months to finally reach the Red River in August 1870. Fearing for their lives, Riel and his followers fled to the United States. On 15 July 1870, Manitoba became a tiny province of Canada with an area of about 2,600 square kilometres (1,000 square miles). The Métis had obtained most of their demands (the use of English and French languages within the government and a dual system of Protestant and Roman Catholic schools); at the same time, Prime Minister Macdonald had begun to ensure Canadian control over Western Canada.

Second Clash: Making Treaty

After Manitoba became part of the Dominion, Ottawa sought to expand its control into the empty Prairie lands. Making treaty with the Indians of this "empty land" was essential before these potential farmlands were filled with homesteaders from Canada, the United States, and

THINK ABOUT IT
Why do you think the Métis chose to negotiate with Ottawa rather than to declare their independence from Canada?

Great Britain. From 1871 to 1877, seven treaties—the so-called numbered treaties 1 to 7—were negotiated to open the West to settlement.

The objective of Ottawa was to extinguish Indian rights to the land, as it had in Ontario with the **Robinson treaties** in 1850, and to promote the assimilation of Plains Indians into Canadian society. The formula was simple—cash, a small annual payment, and land for the exclusive use of Indians (now known as reserves). The assimilation goal soon enough took the form of Indian residential schools. But what were the goals of the various Indian tribes? While they did not speak with one voice, they were aware of the Robinson treaties, the Indian Wars in the United States, and the threat of agricultural settlement on their way of life. More importantly, their main source of food, the buffalo, was disappearing. Word from their cousins in the United States made them very aware of the impact of settlement and railways on the Indian way of life. Few options remained and their main goal was to survive as a people, but the path was not clear. However, treaty negotiations provided a small window of opportunity to improve the terms over the Robinson treaties. Of course, Indian tribes were acutely aware of earlier settlements, and by Treaty 3 they knew all the cards played by the federal negotiators and were able to use this information to gain additional concessions. In this way, the Indians forced the federal government to consider issues far beyond the Robinson treaties. For instance, some Indian leaders hoped that agriculture might provide the basis for a new economy and they were able to have training in farming/ranching plus supplies and tools included in the treaties. Then, too, Indian leaders were able to include "the medicine chest" in Treaty 6, which became the basis for subsequent free health care for First Nations people.

While both Canada and the Prairie Indians agreed to these seven treaties, the federal government and the First Nations saw treaties as necessary elements in achieving their very different goals. Ottawa, for instance, gained ownership of the land but it was not happy with the cost of the "unanticipated concessions" granted to the Indian tribes by federal negotiators. Indians felt the fulfillment of their treaty rights, especially with regard to help in developing an agricultural base on reserve lands, was not forthcoming. Matters turned from bad to worse, culminating in the 1885 Northwest Rebellion.

The Third Clash: The Northwest Rebellion of 1885

While treaties had been signed, Indians faced desperate conditions, and those bands that were not docile in the face of drought and starvation found their meagre treaty provisions cut by federal officials. At the same time, the many Métis from Red River who had migrated north and west into present-day Saskatchewan in the years following the 1869–70 rebellion felt threatened once more by the advancing wave of settlers and by difficult conditions. A delegation went to Montana in 1884 and convinced Louis Riel, in exile as a schoolteacher and an American citizen, to return to Canada to lead their quest for their rights. Late in 1884, Riel sent a petition to Ottawa with various demands for all the inhabitants of the North-West—Indians, Métis, and whites—effectively asking that they be treated with the dignity deserving of loyal British subjects. Eventually, when no remotely supportive government response was forthcoming, Métis soldiers, with a few Indian warriors, ambushed a NWMP contingent at Duck Lake on 26 March 1885, killing 12 men and losing six of their own. Big Bear, a Plains Cree chief, was seeking a peaceful solution to the plight of his people, but a few of his warriors, too, went on the warpath. On 2 April 1885 Cree warriors led by Wandering Spirit rode to Frog Lake to demand food. When the local Indian agent refused them, he was shot. The warriors then looted the settlement and left nine dead.

The Métis, under the leadership of Riel but led militarily by Gabriel Dumont, were prepared to fight the advancing Canadian army, which had arrived quickly from Ontario by means of the Canadian Pacific Railway. Attempts to unite with the Cree failed. Still, the Métis and a few Indians from nearby reserves were successful in surprising the Canadian troops, led by Major-General Frederick D. Middleton, at Fish Creek, but the larger and well-equipped Canadian army eventually wore down the smaller and less well-equipped Métis and Indian forces at Batoche (see Figure 3.12). From Ottawa's perspective, the Northwest Rebellion was crushed. Louis Riel and eight Indian leaders were hung while **Big Bear** and **Poundmaker** were sent to prison. But the uprising had an enduring effect on the Prairie tribes and the Métis, and soured Ottawa's relations with Québec.

CANADA'S HISTORICAL GEOGRAPHY

Figure 3.12 **Western Canada and the Northwest Rebellion of 1885**
The Canadian Pacific Railway played a key role in the Northwest Rebellion by transporting the Canadian forces quickly from Ontario to Qu'Appelle, Saskatchewan. In 1885, the political boundaries in Western Canada (except for Manitoba) still were part of the North-West Territories. The provinces of Alberta and Saskatchewan were formed in 1905 while Manitoba reached its current size in 1912. (*Source:* Based on Rattlesnake Jack's Old West Clip Art Parlour, Font Gallery, North West Rebellion Emporium and Rocky Mountain Ranger Patrol, at: <members.memlane.com/gromboug/P5NWReb.htm>.)

 For more on the reasons for the souring of relations between Ottawa and Québec, see "Strained Relations," pages 113–114.

The Making of Canada

Canada became a different country with the settling of its western territories. Not only was the migration to Western Canada one of the greatest world immigrations, it also placed a "British/Canadian" brand on the landscape and, with Indian treaties and the dispersal of the Métis, Canada relegated the earlier occupants to the margins. The emerging cultural landscape of Western Canada took three forms—the rectangular appearance of its rural landholdings, the orientation of villages and cities to the railways, and the symbols of ethnic/religious diversity as expressed by farm buildings and churches. By 1895, Western Canada had a predominantly British population that had established its own land survey and ownership system, local governments and police to ensure "law and order," and a variety of social institutions.

As evident from the experience of the Métis pattern of landownership, elements of the landscape that did not conform ran into serious problems. During a 10-year span from 1870 to 1880, the Métis lost their majority due to an influx of immigrants from Ontario, many of whom either belonged to or supported the views of the Orange Order, a Protestant fraternal organization with strong anti-Catholic beliefs. Some newcomers saw no place for the Métis and Indians in the emerging society, thus creating tensions between the existing population and the newcomers. From 1871 to 1881, Manitoba's population increased from 25,228 to 62,260, with most immigrants coming from Ontario, the British Isles, and the United States (Table 3.9). At this time, those of British ancestry formed 54 per cent of the population, other Europeans made up 17 per cent, Métis, 17 per cent, and Indians, 11 per cent (Canada, 1882: Table III). The newly formed English-speaking majority focused their attention on the dual school system. By 1891, Manitoba's population exceeded 150,000 (Table 3.9). In an example of the tyranny of the majority, the

THINK ABOUT IT
John Ralston Saul calls Canada a "soft country" where middle ground to conflicts is sought. Was this true in the late nineteenth century when the faultline between Ottawa and Québec widened over events surrounding the Métis and the hanging of Riel?

Table 3.9 **Population in Western Canada by Province, 1871–1911**

Year	Manitoba	Saskatchewan	Alberta
1871	25,228		
1881	62,260	21,652	9,875
1891	152,506	40,206	26,593
1901	255,211	91,279	73,022
1911	461,394	492,432	374,295

The boundaries of Manitoba did not reach their present limits until 1912, and Saskatchewan and Alberta became provinces in 1905. Their populations for 1881, 1891, and 1901 have been calculated from the censuses of Canada for 1881 and 1891.

Source: Canada (1882: 93–6; 1892: 112–13); Statistics Canada (2003).

English-speaking population argued that with so few French-speaking students, funding for the Catholic school was not warranted. In 1890, the government of Manitoba abolished public funding for Catholic schools. This decision took on national significance by becoming a critical issue between Québec and the rest of the country.

What caused this initial influx of settlers? One reason was that Ontario no longer had a surplus of agricultural land and sons of farmers looked to the unsettled lands on the Great Plains of the United States and to Manitoba. A second reason was that the promise of a railway would make farming in Manitoba more viable. With completion of the CPR line from Fort William on Lake Superior to Selkirk, Manitoba, in 1882, grain could be transported by rail and ship to eastern Canada and Great Britain rather than by the more circuitous steamship route to St Paul and then by rail to New York. Wheat farming in Manitoba had become a profitable business because of advances in agricultural machinery and farming techniques, and rising prices for grain. Equally important, new strains of wheat, first Red Fife and then Marquis, both of which ripened more quickly than previous varieties, lessened the danger of crop loss due to frost. Marquis wheat, which matured seven days earlier than Red Fife, allowed wheat cultivation to take place in the parkland belt of Saskatchewan and Alberta where the frost-free period was shorter than in southern Manitoba.

Sifton Widens the Net

By the end of the nineteenth century, Canada's West still needed more settlers. Clifford Sifton of Manitoba, the federal minister responsible for finding settlers, realized that he had to expand his recruitment area beyond Great Britain into Central Europe and Russia. Even though this ran against the creation of a British-populated Western Canada, Sifton (1922) took a pragmatic approach, which he summed up in later years: "I think a stalwart peasant in a sheep-skin coat, born on the soil, whose forefathers have been farmers for ten generations, with a stout wife and a half-dozen children is good quality." Under Sifton, the pattern of immigration took a sharp turn from the main sources of immigrants to the West, namely Canada, the British Isles, and the United States. Within two decades of entering Confederation, Manitoba's population had increased by just over 600 per cent (Table 3.9). Most were of British stock, but substantial numbers of Mennonites and Icelanders had also come to Manitoba. At the same time, few settlers had reached Saskatchewan and Alberta, though the Métis had relocated in Saskatchewan, primarily around the settlement of Batoche on the South Saskatchewan River just north of Saskatoon. In the next decade, the volume of immigrants from Central Europe, Scandinavia, and Russia increased substantially. As peasants, they were prepared for the harsh physical conditions associated with breaking the virgin prairie land and were willing to deal with the psychological stress of living on isolated farmsteads in a foreign country where their native tongue was not accepted. As the numbers of these European immigrants grew, the anglophone majority became concerned about the newcomers and their possible effect on the existing social structure. The demographic impact of the non-British migration to Western Canada is shown in the 1916 census (Table 3.10).

Table 3.10 **Population of Western Canada by Ethnic Group, 1916**

Ethnic Group	W. Canada Population	% of W. Canada Population	% of Manitoba Population	% of Sask. Population	% of Alberta Population
British	971,830	57.2	57.7	54.5	60.2
German	136,968	8.1	4.7	11.9	6.8
Austro-Hungarian	136,250	8.0	8.2	9.1	6.4
French	89,987	5.3	6.1	4.9	4.9
Russian	63,735	3.7	2.9	4.5	3.8
Norwegian	47,449	2.8	0.6	4.2	3.4
Indian (Aboriginal)	39,147	2.3	2.5	1.7	2.9
Ukrainian	39,103	2.3	4.1	0.7	1.8
Swedish	37,220	2.2	1.4	2.5	2.7
Polish	27,790	1.6	3.0	1.0	0.9
Jewish	23,381	1.4	3.0	0.6	0.6
Dutch	22,353	1.3	1.3	1.4	1.3
Icelandic	15,800	0.9	2.2	0.5	0.1
Danish	9,556	0.6	0.3	0.5	0.9
Belgian	9,084	0.5	0.8	0.4	0.4
Italian	5,348	0.3	0.3	1.0	0.9
Other	26,219	1.5	0.9	1.5	2.3
Total	1,701,220	100.0	100.0	100.0	100.0

Source: Census of Prairie Provinces, 1916, Table 7. Data adapted from Statistics Canada, at: <www12.statcan.ca/English/census01/products/analytic/companion/age/provpymds.cfm>.

This wave of Central Europeans had tremendous implications for Western Canada. While most newcomers assimilated into the English-speaking society, a few did not. Often these ethnic groups settled in one area where they were somewhat insulated from the larger society and where they attempted to maintain their traditional customs, language, and religion. The federal government, by providing land reserves for ethnic groups such as the Mennonites and Doukhobors, reinforced this tendency.

While they were successful farmers, the cultural differences between the more conservative Doukhobors and Canadian society were too great for the majority society to accept. Some Doukhobors were able to integrate into local society, but the Community Doukhobors simply were unable to adapt. They remained faithful to their religious beliefs that emphasized communal living. In choosing to settle in Canada, they were granted blocks of land and exemption from military service.

Through negotiations with the Canadian government, Doukhobor leaders had obtained four large blocks of land totalling 750,000 acres. In 1899, the Doukhobors—7,500 in total—arrived in Canada and took possession of pre-selected lands in Saskatchewan where they built 57 villages. The four colonies were located just west of Swan River, Manitoba (North Colony), and at Prince Albert (Saskatchewan Colony) and Yorkton, Saskatchewan (South Colony and Good Spirit Lake Annex).

Farming was not only an economic activity, but it was also central to their religious beliefs, which emphasized the value of a simple, communal life. For example, Doukhobors shared in the returns from farming, and no one person owned the land or the tools. In a land of individual landholdings and the pursuit of profit, the Doukhobors were

THINK ABOUT IT
Does the open discrimination against Aboriginal peoples and non-British newcomers that was common in the nineteenth century and much of the twentieth century reflect Saul's concept of a "hard" country? If so, why?

seen as "out of step" with the surrounding community. As public resentment increased, the federal government took action. In 1905, Frank Oliver succeeded Clifford Sifton as Minister of the Interior. Oliver decided to enforce the Dominion Lands Act, so when the Doukhobors refused to swear an oath of allegiance to the King, Oliver had his excuse to deny them homestead lands.

Failure to take such an oath had two implications. First, it suggested that these people were disloyal to the Queen. Second, it meant that the Doukhobors could not obtain title to their homestead lands. Under this pretext, Oliver used the Dominion Lands Act to cancel their right to land. A hard core of Doukhobors remained committed to the communal way of life, most of whom eventually moved to British Columbia; others abandoned the village life and took title to homesteads. The villages gradually lost members and lands. The South Colony just north of Yorkton, Saskatchewan, was the last holdout, but by 1918 it ceased to exist on Crown land. It persisted in a much reduced area on purchased land until 1938, as did other communal settlements established in the Kylemore and Kelvington areas.

One explanation for the ultimate failure of the Doukhobor experiment was that Canada's model of individual settlement was simply too rigid to accept a communal one. Primarily for that reason, the Community Doukhobors were unable to find a place in Western Canada. They represent a classic example of a people being too different—too "other"—from the majority to be allowed a comfortable space within the predominately British landscape. Ironically, the village model of settlement was perhaps the most effective way of settling the Prairies in the late nineteenth and early twentieth centuries. Professor Carl Tracie (1996: xii) puts it this way:

> At the very time when the individual homesteader was struggling with the very real problems of isolation and loneliness, the Doukhobor settlements, whose compact form allayed these problems, were being dismantled by forces which could not accommodate the communal aspects of the group. Also, although the initial government concern was the survival of the Doukhobors, their very prosperity, based as it was on communal effort, may have worked against them since it illustrated the success of a system diametrically opposed to the individualistic system dictated by government policy and assumed by mainstream society.

The French/English Faultline

Although the ancestors of Aboriginal peoples were the first to occupy North America, the colonization of the continent pushed them to the margins, leaving the two European powers—the French and the British—to place their mark on the land. Following the British military victory at Québec in 1759, the Treaty of Paris (1763) confirmed British hegemony over a French-Canadian majority and its control over the lands of New France. This historic fact underscores the dominant position of the British and their impact on later Canadian institutions and governments. Also, the British way of life was established across Canada, though rural Québec retained its French character, the seigneurial agricultural system, and Roman Catholic religion. Differences between these two cultures have come to represent a major faultline in Canadian society. However, the union of Lower Canada and Upper Canada (Ontario) in 1841 meant that French and English had to work together in a single parliament, which made each dependent on the other and was instrumental in the 1867 Confederation. This interaction of French and English has done much to shape the cultural and political nature of Canada. Over the years, they have accomplished much together. Nevertheless, significant differences between the two communities exist, and from time to time these differences flare into serious misunderstandings. Without a doubt, Canadian unity depends on the continuation of this relationship and the need for compromise, which has become a feature of modern Canadian political life and is a basic aspect of Canadian tolerance between the two official language groups and towards newcomers.

The serious nature of the French/English rift has profound geopolitical consequences for Canada. While blowing hot and cold, the rift was boiling hot in the last three decades of the twentieth century. In 1993, a well-known Canadian political columnist, Jeffrey Simpson, captured this moment:

> We can also hope that, in the 1980s, Canadians gained a deeper understanding

of the faultlines running through their society, and that they will avoid measures that widen them, thereby concentrating on making new arrangements and reforming old ones, so that what the rest of the world rightly believes to be a successful experiment in managing diversity will endure and prosper. (Simpson, 1993: 368)

Origin of the French/English Faultline

The British Conquest of the French on the Plains of Abraham in 1759 marks the origin of this faultline. An event that remains a dark page in French-Canadian history culminated in the battered remnants of the French army and the French colonial elite boarding ships to return to their mother country. The French Canadians had no thought of leaving, but what would happen to them under British military rule? Would they, like their Acadian brethren, be deported to other British colonies? Britain did not need to take such drastic action by this time—the Acadian deportations had occurred in the 1750s when there still was a French threat to Britain's North American possessions. In the Treaty of Paris, France ceded New France to Britain, which placed the French-Canadian majority under the British monarchy. While the English lived in cities in Québec and dominated the Québec economy and politics, French Canadians lived mostly in rural areas where they successfully maintained their culture within a British North America. This relationship between British rulers and the French Canadians would be strengthened with the Québec Act of 1774.

The Québec Act, 1774

With the Québec Act of 1774, the unique nature and separateness of Québec were recognized, thus affirming its place in British North America. This Act is sometimes described as the Magna Carta for French Canada.[5] Its main provisions ensured the continuation of the aristocratic seigneurial landholding system and guaranteed religious freedom for the colony's Roman Catholic majority and, by implication, their right to retain their native language.[6] This gave the most powerful people in New France a good reason to support the new rulers. The Roman Catholic Church was placed in a particularly strong position. Not only was the Church allowed to collect tithes and dues but its role as the protector of French culture went unchallenged. Therefore, the clergy played an extremely important role in directing and maintaining a rural French-Canadian society, a role further enhanced by the Church's control of the education system. The habitants (farmers) were at the bottom of French-Canadian society's hierarchy. They formed the vast majority of the population and continued to cultivate their land on seigneuries, paying their dues to their lord (seigneur) and faithfully obeying the local priest and bishop. The British granted another important concession, namely, that civil suits would be tried under French law. Criminal cases, however, fell under English law.

The seigneurial system formed the basis of rural life in New France and, later, in Québec. In 1774, there were about 200 seigneuries in the St Lawrence Lowland. This type of land settlement left its mark on the landscape (the long, narrow landholdings and the vast estate of the seigneur) and on the mentality of rural French Canadians (close family ties, a strong sense of togetherness with neighbouring rural families, and staunch support for the Church). A habitant's landholding, though small, was the key to his family's prosperity, and by bequeathing the farm to his eldest son the habitant ensured the continuation of this rural way of life. In 1854, the habitant was allowed to purchase his small plot of land from his seigneur, but the last vestiges of this seigneurial system did not disappear until a century later. Even today, the landscape along the St Lawrence shows many signs of this type of landholding.

While the heart of this new British territory was the settled land of the St Lawrence Lowland, its full geographic extent was immense. Essentially, the Québec Act of 1774 recognized the geographic area of former French territories in North America. Québec's territory in 1774 was extended from the Labrador coast to the St Lawrence Lowland and beyond to the sparsely settled Great Lakes Lowland and the Indian lands of the Ohio Basin. After the British defeat, the geographic size of Québec shrunk with the southern part of the Great Lakes Lowland and the Indian lands of the Ohio Basin ceded to the Americans.

The Loyalists

The American War of Independence changed the political landscape of North America. Within the newly formed United States, a number of Americans, known as the **Loyalists**, remained loyal to Britain. Like the French-speaking people in North America, most of these Loyalists were born and raised in the New World. For them, North America was their homeland. During the revolution, they had sided with the British. They were hounded by the American revolutionaries and many lost their homes and property. Most resettled in the remaining British colonies in North America, where Britain offered them land. The majority (about 40,000 Loyalists) settled in the Maritimes, particularly in Nova Scotia. About 5,000 relocated in the forested Appalachian Uplands of the Eastern Townships of Québec. A few thousand, including Indians led by Joseph Brant, took up land in the Great Lakes Lowland in present-day Ontario. The Six Nations of the Iroquois Confederacy, who had been loyal to the British, settled along the Grand River in southwestern Ontario on a vast tract—the Haldimand Grant—given to them in 1784 by the Governor of Québec, Lord Haldimand.

In 2006, the Six Nations became tangled in a nasty 40-acre land development dispute. See Vignette 5.7, "Historical Timeline of the Caledonia Dispute," and Figure 5.6, "The Haldimand Tract," page 184, for more details about the Haldimand Grant.

Within a few decades, the English-speaking settlers in the Great Lakes region grew in number. Soon they sought to control their own affairs so they could have a more British government with British civil law, British institutions, and an elected assembly. In the Constitutional Act of 1791, Québec was split into Upper and Lower Canada.

The Constitutional Act, 1791

The Constitutional Act of 1791 represented an attempt by the British Parliament to satisfy the political needs of the French- and English-speaking inhabitants of Québec. These were the main provisions of the Act: (1) the British colony of Québec was divided into the provinces of Upper and Lower Canada, with the Ottawa River as the dividing line, except for two seigneuries located just southwest of the Ottawa River; and (2) each province was governed by a British lieutenant-governor appointed by Britain. From time to time, the lieutenant-governor would consult with his executive council and acknowledge legislation passed by an elected legislative assembly.

In 1791, Lower Canada had a much larger population than Upper Canada. At that time, about 15,000 colonists lived in Upper Canada, most of whom were of Loyalist extraction, plus about 10,000 Indians, some of whom had fled northward after the American Revolution. Lower Canada's population consisted of about 140,000 French Canadians, 10,000 English Canadians, and perhaps as many as 5,000 Indians.

Following the Constitutional Act, Upper and Lower Canada each had an elected assembly, but the real power remained in the hands of the British-appointed lieutenant-governors. In Lower Canada the lieutenant-governor had the support of the Roman Catholic Church, the seigneurs, and the **Château Clique**. The Château Clique, a group consisting mostly of anglophone merchants, controlled most business enterprises and, as they were favoured by the lieutenant-governor, wielded much political power. In Upper Canada the **Family Compact**—a small group of officials who dominated senior bureaucratic positions, the executive and legislative councils, and the judiciary—held similar positions in commercial and political circles. While these two elite groups promoted their own political and financial well-being, the rest of the population grew more and more dissatisfied with blatant political abuses, which included patronage and unpopular policies that favoured these two groups. Attempts to obtain political reforms leading to a more democratic political system failed. Under these circumstances, social unrest was widespread.

In 1837 and 1838, rebellions broke out. In Lower Canada Louis-Joseph Papineau led the rebels, while William Lyon Mackenzie headed the rebels in Upper Canada. Both uprisings were ruthlessly suppressed by British troops. The goal of both insurrections was to take control by wresting power from the colonial governments in Toronto and Québec and putting government in the hands of the popularly elected assemblies. In Lower Canada the rebellion was also an expression of Anglo-French animosity. While both uprisings were unsuccessful, the British government nevertheless sent Lord Durham to Canada as Governor General to investigate the rebels' grievances. He recommended a form of responsible government and the union of the two Canadas. Once the two colonies were unified, the next step, according to Durham, would be the assimilation of the French Canadians into British culture.

The Act of Union, 1841

In response to Durham's report, in 1841 the two largest colonies in British North America, Upper and Lower Canada, were united into the Province of Canada. This Act of Union gave substance to the geographic and political realities of British North America. The geographic reality was that a large French-speaking population existed in Lower Canada, while a smaller English-speaking population was concentrated in Upper Canada (Table 3.11). The political reality was twofold. Both groups had to work together to accomplish their political goals and neither group could achieve all of its goals without some form of compromise. When the two cultures were forced to work together in a single legislative assembly, a new beginning to the French/English faultline surfaced.

Demographic Shifts

In a democracy, political power is based on population numbers. During the early days of the Act of Union most people lived in Lower Canada, but by 1851 the reverse was true (Table 3.11). In this way, the balance of power shifted to Upper Canada and this shift continues. From 1841 to 1871, for instance, Québec's population had dropped from 45 per cent of Canada's population to 34 per cent.

Strained Relations

During these formative years, several events seriously strained relations between the Dominion's two founding peoples:

- the Red River Rebellion, 1869–70;
- the Northwest Rebellion, 1885;
- the Manitoba Schools Question, 1890.

The Red River Rebellion, 1869–70

The Métis rebellion, led by Louis Riel, soon became a national issue, reopening differences between English, Protestant Ontario and French, Roman Catholic Québec.[7] Québec considered Riel a French-Canadian hero who was defending the Métis, a people of mixed blood who spoke French and followed the Catholic religion. Protestant Ontario, on the other hand, considered Riel a traitor and a murderer. For Canada, the larger issue was the place of French Canadians in the West. A compromise was achieved in the Manitoba Act of 1870. Accordingly, the District of Assiniboia became the province of Manitoba. Under this Act, land was set aside for the Métis, although a number of them sold their entitlements (scrip) to land allotments for a cheap price to incoming settlers and moved further west or sought to continue their former hunting lifestyle in Manitoba. The elected legislative assembly of Manitoba provided a balance between the two ethnic groups with 12 English and 12 French electoral districts. Equally important, Manitoba

THINK ABOUT IT
Saul's premise of a "soft" Canada begins with the Act of Union, when the French and English were forced to work together at the political level, thus requiring an accommodation of the differing interests of the two founding peoples. Still, the raw edge remained in the general population. What events in the early days of the province of Manitoba support the argument of intolerance in Canadian society?

Table 3.11 **Population by Colony or Province, 1841–1871 (%)**

Colony/Province	1841	1851	1861	1871
Ontario	33.0	41.1	45.2	46.5
Québec	45.0	38.5	36.0	34.2
Nova Scotia	13.0	12.0	10.7	11.1
New Brunswick	9.0	8.4	8.1	8.2
Manitoba				< 0.1
British Columbia				< 0.8
Total per cent	100.0	100.0	100.0	100.0

Source: McVey and Kalbach (1995: 38). © 1995 Nelson Education Ltd. Reproduced by permission. www.cengage.com/permissions

had two official languages (French and English) and two religious school systems (Catholic and Protestant) financed by public funds.

The background to the Red River Rebellion is presented earlier in "The First Clash: Red River Rebellion of 1869–70," page 104.

The Northwest Rebellion, 1885

During the 1870s, many Ontarians settled in Manitoba while some Métis sought a new home on the open prairie. Seeking to remain hunters, one group settled along the South Saskatchewan River where they established a Métis colony around Batoche, about 60 km northeast of present-day Saskatoon. Batoche became the new centre of the French-speaking Métis in Western Canada. As settlers spread into Saskatchewan, the Métis again feared for their future. In 1884, when a party of Métis went to Montana to plead with Louis Riel to return to Batoche and lead them again, Riel, convinced of his destiny, accepted this challenge. As we have seen, this uprising ended in failure for the Métis and their Indian allies. For Québec, the defeat of the Métis and the subsequent hanging of their leader not only represented a defeat for a French presence in the West but also widened the gulf between French and English Canadians. Riel's link to Québec and the Roman Catholic Church made him a powerful symbol of language, religious, and racial divisions for over 100 years. Indeed, to this day historians remain divided about Louis Riel's legacy, his place in the story of Canada, and even his sanity as the messianic leader of a doomed rebellion.

The background to the Northwest Rebellion is presented earlier in "The Third Clash: The Northwest Rebellion of 1885," page 106.

The Manitoba Schools Question, 1890

The British North America Act of 1867 established English and French as legislative and judicial languages in federal and Québec institutions. The remaining three provinces (New Brunswick, Nova Scotia, and Ontario) had only English as the official language. The question of French language and religious rights in acquired western territories first arose in Manitoba.

The French/English issue became the focal point for the entry of the Red River Settlement (now Manitoba) into Confederation. Local inhabitants—mostly French-speaking Roman Catholic Métis and the less numerous English-speaking Métis—were determined to have some influence over the terms that would include their community as part of Canada. One of their concerns was language rights, an issue that was ultimately resolved when a list of rights drafted by the Riel's Provisional Government became the basis of federal legislation. When the settlement and surrounding territory of Red River entered Confederation in 1870 as the province of Manitoba, it did so with the assurance that English- and French-language rights, as well as the right to be educated in Protestant or Roman Catholic schools, were protected by provincial legislation.

During the 1870s and 1880s, with the influx of a large number of Anglo-Protestant settlers from Ontario, the proportion of Anglo-Protestants in the Manitoba population increased and the proportion of French and Roman Catholic inhabitants decreased. This demographic change created a stronger Anglo-Protestant culture in Manitoba. In 1890, the provincial government ended public funding of Catholic schools. From Québec's perspective, this legislation shook the very foundations of Confederation. Sir Wilfrid Laurier became Prime Minister in 1896 and, in the following year, Laurier negotiated a compromise agreement with the government of Manitoba. The compromise allowed for the teaching of Catholic religion in a public school when there were sufficient Catholic students. Similarly, if there were sufficient French-speaking students, classes could be taught in French.

One Country, Two Visions

The greatest challenge to Canadian unity comes from the cultural divide that separates French- and English-speaking Canadians and their respective visions of the country. The two predominate visions are (1) a partnership between French and English Canada, and (2) 10 equal provinces.

In the early years of Confederation, events such as those outlined above widened the French/English faultline. For French Canadians these events demonstrated the "power" of the English-speaking majority and their unwillingness to accept a vision of Canada as a partnership between the two founding peoples. The root of each vision lies in the history of Canada and the division of powers by the Fathers of Confederation.

Partnership Vision

One vision of Canada is based on the principle of two founding peoples. This vision originated in French-Canadian historical experiences and compromises that were necessary for the sharing of political power between the two partners. This vision began with the Conquest of New France, but its true foundation lies in the formation of the Province of Canada in 1841. From 1841 onward, the experience of working together resulted in a Canadian version of cultural dualism.

Henri Bourassa, a French-Canadian politician and journalist (and Canadian nationalist) in the early twentieth century, was a strong advocate of cultural dualism. He wrote, "My native land is all of Canada, a federation of separate races and autonomous provinces. The nation I wish to see grow up is the Canadian nation, made up of French Canadians and English Canadians" (quoted in Bumsted, 2007: 307). Bourassa argued that a "double contract" existed within Confederation. Even today, Bourassa's "double contract" is an essential element in the concept of two founding peoples. He based the notion of a double contract on a liberal interpretation of section 93 of the BNA Act, which guarantees denominational schools. Bourassa expanded the interpretation of the religious rights to include cultural rights for French- and English-speaking Canadians. In more practical terms, Bourassa regarded Confederation as a moral contract that guaranteed French/English duality, the preservation of French-speaking Québec, and the protection of the language and religious rights of French-speaking Canadians in other provinces.

From a geopolitical perspective, Canada is a bicultural country. In one part the majority of Canadians speak English, and in another part French is the majority language. Thus, French culture predominates in Québec and has a strong position in New Brunswick. In addition to provincial control over culture, two other geopolitical factors ensure the dynamism of French in those provinces. One factor is the large size of Québec's population—the vitality of Québécois culture is one indication of its success. The second factor is the geographic concentration of French-speaking Canadians in Québec and adjacent parts of Ontario and New Brunswick. In New Brunswick the French-speaking residents, known as Acadians, constitute over one-third of the population.

The Royal Commission on Bilingualism and Biculturalism was designed to bridge the gap between English and French Canadians. This Commission, set up in 1963, examined the issue of cultural dualism, that is, an equal partnership between the two cultural groups. But by the 1960s, Canada's demographics revealed a third ethnic force and the concept of duality no longer reflected reality. English-speaking Canada had changed. English-speaking Canada had evolved from a predominantly British population to a more diverse one with several large minority groups who also spoke other languages besides English, especially German and Ukrainian. Ottawa, in searching for a compromise, established two policies, bilingualism (1969) and multiculturalism (1971).

The Vision of Equal Provinces

In the second vision, Canada consists of 10 equal provinces—yet this, too, is misleading. On the one hand, it represents the simple notion based on provincial powers granted under the British North America Act, which ensured that Canada consists of a union of equal provinces, all of which have the same powers of government. Nonetheless, by assigning provinces powers over education, language, and other cultural matters within their provincial jurisdictions, the BNA Act ensured that Québec's French culture was secure from political tampering by the anglophone majority in the rest of Canada. Thus, Confederation provided a form of collective rights for French culture within Québec. Under Canada's federal system, the powers of government are shared between the federal government and 10 provincial governments. But are all provinces really equal? As noted earlier, population size, geographic extent, and financial strength vary considerably, which is reflected in the need for equalization payments.

The vision of 10 equal provinces may reflect English-Canadian nationalism. For some time, English-speaking Canadians have been searching for their cultural identity. Before World War I, Canadians saw themselves as part of the British Empire. By the end of World War II, this perspective began to change. The Maple Leaf flag, adopted by Parliament in 1964, and "O Canada," the new national anthem approved by Parliament in 1967 and officially adopted in 1980, were signs of this cultural change. While the Québécois

THINK ABOUT IT
Is the demand for Québécois culture within Québec driven by the threat of drowning in a sea of American culture?

culture was flourishing, thanks in part to generous provincial funding for the arts, English-speaking Canadians continued to lean heavily on American culture. Some looked with envy at the cultural accomplishments of the Québécois and wondered aloud if similar achievements in English-speaking Canada were possible. The answer could be yes, providing the provincial governments offered similar financial support for the arts, and providing that English-speaking Canadians supported their artists at the same level as the Québécois public supported francophone artists and cultural producers.

Compromise

Given the incompatibility of the two visions—two founding peoples versus 10 equal provinces—and the historical development of the country, Canadian politicians have had the unenviable task of trying to accommodate demands from different groups—especially French Canadians, new immigrants, and Aboriginal peoples—and from different regions without offending other groups or regions. As in the past, politicians have continued to struggle with this Canadian dilemma, but in reality there is no perfect solution, only compromise. With this object in mind the federal government has made many efforts in search of the elusive middle ground.[8] It seems the search for an acceptable compromise between the two opposing visions of Canada will never end, and perhaps that is a good thing because the process is more important than the end result. To understand the current struggle for compromise, it is important to understand the political, economic, and cultural developments that have taken place in Québec over the past five decades.

Resurgence of Québec Nationalism

After World War II, Québec broke with its past. A rise of Québec nationalism had begun much earlier but gained political momentum during the Quiet Revolution of the early 1960s. This development was the result of four major events. The most important was the resurgence of ethnic nationalism, that is, a pride in being Québécois. The second was Québec's joining the urban/industrial world of North America and the subsequent expansion in the size of its industrial labour force and business class. The third was the removal of the old elite. This reform movement was profoundly anticlerical in its opposition to the entrenched role of the Church in Québec society, particularly the Church's control over education. In many ways, this reform was based on the aspirations of the working and middle classes in the new Québec economy. The fourth was the state's aggressive role in the province's affairs.

With the election of Jean Lesage's Liberal government in 1960, which held power until 1966, the province moved forcefully in a new direction. It created a more powerful civil service that allowed francophones access to middle and senior positions often denied them in the private sector of the Québec economy, which was controlled by English-speaking Quebecers and American companies. It nationalized the province's electric system, thereby creating the industrial giant known as Hydro-Québec, now a powerful symbol of Québec's revitalized economy and society. In turn, Hydro-Québec built a number of huge energy projects that demonstrated the province's industrial strength. By 1968, this Crown corporation had constructed one of the largest dams in the world on the Manicouagan River. Called Manic 5, this dam demonstrated Hydro-Québec's engineering and construction capabilities. To Quebecers, Hydro-Québec was a symbol of Québec's economic liberation from the years of suffocation associated with Maurice Duplessis and his Union Nationale government, which had been closely tied to big businesses owned by English-speaking Canadians and Americans. Clearly, Lesage's political goal of becoming "maîtres chez nous" (masters in our own house) had materialized with the success of Hydro-Québec, thus sparking a growth in Québec nationalism. Québec's desire for more autonomy in its own affairs intensified with increased confidence. In short, a new society had arisen in Québec, a society that wanted to chart its destiny. Charles Taylor (1993: 4) summed up this new feeling as "a French Canada which, after a couple of centuries of enforced incubation [under London and then Ottawa], was ready to take control once more of its history." The political question Taylor raised is a simple one: Would this "control" take place within the framework of Canada's political system or outside it?

Separatism

Separatism—the desire for an independent francophone nation in North America—grew out of the Quiet Revolution. The embers of nationalism were ignited in 1967 by French President Charles de Gaulle, who, when visiting the province for Expo 67, uttered the incendiary words, *"Vive le Québec. Vive le Québec libre"* (photo 3.7), during a speech from a hotel balcony in Montréal. Soon, René Lévesque, who as a member of the Lesage government had been the architect of the nationalization of electricity generation in the province, had formed a new separatist political party, the Parti Québécois. By 1976, the PQ had won a stunning election victory, and since that time separatism, though waxing and waning in public support, has taken on a mainstream political form. By the time of the first referendum on independence in 1980, the separatists made up a substantial minority within Québec's population, with perhaps 20 per cent dedicated separatists and another 40 per cent strongly dissatisfied with their place within Canada.

In the first referendum in 1980, Québec voters rejected the **sovereignty-association** option, with almost 60 per cent voting to remain in Canada, which suggests that just over half of the francophone voters stood with the "Non" side, along with almost all of the English-speaking residents. The rest of Canada responded with a collective sigh of relief, but separatism was far from dead.

The dream of an independent Québec remained a strong political force. In fact, the 1995 referendum vote on independence almost succeeded. "No—by a Whisker!" screamed the headline of the *Globe and Mail* on the morning after the referendum of 30 October 1995. Québec came within 40,000 votes of approving the separatist dream of becoming an independent state (Vignette 3.10).

Photo 3.7
French President Charles de Gaulle during his incendiary "Vive le Québec libre" speech in Montréal, 24 July 1967.

For historical background on the French/English faultline in Québec, see Chapter 6, "British Colony, 1760–1867," page 231.

Moving Forward

The 1995 referendum was a low point in French–English relations, and its after-effects were many

VIGNETTE 3.10

The Results of the 30 October 1995 Referendum

The Question: "Do you agree that Québec should become sovereign, after having made a formal offer to Canada for a new Economic and Political Partnership within the scope of the Bill respecting the future of Québec and of the agreement signed on June 12, 1995?"

The Answer (at 10:30 p.m. Eastern Time, 21,907 of 22,427 polls):

	Number	Per cent
No	2,294,162	49.5
Yes	2,254,496	48.7
Rejected	83,340	1.8
Total	4,631,998	100.0

Source: Globe and Mail (1995).

THINK ABOUT IT
Why won't separatism go away?

and varied. English Canada was dazed by the outcome, but the separatists appeared to be a spent force. In 1996, provincial premiers added their voice to the discussion in the Calgary Declaration, stating: "the unique character of Québec society with its French-speaking majority, its culture and its tradition of civil law is fundamental to the well-being of Canada." In the typical fashion of Canadian provincial leaders, the premiers remained clearly in the camp of 10 equal provinces by adding to their conciliatory Declaration that "any power conferred to one province in the future must be available to all." This Declaration was the third attempt at reconciliation with Québec since the patriation of the Constitution in 1982.[9] The next step was for each provincial government to pass the appropriate legislation, giving the Calgary Declaration legal status. By July 1998, all provinces (except Québec) and territories had passed this resolution in their legislatures.

Since the 1995 referendum, separatism, while not gone, has lost its spark, in part because the threat of being absorbed by English-speaking North America has subsided for Quebecers, who are more confident in the security of their language and culture than had been the case in the immediate post-World War II period. Equally important, Ottawa is more comfortable with the idea of Québécois being recognized as a "distinct cultural group" or nation within Canada. In November 2006, the House of Commons overwhelmingly passed a motion by Prime Minister Harper that recognized Québécois as a nation within Canada. But Québécois nationalism, while somewhat dormant for now, remains deep inside Québécois culture and language, and in September 2012 a minority Parti Québécois government was returned to power after nearly a decade of Liberal rule in the province.

SUMMARY

History and geography explain the nature and complexity of contemporary Canada. Canada is both a young and an old country. Complexities are reflected in its four faultlines. The newcomer/old-timer faultline hinges on the "accommodation" issue. The centralist/decentralist argument has taken a twist with Ontario now declared a "have-not" province. Aboriginal peoples are settling their outstanding issues with the Crown—the land claim settlement process continues to function and the deep sores caused by Indian residential schools, one can hope, have begun to heal. History teaches Canadians that differences will continue to emerge but compromises are necessary for national unity, regional harmony, and social justice.

Over the course of its short history as a nation of regions, Canada, has learned "tolerance" the hard way, and, it would seem, has chosen a "soft" path into the twenty-first century.

CHALLENGE QUESTIONS

1. Why did the doctrine of "terra nullius" allow Europeans to consider North America "unoccupied" and therefore open to European ownership and settlement?

2. What historic events support the concept that Canada has become a "soft" nation that has learned the benefit of compromise and tolerance?

3. Do you believe that political reality forces federal governments to favour Ontario and Québec over other parts of the country? Can you supply an example?

4. Why does Québec support the concept of Canada as "two founding peoples" rather than the concept of Canada as "10 equal provinces"?

5. If the separatists had won the 1995 referendum, would a geographically split country inevitably drift into the political orbit of the United States?

FURTHER READING

Harris, R. Cole, ed. 1987. *Historical Atlas of Canada, Volume I: From the Beginning to 1800*. Toronto: University of Toronto Press.

Gentilcore, R. Louis, ed. 1993. *Historical Atlas of Canada: Volume II: The Land Transformed 1800–1891*. Toronto: University of Toronto Press.

Kerr, Donald, and Deryck W. Holdsworth, eds. 1990. *Historical Atlas of Canada, Volume III: Addressing the Twentieth Century 1891–1961*. Toronto: University of Toronto Press.

The historical geography of Canada recalls past events. Maps play a large role in this rediscovery of Canada's past. In 1970, several geographers and historians explored the idea of preparing a major Canadian historical atlas focused on social and economic themes. These three volumes, which parallel the discussion in this chapter, are the successful outcome. The editors weave together the various strands that constitute Canada's historical geography and provide a rich legacy for Canadian scholars and students. Four more recent major events, however, are not covered—the rise of Aboriginal political power; the threat of separation of Québec from the rest of Canada; the influx of non-European immigrants; and the Canada–US Free Trade Agreement. These issues are discussed in Chapter 4.

CANADA'S HUMAN FACE

Introduction

Canada is home to over 35 million people. The country's population and economy continue to grow, thanks in large measure to the continuous flow of newcomers and to the increasing demand for its natural resources. Canada's declining rate of natural increase provides the rationale for high levels of immigration while the rapidly expanding economies of China and other Asian countries have had a double-edge impact on Canada. On the one hand, Asian demand for Canadian resources has created a booming economy in Western Canada and, to a lesser degree, in British Columbia. On the other hand, Asian but particularly Chinese manufactured goods have flooded into Canada and, in doing so, have replaced much domestic production.

As a result of these demographic and economic forces, the Canadian population and economy are undergoing significant adjustments and these adjustments are resonating differently across the country. For instance, until the global economic crisis that began in late 2008, Ontario generated the greatest economic gains and most newcomers flocked to Toronto, with a smaller number settling in Montréal and Vancouver. Since then, a growing number of immigrants, lured by employment and business opportunities, have settled in the major cities of Western Canada. At the same time, the face of Canada's economy is changing, primarily as a result of globalization. While Canada's resource-rich regions are benefiting from the strong demand and record high prices for energy and mineral products, its manufacturing sector faces ever-increasing competition from low-wage foreign competitors, causing serious restructuring, relocation to offshore countries, and, in too many cases, loss of jobs and lower wages. Ontario and Québec have felt the brunt of the global economic downturn while Western Canada, British Columbia, and even that former "have-not" province, Newfoundland and Labrador, are enjoying economic good times.

Demography plays a critical role in fuelling economic growth. Skilled labour shortages are especially acute in the fast-growing areas of Western Canada. Immigration is one solution. Canada, while enriched by the global wave of diverse peoples to its shores, faces certain challenges. The cleavage between old-timers and newcomers has softened over time and the acceptance of newcomers and their cultures has

CHAPTER OVERVIEW

Issues examined in this chapter are:
- Factors causing Canada's population to increase.
- Implications of immigration for Canadian society.
- Multiculturalism and the "accommodation" of newcomers.
- The link between Canada's aging population and its rapidly increasing Aboriginal population.
- The clash between Canada's traditional continental economic strategy and a global strategy.
- The effect of the global economy on Canada's regional economies.
- Demography and Canada's four faultlines.

The St Jacobs Farmer's Market, St Jacobs, Ontario. The St Jacobs Market is Canada's largest farmer's market, hosting hundreds of food and craft vendors from across Waterloo region including local Old Order Mennonite farmers. Photo: Courtesy of St Jacobs Country.

become almost routine. For the most part, newcomers, but especially the second generation, find their place within an adaptable, pluralistic, and dynamic Canadian society. Even so, those newcomers who do not speak one of Canada's two official languages and whose cultural homeland has sharply different traditions, religions, and social values from those of Canadian society must make the greatest adjustment. Most Canadians welcome the adjustment to an open and democratic society. While the accommodation of newcomers remains a challenge for them and for the rest of Canadian society, Canada has grown over the centuries into what John Ralston Saul calls a "soft" country able to accept and integrate newcomers. This soft country did not appear overnight but has grown from a long history and sometimes hard experiences that moulded the country into a relatively harmonious pluralistic society. This soft faultline distinguishes Canada from many other Western nations and may well a harbinger of what future secular societies should and will look like.

Canada's Population

Canada's population continues to grow. By 2012, the population exceeded 35 million and, with a projected average annual population increase of 500,000 over the next eight years, should approach 40 million by 2020. In comparison to its population in 1851, Canada has increased in size by more than fourteen-fold (Figure 4.1). In spite of its population growth, Canada's society is also aging, which affects the size of the labour force and the age dependency ratio. At the same time, the demographic direction for the Aboriginal population is the opposite—a rapidly expanding and very youthful population. Under normal circumstances, young Aboriginal workers would be flooding into the labour force, but such is not the case because of a mismatch of skills and geography.

Immigration continues to play a key role in Canada's population increase. The country is receiving record numbers of immigrants each year and many are from non-European and

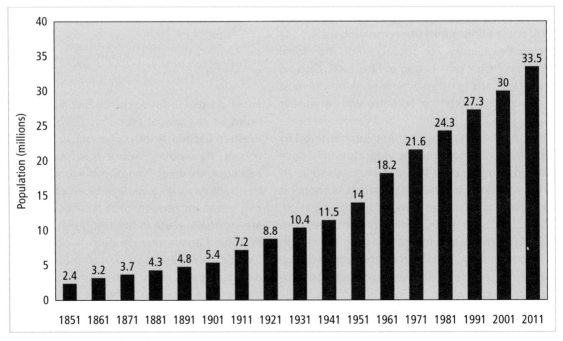

Figure 4.1   **Population of Canada, 1851–2011**
Source: Statistics Canada (2012d).

non-Christian countries. On top of these demographic changes, Canada's economy was buffeted by the global financial crisis that emanated from the United States. Within Canadian society, social issues such as homelessness, urban gang warfare, and sprawling urban centres came into focus in recent years, while the principal economic issue has become the widening income gap between the poor and the rich. All of these issues have caught the public's attention. How they will play out in the second decade of the twenty-first century is anyone's guess, but what is known is that Canada is sailing into uncharted waters.

A New Look

Canada's **demography** has a new look—ethnic composition and cultural diversity now more strongly than ever reflect fresh elements in our society. Canada has a substantial—and growing—number of adherents to Islam, Sikhism, and Hinduism, its larger cities are home to many so-called visible minorities, and, after English and French, the Chinese language (Cantonese and Mandarin) is the third most commonly spoken language in Canada. Clearly, immigration plays a major role in these demographic changes. Other demographic trends are cause for concern: the rapid increase in Canada's Aboriginal population and the declining place of the French language within the nation. Could Canada's new demographic face ignite new faultlines based on the ever widening gap between the rich and poor? A broader question is whether public protests against inequalities within Western societies—such as the Occupy movement in United States, the anti-globalization movement against economic inequality, the Québec student protests against higher tuition fees, and the Idle No More movement of grassroots Aboriginal activists and sympathizers against unilateral actions of the Harper government—represent a fresh wave of social unrest against the "establishment" or are just one-off events.

Population Size

With a population of 35 million in 2012, Canada can be considered a medium-sized country on the world stage—only one-quarter of the nations of the world have larger populations than Canada. From a historical perspective, this population size represents a tenfold increase since Confederation. Three primary factors account for this growth over the last 145 years: natural increase (births minus deaths); population gained from territorial expansion (nearly 320,000, for example, when Newfoundland joined Confederation in 1949); and immigration, which now approaches 300,000 annually.

Population Density

In spite of continuous population growth, Canada's population density is one of the lowest in the world. The explanation is simple—as the second-largest country in the world by geographic area, relatively few people inhabit this northern land. But is Canada underpopulated? Population density figures are more meaningful to answer this question if they are expressed as the amount of arable land per person. This measure is called physiological density. By this measure, Canada's physiological density is similar to that found in the United States.

Population density is determined by dividing the number of people by the land area. Canada has a population density of 3.7 persons per square kilometre, which means the country has an extremely low population density (but not the lowest in the world). Australia, a country with much of its dry lands unsuitable for settlement, has only 3 persons per km^2. Mongolia, another large country with little land suitable for settlement, has an even lower density of 1.7. All other countries of the world have higher population densities. The United States, for example, has 32 persons per km^2, while Bangladesh is one of the most densely populated countries with a density of just under 1,000 people per km^2.

The explanation for these variations between countries is that land varies greatly in its capacity to support human settlement. Most of Canada lies beyond the northern limits of agriculture. The Territorial North, for example, has an exceptionally low capacity to support human life. As of the 2011 census, the Territorial North's population density was only 0.03 people per km^2 (Table 4.1). In comparison, Ontario had a population density of 14.1 people per km^2, and this includes the vast reaches of northern Ontario with very little population.

Population Distribution

Population distribution is the dispersal of people within a geographic area. Canada's population is extremely unevenly distributed across the country (Figure 4.2 and Table 4.1). In fact,

THINK ABOUT IT
Does the export of food from surplus-producing countries to food-short countries make the physiological density argument somewhat irrelevant, or, given global warming, is national food security a critical issue for the twenty-first century?

Table 4.1 Population Size, Percentage, and Density, Canada and Regions, 2011

Geographic Region	Population	Population (%)	% Change, 2006–11	Population Density (per km²)
Territorial North	107,265	0.3	5.9	0.03
Atlantic Canada	2,327,638	7.0	1.9	4.7
British Columbia	4,400,057	13.1	7.0	4.8
Western Canada	5,886,906	17.6	8.9	3.2
Québec	7,903,001	23.6	4.7	5.8
Ontario	12,851,821	38.4	5.7	14.1
Canada	33,476,688	100.0	5.9	3.7

Source: Adapted from Statistics Canada (2012a).

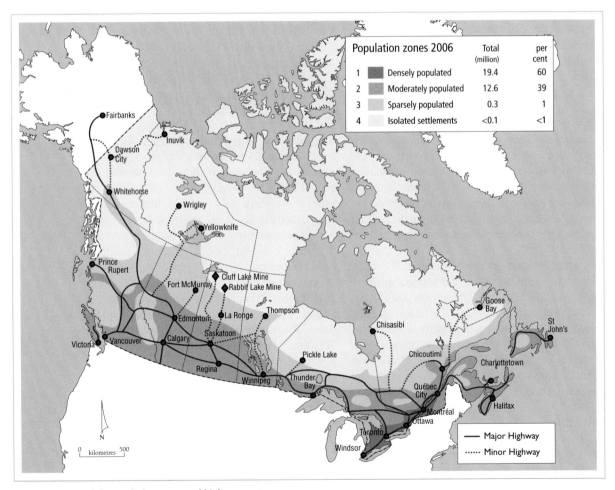

Figure 4.2 **Canada's population zones and highway system**

Canada's population is heavily concentrated in southern Ontario and southern Québec, where a favourable physical geography and an advantageous geographic location have resulted in a dense population. A secondary belt of population spans a southern strip of Canada. Together, the densely and moderately populated zones account for 99 per cent of Canada's population. The core highway system connects cities and towns in Canada's ecumene, which consists of population zones 1 and 2. Outliers of the highway system extend into population zones 3 and 4. Nunavut is the only political territory not connected to the national highway system.

Photo 4.1

Founded in 1642, Montréal, Québec, is one of Canada's oldest cities. Located on the St Lawrence River, Montréal is a transportation hub for international shipping. With a population of 3.6 million (2012), the city has the largest francophone urban population in North America and falls just behind Paris for the largest French-speaking city in the world.

few nations have so much of their population concentrated in such a relatively small area of the country, while the rest of the country is almost vacant. So pronounced is this uneven population distribution that an American geography text described Canada's population as if it were "drawn by a magnet toward the giant neighbour on the south, for they [Canada's

Photo 4.2

Toronto, Ontario, with a population of 5.6 million in 2012, is Canada's most populous city and serves as the economic engine for Ontario and as the financial capital for Canada. Toronto has also become the nation's most culturally diverse city because of its capacity to attract new Canadians.

Photo 4.3

Vancouver, British Columbia, is Canada's leading ocean port, with most goods coming and going to China and other Asian countries. The third-largest Canadian city at 2.1 million in 2012, Vancouver was home to the 2010 Winter Olympic Games. In the foreground, the Cambie Street Bridge spans False Creek and leads to BC Place; to the west of the bridge, at the bottom of the photo, is Granville Island with its public market.

inhabitants] are strikingly concentrated along the United States border" (Trewartha et al., 1967: 542). Canadian scholars often view this same distribution as consisting of a national population core surrounded by a sparsely populated hinterland. The population core is sometimes described as Canada's national **ecumene** (inhabited area). Within this ecumene lies the core of Canada's highway system (see Figure 4.2).

As Table 4.1 shows, Ontario has the highest density at 14.1 persons per km², followed by Québec at 5.8 persons per km². These two geographic regions form a demographic core with 62 per cent of Canada's population. Combined or individually, these geographic regions exert considerable political force within the Canadian federation and, as a result, they are a source of much regional alienation. Their dominant demographic position is enhanced by their geographic situation in the Great Lakes–St Lawrence Lowlands, proximity to the manufacturing heartland of the United States, and their economic/financial strength.

Population zones provide a more exact geographic picture of Canada's population distribution.

As shown in Figure 4.2 and Table 4.2, four population zones vary in population size from very large (59 per cent of Canada's population) to very small (less than 1 per cent of Canada's population). Similarly, the four zones vary considerably in population density. The overall spatial pattern reinforces the image of a highly concentrated population core surrounded by more thinly populated zones.

Canada's core population zone lies in the Great Lakes–St Lawrence Lowlands. As the most naturally favoured physiographic region, the Great Lakes–St Lawrence Lowlands contains 19.5 million people and almost three-quarters of Canada's major cities. This population core includes Toronto, Montréal, Ottawa–Gatineau, Québec City, Hamilton, Oshawa, London, and Windsor, to name only some of the largest cities in the region. As Canada's most densely populated area, its economy is based on manufacturing and its agriculture lands contain the most fertile farmlands in Canada.

The secondary core zone extends in a narrow band across southern Canada. In general, its northern boundary corresponds with the polar edge of arable land. As the second-most favoured zone, it

Table 4.2 Population Zones, 2011

Zone	Population (millions)	Percentage of Canada's population	Major City	Population of Major City
1. Core zone: densely populated	19.5	59	Toronto	5,838,800
2. Secondary zone: moderately populated	13.2	40	Vancouver	2,419,700
3. Sparsely populated zone	0.3	1	Fort McMurray*	61,374
4. Empty zone: isolated settlements	<0.1	<1	Labrador City	9,228

*Wood Buffalo Regional Municipality.

Source: Statistics Canada (2012a, 2012b, 2012c).

occupies the more southerly portions of the Appalachian Uplands, the Canadian Shield, the Interior Plains, and the Cordillera. About 13.2 million Canadians (over one-third of the country's total population) live in this moderately populated zone. Canada's remaining major cities are located within this zone, including Vancouver, Edmonton, Calgary, Winnipeg, and Halifax. Within the secondary zone, some areas, such as southern Alberta and British Columbia, are growing quickly while other areas, such as Newfoundland and Labrador, have experienced much slower growth and even population losses. As a result, the population of the secondary zone is increasing slowly and unevenly.

The third zone, characterized by sparse population, is associated with a narrow band of the boreal forest that stretches across mid-Canada. Three physiographic regions are found here—the Canadian Shield, the Interior Plains, and the Cordillera. As the third-most populous zone, less than 1 per cent of all Canadians (about 300,000) live here. Only one of Canada's major cities, Fort McMurray, Alberta, is in this zone. Fort McMurray is an outstanding example of a booming **resource town**. As the hub of northern Alberta's oil sands extraction and exploration, Fort McMurray (which is within the Wood Buffalo Regional Municipality) is the largest city in the tertiary zone with a population exceeding 61,000. Other larger urban centres range in size from 10,000 to 20,000. Whitehorse and Yellowknife, as the capital cities of Yukon and the Northwest Territories, are administrative centres and **regional service centres**, since they also provide most of the service functions for their areas (Figure 4.3). These two cities, with populations of 20,562 and 18,352, respectively, in 2011, also anchor the poleward edge of zone 3.

Photo 4.4
With the Parliament Buildings (left) and the Château Laurier in the background, the Rideau Canal provides a winter skating experience in Ottawa, Canada's capital. The canal, completed in 1832, was originally built as a military supply route between Kingston and Ottawa. Metropolitan Ottawa, in 2011, accounted for 1.3 million residents.

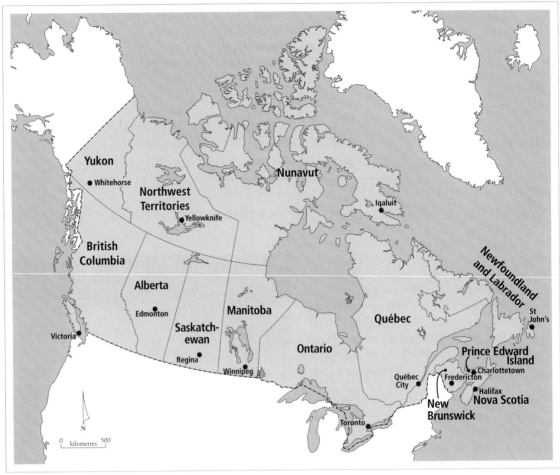

Figure 4.3 Capital cities
With five exceptions, the capital cities of the 10 provinces and three territories are the largest urban centres in each political jurisdiction. The exceptions are: New Brunswick (Moncton); Québec (Montréal); Saskatchewan (Saskatoon); Alberta (Calgary); and British Columbia (Vancouver).

Most of Canada's expansive territory is found in the last, almost uninhabited lands of the North. Here, because of the extremely cold climate, the presence of permafrost, and the polar ice pack, the land is inhospitable for settlement. This zone includes fewer than 100,000 inhabitants. Most reside in small, isolated **Native settlements**, which, sadly, have no economic base and offer little hope for a better life. As described in the "Contested Terrain" box below, these settlements resulted from an initiative of the federal government back in the 1950s. With a small population and a large geographic area, this population zone has the lowest population density. From a business perspective, this sparsely populated zone is the least productive area in the country. Most commercial activities centre on exploitation of non-renewable resources, including mineral bodies and petroleum deposits. Unlike in the other zones in Canada, Aboriginal peoples form the majority in this zone, but few adults are engaged in the mining economy because most settlements are far from mine sites. Even when a First Nation settlement is close to a mine, few have a high school diploma and, without such certification, well-paying jobs are out of reach. In spite of a high rate of natural increase among the Aboriginal population, the quaternary zone is affected by a net out-migration. Urban centres are small, most with populations under 5,000. In 2011, the iron-mining town of Labrador City had the largest population in the fourth zone with 9,228 inhabitants, followed by the rapidly growing Iqaluit, the capital city of Nunavut, with 6,184. Unlike the booming resource town of Fort

> **CONTESTED TERRAIN 4.1**
>
> **Social Engineering Often Backfires**
>
> In the 1950s, the federal government faced a dilemma—should it begin to provide basic public services to Aboriginal people living on the land in a hunting and trapping lifestyle? When Canadians and the outside world became aware of a few hunting families starving to death, Ottawa's hand was forced. The government chose the relocation solution that created Native settlements and thus ended the threat of starvation. Ottawa hoped the settlements would lead to integration into Canadian society, but in fact the settlements were far removed from the rest of society, with few or no employment opportunities, and led to the loss of a traditional way of life, to an unhealthy change in diet, and to a sedentary lifestyle. Other options aimed at keeping hunters and trappers on the land and left the pace of integration in the hands of the people. One option was a fur subsidy that would have provided sufficient cash income to support the hunting/trapping lifestyle. A second one could have been a direct payment to hunters/trappers and their families, similar to the Quebec Income Security Program for Cree Hunters and Trappers. All three approaches have their advantages and disadvantages. Given what you know now, if you were the Prime Minister back in the 1950s, would you still choose the "relocation" strategy?

McMurray, with a population increase of 28.7 per cent from 2006 to 2011, Labrador City's population has remained largely static. From 2006 to 2011, its population increased by only 2.8 per cent. Iqaluit, on the other hand, with a growing public sector, had a remarkable jump in population of 30.4 per cent over the same five-year period.

Urban Population

Canada is an urban country with the vast majority of its population living in cities, towns, and villages. In 2011, 82 per cent of Canadians lived in urban areas with a population of at least 1,000 residents (Statistics Canada, 2012c: Table 2). Canada's urban population has grown steadily over the past century, from 37 per cent in 1901 to 82 per cent (Figure 4.4). The greatest increase has taken place in the largest cities, the so-called census metropolitan areas (CMAs). From 2001 to 2011, these 33 cities grew at a rate of nearly 15 per cent compared to 11 per cent for Canada as a whole. The source of this population increase is due to three factors. First, the arrival of immigrants from overseas has greatly added to urban growth, especially of the larger cities. Second, the stream of rural Canadians abandoning the countryside for urban places remains a powerful factor. In this case, the movement is from farms to small towns and nearby cities. Third, urban centres grow due to the natural increase taking place among the urban population.

Census Metropolitan Areas

The emergence of large cities across Canada is the latest outcome of urbanization. These cities serve as the economic and cultural anchors of their hinterlands. Statistics Canada defines **census metropolitan area** (CMA) as an urban area (known as the urban core) together with adjacent urban and rural areas that have a high degree of social and economic integration with the urban core. The urban core population of a CMA must be at least 100,000 based on the previous census.

The proportion of Canada's population residing in census metropolitan areas has increased from 30.3 per cent in 1931 to 69.1 per cent, or 23.1 million Canadians, in 2011. In 1931, there were 10 CMAs: Halifax, Hamilton, Montréal, Ottawa, Québec, Saint John, Toronto, Vancouver, Windsor, and Winnipeg. By 2011, Canada had 33 census metropolitan areas (Table 4.3). Of the 23.1 million people residing in these CMAs, 15.9 million lived in the six metropolitan areas each with a population of more than 1 million: Toronto, Montréal, Vancouver, Calgary, Ottawa–Gatineau, and Edmonton.

Since 1951, the population growth of the five largest CMAs has clearly outstripped the national

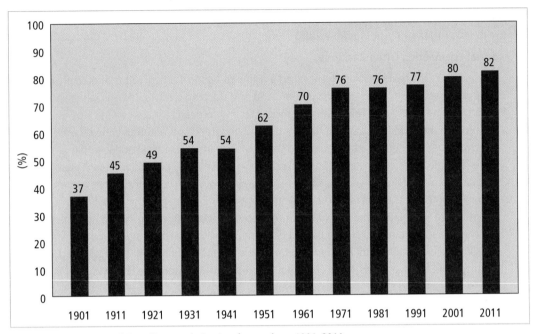

Figure 4.4   **Percentage of Canadian population in urban regions, 1901–2011**
Source: Adapted from Statistics Canada (2007c, 2013a).

rate of increase. Over the period 1951–2011, Toronto has gained the greatest number of people and has consistently ranked in the top five cities by growth rate. By 1971, Toronto surpassed Montréal as Canada's largest metropolitan centre, and Toronto, with a higher growth rate, continues to outdistance Montréal in population size. From 2006 to 2011, for instance, Toronto's growth rate was 9.2 per cent while that of Montréal was 5.2 per cent. Yet, the greatest rate of increase among the census metropolitan areas—a phenomenal 12 plus per cent—took place in Calgary and Edmonton, followed closely by Saskatoon at 11.4 per cent and Kelowna at 10.8 per cent. Only two CMAs suffered a decline—Windsor at 1.3 per cent and Thunder Bay at 1.1 per cent.

What is the attraction of cities? Of greatest importance, most business and employment opportunities are found in cities, especially large cities. As well, Canadians prefer to live in an urban setting where amenities are readily available. Cities are also important for other reasons. Major urban centres are at the cutting edge of technological innovation and capital accumulation. In the new world of the knowledge economy, manufacturing does not determine a city's prosperity; rather, the creativity of its business and university communities is the determining factor.

Despite the steady and in some instances remarkable growth of Canadian cities, all is not well in metropolitan Canada. Urban sprawl affects all cities, forcing cities to spend heavily on infrastructure in these outlying areas while their downtowns lose their raison d'être. Competition from malls and big-box stores in the suburbs has hurt downtown retail areas. Also, automobiles are ill-suited for the downtown while generously accommodated in shopping areas on the urban perimeter. As a result, downtown cores have lost their focus of commercial and social activities and some even seem deserted at night. In the twenty-first century, cities face the daunting task of finding solutions to these problems associated with the growth of suburbia and the shift of commercial activities from downtown to the outlying areas. On the one hand, efforts to curtail urban sprawl by making downtowns more pedestrian- and bicycle-friendly, turning inner-city residential areas into a much denser form of residential housing, and adding more "urban parks" could signal a new direction. On the other hand, the ever-expanding suburban nature of cities that accommodates our

Table 4.3 **Population of Census Metropolitan Areas, 2006 and 2011**

CMA	Population 2011	Population 2006	% Change
Toronto (Ont.)	5,583,064	5,113,149	9.2
Montréal (Que.)	3,824,221	3,635,556	5.2
Vancouver (BC)	2,313,328	2,116,581	9.3
Ottawa–Gatineau (Ont./Que.)	1,236,324	1,133,633	9.1
Calgary (Alta)	1,214,839	1,079,310	12.6
Edmonton (Alta)	1,159,869	1,034,945	12.1
Québec (Que.)	765,706	719,153	6.5
Winnipeg (Man.)	730,018	694,668	5.1
Hamilton (Ont.)	721,053	692,911	4.1
Kitchener–Cambridge–Waterloo (Ont.)	477,160	451,235	5.7
London (Ont.)	474,786	457,720	3.7
St Catharines–Niagara (Ont.)	392,184	390,317	0.5
Halifax (NS)	390,328	372,858	4.7
Oshawa (Ont.)	356,177	330,594	7.7
Victoria (BC)	344,615	330,088	4.4
Windsor (Ont.)	319,246	323,342	−1.3
Saskatoon (Sask.)	260,600	233,923	11.4
Regina (Sask.)	210,556	194,971	8.0
Sherbrooke (Que.)	201,890	191,410	5.5
St John's (NL)	196,966	181,113	8.8
Barrie (Ont.)	187,013	177,061	5.6
Kelowna (BC)	179,839	162,276	10.8
Abbotsford–Mission (BC)	170,191	159,020	7.0
Greater Sudbury/Grand Sudbury (Ont.)	160,770	158,258	1.6
Kingston (Ont.)	159,561	152,358	4.7
Saguenay (Que.)	157,790	156,305	1.0
Trois-Rivières (Que.)	151,773	144,713	4.9
Guelph (Ont.)	141,097	133,698	5.5
Moncton (NB)	138,644	126,424	9.7
Brantford (Ont.)	135,501	124,607	8.7
Saint John (NB)	127,761	122,389	4.4
Thunder Bay (Ont.)	121,596	122,907	−1.1
Peterborough (Ont.)	118,975	116,570	2.1

Source: Statistics Canada (2012c).

automobile-oriented public represents a threat in two ways. First, the costs of providing urban services to new suburbs are taxing city budgets for new roads, schools, fire halls and trucks, parks, transit services, and water/sewer systems, and these costs may force cities to greatly increase the cost of suburban lots for developers. For instance, cities could limit additional suburban development by

Photo 4.5
Calgary's downtown is dominated by skyscrapers, many of which are associated with the petroleum industry. Beyond the skyscrapers, Calgary, like other major cities, faces several challenges, including urban sprawl, homelessness, and inadequate revenue-sharing from provincial and federal governments.

THINK ABOUT IT
Do Canadian cities need more fiscal power to meet their growing needs? As Calgary's mayor, Naheed Neshi, stated, "I am the mayor of a city that has more people in it than five provinces, yet I have the exact same legislative authority as any village of 30 or 40 people. And that has to change" (Agrell, 2011: A6).

substantially increasing the price of city-owned land to companies, to the point where developers would turn to purchasing properties in old neighbourhoods and replace single-unit housing with denser forms of residential development, thus curtailing suburban sprawl. Second, the well-being of central business districts is threatened and already many downtown areas are virtually "dead" in the evenings. Other challenges for cities are smog, traffic congestion, the high cost of rapid transit, the homeless, deteriorating infrastructure, and garbage disposal. The complexity of urban Canada is often the subject of study by geographers.

Variation in Urban Population by Geographic Region

Urbanization is linked to economic development. For that reason, the rate of urbanization has varied across Canada. Not surprisingly, Ontario and British Columbia have the highest percentages of their populations classified as urban, while Atlantic Canada and the Territorial North have the lowest. For years, however, Ontario led all other geographic regions (Table 4.4).

By 2011, as shown in Table 4.4, British Columbia had a higher percentage of urban population than Ontario. The increase in urban population is most obvious in Western Canada where, over the past 100 years or so, the rural nature of the region has declined sharply. In 1901, less than 20 per cent of Western Canada's population lived in urban places. By 2011, the figure had jumped to almost 77 per cent. The shift from rural to urban communities, although most obvious in Western Canada, is a national phenomenon and is associated with two factors: the declining numbers involved in agriculture due to mechanization, and the increase in job opportunities in urban places.

Population Change

Canada's population increased by 1.9 million (5.9 per cent) from 2006 to 2011 (Table 4.1). Population change has three components: births,

Table 4.4 Percentage of Urban Population by Region, 1901–2011

Region	1901	1921	1941	1961	1981	2001	2011
Ontario	40.3	58.8	67.5	77.3	81.7	84.7	85.1
British Columbia	46.4	50.9	64.0	72.6	78.0	84.7	85.4
Québec	36.1	51.8	61.2	74.3	77.6	80.4	80.2
Western Canada	19.3	28.7	32.4	57.6	71.4	75.7	76.7
Atlantic Canada*	24.5	38.8	44.1	50.1	54.9	53.9	54.1
Canada	34.9	47.4	55.7	70.2	76.2	79.7	82.0

Source: McVey and Kalbach (1995: 149). © 1995 Nelson Education Ltd. Reproduced by permission. Statistics Canada (1997b, 2002a, 2007f, 2008a, 2012c).

*Newfoundland is not included in Atlantic Canada's figures until 1961.

While comparable statistics are not available for the Territorial North, two observations are possible. (1) Prior to the 1950s, few people in this region lived in settlements. (2) By 2006, Statistics Canada (2007f) classified nine centres in the Territorial North as urban areas: Whitehorse in Yukon; Hay River, Inuvik, and Yellowknife in the Northwest Territories; and Iqaluit, Pangnirtung, Rankin Inlet, Cambridge Bay, and Arviat in Nunavut. Their total population in 2006 was 55,137, resulting in 54 per cent of the Territorial North's population defined as urban.

deaths, and migration. **Population increase** is the sum of natural increase and net migration over a given period. The term **population growth** is used when this increase is expressed as a rate, that is, as a percentage change over time. The **rate of natural increase** is the difference between the **crude birth rate** (CBR) and the **crude death rate** (CDR). CBR is the number of live births per 1,000 people in a given year. CDR is the number of deaths per 1,000 people in a given year. **Net migration** is the difference between in- and out-migration. One theoretical explanation lies in the **push-pull model**, which, in its most simple form, sees adverse factors at home "pushing" people out, making them want to emigrate. Attractive factors abroad can "pull" people in.

Since 1851, Canada has enjoyed continuous population growth. At first, high rates of

THINK ABOUT IT
Figure 4.5 projects an ever-decreasing natural increase. In fact, from an increase of 134,000 in 2010, the annual figure for natural increase falls to zero by 2031 and then into negative annual numbers, falling to negative 150,000 in 2056. Two questions: How can a population have a negative natural increase? Is this pattern of natural increase reflected in the demographic transition theory (discussed below)?

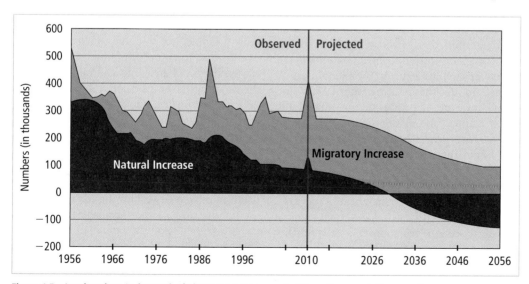

Figure 4.5 **Immigration: An increasingly important component of Canadian population growth**
Source: Statistics Canada (2009k, 2012a).

Photo 4.6
The North Saskatchewan River frames Edmonton's downtown and provincial legislative buildings. Like Calgary, Edmonton is experiencing rapid population growth and this growth is pushing the residential areas further and further from the downtown.

THINK ABOUT IT
While Québec's fertility rate is among the highest in Canada's provinces, its rate of population increase is below the national average. What other element of population change is lagging in Québec compared to other provinces?

natural increase and high levels of immigration propelled this growth. The highest rate of population growth occurred over the 1901–11 period, with a remarkable increase over 10 years of 34 per cent, spurred by the large influx of immigration to the Canadian Prairies. As more Canadians moved to cities, parents opted for smaller families, causing the rate of natural increase to decline. During the 1950s and 1960s, however, the baby boom pushed the average annual rate of population increase to nearly 3 per cent (Vignette 4.1). After that short burst, the rate fell. From 2001 to 2011 the annual average rate was just over 1 per cent. Most population increase is now related to immigration rather than natural increase. However, the **fertility rate** has shown signs of growing in recent years, from a low fertility rate of 1.49 births per 1,000 women between the ages 15 and 44 in 2000 to 1.68 in 2010, and this may signal a modest resurgence of another baby boom (Statistics Canada, 2011a).

Natural Increase

Natural increase is determined by the number of births minus the number of deaths. Until 1986, the greater portion of Canada's population growth (Table 4.5) was due to natural increase. At that time, the number of immigrants began to exceed the number due to natural increase, i.e., the net of births minus deaths (Statistics Canada, 2007g). Until 1921, the majority of Canadians lived in rural settings where large families were the rule. In the 1870s, for example, natural increase exceeded 2 per cent per year. At that rate, Canada's population would double every 30 to 35 years. Birth rates were extremely high, while death rates were low, allowing for a high rate of natural increase. As most available farmland became occupied, however, children of farm parents had no choice but to seek their fortunes elsewhere. Many went to the cities in search of work.

Every industrial country has followed the same demographic path. As a country industrializes,

Table 4.5 **Population Increase, 1951–2011**

Year	Population (000s)	Percentage Change	Average Annual Rate
1951	14,009.4	21.8	1.7
1961	18,238.2	13.4	2.5
1971	21,568.3	7.8	1.5
1981	24,343.2	5.9	1.2
1991	27,296.9	7.9	1.6
2001	30,007.1	4.0	0.8
2011	33,476.7	5.9	1.2

Source: McVey and Kalbach (1995: 42); Statistics Canada (1997b, 2002a, 2007a, 2012a).

THINK ABOUT IT
If more women are having children, why is the birth rate not increasing more rapidly? The answer lies in the nature of measuring birth and fertility rates. See "fertility rate" in the Glossary for the answer to this question.

birth rates decline. The exception in Canada was the baby boom (see Vignette 4.1). The figures for Canada, as shown in Table 4.6, indicate that the crude birth rate was 45 in 1851, but by 2009–10 it had dropped to 11. Over the same period, the crude death rate dropped from about 20 deaths per 1,000 people per year to about seven.

What are the factors behind the decline in both fertility and mortality? The answer lies in the benefits brought to society by industrialization.

VIGNETTE 4.1

The Baby Boom

During the Great Depression and World War II, economic and social conditions did not favour large families. By 1941, the birth rate had slipped to 22 births per 1,000 persons. With the return of Canada's servicemen and women, a stable political world, and improving economic conditions, couples' attitude towards family formation took a positive turn. Demographers refer to this aberration in the downward movement of the crude birth rate as the baby boom. By 1951, the birth rate reached 27 births per 1,000 people and it remained at or near that level for over 10 years. However, during the late 1960s, the crude birth rate began to decline, dropping to 17 per 1,000 persons by 1971 and bottoming out at a rate of 11 by 2001. Since then, a so-called echo effect occurred and the crude birth rate has edged up slightly. By 2010, it was just over 11 (Statistics Canada, 2011b).

The baby boom lasted for 20 years. During that time of high fertility, there were almost 10 million births, which created a bulge in the age structure of Canadian society that continues to have both economic and social implications. Thus, while this demographic phenomenon was short-lived, it has left its mark on Canadian society (Foot with Stoffman, 1996). As consumers of goods and services, baby boomers have had a decided impact on the economy as they move through their life cycle. Companies have geared their products to meet the strong demand created by baby boomers. In the early 1950s, the emphasis was on baby products and larger houses. In the 1960s, a similar age-related pressure was exerted on school facilities, creating a demand for more schools and teachers. As the baby boomers approach old age, the demand for health-care services is expected to rise. Companies are already targeting their advertisements at the growing number of retirees from the baby boom generation. Governments, on the other hand, are concerned about rising health-care costs associated with the expected increase in senior citizens. By 2000, demographers detected an echo effect, namely that the children of the baby boomers were having larger families, thus pushing the birth rate higher.

Table 4.6 Canada's Rate of Natural Increase, 1851–2011

Year	Crude Birth Rate	Crude Death Rate	Natural Increase (%)	Natural Increase (000s)
1851	45	20	2.5	61
1871	42	20	2.2	81
1891	38	18	2.0	97
1911	32	14	1.8	129
1921	29.3	11.6	1.8	160
1941	22.4	10.1	1.2	145
1961	26.1	7.7	1.8	335
1981	15.2	7.0	0.8	200
2001	10.5	7.1	0.3	108
2011	11.3	7.2	0.4	134

Sources: Adapted from Statistics Canada (1997c, 2003b, 2006b, 2007h, 2007i, 2012a); McVey and Kalbach (1995: 268, 270).

THINK ABOUT IT
Table 4.6 shows that both the crude birth and death rates increased from 2001 to 2011. Can you explain this seemingly contradictory situation?

On the one hand, the enormous improvements in public health, such as water purification, improved nutrition, medical advances, and the development of public health-care systems largely explain why **mortality rates** have declined. On the other hand, the explanation for the decline in the birth rate is more complicated and involves various aspects of human behaviour and our perception of the future. Accordingly, the birth rate is driven by a series of social and economic factors that caused parents to opt for smaller or larger families. Among the changes resulting in smaller families, three stand out: the shift of people from rural areas to towns and cities; the sharp increase in the number of women in the labour force; and the widespread acceptance of family planning. Family planning was greatly helped by the birth control pill. Introduced in 1960, the birth control pill has played an important role in reducing the birth rate and was a factor in ending the baby boom. By 2011, both crude birth and death rates increased. (Table 4.6).

By the beginning of the twenty-first century, Canada's crude birth and death rates were extremely low. While these vital rates are similar to those found in other industrial countries, Canada's low rate of natural increase slowly increased in the following decade. Will this rate

VIGNETTE 4.2

The Concept of Replacement Fertility and Canada's Sex Ratio

The concept of replacement fertility refers to the level of fertility at which women have enough daughters to replace themselves. If women have an average of 2.1 births in their lifetime, then each woman, on average, will have given birth to a daughter and a son. The number 2.1 was determined to represent the minimum level of replacement fertility because, on average, slightly more boys than girls are born. In 1961, the Canadian total fertility rate was 3.8 births per woman of child-bearing age (15–49). By 2000, it had dropped to 1.51 but since then has slowly increased, reaching 1.67 in 2009 (Statistics Canada, 2011a).

Nature has ensured that slightly more male than female babies are born. However, male mortality rates are higher than those for females. The net result is a population with more females than males. In 2011, females totalled 17.1 million compared to males at 16.4 million. The **sex ratio**, defined as the number of males per 1,000 females in Canada's population times 100, was 96.2 in 2011, meaning that there were nearly 650,000 fewer males than females in that year (Statistics Canada, 2012e).

Table 4.7 **Population Increase, Selected Years**

Year	Births	Deaths	Natural Increase	Immigration	Total	% Immigrants
1993–4	389,286	207,528	181,758	227,860	409,618	56
2000–1	327,187	231,232	95,955	255,999	351,954	73
2007–8	364,085	237,202	126,883	249,603	376,486	66
2010–11	386,013	252,561	133,452	280,681	414,133	68

Sources: Adapted from Statistics Canada (2003b, 2006a, 2009a, 2011c) and Citizenship and Immigration Canada (2011a).

continue to increase? Based on the experience of other industrialized countries, Canada's birth rate and therefore its rate of natural increase should decline, perhaps both dropping to zero. The echo effect may make Canada an exception to this rule in the short run. In the long run, these two rates are likely to fall as predicted by the demographic transition theory. Population growth remains strong due to the flow of newcomers to Canada's shores.

Immigration

Immigration keeps Canada growing. According to the 2011 National Household Survey, Canada's population contains 6.8 million foreign-born individuals. Stated differently, one out of five persons in Canada is foreign-born (Statistics Canada, 2013b). As shown in Table 4.7, immigration drives Canada's population increase and accounts for more of this increase than does natural increase. For 2010–11, immigration accounted for 68 per cent of Canada's population increase. Newcomers also account for Canada's ethnic diversity and its pluralistic society. This remarkable flow of people to Canada is central to understanding Canada's identity. Each immigrant has added a fresh element to Canadian identity and, over time, each becomes an old-timer. The process of cultural rejuvenation, with the arrival of different cultural groups and their integration into Canadian society within a generation or so, is a remarkable cultural feat and speaks well of Saul's concept of a soft Canada. In the late nineteenth century, for instance, newcomers from Central Europe and czarist Russia brought their customs, language, and religion with them. Adjusting to the existing "British" way of life in Canada and being accepted by their neighbours was not always easy, but the vast majority of the second generation melded into the Canadian fabric and, in their own way, remade Canada's identity. Fast forward to the twenty-first century when Canadian Muslims are growing in numbers through immigration and high birth rates among landed immigrants; at the same time, they are becoming established within Canadian society. Of course, there are bumps along the road of accommodation/acculturation/integration, but this process of adjusting takes time—at least one or two generations—and is a normal aspect of Canadian nation-building. This process includes changing fertility rates (Bélanger and Gilbert, 2006). In the popular press, sharply different interpretations of this topic exist. No wider gap exists than that between Mark Steyn and Doug Saunders. In contrast to Steyn's *America Alone: The End of the World As We Know It*, where he predicts that high birth rates among Muslim immigrant families will create an America dominated by Muslims, Saunders, in *The Myth of the Muslim Tide*, recognizes that large Muslim families are a temporary phenomenon and that second-generation families tend to be of similar size to other Canadian families.

The flow of immigrants to Canada varied from decade to decade due to economics and wars. The greatest wave of newcomers—especially as a proportion of the Canadian population—took place in early years of the twentieth century when Canada needed to settle the vast interior of the country. With the outbreak of World War I, immigration almost ceased until the end of World War II. From then on, the flow of newcomers has increased, playing a critical role in Canada's population growth and its cultural identity (Figure 4.6). Ottawa encourages immigration for three reasons: (1) newcomers keep Canada's population increasing, which is believed necessary for economic growth; (2) newcomers add valuable members to Canada's workforce; (3) Canada takes in refugees who are fleeing oppressive socio-political conditions in their homelands.

THINK ABOUT IT
The late phases of the demographic transition theory predict a falling birth rate. Does this theory help buttress Saunders's or Steyn's argument?

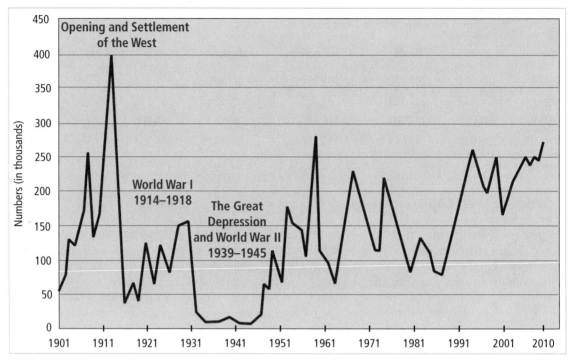

Figure 4.6 Annual number of immigrants admitted to Canada, 1901–2010
Source: Statistics Canada (2003a: 2; 2009b); Citizenship and Immigration Canada (2011a).

THINK ABOUT IT
As pointed out in the discussion of "One Country, Two Visions" in Chapter 3, Québec saw Canada as a country of two founding peoples. The question is, has immigration strengthened or weakened the case for this historic position?

A key social issue is how well immigrants enrich and add to the cohesiveness of Canadian society. Until the 1970s, most came from Western countries. Today, most come from Asia (Figure 4.7). As a result, the adjustment process has become more challenging for newly arrived immigrants and for other Canadians. For those belonging to non-Christian religions and those who are "visible minorities," social barriers into the mainstream society often exist. These barriers include language and education as well as discrimination and exclusion.

Geography plays a role in the destination within Canada of newcomers. Driven by economic considerations, most newcomers relocate in Canada's largest cities. This concentration of immigrants in Canada's major cities has both advantages and disadvantages. Newcomers have more cultural anchors to support them, such as family and friends who speak their native tongue, restaurants that serve traditional foods, and religious institutions that provide both spiritual and community support. A disadvantage may be a sense of isolation from other Canadians. But adjustment to Canadian life and the elusive Canadian identity come with the next generation, whose attachment to the "new country" is stronger and whose roots to the "old country" are much shallower than those of their parents. This process of adaptation may be more difficult for visible minorities. For example, some visible minorities born in Canada do not feel connected to Canadian society (Lewington, 2007).

The issue of cultural adjustment is discussed in Chapter 1 under "Newcomers and Old-Timers," page 15.

From 1971 to 2011, the proportion of immigrants to Canada born in Europe and Asia has reversed. In 2010, the leading countries providing immigrants to Canada were the Philippines, India, and China, followed by the United Kingdom, United States, and France.

The Demographic Transition Theory

The **demographic transition theory** is the most widely accepted theory that describes population change in industrial societies. This theory is based on the experiences of European countries. From the European experience, the assumption is that birth and death rates decline as a society

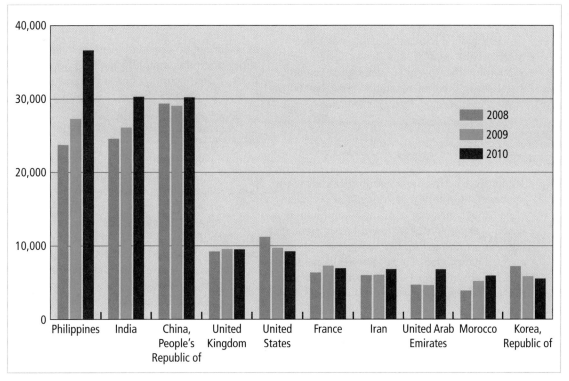

Figure 4.7   **New Canadians: Permanent residents by top 10 source countries, 2008–2010**
Source: Citizenship and Immigration Canada (2011b). Reproduced with permission of the Minister of Public Works and Government Services Canada, 2013.

moves from a pre-industrial to an industrial economy. Most significantly, this theory calls for the death rate to decline well before the birth rate, resulting in a population explosion. The next phase calls for both rates to reach a low level, resulting in population decline.

According to this theory, these demographic changes occur in five phases, each of which has a distinct set of vital rates that coincide with the phases in the process of industrialization (Table 4.8).

While applicable for most European countries, does this theory have relevance for Canada and its multicultural society? More specifically, does this very broad theory have within it elements of **acculturation** whereby immigrants tend to take on the demographic characteristics of their host country? (Bélanger and Gilbert, 2006).

A cursory examination of Canada's birth and death rates over the last 150 years reveals strong similarities to the early industrial, late industrial, and post-industrial phases of this theory. Assuming

Table 4.8 **Phases in the Demographic Transition Theory**

Phase	Birth and Death Rates	Rate of Natural Increase
Late pre-industrial	High birth and death rates	Little or no natural increase but possible fluctuations because of variations in the death rate
Early industrial	Falling death rates	Extremely high rates of natural increase
Late industrial	Falling birth rates	High but declining rates of natural increase
Early post-industrial	Low birth and death rates	Little or no natural increase; stable population
Late post-industrial	Birth rate at or below zero	Declining population

that Canada is now in the early post-industrial phase, the theory makes sense when applied to Canada's natural increase (i.e., the difference between births and deaths for a given year). Supporting that position, demographers argue that Canada's rate of natural increase has fallen below its replacement level (Vignette 4.2).

Canada's Aging Population

A country where seniors outnumber children is uncharted territory. This scenario is facing Canada, with an aging population representing a serious demographic event with implications for the labour force and for the working-age taxpayers who have to foot the bill for public pensions and a variety of social costs ranging from more nursing homes to higher health-care costs. This demographic event is expected to take place in the first half of the twenty-first century. The predicted trend is driven by three factors:

- a decreased fertility rate;
- an increase in life expectancy;
- the movement of the baby boom generation into retirement and old age.

In Figure 4.8, Canada's demographic picture clearly indicates a rapidly "greying population." In less than a century, Canada's seniors increased from 8 per cent of the population in 1971 to a projected 25 per cent in 2061.

What are the implications of an older Canada? First, Canada's population structure will change with a smaller proportion of children (under 15 years of age), a smaller proportion of the population in the workforce (ages 15–64), and a much larger percentage over 64 years of age. However, it is expected that many Canadian seniors will remain in the workforce past the age of 64, either because they need to financially or because they want to continue working. It is not known how these working seniors might affect current projections. Second, Canadians are living longer, which not only augments the senior age category but also adds economic costs for this group in the form of greater drug and health costs, creating a larger tax burden on those in the productive age group and rising costs to the federal treasury to

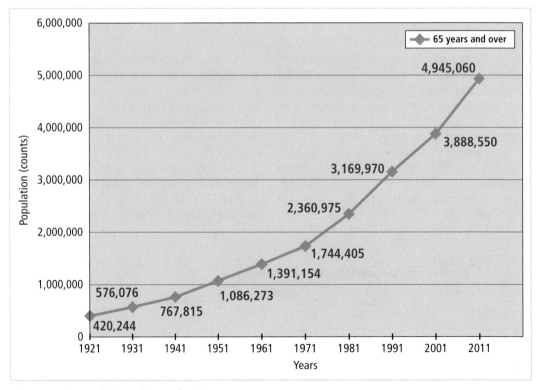

Figure 4.8 **The unrelenting advance of seniors**
Source: Adapted from Statistics Canada (2012f).

pay for Canada's public pensions—Old Age Security and the Canada Pension Plan. The burden for the provinces and territories may be unsustainable because health costs for the growing number of senior Canadians are projected to turn sharply upward. In this discussion about the burden of the elderly, a key point is that health needs peak in the last few months of life (Gee and Gutman, 2000). During this short span, the burden on the remaining family members—often intergenerational members—can be overwhelming emotionally and financially. A gap exists for extended care/nursing within the home and/or care facilities that would provide a more dignified closure and a less expensive one than occupying a hospital bed. Unfortunately, while hospital care is fully funded by the state, the cost of alternative but less expensive forms of care often falls on the family.

Already, health costs make up the major component of provincial/territorial budgets, at around 40 per cent. Could these costs reach 50 per cent, and thus squeeze funds from post-secondary education and social services budgets?

Another measure of the aging process is age dependency ratio. This innocent-sounding term provides another road map to Canada's future demographic destination. Age dependency ratio is the ratio of persons in the "dependent" age groups (under 15 and over 64 years) to those in the "economically productive" age group (between 15 and 64 years). The assumption is that productive members of society are those between the ages of 15 and 64, while unproductive members are either too young (under 15) or too old (over 64) to make an economic contribution. The purpose of the age dependency ratio is to compare the number of dependents with the number of economically productive members of society, thus giving a rough measure of the economic burden on those in the economically productive age group.

In 2011, the age dependency ratio was 46. This means that there were 46 persons in the dependent ages for every 100 persons in the working ages. In the years to come, however, as the baby boom generation continues to reach retirement age, the number of seniors is expected to jump significantly, thus increasing the number of Canadians making up the dependent age group. Added to that concern, many couples are having children at older ages, so that they are responsible for young children and elderly parents at the same time. This small but growing demographic phenomenon is called the "sandwich generation." Placed within a broader social context of the unfolding twenty-first century, an intergenerational care relationship within family units appears as a new but necessary fallout from the demographic trend of seniors living well beyond their seventies.

The 2011 census numbers underscore regional tensions over dwindling revenues that can be attributed to variations in age dependency ratio. On the one hand, an increasingly aging population in Atlantic Canada means fewer taxpayers to support provincial social programs. On the other hand, Western Canada's population has an increasing number of taxpayers because its booming economy is attracting more workers to fill labour shortages.

Ethnicity

An **ethnic group** is made up of members of a population who share a culture that is distinct from that of other groups. Each group has a common identity, shared values, and cultural/linguistic/religious bonds and symbols. **Culture** is the learned collective behaviour of a group of people. Canadian society, historically, was based on British, French, and other European peoples, as well as the Aboriginal peoples, but it is now composed of many ethnic groups[1] from around the world with more than 200 different **ethnic origins** reported in the 2011 National Household Survey. By contrast, the 1901 census identified only 25 ethnic origins. Statistics Canada (2011d) relates ethnic origin to "the ethnic or cultural origins of the respondent's ancestors. An ancestor is someone from whom a person is descended and is usually more distant than a grandparent."

While the National Household Survey did not published detailed ethnic origins tables, their written account indicated little change from the 2006 census (Statistics Canada, 2013c). The leading ethnic group remains "Canadian" at 32 per cent while the ranking of the next nine ethnic groups remains unchanged from the 2006 census (Table 4.9).

In spite of immigrants coming from different parts of the world, time erodes the ethnic connection with the country of origin. For Canadians born and raised in this country, the connection

THINK ABOUT IT
Do seniors without intergenerational family support simply fall between the cracks or is there a role for our public health system to fill?

Table 4.9 **Ethnic Origins of Canadians, 1996 and 2006**

	1996			2006	
Ethnicity	**Number**	**%**	**Ethnicity**	**Number**	**%**
Total population	28,528,125	100.0	Total population	31,241,030	100.0
Canadian	8,806,275	30.9	Canadian	10,066,290	32.2
English	6,832,095	23.9	English	6,570,015	21.0
French	5,597,845	19.6	French	4,941,210	15.8
Scottish	4,260,840	14.9	Scottish	4,719,850	15.1
Irish	3,767,610	13.2	Irish	4,354,155	13.9
German	2,757,140	9.7	German	3,179,425	10.2
Italian	1,207,475	4.2	Italian	1,445,335	4.6
Ukrainian	1,026,475	3.6	Chinese	1,346,510	4.3
Chinese	921,585	3.2	North American Indian	1,253,615	4.0
Dutch (Netherlands)	916,215	3.2	Ukrainian	1,209,085	3.9

Notes: (1) Table shows total responses. Because some respondents reported more than one ethnic origin, the sum is greater than 100 per cent. (2) Figures referring to North American Indian are based on Aboriginal ancestry population, i.e., those persons who reported at least one Aboriginal ancestry (North American Indian, Métis, or Inuit) to the ethnic origin question. "Ethnic origin" refers to the ethnic or cultural origins of a person's ancestors.

Sources: Statistics Canada (2003a, 2010a).

with their ethnic homeland can be tenuous at best (Beaujot, 1991: 297). Place, as cultural geographers insist, plays a critical role in the development of a regional/national identity. This phenomenon is well known. Consider, for example, the attachment of the French born in New France who had no interest in returning to France in 1763 because their lives were centred on the New World. They were no longer French but had become French Canadian. Not surprisingly, then, the ethnic selection of "Canadian" by nearly one-third of the population is attributed to the geographic notion of place overriding the concept of ethnicity. To put it differently, the resettlement of people in a new place causes their ethnicity to fade with time, and, as the roots of the second generation sink into Canadian soil, commitment to their parents' homeland fades while their new identity takes hold. This process of putting down cultural roots is normal, especially if the newcomers are welcomed into the mainstream society. Of course, there are cases of newcomers finding it difficult to put down roots in their new country and they often return to their country of origin.

Immigration is a key factor in population increases. But for Québec, immigration can be a double-edged sword. While immigrants add to the Québec population, some of these new immigrants bring unfamiliar ways that, for some Québec natives, have seemed to threaten the existing way of life. Matters came to a boil in 2007 when Hérouxville's town council announced an Islamophobic "code of standards" for newcomers (see Vignette 1.3).

Language

Language is a key component of ethnicity. Indeed, language is the most durable link to the past and the tool for maintaining a culture. Canada's two official languages are English and French. Language provides a measure of the strength of the two founding peoples, with 80 per cent of Canadians declaring that English or French is their first language. However, this figure is expected to drop to 70 per cent by 2031 (Statistics Canada, 2010b: Table 6), with only 22 per cent claiming French. At the same time, virtually all Canadians can speak one or both official languages (Statistics Canada, 2008b). Yet why is a non-official language the language spoken at home for so many Canadians? The answer is the number of immigrants who speak neither official language. Such Canadians are known as allophones in Québec and, to a lesser degree, in the rest of Canada. Within Canada, the share of the allophone population grew from 18 per cent in 2001 to 21 per cent in 2011 (Statistics Canada, 2012j), whereas in Quebec only 12.3 per cent reported a non-official language and 4.7 per cent spoke English only (Statistics Canada, 2013d).

Religion

Religion is another key element of culture. The changing nature of Canada's religious composition is indicative of the changing cultural mix of people, especially the large number of immigrants arriving from non-Christian countries and the increasing number of Canadians who declare "no religious beliefs" and who have no association—except perhaps at birth, marriage, and death—with the religious institution of their parents. On the other hand, the numbers of Canadians who subscribe to religions such as Islam, Hinduism, Sikhism, and Buddhism have increased substantially.

Culture is not only a link to the past but also provides the institutional organization to preserve ethnicity. Religious organizations provide an institutional structure that consolidates people of similar beliefs. One example is the Roman Catholic Church. In the early history of Canada, the Catholic Church helped sustain Catholicism, the French language, and the French-Canadian way of life. While its role has diminished in recent decades, the Church was the dominant cultural force among francophones from the conquest of New France to World War II. Not only did it give spiritual direction to French Canadians within Québec but it organized its parishes to sponsor group immigration to more isolated areas of Québec (such as the Clay Belt in northwestern Québec), to other provinces, and, after initial opposition, to New England. In these group migrations, priests provided the leadership for the move and for organizing the new settlement, including its educational, social, and religious structures.

VIGNETTE 4.3

Dual Loyalties and Dual Passports

Dual loyalties exist in various forms in Canada, ranging from loyalty to a province and to country of birth. Loyalty to a province is strongest among Québécois, partly because they look to their provincial government to protect their interests rather than to the federal government. Dual loyalties also exist among **recent immigrants** whose ties to their home country remain strong and who may retain a passport from their former country. In 1977, Canada made dual citizenship legal. Dual passports have the advantage of making travel for pleasure or business more convenient. Dual passports are also a reflection of the transnational reality of the global economy. As Canadian novelist Yann Martel said in accepting the 2002 Booker Prize, Canada is "the greatest hotel on earth, the kind of place people can pass through at their leisure, claiming citizenship as if it were a room key" (Brean, 2012). But dual passports may also be a "convenience" or safeguard for people living outside of Canada for a long period of time. For instance, thousands of Lebanese-Canadian citizens were residing in Lebanon on what seems to have been a permanent basis when, in 2006, the Israel army invaded Lebanon, thereby forcing Canada to evacuate those holding dual Lebanese and Canadian passports to Canada. The former Minister of Citizenship and Immigration, Judy Sgro, has stated, "We need to be loyal to one country as far as your citizenship. Your heart can be where you were born, but I think the commitment to Canada has to be strong and I think dual citizenship weakens that" (Woods, 2006). But does it? Many Canadians hold more than one passport. Former Prime Minister Turner holds dual citizenship, as does Opposition Leader Thomas Mulcair. And few would question their loyalty.

THINK ABOUT IT
Did the recommendation of the Royal Commission on Bilingualism and Biculturalism to recognize the multiplicity of Canada's population inadvertently water down biculturalism and thus weaken Québec's place in Canada?

Religious freedom is the right of all Canadians. This right was inscribed in the British North America Act and is contained in the Constitution Act, 1982. In fact, many immigrants have come to Canada seeking religious freedom. The Hutterite Brethren, for example, are a small Christian group who live in agricultural colonies. Like the Amish and Mennonites, they place pacifism at the core of their faith. In 1899, Hutterites fled Europe to the United States. In 1918, they immigrated to Canada because the US government refused to exempt them from military service. Initially, Hutterite colonies were formed in Manitoba and Alberta. Later, colonies were established in Saskatchewan. Hutterites number about 30,000 and the majority live and work in one of 300 agricultural colonies where they remain committed to their religion, a communal lifestyle, and pacifism (Statistics Canada, 2003d).

Canada is thought of as a Christian country. This image was certainly true in 1867, but today Canada is much more religiously diverse, plus, as noted above, a significant number claim no religious affiliation. As recently as the 1960s nearly 90 per cent of Canadians declared themselves to be Christian (though some may not have been active church members). By 2001, this figure had dropped to 77 per cent, while 16 per cent claimed to have no religious beliefs (ibid.). Statistics Canada used to ask Canadians about their religion every 10 years, but this procedure ended in 2011. Statistics Canada (2010b: Table 5) estimated that the percentage of those following Christian religions would fall from 74 per cent to 65 per cent by 2031. The numbers of Muslims, Hindus, Buddhists, and Sikhs more than doubled from 1991 to 2011, to reach 2.5 million, and were expected to exceed 6 million by 2031, making up 14 per cent of Canada's population (ibid.). Those Canadians reporting no religion would comprise the remaining 21 per cent.

Multiculturalism

Multiculturalism is the cornerstone of Canada's social policy towards newcomers. This federal policy was a direct result of the work of the Royal Commission on Bilingualism and Biculturalism (1970), which recommended that Ottawa recognize

Photo 4.7
The Basilica of Notre Dame, opened in 1829, is the principal Roman Catholic Church in Montréal, and is a reminder of the Church's powerful role in the history of French-Canadian society. Today, however, the Church has lost not only its role but also much of its active church membership.

the multiplicity of Canada's population. The following year, in 1971, the federal Liberal government made multiculturalism official policy and in 1972 the cabinet position of Minister of State for Multiculturalism was created. Government funding to ethnic organizations soon followed. In 1988, the federal government passed the Canadian Multiculturalism Act, which is designed to encourage greater human understanding and stronger bonds among Canadians of different cultural backgrounds and ethnic origins, and today official multiculturalism is under the purview of Citizenship and Immigration Canada.

For Charles Taylor, multiculturalism is a way for the Canadian government and society to recognize the worth of newcomers' distinctive cultural traditions without compromising Canada's basic political principles (Vignette 4.4). The outer boundary of multiculturalism is where it rubs against the edge of conventional values held by mainstream Canadians. An example was the rejection of shariah law to settle family disputes within the Muslim community in Ontario. On the other hand, subtle adjustments are occurring in a wide range of areas. For example, the Toronto Stock Exchange has sought to accommodate Muslim investors by launching a Canadian stock index (the S&P/TSX 60 Shariah) in May 2009. This index excludes banks, pork producers, and entertainment and gambling stocks, thus allowing Islamic investors to abide by Islamic law and still participate in equity investment strategy (www.theglobeandmail.com/globe-investor/funds-and-etfs/sp-launches-canadian-sharia-compliant-index/article1198352/).

Photo 4.8

The growing Muslim population in Alberta is reflected in Canada's largest mosque, which serves the Ahmadiyya Muslim community. Based on projections by Statistics Canada, the Muslim population comprised a record 2.7 per cent of Canada's population in 2011 and could reach 6.8 per cent by 2031 (Statistics Canada, 2010b: Table 5).

THINK ABOUT IT

Canada's multicultural programs recognize the need for public support for newcomers to adjust to their new world. From this perspective, why shouldn't the same level of support and programs be extended to urban Aboriginal people who, as new arrivals to cities, are struggling to adjust to this urban world?

In many ways, multiculturalism is the opposite of **ethnocentricity**. But tolerance and respect of others are not an automatic outcome of life in pluralistic societies. Canadian tolerance and respect were learned the hard way, going back centuries to often bitter (and sometimes violent) conflicts

---

VIGNETTE 4.4

Charles Taylor on Multiculturalism

One of Canada's leading philosophers, Charles Taylor (1994: 63), argues in *Multiculturalism: Examining the Politics of Recognition*, that:

> . . . all societies are becoming increasingly multicultural, while at the same time becoming more porous. Indeed, these two developments go together. Their porousness means that they are more open to multi-national migration; more of their members live the life of diaspora, whose centre is elsewhere. In these circumstances, there is something awkward about replying simply, "This is how we do things here." This reply must be made in cases like the Rushdie controversy, where "how we do things" covers issues such as the right to life and to freedom of speech. The awkwardness arises from the fact that there are substantial numbers of people who are citizens and also belong to the culture that calls into question our philosophical boundaries. The challenge is to deal with their sense of marginalization without compromising our basic political principles.

between French- and English-speaking Canadians. For Canada to survive as a nation, the two antagonists had no choice but to become partners. One product of this so-called partnership was biculturalism. The other path to nation-building—a classic European-style nation-state founded on a single common ethnicity and language—was not possible in the northern half of North America. Reaching an accord (not a solution) between the two founding peoples was not a simple task and disputes continue to emerge, as discussed in Chapter 3. Since dominance is not feasible in the long term, then compromise becomes a political necessity and eventually the search for compromise becomes ingrained as a national trait. In *Reflections of a Siamese Twin: Canada at the End of the Twentieth Century* (1997), John Ralston Saul described Canada as a "soft" country to capture this flexible trait.

 See the section "Faultlines within Canada" in Chapter 1, page 10, for more on the subject of Canada as a "soft" country.

Although multiculturalism is a distinctively Canadian approach to social equality in nation-building and an official way to encourage respect for cultural diversity, does it rank as a core value? Outside of Canada, multiculturalism is under attack, and some leading Canadians see a dark side to multiculturalism. In 2005, Bernard Ostry, who was responsible for drafting and implementing federal multicultural policy in the 1970s, noted the concerns that some have expressed about the direction of multiculturalism:

> The multiculturalism policy of the 1970s is under increasing criticism. This policy, developed from grassroots, was intended to assure new citizens full participation in the nation's life. Some now claim that it has had the opposite effect, that it has encouraged minorities to retreat to their own corners. (Ostry, 2005)

The concentration of ethnic minorities into specific areas of Canadian cities may be a natural short-term phenomenon rather than a fixed geographic feature of ethnic enclaves (Walks and Bourne, 2006). Two factors may be at play in the preference of ethnic minorities to congregate in certain parts of a city. First, economic factors such as low incomes may keep visible minorities in certain neighbourhoods where rents are low. Second, social cohesiveness may cause minorities to live in the same neighbourhoods for the comfort and security of living near others of the same socio-cultural background. Third, religion plays a key role and many people wish to locate near their place of worship.

Over time, immigrant children and grandchildren are more likely to find a place in mainstream Canada. Upward mobility is associated with education and perhaps interracial marriages. But the question remains: "Might multiculturalism increase ethnic group identification at the expense of Canadian social cohesion?" Without doubt, tensions have arisen from time to time in Canada as people of different cultures, languages, and racial origins have chafed against what they perceive as barriers within Canadian society, but peaceful discord is not in itself a failure of multiculturalism but part of a process of social interplay necessary to expose and resolve differences. Canadian history is on the side of both multiculturalism and immigration because immigrants—particularly their children—have found a place within Canadian society. While this process has not always been easy, the challenge for visible minorities is even greater due to racism.

VIGNETTE 4.5

Canadian Identity

Canadian identity, flexible and porous as it is, includes core values that are defined by Canada's history and geography. Four core values are: (1) government is based on British parliamentary institutions and the rule of law; (2) two official languages ensure a place for French as well as English, which also means that other languages have no standing in the political and public affairs of Canadian society (except in Nunavut, where Inuktitut also is recognized as an official language); (3) Aboriginal peoples, but especially status Indians, have special rights, which flow out of historic treaties, modern land claim agreements, recognition in Canada's Constitution, and numerous court cases, as we have seen in Chapter 3; and (4) tradition and law are encapsulated in the Canadian Constitution, which includes the Charter of Rights and Freedoms.

CONTESTED TERRAIN 4.2

Immigration and Multiculturalism

Immigration dictates that Canada will become an increasingly more pluralistic society and therefore multiculturalism will continue to play a role not only as an adjustment mechanism for newcomers but also as a core Canadian value. On the other hand, the jury is still out on whether multiculturalism is a lasting feature of Canada's social policy and, therefore, whether it should be considered a core value.

Sense of Place: Now and Then

Canada's sense of place has changed remarkably over time. In the last 60 years, Canada has remade itself. In the early twenty-first century, population trends and faultlines signal the direction of a future Canada with another, yet-to-be-determined sense of place. Looking back, the Canada described by Kenneth Hare in 1968 is totally unrecognizable today. Hare's Canada had just experienced an unprecedented baby boom and record-breaking economic expansion, and the country was heading into trade agreements with the United States that would effectively replace the east–west-based national economy with a continental one, starting with the Auto Pact of 1965. Immigration, while at a far lower annual volume, was primarily from Europe and the United States.

Photo 4.9
How do Canadians, whether newcomers or old-timers, vent their anger about affairs in their former homelands or in the world? Public demonstrations are one way and such demonstrations allow "steam" to be released and for the general public to be made aware of their concerns. Almost all end peacefully. Here, Tamil supporters block the Gardiner Expressway in downtown Toronto, 10 May 2009, to call attention to the escalating civilian death toll in the Sri Lankan civil war.

The divide between rural and urban Canada continued to see rural Canada lose its grip, but rural Canadians still formed 30 per cent of Canada's total population. Urban Canada, from big cities to small towns, was on the upswing. At the city level, the sense of place is perhaps best expressed by Toronto, for no one would recognize Toronto "the Good" of the 1950s or even 1960s as the cosmopolitan Toronto of today. Toronto, Montréal, and other cities have undergone dramatic transformations of their cityscapes, their size and functions, and their ethnic/racial makeup. Indeed, "visible minorities" were predicted to comprise the majority of Toronto's population by 2017.

Sense of place has changed at the regional level, too. From 2006 to 2012, the internal shifting of Canada's population and economy was responding to growing economic opportunities, especially those in Western Canada. Over this six-year period, all regions saw their population increase, but the greatest percentage increases took place in Western Canada, Ontario, and British Columbia (Statistics Canada, 2012a). Within Western Canada, Saskatchewan, for long a "have-not" province, has enjoyed an economic boom that seems to be immune to the current global crisis and has resulted in an increase in population, which now exceeds one million. Why? First, Saskatchewan's resources base—agriculture, natural gas, oil, and potash—has had a run of "good" years, starting in 2006. Spinoff demand from this boom has resulted in specialized products from local manufacturers. Second, knowledge-based industries have taken hold, especially in the bio-agricultural area, encouraged by the research parks at the two universities, in Regina and Saskatoon, and by provincial and federal funding. Third, with major urban centres expanding, the construction industry has been going full blast, creating a demand for more trades workers.

At the other end of Canada, Atlantic Canada, but especially Newfoundland and Labrador, saw massive out-migration and the emergence of commuting to the oil sands and other western sites. Former federal cabinet minister John Crosbie put it this way:

> Christ, everywhere you turn around there's somebody leaving for Alberta. That they've got somewhere to go to better themselves and get work is not looked upon negatively. But we're exporting the ones that have the most initiative and are the most adventuresome and this weakens us. (Martin, 2005: A18)

What Canada, its regions, and the urban landscape will look like by the mid-twenty-first century remains uncertain. But what is almost assured is a realigned sense of place and purposes for both Canada's regions and urban centres. The four faultlines present particular perspectives into that nebulous and fluid concept: sense of place. Canada's population likely will become significantly more diverse in the twenty-first century. For that reason, we turn next to the impact of newcomers on Canada's growing diversity.

For the thousands of Newfoundlanders working in the Alberta oil fields, see Vignette 9.11, "The Big Commute," page 396.

Newcomers and Old-Timers

Everyone is a newcomer—at first. As shown in Figure 4.9, newcomers, defined as foreign-born, have always formed a significant proportion of Canada's population. This proportion is projected to increase to 22 per cent by 2011 and to 26 per cent by 2031 (Statistics Canada, 2010b: Chart 1). Within less than 20 years, then—assuming these demographic assumptions hold true—at least one person in four living in Canada will be foreign-born. At this level, the proportion of foreign-born will reach the highest point in the last 100 years. Clearly, the potential impact of immigration on the makeup of Canada's future population will be profound.

For the most part, newcomers grab the opportunity to succeed in their new country. As Nahlah Ayed, CBC Middle East reporter, said in an interview, "achievement is [the] ethos common to first generation immigrants" (Hampson, 2012: L3). Interestingly, the issue of hard-working and high-performing Asian-Canadian students was injected into the 2012 Québec election when François Legault, the leader of Coalition Avenir Québec (CAQ), lamented that Québec students seek the good life (*la belle vie*) rather than working hard at their studies like Asian-Canadian students (Canadian Press, 2012). Much of the reason for this difference is that immigrant parents often make personal sacrifices to ensure that their children succeed in school, universities, and business, and these parents expect their sons

---

THINK ABOUT IT

In 2013, the federal government introduced new rules for unemployment insurance with the objective of requiring unemployed Canadians but especially seasonal workers to seek employment beyond their immediate location or lose their unemployment insurance payments. Since most seasonal workers reside in rural Québec and Atlantic Canada, pressure on these workers and their families to relocate to urban cities within their provinces or to other regions of Canada is expected to increase. What would such "forced" relocation do to people's sense of place?

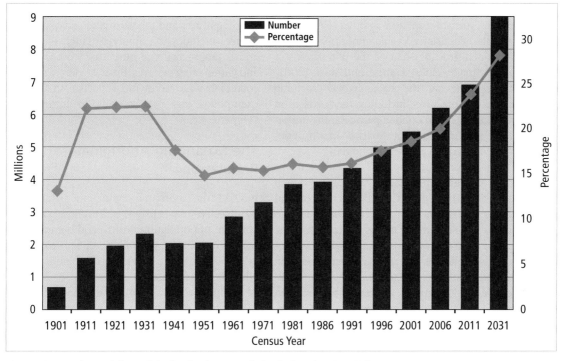

Figure 4.9 **Number and share of the foreign-born population in Canada, 1901–2031**
Source: Chui et al. (2008); Statistics Canada (2010b).

and daughters to excel at school. A few immigrant parents, especially those from non-Western countries, sometimes have trouble accepting the acculturation of their children into Canadian society and, in a very few extreme cases, have actually murdered their children as a consequence of the clash in values.

Cultural differences can rub old-timers the wrong way. For that reason, not everyone welcomes newcomers and their different ways. Resistance to cultural change is common to all societies, particularly those defined as "hard" countries. Canada, as a soft country, is perhaps as open to immigrants as any other country in the world. Still, new people and new ways create ripples in the existing society. The greatest reaction to cultural change has taken place in Québec where concern about maintaining the French language and customs remains a lightning rod. The issue of accommodation runs across Canada but is most strongly expressed by the old-timers in Québec. In Chapter 6, this topic is discussed more fully.

Historically, Canada has needed newcomers and they, in turn, have sought a more secure life in a more open and affluent society. Children of newcomers become old-timers. This pattern of inclusiveness—some would say integration—forms the basis of the evolving Canadian society. Inclusiveness still leaves geographic space for cultural differences, as expressed by the province of Québec, by the territory of Nunavut, and by cultural enclaves in Canadian cities. Aboriginal peoples are a special case. As Peter Li (2003: 9) points out, "Over time, earlier immigrants and their descendants become old-timers of the land, and their charter status places them in a privileged position vis-à-vis the newcomers who must accept the conditions of entry and rules of accommodation laid down by those who came before them."

Since newcomers often come from a different cultural world, they may bring with them different languages, customs, foods, and religious beliefs from those found in mainstream Canada. While these cultural imports often enrich Canadian society (Vignette 4.6), a few have caused a cultural clash between some members of the two groups. On top of these cultural differences, since many recent immigrants come from non-European countries, racism can add another element to these cultural tensions. Visible minorities make up a major and growing share of the populations of Canada's major cities. Bourne and Rose (2001: 115–16) argue that "the challenge of overcoming exclusionary practices, including those based on racism in all its forms, and in particular the impact of these practices on labour market opportunities for youth, becomes all the more urgent."

Immigration brings benefits, including maintaining Canada's population, supplying much needed skilled labour to Canada's workforce, and bringing investment capital. Yet, another side to immigration exists in the minds of the old-timers. The two primary concerns are that the newcomers will somehow lessen their economic position in Canadian society, and that, coming from a different cultural, linguistic, and racial background, they will remake Canadian society in their image. This latter fear is reinforced by the concentration of immigrants in Canada's three major cities, where they may create cultural enclaves that prevent their integration into Canadian society.

The history of immigration has shown two remarkable developments over time. First, Canadian society absorbs some of the cultural imports and multiculturalism fosters a social expression of the cultural contributions of newcomers. Second, recent immigrants, especially their children, have shown a capacity not only to integrate into Canadian society, but also to reshape it. Charles Taylor would consider such reshaping as part of the flexible and porous nature of Canadian identity (Vignette 4.4). The cohesiveness of Canadian society depends on such integration and on an acceptance by all Canadians of cultural adaptation within the society to Old World cultures and religions. Such adaptation, of course, includes the ability to laugh at oneself and to appreciate the humour of other people and within other cultures. Russell Peters, a standup comedian who pokes fun at racism in Canada, exemplifies this positive side to immigration (Vignette 4.6).

THINK ABOUT IT
Why do religious institutions play such a prominent role for newcomers compared to other Canadians, many of whom have no religious affiliation and who are not active in an institution-based faith?

The Expanding Aboriginal Population

In 2011, the Aboriginal population, as measured by ancestry, was just short of 2 million, a remarkable rate of population growth and demographic recovery from the low point of approximately 100,000 in the late nineteenth and early twentieth centuries (Statistics Canada, 2013c). In 1951, Aboriginal people comprised less than 2 per cent of Canada's population and now approximates close to 6 per cent of Canada's population. By 2011, the Aboriginal population as measured by ancestry, totalled 1,889,400 with North American Indian ancestry accounting for 1,369,100; Métis, 447,700 and Inuit, 72,600 (Statistics Canada, 2013c). Clearly, the early twentieth-century myth of the vanishing Indian has been put to rest (Figure 4.10).

Within the Aboriginal population considerable cultural diversity exists well beyond the three constitutionally recognized groups—North American Indians (about 62 per cent of total Aboriginal population), Inuit (about 4 per cent), and Métis (about 34 per cent). First Nations alone represent many former tribes, different economies, and native languages. Even so, two prominent features are common within that cultural

VIGNETTE 4.6

Russell Peters, a Premier South Asian-Canadian Comic

Russell Peters is a Canadian comedian who explores attitudes towards race (his and others) in a way that is challenging and cheeky. Born and raised in Ontario by parents who moved to Canada from India, Peters combines his experiences of growing up "as the only brown child on the block" with insightful observations on race relations today. Now a North American star who has toured globally, Peters delights in exposing the humorous side of clashing cultures in Canada and the United States.

David Johns

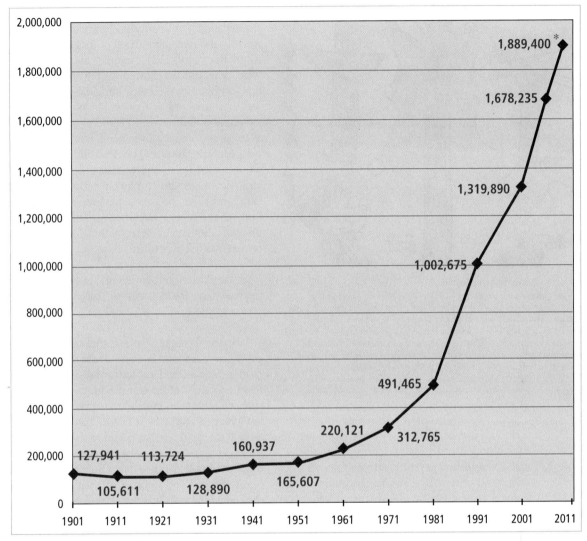

Figure 4.10 **Aboriginal population by ancestry, 1901–2011**
Note: For discussion of the difference between "Aboriginal identity" and "Aboriginal ancestry," see Table 1.2.
*In the 2011 census, many questions were no longer classified as "mandatory," including questions regarding the Aboriginal population. The National Household Survey questionnaire was mailed to some 4.5 million households requesting information on social and economic subjects, including Aboriginal population. *Source:* Adapted from Statistics Canada (2003c, 2013b).

diversity. A key element dividing Aboriginal peoples from the rest of the population is the level of education. The lack of a high school diploma, so common among the Aboriginal population, represents a major barrier barring Aboriginal youth from access to jobs and the possibility of breaking out of the dependency trap. The second element is the urban/homeland divide. This divide is widening, with more and more Aboriginal people voting with their feet to live in urban Canada. This cultural and economic division becomes more apparent with the emergence of a second generation of urban Aboriginals and with some families living a middle-class lifestyle.

In the twenty-first century, the challenge is for Aboriginal people to find their path to modernity, namely, Aboriginal-defined places within Canadian society. Such a path is a long and winding one that may branch out to take different directions to accommodate the diversity of Aboriginal peoples. Even with self-government at the high political level of a territorial government in the case of Nunavut, the gap between existing education levels and those levels demanded by companies and governments—

Figure 4.11 Reset the relationship
While Ottawa sees jobs as key to a better future for First Nations, the Grand Chief of Nishnawbe Aski Nation in Ontario has a different view. Grand Chief Stan Beardy feels that rather than his people being trained to work for somebody else, "we want to develop [the land] so somebody would work for us and make money for us" (Curry and Galloway, 2012). Meanwhile, Ottawa is aiming for fundamental changes to relations between the federal government and First Nations—on property rights, on matrimonial rights, on financial transparency, and on education (Ibbitson, 2012). Some Aboriginal leaders and Aboriginal activists are wary of any Ottawa-initiated changes.

peoples, who for the most part were excluded from participating in mainstream Canadian society until recently and remain hindered by locational and institutional factors. For example, only after 1960 were status Indians allowed to vote in federal elections without losing their Aboriginal status, and it was not until 1958 that an Indian band—the Tyendinaga Reserve near Belleville, Ontario—was allowed any control over band funds (Dickason with McNab, 2009: 371). The Inuit and Métis, while not excluded legally, had little opportunity and encouragement to participate in political affairs, and, like the status Indians, sought some form of autonomous existence within Canada. The place of Aboriginal peoples within Canadian society has evolved over time.

For a more complete discussion of Aboriginal peoples in Canadian society, see "The Aboriginal/Non-Aboriginal Faultline" in Chapter 3, pages 92–103.

Aboriginal people have gone through great changes since European contact. These demographic changes are classified into four phases in Table 4.10. Early contact was associated with population decline while population stabilized in late contact. The last phase shows a high rate of natural increase, though the birth rate has begun to decline. During the most recent phase, Aboriginal population has increased at a higher rate than Canada's population (see Table 4.11 and Figure 4.10). Statistics Canada, through two census questions, produces two measures of Aboriginal population. One is based on identity and the other on ethnic origin (or ancestry); by ethnic origin, the figure is 1.9 million Aboriginal people, forming 5.8 per cent of Canada's population in 2011 while the identity census question accounts for 1.4 million or 4.3 per cent.

including that of Nunavut—is not shrinking. As well, history has tied Aboriginal peoples to the federal government and that relationship is ready for review, not merely by Canadian governments and political leaders but by the Aboriginal leadership, with strong input from outspoken young indigenous activists who speak from outside of the existing Aboriginal circle of political power. In 2012, a few activists from Saskatoon galvanized those outside of this circle of power into a populist movement known as Idle No More (Figure 4.11).

Vignette 3.9 explains why the "newcomer to old-timer" model is not applicable to Aboriginal

With this rapidly growing population, the economic challenge—for Aboriginal people, their

VIGNETTE 4.7

The Assembly of First Nations

The Assembly of First Nations (AFN) is the national representative organization of the 617 First Nations in Canada. Its members consist of the chiefs of these First Nations. These chiefs vote to select their National Chief for a three-year term. The present National Chief is Shawn Atleo. The federal government recognizes the AFN as the official body with which Ottawa interacts on the business of First Nations. The federal government funds the operations of the AFN. Some First Nations band members fear that the AFN has been co-opted by the federal government. For instance, why did the Idle No More movement, rather than the AFN, initiate the protest against the Harper government's omnibus Bill C-45 that affects the environment and First Nation sovereignty?

Table 4.10 Major Phases for the Aboriginal Population in Canada

Phase	Characteristics
Pre-Contact	The Aboriginal population in Canada in the centuries preceding European contact and settlement was at least 200,000 and possibly as large as 500,000. This population may have varied in size due to the carrying capacity of the land, which, in a hunting society, is controlled by the availability of game (food). For instance, natural conditions, especially weather, could affect the size and migration routes of animal populations.
Early Contact (1500–1940)	Aboriginal peoples who came into contact with Europeans were exposed to new diseases, and these new diseases often spread across the land prior to the arrival of Europeans in a particular place. Population losses were heavy. Loss of hunting lands also added to their demise. By the end of the nineteenth century, the Aboriginal population was just over 100,000.
Late Contact (1940–1960)	Rising fertility rates coupled with high mortality rates resulted in the stabilization of the Aboriginal population. Towards the end of this phase, fertility rates were high and mortality rates declined. The result was the start of rapid population increase.
Post-Contact (1960 to present)	High fertility and low mortality account for remarkable population rebound. The net result has been a population explosion. While the Aboriginal population is likely to increase at a rate well above the national average in the coming decades, its natural rate of increase is expected to diminish due to a declining fertility rate. The Aboriginal population now approaches 2 million with a growing number living in cities.

THINK ABOUT IT
Is there evidence from the Aboriginal community that their vision of the future is to match the education and income levels of the rest of the Canadian population and, at the same time, to retain the key values of their cultures?

leaders, and governments—is to facilitate their participation in the economy. This challenge is especially evident in Manitoba and Saskatchewan. Aboriginal political power is, in part, linked to population size, so with the percentage of Aboriginal people in Canada increasing, the federal, provincial, and territorial governments, but especially western provinces, should heed the political voice of Aboriginal leaders more carefully. But what is their voice? In Nunavut, Inuit political voices are loud and clear as they pertain to that territory. At the national level, Shawn Atleo, the current National Chief of the Assembly of First Nations, speaks for all First Nations people and Chief Atleo has called for a resetting of the relationship with Ottawa.

Population increase within Aboriginal communities places pressure on limited resources.

THINK ABOUT IT
What does the urban/homeland divide among Aboriginal peoples reflect?

Table 4.11 Aboriginal Population by Identity, Canada and Regions, 2001 and 2011

Region	Aboriginal Population 2001	Aboriginal Population 2011	Increase 2001–11	Increase 2001–11 (%)	% by Region
Territorial North	47,990	56,225	8,235	17.2	4.1
Atlantic Canada	54,120	94,490	40,370	74.6	6.8
Québec	79,400	141,915	62,515	78.7	10.1
British Columbia	170,025	232,290	62,265	36.6	16.6
Ontario	188,315	301,425	113,110	60.1	21.5
Western Canada	436,455	574,335	137,880	31.6	41.1
Canada	976,305	1,400,685	227,890	43.5	100

Note: For more information on Aboriginal populations by identity and ancestry, see Table 1.2 and Figure 4.10. The uneven regional increases over the ten year period probably indicate that more Canadians are identifying themselves as Aboriginal, especially in regions where Métis reside.

Source: Statistics Canada (2013e).

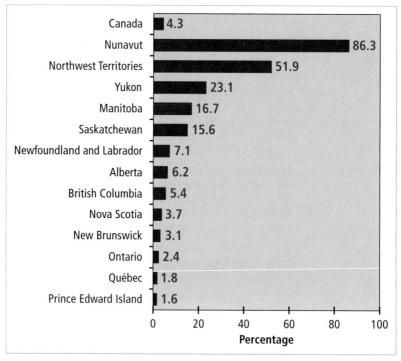

Figure 4.12 Percentage of Aboriginal people in the population, Canada, provinces, and territories, 2011
Source: Statistics Canada, 2013e:Table 2.

VIGNETTE 4.8

Early Estimates of Aboriginal Population

When Jacques Cartier sailed into Baie de Chaleur in 1541, the Indian and Inuit population in Canada may have been as high as 500,000. The exact figure will never be known. What we have are only estimates. By reconstructing the land's capacity to support wildlife and therefore also hunting societies, anthropologist James Mooney (1928: 7) estimated that about 220,000 Indians and Inuit lived in Canada at the time of contact. More recently, scholars have revised this figure upward. Dickason (Dickason with McNab, 2009: 40) and Denevan (1992: 370) estimate that the number of Aboriginal peoples living in Canada was closer to half a million. Whatever the exact figure, initial contact with Europeans resulted in a rapid depopulation of Aboriginal peoples. Factors include loss of hunting grounds and therefore food shortages, increased warfare, the spread of new diseases from Europe among the Indian tribes, and, in some instances, the intentional slaughter of Native people by the European newcomers. Communicable diseases, such as smallpox, caused great suffering and many deaths among Indian tribes. Epidemics sometimes quickly reduced the size of tribes by half. Depopulation did not take place across British North America at once but in a series of regional depopulations associated with the arrival of British settlers, although European epidemic diseases, spread through Aboriginal trade networks, often preceded the actual appearance of Europeans. In 1857, the first comprehensive counting of the Aboriginal population for British North America, undertaken by the Hudson's Bay Company at the request of the British House of Commons, totalled 139,000 (Bone, 2012: 66). By 1881, the census of Canada recorded 108,000 Aboriginal people (Canada, 1884: Table 3.1). As shown in Figure 4.10, the low point was reached in 1911 when the Aboriginal population was recorded as 105,611.

For example, Aboriginal communities face a serious housing shortage. Funds from Ottawa for more housing are never enough to keep up with the demand and the losses to the housing stock through fires and rapid deterioration of substandard dwellings. One solution has been to "double up," that is, to have several families living in the same house. The consequences of overcrowding result in social problems, health issues, and deterioration of the physical dwelling. Trapped in a poverty cycle, few have sufficient incomes to pay rent, let alone cash to maintain their dwellings.

Education is an essential key to upward mobility. From the perspective of public policy, perhaps the most important federal program to assist First Nations students to enter post-secondary institutions and thereby equip themselves for a more productive role in society has been the Post-Secondary Student Support Program (PSSSP), which covers tuition, living expenses, and travel. Unfortunately, this program, administered by First Nation band governments, is unable to fund all students who apply. The federal government views this program as special help for an unprivileged group within Canada society. First Nations take a different view and maintain that post-secondary education is a treaty right.

Photo 4.10
Elder Siipa Isullatak teaches children how to sew traditional Inuit clothing at Nakasuk Elementary School in Iqaluit, Nunavut, on 1 April 2009, the tenth anniversary of Nunavut becoming a separate territory.

VIGNETTE 4.9

Culture, Stability, and Aboriginal Families

All cultures are rooted in the concept of family structure. Aboriginal people are no exception. Yet, in its early efforts to assimilate Aboriginal people, the Canadian government destroyed many core elements of Aboriginal cultures, such as language, and, at the same time, negatively affected Aboriginal families, especially adult males, many of whom are struggling to find their role in the modern Aboriginal and Canadian societies. The sad result is that many Aboriginal families do not have a father. While the reasons are complex, the outcomes for the children, and hence for the future of the Aboriginal communities, appear bleak because the cycle of poverty, family breakup, and a dysfunctional lifestyle will continue. The residual effect of residential schools may be one factor, but the cycle of poverty is the more apparent factor.

A 2012 CBC television production—*Blind Spot: What Happened to Canada's Aboriginal Fathers?*—explores the subject of the "missing" fathers by interviewing a small number of Aboriginal males. Some fathers, for one reason or another, have abandoned their partners and children while others are trying to make family life and parenting work. The producer of *Blind Spot*, Geoff Leo, suggests that African-American society, where fathers for one reason or another are absent—some in jail, some hooked on drugs, and others leading lives of petty crime—may provide an insight into the fatherless Aboriginal families. According to the CBC documentary, many of the social problems facing Aboriginal youth can be traced to dysfunctional, fatherless families. Leo looked into the literature for information on the incidence of fatherless Aboriginal families. He found much literature on fatherless African-American families and the negative impact on these children, but found nothing in the academic or popular literature on fatherless Aboriginal families. Why, Leo asks, have Canadian and Aboriginal leaders ignored this hurtful social issue? He further questions why academics, Aboriginal Affairs Canada, and Aboriginal organizations have not recognized this problem and offered solutions. Leo challenges them to address the issue and take appropriate action. You can view the trailer at: <youtube.com/watch?v=rLn1-tTVjXs>.

Table 4.12 **Population with French Mother Tongue, 1951–2011**

Year	Canada (000s)	Canada (%)	Québec (000s)	Québec (%)	Rest of Canada (000s)	Rest of Canada (%)
1951	4,069	29.0	3,347	82.5	722	7.3
1961	5,123	28.1	4,270	81.2	854	6.6
1971	5,794	26.9	4,867	80.7	926	6.0
1981	6,178	25.7	5,254	82.5	924	5.2
1991	6,562	24.3	5,586	82.0	976	4.8
1996	6,637	23.5	5,747	81.5	970	4.5
2001	6,782	22.9	5,802	81.4	980	4.4
2006	6,892	22.1	5,917	79.6	975	4.1
2011	7,055	21.3	6,102	78.1	953	3.8

Source: Adapted from Harrison and Marmen (1994: Table 2); Statistics Canada (1997a, 2002b, 2008d, 2012k).

THINK ABOUT IT

Is bilingualism of the Supreme Court of Canada under fire because the last two appointments, Marshall Rothstein and Michael Moldaver, do not speak French?

French/English Language Imbalance

Canada is a bilingual country.[2] Yet, the weight of numbers is working against French-speaking Canadians. Since Confederation, even though their population size has increased, the percentage of French-speaking Canadians in Canada has slipped from 31.1 per cent in 1871 to 22 per cent in 2006 (Table 4.12) and to 21 per cent in 2011. With this decline, the place of French-speaking Canadians within Canada has weakened. Only in Québec has the French language remained strong and vibrant.

Why has the language imbalance increased over time? The answer remains the faster rate of increase of the English-speaking population, which is largely due to the number of new Canadians arriving each year who adopt English over French. For all regions except Québec, newcomers learn English.

French/English dualism is a fundamental aspect of Canada. As Jacques Bernier (1991: 79) of Université Laval stated: "Canada's duality is intrinsic, and as long as it is not clearly recognized and dealt with, the issue of Canadian unity will remain." This duality is a political concept embedded in the historical relationship between the two cultures. The main indicator of the stability of this dualism is language; in other words, the stability of this concept depends on a relatively constant number of Canadians speaking each language. But how should we measure duality? Should mother tongue or household language hold the key? Or is the number of bilingual Canadians the most important criterion? Outside of Québec and New Brunswick, the French language is losing ground.[3] Anne Gilbert (2001: 173) points to the decision of the government of Ontario to reject the concept of recognizing both English and French as official languages as a missed opportunity to showcase Canada's serious commitment to its dual languages.

Geography of Canada's Population

Central Canada continues to hold most of Canada's population. From 2001 to 2011, Ontario and Québec saw their combined population jump by just over 2.1 million people. Still, Canada's centre of population gravity has shifted westward because of rapid population growth in Western Canada that outpaced the other five regions with a percentage gain of 16 per cent (Table 4.13). British Columbia also contributed to this western shift but at a lower rate (12.6 per cent). Ontario has continued to grow but no longer occupies the lead position, with a percentage change of 12.6 per cent. Québec's rate of population increase remains below the national average while Atlantic Canada remains close to zero.

Economic growth in Western Canada and British Columbia offers the best explanation for the population shift. The global economic slowdown that began in 2008 has dramatically affected Canada's manufacturing regions, causing major labour force declines. On the other hand, the resource-based economies in British Columbia and Western Canada have fared much better and the demand for more workers remains strong. The push-pull migration model frames this discussion. The "pull" reality driving this process sees an irresistible lure of high wages and jobs galore for

Table 4.13 Changes in Regional Populations, 2001–2011

Geographic Region	2001	2011	Increase	% Change
Atlantic Canada	2,285,729	2,327,638	41,909	0.2
Québec	7,237,479	7,903,001	665,522	9.2
Ontario	11,410,046	12,851,821	1,441,775	12.6
Western Canada	5,073,323	5,886,906	813,583	16.0
British Columbia	3,907,738	4,400,057	492,319	12.6
Territorial North	92,779	107,265	14,486	15.6
Canada	30,007,094	33,476,688	3,469,594	11.6

Source: Adapted from Statistics Canada (2007a, 2012a).

skilled and unskilled labour in Alberta and Saskatchewan. The "push" reality is the dismal employment situation in Central Canada and Atlantic Canada. A side note to this migration is the phenomenon of air commuting from distant provinces, including Newfoundland and Labrador, to Canada's main industrial node—the Alberta oil sands.

Has the pendulum swung to the West? By the next census in 2016, we should know if these regional demographic shifts are just short-term blips or if they represent an irreversible trend. Already, the combined population of the four western provinces exceeds that of Québec and Atlantic Canada. Could they catch Ontario by 2016?

Canada's Economic Face

The global economic crisis that began in 2008 seems to have no end. By 2012, the European Union was teetering on the brink of economic and political collapse and the recovery of the economy of the United States remained stalled (Figure 4.16). On the other hand, Canada has weathered this global storm relatively well and is in a much better position than its major trading partner, the United States. This economic performance is remarkable because, given the dependency on the US market, Canada's economy has always closely followed that of the United States. Even so, Canada's economic performance varies across the country. Central Canada, with its emphasis on manufacturing, is struggling with unemployment rates well above the national average. On the other hand, the resource-based regions are growing rapidly and, in the process, are drawing more Canadians and immigrants. In sum, Canada now faces a new economic divide between "have" and "have-not" provinces.

The global economy has accelerated this process of the **hollowing-out** of Canada's manufacturing base (see Figures 4.13 and 4.15). Even the

THINK ABOUT IT
What would happen to Ontario's population if the final step in the hollowing-out process sees Ontario's automobile manufacturing sector relocated offshore in China, India, and other low-wage countries?

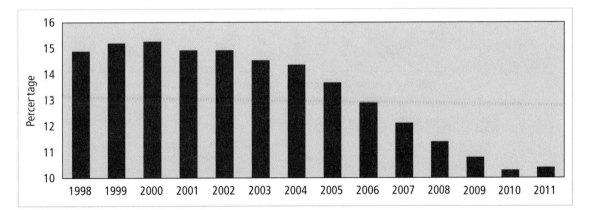

Figure 4.13 Manufacturing's declining share of employment
Source: Bernard (2009: Chart B); Statistics Canada (2012h).

high dollar (which has hovered around parity with the US dollar for the past five years) has taken its toll in this process. Plants that close and relocate offshore because they are unprofitable at current exchange rates are highly unlikely to return even if the Canadian currency declines.

In previous world downturns, Canada's economy has sunk well below that of the United States and has been much slower to recover. While this time is different, the recovery of the world's largest economy augurs well for Canada, especially Ontario and Québec. But just when Canada's manufacturing sector will recover—and to what extent—is not clear because the timing of the cyclical nature of the global economy is not well understood and certainly not predictable (Figure 4.14). Equally important for Canada is the shape of its economic structure in the post-2008 economic meltdown. For the Canadian economy, a major caveat is the US economy. With such a heavy dependence on exports to the US, Canada's economy, but particularly its manufacturing and forestry sectors, will not hit stride until the US economy recovers—and that may still take some time.

Economic Structure

Like other modern industrial countries, Canada's **economic structure** has evolved from an agrarian

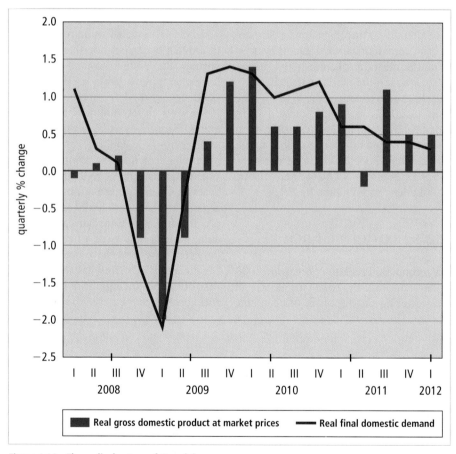

Figure 4.14 The cyclical nature of Canada's economy
Canada's economy is strongly affected by outside forces, especially trade with the United States. This graph indicates the effect of the global economy, especially the depressed US economy and the resulting decline in exports to the United States. Diversifying Canada's trade with other countries has become a federal priority. With oil representing the largest single export, access to foreign markets is critical. Two pipeline projects, the Keystone XL (south to the central US to refineries on the Gulf coast of Texas) and the Northern Gateway (to the British Columbia coast for transshipment to Asian, chiefly Chinese, markets), if constructed, would provide such access. Yet, fierce opposition from the Aboriginal and environmental movements, as well lack of support from the BC population, may halt these projects. *Source:* Statistics Canada (2012i).

Table 4.14 Traditional View of Economic Structure

Economic Sector	Characteristics
Primary	Economic activities, such as fishing, farming, forestry, mining, and trapping, concerned with the extraction of natural resources.
Secondary	Economic activities that process, transform, fabricate, or assemble the raw materials derived from primary activities, or that are high-tech knowledge-based activities that reassemble, refinish, or package manufactured goods. Examples include automobile manufacturing, meat-packing, pulp and papermaking, and software manufacturing.
Tertiary	Economic activities involving the sale and exchange of goods and services, including professional services, retail sales, and education. The service economy is extremely bifurcated with some high-end careers (accountants, bankers, doctors, and lawyers) and many more poorly paid jobs, such as in food services and retail (McJobs), that often offer only part-time employment at minimum wage.
Quaternary	Economic activities that deal with the handling and processing of knowledge and information that leads to decision-making by companies and governments. The leading edge of the knowledge-based economy is found at research centres located at universities and within governments and corporations. Innovative research clusters form around these centres with companies transforming the innovations into commercial production.

to a post-industrial economy. By structure, we mean the relative share of activity in the **primary**, **secondary**, and **tertiary/quaternary sectors** of the economy (Tables 4.14 and 4.15). With the shift of manufacturing to China and other low-wage countries, the economic structures of developed and developing countries are changing. The implications for Canada, especially its secondary or manufacturing sector, are profound. This global shift involves developing countries expanding their manufacturing based on low-cost labour and, with few trade restrictions between countries, exporting their manufactured products to wealthy industrialized countries. In turn, these exports have eroded much of the manufacturing base of industrialized countries, forcing them to turn to more sophisticated forms of manufacturing that rely heavily on research and innovations. High-speed train technology developed by Bombardier, which has garnered worldwide sales, including in China, is but one example. Perhaps the future for manufacturing lies in highly specialized clusters of enterprises that are closely connected with research institutions and universities where innovations can be turned into new manufactured products. Current examples are energy-related manufacturing in Calgary; aerospace in Montréal; computer

VIGNETTE 4.10

Canada's Growing Age/Wealth Divide

The opportunities that allowed older Canadians to gain wealth seem no longer available to the younger generation. The *Globe and Mail* financial columnist, Rob Carrick (2012a, 2012b), supports that view and suggests that graduating students today have it much harder than those who graduated from university three decades ago. Carrick points to much higher tuition fees, crippling student debt, sky-high housing prices, and a dismal job market compared to what he faced as a graduate in 1984. Tuition fees, for instance, have increased from 2001 to 2011 by 45 per cent, while inflation has only increased by 20 per cent. This age/wealth gap is best expressed by the average house prices. In 1984, the average house price in Canada was $76,214; by 2012, house prices had increased almost five times, to an average of $370,000. No doubt, this is not only an expression of the income inequality found in Canadian society, but also it presents an unrelenting force shaping the wealth distribution for the next generation.

Table 4.15 **Sectoral Changes in Canada's Labour Force, 1881–2011**

Development Stage	Year	Primary (%)	Secondary (%)	Tertiary* (%)
Agricultural	1881	51	29	19
Early industrial	1901	44	28	28
Late industrial	1961	14	32	54
Post-industrial	2011	4	19	77

*Because data for the quaternary sector are not available, percentages for the tertiary sector conventionally include both tertiary and quaternary jobs.

Source: Adapted from McVey and Kalbach (1995: Table 10.3); Statistics Canada (2006c, 2012h).

THINK ABOUT IT
Does the city where you reside show any signs of private/public research activity and, if so, is it associated with the local university and its research park?

technology in Waterloo, Ontario; and plant biotechnology in Saskatoon.

More information on Bombardier's train technology can be found in Chapter 6, page 224.

Provincial and Canadian unemployment data for 2007, 2009, and 2012 are shown in Table 5.1, page 166.

Economists Peter Drucker (1969) and Daniel Bell (1976) and geographer David Harvey (1989) were among the first scholars to recognize this global-driven adjustment leading to an information society and the growing importance of a knowledge-based economy. Unlike Bell, who describes this change through the rose-coloured glasses of capitalism, Harvey looks at this historic process through darker-tinted glasses of Marxist analysis. The **knowledge-based economy** offers hope for advanced industrial countries, like Canada, to offset a decline in traditional manufacturing activities. Success depends on integrating scientific knowledge, in the form of technology and innovation, into new products and services that can successfully compete on the international stage. Investments by both the public and private sectors are essential. Each region of Canada has a different set of possibilities in the knowledge-based economy, and these possibilities are discussed in the regional chapters.

The Information Society

An information society has a large portion of highly educated citizens concentrated in major cities. These workers function within the digitized world, and a significant number are employed in scientific research. Richard Florida's theory of the creative class fits into the concept of the information society, and his theory supports both the notion that creative people are the drivers of urban and regional growth and the concept of "clustering," namely, that the creative class is attracted to cities with a cultural soul, where a wide variety of artistic and cultural events flourish and where ethnic diversity and tolerance to non-mainstream lifestyles are well established (Florida, 2002a, 2002b, 2005, 2008, 2012). Florida further claims that innovative firms, public research centres, and universities situated in culturally rich cities have an advantage in recruiting and retaining a highly creative labour force.

In addition, an information society is supported by a knowledge-based economy that places

Figure 4.15 **Global competition and hollowing-out**
Is the offshore relocation of Canada's manufacturing industries an inevitable consequence of the global economy?

a high priority on scientific research. Such research creates new products that transform society: the potential impact of an electric car on city design and greenhouse emissions; biotechnical innovations that have led to higher-yielding grains; application of green technology (e.g., green roofs) to city buildings, thus making large cities more livable places. The role of Canadian universities is crucial. Canada's most recognized technological cluster, the Technology Triangle, consists of Waterloo, Kitchener, and Cambridge, Ontario, and in that cluster the University of Waterloo is an essential catalyst both in research and in training scientists. Universities, by establishing research parks, not only foster partnerships with federal research institutions and private firms, but also provide a link between pure and applied research. In this process, the **National Research Council of Canada** plays an important role in conducting research and in providing funding through various programs. These innovative incubators create scientific research and innovation zones like the Waterloo region, thus demonstrating the power of place. Such scientific nodes are one expression of the clustering approach where the search for innovations in science and technology flourishes. Innovation Place on the grounds of the University of Saskatchewan is one instance of a Canadian university fostering innovation, commercialization, and economic growth through university, industry, and government partnerships. In 2009, Saskatoon's Innovation Place received the "2009 Outstanding Research-Science Park Award" from the Association of University Research Parks, based in Tucson, Arizona (Association of University Research Parks, 2009).

Figure 4.16 Economic bunker

The global economy is highly integrated. The financial crisis that struck the United States in late 2008 caught a lot of people by surprise, although many of the financial firms on Wall Street that were at the centre of crisis received a multi-billion dollar bailout from the US government. By 2012, world attention had focused away from the US and to the crippling economic situation in the European Union, where the so-called PIGS (Portugal, Italy, Greece, and Spain) are on the edge of bankruptcy. The consequences of a collapse of the European Union for the world economy are extremely bleak.

More on innovation zones and the clustering of high-technology firms and related research is found in Vignette 5.10, "The Wave of the Future: Clusters and High Technology," page 192.

VIGNETTE 4.11

Public Support for Scientific Research and Knowledge-based Nodes

Scientific research is expensive but essential in post-industrial economies. Basically, scientific research to produce innovative products and services is a key factor in transforming Canada's economy from an industrial one to a post-industrial economy. Public support is essential and the federal Scientific Research and Experimental Development (SRED) program is one example. SRED provides tax rebates and credits for companies engaged in scientific research for product/service development. Provinces also support such research, with Québec providing the most generous support with a tax credit of 37.5 per cent compared to Ontario's similar credit of 10 per cent (PriceWaterhouseCoopers, 2009). The impact of federal support on one software company in Saskatoon is revealed in comments by its CEO, Amit Gupta of Solido Design: "The big trend right now is to export R&D operations to China and India. As a result of this SRED program, it becomes cost effective for us not to export those jobs and keep those jobs here locally in Saskatoon" (Kyle, 2009).

SUMMARY

Canada's population continues to grow, thanks in large part to immigration, and population distribution remains concentrated along the southern border with the United States. Urbanization remains a dominant driver of Canada's population geography, with more and more people living in urban settings. Given Canada's increasingly pluralistic society, multiculturalism remains a key plank in Ottawa's effort to accommodate newcomers. The search for cultural accommodation without jeopardizing Canada's traditions remains an ongoing process, reflecting the dynamic nature of Canadian culture.

Canada's economy is under pressure to find its place within the global economy. This search is made more challenging by the worldwide economic crisis that began in late 2008. Given Canada's strong trade dependency on US markets, efforts by Ottawa to diversify the Canadian economy have resulted in more trade agreements with other countries and an aim to provide access to Asian markets for Canadian producers. When a full recovery of the US economy takes place, Canada—especially Central Canada—will benefit. But until then, the so-called "resource curse," that is, the fear that the wealth generated by resources is smothering Canadian manufacturing, remains popular in certain political circles in Ottawa and Toronto. Earlier in this book, another interpretation of the adverse effect of a resource boom on manufacturing was described as the Dutch disease. The counter-argument to the resource curse is that manufacturers must focus on productivity, invest in training and research, produce high-value products, and, in the case of Canada, join the supply chain for the expanding resource industries.

Stresses within Canadian society take the form of four faultlines: Aboriginal/non-Aboriginal, core/periphery, French/English, and newcomers/old-timers. Key demographic changes are occurring along these divides, reflecting internal forces and revealing population and political shifts. The first faultline is characterized by a much higher rate of natural increase among the Aboriginal population than the general population. The second faultline involves changes in the sizes of Canada's regional populations. The third faultline focuses on the growing imbalance in the numbers of English- and French-speaking Canadians The fourth faultline—between new Canadians and those born in Canada—is reflected in the federal policy of multiculturalism and is played out daily in Canada's major urban centres, where the vast majority of new Canadians settle.

The following chapters, 5 through 10, focus on the regional nature of Canada, with each chapter devoted to exploring one of Canada's six geographic regions. Friedmann's version of the core/periphery model provides a guide for the ordering of these six regions (Table 1.4). The regional discussion therefore begins with Ontario and Québec, which represent Canada's economic core. These two chapters are followed by chapters on British Columbia, Western Canada, Atlantic Canada, and the Territorial North. While many of the same topics and issues appear in each of the six geographic regions, each is examined from the perspective of the region under discussion. Each regional chapter will illustrate the physical diversity within the region and show how this diversity affects human activities and settlement. Each chapter also has a section on the region's historical development to enrich our sense of the present by providing a link with the past. One recurring theme is the relationship among the six regions. This relationship is cast in the core/periphery model, but it is changing due to continental economic integration (NAFTA), global trade (the WTO), the restructuring of the world economy and the North American/Canadian economies, plus the shift of power—demographically and economically—beyond the core regions of Ontario and Québec. Within this discussion, the shift of regional economies towards a greater emphasis on knowledge-based activities is highlighted.

Finally, Canada's regions are witnessing radical changes in their economic circumstances that challenge the core/periphery theory. This "grand" theory assumed that manufacturing prices would always increase at a greater rate than foodstuffs and commodities; hence, the manufacturing centres of the world would gain more and more over time. By the twenty-first century, this assumption proved incorrect due to two watershed events—the globalization of markets and the industrialization of Asia. These changes have tipped the world economy towards higher prices for foodstuffs and commodities. If this price shift is permanent, the economic and demographic consequences for Canada and its six regions will vary. The implications for each region and these differences are noted in each regional chapter.

CHALLENGE QUESTIONS

1. From 2001 to 2011, some geographic regions increased their share of Canada's population while others did not. Is this just a short-term blip on the demographic radar screen or are the fundamentals in place for this demographic shift to continue for the foreseeable future?

2. Is Calgary on the right track to curb urban sprawl on its outer edges by increasing the cost of residential lots to developers?

3. Since the Aboriginal population is rapidly expanding, why is it not playing a larger role in meeting Canada's labour shortages?

4. What could slow and even reverse the process of hollowing-out, where far too often Canadian manufacturers have relocated to low wage countries?

5. What is the counter-argument to the "resource curse"?

FURTHER READING

Bunting, Trudi, Pierre Filion, and Ryan Walker, eds. 2010. *Canadian Cities in Transition: New Directions in the Twenty-First Century*, 4th edn. Toronto: Oxford University Press.

Canada is a highly urban society. Its cities are complex, dynamic, and ever-changing. In this most recent edition of *Canadian Cities in Transition*, the editors have brought together many of Canada's leading urban geographers and planners to discuss why and how Canadian cities work or don't work. The underlying theme of the 25 chapters is transformation, as cities today are in the midst of remarkable change and new adaptations brought about by such factors as environmental degradation, traffic congestion, immigration and internal migration, and the new economy.

The range of subjects covered in this volume is broad, and includes such new topics as obesity and the built environment, food systems and food "deserts," Aboriginal people in urban Canada, gentrification, cities with declining populations, and social polarization and neighbourhood inequality. Of special interest for those concerned about the future of Canadian cities in a changing world is Chapter 5, by William E. Rees, "Getting Serious about Urban Sustainability: Eco-Footprints and the Vulnerability of Twenty-First-Century Cities," in which the author introduces the startling concept of cities as "human feedlots" and outlines why cities must become "green" to survive and thrive. The final chapter, by Larry S. Bourne and R. Alan Walks, "Challenges and Opportunities in the Twenty-First-Century City," examines the issue of homelessness and the tensions between the reality of urban sprawl and the goal of contemporary planners and politicians to achieve the "compact city." Chapters on the history of Canada's cities, as well as on transportation, housing, economic restructuring, inner cities and the creative class, and governance also are of special interest.

5

ONTARIO

Introduction

Ontario's prominent position within Canada and North America has been badly shaken by the economic downturn that began in 2008 and shows no sign of ending. Manufacturing has been hit hard, especially the automobile industry. Yet, the basic factors that propelled Ontario to the lead position within Canada and its prominent role in North America have not changed. Ontario remains the heartland of Canada's manufacturing industry, continues to play a dominant role in the financial industry, and serves as the cultural centre for English-speaking Canada. While the old economy is slowly going, the shape of the new, global-driven economy is uncertain. For Ontario to prosper and progress into the next decade as well as to retain its title as the economic engine of Canada, bold and innovative measures will be necessary. Two key components of its economy—forestry and manufacturing—are stalled in the old economy and they need two things: (1) fresh directions and investments to get back on the domestic and global economic highways; and (2) recovery of the US economy and the resumption of high levels of automobile and lumber exports to the US.

CHAPTER OVERVIEW

The upheaval in the North American automobile industry makes our "Key Topic" the restructuring of the automobile industry. As well, this chapter will consider the following topics:

- Ontario's physical geography and historical background.
- The downsizing of the manufacturing sector.
- The basic characteristics of Ontario's population and resources.
- The economic and environmental challenges facing Ontario.
- The significance to Ontario of several important trade agreements.
- Ontario's dichotomy—an industrial core in the south and a resource hinterland in the north.
- Ontario's changing role within Canada, North America, and the world.

The Parliament Buildings as seen from across the Ottawa River, Ottawa, Ontario. The Peace Tower situated at the centre of the parliament complex is 92.2 metres tall and houses 53 bells. Photo: Alan Copson/Getty Images.

Ontario within Canada

The first decade of the twenty-first century has not been kind to Ontario. Its economy took an economic jolt, especially its forest and manufacturing industries. While signs of an economic decline appeared in the first phase of globalization, the hardest blow came in 2008 with the collapse of two of its principal automobile manufacturers, Chrysler and General Motors. The negative spinoff rippled through the manufacturing sector, causing massive layoffs. Ontario's economy fell off the rails and the province received its first equalization payments in 2009, making it a "have-not" province. In a short space of time, Ontario's economy has slipped badly and, as a consequence, so has its population growth. As shown in the Figure 5.1, prepared by two of Canada's leading economists, Don Drummond and Derek Burleton (2008), Ontario's economy could shift either way. As they put it:

> With much of Ontario's economic success driven by advantages that no longer exist, a new direction is required. We look to the provincial government to take leadership on this front by developing a vision on where it plans to take the economy down the road.

What happened to the strongest partner in Confederation? For certain, Ontario's place has slipped for four reasons. First, the principal market for Ontario products, the United States, dried up following the 2008 economic crisis. Second, in 2002 when China joined the World Trade Organization, low-cost Chinese manufactured goods flooded the Canadian market, thus displacing local manufacturers. Third, within Canada, western provinces are enjoying the commodity boom, allowing those provinces to outpace the economic and population growth rates of Ontario. At the same time, this commodity boom has increased the value of the Canadian dollar, creating the conditions for the Dutch disease, i.e., making other exports more costly. As a result, Western Canada and British Columbia have sharply increased their share of the national GDP and population. Fourth, the Ontario government has little latitude to lead the province out of the economic doldrums because of its heavy debt. The impact of these four economic factors resulted in unusually high unemployment rates that soared to 9 per cent in 2009—well above the national average of 8.3 percent and, for the first time, above Québec's rate of 8.5 per cent (Table 5.1).

The foremost challenge to Ontario's economy has been the sharp fall in exports to its principal

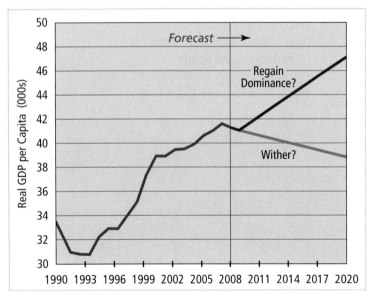

Figure 5.1 Ontario's economy to 2020: Which way?
Ontario's manufacturing sector could suffer from deindustrialization and become part of the adjacent US Rust Belt; or Ontario could reinvent itself. *Source:* Drummond and Burleton (2008: 1).

Table 5.1 Before and After the Crash: Unemployment Rates by Province, 2007, 2009, and 2012

Province	2007	2009	2012 (July)
Alberta	3.5	6.6	4.6
Saskatchewan	4.2	4.8	4.9
Manitoba	4.4	5.2	5.2
British Columbia	4.2	7.6	6.6
Ontario	6.4	9.0	7.7
Québec	7.2	8.5	7.7
New Brunswick	7.5	8.9	9.5
Nova Scotia	8.0	9.2	9.6
Prince Edward Island	10.3	12.0	11.3
Newfoundland and Labrador	13.6	15.5	13.0
Canada	6.0	8.3	7.3

Source: Statistics Canada (2009c, 2010, 2012a).

market, the United States. International exports—80 per cent going to the US—hit rock bottom in 2009 at $125.6 billion, and they struggled to reach $155.4 billion in 2011, which remains far below the 2008 figure of $163 billion (Ontario, 2012).

The problem was (and remains) that Ontario has tied its economic wagon to the United States. Back in 1996, Professor Glen Norcliffe (1996: 44) warned about the risks of too heavy a dependence on trade with one country, meaning the United States. The unexpected collapse of the US market in late 2008 proved that Norcliffe's point was valid. But what has been surprising is the length of the US recession, which has put a serious dent in Ontario's exports. The good news is that the US economy shows very modest signs of recovery and Ontario exports to the US increased in 2012. In fact, automobile production for 2012 is on track to reach the pre-2008 levels (Scotiabank, 2013). However, at this juncture, many problems still face the United States and a full recovery—so necessary for Ontario's economic well-being—is not yet in sight.

The second blow to Ontario's economy was caused by the globalization of trade. This ongoing erosion of Ontario's manufacturing base began well before 2008. The problem is a shift of manufacturing to low-cost countries. For instance, many Ontario manufacturers cannot match the production costs found in China and other low-wage countries. Manufacturing plants in Canada and the United States continue to close and the number of workers in the manufacturing sector continues to decline. Economists have termed this effect of globalization on the manufacturing sector found in developed countries as the **"China Syndrome."**

The third blow was the loss of the low Canadian dollar, which kept the price of exports to the United States at low levels. In 2002, Canada's dollar was valued at 64 cents on the US dollar. By April 2010, the Canadian dollar had reached parity with the US dollar and, as of March 2013, it remained slightly below par. Whether you call it the "petro-dollar impact" or the Dutch disease, the days of a low Canadian dollar are gone, making cross-border manufacturing more costly and less attractive for locating new plants in Ontario.

The last factor reveals a missed opportunity. Put it this way: has Ontario missed the economic boom in Western Canada and British Columbia? At the beginning of the twenty-first century, the western economic boom was underway, and since then it has only gotten stronger. Yet, why do foreign firms supply the largest share of the high-value manufactured products needed by the huge resource projects in Alberta, Saskatchewan, and British Columbia? Can Ontario reset its economy to meet the needs of its western neighbours?

Despite Ontario's new status as a "have-not" province, its position within Canada goes beyond its economic role. Ontario remains the political linchpin in Confederation and is a cultural sparkplug for English-speaking Canada. Its cultural and political role can be attributed largely to Ontario's large and affluent population. Ontario, with nearly 13 million people, has by far the largest population of the six regions. For that reason, Ontario sends more representatives to the House of Commons than any other province. Its large market has the capacity to support the arts, including the Toronto Symphony Orchestra, the Hamilton Symphony, the Canadian Opera Company, the National Ballet, the Stratford Shakespearean Festival, the Shaw Festival, many other venues for staging theatrical works and popular music concerts, the Art Gallery of Ontario, and the Royal Ontario Museum; an

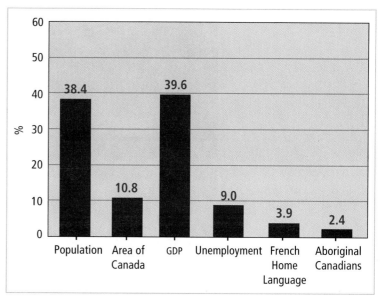

Figure 5.2 Ontario basic statistics, 2011

Ontario's economic and political strength within Canada is revealed by its share of the nation's GDP and population. Ontario comprises the largest single market in Canada, forming 38.4 per cent of Canada's population and accounting for nearly 40 per cent of its GDP. However, the province's economy is sputtering as illustrated by much higher unemployment rates than those found in Western Canada and BC (Table 5.1). Franco-Ontarians and Aboriginal people form small minorities, but the Aboriginal population is growing in size and percentage of Ontario's population while the reverse is true for the Franco-Ontarians. *Source:* Statistics Canada (2013: Tables 1.1 and 1.2).

THINK ABOUT IT
If the closure of the Caterpillar plant in London in 2012 was a result of the seemingly relentless push by global companies to match costs to those countries with low wages and few benefits, what message does it send to Ontario's manufacturing industry and its workers?

Photo 5.1
Toronto's high-order cultural facilities include the Four Seasons Centre designed specifically for opera and ballet. Opened in August 2006, the performance of Wagner's Ring Cycle attracted Wagner enthusiasts from around the world.

extensive network of post-secondary education institutions; and professional sports teams not found elsewhere in Canada—the Toronto Blue Jays (baseball), the Toronto Raptors (basketball), and Toronto FC (soccer).

Ontario's Physical Geography

Ontario is larger than most countries (Figure 5.3), stretching out over 1 million km². Extending over Ontario are three of Canada's physiographic regions (Great Lakes–St Lawrence Lowlands, Canadian Shield, and Hudson Bay Lowlands), and three of the country's climatic zones (Arctic, Subarctic, and Great Lakes–St Lawrence) (Figures 2.1 and 2.6). Manitoba lies to its west, Hudson and James bays to its north, while Québec, on its eastern boundary, is bordered in part by the Ottawa River. This central location within Canada and its close proximity to the industrial heartland of the US have facilitated Ontario's economic development.

Ontario is not a homogeneous natural region. For that reason, it is divided into two sub-regions (northern and southern Ontario). This division is based on its physiographic regions. Accordingly, northern Ontario consists of the Canadian Shield and Hudson Bay Lowlands while southern Ontario matches the geographic extent of the Great Lakes–St Lawrence Lowlands that fall within Ontario. Each sub-region has a different economy. Northern Ontario has the characteristics of a resource hinterland, while southern Ontario is the epitome of an agricultural-industrial core (Vignette 5.2).

Northern Ontario stretches across 10 degrees of latitude—from 46° N to nearly 57° N. Fort Severn First Nation is located on the shores of Hudson Bay at 56°37′ N while Sudbury is at 46°30′ N. The Subarctic climate of northern Ontario has longer and colder winters as well as shorter and cooler summers than those occurring in southern Ontario. Even along its southern edge at 46° N, short summers make crop agriculture vulnerable to frost damage. In addition to a difficult climate for agriculture, the rocky Canadian Shield has very little agricultural land while the Hudson Bay Lowlands has none. The rugged, rocky terrain of the Canadian Shield has only a

ONTARIO 169

Figure 5.3 **The province of Ontario**
Ontario has the largest economy and population of Canada's six regions, but several factors—the high Canadian dollar, the loss of manufacturing jobs, rising unemployment rates, and the slow recovery of its major US market from the 2008–9 global economic crisis—have placed Ontario on a lower rung. A measure of its demise is that it now qualifies for equalization payments from the federal government. *Source: Atlas of Canada*, 2006, "Ontario," at: <atlas.nrcan.gc.ca/site/english/maps/reference/provincesterritories/ontario>.

few pockets of agriculture where former lakebeds provide the basis for soil development. Climate, soils, and physiography combine to limit agriculture in northern Ontario.

Unlike Québec, the Ontario section of the Canadian Shield has relatively low elevations, limiting the opportunities for hydroelectric power developments. However, this sub-region does have vast forests, superb scenery (which supports tourism), and extensive mineral wealth. The Hudson Bay Lowlands consists of a poorly drained plain associated with muskeg and permafrost. Few opportunities for resource development exist and it remains an area for hunting and trapping inhabited mainly by Cree Indians. As a resource hinterland, northern Ontario is limited economically to the Canadian Shield almost entirely, and its development is dependent on forestry, tourism, and mining.

Southern Ontario, located in the southernmost part of Canada, has Canada's longest growing season. Windsor, for example, is at latitude 42° N and

THINK ABOUT IT
While a distance of 1,200 km separates Fort Severn from Sudbury, why is the cultural distance between the two communities even greater?

THE REGIONAL GEOGRAPHY OF CANADA

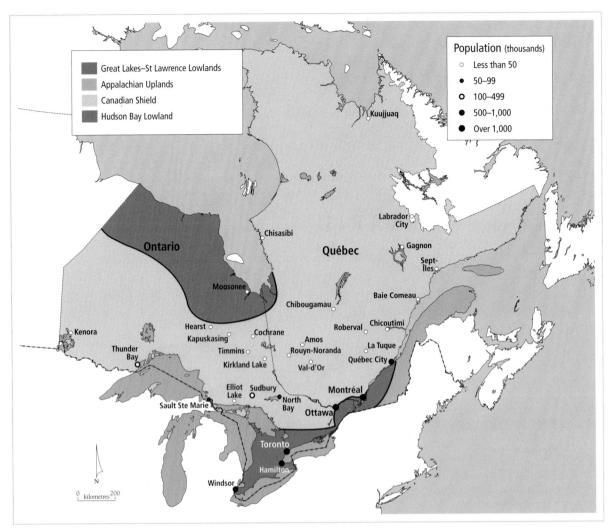

Figure 5.4 **Central Canada**
Central Canada consists of Ontario and Québec. Three physiographic regions are found in Ontario and four in Québec. Canada's most productive agricultural lands, its manufacturing belt, and its core population zone all are located in one physiographic region—the Great Lakes–St Lawrence Lowlands.

VIGNETTE 5.1

The Burden of Confederation

One sign of Ontario's "Burden of Confederation" is that Canadians living in provinces less prosperous than Ontario have benefited from equalization payments. Ontario and Alberta have been the major source of tax revenue for Ottawa and these revenues have provided much of the funds for equalization payments. In 2009–10, however, Ontario, for the first time, received an equalization payment while Saskatchewan and Newfoundland and Labrador joined Alberta in shouldering the "Burden of Confederation." Ontario was expected to bounce back quickly, but after four more years the province remains in the "have-not" category.

Toronto is close to 44° N. Southern Ontario has a moderate continental climate. This climate is noted for long, hot, and humid summers, warm autumns, short but cold winters, and cool springs. Annual precipitation is about 1,000 mm. The greatest amounts of precipitation occur in the lee of the Great Lakes, where winter snowfall is particularly heavy (see Vignette 5.3). The Great Lakes modify temperatures and funnel winter storms into this region. During the winter, this region experiences a great variety of weather conditions.

Southern Ontario is the most favoured physical area in Canada. Not surprisingly, the vast majority of the province's population, nearly 12 million, or 93 per cent, live in southern Ontario. The area is underlain by slightly tilted sedimentary rocks, which are covered by a thin deposit of glacial till. Except for the Niagara Escarpment, there is little relief topography. Formed since deglaciation, the Niagara Escarpment is an erosional remnant of the more resistant sedimentary rock that extends from the Bruce Peninsula to Niagara Falls. The Escarpment provides the most spectacular scenery in southern Ontario and a nature trail known as the Bruce Trail extends some 725 kilometres along this limestone topographic feature. Prior to settlement, a mixed forest vegetation flourished in this temperate continental climate (Figure 2.7). With a long growing season, ample precipitation, and fertile soils, the southern Ontario lowland has the most productive agricultural lands in Canada.

The section on "Physiographic Regions" in Chapter 2, pages 32–44, especially the figures and tables, provides important background information for Ontario.

Climate and Agriculture

Southern Ontario's climate is dominated by its long, warm summer that extends from May to September. Tropical air masses that originate in the Gulf of Mexico extend over this area, resulting in hot and humid weather. Winter, on the other hand, takes hold for three to four months from mid-November to March when occasional invasions of Arctic air masses bring exceptionally cold weather. Such weather is associated with cold, sunny days and frigid nights. During the early spring and late fall, unsettled, cloudy weather dominates, with minimum temperatures falling below the freezing point. Proximity to the Great Lakes affects the local weather by funnelling air masses into this region and by increasing local precipitation from air masses absorbing moisture from the surface of the Great Lakes. During the winter, high winds funnelled along the Great Lakes storm track often accompany cold spells, driving the wind chill to a dangerous level. Storms are short-lived, followed by relatively mild weather. Annual precipitation ranges around 100 cm. Warm, moist air masses from the Gulf of Mexico provide most of the moisture for the region, which falls as convectional and frontal precipitation.

Because of its mild climate and fertile soils, southern Ontario accounts for much of Canada's agricultural output by value. Southern Ontario has over half of the highest-quality agricultural land (known as Class 1) in Canada. With a combination of fertile soils and a warm summer, farmers in southern Ontario accounted for most of the $11.7 billion of agricultural produce in

VIGNETTE 5.2

Contrasting Sub-Regions: Northern and Southern Ontario

Ontario is a geographic paradox. The contrasts between southern Ontario and northern Ontario are extreme. Each has very different physical conditions and geographic situations. Over time, two distinct economies have emerged, each with its own spatial pattern of population distribution. While southern Ontario forms the industrial and population heartland of the province, northern Ontario is an old resource hinterland where economic and population growth have stalled. Because so many young people are moving to more prosperous areas in Canada, the population of northern Ontario is losing a valuable segment of its labour force, leaving its cities and towns with disproportionately large numbers of older and very young people. By 2011, northern Ontario had a population of just under 850,000, while southern Ontario had a population of 12.7 million (Statistics Canada, 2012c).

Photo 5.2
The Niagara Escarpment, stretching over 700 km from Queenston on the Niagara River to Tobermory on the Bruce Peninsula, is the most significant landform in southern Ontario. The Escarpment reaches over 350 metres above the surrounding land at various points.

2010 (Statistics Canada, 2012d). Yet, the number of farms and the amount of acreage keep declining. In 2011 Ontario had 52,000 farms, compared to 57,000 in 2006. Part of the explanation is larger farm size, but another explanation is a decrease in total acreage (ibid.).

Located south of 44° N, southern Ontario has abundant moisture and a long, warm-to-hot summer permitting the production of highly specialized crops such as grain corn, soybeans, and sugar beets as well as a wide variety of fruits, grapes, and vegetables that cannot be grown in

VIGNETTE 5.3

Ontario's Snowbelts

Ontario's snowbelts are legendary. On the upland slopes facing Lakes Huron and Superior and Georgian Bay, huge snowfalls totalling in the 300–400-cm range occur each winter from November to late March. The uplands on the northeastern shores of Lake Superior receive the greatest total snowfall amounts of any area in Ontario, exceeding 400 cm annually. Much of the snowfall in snowbelt areas can be attributed to cold northwesterly to westerly winds blowing off the lakes and ascending the highlands. As the Arctic air travels across the relatively warmer Great Lakes, it is warmed and moistened. Snow clouds form over the lakes and, once onshore, intensify as the air is then forced to ascend the hills to the lee of the lakes, triggering heavy snowfalls. Areas on the downslope side of the higher ground to the lee of the lakes receive less than half the annual snow totals of the upslope snowbelt areas. For example, Toronto, Hamilton, and other places to the lee of the Niagara Escarpment are snow-shadow regions with winter amounts of 100 to 140 cm. In the snowbelt regions, snowfall accounts for about 32 per cent of the year's total precipitation; but in the snow-sparse area of southwestern Ontario around Windsor and Chatham, the snow contribution is only about 13 per cent.

CONTESTED TERRAIN 5.1

Urban Sprawl and Farmland

Urban Ontario is growing and requires more land for housing, roads, and retail outlets. On the other hand, high-quality agricultural land is a scarce commodity but its market value is low compared to the same land used for urban uses. Would you, as Premier of Ontario, consider protecting Niagara Fruit Belt farmland (see Vignettes 5.4 and 5.5) as a priority and thereby limit urban sprawl? If so, how would you justify such a policy to landholders and developers?

other parts of Canada. In the Niagara Fruit Belt, a unique set of growing conditions has provided the basis for grape production, resulting in high-quality wine production (Vignette 5.4). Equally important, the large urban population provides a stable local market for its dairy, livestock, soft-fruit, and vegetable products. Although the amount of cropland is relatively small, farming operations in southern Ontario are much more intense than in Western Canada, where extensive grain crops dominate farming operations. The difference between the two agricultural areas is revealed in the size of farms and the types of crops. In 2010, the average size of farms in Ontario, 244 acres, compared to an average of 1,668 acres for farms in Saskatchewan (ibid.). Grain farming and cattle ranching dominate in Western Canada, while a much more diversified use of agricultural land prevails in southern Ontario, where corn, barley, and winter wheat are important crops along with highly specialized ones such as tobacco and vegetables. Corn and grain often serve as fodder for the hog, dairy, and beef livestock farms that operate in Ontario.

Agricultural land use varies within southern Ontario due to subtle but important physical differences. For example, soils are less fertile and the growing season is shorter east of Toronto than southwest of Toronto. There are three highly specialized agriculture zones west of Toronto: the Essex–Kent vegetable area, the Norfolk Tobacco Belt, and the Niagara Fruit Belt. All lie in southwestern Ontario where the growing season is relatively long. These zones, located south of 43° N, are the southernmost lands in Canada.

Unfortunately, the expansion of cities and towns in southern Ontario has spread into these farmlands. Valuable farmland is now part of the urban landscape. Until the price of agricultural land is equal to or surpasses that of urban/industrial land prices, this process of losing irreplaceable arable land will continue. How to resolve this land-use conflict is beyond the market economy because the price of farmland is well below that of urban land. Ontario's government has recognized this dilemma but state intervention in the real estate market has been ineffective. A perfect example of the mismatch between the value of farmland and other commercial uses is revealed in the proposal of Highland Companies to convert Ontario's most valuable soil, rated as Class I, into a quarry. This agricultural Garden of Eden, with some of the best farmland in Canada, is an hour northwest of Toronto. Within this rich agriculture land, Highland Companies is seeking to establish a large quarry for high-quality Amabel dolostone bedrock on a portion of its landholdings in the Township of Melancthon, in Dufferin County. The potential value of the Amabel dolostone exceeds $6 billion. When Highland purchased the land from farmers at or above market value, little was said about a quarry. Rather, the company's credible sales pitch was that by assembling these small farms into larger ones, economies of scale could result in a more profitable farming operation and ensure the long-term viability of farming in the area. Besides the cash in their pockets, the farmers felt good about the company continuing the tradition of farming. On the credit side of this deal, Highland Companies has become the largest potato operation in the province. On the negative side, the company, according to local folks, was not forthcoming about its plans for a quarry. Some go so far as to call the Highland deal "an underhanded bait-and-switch" trick (Leeder, 2012: A5). Added to the idea of "underhandedness," the plan is funded by Baupost, a US hedge fund company based in Boston. Not surprisingly, then, a storm of protest erupted among residents, and Ontario's Ministry of the Environment was flooded with letters calling for an environment impact review. On 1 September 2011, the province established such an inquiry. As this review plays out, Ontarians will see the arguments for and against the quarry. The quarry, if given approval, will supply much needed aggregate for building highways and other infrastructure. On the other hand, the land could feed people forever. Given the pricing of potatoes compared to that generated by the sale of aggregate, the quarry proposal wins hands down, that is, until food becomes such a scarce commodity that its price rises to that assigned by market forces to the aggregate.

VIGNETTE 5.4

The Niagara Fruit Belt

The Niagara Fruit Belt lies on the narrow Ontario Plain that extends from Hamilton to Niagara-on-the-Lake. The Ontario Plain has fertile, sandy soils suitable for most types of agriculture. Lake Ontario marks its northern limit while the Niagara Escarpment forms its southern edge. This small agricultural zone contains the best grape and soft-fruit growing lands in Canada, and is home to vineyards that account for most of Canada's quality wines, including the unique ice wine.

While the Niagara Fruit Belt occupies a northerly location for grapevines and soft-fruit trees, local factors have more than offset the threat of frost at 43° N latitude. These factors are:

- **Air drainage** from the Niagara Escarpment to Lake Ontario reduces the danger of both spring and fall frosts.
- The water of Lake Ontario is warm in autumn and its proximity to the Niagara Fruit Belt helps to moderate advancing cold air masses.
- In early spring, the cool waters of Lake Ontario keep air temperatures low, thereby delaying the opening of the fruit blossoms until late spring when the risk of frost is much lower.

Photo 5.3
The Niagara Fruit Belt extends about 65 km between Hamilton and Niagara-on-the-Lake, and is one of the major soft-fruit and grape-producing areas in Canada. Most vineyards are located on the slopes of, or below, the Niagara Escarpment. For many years, hardy vines that produced low-quality grapes resulted in a poor-quality (but inexpensive) wine. With the Free Trade Agreement, Canada had to remove its tariffs, making it difficult to compete with foreign wines. Since that time, farmers in the Niagara Fruit Belt have been successful in growing the finest varieties of grapes and have been able to make some of the finest wines in the world.

Environmental Challenges

Ontario faces two major environmental challenges—air pollution and water pollution. Solutions are costly and require lifestyle changes. Nevertheless, action is needed and without more robust action by governments, these two problems will only get worse, causing serious harm to the environment and to the health of citizens. Both types of pollution represent the hidden costs of our industrial world.

Ontario's cities, especially Toronto, face two hidden costs associated with urban pollution, namely disposing of waste products and paying for the health cost of **air pollution** from smog caused mainly by vehicular traffic. City governments have had more success in convincing their residents to reduce their waste products through recycling programs and composting than they have in getting people to shift from the automobile to public transportation. The solution to both challenges lies in a behaviour change, but where one issue has met with some success, the other has not.

Toronto used to export most of its garbage south of the border to Michigan but that arrangement ended in 2010. Now all garbage is sent some 200 km to the city-owned Green Lane Landfill site south of London, Ontario. Advances in urban garbage disposal lie in converting garbage into energy. Gasification, a low-emission method of extracting energy from a variety of waste materials, may offer a solution. While the landscape is littered with garbage-to-energy technologies that promised breakthrough advances, none have cleared the commercial hurdles; the cost of moving from a demonstration plant to a commercial one is high. **Sustainable Development Technology Canada** has provided base funding for promising technologies such as the Plasco Energy Group's test facility in Ottawa that converts urban wastes (garbage) into reusable products (Cleroux and Woods, 2012).

For more on the high cost of developing new technologies in Ontario's knowledge-based economy, see "Ontario Today," page 185. For more on a knowledge-based economy, see "Economic Structure" in Chapter 4, page 158.

Pollution has a health cost. The densely populated area of the Golden Horseshoe—which includes Toronto and Hamilton—has significant health problems caused by smog. Smog becomes a serious health hazard in the summer months when high temperatures and air inversions combine to keep the smog hanging over the city as a brownish-yellow haze. The word "smog" is actually a combination of "smoke" and "fog." Smog is the most visible form of air pollution, and is caused when heat and sunlight react with various pollutants emitted by industry, vehicle exhaust, pesticides, and oil-based home products. A study by the Ontario Medical Association estimated that illnesses caused by smog cost Ontario more than $1 billion a year in hospital admissions, emergency room visits, and absenteeism. Even more startling, the OMA study (2005) concluded that about 5,800 people die prematurely in Ontario each year because of smog-related respiratory problems.

VIGNETTE 5.5

The Tragedy of Urban Sprawl

Urban sprawl devours valuable but underpriced agricultural land. Since the economic value of urban/industrial classified land is higher than land designated for agricultural uses, urban/industrial encroachment continues to spread into the rural landscape. Although Ontario has over half of the best agriculture land in Canada, urban encroachment and other non-agricultural interests have cut deeply into the amount of farmland. For instance, over the three decades between 1976 and 2006, nearly 200,000 acres or 25 per cent of the province's Class 1 agricultural land was lost. Even the extremely valuable lands in the Niagara Fruit Belt remain under pressure from residential and commercial land developers. An indirect impact of urban sprawl is the release of more greenhouse gases because of the longer drives from suburbia to places of work. Ontario has put into place measures like the Greenbelt Plan, the Places to Grow Act, and the Growth Plan for the Greater Golden Horseshoe Area, but these efforts fall far short of a comprehensive land-use plan to regulate urban sprawl and to convince the public to switch from the automobile to public transportation.

In Ontario, most air pollution comes from automobile exhaust, followed by coal-burning thermoelectric plants. While the number of vehicles continues to rise, the good news is that more efficient engines have limited the increase in exhaust fumes. While this is a step in the right direction, until Canadians' love affair with the automobile ends, little relief from automobile and truck pollution can be expected. On the other hand, Ontario is committed to closing by 2014 its five coal-burning plants, which have contributed 16 per cent of Ontario's electrical needs, and replacing the energy they have produced through other, less polluting sources. Progress in moving away from coal has been slow for two reasons. First, coal plants provide the cheapest form of electrical generation. Second, provincial and then local approval takes time before construction of alternative generating plants begins—and political backtracking can take place. For instance, in 2010 TransCanada completed a gas-fired plant at Halton (just north of Oakville) but its proposed plant for Oakville was cancelled due to local opposition.[1] Since 2003, the Lakeview generating station near Toronto, two units at Nanticoke, and a similar number at Lambton have closed, and the Thunder Bay coal plant in northwestern Ontario was to be converted to natural gas but those plans are in limbo. By 2014, the conversion of the Atikokan coal plant into a biomass-fuelled plant is expected to be completed. In sum, Ontario remains committed to ending the production of coal-produced electrical power, perhaps as soon as December 2013 (Spears, 2013).

Under the Ontario government's plan, additional nuclear plants will provide more and more electrical power. Yet construction time could take between five and 10 years, which means that new nuclear electric power may not be available until 2020 or later. With the removal of coal-produced electricity by 2014, the Ontario plan calls for natural gas to make up most of the shortfall. To date, Ontario's nuclear plants have not experienced major problems, but the risk of a serious accident remains a possibility. Then, too, the issue of the disposal of nuclear waste remains an environmental concern. At the same time, from a financial point of view, who is to say that past cost overruns that have saddled Ontario with a massive debt will not be repeated in the planned expansion of its nuclear industry? Here, the consumer pays, as the Ontario government in 2012 authorized an additional 7 cents/kWh to every customer's monthly electric bill to pay down the debt incurred by past cost overruns related to the province's nuclear plants.[2] Alternative energy sources, such as wind farms, have begun to feed into Ontario's electrical grid, but these provide a minuscule amount of total energy and at a higher cost than conventionally produced electricity.

Although the plan calls for the end to coal-produced electricity by 2014, this date might not be met if construction delays occur on the replacement sources—and construction will take time. Here are three examples:

- Two natural-gas generating stations that will replace the electrical output from the Lambton coal plant are not expected to be operative until 2014.
- The completion date is now 2014 for the Niagara Tunnel Project, a huge undertaking that will divert water from the Niagara River downstream to the Sir Adam Beck Generating Station via a tunnel, bored by an Austrian firm, underneath the city of Niagara Falls. The tunnel is over 14 m wide and over 10 km in length.
- The approval of an expansion of the Darlington nuclear plant was made in July 2012 but planning and construction may take a decade to complete. The end result will see this new facility push nuclear power's share of electrical output more than 60 per cent.

Water pollution constitutes a serious problem across Ontario. Industrial effluents, farm chemicals, and livestock waste are the main culprits. At the local level, in May 2000, the contamination of drinking water with E. coli at Walkerton in southwestern Ontario provided a deadly example of farm-waste hazards with seven deaths and many illnesses. At the regional level, the integrity of the Great Lakes has been compromised by human activities. Over 35 million people, many of whom live in large urban cities such as Chicago, Detroit, Cleveland, Toronto, and Montréal, depend on the Great Lakes/St Lawrence for water supply and sewage disposal. The area surrounding Lake Michigan, Lake Erie, and Lake Ontario contains many industries and factories that, in the past, dumped their toxic wastes into the Great Lakes and buried their chemical wastes along the shores. Later, these chemical wastes

CONTESTED TERRAIN 5.2

Cleaner Air versus Higher Electrical Costs

The good news in Ontario's energy plan is that it leads to less pollution and a healthier environment, especially in the heavily population Golden Triangle of southern Ontario. As stated earlier, by reducing air pollution, the hopeful outcome is fewer respiratory problems and lower health costs. The bad news is that, as Ontario closes its low-cost coal plants and employs other forms of energy production, higher electrical costs will result. Not only will the consumer pay more in the years to come, but the already endangered Ontario's manufacturing industry will face another challenge. As Adam White, president of the Association of Major Power Consumers of Ontario, stated (McCarthy, 2013):

"Ontario has the highest delivered prices for industry in North America, so we need to have a reality check." Industrial customers now pay roughly $85 per megawatt hour, given all the various components of the price, compared to roughly $40 in Quebec, Manitoba, and Michigan.

Photo 5.4
The Niagara Tunnel is nearly 15 metres high and 10.2 km long. The tunnel boring machine, nicknamed "Big Becky," was assembled on site. Located deep in the bedrock beneath the city of Niagara Falls, this huge tunnel diverts water from the Niagara River, sending it downstream to the Sir Adam Beck Generating Station where it produces clean electrical energy. In this photo the tunnel is nearing completion; the tunnel is lined with a concrete-like substance, called "shotcrete," that is sprayed on the interior tunnel wall robotically; the yellow tube is a tunnel ventilation duct; the movable platform structure allows workers to access the tunnel roof.

began to leak into the lakes. Measures were taken to regulate these polluters, but pollution from the runoff from agricultural lands, the waste from cities, and toxic discharges from industry continue to affect the Great Lakes.

Canada and the United States have recognized the need to regulate the water levels and the discharge of water and toxic wastes into the Great Lakes. In spite of the fact that the hydrological cycle can increase the volume of water in the Great Lakes over a short time span, the long-term trend has seen the water levels of the Great Lakes slowly declining due primarily to the warming climate (International Great Lakes Study, 2012), with the water levels dropping over the last 50 years. This decline translates into retreating shorelines. In the case of Lakes Huron and Michigan, which are, in reality, one lake, an increase in water level would benefit Lake Huron, which faces stranded cottage docks, the loss of wetlands, and receding shorelines, but at the same time this would threaten the exposed sandy beaches and dunes along the shores of Lake Michigan with water erosion.

Toxic waters represent another challenge. The Great Lakes Water Quality Agreement, first signed in 1972 and renewed in 1978, expresses the commitment of the two countries to restore and maintain the chemical, physical, and biological integrity of the Great Lakes Basin (see Figure 5.5). By the beginning of the 1990s, cleaner water and increasing fish populations indicated that these joint efforts were making solid progress. However, industrial discharges of pollutants remain a problem. Growing levels of phosphorus, primarily from waste water containing detergents and from agricultural runoff containing high levels of chemical fertilizers, are contributing to the creation of a "green slime": a dead zone where only toxic organisms can survive. In addition, the careless introduction of exotic species is resulting in sea lampreys, Asian carp, and zebra mussels squeezing out native species and thereby radically changing Great Lakes ecosystems. Ocean-going ships have carried unwanted passengers such as zebra mussels and sea lampreys into the Great Lakes via the St Lawrence Seaway. Both have had a negative impact on domestic fisheries.

 Under "Environmental Challenges" in Chapter 6, page 228, the impact of the zebra mussel on the St Lawrence River and Great Lakes is presented.

Ontario's Historical Geography

When the American Revolution began in 1775, Ontario—except for the French settlement around Detroit (established 1701)—was a densely forested wilderness inhabited by a few fur traders and Indians. By 1750, Detroit had a French population of 600, of whom 150 lived across the Detroit River at Petite Côte. In 1760, British troops took control of Fort Detroit and the surrounding settled area.

After losing the American colonies, loyal British subjects returned to England or relocated in one of Britain's colonies. In 1782–3, Loyalists moved north to Nova Scotia and New Brunswick, while others resettled in the Eastern Townships of Québec. A smaller number, perhaps as many as 10,000 Loyalists and members of the Six Nations who had fought alongside British troops, settled on land along the St Lawrence River that was upstream from French-speaking habitants, thus colonizing the wilderness area that later became known as Upper (relative to the mouth of the St Lawrence) Canada. In 1784, Britain rewarded its Indian military allies, the Six Nations, with the Haldimand Tract, which consisted of a strip of land on each side of the Grand River totalling 385,000 hectares (Darling, 2007). These Loyalists were followed by American, British, and European newcomers, who flooded into Upper Canada over the next half-century in search of land. In less than 100 years, the natural forest landscape was transformed into a British agricultural colony.

Until the War of 1812, many settlers had come from the United States. They, like most other colonists, sought land in the fertile lowlands of the Great Lakes. When hostilities between Britain and the United States escalated, the Americans launched an attack on British North America. The War of 1812 effectively ended the influx of American settlers into Upper Canada. After the war ended with the signing in December 1814 of a treaty of stalemate at Ghent, an increasing number of settlers came from the British Isles, especially Ireland and Scotland. By 1851, Canada West (as Upper Canada had become with the Act of Union of 1841) had reached a population of 952,004, 86 per cent of whom lived in rural settings (Statistics Canada, 2007a). By Confederation, virtually all the

Figure 5.5 **The Great Lakes Basin**
Created 10,000 years ago at the end of the last continental glaciation, the Great Lakes form the largest freshwater system on the planet. Except for the state of Michigan, much of the Great Lakes Basin is located in Ontario. The watersheds of the Mississippi River and Ohio River in the United States are just south and west of the Great Lakes and form part of the Gulf of Mexico Basin. *Source:* Based on *Atlas of Canada,* 2004, "Great Lakes Basin," at: <atlas.nrcan.gc.ca/site/english/maps/reference/provincesterrritories/gr_lks/index.html>.

arable land in the Great Lakes Lowland had been cleared of forest by settlers. With a severe land shortage, some turned further north to try their luck in the few pockets of arable land within the Canadian Shield, but few were successful. Others migrated to towns and cities and were absorbed into the urban workforce, and still others moved west to what would soon become Manitoba. Many more took their chances in the last great American land rush, west of the Mississippi River. By 1867, the demographic balance of power remained in the rural community—urban areas, such as villages and towns, contained less than 20 per cent of the population of Ontario.

When Canada West joined Confederation in 1867, it was renamed Ontario (Figure 3.4). At that time, the geographic extent of Ontario was about 100,000 km² — a fraction of its present size — but as Canada acquired more territory from Great Britain, Ontario and Québec obtained some of these new lands (Figures 3.5–3.7). They had little immediate value for economic development and settlement, however, because they were carved from two physiographic regions (the Canadian Shield and the Hudson Bay Lowlands) that were far from markets and had little or no agricultural potential. Since Confederation, the borders of Ontario have been extended three times, greatly increasing the geographic size of the province, but not its agricultural lands. The first expansion occurred in 1874 when Ontario's boundaries were pushed northward to about 51° N and westward towards the Lake of the Woods. Ontario's second expansion, in 1889, ended the bitter contest between Manitoba and Ontario for the land around the Lake of the Woods. At the same time, Ontario's northwest boundary was adjusted to the Albany River, which flows into James Bay, gaining Ontario access to James Bay. In 1912, the final boundary modification occurred when the District of Keewatin south of 60° N was assigned to Ontario and Manitoba. As a result, Ontario extended its political boundary to the northwest, stretching from Manitoba to Hudson Bay at the latitude of 56°51′ N. Through these boundary adjustments, Ontario reached its present geographic extent of 1 million km².

The Birth of an Industrial Core

The economic essence of Ontario is its industrial base. Transportation routes played a key role, especially the Welland Canal, which facilitated low-cost transportation (Vignette 5.6). Manufacturing in southern Ontario (then Upper Canada) began in the nineteenth century and received a boost in 1854, when Britain and the United States signed the Reciprocity Treaty, which allowed trade between the British North American colonies and the United States. By 1867, Ontario had the largest population of Canada's four founding provinces and a fledgling industrial base. Small manufacturing outfits in small villages often consisted of less than five employees (e.g., a village blacksmith; a miller). Larger manufacturing took place in towns and cities, where sawmill, gristmill, and distillery

VIGNETTE 5.6

The Welland Canal

The Welland Canal connects Lake Ontario and Lake Erie, allowing ocean-going ships to enter the heart of North America. To avoid the Niagara River and its huge falls, the first canal-builders faced the daunting task of constructing a canal across the Niagara Peninsula, a distance of some 44 km. The first canal, opened in 1829, was dug by hand. A series of locks made from hand-hewn timbers connected a series of creeks and lakes. As the size of ships increased, the original canal proved inadequate and a new canal was built in 1845. Within 40 years, even larger ships required a third renovation, which was opened in 1887. The present canal was completed in 1932. In 1973, to bypass the city of Welland, a new channel was constructed, for which a series of lift locks were needed to overcome a difference in elevation of nearly 100 metres between Lake Ontario and Lake Erie. The Welland Canal has been part of the St Lawrence Seaway since 1959 and is operated by the St Lawrence Seaway Management Corporation.

Photo 5.5

The Welland Canal is a strategic link between Lake Ontario and Lake Erie that provides a water route around Niagara Falls. To accommodate ever-increasing traffic, the lock system was divided into two at several places, as shown in this aerial photograph. In 2006, approximately 40 million tonnes of goods passed through these locks. The three leading products were grain, iron ore, and coal.

operations were commonplace. At that time, most manufacturing activities depended on water power, so most industries were located near a stream or river.

The Americans allowed the Reciprocity Treaty to lapse in 1866, cutting off access to the American market for Canada West. However, Confederation and the 1879 National Policy of Prime Minister Macdonald enabled what was now Ontario to secure the Canadian market. Under the National Policy high tariffs were imposed on imported manufactured goods, which allowed manufacturing in southern Ontario to flourish. The consequences of Confederation and protective tariffs for Ontario were threefold: (1) the creation of a Canadian market for Ontario products; (2) an increase in the size of the more successful manufacturing companies; and (3) the growth of the industrial workforce in Ontario. In the rest of the country, however, prices for manufactured goods (made in Ontario) were generally higher than in the adjacent areas of the United States. The regional price variations reflected two factors: Canadian transportation costs and Ontario's more limited economies of scale compared to those of US manufacturers.

The National Policy of high tariffs provided the impetus for a national industrial core in southern Ontario, while the rest of the country—except for Québec—was relegated to a domestic market for these manufactured goods. This economic arrangement made necessary the east–west transportation axis and, in doing so, favoured Ontario and Québec, thus laying the groundwork for western alienation and Maritime dissatisfaction with Confederation. Not until the Auto Pact of 1965 and then the 1989 Free Trade Agreement did this national core/periphery relationship undergo significant change and a north–south transportation axis emerge as a prominent aspect of Canada's manufacturing trade.

See Table 1.5, "Changing Trade Direction: From Continental to Global Trade," page 19, for more on the Auto Pact and the FTA

By the early twentieth century, four factors had led to successful manufacturing in southern Ontario, which can be illustrated by the automobile industry. The first factor was Ontario's

Photo 5.6

In 1794, York became the capital of Upper Canada. Despite its political status, this frontier village remained on the western edge of British settlement that stretched westward from Lower Canada along the north shore of Lake Ontario. By 1812, York had only 700 residents. This painting (dated 1804) illustrates a group of houses strung along the shore of Toronto Bay. Beyond this narrow strip of cleared land lies the original forest of southern Ontario.

geographic advantage: close proximity to America's manufacturing belt led American industries to locate branch plants in Canada early in the twentieth century. This spillover effect first took place in the Windsor–Detroit area in 1904, when the Ford Motor Company established an automobile assembly plant in Windsor. The second factor was trade restriction on foreign manufactured goods (imposed by the National Policy in 1879). Automobile parts from Ford's plant in Detroit had to be ferried across the Detroit River and assembled in an old carriage factory in Windsor. The third advantage was access by American branch plants in Canada to the lower tariffs for Canadian-made products in the British Empire. As a result, Canadian-built Ford cars were sold not only in Canada but in various places in the British Empire. General Motors followed the same strategy by opening a branch plant in Oshawa. The size of its domestic market provided southern Ontario with its fourth location advantage for the automobile industry: most GM cars were sold in southern Ontario, thus minimizing transportation costs.

With the liberalization of trade, manufacturing in Canada escalated. First came the Auto Pact of 1965. The Big Three automobile companies in the early 1960s had proposed a common market between Canada and the US for the production and marketing of their products. In the year Ottawa and Washington passed the appropriate legislation, Canada produced nearly 900,000 vehicles. By 2000, the figure had reached 2.5 million. Until 2008, annual production hovered around 2.5 million cars and trucks, most of which were exported to the US market. At this level of production, Canada was accounting for 14 per cent of North America's vehicle output. Ontario's automotive assembly and parts firms remained the anchor of manufacturing in the province, employing about 150,000 people. These well-paid workers produce high-quality motor vehicles that account for 12 per cent of the country's **gross domestic product** (GDP).

The second step took place in 1989 when the Canada–US Free Trade Agreement brought other components of the economy under a single North American market, thus removing tariffs

Photo 5.7
"View of King Street [Toronto], Looking East" (1835) by Thomas Young. At the time of this painting, York had just been renamed Toronto and had a population of nearly 10,000.

(some high) between the two countries. Since American firms had already benefited from economies of scale, their costs of production were considerably lower than those of similar Canadian firms. As well, the American firms were well established in the marketplace and dislodging them was a daunting task. Canadian-based manufacturing firms were forced to play catch-up and many did not survive. American firms under stress often closed their smaller, less efficient branch plants and served their Canadian customers from their American factories. Some Canadian firms merged to attain a size deemed necessary to compete in the North American and global markets and to avoid takeovers by large foreign companies. Mergers, however, resulted in the closing of redundant offices, thus reducing the number of employees.

The golden days of Ontario's manufacturing industry ended with the financial crisis of 2008–9. As a result, Ontario's manufacturing industry and its workers are caught in a global price/wages squeeze. This squeeze is particularly hurtful to those in the automobile industry. These issues and challenges are discussed later in this chapter under "Key Topic: The Automobile Industry."

Aboriginal Territory within Ontario

Ontario has 126 First Nations holding Aboriginal territory, known as reserves, that was determined through treaty negotiations between government and individual tribes. Most were classified as "unnumbered" and took place before 1923. Three numbered treaties—3, 5, and 9—cover northwestern and northern Ontario. The first land grants to Indians took place during the days of British North America, beginning with the Haldimand Proclamation of 1784, which assigned land along the Grand River to the Iroquois who fought alongside the British in the American Revolution (Figure 5.6). In 1850, the two Robinson treaties were completed; they covered large areas east and north of Lake Huron and north of Lake Superior.

Figure 3.10, "Historic treaties," page 98, shows the location of the numbered and unnumbered treaties that blanket Ontario.

Photo 5.8
With an ideal climate and proximity to large urban markets, dairy farming dominates the agricultural landscape of much of Ontario.

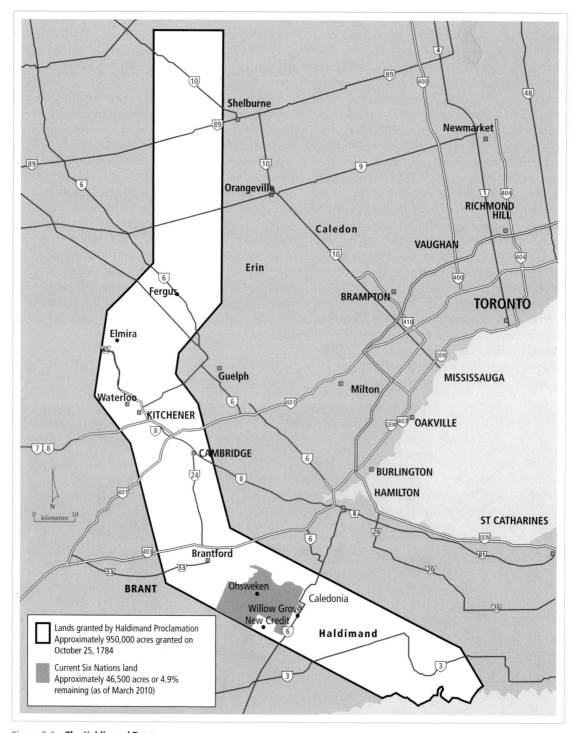

Figure 5.6 **The Haldimand Tract**
Source: Six Nations Land Resources. Based on map at: <www.sixnations.ca/LandsResources/HaldProc.htm>.

In Ontario, Indians were granted land hundreds of years ago. When disputes arise today, reaching an agreement is challenging because the "facts" are buried in time (see Vignette 5.7). The slow pace of resolution is frustrating to Native Canadians and two land disputes (Ipperwash and Caledonia) have led to violent protests by Indians from the Kettle and Stony Point and Six Nations reserves. In the Ipperwash dispute, the facts are relatively clear: land from Stony Point was taken in 1942 to serve as a military training camp, named Camp Ipperwash. After the war, the land was supposed to have been returned but the Department of National Defence decided to keep the camp—adjoining Ipperwash Provincial Park, established in 1936—to train cadets. Promises from Ottawa to return the land were not fulfilled. By 1993, the Stony Point Indians were utterly frustrated with the repeated failure of Ottawa to act on its promises and in September 1995 Indian protesters moved into Ipperwash Provincial Park, the alleged site of sacred burial grounds, where a confrontation with the Ontario Provincial Police took place, ending in the shooting of an unarmed Native protester, Dudley George, by an Ontario police sniper. In 2003, the Ontario government asked Justice Sidney Linden to conduct a public inquiry into the circumstances surrounding the 1995 death of Dudley George, including the role of Premier Mike Harris. In May 2007, Justice Linden issued the Ipperwash Inquiry Report, in which he concluded that Harris had not explicitly ordered the Ontario police into the Ipperwash Provincial Park to remove the Indian protesters. At the same time, Justice Linden called for the immediate return of Camp Ipperwash to the Kettle and Stony Point First Nation. Jim Prentice, the federal Minister for Indian Affairs and Northern Development, responded by stating that, "We'll do something immediately" (*National Post*, 2007). Camp Ipperwash was returned to the Chippewas of Kettle and Stony Point in 2007, and in 2010 the process began of signing over to the Chippewas the 56-hectare site of the provincial park, which has been closed since 1995. This is complicated by the fact that Ottawa determines reserve lands and has not acted, and currently the former park is under the control of the Kettle and Stony Point First Nation on a basis of co-management with the province and in consultation with the surrounding community. Ownership of the land remains in limbo.

Specific land claims by Native groups are not usually as straightforward as the Ipperwash claim—and even that claim is unresolved. The Caledonia dispute exemplifies the complexity of some claims. An outline of the historical evolution of the Six Nations claim to a 40-hectare parcel owned by a land developer at Caledonia, Ontario, near Hamilton, suggests why, in many instances, settlements have been achieved at such a slow rate (Vignette 5.7). The basis of the Six Nations claim goes back to the original Haldimand Grant of 1784 and land surrenders in the eighteenth and nineteenth centuries. History is not clear on these issues. While the protest finally ended, negotiations between the Six Nations and the federal government are at an impasse since the federal government, in 2009, rejected the $500 million claim of the Haudenosaunee/Six Nations, leaving its offer of $125 million on the table.

Ontario Today

Canada's centre of gravity—as measured by economic performance and population size—continues to shift westward. Yet, the sheer size of Ontario's economy and population keeps that centre anchored in Ontario. This westward shift is due more to the weak performance of Québec and Atlantic Canada, though Ontario's post-2008 performance—as indicated by its 2007, 2009, and 2012 unemployment rates (Table 5.1)—also made a contribution to this shift. Yet, long-term statistical indicators of the powerful position of Ontario within Confederation include:

- Largest economy and population of the six regions.
- Average personal income well above the national average.
- Greatest cluster of major cities, universities, and technological centres of any region.
- Elects more members of Parliament than any other region.
- Central location within North America further facilitated by its hub position in the east–west and north–south transportation systems.

While Ontario enjoys its paramount position within the nation, the road has recently become bumpy and the future less certain. Ontario needs a vision and investment strategy based on the so-called knowledge-based economy. The hope is

THINK ABOUT IT
What is a fair price for land? Canada's offer of $125 million as a financial settlement for the four outstanding Six Nations claims falls short of the $500 million counter-offer by the Six Nations. Should the two sides simply split the difference and move on?

that innovation and technology will move developed regions, like Ontario, to a more sophisticated information society where cost of labour is less critical to determining the success of businesses and where exports can penetrate global markets and, in the case of Canada, supply the rapidly expanding resource sector. Labour is critical, however, because bright workers are the key to the knowledge-based industry. Indeed, if Richard Florida (2002 and 2012) is correct in thinking that the creative class (which includes the so-called knowledge-based workers) want to live in "interesting" cities, then that relationship reinforces the concentration of these firms in such urban places. Because of the inventive nature of this type of economy, small as well as large firms can be engaged in the search for new and better ways of production. However, firms with a global reach, while vulnerable to the fierce global competition, represent the principal drivers of innovation and they have important spinoff effects for the surrounding scientific community. For example, because these global "technology" firms are at the cutting edge of world technology, they push local businesses and university researchers to higher levels of "applied" scientific investigations, which leads to new forms of technology. Unfortunately, the loss of Nortel and the recent loss of market share by Research In Motion (RIM, renamed BlackBerry in early 2013) has had

VIGNETTE 5.7

Historical Timeline of the Caledonia Dispute

Eighteenth Century

1784

The British Crown allows the Six Nations (Iroquois Confederacy) to "take possession of and settle" a strip of land nearly 20 kilometres wide along the Grand River, from its source to Lake Erie, totalling about 385,000 hectares; called the Haldimand Grant, it is named after the governor of Québec, Frederick Haldimand, who previously had negotiated the purchase from the Mississauga Indians of over 1.2 million hectares on the Niagara Peninsula, for a little over £1,000. Haldimand's intent in this purchase was to provide land for settlement to loyal Iroquois who had fought beside the British in the American Revolution.

1792

Upper Canada's lieutenant-governor, John Graves Simcoe, reduces the grant to the Six Nations by two-thirds, to 111,000 hectares.

1796

The Six Nations Confederacy grants its chief, Joseph Brant, the power of attorney to sell some of the land and invest the proceeds. The Crown opposes the sales but eventually concedes.

Nineteenth Century

1840

The government of Upper Canada recommends that a reserve of 8,000 hectares be established on the south side of the Grand River and the rest sold or leased. Six Nations council agrees to surrender for sale all lands outside those set aside for a reserve, on the agreement the government would sell the land and invest the money for them. A faction of Six Nations members petitions against the surrender, saying the chiefs were deceived and intimidated. Six Nations would challenge the 1840 surrender in a 1995 lawsuit and it is part of the basis for the current protest.

1850

The Crown passes a proclamation setting out the extent of reserve lands on both sides of the Grand River—about 19,000 hectares agreed to by the Six Nations chiefs.

the opposite effect. Nevertheless, Ontario, with a majority of Canada's creative wealth as reflected in its universities and provincial/federal research facilities, is well placed to lead; interestingly, Québec devotes more public funds to its research efforts.

 In "The Information Society," Chapter 4, page 160, Florida's theory is discussed within the broader context of the information/knowledge society and his "creativity" thesis is examined in Further Reading in this chapter.

Florida's vision of the creative class pushing Ontario into the new economy is perhaps best reflected in two sectors. First, the entertainment industry is flourishing. As the centre of the Canada's music and film industry, Toronto's reach goes beyond Canada (Florida and Jackson, 2010; Florida et al., 2010, Hracs et al., 2011). The best example is the Toronto International Film Festival, which has rocketed Toronto onto the world stage. Another is the Pride Parade that draws tourists and participants from around the world. Second, Ontario universities play a key role in fostering collaborative innovations, while research-connected companies supply the applied component. Ryerson University's Digital Media Zone, a high-tech incubator for students, provides one example while the much older "cluster" of knowledge-based companies, the University of Waterloo, and public research institutions provides another example. This raises the

Twentieth Century

1992

Henco Industries Ltd purchases 40 hectares of land near Caledonia and names it the Douglas Creek Estates.

1995

The Six Nations disputes the ownership of this Crown land.

Twenty-First Century

2005–6

Henco Industries' subdivision plan for Douglas Creek Estates is registered with the province of Ontario. The following year, a group of Six Nations members occupies the housing project, erecting tents, a teepee, and a wooden building. Court challenges and non-Native counter-protests take place. Several hundred non-Native demonstrators gather in Caledonia to protest occupation and what they call police inaction. Dozens of police officers form lines between Native and non-Native protestors. Three people arrested. In a search for a compromise, the Ontario government buys out Henco's interest in the disputed property for $15.7 million, thus maintaining Crown ownership of this disputed land.

2007

The federal government enters negotiations with the Six Nations to resolve the historic and current land claim disputes. Canada makes an offer of $125 million to compensate the Six Nations for four outstanding historic claims based on nineteenth-century land surrenders known as Grand River Navigation Company investment; Block 5 (Moulton Township); Welland Canal flooding; and the Burtch Tract. As well, Ottawa compensates Ontario for $26.4 million for the province's costs incurred as a result of the occupation near Caledonia and the province's purchase of the land.

2008 to present

Canada receives a formal counter-offer of $500 million from the Six Nations in 2008. The following year, Canada rejects the claim for $500 million and restates its offer of $125 million. As a result of the inability of the two parties to breach this gap, negotiations ceased.

Adapted from CBC News (2006a, 2006b); INAC (2009).

question: Is Massachusetts Institute of Technology recognition of Hossein Rahnama, associate director of Ryerson's Digital Media Zone, and of Joyce Poon of the University of Toronto as "rising stars in a new international high-tech ranking" a reflection of Toronto's growing creative class or simply a one-off event (Church, 2012: A5).

Ontario Advantage: Trade with the United States

Ontario is geographically positioned to engage in trade, both domestically and internationally. Yet, over 80 per cent of Ontario's exports go to the United States. The importance of the US market to Ontario firms is critical. Just how critical is revealed in the recent drop in Ontario manufacturing exports to the United States. For nearly five years, the United States economy has sputtered and, as a result, Ontario exports to its neighbour have also slipped. As Stephen Beatty (2009), managing director of Toyota Canada, stated:

> I want to emphasize, though, that we could not do this—the business case would not exist for assembly plants in Canada—if we did not have open access to the North American market. Even though our Ontario plants build such a high proportion of vehicles destined for Canadian driveways, the vehicles we export are crucial to supporting our investments here in Canada.

Production of forest, mineral, and other products in northern Ontario follows the same pattern: production is far greater than the Canadian market can absorb. The advantage has to do with costs of production. At a high level of output, economies of scale are achieved, keeping the cost per unit of output low. Forestry companies, like automakers, sell much of their lumber, pulp, paper, and other products to the US market. In southern Ontario, manufactured goods ranging from aircraft to steel products are produced for both these markets. **Restructuring** of Ontario's manufacturing industry took place after 1989 when the FTA was signed. Canadian firms, especially those with high labour costs in their manufacturing process, were faced with a highly

VIGNETTE 5.8

High-Tech Stars and the Creative Class

Canada is used to the "brain drain," but in the past this meant leading scholars and innovators immigrating to the US. Perhaps this pattern is changing. For example, Professors Hossein Rahnama and Joyce Poon are two of 35 young innovators named by MIT's *Technology Review* for tackling research challenges in fresh, unconventional, and innovative ways. Both came to Toronto as children of immigrants and both completed their undergraduate education where they now are faculty members—Ryerson University and University of Toronto, respectively. Both studied abroad and then decided to return to Toronto "as the best place to conduct their cutting-edge research, turning down the chance to work in other high-tech hubs, including Silicon Valley."

For years, the University of Waterloo has focused on entrepreneurship and innovation. The Conrad Business Entrepreneurship and Technology Centre works closely with local technology firms. Anchored by BlackBerry, Canada's leading high-technology company, the Waterloo Region has one of the most diverse economies in Canada, with strengths in advanced manufacturing and communications technology. While BlackBerry has hit a rough patch in the global technological world, this advanced technology cluster consists of much more than one firm and boasts a healthy mix of small-, medium-, and large-sized innovative firms that often leads to joint innovation efforts and commercial breakthroughs. Public funds often kick-start projects. For example, Communitech, a digital media hub in Waterloo that brings multinational companies, academic institutions, and entrepreneurs together to create digital media companies with a global reach, received $26.4 million from the province of Ontario in 2009 for its $107 million digital media strategy. In just two years, high-tech jobs, innovations, and numerous new companies have flowed from this strategy and the future for Communitech Hub and its innovators looks bright (Chartrand, 2012).

competitive market that often resulted in plant closure and/or relocation to the US or Mexico.

The FTA and then NAFTA, along with Ontario's geographic location within North America, have allowed the region's business firms to penetrate the huge US market and that has led to greater integration into the North American economy. For example, Ontario's exports to the United States in the 1980s were of about the same value as those to the rest of Canada; but by 1998 Ontario's exports to the US soared to two-and-a-half times the value of exports to the rest of Canada (Courchene with Telmer, 1998: 276–7). Driven by automobile shipments to the US market, this trade gap continued to widen in the first years of the twenty-first century as the sheer size of the US market has tilted the province's trade in that direction.

Once the Free Trade Agreement was in effect, southern Ontario saw itself and its future as a northern extension of the American manufacturing belt. Trade with the rest of Canada was viewed more as an add-on rather than as a core element. The reason was a matter of market size and easy access. When the Canada–US Auto Pact came into play in the mid-1960s, it was designed to integrate Canada's automobile industry into the North American market. At that time, over 90 per cent of Canadian automobile production was controlled by the Canadian branch plants of the Big Three automobile companies. These companies wanted to rationalize their production and marketing of automobiles within North America to be prepared for competition from foreign producers. While the Auto Pact ended in 2001 as a consequence of a WTO ruling, by then this Canada–US agreement not only had altered the nature and purpose of automobile manufacturing in Ontario, but it also had created a shift in federal policy towards more liberal trade arrangements and towards economic continentalism in North America. Automobile trade is a critical factor in bilateral Canada–US trade and accounts for 30 per cent of Canada's trade with the US. Economic hard times, however, can alter this trade significantly. Canadian auto sales (mainly to the US market) were $61 billion in 2007 and $47 billion in 2008, a drop of 22 per cent (Kowaluk and Larmour, 2009: Table 1).

Ontario's Economy

Across Canada, regional economies vary. Ontario is seen as the manufacturing and financial centre of Canada. More precisely, an economy contains an industrial structure. Economists measure this structure in several ways, but the most common method is by the number of workers in an industrial sector. Each sector describes a particular set of economic activities, such as agriculture, forestry, manufacturing, education, and trade. To simplify matters, these sectors are grouped together into three broad categories: primary, secondary, and tertiary activities.

THINK ABOUT IT
Should Ontario's manufacturers look to Western Canada for increasing sales of their manufactured goods or wait for a rebounding US economy?

VIGNETTE 5.9

Importance of Central Location to Ontario

The role of proximity cannot be underestimated. For example, trucks can deliver Ontario goods to firms in Michigan, Ohio, Pennsylvania, and New York in a matter of hours, allowing a just-in-time production system to flourish on both sides of the border. Backups at the major border crossings—especially between Windsor and Detroit—caused by heightened concerns about security in the US have become a major problem for "just-in-time" delivery. With much of Canada's $1.7 billion worth of daily exports to the US coming from Ontario, hampering the flow of people and goods across the border to satisfy Washington's concerns about security and terrorism has a negative impact on Ontario—the only question is "how much?" The answer to that question is still unknown, largely because of the difficulty of collecting information on the cost. A 2004 report to the Ontario Chamber of Commerce estimated the slowdown at the border cost Ontario $5.25 billion the previous year. More recently, Len Crispino, chief executive of the Ontario Chamber of Commerce, stated: "That border now has become more of a choke point, rather than a conduit for trade" (French, 2007). He estimated that Canada loses as much as Cdn$8 billion every year to border delays.

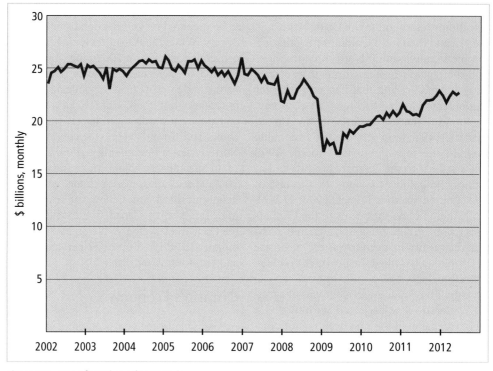

Figure 5.7 Manufacturing sales: Ontario
Ontario's manufacturing sector is slowly climbing out of the 2008–9 economic hole. However, until the US economy recovers, Ontario's manufacturing robustness is unlikely to return to pre-2008 levels. *Source:* Royal Bank of Canada (2012) and Statistics Canada.

Economic structure is discussed in Chapter 4. In particular, see Table 4.14, "Traditional View of Economic Structure," page 159, and Table 4.15, "Sectoral Changes in Canada's Labour Force, 1881–2011," page 160.

Ontario accounts for almost half of all manufacturing jobs in Canada. Several factors explain this concentration but paramount is the availability of a skilled and hard-working labour force, which is essential to producing quality products. Since the FTA, the face of manufacturing has changed, and so has southern Ontario's mix of manufacturing activities. Textile firms and other manufacturers that required semi-skilled labour working at or near the minimum wage have largely disappeared. New to the scene are high-technology companies that depend on highly skilled workers who can command high wages. Ontario is at the leading edge for Canada's high-technology industry (Vignette 5.8).

Globalization has caused major problems for Ontario's economy. In addition, the weak US housing market has crippled the forest industry while the high Canadian "petro-dollar" has added to the woes of Ontario manufacturers. Coupled with this downturn in manufacturing, many workers—those still employed in the secondary sector—face wage and benefit cuts. Unionized workers at the now-closed Caterpillar plant in London were asked to accept a previously unheard of 50 per cent cut in their wages. Even unionized workers may see their wages reduced in the next round of negotiations or, as the automobile companies want, have wages based on profit-sharing instead of annual percentage wage increases (Keenan, 2012: B3).

Until 2008, Ontario's economic structure was changing relatively slowly, though its manufacturing and forestry industries were losing plants, mills and, worst of all, employees. Manufacturing sales remained steady, however, until this time (Figure 5.7). This process resulted in the tertiary sector gradually but steadily gaining ground and the primary and secondary sectors losing their share of the labour market. The catalyst that accelerated this decline was the global economic crisis, which hit Ontario's manufacturing sector particularly hard. While the total number of workers in Ontario increased from 6.4 million to 6.7 million from 2005 to 2011, only the tertiary

Table 5.2 **Ontario Employment by Industrial Sector, Number of Workers, 2005 and 2011**

Industrial Sector	Workers, 2005 (000s)	Workers, 2011 (000s)	Difference
Primary	128	129	1
Secondary	1,509	1,292	−217
Tertiary	4,761	5,310	549
Total	6,398	6,731	333

Source: Statistics Canada (2006b, 2009c, 2012e).

sector recorded a significant increase in the number of workers (Table 5.2). By 2011, the tertiary sector had 79 per cent of the workforce, up 4.5 percentage points from 2005 (Table 5.3). Most job losses in the secondary sector were associated with manufacturing, especially in automobile manufacturing and parts production. The figures in Tables 5.2 and 5.3 demonstrate the profound shift in Ontario's economy and its labour force structure. Yet, resilience can be found in the manufacturing sector. One example is Russel Metals, a Mississauga-based firm that barely weathered the 2008–9 downturn, but its approach of acquiring local businesses and delegating authority to make decisions quickly worked well, particularly in the Alberta oil patch (Preville, 2012: B8). In this way, an Ontario firm kept its sales growing by tapping into the booming Alberta market.

Energy Crisis

Ontario faces a challenging energy situation. Pollution is a problem and green energy solutions are costly. Even though Ontario produces much power from hydroelectric installations at Niagara Falls and along the Ottawa River, from four thermal coal-fired plants using coal from Pennsylvania and West Virginia, and from nuclear generators in plants at Pickering and Darlington along Lake Ontario east of Toronto and at Tiverton on the Bruce Peninsula, the province faces both the threat of an energy shortage and higher energy prices. To meet that demand, additional electricity is imported from Québec and produced within Ontario by natural gas from Alberta. While Manitoba has the capacity to produce more low-cost power along its Nelson River, its cost delivered to southern Ontario is around 10 cents per kilowatt hour, about double the current price to residential consumers. Still, what is Ontario to do? With the record high prices of coal and uranium expected to continue into the next decade, Ontario faces both energy shortage and higher energy costs. Green energy provides one alternative, but an expensive one; natural gas has relatively low prices now, but will they stay low?

The provincial government turned to green energy with its 2009 **Green Energy Act**, which was designed to stimulate green construction projects and to produce equipment for these anticipated projects. Ontario hoped to become a global leader in clean, renewable energy and conservation, and in so doing to create thousands of jobs, economic prosperity, energy security, and climate protection. The intent also, of course, is to export the

THINK ABOUT IT
Does the shift of workers to the tertiary sector fit with Florida's thesis about the rise of the creative class or does it say more about the slide into low-wage service jobs?

Table 5.3 **Ontario Employment by Industrial Sector, 2005 and 2011 (%)**

Industrial Sector	Workforce, 2005	Workforce, 2011	Percentage Difference
Primary	2.0	1.9	−0.1
Secondary	23.6	19.2	−4.4
Tertiary	74.4	78.9	4.5
Total	100.0	100.0	

Source: Statistics Canada (2006b, 2009d, 2012e).

resulting technology around the world. While the emphasis on a green economy could reduce greenhouse gas emissions and provide renewable energy sources, Ontario has had to provide large subsidies in the form of a "feed-in tariff"; that is, green energy will be marketed at relatively high rates—and certainly much higher than for coal energy—for electricity sold to Ontario Power Authority. These costs have a negative impact by increasing power rates for consumers and companies, pushing Ontario into a high-cost energy region.

The stakes are high. The real question for Ontario is whether it can position itself for a recovery in the next upturn of the US **business cycle** and regain its export position to the US, as well as tap into the huge market in Western Canada for high-end manufactured goods. Or is Ontario, like its American neighbours Michigan and Ohio, destined to slide into a depressed state, characterized as the Rust Belt?

 Table 1.4, "Economic Structure of Canada and Its Six Geographic Regions, 2011," page 18, provides an overview of each region's employment pattern.

Key Topic: The Automobile Industry

The automobile industry remains the heart and soul of manufacturing activity in Ontario. Yet, this industry is feeling the cold chill of the dark side of globalization. Canada's formerly sheltered manufacturing industry within the North American Auto Pact is now exposed to competition from countries with much lower wage rates. Even substantial federal and provincial grants and loans and research links with universities cannot turn the tide.[3]

In 2011, for example, the federal and Ontario governments provided $500 million to Toyota as an incentive to expand production at its Woodstock, Ontario, plant. By 2013, manufacturing of the RAV4, a crossover utility vehicle, will expand from 150,000 units per year to 200,000 units, with the bulk exported to the expanding market in the United States. Employment at the Woodstock plant will increase from 2,000 to 2,400 (Keenan, 2012: B3). Yet, these gains are small potatoes compared to those taking place in Mexico, where low wages attract fresh investment. As a result, Mexico surpassed Canada's annual automobile production in 2008 and this production gap between the two countries is widening. Canadian employment and production data for the last 10 years is not encouraging. As Figure 5.8 indicates, employment has dropped significantly over the last 20 years while the output of vehicles has fallen (Table 5.4).

Automobile manufacturing in Ontario expanded greatly following the Auto Pact (the Canada–US Automotive Products Agreement of 1965). As a result, Ontario became a key player in the North American automobile industry as

VIGNETTE 5.10

The Wave of the Future: Clusters and High Technology

High-technology companies are found in many places across Canada but particularly in Canada's major urban centres. In fact, Britton (1996: 266) states that "Canadian urbanization is the key to understanding the location of technology-intensive activities." Marc Busch takes this idea further. His answer is "clusters" (Atkin, 2000: C1). A cluster is a place where institutions, companies, and individuals have a commitment and enthusiasm for innovative technological research along with the capital necessary to develop and market the product. High-tech clusters often are anchored around a university or a public research agency like the National Research Council. Most high-technology companies are found in Toronto, Montréal, and Ottawa rather than in Saskatoon, Halifax, or Victoria. The latter do have high-tech operations, but they are on the edge of the high-tech world and depend on a niche operation. Saskatoon's niche is agricultural biotechnology. For St John's, it is marine technology. Can technological innovations rejuvenate Ontario's struggling economy? Waterloo has already experienced the technological boom. The Waterloo region hosts the Technology Triangle of Canada and boasts 550 technology companies, including Open Text, the largest software company in the country, and Christie Digital Systems Inc., maker of a high-end projection system.

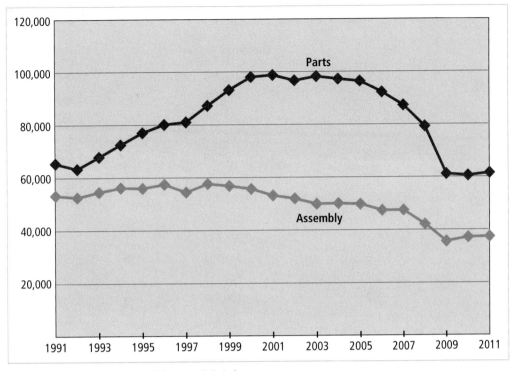

Figure 5.8 Employment in Canada's automobile industry
In the 1990s, employment in the automobile industry reached its peak. Its decline in the first decade of the twenty-first century is a remarkable reversal of good fortune for auto workers and their industry. In the next decade, further contractions are possible—unless, of course, workers' wages adjust to the global reality. *Source:* Canadian Auto Workers Union (2012: Figure 1), from Statistics Canada, CANSIM Table 281–0024. Reprinted with permission from Canadian Auto Workers Union (CAW-Canada).

80 per cent of Canadian manufactured vehicles were exported to the United States. The switch from a domestic market to a North American one was dramatic and transformed the auto industry in Canada from one producing a wide variety of cars for the Canadian market into one where large, low-cost plants achieved economies of scale by manufacturing only a few types of cars for the entire North American market. As such, the Auto Pact was a precursor to the FTA and NAFTA. From Canada's perspective, the Auto Pact served four objectives:

1. It secured guarantees that Canadian automobile plants would not close.
2. It brought to Canadian plants the advantage of economies of scale by allowing them to specialize in a few types of automobiles that would supply the North American market.
3. It reduced the price of cars for Canadian customers.
4. It resulted in higher wages and benefits for workers.

The Auto Pact came to an end due to a ruling of the World Trade Organization in 2001.[4]

The demise of the Auto Pact reduced the grip of the Big Three on the North American market and helped Toyota and Honda expand their market share in North America. More importantly, this WTO ruling left the door open for foreign countries to ship their automobiles and parts to Canada. Germany and South Korea soon penetrated the Canadian market. In 2012, the effect of globalization, namely low-cost manufacturing in China, saw Honda shipping its subcompact cars to Canada from China. Clearly, globalization of the automobile industry threatens Ontario and its automobile employees—unless, of course, the Canadian costs of production are sharply reduced.

The 2008–9 economic crisis hurt the North American automobile industry. The demand for cars and other vehicles dropped dramatically in the United States. Consequently, Ontario's principal export market dried up, causing a decrease in Canada's annual production by over 20 per cent, to 1.4 million vehicles, from 2007 to 2009.

THINK ABOUT IT
The closure of the profitable Caterpillar locomotive plant in London, Ontario, in February 2012 highlights the risks Canadian workers face in a world where global corporations can cut and run. Is this corporate strategy the opposite of the concept of a sense of place?

Table 5.4 **Canadian Motor Vehicle Production, 1999–2011**

Year	Cars	Commercial Vehicles*	Total
1999	1,626,316	1,432,497	3,058,813
2000	1,550,500	1,411,136	2,961,636
2001	1,274,853	1,257,889	2,532,742
2002	1,369,042	1,260,395	2,629,437
2003	1,340,175	1,212,687	2,552,862
2004	1,335,516	1,376,020	2,711,536
2005	1,356,197	1,332,165	2,688,362
2006	1,389,536	1,182,756	2,572,292
2007	1,342,133	1,236,657	2,578,790
2008	1,195,426	882,153	2,077,579
2009	822,267	668,215	1,490,482
2010	967,077	1,101,112	2,068,189
2011	509,398	575,398	1,085,211

*A commercial vehicle is a larger type of motor vehicle used for transporting goods or 10 or more passengers; thus, a taxi would be classified as an automobile while a city bus would be a commercial vehicle.

Source: OICA (2009); Industry Canada (2012).

Ironically, sales in Canada remained strong but those in the United States hit rock bottom. Canadian vehicle production in 2010 indicated that the North American automobile industry was on the road to recovery, with much stronger sales in the United States where most of Ontario's production is sold, but in 2011 production fell to the lowest level in 25 years (Table 5.4). This decline was largely due to the closing of Toyota and Honda plants for several weeks because certain components from Japan could not be shipped to Canada because of the tsunami and ensuing nuclear disaster in Japan in March of that year. A slowdown in sales in the United States also affected Canadian auto exports in 2011.

The Importance of Ontario's Automobile Industry

The automobile industry drives the Ontario economy. Yet, the industry—and, consequently,

VIGNETTE 5.11

Automobile Production: Industrial Core or Dead End?

Can Canada afford to lose its automobile industry? Put differently, is this industry a key anchor with spinoffs to "numerous other industries" that warranted the huge bailout in 2009? According to geographer Peter Dicken (1992: 268), "[t]he motor vehicle industry came to be regarded as a vital ingredient in national economic development strategies." But in the age of globalization, is that premise still true? Global events raise a critical question for the future of Ontario's economy and its workers, namely: Is the automobile industry in Canada going the way of the textile industry? Some 30 years ago, low-cost labour in other countries plus easy access to the Canadian market spelled the end of this manufacturing enterprise because it encouraged Canadian textile producers to move offshore where much lower labour costs more than offset higher transportation costs. Other textile manufacturers simply ceased to operate.

Ontario—is hurting. One sign is the loss of nearly one-third of those employed in this industry over the last decade (Figure 5.8). Wages in the unionized assembly plants are at the high end for semi-skilled workers and companies are demanding lower wages. Negotiations for new contracts with the Big Three auto companies reached a conclusion in September 2012. The result was a new four-year collective agreement with mixed results for workers and lower labour costs for the Big Three auto companies. While auto workers obtain job security, their take-home pay was reduced by freezing wages and downgrading benefits. For instance, instead of an hourly pay increase, workers will receive an annual bonus of $3,000. As well, new employees face even darker prospects, with lower starting wages ($20/hr rather than the existing $24/hr) and they will take 10 instead of six years to reach the maximum rate of $34/hr (Flavelle, 2012).

Nevertheless, with these relatively high incomes, workers are able to make substantial purchases, thereby stimulating southern Ontario's retail sector. Another sign of the weak state of the automobile industry is revealed by the decline in its share of exports. In 2007, automotive products comprised nearly 17 per cent of Canada's exports, but by 2011 this had fallen to 13 per cent (Statistics Canada, 2012f).

The tide that once favoured Ontario has turned. Ten years ago, Ontario had three advantages over its American counterparts:

- Ontario assembly plants had a higher productivity than comparable US plants. In 2001, Canadian workers had a 7 per cent advantage over American workers in the time required to produce an automobile (Brent, 2002).
- From 1976 to 2007, the much lower Canadian dollar allowed Canadian exports of cars and trucks to cost less than similar vehicles produced in the United States, with the floating Canadian dollar reaching its low exchange point of 62 US cents on 21 January 2002.
- Health-care costs are covered by Ontario while these costs are part of the wage package in the United States.

By 2012, Ontario still had an edge of productivity but the high Canadian dollar has more than eliminated that advantage. Canada's universal health-care system remains an important cost advantage, especially since the US automobile companies are reducing their contribution to workers' health plans and have left retired workers out on a limb by

VIGNETTE 5.12

The Bailout of Chrysler and GM: Sound Public Policy?

Governments are loath to see the bellwethers of their manufacturing industries—motor vehicles—die, for then, it is asked, who can lead the manufacturing sector? In 2009, Canada and Ontario provided financial assistance to Chrysler Canada (US$3.8 billion) and General Motors Canada (US$10.6 billion) to keep the automobile industry in Canada at its historical share of North American production (roughly 20 per cent). In 2009, Fiat's successful buyout of Chrysler marked a new beginning and direction for Chrysler Canada, with Fiat's Sergio Marchionne taking over as Chrysler's CEO. The federal and Ontario governments secured a 2 per cent stake in Chrysler. In the US, a $25 billion government bailout of GM and Chrysler went through, although both of these companies were in bankruptcy protection by mid-2009 and many auto dealership franchises had been forced to close.

Was this sound public policy by the federal and Ontario governments? Fast forward to 2012. Shiell and Somerville (2012) look back over the bailout and conclude that the 2009 federal–provincial support for GM Canada and Chrysler Canada was a smart one-time move, allowing time for the companies to restructure, return to profitability, and begin the process of repaying a significant portion—but not all—of the bailout funds. Shiell and Somerville estimate that the alternative cost associated with letting the two companies fail at about $20 billion, most of which would have been borne by Ontario.

Additional proof that Ottawa has bought into automobile subsidies was revealed in the 4 January 2013 announcement of its continued financial support of the Automotive Innovation Fund (Keenan and Chase, 2013).

turning over the under-funded retired workers' plan to their union, the United Auto Workers (UAW).[5]

The Canadian automotive industry has seen a shift from the domination of the Big Three to the rising power of Japanese-based firms. Some 25 years ago, the Big Three accounted for nearly all cars and trucks manufactured in Canada. By 2011, foreign firms, led by the Japanese automobile manufacturers, had cut into that dominant position and the Big Three accounted for 64 per cent of automobiles and 79 per cent of trucks.

Is there another approach to automobile manufacturing for Ontario? One is the electric car. The Ontario government is promoting both a green environment and the production of electric automobiles. To that end, former Premier Dalton McGuinty encouraged Ontarians to purchase electric cars. Besides offering consumers an incentive of up to $8,500, Ontario will provide $80 million to increase the number of power stations, thus allowing operators of electric cars more geographic options to recharge. As stated in 2009, electric cars are but one element in Ontario's effort to create a greener environment (CBC News, 2009). By 2020, it was hoped that one in 20 cars would be electric-operated. However, electric cars are expensive: General Motors' Chevrolet Volt is priced at over $40,000. As well, lower-priced cars now have much more efficient engines and so the lure of electric cars has not had much traction.

A second approach could be to produce specialized vehicles needed by the mining industry, such as the oil sands companies. Already Hitachi, under the name of Euclid-Hitachi Heavy Equipment, is producing a range of huge trucks in Guelph and some have found a home in the oil sands. While trade missions to China are popular with political and business leaders, in the same spirit of encouraging trade, Ontario and its automotive leaders might benefit from a visit or two with the Alberta Premier and the oil sands company officials.

Automobile Assembly Plants

Automobile assembly plants are concentrated in southern Ontario where transportation links to the major markets of Canada and the United States are readily available and driving distances are short (Figure 5.9 and Table 5.5). In recent years, however, some plants have shut down: the Sainte-Thérèse

Table 5.5 **Ontario Automobile Assembly Plants, 2012**

Location	Products
CAMI Automotive Inc.	
Ingersoll, Ontario	Chevrolet Equinox; GMC Terrain
Chrysler Canada Inc.	
Brampton, Ontario	Chrysler 300; Dodge Challenger; Dodge Charger; Lancia Thema
Windsor, Ontario	Dodge Grand Caravan; Chrysler Town & Country; Volkswagen Routan; Lancia Voyager
Ford of Canada Ltd	
Oakville, Ontario	Ford Edge; Ford Flex; Lincoln MKT; Lincoln MKX
General Motors of Canada Ltd	
Oshawa, Ontario	Chevrolet Equinox; Chevrolet Impala
Oshawa, Ontario	Buick Regal; Chevrolet Camaro
Honda Canada Manufacturing Inc.	
Alliston, Ontario	Honda Civic; Acura CSX
Alliston, Ontario	Acura MDX; Acura ZDX; Honda Civic; CR-V
Toyota, Canada	
Cambridge, Ontario	Toyota Corolla; Toyota Matrix; Lexus RX350
Woodstock, Ontario	Toyota RAV4

Source: Industry Canada (2012). Reproduced with the permission of the Minister of Public Works and Government Services Canada, 2013.

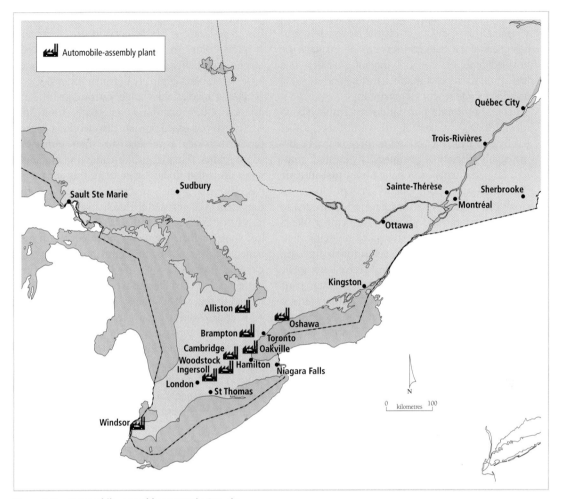

Figure 5.9 **Automobile assembly centres in Ontario**

GM plant north of Montréal closed in 2002 and the Ford plant in St Thomas, Ontario, closed in 2011. While Japanese auto factories are located in the United States, the Canadian advantages mentioned earlier—a highly motivated workforce, the lower value until recently of the Canadian dollar, and the public medical system—have attracted substantial Japanese investment in Ontario. In American-based assembly plants, the cost of providing medical insurance is often built into the wage agreement.

Asian automakers are capturing more and more of the North American market. Honda and Toyota are scouting for new manufacturing sites in North America, especially in Mexico. In Ontario, Honda and Toyota are expanding their production capacity, thus creating more jobs and more demand for parts. Toyota, in 2008, built an assembly and parts plants at Woodstock to support production of the RAV4 sports utility vehicle. Toyota plans to expand production at Woodstock, bumping the number of RAV4 from 150,000 to 200,000. In 2012, Honda began manufacturing a new SUV, the Honda CR-V. These two Japanese firms expanded their Ontario manufacturing primarily because of strong demand for their automobiles in the US. Ontario is an attractive production site in North America for the Japanese manufacturers as a consequence of support from both the federal and provincial governments, easy access to the US market, and a productive, "small-town" labour force.

Automobile Parts Firms

The automobile industry consists of two separate operations: the assembly of automobiles and trucks and the production of their parts. In addition, some manufacturing firms supply semi-processed materials. In southern Ontario, fabricating firms produce steel, rubber, plastics, aluminum, and glass parts for automobile assembly and parts plants in Canada and the United States. Finally, service firms, ranging from the advertisers and designers to the sales and service staff, manage the finished product. In short, the auto industry is a final-product type of manufacturing, and as such, its added value reaches a maximum.

The automobile parts industry employed almost 130,000 workers in 2005 but this number fell by half over the following three years. By being highly efficient and strategically located, automobile parts firms can operate on a **just-in-time** principle—auto components are produced in small batches and quickly delivered as needed to their customers. This allows the assembly plants to achieve considerable savings by reducing their inventories, warehousing space, and labour costs.

In the late 1980s, the Big Three automobile manufacturers decided to subcontract their parts business. This practice of subcontracting parts is called outsourcing. **Outsourcing** had two advantages for automobile companies. First, it allowed manufacturers to concentrate on assembling automobiles, thereby reducing their costs and improving the quality of their product. Second, parts companies were not unionized and therefore had lower wages. This wage differential—and the savings it provides—is the main reason why General Motors, Ford, and DaimlerChrysler continue to divert work to parts firms. In Ontario, the Canadian-based Magna International has grown into the third-largest auto parts company in North America.

The "Canadian advantage" encouraged the expansion of Canadian-based auto parts firms. This advantage was based on Canadian workers' higher productivity as compared to their American counterparts, Canada's health-care program, and a weak Canadian dollar. Since then, this advantage has slipped. First, the strengthening of the Canadian dollar caused the exchange advantage to largely disappear. Second, US autoworkers have lost some of their company-provided health-care benefits, thus relieving the companies of that obligation. Third, access to the US market became more complicated because of security concerns related to the 11 September 2001 attacks on New York and Washington. Delays at the border increased costs for truckers and compromised the "just-in-time" system for the parts industry. Added to security issues, the Windsor–Detroit bridge and tunnel crossings are insufficient for the volume of traffic, and this jeopardizes the "just-in-time" system used in the automobile industry.

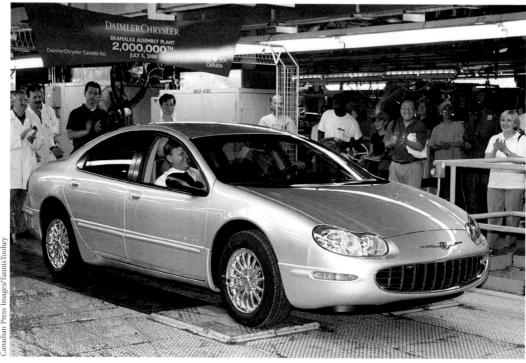

Photo 5.9
A new car rolls off the assembly line at the Chrysler plant in Brampton, Ontario, but will Chrysler survive?

Ottawa and Washington currently are seeking a solution to the traffic congestion that hampers trade between the two countries, with plans well underway for a new international bridge called the Detroit River International Crossing (DRIC) that would open in 2015. This new border crossing is expected to tie in directly with Highway 401 as it approaches Windsor and thereby avoid the present crawl through the city streets of Windsor to reach the Ambassador Bridge, which lies north of the proposed DRIC. In 2009, the project was approved by both Canadian and American environmental assessment agencies, and in April 2010 agreement was reached between Windsor and the province on a $2 billion project, involving 12,000 construction jobs, for building the freeway connection to the new bridge (CBC News, 2010). Funding for Canada's portion of the multi-billion dollar project will come from Ottawa's infrastructure funds, making Canada's section publicly owned. With Michigan unable to afford its share, Canada has agreed to pay for the Michigan share and then recoup these costs in future toll revenues (Detroit River International Crossing, 2009; Transport Canada, 2009; Potter, 2010; CBC News, 2012). Approval by President Obama is the final step before breaking ground for the new bridge.

Northern Ontario

The physical base of northern Ontario consists of the Canadian Shield and the Hudson Bay Lowlands. While both have disadvantages for settlement, development has taken place along the two transportation corridors between Ontario and Manitoba. One corridor follows the Trans-Canada Highway (Ontario Highway 17) and the Canadian Pacific rail line while the more northern corridor parallels the Canadian National rail line and the northern link of the Trans-Canada Highway (Ontario Highway 11).

Figure 2.1, "Physiographic regions and continental shelves in Canada," page 33, and Figure 5.4, "Central Canada," page 170, delineate Ontario's physiographic regions. Also see "Physiographic Regions" in Chapter 2, page 32.

Northern Ontario is an old resource hinterland, and a sluggish economy, a declining population base, and high unemployment rates keep northern Ontario's economy in the doldrums. The Rim of Fire, a highly mineralized zone, represents the main bright spot in an otherwise dismal economic future. Until 2012, the seemingly never-ending housing crisis in the US crippled lumber exports, while pulp and paper products are facing an ever-diminishing demand: to some extent, computers have replaced newspapers, which have gone out of business or shrunk in size, and also have largely replaced paper in such business applications as billing, accounting, and banking; retailers use less paper; packaging of food and other products tends to use plastic (or consumer-purchased cloth bags) in preference to paper; recycling programs have replaced a small percentage of new paper products. Even northern Ontario's transshipment role between Ontario and Western Canada has weakened as more products,

---

VIGNETTE 5.13

The Threat of Hollowing-Out

In the coming decade, Ontario's principal challenge will be its manufacturing base. How can Ontario avoid the threat of hollowing-out? China, now the manufacturing centre of the world, presents intense competition: its firms can produce manufactured goods at a lower cost because of the wage differential between China and Ontario. Many Ontario firms have responded by relocating offshore where wages are much lower, and even with the cost of shipping the products back to Canada, production and transportation costs still are lower than they would be for domestically manufactured goods. This geographic shift of manufacturing from Canada to other countries began with the textile industry and is now affecting more sophisticated forms of manufacturing. With Chinese-based auto manufacturers now producing automobiles for domestic customers in China, additional production has begun to be exported to North America. This ongoing erosion of Canada's industrial base is a result of open world markets. For Canada, its vulnerability has serious consequences for its two core manufacturing regions—Ontario and Québec.

such as grain and potash, are shipped to Vancouver for sale in Asian countries. With a faltering economy, northern Ontario's population has taken on four demographic characteristics that are strikingly different from southern Ontario:

- an aging population;
- a net out-migration, especially of younger members of its population;
- few immigrants;
- a small but rapidly expanding Cree and Métis population.

The physical base of northern Ontario is considerable but its population consists of fewer than 850,000 people. In percentage terms, northern Ontario comprises 87 per cent of the geographic area of Ontario but has less than 7 per cent of Ontario's population. With virtually no agricultural base, most people live in towns and cities along the two transportation routes that link Montréal and Toronto with Winnipeg. Both of these trans-Canada routes cross the rugged Canadian Shield. The southern route, defined by the CP rail line and the Trans-Canada Highway, connects North Bay with Sudbury, Sault Ste Marie, Nipigon, Thunder Bay, Dryden, and Kenora. The northern route, defined by the CN line and a major highway, connects North Bay, Timmins, Kirkland Lake, Cochrane, Kapuskasing, and Nipigon to Thunder Bay.

Mining, forestry, and tourism are the major economic activities, although many people are employed in the public sector. Like other resource hinterlands, the development of northern Ontario's economy was linked to external markets. Resource exploitation in northern Ontario began after the construction of railways in the late nineteenth century. Soon thereafter, resource development took place along the two east–west rail lines. The same pattern of resource development took place along the Temiskaming and Northern Ontario Railway (also known as the Ontario Northland Railway).[6] The railway extends northward from North Bay to New Liskeard, Kirkland Lake, and Cochrane, which is on the CN main line. Northern Ontario exports minerals and forest products, including gold, nickel, newsprint, and lumber. These resource products contribute less than 10 per cent of the value of Ontario's exports to foreign countries.

Unlike the northern hinterland of Québec, the natural conditions in northern Ontario are not conducive to major hydroelectric developments. While a number of rivers flow across northern Ontario to James Bay and Hudson Bay, the gentle slope of the land, especially in the Hudson Bay Lowlands, does not make the construction of hydroelectric dams feasible. A few established sites for hydroelectric production, such as along the Ottawa River and at Abitibi Canyon about 100 km north of Cochrane, are important, but they cannot meet the energy demands in Ontario. The much higher elevations in the Canadian Shield area of Québec provide the necessary natural drop to drive the turbines that produce hydroelectric power. Since Québec has a surplus of electrical power, some of it is transmitted to southern Ontario.

Northern Ontario has a strikingly different settlement pattern from that of southern Ontario. Long distances separate the four major cities. For example, the distance from North Bay to Sudbury is over 100 km; from Sudbury to Sault Ste Marie is about 300 km; and from Sault Ste Marie to Thunder Bay is about 500 km. The explanation for this oasis-like settlement pattern is due to northern Ontario's physical geography. The rocky terrain of the Canadian Shield discourages continuous settlement; a series of isolated settlements are located at mining sites, pulp plants, and key transportation junctions. Similar to other old hinterlands, the urban centres of northern Ontario are declining and the populations of single-industry towns like Cobalt, Kirkland Lake, and Porcupine rose and fell as the cycle of resource exploitation ran its course.

Forest Industry

The boreal forest stretches across northern Ontario, providing the province with most of its 57 million hectares of productive forest area, most of which is classified as softwood. The main species are black spruce, poplar, and Jack pine. From this resource, the forest industry of Ontario produces approximately $12 billion of products each year, with most exported to markets in the US. Access to the American market is critical, and is controlled by agreements between Canada and the US. The most recent agreement, the 2006 Softwood Lumber Agreement (SLA), limits Canadian softwood exports. Under this agreement, Canadian lumber firms are allocated 34 per cent of softwood lumber sales in the US market. Each province is allocated a share of those exports based on 2004–5 exports, so that Ontario gets around 9 per cent of the Canadian softwood lumber exports to the US (CTV, 2006).

For a fuller discussion of the Softwood Lumber Agreement, see Chapter 7, "North American Free Trade Agreement," page 302.

The harvest of this vast forest is a central component in the economy of northern Ontario. Yet, the harvesting of the boreal forest is not without challenges. From a sustainable forest perspective, the main challenge is to maintain a balance between logging and the regeneration of the forest. Since over 90 per cent of Ontario's forest lands are owned by the provincial government, private companies must obtain forest leases. The Ontario government, like most other provincial governments, has insisted that these private companies take on the responsibility for restoring trees to logged areas through forest management agreements. Efforts to hand-plant seedlings are helping to speed up the process of reforestation, but how successful these efforts have been can only be known 20 years from now. The practice of granting forest companies long-term timber leases is based on the assumption that it is in the companies' self-interest to manage the forests well. It remains to be seen whether this assumption and the corresponding leasing policy will ensure the protection and regeneration of Ontario's forests.

From an economic perspective, the forestry industry remains by far the most important primary industry in northern Ontario. Some 50 communities in the boreal forest region depend on forestry. Within Canada, Ontario ranks just behind British Columbia and Québec in forestry production. Ontario's mills produce pulp and paper, lumber, fence posts, and plywood. The pulp and paper industry in northern Ontario accounts for about 25 per cent of the national production, and Ontario is a leading exporter of newsprint and pulpwood to the US. In fact, most pulp and paper firms operating in northern Ontario are American-owned companies.

The forest industry continues to face tough times. The primary reason is that its principal export market is the United States and demand from the US has dropped dramatically and only showed signs of recovery in 2012. In 2007, Canadian exports of softwood lumber reached $12.6 billion, but four years later they had fallen to $8.5 billion (Statistics Canada, 2012f). Since Ontario accounts for just under 10 per cent of softwood lumber exports, its export figures were likely $1.2 billion in 2007 and $0.8 billion in 2011. A further measure of the decline is revealed in the value of production—in 2009 it was $12 billion compared to $21 billion in 2008 (Ontario, Ministry of Natural Resources, 2012).

The collapse of the US market affected at least a dozen communities in northern Ontario. Shutdowns put thousands out of work. In Kenora, Red Rock, Dryden, Thunder Bay, Terrace Bay, Kapuskasing, and other centres in Ontario's north, workers in the forest industry faced layoffs, closures, and limited prospects. The negative spinoff effects reverberate through these communities and across northern Ontario. While plant closures have taken place across the country, the high cost of energy is an added factor affecting mills in Ontario. Electricity prices have skyrocketed in the province, to the point where energy makes up 30 to 40 per cent of the cost of getting wood from the forest to the mill and then to the market. For many companies in northern Ontario, energy has become the make-or-break number on the balance sheet.

Today, the forestry industry faces two issues. First, many mills were built before World War II and continue to use old technology, which results in much higher discharges of toxic wastes into the environment. The consequences of such practices can be life-threatening. The worst cases of massive toxic discharges into streams and rivers occurred in the 1960s. At that time, a chemical plant that prepared the bleaching solution for the pulp plant at Dryden was releasing effluents containing

Photo 5.10
The boreal forest stretches across most of northern Ontario, providing the basis for a forest industry. It is too early to know if the Canadian Boreal Forest Agreement, signed 18 May 2010 between forest companies and environmental groups, will prove to be both environmentally sound and economically attractive.

mercury into the English–Wabigoon river system.[7] At considerable cost, most forestry mills have updated their operations to ensure a cleaner and safer natural environment, but some have not.

The second issue is fluctuations in the US demand for softwood lumber. Since Ontario's softwood producers are so dependent on the US market, low American demand means low Canadian prices. In 2006, the sharp downturn in the US housing market produced a double whammy: fewer exports to the US and a tax on Canadian softwood lumber exports. The tax, which comes into effect when the price drops below US$355 per thousand board feet, is part of the new softwood lumber agreement. As US demand for lumber fluctuates, so does the price of lumber. From May 2006 to May 2007, the price of softwood lumber had dropped from US$367 to US$224, thus triggering the maximum export tax of 15 per cent, which is collected by Ottawa (Anderson, 2006; Export Development Canada, 2007). By 2009, softwood lumber prices remained low, falling below US$200, but a slow (and hopefully steady) rise in prices that reached $291 in July 2012 may be the harbinger of better times for Ontario's softwood lumber industry (Natural Resources Canada, 2012a: Table 1).

Mining Industry

The mining industry is northern Ontario's second economic anchor. Based on non-renewable resources, the major drawback of such development is its limited lifespan. While a few deposits, such as the Sudbury nickel deposit, last for more than 100 years, most have a much shorter lifespan, often less than 20 years.

The centre of mining is the Canadian Shield, which provided ideal geological conditions for the formation of hard-rock minerals such as diamonds, gold, nickel, and copper. In 2011, metallic minerals in the Canadian Shield in Ontario had a value of $7.5 billion, with gold, nickel, and copper leading the way. Including non-metallic minerals, many of which come from the sedimentary strata in southern Ontario, the total value of mineral production in 2011 was $10.7 billion (Natural Resources Canada, 2012b). Ontario was the leading Canadian producer of gold and nickel, and was the leading producer of minerals among the provinces and territories, accounting for 21 per cent of the total Canadian value (Table 5.6). Most production comes from Red Lake (gold), Hemlo (gold), Wawa (gold), Manitouwage (gold), Marathon

Table 5.6 Mineral Production by Province and Territory, 2011 ($ millions)*

Province/Territory	Metallics	Non-metallics	Coal	Total	% Share of Production
NL	5,112	78	0	5,190	10
PEI	0	3	0	3	0
NS	0	247	0	247	1
NB	817	491	0	1,308	3
QC	6,052	1,698	0	7,750	15
ON	7,505	3,159	0	10,664	21
MB	1,646	188	0	1,834	4
SK	1,157	8,057	**	9,214	18
AB	1	1,060	**	2,015	5
BC	2,096	806	5,691	8,593	17
YT	395	7	0	402	1
NWT	65	2,080	0	2,145	5
NU	414	0	0	414	1
Totals	25,260	18,038	7,050	50,348	100

*Preliminary data.
**Confidential coal data, but Alberta and Saskatchewan accounted for $1.4 billion in 2011.

Source: Adapted from Natural Resources Canada (2012b).

(gold), Thunder Bay (gold, copper, zinc), and Sudbury (nickel, copper). Ontario's first diamond mine (the Victor Mine), located in the Hudson Bay Lowlands some 500 kilometres north of Timmins, came into production in 2008. The promise of a chromite mine widens the range of minerals flowing from the Canadian Shield. The Ring of Fire mineralized belt in northern Ontario is one of the most promising mineral development opportunities in Ontario in almost a century (Figure 5.10). Deposits include nickel, copper, and platinum, but huge deposits of chromite known as Black Thor have attracted most attention. An American firm, Cliffs Natural Resources, intends to invest $5.1 billion to establish a chromite mine west of James Bay and build a smelter near Sudbury (Shufelt, 2012). Since this area is inaccessible by land transportation, the Ontario government has agreed to build an all-season road to the mining site that will allow two-way traffic—ingoing supplies to construct the mine and then outgoing trucks to carry the chromite ore to the refinery near Sudbury. Chromite is the principal component in stainless steel and the leading global steel producer is China, which accounts for 70 per cent of world production. For the time being, however, Cliffs Natural Resources remains in the early planning stages of its Ring of Fire endeavour, is apparently seeking partners for the venture, and in November 2012 announced a delay in becoming operational until at least 2017.

THINK ABOUT IT

Is it a reasonable gamble for Ontario to invest millions in constructing a highway to the proposed chromite mine in northern Ontario when the major importer, China, might well find new chromite deposits closer to home?

Figure 5.10 Northern Ontario's Ring of Fire

The Ring of Fire mineral region in northern Ontario was allegedly so named by a prospector and former chairman of a Toronto-based mining company, Noront Resources, who is a Johnny Cash fan—one of Cash's best-known songs is titled "Ring of Fire." The economic potential of the Ring of Fire mineral deposits is about as limitless as non-renewable resources can be, but the exploitation of these resources will require careful planning and co-operation among the major players: the provincial government, the mining companies, and more than a dozen First Nations whose traditional homelands are in this region of Ontario's north. *Source:* Shufelt (2012). Material reprinted with the express permission of: National Post, a division of Postmedia Network Inc.

Canada's leading minerals by value of production for 2011 are shown in Table 7.3, page 294.

Without an agricultural land base, the search for mineral wealth in the Canadian Shield was crucial in the development of northern Ontario. The first important mineral discovery took place in 1883 when the copper-nickel ores of the Sudbury area were discovered during the building of the Canadian Pacific Railway. In 1903, a rich silver deposit near Cobalt was detected during the construction of the Temiskaming and Northern Ontario Railway. Rich gold deposits were uncovered near Timmins in 1909 and at Kirkland Lake in 1911. While new mineral finds kept the mining economy of Ontario going, the overall pattern was one of boom and bust. During the 1960s, a large lead-zinc deposit was discovered in Timmins. A decade later, rich gold deposits were found at Marathon (the Hemlo gold mine, now consisting of the David Bell and Williams mines) and at Pickle Lake. The Victor diamond mine, owned by the South African mining company De Beers, began production in 2009, and, in an agreement with Ontario, has committed 10 per cent of its production to the Crossworks diamond factory in Sudbury.[8] In 2011, Crossworks was awarded the De Beers allocation for 2012–15, ensuring that Sudbury remains the leading diamond-cutting centre in North America (Ontario, 2012). Ironically, while Sudbury's unemployment rate hovered around 10 per cent in 2009, no worker in Sudbury had the required skill set or was trained in diamond cutting. Instead, 27 highly skilled diamond cutters and polishers were flown in from Vietnam (Hoffman, 2009: B1).

Ontario's Urban Geography

Ontario is the most highly urbanized province in Canada. Almost 11 million people—nearly 85 per cent of the province's total population—live in towns and cities. As well, 10 of Canada's 25 largest cities are in Ontario. With the exceptions of Sudbury and Thunder Bay, these census metropolitan areas (CMAs) are located in southern Ontario. Most importantly, Toronto, with a population of 5.5 million, is by far the largest city in Canada and, as a result of its size, contains many higher-order businesses, cultural attractions, and services not found in smaller cities.

Since cities are the engines of the Canadian economy, Ontario's urban geography provides an enormous economic advantage. Cities are self-perpetuating: employers set up for business and that is where most employment is located. The pattern of urban growth from 2001 to 2011 varied considerably for these CMAs (Table 5.7). The fastest-growing cities in Ontario were Guelph, Oshawa, and Toronto. Ottawa–Gatineau ranked fourth but part of its residents live across the Ottawa River in

Photo 5.11
Mining forms a major part of northern Ontario's economy. The Williams gold mine near Marathon, owned by Barrick, remains the largest gold-producing mine in Canada. In 2011, Ontario accounted for 53 per cent of Canada's gold production by value, followed by Québec at 28 per cent, Nunavut at 9 per cent, and Manitoba at 8 per cent (Natural Resources Canada, 2012b).

Table 5.7 Census Metropolitan Areas in Ontario, 2001–2011

Census Metropolitan Area	Population 2001 (000s)	Population 2011 (000s)	Change (%)
Peterborough	110.9	119.0	7.3
Thunder Bay	122.0	121.6	0.0
Brantford	118.1	135.5	14.7
Guelph	117.3	141.1	20.3
Kingston	146.8	159.6	8.7
Sudbury	155.6	160.8	3.3
Windsor	307.9	319.2	3.7
Oshawa	296.3	356.2	20.2
St Catharines–Niagara*	377.9	390.3	3.3
London	432.5	474.8	9.8
Kitchener*	414.3	477.2	15.2
Hamilton	662.4	721.1	8.9
Ottawa–Gatineau*	1,067.8	1,236.3	15.8
Toronto	4,682.9	5,583.1	19.2
Total	9,012.7	9,691.8	7.2

*Statistics Canada has combined St Catharines and Niagara Falls as a single census metropolitan area although these cities still exist as separate political jurisdictions, as, of course, do Ottawa, Ontario, and Gatineau, Québec. Likewise, Statistics Canada includes Waterloo and Cambridge with Kitchener as a single CMA.

Source: Statistics Canada (2007c, 2012g).

Québec. In contrast, the three slowest-growing cities were St Catharines–Niagara, Sudbury, and Thunder Bay, with the latter two located in the resource hinterland of the Canadian Shield.

The eight largest cities in Ontario are located in southern Ontario (Figure 5.11). The four largest CMAs (Toronto, Ottawa, Hamilton, and Kitchener) form the core of the three urban clusters in southern Ontario, namely, the Golden Horseshoe, southwestern Ontario, and the Ottawa Valley. Each of southern Ontario's three major urban clusters has a population of at least 1 million. As mentioned earlier, northern Ontario has fewer and smaller cities, each separated by long distances. It has two major cities (Sudbury and Thunder Bay) and no single urban cluster. Sudbury, the largest centre, had 160,770 residents in 2011, while Thunder Bay's population was 121,596.

The Golden Horseshoe

The Golden Horseshoe obtained its name because of its horseshoe-like shape around the western end of Lake Ontario and its outstanding economic performance over the years. This tiny area of the Great Lakes Lowland forms the most densely populated area of Canada. The Golden Horseshoe extends from the US border at Niagara Falls northward to Hamilton, Toronto, and then on to Oshawa. Nearly 8 million Canadians live, work, and play in Canada's largest population cluster, and many visitors come either as tourists or on business trips. Accounting for nearly one-quarter of Canada's population, the Golden Horseshoe contains numerous towns and cities, including Toronto, Hamilton, Oshawa, St Catharines, Niagara Falls, Burlington, Oakville, Pickering, Ajax, and Whitby. Toronto, the largest city in Canada, is its urban anchor, while Hamilton, with its steel plants, is the focus of heavy industry (Vignette 5.14), and Oshawa is Canada's leading automobile-manufacturing city.

Toronto

Toronto drives the province's cultural, demographic, and economic growth and dominates the province's urban landscape. It also plays a key role in Canada's urban hierarchy by offering unique "high-end" services, ranging from finance

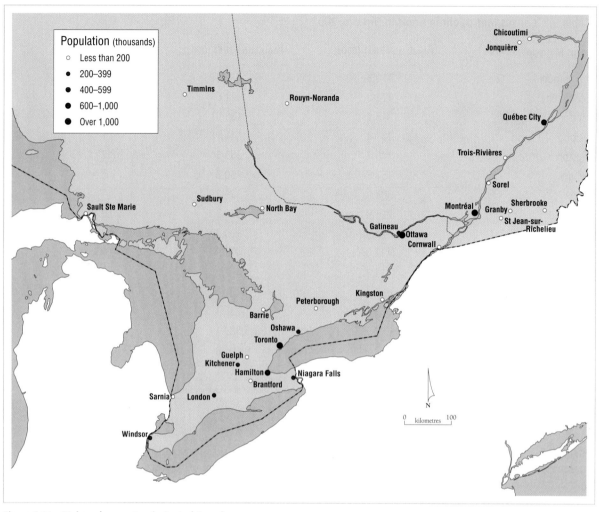

Figure 5.11 **Major urban centres in Central Canada**
Most large urban centres in Central Canada are located in the Great Lakes–St Lawrence Lowlands, especially in southern Ontario.

to opera, Major League Baseball, and other major league sports franchises. Toronto is the financial capital of Canada, housing the main offices of national and international banks and investment firms, as well as the Toronto Stock Exchange. It is the site of the famous Roy Thomson Hall concert venue and the Toronto Blue Jays' multipurpose stadium, the Rogers Centre (formerly called SkyDome).

One measure of Toronto's key role in Ontario's growth is population increase. For decades, Toronto's population growth has outpaced that of most other cities in Ontario, and by 2011 the population of the Greater Toronto Area (GTA) had reached 5.6 million. Over the last decade, the main factor driving Toronto's growth has been immigration from foreign countries. At the same time, people and businesses have spilled over Toronto's boundaries, causing a geographic expansion of the GTA. The spread of Toronto's population and businesses into neighbouring areas is due partly to lower land values and rents. Land values and rents are highest in downtown Toronto. Cities surrounding Toronto have significantly lower land values and can therefore attract businesses by offering lower office rents.

As the province's major cultural and entertainment centre, Toronto is a hub for the entertainment industry, which ranges from music and drama to professional sports, and plays an essential role in shaping Canadian culture. Just how dependent the cultural and service industries are on tourists was revealed by the deadly outbreak of SARS in April 2003; as a result of a travel advisory issued by the World Health Organization, tourism dropped precipitously for a brief period, resulting in an estimated loss to the city of $1 billion in one month alone (Brean, 2003).

Like other major cities in North America, Toronto is trying to cope with a rapidly expanding population, much of which has moved into adjoining urban areas. One effort to deal with the administration of this urban area was to create a super-city known as the Greater Toronto Area (GTA). In 1998, the municipalities of the former Metropolitan Toronto (Toronto, North York, Etobicoke, Scarborough, East York, and York) officially formed a single city government with the hope that solutions to problems arising from rapid population growth and geographic expansion of the urban population could be found.[9] While many opposed this approach to mega-urban government, the objective is to reduce administrative overlap and thereby create a more efficient urban government. Whether or not this single administration approach is succeeding in serving such a large population remains to be seen. However, many problems remain, including traffic congestion. With so many people living outside of downtown Toronto but working in the city centre, massive numbers of them commute daily, creating traffic jams during the morning and afternoon rush hours. Traffic congestion leads to time lost travelling to and from work. A radical solution would be charging a toll on automobile traffic into and out of Toronto. London, England, imposed such a tax and it resulted in a reduction in traffic flow and a decrease in travel time.[10]

As the most popular destination for immigrants to Canada, Toronto has experienced remarkable ethnic diversification over the past 35 years, to the point where visible minorities are hardly in the minority. Visible minorities comprised 37 per cent of Toronto's population in 2001 compared to 26 per cent in 1991 and are projected to be in the majority by 2017.

Wave after wave of immigrants has come to Toronto, first from Europe, especially from Italy and the United Kingdom. More recent newcomers have come from Asian countries. According to the 2006 census, Toronto had 2.3 million immigrants who accounted for 46 per cent of the city's population; of these immigrants, nearly 14 per cent came from China and Hong Kong, 10 per cent from India, and 6 per cent from Italy and the Philippines (Table 5.8).

Toronto is known as a city of neighbourhoods, partly because immigrant groups have clustered in certain areas. Little Italy and Little Portugal are older, well-established ethnic neighbourhoods. More recent Asian, African, and Caribbean neighbourhoods have emerged. Recent immigration has had an impact on Toronto's cityscape, including "foreign" architecture and commercial activities designed to meet the demands of these

VIGNETTE 5.14

Hamilton: Steel City or Rust Town?

Hamilton, situated at the west end of Lake Ontario only 50 km from Toronto, is known as Steel City. Unfortunately, the North American steel industry has fallen on hard times and Hamilton is hurting. Canada's largest steel firms (Stelco and Dofasco) were purchased by US Steel and ArcelorMittal (a transnational from India with headquarters in Luxembourg) in 2007 and 2008, respectively. The companies were renamed US Steel Canada Inc. and ArcelorMittal Dofasco. Both steel companies are struggling and economic growth is not on Hamilton's horizon. Trevor Cole's dark perspective may still hit the mark: "As prosperity plumped nearby rivals such as Burlington, Oakville, Mississauga, Kitchener–Waterloo and—especially—Toronto, it skipped Hamilton completely, cruelly, until most of its big-name companies were gone, the stores along Barton Street deteriorated into dark and crumbling shells, downtown became a kind of forbidden zone, and even the Mafia couldn't make any money" (Cole, 2009). Manufacturing is hurting, leaving Hamilton's future in doubt unless the US economy recovers quickly to create demand for Hamilton's manufactured products. In a sombre tone, the Conference Board of Canada's *Metropolitan Outlook* study (Arcand et al., 2012) calls for very modest growth of 2 per cent in 2012. What can be done? A bold shift to the knowledge-based economy may drag Hamilton out of the old economy and slow growth by using McMaster University as an anchor and collaborating with leaders in the high-tech field in Toronto, such as Ryerson University.

Photo 5.12
The Eaton Centre, first opened in 1977, is Toronto's most popular tourist attraction and a major shopping centre, with about one million visitors each week.

new Canadians. Asian theme malls, which consist of a large number of small retail outlets, many of which are either Chinese restaurants or specialty stores, represent a dramatic change to Toronto's space. Unlike other suburban malls, no anchor store exists and, rather than a single developer who leases space, each retail unit is owned by the operator of that retail space. The failed development of an Asian theme mall in Richmond Hill indicates the resistance by local residents to a changing urban space with a new ethnic identity (Preston and Lo, 2000: 182–90). Toronto Island also represents a distinct neighbourhood that has survived confrontation with city planners who had decided to transform the area, accessible by a short ferry ride from downtown, into public open space. Its geography—essentially a sandbar extending into Lake Ontario that at one time was attached to the mainland—consists of a series of islands, most of which now is dedicated to parkland. However, residential communities exist on two islands (Ward's Island and Algonquin Island).

Ottawa Valley

Ottawa–Gatineau is not only a major population cluster in Canada, but is distinctive as an area where both official languages are used. As the nation's capital, federal government operations are found on both sides of the Ottawa River. In 2001, Ottawa had a population of just over 800,000 while Hull (now Gatineau) in Québec had nearly 260,000 citizens. Together, Ottawa–Gatineau is the fourth largest metropolitan area in Canada, and in 2011 its population totalled 1,236,324. On the Ontario side, Greater Ottawa is the third-largest urban cluster in Ontario. Much of its growth has come from in-migration from other parts of Canada and from foreign countries. Many are attracted by the employment opportunities offered by the federal government and the business community. In 2001, 18 per cent of its population was foreign-born compared to 15 per cent in 1991. Visible minorities accounted for 14 per cent of the population, with the three leading groups being blacks, Chinese, and Arabs and West Asians.

The federal government, located in the national capital, is the major employer, followed by the high-technology sector. The federal government requires a wide variety of goods and services in its daily operations. This demand provides an opportunity for many small and medium-sized firms in the Ottawa Valley. By locating its departments and agencies in both Ottawa and Gatineau, the federal government has ensured that Ottawa's economic orbit extends to a number of small towns on both the Ontario and Québec sides of the Ottawa River. Most people live on the Ontario side, especially in Ottawa. Other urban centres on the Ontario side include Nepean, Gloucester, Kanata, and the municipalities of Rockcliffe and Vanier. Around Ottawa, in its periphery, are a number of smaller centres, including Pembroke, Cornwall, Brockville, and Kingston. However, with the exception of Pembroke, these cities are more closely tied to economic and social activities in Toronto and Montréal.

As the capital of Canada, Ottawa is the focus of national and international affairs and has subsequently developed into a national administrative centre with a large federal civil service. In its early days, the forest industry played a strong economic

Table 5.8 **Top 10 Countries by Birth of Immigrants* to Toronto, 2006**

Country of Birth	Number of Immigrants	% of Immigrants
India	221,930	9.6
China, People's Republic of	191,120	8.2
Italy	130,685	5.6
Philippines	130,315	5.6
United Kingdom	125,980	5.4
Hong Kong, Special Administrative Region	103,090	4.4
Jamaica	93,845	4.0
Pakistan	85,635	3.7
Sri Lanka	84,225	3.6
Portugal	76,304	3.3
Total of all immigrants to Toronto	2,320,165	100.0

*Immigrants are foreign-born persons who are, or have ever been, landed immigrants in Canada. A landed immigrant is a person who has been granted the right to live in Canada permanently by immigration authorities. Some immigrants have resided in Canada for a number of years, while others are more recent arrivals.

Source: Statistics Canada (2009f).

role, but it was eventually overshadowed by the activities of the federal government and then the high-tech industry. Still, the geographic location made this an ideal place for a pulp and paper mill: pulpwood was easily harvested from the interior forest lands and then transported along the Ottawa and Gatineau rivers to the Eddy Pulp and Paper Mill in Hull, Québec. An added advantage of this site is the availability of low-cost electrical power from the hydroelectric installation on the Gatineau River near Chaudière Falls. Today, the E.B. Eddy plant produces fine and coated paper.

Described in the press as the "Silicon Valley of the North," Ottawa has become an industrial leader in high technology. Ottawa firms specialize in telecommunications, computers, software and computer services, and electronics (Britton, 1996: 266). Northern Telecommunications (Nortel) was the leading high-tech company, specializing in fibre optics. In 2009, Nortel declared bankruptcy and its assets were sold to foreign buyers, led by Ericsson, headquartered in Sweden. The loss of Nortel has seriously hurt Canada's high-tech industry. It remains to be seen if Canada can produce another Nortel. Other leading figures in Ottawa's high-tech world failed and some, like Nortel, have been purchased by larger companies outside of Canada. As James Bagnall (2003) reported, "With software-maker Corel now in the hands of a California-based venture capital firm and fibre-optics star JDS Uniphase set to shift its headquarters next month to San Jose, Ottawa is swiftly reverting to its former role as a technology R&D centre, serving the whims of multinationals based elsewhere."

Southwestern Ontario

Southwestern Ontario is the third major urban cluster in southern Ontario. About 1 million people live in this area, which extends from Lake Erie to Lake Huron. Although this region does not have a major metropolitan city, the Kitchener CMA (Cambridge, Kitchener, and Waterloo), with a 2011 population of 477, 160, and London (population 474,786) are the largest cities in southwestern Ontario. These urban centres are close to Toronto, Buffalo, and Detroit. Windsor and Sarnia are on the edge of the London/Kitchener urban cluster and fall into the orbit of Detroit.

London provides administrative, commercial, and cultural services for the larger region. It is also the headquarters of several insurance companies, including London Life Insurance Company. Among its economic activities, London is noted for manufacturing, including the production of armoured personnel carriers and diesel locomotives by General Dynamics. London, therefore, has a sound and growing industrial foundation based on insurance, manufacturing, and high-tech industries.

THINK ABOUT IT
Can Canada grab a share of the global high-technology business without substantial support similar to that received by the automobile and aircraft manufacturers?

Such manufacturing firms pay relatively high wages to their employees, and consumer spending by these employees supports a strong retail sector.

Automobile assembly plants are located in Cambridge, Ingersoll, Alliston, Woodstock, and Windsor (Table 5.5 and Figure 5.9). Hitachi produces Euclid-Hitachi trucks at its plant near Guelph. Auto-parts plants play an important role in the economy of southwestern Ontario while a number of high-tech firms, particularly in the Kitchener CMA, add to the region's economic diversity. This area comprising Kitchener, Waterloo, and Cambridge has been known for more than 20 years as Canada's "Technology Triangle" where innovative technology is often developed in research institutes or universities. Such technology is then used to create commercial products.

Cities of Northern Ontario

In sharp contrast to cities in southern Ontario, the economies of urban centres in northern Ontario are stalled and their populations static or declining, so that many northern Ontarians are dissatisfied with their lot. Such grumbling is common in resource hinterlands. Claiming to be isolated and ignored by the provincial government, some of its citizens have called for a new province called Mantario (Matteo, 2006). Like most downward transitional regions, the resource base is losing its economic strength for three reasons: (1) the most accessible mineral and timber resources have been exploited, so resource production is now more costly; (2) depressed prices for lumber have resulted in mill closures; and (3) when companies introduce more technology, fewer workers are required.

Timmins, located in the mineral-rich Canadian Shield, is in the centre of a gold belt, while Sudbury is in a world-famous nickel belt, where the mining and smelting of nickel and copper ores are core functions. Both cities began as single-industry towns, but over time they have become important regional centres with many service industries and more diversified economies, factors that have supported community stability. Even with these efforts, however, the two communities have been unable to maintain their populations over the past 15 years. For example, from 1996 to 2011, the populations of Sudbury and Timmins declined. Sudbury's population dropped from 165,336 in 1996 to 160,770 while that of Timmins fell from 47,499 in 1996 to 43,165 in 2011 (Statistics Canada, 2002, 2007c, 2012g). For the past 10 years, Sudbury has enjoyed a slight increase in population; the Timmins population has stabilized in the last five years (Table 5.9).

Sault Ste Marie is a steel town located on the Great Lakes between Lake Superior and Lake Huron. Like most other centres in northern Ontario, its population trend over the last 15 years has been downward, declining from 83,619

Table 5.9 Urban Centres in Northern Ontario, 2001–2011

Centre	Population 2001	Population 2011	% Change 2001–11
Sudbury	155,601	160,770	3.3
Thunder Bay	121,986	121,596	0.0
Sault Ste Marie	78,908	79,800	1.1
North Bay	62,303	64,043	2.8
Timmins	43,686	43,165	−1.2
Kenora	15,838	15,348	−3.1
Temiskaming Shores*	12,904	13,566	5.1
Elliot Lake	11,956	11,348	−5.1
Kapuskasing	9,238	8,196	−11.3
Kirkland Lake	8,616	7,334	−14.8
Total	521,036	525,166	0.7

*Temiskaming Shores is the newly restructured area comprising the three former municipalities of New Liskeard, Haileybury, and Dymond Township.

Source: Statistics Canada (2007c, 2012g).

in 1996 to 79,800 in 2011 (Statistics Canada, 2002, 2007c, 2012g). The town's major firm is the former Algoma steel mill, now called Essar Algoma. The mill has struggled to succeed in a highly competitive marketplace. Geography poses a major disadvantage because the steel plant is located at some distance from its major industrial customers in southern Ontario and the United States. With high transportation costs, the Algoma mill is on the edge of failure. While Essar Algoma maximizes its use of local railway and seaway connections to secure raw materials and to market its steel products, the mill has struggled over the years to keep afloat. After Essar Steel Holdings Ltd purchased the firm in June 2007, the market for steel plummeted because of the start of the global financial crisis in the following year. After a series of difficult years, the firm managed to turn a profit in its 2012 fiscal year largely because sales to the US increased by nearly 10 per cent.

Thunder Bay, at the western end of Lake Superior, is a major transshipment point and a key element in the east–west transportation system. Bulky products—particularly grain, iron, and coal—are shipped to Thunder Bay for loading onto lake vessels. In recent years, however, this function has diminished. In the past, for example, iron from mines at Atikokan passed through Thunder Bay on its way to steel mills located along the shores of the Great Lakes. This mining operation is now defunct. Grain and potash from the Canadian Prairies are the major products handled at Thunder Bay. Until 1996, Ottawa heavily subsidized grain shipments under the Crow Benefit. With the elimination of that subsidy, the advantage of sending grain by rail to Thunder Bay has diminished, thus weakening its grain transportation role. In addition, the rise of Asian markets for resources from Western Canada has cut into Thunder Bay's importance as a transshipment centre.

For more on the Crow Benefit, see Chapter 8, page 327.

As a forest and transportation centre, Thunder Bay grew rapidly after World War II. However, new economic conditions slowed the growth of its economy and population. The loss of the American softwood lumber market hurt, as did the decline in grain moving from the Prairies through Thunder Bay. Its population dropped from 126,643 in 1996 to 121,986 in 2001 (Statistics Canada, 2002, 2007c, 2012g). In the last decade, however, Thunder Bay's population has stabilized at just under 122,000 (Table 5.9).

Overall, thanks to modest population increases in the larger centres—Sudbury, Sault Ste Marie, and North Bay—the cities in northern Ontario just held their own over the most recent census period (2001–11), but the smaller centres—Timmins, Kenora, Elliot Lake, Kapuskasing, and Kirkland Lake—all lost population. The only exception was Temiskaming Shores. Yet, cities and towns in northern Ontario are struggling to diversify. One option being pursued is tourism. Many small private firms—motels, summer camps, and fishing and hunting lodges—have sprung up (especially in smaller communities) to meet the demand from the growing number of tourists coming to experience the wilderness of northern Ontario. The urgency to diversify is driven by the continuing reduction in the size of the primary labour force. For example, the workforce at the nickel mines in Sudbury has

VIGNETTE 5.15

Population Distribution in Northwestern Ontario

Northwestern Ontario stretches from Lake Superior to Hudson Bay and James Bay. This vast area has less than 225,000 people and consists of three census divisions—Kenora (57,607 in 2011), Rainy River (20,370), and Thunder Bay (146,057). Most people live in a few large urban centres while the remainder reside in mining towns and First Nation communities. Northwestern Ontario's population geography has three distinct zones:

- Zone 1: Most people (67 per cent) reside in Thunder Bay. In 2011, this city had 121,596 people.
- Zone 2: Most of the remainder (27 per cent) live south of the CN rail line, where a number of small towns, villages, and First Nation reserves are located.
- Zone 3: North of the CN rail line is the "empty" population zone, which is beyond the reach of the national highway system and has only 2 per cent of the region's population, consisting of a few mining centres and First Nation settlements.

fallen by half over the last 30 years. Sudbury has reacted by aggressively seeking to expand its service industry. The city has met with some success, partly because of its strategic location at the junction of three highways—the Trans-Canada Highway, Highway 69 (which leads to Toronto), and Highway 144 (which connects Sudbury with Timmins)—and partly because of its efforts to have federal and provincial agencies (the federal National Tax Data Centre and provincial Department of Mines) and institutions (Laurentian University and the Science North Museum) locate in Sudbury.

Many First Nations are found in northern Ontario. Some are situated in remote locations, such as the Kashechewan First Nation reserve. The residents of Kashechewan, located near James Bay some 450 km north of Timmins, were evacuated three times in 2005–6 because of spring floods and polluted drinking water. The federal government was considering relocating the reserve to higher ground, but another option was to move the reserve to the Timmins area where access to drinking water, health, housing, schools, and employment would be greatly improved. Such an integrationist approach has some merit, but social engineering strategies also involve certain dangers. The mayor of Smooth Stone, near Timmins, was not consulted about the proposed relocation to his community and was not enthusiastic. With the closing in July 2006 of a timber mill, 230 employees lost their jobs. As the mayor stated: "There's no work for my own people. You don't take a community in crisis and dump them on another community in crisis. This has to be a win-win situation. Our community has to be respected" (Campion-Smith, 2006). However, culture trumped economics. On 30 July 2007 Kashechewan Chief Jonathan Solomon signed an agreement with Ottawa to redevelop the community at its present location, although in 2012 a new chief, who has expressed interest in relocation, was elected.

Beyond the towns and cities of northern Ontario lies a sparsely populated area where fewer than 10,000 people live. Cree and Ojibwa Indians live on isolated reserves or in small Native settlements in this vast region of scrub forest and muskeg. Only a few gold mines, trapping cabins, and fishing lodges dot the landscape. For the Indian residents, fishing and hunting are an important source of food, while trapping and guiding provide cash income. Since this cash income usually does not meet their basic needs, transfer payments supplement their income. The development of the Victor diamond deposit near the Attawaspiskat First Nation reserve provides a potential economic boost for the band because De Beers agreed to hire local workers for both its construction and production operations. Two other proximate Aboriginal groups, Moose Cree First Nation and Taykwa Tagamou Nation, also signed agreements with De Beers for education, training, and compensation (Bone, 2012: 245). This open-pit mine opened in July 2008, under budget and ahead of schedule, and in 2009 was named "Mine of the Year" by readers of the international trade publication *Mining Magazine*.

SUMMARY

Ontario, located in Canada's most favoured physiographic region, faces serious economic challenges for two reasons. First, the shrinking of the US market has had an enormous negative impact on Ontario, especially for its manufacturing and forest industries. Second, the global economy has seen manufacturing shift to China and other low-wage countries, leaving Ontario's manufacturing sector extremely vulnerable. Over the last 10 years, a declining US market for softwood lumber and paper products has devastated northern Ontario. Sawmills and pulp and paper plants have closed and those still operating confront a gloomy future with more shutdowns possible. Southern Ontario has its own challenges. Its manufacturers also are looking at a bleak future. Fierce competition from foreign countries with low labour costs will continue to dog Ontario-based firms, forcing them to make hard decisions, such as to reduce their labour costs, search for niche production, and, if all else fails, either relocate to low-wage countries or close shop. Within North America's automobile manufacturing sector, Canada is lagging behind Mexico and the United States in securing new investment in the automobile industry. The primary reason is much lower wages in these two countries, particularly Mexico. In a broader context, globalization translates into lower wages for manufacturing workers in developed countries. The loss of manufacturing jobs in Ontario began over a decade ago. In 2012, the shocking ultimatum by Caterpillar officials to its Ontario workers—take a 50 per cent pay cut or lose your job—was not a pleasant trade-off, but that is the harsh face of the global corporate world.

Historically, Ontario's economy has been oriented to continental trade, and in a globalizing world fresh directions and investments are needed to get back on the domestic and global economic highways. After five years and counting, the strategy of waiting for the recovery of the US economy and the resumption of high levels of automobile and lumber exports to the US seems misplaced. For Ontario to prosper and progress into the next decade as well as to retain its title as the economic engine of Canada, this dominant core region must turn to the new economy more quickly and accelerate the development of knowledge-based firms and a highly skilled workforce by investing heavily in innovation and research partnerships among universities, federal research institutions, and private industry. Ottawa can help, too—perhaps by shifting a portion of its financial support for the automobile industry to high-tech businesses. By succeeding in these areas, Ontario can hope to again secure its place within the North American market by developing a stronger north–south economic axis while maintaining its traditional centrality as the core in the east–west axis.

CHALLENGE QUESTIONS

1. For much of its recent history, Ontario has fashioned its prosperity on trade with the United States. Is this "all eggs in one basket" strategy still valid?
2. Has globalization doomed Ontario manufacturing workers to fewer jobs at lower wages?
3. What is needed to pull the forest economy of northern Ontario out of its tailspin?
4. Is the Ontario government's "green strategy" for electric production and for electric cars a move in the right direction, or is it going to add to Ontario debt?
5. Hobson's choice: If the knowledge-based economy represents the future of Ontario, would you have "saved" Nortel rather than General Motors and Chrysler?

FURTHER READING

Florida, Richard. 2012. *The Rise of the Creative Class—Revisited*. New York: Basic Books.

Without a doubt, Richard Florida has turned the academic world on its head with his creative class thesis whereby urban economic development is driven by urban amenities. For Toronto and other metropolitan cities in Canada, Florida sees the emergence of a new social class based on creativity transforming the very nature of city life. His argument is that creativity is the driving force behind the emerging urban society, where more and more emphasis is placed on cultural amenities such as the arts and sports. As Florida (xxiii) puts it, "the role of creativity as the fundamental source of economic growth and the rise of the new Creative Class" represent the common thread underlying the broad social changes taking place in major cities. Companies need talented people and talented people want urban amenities.

Yet, Florida presents a puzzle, claiming that unlike past emerging classes of people with common traits and concerns who steered the direction of their society, the creative class is unaware of its own existence and thus unable to consciously influence the course of the society it leads. Perhaps the answer lies in applying the biological notion of osmosis whereby social change gradually seeps into the urban society, thus modifying society without a direct plan of action?

Urban amenities take many forms. Nowhere is urban transformation more apparent than in the arts. Hracs et al. (2011) have observed how musicians—a leading edge of the creative class—fashion a vibrant and innovative music scene. Like other highly educated and mobile creative people, musicians have clustered together in major cities and, like technology firms, benefit from the interaction among themselves. Toronto's award-winning Broken Social Scene, a pop music collective of ever-changing membership, is a case in point. Both technologists and musicians love the life they are pursuing, though in most instances the material reward for musicians falls far below that for those involved in high-technology innovation.

6

QUÉBEC

Introduction

"La Belle Province" retains its place as a cultural, economic, and demographic power within Canada. Of the six geographic regions, Québec alone has a dominant francophone culture, thus distinguishing it from the rest of Canada, but also from the rest of North America. For that reason, unlike the rest of Canada, culture and identity remain critical, even explosive, political elements within Québec society. Equally challenging for Québec, its power base within Canada is slowly eroding. While Québec continues to rank second among the six regions in terms of economic output and population size, this ranking is less secure than ever before because of the demographic and economic growth in Western Canada and British Columbia.

Québec's culture adds a powerful third dimension to its economic and demographic contribution to Canada. By virtue of its geography, language, and history, Québec occupies a unique place in Canada and North America. As a French-speaking region making up about 2 per cent of the population of English-speaking North America, Québec worries about its vulnerable situation, especially the well-being of its French language. This concern is compounded by the close links of its economy to the rest of Canada and to the United States, which dictates that English is the language of business and tourism.

Québec is by far the largest province in Canada. The St Lawrence River provides an entry to the heart of North America while the vast mineral and water resources of northern Québec supplement its economy and serve as the homelands of the **Eeyou Istchee** (the James Bay Cree) and the Inuit of **Nunavik**.

CHAPTER OVERVIEW

This chapter will examine the following topics:

- Québec's physical geography and historical roots.
- The basic elements of Québec's population and economy in the context of the region's physiography.
- Québec's cultural and economic development within the context of the core/periphery model.
- Québec's position within Canada, its francophone culture, and its separatist movement.
- The long-term effects of the James Bay and Northern Québec Agreement and the Paix des Braves for the Cree.
- Hydro-Québec's strategy of developing low-cost electrical power in northern Québec to stimulate industrial activities in southern Québec.
- Québec's changing role within Canada and North America.

Given the vast hydroelectric resources of northern Québec, the "Key Topic" in this chapter is the harnessing of these waters by Hydro-Québec, with special emphasis on the James Bay Hydroelectric Project. This project—only partially completed—has placed Québec in the forefront of hydroelectric production,

The Cap-des-Rosiers Lighthouse, Gaspé, Québec. The Cap-des-Rosiers Lighthouse is in the village of Cap-des-Rosiers near the tip of the Gaspé Peninsula. It was erected in 1858 and is Canada's tallest lighthouse at 34.1 metres. Photo: Claude Bouchard/Corbis.

stimulated its economy, and made Québec an essential source of power for New England. At the same time, the James Bay Project led to the first modern land claim agreement, the 1975 James Bay and Northern Québec Agreement (JBNQA), which was followed 30 years later by the Paix des Braves. The JBNQA also laid the groundwork for the political evolution of the Inuit of Québec, who are close to achieving a new form of public self-government within a province.

Québec's Culture, Identity, and Language

Following in the footsteps of Samuel de Champlain, settlers came from France to the St Lawrence Valley more than 400 years ago. Over the centuries, by clinging tenaciously to the land through many adversities, their ancestors made this corner of North America a homeland for the French-speaking people. La Fête nationale du Québec (the national holiday of Québec) celebrates this accomplishment on 24 June each year. However, the road has not always been easy or pleasant, and the collective memory often recalls humiliations and resistance as well as hard-fought victories. Flowing through this collective memory is the fear of a loss of Québec's culture, identity, and language. Separatism is one response to that fear while another is Confederation, which allocates some powers to Québec.

In spite of all that has passed, however, Québec remains a vital part of Canada with its distinct culture and language recognized and respected by the rest of Canada more than ever before.

The French language, Québécois culture, and a francophone identity have generated a strong sense of belonging among the majority of Québec citizens, forming the basis of ethnic pride, loyalty, and nationalism, which from time to time fuels the desire for an independent political state. Such feelings are particularly strong among the "old stock"—the descendants of some 10,000 settlers who migrated from France in the seventeenth and eighteenth centuries—but are less strong among anglophones and allophones. While their first loyalty is often to Québec, most Québécois also have a strong attachment to Canada. This sense of dual loyalty is unique in Canada, but it underscores the important role of the Québec government as the protector of the French language, Québécois culture, and the francophone identity. As well, it explains Québec's interest in the concept of a political partnership within Confederation. In the past, federal politicians have avoided referring to Québec as a nation because they feared the acceptance of the word "nation" would encourage the separatist movement as well as alienate the other provincial governments, which subscribe to the concept of 10 equal provinces. Times have changed, however, and in November 2006 the Canadian Parliament unanimously passed a motion by Prime Minister Harper that recognized the Québécois as a nation within Canada.

For a more complete discussion of the historic origins of the concept of a political partnership, see "One Country, Two Visions" in Chapter 3, page 114.

Québec's culture is largely derived from the historical experience of **francophones** living in North America for over 400 years and of their being part of Canada for more than 145 years. Clearly, Québec's sense of place and identity is based on its struggle to survive within an English-speaking North America. The decision to join Confederation has had both benefits and costs. Confederation provides a safe place within North America, though relations between French- and English-speaking Québecers have been strained from time to time. The resolution of these strains has led to a type of federation built on compromise. This complex concept of compromise—as identified by Saul in his book, *Reflections of a Siamese Twin*—helps defuse tensions between French- and English-speaking Québecers. As a living, dynamic concept, "compromising" federalism continues to evolve and survive.

For a broader historic discussion of two competing visions of Canada, see "The French/English Faultline" in Chapter 3, page 110, and "French/English Language Imbalance" in Chapter 4, page 156. For a brief account of John Ralston Saul's concept of Canada, see "Faultlines within Canada" in Chapter 1, page 10.

Québec, while home to the largest francophone population outside of France, contains a

> **THINK ABOUT IT**
> Are ideologies like railway tracks? If so, does the concept of two founding peoples—if accepted by Ottawa over the model of 10 equal provinces—inevitably lead to two independent states, Québec and the rest of Canada?

> **THINK ABOUT IT**
> Since the paramount role of the government of Québec is to protect the French language, Québécois culture, and the francophone identity, does Québec have enough powers under Confederation to fulfill this role or is it necessary to form an independent country?

minority composed of **anglophones**, whose mother tongue is English, and **allophones**, immigrants whose mother tongue is neither English nor French. This minority is concentrated in the Montréal and Gatineau areas. The **Québécois** whose first language is French represent the dominant demographic force in the province. The remaining Québecers include Aboriginal peoples (whose mother tongues include Cree, Inuktitut, and Innu-aimun). Yet, from a linguistic perspective, almost all speak some French. In northern Québec, Cree, Inuit, and Innu (Naskapi) form the majority. From time to time, social tensions surface between the French-speaking majority and the province's minority groups as the Québécois continue to assert their desire to be "*maîtres chez nous.*"

In the 2012 election, the perceived increase in the use of English, especially in Montréal, triggered a vigorous debate, with Parti Québécois leader Pauline Marois, who would win a minority government, playing the "identity card" by promising to prevent newcomers whose French is limited from running for public office. Once in office, the Marois government, to provide more protection to the French language, introduced Bill 14, which is designed to further limit English and thus strengthen the use of French. Under Bill 14, small companies will join larger companies that must operate in French. The new PQ government also has demanded that federal institutions in Québec function in French and has restricted the access of francophone and ethnic students to English-language colleges. In addition, the PQ Minister for Education, Marie Malavoy, plans to end an intensive English-language program for francophone students, a program the previous Liberal government introduced in francophone schools to create more career opportunities for students (Séguin, 2013).

Even though the number of Québec residents who speak only English declined from 6.6 per cent to 6.2 per cent from 2006 to 2011 (Statistics Canada, 2013: Table 5), reopening the language debate and the more aggressive activities of the "language police" from the Office québécois de la langue française have created an uneasiness among non-francophones in Montréal. The actions of a language inspector who ordered a restaurant owner to remove the word "pasta" from his menu is an extreme example of common sense running aground (Figure 6.1). But more

Figure 6.1 **Pastagate: Language inspector rejects "pasta" on Italian restaurant menu**
The maelstrom that erupted over Pastagate may have reduced the chances of Bill 14 passing in the National Assembly. In any case, one casualty was Louise Marchand, head of the Office québécois de la langue française, who was forced to resign because one of her language inspectors told a Montréal Italian restaurant owner that words on the menu such "botiglia," "pasta," and "antipasto" must have French translations.

CONTESTED TERRAIN 6.1

Fortress Québec?

The Québec government passed legislation (Bill 101) in 1977 limiting the use of English in order to preserve the French language and has continued to revise and tighten the legislation. Even so, from the perspective of Martin Lemay, the Parti Québécois MNA for St Marie–St Jacques, Québec needs to continue to strengthen the use of French, a point of view he expressed in an article titled "Francophones Have Reason to Be Paranoid" (Lemay, 2010). On the other hand, a leading Montréal businessman, Stephen Jarislowsky[1] (2012: A19), argues that language policy is turning Québec into a "hermit state," isolating Québec from Canada and the rest of the world.

importantly, is Québec shooting itself in the foot over the language issue? Aubin (2013) seems to think so. As an unnamed businessman interviewed by Aubin said: "The problem is not just that anglos are leaving Quebec—they've been leaving for years and years. The problem is also that we've built a great big fence around Québec that effectively keeps outside talent out. Any dynamic economy has to cross-fertilize with other cities and bring in new talent." This raises the question of whether, in its enthusiasm to strengthen French, the Québec government has slipped into an anti-creative class mentality, which, according to Florida, would keep talented people from moving to Montréal and thus harm economic growth in Montréal and, to a lesser degree, in the rest of the province.

VIGNETTE 6.1

Sovereignty: Now or Tomorrow?

Sovereignty remains a powerful undercurrent in Québec society. Still, Pauline Marois's bold declaration in her victory speech after the September 2012 election—"we want a country and we will have it"—is not supported by the results from the election. In fact, with only 31.9 per cent of Québecers voting for the Parti Québécois (Table 6.1), another referendum is highly unlikely in the foreseeable future. Yet, the actions of the Marois government have stirred the French/English language faultline in Montréal and, to a lesser degree, in Québec as a whole.

Table 6.1 **Québec Election Results, 4 September 2012**

Party	Number of Seats	Popular Vote (%)
Parti Québécois	54	31.9
Liberals	50	31.2
Coalition Avenir Québec	19	27.1
Québec Solidaire	2	6.0
Others	0	3.8
Totals	125	100.0

Source: La Presse, Élections Québec 2012.

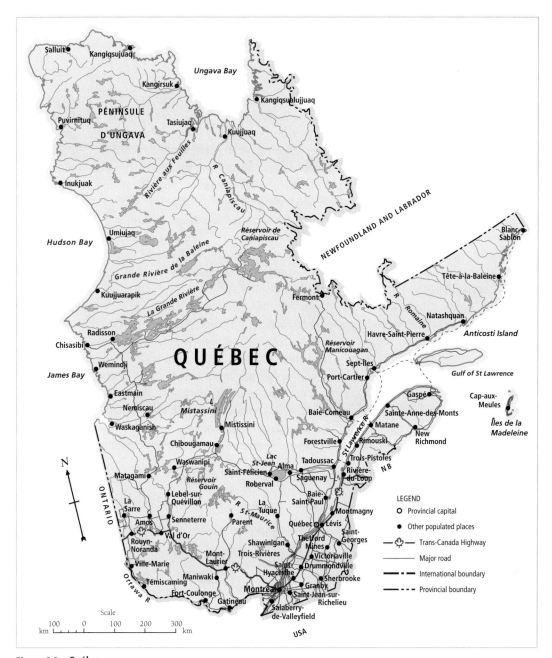

Figure 6.2 **Québec**
Source: Adapted with the permission of Natural Resources Canada 2013, courtesy of the Atlas of Canada. At: <atlas.nrcan.gc.ca/site/english/maps/reference/provincesterritories/quebec>.

Québec's Place within Canada

Québec's economic and demographic position within Canada remains just behind that of Ontario. Yet, Québec's economy and population are losing ground to the rest of Canada. While still ranked second, Québec production dropped from 21.1 per cent of GDP in 2001 to 19.6 per cent in 2010.

Its population fared better, only slipping from 24 per cent in 2001 to 23.6 per cent in 2011 (Figure 6.4). Population increase in Québec was due to an unexpected boost in the province's fertility rate—in defiance of the demographic transition theory. In 2005, the birth rate was 1.52 compared to 1.54 for Canada as a whole, but by 2009 Québec's birth rate had reached its highest point

VIGNETTE 6.2

The St Lawrence River

The St Lawrence River provides a natural waterway into the interior of North America and, consequently, played a key role in the history of New France. Today, the St Lawrence is an essential part of North America's transportation system. Cities along its shores, particularly Québec City and Montréal, have benefited greatly from the waterway's role as a major trading route. Montréal's favourable location on the St Lawrence gives it an economic advantage and fuels the city's growth. When the era of supertankers began in the 1950s, these ships required a much greater depth of water to reach the Great Lakes. For around a decade, Montréal served as a transshipment centre, transferring the sea cargo to rail cars.

After the construction of the St Lawrence Seaway in 1959 Montréal lost its natural advantage as a transshipment point. Still, the city remains an important port, especially for its growing container traffic. From 2001 to 2011, container tonnage at the Port of Montréal increased from 8.7 million to 12.5 million metric tonnes. In 2012, 28.4 million metric tonnes of cargo—categorized as liquid, dry, containerized, and non-containerized—reached the docks at Montréal (Port of Montréal, 2012).

As the second decade of the twenty-first century unfolds the Seaway and the Port of Montréal (photo 6.2) face major challenges. First, the anticipated trade agreement with the European Union could boost trade through the Port of Montréal. Second, a shift of Western Canada grain cargo to Pacific ports and Asian markets has greatly reduced that traffic through the Port of Montréal. At the same time, the downsizing of the North American steel industry saw the demand for iron ore shipments from northern Québec/Labrador decline sharply. Third, the long-term drop has meant that, for the past two decades, the Seaway has operated at about 60 per cent capacity. The magnitude of the decline in traffic, notably in grain and iron ore, is revealed by the drop in shipments from a high annual average tonnage of around 75 million in the late 1970s to 38 million tonnes in 2011 (Jenish, 2009: 41; St Lawrence Seaway Management Corporation, 2012: 4).

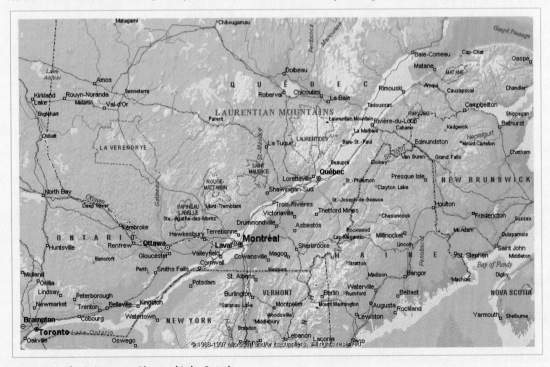

Figure 6.3 **The St Lawrence River and Lake Ontario**
Source: Paul Hebert, Biodiversity Institute of Ontario. At: <www.aquatic.uoguelph.ca/rivers/stlawmap.htm>.

VIGNETTE 6.3

Québec Students March

Where else in Canada would thousands of students demonstrate in the streets so passionately, day after day, for over five months? The rest of Canada stood by, dumbfounded by this seemingly irrational act. How could students who enjoy the lowest tuition fees in Canada be annoyed with a relatively modest increase?

Vive la difference; c'est Québec, pas Ontario, where students face nearly double the Québec tuition fees. During the demonstrations, many protestors wore red-square or *carré rouge* emblems, symbolizing the crushing debts faced by students that place them "squarely in the red." And yet, the proposed tuition increases in the provincial Liberal government's Bill 78 would still have left Québec students with the lowest tuition fees in Canada. So what motivated the daily marches? While the spark igniting the students back in February 2012 was the government's announcement of tuition fee increases, a complex mix of factors sustained the protest, including a desire to challenge the Liberal government on a number of issues and a rationale based on the Quiet Revolution policy in the 1960s of setting low tuition fees to draw students into post-secondary education and thus to catch up with the rest of Canada and the Western world. As former Premier Jacques Parizeau reflected: "In 1962, we had the world's lowest secondary schooling rate in the so-called civilized world, with Portugal" (Lagacé, 2012: A8). Other factors were at play, including (1) student debt, a highly unionized student body, and its charismatic student leadership; and (2) a desire to challenge the Liberal government in its efforts to balance the provincial budget, which, in the students' minds, was on the backs of students and lower- and middle-income Québecers. Soon, the marches morphed into a Québec version of the anti-corporate movement known as Occupy Wall Street. By April, with warm evening weather arriving, thousands marched with hundreds of police following and helicopters overhead. Such demonstrations are a sign of the passionate nature of Québec society rarely witnessed in the rest of Canada.

Photo 6.1

The Québec student strike started on 13 February 2012 and ended shortly after the announcement on 1 August of a provincial election. The first students to take to the streets were social science students at Université Laval. Within a month, hundreds of thousands of students were marching. Daily protests took place and in the summer, evening marches became the order of the day. As shown here, on 22 May the "largest act of civil disobedience in Canadian history" took place in Montréal when between 400,000 and 500,000 people marched (Schonbek, 2012).

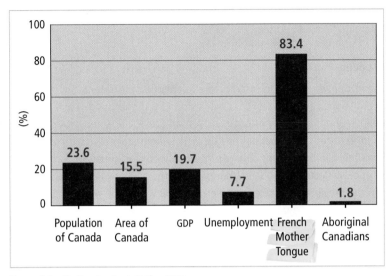

Figure 6.4 Québec basic statistics, 2011
By population size and gross domestic product, Québec remains the second-ranking geographic region in Canada, though Western Canada and British Columbia are narrowing the gap.
Source: Tables 1.1, 1.2, 5.1.

in many decades at 1.74, exceeding the national figure of 1.67 (Statistics Canada, 2011).

 Annual crude birth rates for Canada are found in Table 4.6, page 136.

In both cases, the rate of economic and population gains taking place in British Columbia and Western Canada are outstripping those in Québec. For example, since 2006, the combined 2006 populations of British Columbia and Alberta (7,686,215) exceeded that of Québec (7,546,100) by 140,115; by 2011, this gap had reached 142,313. From another perspective, the percentage gain in population from 2006 to 2011 accentuates this demographic point, with Québec, at 4.7 per cent, far below Alberta at 10.8 per cent, British Columbia at 7 per cent, and Saskatchewan at 6.7 per cent. Two questions are central to understanding Québec's future position within Canada and the future of French in Québec:

THINK ABOUT IT
Has the decline in Québec's share of Canada's population eroded its political influence in Ottawa and with the rest of the provinces?

1. In 1966 Québec's share of the total Canadian population was 28.9 per cent but by 2011 this share had slipped to 23.6 per cent. Will this long-term downward trend continue?
2. If most population gains in Canada come from immigration, will this put more pressure on the issue of "accommodation" of minorities and, at the same time, reduce the use of French in Montréal?

Concern over sluggish economic growth and low birth rates is not new to Québec. In 1995, the separatist leader, Lucien Bouchard, called on Québécoise women to have more children. Eleven years later, Bouchard, who had left politics for the business world, was worried about what he saw as an economic malaise in Québec. He led a group of business leaders in preparing a manifesto, *Manifest—pour un Québec lucide* [Manifesto for a clear-eyed vision of Québec]. The manifesto sees the province slipping behind other provinces and states. It charts a different path for the province's economic future and calls for greater involvement of the private sector. Québecers, these business leaders believe, must work harder because other provinces and states have higher levels of worker productivity (Bouchard et al., 2006). Yet, this call for a different path may not resonate because Québec has always maintained a strong involvement in its economy as exemplified by Hydro-Québec and in the somewhat nationalistic platform of the recently elected Parti Québécois.

Why has Québec's position within Canada weakened since Confederation? First, of course, Québec's geopolitical position within Canada has changed—from being one of four provinces in 1867 to one of 10 provinces and three territories in 2012. Equally important, Québec's demographic clout has shrunk from 32 per cent of Canada's population in 1871 to 23.6 per cent and 2009, culminating with a $4 billion sale of high-speed trains to China. Bombardier, now a well-established global company, produced these trains in China using its Zefiro technology, which boasts speeds of 380 km/h. The Montreal-based manufacturer operates a joint venture, Bombardier Sifang Transportation Ltd, in China and in September 2012 was awarded a $2 billion order for its Zefiro trains (Deveau, 2012). Bombardier has had several high-profile contracts—the Beijing Olympics in 2008 and the World Cup soccer matches in South Africa in 2010—that project its global reach. By mid-2012, Bombardier regained its footing in the US market with a sale to Amtrak for service on its Northeast corridor. Prospects for future US sales appear bright as regional passenger train systems, such as the California High-Speed Rail System, are planning to replace aging rolling stock with high-speed trains.

 For more on Bombardier and SNC-Lavalin, see below, "Québec's Economy," page 237, and Vignette 6.7.

The challenge for Québec is to continue on the path to create a more knowledge-based economy within its industrial core, to advance its global reach, and to ensure an atmosphere of harmony between its francophone majority and its minority-language groups. The rising temperature along the French/English language faultline initiated recently by the separatist Québec government can only heat up the existing atmosphere of language harmony and limit the prospects for an inflow of talented people who are so essential for what Florida calls the creative class. Ironically, the business community, especially the retail sector in Montréal, prefers bilingual employees, thus forcing non-English-speaking immigrants to learn English in order to get a job.

The minority Parti Québécois government faces a huge provincial deficit ($4.5 billion), and so the government has had little choice but to reduce public expenditures. Québec is not alone. Canadian provinces are carrying substantial debt loads, with Ontario leading the way with a debt of nearly $20 billion.

Petroleum refining continues as the second-largest manufacturing enterprise in Québec. The refining business is based on high-cost Middle East oil, which is shipped by tankers to Montréal and Québec City. Shipping lower-cost oil from Alberta could benefit both Alberta and Québec. The demand for gasoline, however, is declining as more drivers own smaller, more fuel-efficient cars. As a result, refineries are closing across North America, with Montréal's Shell refinery having ceased operations in early 2010 (VanderKlippe, 2010). However, lower-cost Alberta oil refined in Montreal and possibly Saint John, New Brunswick, would offer the opportunity to sell the product in the cities along the US eastern seaboard.

The francophone cultural industry is one of Québec's economic strengths. Generously supported by the provincial government and the people of Québec, the arts remain a vibrant element within Québec society and its economy. Québec's film industry has a worldwide reputation and its music community, ranging from popular to classical music, ranks highly within Canada and North America. Montréal is key to the province's cultural renaissance, followed by Québec City. For instance, Québec City's Winter Carnival makes it an international winter tourist destination.

Unlike the rest of Canada, Aboriginal relations with the provincial government are conducted on a nation-to-nation basis. One outcome is the sharing of benefits of natural resource development and the resulting diminishing of dependency on government. For the Cree, a close partnership with the provincial government resulted from a series of agreements beginning with the James Bay and Northern Québec Agreement (1975) over hydroelectric development in the James Bay region. The emergence of Aboriginal regional governments in northern Québec benefited both the province and the Cree and the Inuit, whose traditional homelands are Subarctic and Arctic Québec. In the last days of the Charest Liberal government, the Cree Nation and Québec signed a land management arrangement that grants the Cree more powers over land and resources and provides for a Cree regional government that will replace the municipality of James Bay (Canadian Press, 2012b).

Québec's Physical Geography

Québec, the largest province in Canada, has a wide range of natural conditions and physiographic regions. Its climate varies from the mild continental climate in the St Lawrence Valley to the cold Arctic climate found in Nunavik (Inuit

THINK ABOUT IT
Distance provides a measure of Québec's enormous size. Using the scale in Figure 6.2, what is the distance between the two major cities in the St Lawrence Lowland? And what is the distance from Montréal to Kuujjuaq, the largest community in Arctic Québec?

Photo 6.3
Estrie, formerly called the Eastern Townships, lies in the Appalachian physiographic region. Dairy farms are found in the rolling countryside, which is surrounded by wooded uplands. Hay is the principal crop and is used as feed for dairy cows.

lands lying north of the fifty-fifth parallel). Four of Canada's physiographic regions extend over the province's territory—the Hudson Bay Lowlands, the Canadian Shield, the Appalachian Uplands, and the Great Lakes–St Lawrence Lowlands—and each has a different resource base and settlement pattern (Figure 5.4). The Canadian Shield extends over most of Québec—close to 90 per cent—while the Appalachian Uplands and Great Lakes–St Lawrence Lowlands together form nearly 10 per cent. By far the smallest physiographic region is the Hudson Bay Lowlands, which constitutes around 1 per cent of Québec's land mass.

The heartland of Québec is the St Lawrence Lowland. Formed from the Champlain Sea some 10,000 years ago, this physiographic region provides the best agricultural land in Québec. Settlers from France began farming along the edge of the St Lawrence River some 400 years ago. New France was established within this physiographic region and, by means of the St Lawrence, spread into the interior of North America. It remains the cultural core of Québec. The St Lawrence Lowland offers several natural advantages that give the region its central role in modern Québec. Arable land is one such advantage. By far the best agricultural land is in a small area between Montréal and Québec City. As well, most industrial plants are located in this same area.

The Appalachian Uplands physiographic region is an extension of the Appalachian Mountains, which stretch northward from the state of Georgia. While this ancient geological feature reaches into Québec and Atlantic Canada, its

VIGNETTE 6.4

Variations in Agriculture Resources in Québec's Physiographic Regions

In 2006, Québec had 3.5 million hectares of farmland with just over half in crops (Statistics Canada, 2008). The remaining land was in pasture. Within each physiographic region, climate and soils vary widely. Most agricultural land is found in the St Lawrence Lowland with a secondary core in the western section of the Appalachian Uplands (Estrie). Much farther north in the Clay Belt of the Canadian Shield, the shorter growing season limits crop agriculture to hay and other forage crops. Encouraged by climatic and economic factors, dairy farming dominates, making up more than half of the commercial farms in the province. However, dairy farming is more widespread outside of the St Lawrence Lowland. Over three-quarters of the farms in Estrie and close to 90 per cent in the Clay Belt are devoted to dairying. The predominance of dairy herds and the preference for cheese in the Québécois diet have resulted in a spinoff effect, with numerous local butter and cheese processing firms. Other specialty crops include apples, blueberries, other fruits, vegetables, tobacco, and sugar beets. In the marginal lands of the St Lawrence Lowland and Estrie, stands of maple trees provide the sap for maple syrup.

topography is much subdued and consists more of rugged hills and rolling plains. Most arable land in this region is in Estrie, where dairy farming prevails (photo 6.3). On the rugged Gaspé Peninsula, small communities dot the coastline, reflecting the inhabitants' orientation to the sea and the fishing industry. Because of the area's spectacular scenery, tourism has become an important source of income during the summer on the Gaspé Peninsula. Mining and forestry are other economic activities in this physiographic region. With the exception of the Lake Champlain gap in the Appalachian Uplands, easy road access to the populous parts of New England is blocked by these rugged uplands. The Lake Champlain gap has therefore become a very important north–south transportation link between Montréal and points south in the US, especially New York City.

As the largest physiographic region in Québec, the Canadian Shield occupies nearly 90 per cent of the province's territory (Figure 5.4). The Canadian Shield is noted for its forest products and hydroelectric production—it has most of the hydroelectric sites in Canada because of a combination of heavy precipitation, large rivers, and significant changes in elevation. Near Montréal and Québec City, the Canadian Shield is a recreational playground where the rolling, rugged, forested upland with its numerous lakes has become a popular site for urbanites and tourists. Further north, single-industry towns based on mining and forestry dot the landscape. In the Lac Saint-Jean region of the Canadian Shield is a large flat deposit of clay—the Clay Belt. It is not rich or well drained, and the growing season is short, but in most years it can produce good crops of hay and other forage and sustain dairy and other livestock operations to supply the nearby mining and forest industries.

Beyond the **commercial forest** zone lie the lands of the Cree and Inuit, where the Cree must coexist with the massive La Grande hydroelectric project. The huge hydroelectric reservoirs initially resulted in a serious environmental problem for the Cree—the microbial breakdown of large amounts of submerged vegetation creates high levels of methylmercury contaminants. These contaminants enter the aquatic food chain and eventually find their way into fish. The Cree, who traditionally have consumed large quantities of fish, were confronted with the risk of mercury poisoning, but, as the microbial breakdown eased, the mercury problem has gradually diminished. In Arctic Québec, a number of Inuit settlements are found along the coasts of Hudson Bay and Hudson Strait. Further to the east, near the Québec–Labrador border, upward of 15,000 Innu live in several communities. The Native communities are unable to provide sufficient employment opportunities, and consequently, they have high unemployment rates. The Québec government supports a hunting-and-trapping program, thus encouraging many Cree families to stay on the land. This popular program has important cultural and social values.

The Hudson Bay Lowland extends from Ontario along the southeast edge of James Bay near Rupert River. By far, this lowland is the smallest physiographic region in Québec. Few people live here.

With four climatic zones found in Québec, weather varies greatly. The four zones are Arctic, Subarctic, Atlantic, and Great Lakes–St Lawrence. The Subarctic climatic zone extends over 60 per cent of Québec followed, in extent, by the Arctic, Great Lakes–St Lawrence, and Atlantic climatic zones.

The geographic extents of Québec's four climatic zones are shown in Figure 2.4, page 47, while the characteristics of each zone are discussed in the section "Climatic Types and Zones," Chapter 2, page 49.

For the most part, the weather across Québec falls within a predictable range for each climatic zone, but, like other regions of Canada, Québec has had its share of extreme weather events. While Québec commonly experiences heavy rainfall and freezing rain, two storms, in 1996 and 1998, were especially dramatic. In July 1996, the rainstorms in the Saguenay region caused extensive flooding and resulted in loss of life, extensive damage, and the evacuation of 15,000 people. Seven were killed in the flood, including two children when a mudslide buried their house. The loss of property from this storm was over $600 million. The damage was widespread with the communities of L'Anse Saint Jean, La Baie, Jonquière, Chicoutimi, and Hébertville hardest hit. Two years later, the worst ice storm in Canada's history struck southern Québec and eastern Ontario, causing great damage and disrupting electrical power. The physical damage was enormous, perhaps reaching $500 million. The ice storm began on 5 January 1998 and continued for several more days. Over 4 million people were without electricity, many for several weeks. Estrie was hard hit, including its largest city, Sherbrooke. In the spring of 2011, a severe flood of the Richelieu River forced the evacuation of 3,000 residents whose homes were either flooded or threatened by high waters.

 The section "Extreme Weather Events" is found in Chapter 2, page 52.

Environmental Challenges

Québec, as an old industrial region, is confronted with a series of environmental problems, most of which stem from the discharge of agricultural, industrial, and mining wastes into the atmosphere and water bodies or from the construction of huge hydroelectric dams. Even its so-called pristine North has been affected as mineral exploration companies have failed to clean up their sites, leaving the landscape littered with abandoned machinery, chemicals, and oil barrels (Duhaime et al., 2005: 262). Then, too, the plan to expand the mining industry and thus create more high-paying jobs in this economically depressed area could cause further harm to the northern environment, especially the boreal forest (Moreault, 2009b). But the recently elected PQ government has put a hold on former Premier Jean Charest's Plan Nord (Plan North) and has cancelled a loan to the asbestos mining firm at Asbestos, effectively ending over 100 years of mining of this dangerous mineral still used as an insulating material in construction in some parts of the world. From an environmental/health perspective, the PQ government did the right thing. In a more controversial move, the Parti Québécois government of Pauline Marois intends to close the aging Gentilly-2 nuclear power plant in Bécancour at a cost of $1.3 billion and ban the locally unpopular process of fracking to recover oil and gas deposits (Cousineau, 2012b; Blatchford, 2012). Plan Nord might yet see another day under the PQ as Premier Marois invited French businesspeople to invest in Quebec's North with the lure of tax credits based on the number of employees but requiring the processing of the minerals in Québec (Cousineau, 2012a: B5).

In the heartland of Québec, the greatest concern is for the St Lawrence River. As one of the most populated and industrial areas in Canada, the disposal of sewage and industrial wastes into the river gradually overpowered the river's natural capacity to deal with such wastes. In the 1970s, matters came to a head and Québecers feared for the future of the river and their drinking water. Controls over dumping municipal sewage and industrial waste into the river were virtually non-existent. The result was a river struggling with high levels of toxic chemicals—polychlorinated biphenyls (PCBs) and dichlorodiphenyltrichloroethane (DDT), as well as heavy metals such as lead, mercury, and cadmium—and the introduction of the notorious **zebra mussel**, which has had a devastating impact on local aquatic ecosystems.

Efforts to stem the pollution of the St Lawrence River and, more importantly, to reclaim its water quality and biodiversity began in earnest in 1988 when Québec and Canada joined forces to tackle the problems facing this river (Québec and Canada, 2012). The goals were and remain threefold: (1) biodiversity; (2) improved water quality; and (3) sustainable use. While the water quality has improved, the St Lawrence still rates as one of Canada's most polluted rivers. One indicator of water quality is the state of the famous St Lawrence beluga whales (see Vignette 6.5). So it was not surprising that, in 2011, the two governments launched a long-term effort called *The St. Lawrence Action Plan: 2011–2026* (ibid.).

Québec's Historical Geography

Relative to some other parts of North America, the history of European settlement in Québec is long, rich, and complicated by the period of British rule and then the search for a place within Confederation (see Tables 6.2–6.4). Beneath the surface of this search lies the deeply fractured French/English faultline. Québec's historical geography can be divided into three periods: New France, British occupation, and Confederation.

New France, 1608–1760

The introduction of the French to North America began in 1534 when Jacques Cartier sailed into the Bay of Chaleur and set foot on the shores of

Table 6.2 Timeline: Historical Milestones in New France

Year	Geographic Significance
1534	Jacques Cartier sails into the Bay of Chaleur and claims the land for France. The following year, Cartier discovers the mouth of the St Lawrence River, which provides access to the interior of North America.
1608	Samuel de Champlain, described as the "Father of New France," founds a fur-trading post near the site of Québec City. Champlain was instrumental in the development of the fur economy, which provided the initial economic basis for New France.
1642	Paul de Chomedey de Maisonneuve establishes Ville-Marie on Île de Montréal, which is strategically situated at the confluence of the Ottawa and St Lawrence rivers. Later, Ville-Marie was renamed Montréal.
1759	The struggle between France and England over North America sees the British defeat the French army on the Plains of Abraham at Québec. The final battle between French and English forces ends with the capture of Montréal by the British. In 1763, the formal surrender of New France to England takes place with the Treaty of Paris.

the Gaspé Peninsula. The first permanent French settlement in Québec was established in 1608, when Samuel de Champlain founded a fur-trading post at the site of Québec City, thus establishing a French colony in North America. Over three decades, on foot and by ship and canoe, Champlain explored in the unknown heart of the continent through what are now six Canadian provinces and five American states, and by doing so he created the territorial basis for the French Empire in North America. Although France eventually lost its North American colony, it left a cultural legacy in the form of the French language and the Catholic religion, and a French stamp on the landscape with its unique settlement pattern and arrangement of farms into long lots. Notably, the vision of Champlain to strive for a French settlement in the New World founded on harmony with and respect for the Huron Indians differed sharply from English and Spanish settlements and attitudes towards the First Peoples. David Fischer (2008) argues that the "humanist" values of Champlain and the respect for the values and traditions of the "other" have been important in marking Canada as different from other "New World" countries.

During the seventeenth and eighteenth centuries, France had control over vast areas of North America. Its core, however, was the St Lawrence Valley, from which New France developed a vast

VIGNETTE 6.5

The Future of Beluga Whales in the St Lawrence

Beluga or white whales can live 75 years or more. A Fisheries and Oceans Canada report describes the situation of beluga whales, a species at risk in the St Lawrence River:

> The beluga whale is primarily an arctic species; the St. Lawrence Estuary beluga is at the southernmost limit of the range and is geographically isolated from other populations. Before 1885, there were as many as 10,000 belugas in the St. Lawrence Estuary and Gulf. Today's most recent estimates put the current population at close to 1,000 whales.
>
> Commercial whaling has depleted the population severely. Although whaling for belugas has been banned since 1979, there has been no noticeable recovery in the population.
>
> A number of factors seem to contribute to the lack of recovery of this species in the St. Lawrence. Among them, pollution, reduced food resources, disturbance by humans and habitat degradation are considered to be the main threats to the recovery of the population. Beluga whales can also be the victim of ship strikes and become entangled in fishing gear. (Fisheries and Oceans Canada, 2012)

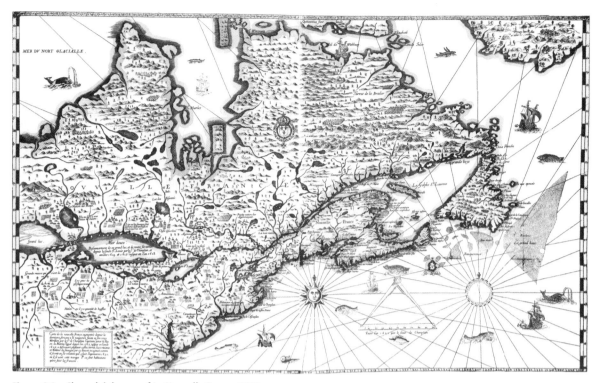

Figure 6.6 Champlain's map of La Nouvelle France, 1632
Source: Bibliothèque nationale du Québec, at: <www.nsexplore.ca/maps1/novascotia1632champlain.jpg>.

fur-trading empire. The wealth from the fur trade was enormous and was the reason for France's interest in the New World. Almost every male French settler wanted to participate in the fur trade, which left only a few to clear the forest and till the land. Indeed, several canoes full of beaver pelts could make a man extremely wealthy compared with the meagre returns obtained from the back-breaking toil of clearing land and breaking the soil. Frenchmen who were coureurs de bois (fur traders) often lived with the Indians. A few were extremely successful and returned to France to enjoy their good fortune. Others remained in the fur trade or settled in New France.

Geography played a part in New France's success both as a fur-trading empire and as an agricultural colony. The St Lawrence River provided a route to the interior, which gave the French explorers and fur traders an advantage over their English rivals, who had to contend with crossing the Appalachian Mountains or, in Canada after 1670, ply the trade further north through Hudson Bay. With exploration of the interior of North America, the French were able to expand their territory and secure more and better fur-trading routes. Thus, by portaging from the Great Lakes to the Ohio River and then to the Mississippi, the famous French explorer, René-Robert Cavelier, Sieur de La Salle, reached the mouth of the Mississippi River at New Orleans in 1682. Early in the eighteenth century, French fur traders, led by Pierre Gaultier de Varennes, Sieur de La Vérendrye, established a series of fur-trading posts in Manitoba. These trading posts, especially Fort Bourbon on the Saskatchewan River at Cedar Lake (just east of the present-day community of The Pas), made it convenient for Indians to trade their furs with French traders rather than travel farther to the British fur-trading posts along the Hudson Bay coast.

New France also established a successful agricultural society. Once the land was cleared, the fertile soils in the St Lawrence Lowland provided a solid basis for essentially feudal agricultural settlement. Farming took hold in New France, particularly following the efforts of Jean Talon (1626–94), the greatest administrator of New France. By the early eighteenth century, the French had turned to farming, leaving the lure of the wilderness to a relatively small number of more daring souls. By then, New France had existed for over 100 years, and many of its people had been born and raised in the New World. They were no longer "French" but Canadiens.

By the middle of the eighteenth century, almost all the lands in the St Lawrence Valley were under cultivation. Farmlands stretched in a continuous belt from Québec City to Montréal. As in the feudal agricultural system in France, peasant farmers (**habitants**) worked the land, paying their lords (**seigneurs**) both in kind and in labour. Through the efforts of the habitants and their seigneurs, New France became a successful agricultural colony, but one with a system of landownership and rural life that was quite different from that of the British colonies along the Atlantic seaboard (a difference that would eventually cause Britain to split Québec into Upper and Lower Canada in 1791; see Chapter 3).

The Seigneurial System

When the first Intendant of New France, Jean Talon, arrived in New France in 1665, he encountered a population of only 3,000 inhabitants, most of whom were men engaged in the fur trade. Talon had been instructed by Louis XIV to create a feudal agricultural society resembling that of rural France in the seventeenth century. Talon undertook three measures to achieve this goal. First, he recruited peasants from France. Second, he sent for young women—orphaned girls and daughters of poor families in France—to provide wives for the men of the colony. Third, he imposed the French feudal system of landownership, known as the seigneurial system. In the seigneurial system, huge tracts of land were granted to those favoured by the king, namely, the nobility, religious institutions of the Roman Catholic Church, military officers, and high-ranking government officials. The seigneur was obliged to swear allegiance to the king and to have his tenants cultivate the lands on his estate. In exchange for use of the land, the tenants owed certain obligations to their seigneur: paying yearly dues (*cens et rentes*) to their seigneur, working the seigneur's land, especially in regard to road maintenance (*corvée*), and paying rent for using the seigneur's grinding mill and bake ovens (*droit de banalité*).

By 1760, there were approximately 200 seigneuries. Seigneuries, which were usually 1 by 3 leagues (5 by 15 km) in size, were generally divided into river lots (rangs). These long, rectangular lots were well adapted to the St Lawrence Valley for several reasons, the most important of which was that each habitant had access to a river, either a tributary of the St Lawrence or the river itself. At that time, most people and goods were transported along the river system in New France. For that reason, river access was vital for each habitant family.

After the British Conquest of New France, the seigneurial system was retained by the British more for political reasons—to gain the support of the principal source of power (the seigneurs and the Roman Catholic Church)—than for economic reasons. By the early nineteenth century, however, agriculture in Lower Canada had become a commercial venture, making the seigneurial system an anachronism. The seigneurial system was abolished in 1854 by the legislature of the Province of Canada. Two of these seigneurial landholdings, now the Vaudreuil-Soulanges municipality, were located southwest of the Ottawa River in a triangle at the confluence of this river and the St Lawrence. Although the Ottawa River generally forms the boundary between Québec and Ontario, what is now the only part of Québec south of the Ottawa River is a consequence of the desire in 1791, with the creation of Upper and Lower Canada, to keep this small French-speaking district in Lower Canada.

British Colony, 1760–1867

Following the defeat of the French in 1760, the British ruled Québec for over 100 years. The British governor was installed at Québec City along with a regiment of British troops, while the fur trade continued to flourish and the agricultural economy went unchanged. Most French Canadians were peasant farmers. After the Conquest, their life on the land remained much the same. Their social and economic lives revolved around the parish church and a landholding system centred on the seigneuries. Life in the towns and cities, however, changed radically due to a massive influx of British immigrants, the powerful political position of English Canadians, and their control of the commercial and industrial sectors of urban places. By 1851, French Canadians accounted for only about half of the population of Montréal. From 1851 to 1951, enormous demographic changes took place. The saving grace for the Canadiens was their high fertility rate, which was known as the "revenge of the cradles" (*revanche des berceaux*). Even though

Table 6.3 **Timeline: Historical Milestones in the British Colony of Lower Canada**

Year	Geographic Significance
1763	The Treaty of Paris awards New France to Great Britain.
1774	The British Parliament passes the Québec Act, which recognizes that Québec, as a British colony, has special rights, including use of the French language, the Catholic religion, and French civil law.
1791	The British Parliament approves the Constitutional Act that creates two colonies in British North America called Upper Canada and Lower Canada.
1841	Based on the report of Lord Durham following the failed rebellions in Lower and Upper Canada of 1837 and 1838, the British Parliament passes the Act of Union that reunites the Canadas into a single colony and makes English the official language of the newly formed Province of Canada.

this rate began to slow, for 100 years Québec's birth rate remained higher than the birth rate in the rest of Canada. Thus, Québec's slower decline in fertility helped to compensate for the province's out-migration and for the growing English-speaking population.

Land hunger forced many **French Canadians** to migrate. By the middle of the nineteenth century, many had left the St Lawrence Lowland due to a land shortage. Birth rates were so high in this rural society that there was not enough land left for the children of farm families. French Canadians migrated in three directions: to the Appalachian Uplands, where they either purchased farms from English-speaking farmers or found jobs in textile mills; to the Canadian Shield, where they tried to exist on extremely marginal agricultural land; or to New England's industrial towns, where most were employed in textile factories. By the early twentieth century large numbers of French Canadians, perhaps as many as one million, had left Québec for the United States, while only a small number settled in the Canadian West that was calling out for homesteaders. Through all of these economic and political changes, the vast majority of French Canadians maintained their language and Catholic religion. They turned to the Roman Catholic Church for both spiritual and political leadership. The Church, in turn, encouraged the people to stay on the land, far from the secularizing influences in towns and cities, where the English Protestants lived.

By the 1830s, political unrest was growing in both Upper and Lower Canada. Each colonial government was headed by a governor who was appointed in England and had absolute power. The governor administered the colony along with leading members of the community. This cozy arrangement not only concentrated power in the hands of a few but also led to blatant abuse by powerful elites. In Upper Canada, the political elite was known as the Family Compact; in Lower Canada it was the Château Clique. In 1837, rebellions broke out in each colony (Vignette 6.6). Both rebellions, which called for political reforms and the curbing of the political power of the elites, were crushed by the British army.

Britain, however, was determined to remedy the political situation in its two colonies. In an attempt to identify the main sources of discontent in Upper and Lower Canada, the British government sent Lord Durham, a politician, diplomat, and colonial administrator, to North America. Durham recognized that the political solution lay in an elected government, where power was dispersed among elected representatives rather than concentrated within an appointed elite. Durham also observed the French/English faultline, which he described as "two nations warring in the bosom of a single state." In Durham's report he recommended responsible (elected) government and the union of English-speaking people in Upper Canada with the French-speaking settlers of Lower Canada.[2] Lord Durham believed that assimilation of the French was desirable and possible, claiming that the French Canadians were "a people with no literature and no history" (Mills, 1988: 637). He recommended that English be the sole language of the new Province of Canada, and that a massive immigration of English-speaking settlers be launched to create an English majority in Lower Canada. In response to Durham's recommendations, the Act of Union was passed by the British Parliament in 1841, uniting the two colonies into the Province of Canada and thus creating a single elected assembly.

Lower Canada was now known as Canada East, and in spite of the political changes and the flood of immigrants from the British Isles, the French Canadians were not assimilated. Under the new form of British administration, the task of maintaining their French culture and language was not easy, but it was achieved because of several factors, the most important being their strong desire to remain Catholic and French-speaking. A second factor was the institutional support provided by the Roman Catholic Church. By providing spiritual guidance and schooling in French, the clergy played an essential role in cultural preservation. Geography and demography were other factors essential to the survival of French Canada in the nineteenth century. The overwhelming number and concentration of French-speaking people in Canada East provided a critical mass necessary for cultural survival. A high birth rate and high rate of natural increase for the Canadiens ensured an expanding population of French Catholics. Other factors were the rural nature of the French-speaking population, which isolated them from English-speaking residents of the major cities, and the emergence of a French-Canadian intellectual elite whose writings preserved the history and created a literature of French Canada. One of the most popular novels in early Québec was *Jean Rivard* (1874), published not long after Confederation. Written

Photo 6.4

Roman Catholic churches are found in most communities across Québec, signifying the dominant role that the Roman Catholic Church once played in the social life and political affairs of the province. In fact, faith was a key pillar of French-Canadian identity, the other being language. By the time of the Quiet Revolution in the 1960s, however, the Church's influence in Québec affairs had greatly diminished and many churches had lost many of their parishioners. Language remained the pillar of distinctiveness of Québec.

VIGNETTE 6.6

The Rebellions of 1837–8

In Lower Canada, Louis-Joseph Papineau, a lawyer, seigneur, and politician, was the leader of the French-speaking majority in the Assembly of Lower Canada. Papineau had fought with the British forces against the invading American army at Chateauguay in 1812. A strong supporter of the Church and the seigneurial system, Papineau was a leading political figure of the Patriotes. In 1834, Papineau issued a list of grievances known as "The Ninety-Two Resolutions." At this time, the economy was depressed and tensions between the French-Canadian majority and the British minority were growing. Papineau sought to shift political power from the British authorities to the elected Assembly of Lower Canada. He planned to use his majority in the Assembly to pass legislation, including tax bills. The British government rejected his resolutions, and it was just a matter of time before an armed uprising broke out. When it did, the British reacted with force. Even with the strong support of rural areas, Papineau and his Patriotes were soundly defeated. Nearly 300 rebels were killed in six battles. Papineau fled to the United States, but in 1845, he was granted amnesty and returned to Québec. Following a second uprising in Lower Canada in November 1838, the British captured hundreds of rebels and ultimately 12 men were sentenced and executed and another 58 were exiled to Australia (Bumsted, 2007: 169). A rebellion based on the same popular objections to elite rule also took place in Upper Canada at the same time and it, too, was crushed. The British government sought to remedy the unrest in both of its colonies. It began this process with a fact-finding mission headed by Lord Durham. The result and Britain's solution was the Act of Union in 1841.

Photo 6.5

The Clay Belt stretches across the Canadian Shield of northern Ontario and Québec. This ancient lake bottom contains the major soil deposit suitable for dairy farming in the Canadian Shield. With the encouragement of the Roman Catholic Church, French-Canadian farmers began to settle these lands in the late nineteenth century. Life was hard and many farms were abandoned. The process of farm consolidation—larger farms but fewer farmers—was one answer, but even then the Clay Belt is not a farmer-friendly area.

by Antoine Gérin-Lajoie, *Jean Rivard* is the story of a young French Canadian in Canada East who is advised by his *curé* on the advantages of becoming a farmer rather than a lawyer. The novel promotes the virtue of living a subsistence rural lifestyle in a remote area of Québec's Clay Belt, far from Montréal with its many worldly temptations. Here, the French language, religion, and heritage could flourish.

Confederation, 1867–Present

Confederation, achieved in 1867, sought to unite two cultures—English and French—within a British parliamentary system. For Québec, Confederation provided a political framework offering three benefits: an economic union with Ontario, Nova Scotia, and New Brunswick; a political environment where Roman Catholicism and, to a lesser degree, the French language were guaranteed protection by Ottawa; and provincial control over education and language. George-Étienne Cartier, one of the Fathers of Confederation and a French-Canadian leader, viewed these provincial powers as a way for Québec to shape its own destiny within Confederation. Cartier may have identified a fourth benefit—since Québec and Ontario often had mutual economic interests, they could, by working together, influence federal policies and thereby shape the future of Canada.

Confederation also led to the expansion of the geographic size of Québec (Figures 3.4 and 3.6). Since Confederation, Québec's geographic size has increased greatly. It is now 1.5 million km². As Canada acquired more territory from the British government, Ottawa assigned to Québec parts of Rupert's Land lying north of the St Lawrence drainage basin. In 1898, the Québec government received the first block of Rupert's Land. The second was obtained in 1912.[3] Some of this land, however, was claimed by the British colony of Newfoundland. In its argument, Newfoundland demanded all of Québec's territory that drained into the Atlantic Ocean. Though the British Privy Council awarded this land, known as Labrador, to Newfoundland in 1927

Table 6.4 Timeline: Historical Milestones for Québec in Confederation

Year	Geographic Significance
1867	The Dominion of Canada is formed, the new state consisting of Québec, Ontario, New Brunswick, and Nova Scotia.
1898	Ottawa extends Québec's northern boundary to the Eastmain River, thus expanding Québec's territory well beyond its core area of the St Lawrence Lowland into the Cree lands of James Bay in the Canadian Shield.
1912	Ottawa adds the Territory of Ungava to Québec, thus extending Québec to Inuit lands of Nunavik. With the addition of these two northern lands in 1898 and 1912, Québec's territory more than doubles.
1927	In settling a dispute between Canada and Newfoundland, Britain rejects Canada's claim that the boundary should be placed just inland from the shore. Instead, Britain declares that the boundary is to follow the Hudson Bay and Atlantic Ocean watersheds. Québec does not recognize this boundary.

THINK ABOUT IT
Why do Québec governments, regardless of political affiliation, advocate a vision of Canada as a "partnership" between founding peoples?

(Figure 3.7), to this day the Québec government does not recognize the decision. As Premier Jean Charest said in September 2008, "This is a traditional position that all governments have reiterated. There is a boundary line on which there is no agreement" (Robitaille, 2008).

Prior to World War II, Québec continued to project an image of a rural, inward-looking, Church-dominated society. In 1960, the Quiet Revolution unleashed the force of change that drove Québec into a modern industrial state. As with other social revolutions, its origin of changes began earlier. Yet, the election of the Liberal government of Jean Lesage marked a dramatic transformation in government, French-Canadian society, and the place of French within Québec. His government initiated major political innovations that accelerated the process of social and economic change. In effect, the provincial government replaced the Catholic Church as the leader and protector of French culture and language in Québec. The Quiet Revolution instilled a sense of pride and accomplishment among Québecers. The main reforms of the Lesage government were hinged on state intervention in the Québec economy through Crown corporations, and on the expansion of a French-speaking provincial civil service. The government's principal achievements were:

- nationalization of private electrical companies under Hydro-Québec;
- modernization and secularization of the education system, making it accessible to all;
- investment of Québec Pension Plan funds in Québec firms, thereby stimulating the francophone business sector;
- establishment of Maisons du Québec (quasi-embassies) in Paris, London, and New York, thus signalling to Ottawa that the Québec government wanted to represent Québec interests to the rest of the world.

For more on the Quiet Revolution, see Chapter 3, page 116.

With these accomplishments behind them, Québecers felt confident about their future. Lesage's 1963 campaign slogan, "*Maîtres chez nous*" (Masters in our own house), became a reality. For federalists in Québec, these achievements proved that a strong Québec could function within Canada, but for separatists they were not enough. The rise of separatism in Québec signalled that some Québécois felt only an independent Québec could adequately represent French-Canadian interests. For them the slogan became "*Le Québec aux Québécois*" (Québec for the Québécois). After two referendums, Québecers have, for the time being, turned away from pursuing political separation and are focusing more on economic and social concerns. Then, too, Québec nationalism, which is at the core of the separatist movement, has shifted somewhat from the goals and values of the "old stock" francophones and has become more inclusive, i.e., embracing all French-speaking Québecers, including immigrants (allophones) and bilingual anglophones. Even the federal government has referred to Québec as a "nation" within Canada. Nonetheless, philosophical and jurisdictional disagreements are never far from the surface in the relationship between Québec and Ottawa.

Discussion of the French/English faultline in Chapter 3, page 110, includes more information on the historic twists and turns to Québec's place within British North America and then within Confederation.

Québec Today

Québec is a modern industrial society operating within a francophone environment but dependent on exports to the rest of Canada, the United States, and global markets. Like Ontario, Québec is a major manufacturing centre, ranking just behind Ontario but well in front of the other four geographic regions. Unfortunately, Québec's economy did not benefit from the Auto Pact, which provided a huge boost to Ontario's manufacturing sector. On the other hand, Québec has many more natural sites for the production of hydroelectricity than does Ontario. Through a Crown corporation, Hydro-Québec, the province developed some sites, creating a surplus of electrical energy that it exports to the United States. Another key geographic advantage lies in Québec's position along the St Lawrence River, which provides low-cost access to the heart of North America. This location provides Québec with a hub position in the east–west and north–south transportation systems and accounts for the leading position of Montréal in container traffic from Europe to North America. Its fledgling knowledge-based economy is moving forward and is centred on innovative firms and technological institutions in Montréal, Québec City, and Sherbrooke. In spite of these achievements, Québec remains a "have-not" province and its citizens are burdened with the largest per capita debt in the country, which stood at just over $40,000 in September 2012 (Canadian Federation of Independent Businesses, 2012).

Photo 6.6
The popular Victoria Square in the heart of downtown Montréal near McGill University is situated at the intersection of Beaver Hall Hill and McGill Street.

One challenge for Québec in the twenty-first century is its relatively slow growth rate compared to Ontario and western regions. As a result, Québec's share of Canada's economy and population has slipped. If the current trend continues, both Western Canada and British Columbia will surpass Québec before mid-century. The political implications are considerable. Even though Québec is guaranteed 75 seats in the House of Commons, Québec's political clout will diminish as the total number of seats increases to accommodate population increases in other geographic regions.[4]

Another challenge, of course, is the maintenance of the province's French language and culture. Attracting immigrants ensures a positive rate of population growth, but some worry that immigrants could weaken the francophone culture and the linguistic balance between French- and English-speaking Québecers. Language laws that require immigrants to send their children to French schools, thereby ensuring the growth of the French-speaking population, appear to have solved the language question,[5] but the PQ government has fuelled the French/English faultline in Québec with its Bill 14, which as of March 2013 had not yet become law.

The last, and perhaps most important, challenge revolves around relations between Québec, Ottawa, and the rest of Canada. Language and culture remain issues in the province's relationship with Ottawa and the other provinces. Most other concerns are framed around language—in other words, most political, social, and economic issues in Québec are seen from the perspective of language and culture. With the election of the Parti Québécois in September 2012, the French language issue returned to the front page with Premier Marois calling language "the centre of my preoccupation" (Séguin and Leblanc, 2012: A1).

Québec's Economy

Québec's economy, though the second largest of the six regions in Canada, is struggling. Its past success was largely based on the manufacturing sector, but the importance of manufacturing has shrunk over the past 40 years. More recently, aggressive competition from China and other countries has hurt Québec firms. Not surprisingly, then, Québec's economic performance is lagging

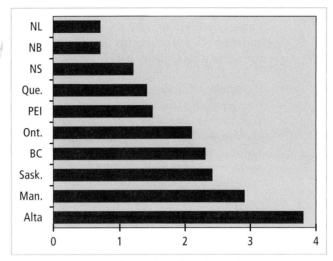

Figure 6.7 **Percentage change* in real GDP by province, 2012**
*Based on 2002 dollar.
Source: Hodgson (2012).

behind that of resource-rich provinces (Figures 6.5 and 6.7). Slow growth remains a fact of life for Québec, with the inevitable consequence of a widening gap with other provinces, especially those in Western Canada. The Conference Board of Canada (Hodgson, 2012) describes Québec's economy in the coming decade:

> For Quebec, future economic growth is expected to decline to around 1.5 per cent annually after 2015, largely due to slowing labour force growth, with little prospect for stronger growth down the road without deep and fundamental changes that could unlock some new growth potential. Real annual economic growth of 1.5 per cent or less will make it very hard for Quebec to sustain its health care and public education systems in their current form.

The liberalization of world trade promised much to nations and their workers. The reality for both Ontario and Québec has been hard and both provinces have found their manufacturing economies floundering. Fierce competition from low-wage countries has placed great pressure on this sector, especially on firms based on semi-skilled labour. After China became a member of the World Trade Organization, Canada had no means of preventing Chinese goods from entering Canada. As a result, Québec and Ontario have seen a sharp decline in the size of their manufacturing employment. From 2001 to 2011, employment in manufacturing industries in Canada declined from 2.2 million to 1.8 million and most job losses occurred in these two provinces (Statistics Canada, 2012c: Table 20).

Equally devastating for Québec, not only did the number of workers in the manufacturing sector decline but the rate of decline was greater than the national average. In 2000, Québec accounted for 28.7 per cent of Canada's manufacturing workers but by 2011 its share had dropped to 27.7 per cent (Human Resources and Skills Development Canada, 2012).

Québec, like Ontario, depends heavily on foreign trade, especially to the United States. Québec accounts for much of Canada's exports: 25 per cent of information technologies, 55 per cent of aerospace production, 30 per cent of pharmaceuticals, 40 per cent of biotechnology, and 45 per cent of high-tech exports. An increase in exports, led by aircraft sales, took place in 2011, hopefully signalling an upsurge in the economy (St-Pierre and Carrier, 2012). For a similar pattern of rising exports in the years to come, Québec is dependent on the recovery of the US economy and, perhaps, gains from the pending trade agreement with the European Union. Québec's high-value exports, if annual sales abroad continue to expand at 2011 rates, offer hope for a brighter economic future and may somewhat offset the rather dismal economic forecast of the Conference Board of Canada.

Within Canada's manufacturing belt, Canada's aerospace industry is concentrated in the Montréal area with Mirabel airport emerging as an important hub. Major players are Bombardier, Pratt & Whitney, Bell Helicopter, Textron, and CAE. Bombardier's extended C model passenger plane with 160 seats will allow the company to expand its sales beyond business jets and regional jets to large commercial aircraft. Though the competition with passenger jets produced by Airbus and Boeing will be a challenge, initial indications from potential customers suggests that the extended C aircraft has a chance of gaining market share (Keenan, 2013).

Over the years, Québec-based firms have specialized in certain industrial sectors, including apparel and textiles, high-tech industries, metal refining, printing, and transport equipment. The apparel and textile industries formed the traditional heart of manufacturing in Québec, but since the FTA these industries have faced stiff competition from abroad because of the reduction in tariffs on apparel and textiles, which has allowed imports from low-wage countries to grab a major share of the North American market. In 2002, the final blow came with China's entry into the World Trade Organization (WTO). As a result, quotas on textiles and garments disappeared. With Canadian quotas on Chinese textiles removed in 2003, imports into Canada soared. In 1995, for example, China accounted for 6 per cent of textile imports, but by 2005 Chinese imports had reached 46 per cent (Wyman, 2006). The resultant decline in such manufacturing in Québec is a common phenomenon found in the rest of Canada and in other countries. By 2007, China was clearly the world's biggest and least expensive producer of quality garments, thus driving Québec and other Canadian garment firms from the marketplace.

High-tech industries have fared much better and provide hope for the future. In fact, many economic gains in Québec have come from this knowledge-based sector. These high-tech firms require a global reach, something that a select number of Québec companies have already achieved. Because Montréal has a critical mass of high-tech companies, several universities, and strong provincial support, it is the most important centre for the new economy in Canada. Leading components are found in aerospace, biotechnology, fibre optics, and computers (both hardware and software). Québec also excels in engineering/construction on the international stage. SNC-Lavalin is such a firm (Vignette 6.7).

Aerospace firms, led by Bombardier, CAE Inc., and Héroux-Devtek, have been particularly successful in recent years. Héroux-Devtek, for instance, has gained global dominance by focusing on its core business—designing and manufacturing airplane landing gear systems—while CAE has achieve the same global position by specializing in flight simulation for civil aviation and defence. Bombardier is a global player in the manufacture of passenger aircraft and high-speed trains. Sales in the past five years indicate its global presence. In 2009, Bombardier secured a contract worth $1.23 billion from Toronto to build the next generation of 70 subway train sets consisting of six cars for a total of 420 cars. Known as the Toronto Rocket, these trains were manufactured at its Thunder Bay plant. The first train set arrived in 2011 and the final is scheduled for delivery in 2014 (Bombardier, 2010; Toronto Transit Commission, 2013).

In the same year, the Montreal-based manufacturer's foothold in China was secured when the company reported that its Chinese joint venture, Bombardier Sifang Transportation Ltd, had a contract to produce 80 of its high-speed Zefiro trains that were featured at the 2008 Beijing Summer Olympics. In 2012, this contract was extended with China buying more high-speed trains. In February 2010, Bombardier won a huge contract worth $11.4 billion from France's national railway. In the meantime, orders for its aircraft, especially its C Series, a family of twin-engine, medium-range jet aircraft, have picked up as the global economic recovery strengthens. This series has undergone revisions to keep the cost of flying low, which appeals to main airlines. Again, Bombardier has succeeded as a global player only because of its strong research and development efforts and its focus on certain types of high-speed trains and passenger aircraft. But the Québec aerospace industry may be at a crossroads as global competition in this industry increases. For instance, Bombardier used to have only one rival (Brazil's Embraer SA) in the regional jet market. Now the company faces stiff competition from Boeing and Airbus as well as from new players in China, Japan, and Russia, which are backed by their governments. Ottawa is so concerned that a sweeping review of federal aerospace and space policies is underway, with the report due in early 2013 (Marotte, 2012: B10).

Founded in 1947, CAE is another anchor in Montréal's transportation sector. The company is a world leader in providing simulation and modelling technologies and integrated training solutions for the civil aviation industry and defence forces around the globe. The company employs about 4,000 highly trained workers in the Montréal area and another 2,000 in other parts of the world. Approximately one-third are engaged in research and in engineering, designing, and testing new products, such as a robot that strips paint from aircraft bodies. Its main product, however, is a flight simulator manufactured at its plant in Saint-Laurent near Montréal. Unlike many manufacturing firms, most of CAE's sales do not go to the United States but to customers in Europe, the Middle East, and Asia.

Like other regions in Canada, Québec is moving quickly towards a knowledge-based economy with global connections. Each region has a different emphasis, with Ontario focusing on automobile research, including the electric car. Québec has already established itself in several areas—large-scale engineering projects, cutting-edge transportation production, pharmaceutical research, and biotechnology. Most of this activity is located in Montréal. However, international exports require branch locations in other countries. With research costs high, governments must invest heavily in universities and public research institutes as well as provide tax incentives for private research firms. Most funds come from the federal government, with minor support from provinces. Of the provincial governments,

VIGNETTE 6.7

SNC-Lavalin: Building a New City and Transport System

SNC-Lavalin has a global reach, which is critical for Canadian export-dependent firms. The company has offices across Canada and in over 35 countries around the world and has projects in about 100 countries. In July 2009, SNC-Lavalin Group Inc., a world-renowned engineering/construction company, won a $508 million contract for a massive urban planning project in Algeria—to build a new city to be called Hassi Messaoud near Algeria's largest oil field. This contract for a future city to house 80,000 people was cancelled in June 2012. However, SNC-Lavalin has the option to enter another bid. In 2011, SNC-Lavalin bought a 48 per cent stake in a Russian engineering company to improve its chances of involvement in Russian oil and gas operations. As a key player in the transportation system for the 2010 Winter Olympics, the company built the rapid transit train from Vancouver's airport to the downtown. SNC-Lavalin's global reach has also encountered problems, for in some instances overseas contracts have been secured through illegal payments. The "Gaddafi disaster" ranks as one of the worst cases of its global business, and a former top executive, Ben Aïssa, sits in a jail cell in Switzerland on suspicion of paying bribes in Libya (McArthur and Smith, 2012).

Table 6.5 **Employment by Industrial Sector in Québec, 2011**

Industrial Sector	Workers (000s)	Workers (%)	Percentage Change from 2005
Primary	90.9	2.3	−0.4
Secondary	756.3	19.1	−3.1
Tertiary	3,106.4	78.6	+3.5
Total	3,953.6*	100.0	6.4

*Number of workers in 2005 was 3.72 million.
Source: Statistics Canada (2012d).

Québec's financial support to provincial-based firms is by far the greatest. Such investments are costly, however, and governments must know what research areas should be emphasized.

Federal and provincial tax credits for scientific research are discussed in Vignette 4.11, "Public Support for Scientific Research and Knowledge-based Nodes," page 161.

Industrial Structure

THINK ABOUT IT
Can you make a case either for or against other provinces providing the same level of tax credits for knowledge-based research and development as Québec?

As a core region of Canada, Québec has the second-largest number of workers—nearly 4 million, or 23 per cent of the country's workers (Table 6.5). Its industrial structure is very similar to that of Canada's other core region, Ontario. First, from 2005 to 2011, both regions saw their workforces increase in size. From 2005 to 2011, Québec's workforce increased from 3.7 million to 4.0 million. Second, the division of Québec's and Ontario's industrial labour forces into the three principal sectors (primary, secondary, and tertiary) indicates that around 2 per cent of workers in each province are engaged in the primary sector (most in agriculture), just over 19 per cent are in the secondary sector (most in manufacturing), and approximately 79 per cent are in tertiary industries. Third, the tertiary sectors of Québec and Ontario increased their share of the total labour force from 2005 to 2011 at the expense of the primary (forestry) and secondary (manufacturing) sectors (Statistics Canada, 2012d).

But what aspect of the Québec and Ontario industrial structures distinguishes them from those of the other geographic regions? The answer lies in their secondary sectors. As core regions, both Québec and Ontario have strong secondary industrial sectors dominated by manufacturing (Tables 6.5 and 6.6). In comparison, the four other geographic regions, the so-called hinterland, had smaller secondary sectors. In 2011, both British Columbia and Western Canada had around 17 per cent of their labour force in the secondary sector, while Atlantic Canada had 16 per cent and the Territorial North approximately 2 per cent. Québec's labour force increased by 237,000 workers from 2005 to 2011, a gain of 6.4 per cent. In comparison, Ontario's labour force grew by 333,000 or 5.2 per cent—surprisingly, lower than the Québec rate.

Key Topic: Hydro-Québec

Since the Quiet Revolution, successive Québec governments have played an active role in shaping the province's industrial economy and its energy export strategy. Hydro-Québec has been central to this strategy. As a Crown corporation of the Québec government, three-quarters of its profits, which are around $2 billion each year, flow to the provincial government (Cousineau, 2012b: A15). Since the Crown corporation manages the Gentilly-2 nuclear plant, Hydro-Québec will be responsible for the cost of decommissioning the plant, which is estimated at $1.3 billion.

Table 6.6 **Shift of Ontario and Québec Industrial Structures, 2005–2011 (%)**

Sector	Ontario, 2011	Ontario, 2005	Québec, 2011	Québec, 2005
Primary	1.9	2.0	2.3	2.7
Secondary	19.2	23.6	19.1	22.2
Tertiary	78.9	74.4	78.6	75.1
Total	100.0	100.0	100.0	100.0
Workers (000s)	6,731	6,398	3,954	3,717

Source: Statistics Canada (2006a, 2012d).

The core of Hydro-Québec's strength lies in the vast water resources found in the Canadian Shield. Once these are developed through dams, reservoirs, diversions, generating stations, and transmission lines, electrical power is produced at a very low cost relative to nuclear or gas-fired plants. A second advantage lies in Québec's geographic situation—its close proximity to energy-short New England. Long-term sales to utilities in New England have provided much of the financial clout needed to construct northern dams and transmission lines (Vignette 6.8). This approach to economic development, initiated by the Lesage government and continued by successive governments, has pursued two goals—one economic, the other political. Its economic objective has been to stimulate economic growth through state intervention in the marketplace. Hydro-Québec has undertaken construction of huge hydroelectric projects, developed high-voltage transmission systems, and offered low electric rates to industrial firms. Its political goal was to increase Québec's public and private ownership of its economy within the francophone business community. This strategy has been highly successful. Hydro-Québec is Canada's largest electric utility, and the expertise gained from huge hydroelectric construction projects has allowed its contractors, such as SNC-Lavalin, to undertake similar projects around the world. Another spinoff was the breakthrough **transmission technology** that enabled Hydro-Québec to transmit electrical power from Churchill Falls[6] and James Bay to markets in southern Québec and New England with acceptable levels of energy leakage. In 2009, Hydro-Québec took a bold step by seeking to purchase New Brunswick Power, a deal that fell through in March 2010. The strategy was threefold: first, to gain control over the Maritimes power grid; second, to block the export of electric energy from Labrador to the Maritimes; and third, to achieve unfettered access to the lucrative New England market.

For discussion of the difficult relationship between Hydro-Québec and Newfoundland and Labrador, see the section "Can the Lower Churchill Hydro Project Mend Atlantic Canada's Fractured Geography?" in Chapter 9, page 377.

While the completion of the next phase of the James Bay Project remains central to Hydro-Québec's plans, its next construction project calls for the harnessing of the Romaine River (Figure 6.8), which flows from the Canadian Shield to the St Lawrence River on the North Shore near Havre-Saint-Pierre, some 200 km east of Sept-Îles (Hydro-Québec, 2009c). The cost is estimated at $6.5 billion, and construction will extend over a 10-year period. The Romaine complex will enable Hydro-Québec to secure Québec's energy future and to increase its exports to markets outside Québec. The commissioning of the first hydroelectric generating station is scheduled for 2014 and the fourth for 2020. By then, the Romaine hydroelectric system will have an installed capacity of 1,550 MW (Canadian Hydropower Association, 2009b). For this project, Hydro-Québec will also erect 500 kilometres of transmission lines and carve out more than 150 kilometres of roads (Yakabuski, 2008). Because the river is sharply incised, the total area of the four reservoirs will be only 280 square kilometres.

Meanwhile, work on the James Bay Project continues with the construction of the Eastmain Diversion Project with a completion date of 2012. By 2011, four dams, 74 dikes, a transfer tunnel allowing water to flow northward to the existing powerhouses on La Grande Basin, and the Eastmain-1-A powerhouse were completed; the last element, the Sarcelle powerhouse, began production in 2013 (Hydro-Québec, 2013). This project entails diverting a portion of the water in the Rupert River watershed into the Eastmain River watershed and then into La Grande Basin. The project consists of a 768-MW generating station—Eastmain-1-A powerhouse—near the existing Eastmain-1 powerhouse, and diverting part of the flow of the Rupert River into these two facilities, then through the future Sarcelle powerhouse, on to existing reservoirs, and then to existing powerhouses Robert Bourassa; LG-2A, and LG-1, before flowing into James Bay.

Québec's Industrial Strategy

With the production of low-cost hydroelectric power in Québec's Canadian Shield (Figure 6.9), an industrial strategy was born. The vast electrical power generated by the first phase of the James Bay Project provided the provincial government with an opportunity to attract energy-hungry industries into southern Québec by offering them special, low electricity rates, and to export surplus power to energy-hungry utilities in New England. Such an industrial strategy takes on the spatial form of the core/periphery model where the hinterland supplies the energy for industrial users in the core. An early version of this strategy took place in 1957 when Reynolds Aluminum built a smelter at

THINK ABOUT IT
On what basis does Québec dispute the boundary with Newfoundland and Labrador?

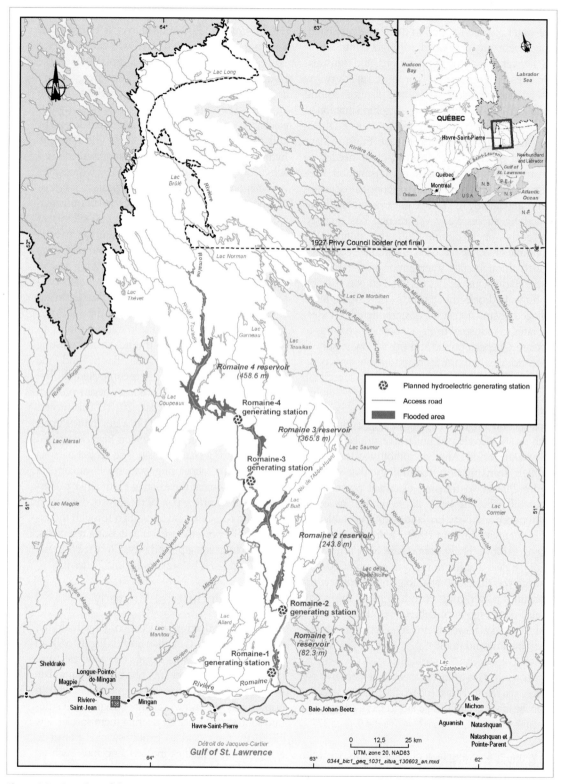

Figure 6.8 **Overview of the Romaine complex**
Hydro-Québec is constructing a huge hydroelectric complex with four power stations with a total capacity of 1,500 MW on the Romaine River in the Lower North Shore region. Note that Hydro-Québec has qualified the boundary with Labrador as "1927 Privy Council boundary (not final)." The explanation for this boundary is found in Table 6.4. *Source:* Hydro-Québec.

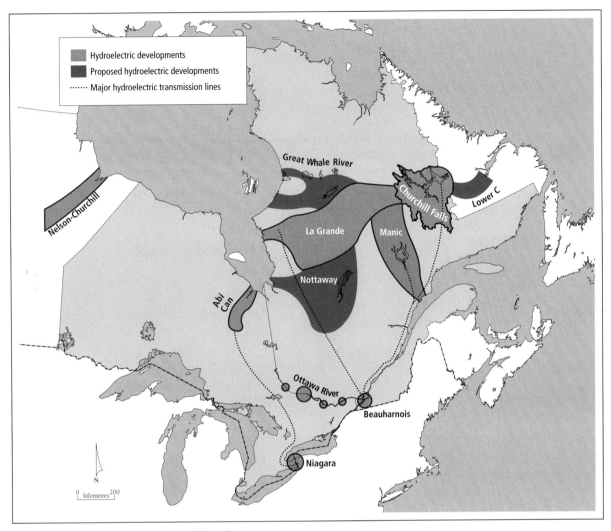

Figure 6.9 **Hydroelectric power in Central Canada**
Québec's dominant role in the production of hydroelectric power in Canada is due to three natural factors found in Québec's Canadian Shield: (1) abundant precipitation, (2) natural reservoirs, and (3) high elevations. Geography dictates that the Churchill Falls hydroelectric facility, while located in Labrador, sells virtually all its electric power to Hydro-Québec.

Baie-Comeau because Hydro-Québec provided it with low-cost power. In return, the company added to the value of production in the province and, more importantly, it provided jobs in an economically depressed area. Today, Reynolds Aluminum, known locally as Société canadienne des métaux Reynolds, employs approximately 1,000 workers. Alcoa provides another example. In November 2011, Alcoa secured a 25-year commitment for preferential electricity rates from Hydro-Québec. On the basis of this agreement, Alcoa announced a five-year, $2.1 billion expansion and modernization program for its three smelters at Baie-Comeau, Deschambault, and Bécancour. In spite of low prices for aluminum products in 2012, Alcoa is proceeding with its plan (Alcoa, 2013).

A recent outcome of this strategy is the decision by the Stockholm-based technology company, Ericsson, to open a global size information and communications centre close to Montreal. The proposed centre at Vaudreuil-Dorion will begin operations in 2015. Since its operations require vast quantities of electrical power, the offer of preferential rates from Hydro-Quebec was a key factor. Placing the Ericsson decision in the larger context of Quebec's economic well-being, the Premier, Madame Marois, declared that "The future of Quebec and the greater Montreal area relies on the knowledge base, the training opportunities and the skills available locally" (Marotte, 213).

Hydro-Québec is able to provide industrial firms with low-cost energy for three reasons:

- Northern Québec can produce vast quantities of low-cost electrical power.
- Hydro-Québec has a long-term contract to buy power from Churchill Falls in Labrador at 1969 prices.
- Hydro-Québec has control over its price structure and can set extremely low power rates for its industrial customers.

Québec's industrial strategy attracted metallurgical firms, but global competition still constitutes a threat. Norsk Hydro provides such an example.[7] In 1988, Norsk Hydro's large magnesium-smelting operation near Trois-Rivières was supplied with electrical power from Hydro-Québec at very low rates. Yet, by 2003, even with such a subsidy, the plant could not complete with lower priced Chinese imports. In 2007, the Québec plant closed.

Hydro-Québec's James Bay Project

The massive James Bay Project in northern Québec began in 1972, and while several projects are completed more hydroelectric development remains. The ambitious project calls for the harnessing of the power from all the rivers that flow into James Bay from Québec. One complicating factor is the fact that these Crown lands are also the homelands of the Québec Cree and Inuit. At first, the concept of

Table 6.7 **Technical Potential Hydro Power by Province and Territory**

Province/Territory	Technical Potential in Megawatts
Québec	44,100
British Columbia	33,137
Yukon	17,664
Alberta	11,775
Northwest Territories	11,524
Ontario	10,270
Manitoba	8,785
Newfoundland and Labrador	8,540
Nova Scotia	8,499
Nunavut	4,307
Saskatchewan	3,955
New Brunswick	614
Prince Edward Island	3
Canada	163,173

Source: Canadian Hydropower Association (2009a).

Aboriginal title to lands traditionally utilized by quasi-nomadic hunting tribes was not recognized by the courts, but with the Supreme Court ruling in the 1973 **Calder** case, the legal door opened. By 1997, the ***Delgamuukw*** ruling defined Aboriginal title.

VIGNETTE 6.8

Natural Advantages for Hydroelectric Developments in Québec

The natural setting of the Canadian Shield in Québec provides numerous technical potential hydroelectric sites, giving the region a clear lead over other provinces (Table 6.7). Hydroelectric developments depend on three factors: precipitation, topography, and access to market. The Canadian Shield provides two of the factors. First, annual precipitation often exceeds 800 mm thus providing a regular source of water for the lakes and rivers. Second, favourable topography with entrenched river valleys that begin at elevations of 1,000 metres; deep lakes and reservoirs store water and thereby ensure a steady flow for the power plants; and steep rock walls blasted out of the Canadian Shield facilitate dam and diversion projects (photo 6.7). Distance to market was a problem but the innovation of high-voltage transmission lines in the mid-1960s provided a solution, making it possible to reach the large but distant markets, such as southern Québec and the surrounding provinces and American states. The main advantages of hydroelectric developments are: the generation of clean, renewable, low-cost power; the long life of the facilities; low operating costs; job creation during the construction phase; and zero air pollution or GHG. However, there are drawbacks: the initial high capital investment; the long construction period; and the extensive and time-consuming environmental studies.

This hydroelectric project, announced in 1971 by Premier Robert Bourassa, targeted three separate river basins (La Grande, Great Whale, and the Nottaway-Broadback-Eastmain-Rupert basins). The project involves about 20 rivers and affects an area one-fifth the size of Québec. Construction of the first phase of the James Bay Project, La Grande, began in 1972 and was completed 10 years later at a cost of $15 billion (Bone, 2012: 167). The La Grande project involved diverting waters from three other rivers (Eastmain, Opinaca, and Caniapiscau) into La Grande Rivière. Electrical energy generated from the three power stations in La Grande Basin is more than 10,000 MW each year. The power is sent to southern markets via transmission lines suspended from large steel towers at a high voltage of 735 kV. At the market, the voltage is reduced to levels suitable for local distribution lines.

The first phase of the James Bay Project raised considerable controversy. It evoked an unprecedented and angry opposition from Aboriginal peoples and environmental organizations. For the Cree, the James Bay Project unleashed social and environmental problems on their traditional way of life, while environmentalists decried the loss of wilderness. Diversion of rivers, creation of huge dams and reservoirs, and the construction of high-voltage transmission lines change the landscape forever. In the process, traplines were flooded and wildlife disturbed, and the fish stocks in huge reservoirs contained a high level of mercury caused by the rotting of trees and other vegetation in the reservoirs. Since fish are a traditional source of food, the presence of mercury posed a serious health problem. In response, Hydro-Québec maintained that the environmental impacts have been mitigated to an acceptable level through modifications to the design of the project. Fortunately, high mercury content in the waters of the reservoirs did diminish over time as the submerged vegetation dissipated.

In 1985, when the second phase of the James Bay Project was announced, opposition from the Quebec Cree and environmental organizations reached a high peak. The Great Whale River project, located just north of La Grande Basin, was to consist of three powerhouses, four reservoirs, and the diversion of two rivers. In addition, another generating station (LG-1) was to be located at the mouth of La Grande Rivière.

Opposition from the Cree and the Sierra Club, who mounted a joint public relations campaign,

Photo 6.7
Construction work for the Sarcelle powerhouse. The Eastmain River diversion will generate more electricity at power stations located on La Grande Rivière. Note the steep rock walls carved out of the Canadian Shield by construction workers.

had an effect on public opinion in New England and New York—the principal export markets for Québec electric power—and when a new natural gas pipeline from Alberta to New England came on stream to provide a low-cost alternative the government of Québec announced in 1994 that the project would not proceed until the demand (price) for electricity in New England improved.

From the very beginning, the Cree opposed the James Bay Project because of its effect on their hunting grounds. The Cree, joined by the Inuit of Arctic Québec, forced a land claim settlement known as the James Bay and Northern Québec Agreement. Over time, however, the interests and objectives of the Cree and Inuit shifted from the land to their emerging settlement economies. The first sign of this shift appeared in 2001 when the province signed the Paix des Braves agreement with the Québec Cree. In the following year, the third phase of the James Bay Project was announced, and the Cree were looking for economic benefits and participation. The $2 billion Eastmain-1 project provided such benefits. Located south of the original development on La Grande Rivière, the three generating units of Eastmain-1 have a total installed capacity of 507 megawatts. Construction began in 2002 following the 2001 Paix des Braves agreement. The main components of the Eastmain-1A Sarcelle-Rupert project, completed in 2013, are the diversion of some water from the Rupert watershed, two powerhouses, the main dam across the

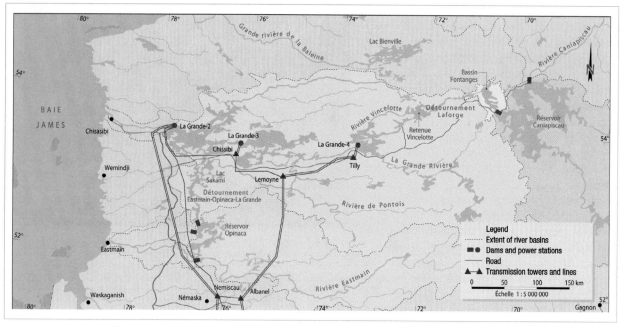

Figure 6.10 **Overview of the Eastmain-1-A/Sarcelle/Rupert project**
La Grande's flooded lands (blue-coloured reservoirs) lie east of the LG-2 dam. Downstream is the newly constructed community of Chissibi. In 2007, waters from the Eastmain River were diverted into La Grande Basin. *Source:* Hydro-Québec.

Eastmain River, the spillway, and the dikes for reservoir closure (photos 6.7, 6.8, and 6.9). A 315-kV transmission line links the powerhouse to Nemiscau substation. With the completion of the dam in 2006, water from the Eastmain River formed a reservoir 35-km long with a total area of 603 km^2. After passing through the Eastmain powerhouse, water flows to Opinaca Reservoir and then to Robert Bourassa Reservoir. The same water is utilized four times—at Eastmain-1 or Eastmain-1A, Sarcelle, Robert Bourassa or La Grande-2A, and La Grande-1 power stations—before flowing into James Bay.

Hydro-Québec's Exports to New England

Hydro-Québec exports close to $2 billion worth of electricity to the nearby New England states of Vermont, New Hampshire, Maine, and New York (Bordeleau, 2011). Geography favours the New England market for two reasons—distance and price. In terms of distance, the length of the transmission line is significant. The major market of Boston is only 400 km from Montréal and New York City is 500 km. In comparison, Toronto is also about 500 km from Montréal. Price differential is significant, with the energy-short New England region having approximately double the electrical rate found in Ontario (the Boston rate is approximately 15 cents/kilowatt hour while Toronto was closer to 8 cents in 2012). With Québec's surplus of hydroelectric power, exports are a key economic factor for building megaprojects in northern Québec. As well, long-term agreements to purchase Québec electricity help pay for the construction costs of these massive projects. For example, in 2010 Vermont's two largest utilities signed a $1.5 billion agreement with Hydro-Québec to purchase 225 megawatts of electricity at 6 cents/kilowatt hour until 2038 (CBC News, 2010). Another example of the interconnection with New England is the high-voltage transmission line that connects the La Grande complex in the James Bay area to Boston.

The James Bay and Northern Québec Agreement

In 1971, 6,000 Cree in northern Québec lived as eight bands scattered across 375,000 km^2 of rivers and forest. They were under the administration of the federal Department of Indian Affairs

Photo 6.8
This photo shows the Eastmain-1 spillway (left), dam (right) and reservoir (background). Diversion of waters from the Eastmain and Rupert rivers to La Grande Basin enabled the increase of electric power from existing generating stations. To facilitate the flow of water and navigation, the land has been cleared of trees. About 45,000 cubic metres of wood having a commercial value were salvaged by the Cree and used in their sawmills.

and Northern Development. The James Bay Project threatened to flood their lands. This threat united the eight bands. When construction began in 1972, the Cree asked the Inuit to join them in taking legal action to halt the construction until the Cree and Inuit land claims were addressed. This action forced the Québec government and the Aboriginal claimants to the bargaining table. The result was the **James Bay and Northern Québec Agreement** (JBNQA). Under this agreement, both the federal and Québec governments became responsible for providing the "treaty" benefits. As the first modern land claim agreement in Canada, this 1975 agreement provided land, cash, and the power to administer cultural matters (education, health, and social services) to Aboriginal peoples. In exchange, the Cree and Inuit surrendered their Aboriginal claims to much of northern Québec and agreed to allow construction of La Grande project to proceed.

In combination, these events—the negotiations, the agreement, and the construction project—have forever altered the lives of the Cree and Inuit. Both groups, now living in settlements (Figure 6.11), are more involved in the modern industrial society than ever before. Many are employed in businesses run by Cree and Inuit organizations, while others work in construction activities in the growing Cree and Inuit settlements and for Hydro-Québec. Still others are involved in the administration of their cultural affairs through the Cree Regional Authority and the **Kativik Regional Government**. In comparison with other Aboriginal peoples in Québec, the economic situation of the Cree and Inuit is much improved and certainly much better than that of those without such an agreement (Simard et al., 1996). Still, the Cree felt that both Ottawa and Québec had failed to honour their responsibilities under the JBNQA. When Québec announced plans for the second phase of the James Bay Project in 1985, relations between the Cree and the Québec government were so confrontational that the Cree actively opposed the Great Whale River project and took the Québec government to court over a number of issues related to the earlier agreement. Without a doubt, tensions between the Cree and the Québec government reached a low point in the

THINK ABOUT IT
If you were the Premier of Ontario, would you seek an agreement for low-cost energy from Québec to solve your energy shortfall or would you build more power facilities? Conversely, if you were the Premier of Québec, why would you prefer to supply New England with electricity rather than Ontario?

Figure 6.11 Cree communities of Québec
In a relatively short time (several decades), the Québec Cree have accepted settlement life. As of 2011, virtually all of the 16,000 Cree resided in eight communities. The largest Cree community is Chisasibi, which had nearly 4,500 inhabitants in 2011. Eastmain, Mistissini, Nemaska, Waskaganish, Waswanipi, Wemindji, and Whapmagoostui are the other Cree communities. *Source:* Based on <www.ottertooth.com/Native_K/jbcree.htm>. Copyright © Brian Back. All rights reserved.

latter part of the twentieth century, marking a particularly strained Québec version of the Aboriginal/non-Aboriginal faultline.

Turnaround

In 2001, the Cree reached an agreement with the Québec government for the economic development of the resources of northern Québec. This agreement, known as the Paix des Braves, opened the door to a major diversion of the Rupert and Eastmain rivers into La Grande Basin, thus adding more water for its hydroelectric plants. The acceptance of such an agreement was an astonishing reversal for the Québec Cree, who had bitterly opposed the project and who had mounted numerous national and international protests against further hydroelectric developments in their traditional lands. Some argue that the Cree now recognize that their participation in northern economic development is their only option. But feelings run high because efforts to protect the land for a hunting and trapping lifestyle have faltered. Paul Dixon, the Cree trapper representative, put it this way: "They [Québec] promised the traditional way of life would continue undisturbed. Today, the whole territory has been slated for development" (Roslin, 2001: FP7).

Some, especially younger Cree, have chosen an urban lifestyle. For them, living on the land is no longer a viable option. Some say that the Cree leaders had to make a deal. Faced with a rapidly growing population, high unemployment rates, a critical shortage of public housing, and a desperate need for sewer and water systems, the Cree leaders had to seek an agreement with the Québec provincial government. Québec wanted to develop the northern resources and the Cree needed revenue to operate their communities and to find work for their people. Under the terms of the agreement, the Cree receive $3.6 billion over 50 years (roughly $70 million a year), but these funds release Québec from its obligations for economic and community development associated with the James Bay and Northern Québec Agreement. A newly created Cree Economic Development Agency will administer these funds with a mandate to foster growth of Cree businesses. The agreement also required the Cree to drop their lawsuits against the Québec government for failure to meet its obligations under the James Bay and Northern Québec Agreement. Whether the Cree will benefit from this model of economic development remains to be seen. What is clear, however, is that similar agreements are taking place across the country and all are designed to allow Aboriginal people to participate in economic development taking place in their traditional lands.

Tourism

Tourism is extremely important to Québec's economy. In fact, Québec has combined its natural beauty, historic past, and francophone

Photo 6.9
The James Bay Project generates huge amounts of electrical energy. At the same time, the project has altered the natural environment by diverting rivers, changing their seasonal flow, and creating enormous reservoirs in Québec's Canadian Shield.

culture to draw more tourists each year for over a decade, and tourist dollars spent in the province also have increased. In 2010, just over 10 million tourists visited different areas in Québec. Approximately half originated in the province. Of those coming from outside of the province, nearly 30 per cent came from the United States, 40 per cent from other provinces, and 30 per cent from other foreign countries, especially France, the UK, Germany, and Japan (Québec, 2012). Since Québec is ideally located to cater to tourists from Ontario, New England, and Western Europe, its future as a world-class tourist destination seems secure, although a Canadian dollar at parity with the US dollar and global economic troubles can affect the tourism.

Montréal and Québec City are major attractions for tourists seeking an urban vacation with a francophone atmosphere, while the Laurentides, a low range of mountains bounded by the Saguenay, St Lawrence, and Ottawa rivers, attract visitors looking for a summer or winter playground in the forests and lakes of the Canadian Shield. In addition to the natural beauty, the Laurentides are but a short distance from Montréal and relatively close to New England and to major American cities such as Boston and New York. The climate in this area is ideal for the tourist business—the summers are hot, and heavy snowfall provides ideal conditions for winter sports. Mont Tremblant Resort, for example, is a world-class tourist destination for both winter and summer activities. Québec City has cashed in on its cold winter climate, holding an annual two week Winter Carnival each February, as well as hosting a wild ice and dangerous skating race called the Red Bull Crashed Ice, which attracts skaters and visitors from around the world. The skaters must navigate a narrow winding course some 430 metres long with a vertical drop of over 60 metres. While Québec has the setting for summer and winter recreation activities, the tourist industry is vulnerable to external events. With weak economies in the European Union and the United States, the tourist industry lost some American and European tourists, but gained others from the rest of Canada.

Southern Québec

Southern Québec is the economic, social, and political core of Québec, while northern Québec is a sparsely populated resource hinterland. The relationship between the two regions is a provincial version of the core/periphery model and overlaying that model are the francophone presence in southern Québec and an Aboriginal majority in northern Québec. Most anglophones and allophones reside in Montréal, giving that city a cosmopolitan atmosphere.

The best farmlands are found in the St Lawrence Lowland. Southern Québec contains two physiographic regions: the Appalachian Uplands and the St Lawrence Lowland. Bordered on the north by the Canadian Shield and on the south by the United States, southern Québec contains over 90 per cent of Québec's population and agricultural land and is the industrial heartland of the province. The human and physical characteristics of each physiographic region in southern Québec are presented below.

Appalachian Uplands

The northern edge of the Appalachian Uplands region faces the St Lawrence Lowland. Because of this geography, the Appalachian Uplands were settled at different times and in different ways. First settlements took place along the Gaspé coast. In the sixteenth century the waters off the Gaspé Peninsula attracted fishers from Spain, Portugal, England, and France. Even today, fishing in this part of Québec provides a way of life, though many supplement their income with farming and wage employment in the small coastal settlements. The main settlements are scattered along the Gaspé coast and the south shore of the lower St Lawrence River, leaving the rugged interior with few settlements. The largest urban centres are along the south shore. Rimouski, with a population of nearly 40,000 in 2011, is by far the largest city in the region. Rivière-du-Loup and Matane are medium-sized towns with populations roughly half the size of Rimouski. Towns are much smaller along the Gaspé coast, as the area has a very limited resource base. Percé, now an important tourist town, is the largest of the small centres along the Gaspé coast with a population of about 4,000. This French-speaking area has place names like New Richmond, New Carlisle, and Chandler, the legacy of early English settlers, some of whom had relocated along the Gaspé coast after the American Revolution. Over the last half-century, the lack of jobs caused many to migrate out of this area, which seriously reduced the size of English-speaking communities. In 2011, English-speaking residents constituted less than 5 per cent of the region's population.

Estrie (the Eastern Townships) is a pocket of communities in the rolling land of the Appalachian Uplands located east of Montréal. Estrie was settled after the American Revolution by British Loyalists. The British organized the surveyed land into rectangular townships rather than the French long lots found in the St Lawrence Valley. Much of the land was ill-suited for cultivation. Within several generations, the more marginal lands were abandoned when English-speaking owners left to look for jobs in Montréal and Boston or to try homesteading on the American frontier. Because of a land shortage in the St Lawrence Valley, French Canadians began to move into the Eastern Townships. French-Canadian migrants, often sons of farmers living in the St Lawrence Lowland, either bought or took over abandoned farms from the original English-speaking settlers.

The physical geography of Estrie is much more conducive to economic development and agricultural settlement than are the Gaspé coast and south shore of the St Lawrence. From the 1870s to the 1970s, mining at Thetford Mines and Asbestos was a key sector of the regional economy. Since the 1970s, asbestos mining has fallen on hard times because this mineral, formerly used as insulation in the construction trade, has proven hazardous to human life. Banned in Canada, most sales were to developing countries. In 2012, the Québec government ceased to support this mining operation, causing it to close. Agriculture and forestry have continued to be durable in this area. Dairy farming is pursued in the broad valleys and logging in the forested uplands. Overall, Estrie is the most prosperous area of the Appalachian Uplands. Sherbrooke, which has grown over the years and is the largest urban centre in the Appalachian Uplands, exemplifies the relative well-being of the region. With a population of 201,890 in 2011, Sherbrooke has become an important regional centre (Table 6.10). From 2006 to 2011, its population increased by 5.5 per cent, making it the third-fastest-growing city in Québec after Gatineau (9.1 per cent) and Québec City (6.5 per cent). Montréal was fourth at 5.2 per cent. This population increase is reflected in its economic growth. Proximity to Montréal has often worked in Sherbrooke's favour, allowing it to engage in the high-technology industries.

St Lawrence Lowland

Most of Québec's agricultural and industrial production is in the St Lawrence Lowland, which functions as the province's core. It contains Québec's largest market and is close to transportation networks and the St Lawrence River, which facilitate trade with foreign countries.

The warm-to-hot summer climate coupled with abundant rainfall makes the St Lawrence Lowland the most favoured region in Québec for agricultural activities. Livestock farms specializing in cattle, hogs, or sheep are common, while some farmers concentrate on dairy, poultry, and egg production. Livestock farmers grow forage crops for winter feed. During the summer, the cattle graze on pastures. Specialized crops, particularly vegetables and fruit, are also popular and are sold mostly in the major urban centres of the province.

Though farmers in this region engage in a variety of agricultural activities, dairy and vegetable farming predominate. With nearly half of Canada's dairy farms, Québec leads in the production of dairy products. While each province has its own milk marketing board, the National Milk Marketing Plan calculates the allocation of milk quota for each province, with Québec receiving nearly 45 per cent of the Canadian market. Moreover, the processing of agricultural products provides an added benefit for the Québec economy. Large butter and cheese firms rely on the dairy farms. Often, the processed food products are designed for the provincial market. Because of a ready market for unpasteurized cheese products in Québec, a few cheese firms began producing *fromage au lait cru*, cheese made with unpasteurized milk. These cheeses compete favourably with European imports, including popular brands from France.

Dairy farmers have fared relatively well under Québec's fluid milk marketing board and the national marketing system. In 2011, cash sales from milk and cream amounted to $2.1 billion in Québec, which represented 37 per cent of Canada's milk sales (Canadian Dairy Information Centre, 2012). Under WTO rules, however, Canada is under pressure to dismantle its marketing boards. Without a regulated dairy industry, dairy farmers in Québec would face stiff competition from American dairy imports. In 2002, the World Trade Organization ruled that Canadian dairy exports are illegally subsidized because the national marketing system's price for Canadian milk production is above market price. Since then, Canada has defended its marketing system, but the pending trade agreement with the European Union may affect Canada's protected dairy industry.

Manufacturing is concentrated in the Montréal area but extends eastward to Sherbrooke and northeastward to Québec City. As part of the Canadian manufacturing axis, this industrial zone serves as the engine driving both the provincial and the national economies. The manufacturing sector is very sensitive to global trade. While aerospace, information technology, pharmaceutical, and biotechnology firms have benefited from the liberalization of world trade and NAFTA, they are now tied to world demand. To stay ahead of foreign competitors, these firms have spent heavily in research, making Montréal a leading research centre in Canada.

Since 1989, these firms have expanded, although in the early years of the twenty-first century some businesses, such as the aerospace industry, have contracted due to a reduction in worldwide demand. These high-technology industries employ

Photo 6.10
The rolling terrain of the Appalachian Uplands in Estrie is well suited for dairy operations.

highly skilled and well-paid workers. Most of Québec's manufacturing sector is in Montréal, where high-tech companies are now the leading edge in manufacturing. These companies were responsible not only for Montréal's impressive economic recovery in the late 1990s but also for transforming the city into one of the leading high-tech centres in North America. This shift from traditional to high-tech manufacturing has three consequences for Montréal's labour force. First, the demand for highly skilled workers is increasing, creating labour shortages and a search for skilled immigrants. Second, the demand for semi-skilled workers is decreasing, thereby contributing to the relatively high unemployment rate in the Montréal area and province. Lastly, a command of both French and English is often required in the high-tech manufacturing industries.

While Montréal remains the focus of Canada's clothing and textile industry, its grip on market share has slipped for two reasons. First, established firms are either shifting their production from Canada to countries with low wages or they are subcontracting various elements of production in these low-wage countries. Second, in accordance with WTO rulings, Ottawa has continued to reduce tariffs on textiles, clothing, and related products from developing countries. While these measures may stimulate economic development in foreign countries, including China, the negative impact on Québec firms is contributing to the collapse of the textile and garment industry in Montréal and other Québec centres.

With the automobile and aerospace industries facing a slump in demand, both industries have had to reduce production capacity. In 2002, Québec lost its only automobile assembly plant when GM closed its plant at Sainte-Thérèse, laying off more than 1,000 workers. Québec's aerospace industry has been a major player on the world scene. Assisted by generous funding from Ottawa and the Québec government, Bombardier is Canada's leading aerospace and high-speed train firm. The firm accounts for around one-quarter of business jet sales in the world and employs nearly 20,000 workers. Bombardier produces passenger jets (the Challenger and the Regional Jet) and smaller business jets (Learjets). The Regional Jet, a stretched version of the Challenger, is the leading aircraft for short-haul markets in North America and Europe. The success of this firm is due largely to its ability to sell its product in the global market. Two international factors have hurt Bombardier's aerospace division. First, the 9/11 terrorist attacks caused a drop in air travel. Second, the 2008 global economic downturn has affected Bombardier and other aerospace firms. They face a difficult future. Bombardier, however, has achieved impressive gains in train technology and production, as noted earlier. Its continued development of its high-speed trains is paying off. In September 2012, Bombardier reported that sales to the United States had jumped substantially, not because the US economy had rebounded but because US transit authorities have had to purchase new transportation equipment as their existing fleets of rail cars are aging. This pent-up demand, coupled with federal funding for state transportation infrastructure, led to a record $5.7 billion train order from Amtrak for Bombardier in the first half of 2012 (McKenna, 2012: B5).

Northern Québec

Northern Québec lies beyond the ecumene of the St Lawrence Valley and in the terrain of two physiographic regions—the Canadian Shield and the Hudson Bay Lowland. Here are the traditional homelands of the Cree, Innu (Naskapi), and Inuit. Its resource-based economy—mining and forestry—depends on foreign markets for much of its production. With the housing crisis in the US, exports have dropped for lumber while pulp and paper products face an ever-diminishing demand. As an old resource hinterland, its forest economy is depressed, the population is declining, and unemployment rates are high. On the other hand, northern Québec is the site of exciting political developments among the Inuit and Cree. One such development is taking place in Nunavik, the political territory of the Inuit. This territory soon will have a public style of regional government, like that of Nunavut, rather than an ethnic one (such as First Nations governance structures), which means that all residents are treated equally. The political significance is twofold. First, public government offers more opportunities for political autonomy within Québec. Second, it allows for closer integration with Québec and therefore better access to its programs. Since the signing of the James Bay and Northern Québec Agreement, **Makivik Corporation** has been responsible for managing the political and economic interests of the Inuit of Nunavik (Vignette 6.10), who are known as Nunavimmiut. Makivik has led negotiations for the Inuit with the governments of Québec and Canada regarding the creation of a Nunavik government (Vignette 6.9).

With a faltering economy, northern Québec's population has taken on four demographic characteristics that are strikingly different from southern Québec:

- an aging population;
- a net out-migration, especially of younger members of its population;
- few immigrants;
- rapidly expanding Inuit and Cree populations.

As a resource hinterland lying in the Canadian Shield, the main economic activities are associated with forestry, mining, and hydroelectric generation. Today, such economic activities require relatively little labour. Consequently, the region is sparsely populated, with a few large mining towns such as Chibougamau. Further north in the James Bay region, most people are Aboriginal Québecers. While tourism flourishes in the Laurentides, most of northern Québec is too remote to attract tourists. Attempts at agricultural settlement have had marginal success in the Lac Saint-Jean area, but the short growing season and thin soils associated with the Canadian Shield prohibit commercial agriculture. Northern Québec's geography is best suited as a resource frontier. Low-cost electricity makes the smelting of bauxite (alumina ore) a major processing industry in the Saguenay region and along the north shore of the St Lawrence. Alcan, the world's second-largest aluminum producer, continues to expand its production capacity in Québec. Low-cost electrical power and an ocean shipping route are important locational factors.

Agriculture

The settlement of the Clay Belt demonstrates the difficulty of farming in the northern area of Québec. The Clay Belt occupies an enormous area of northwestern Québec and northeastern Ontario. The Canadian Shield acquired this relatively thick layer of sand, silt, and clay sediments near the end of the last ice age about 10,000 years ago. The rivers flowing to Hudson and James bays were blocked by the remnants of the Laurentide Ice Sheet. Gradually, a huge glacial lake that lasted for several thousand years, called Lake Barlow-Ojibway, spread over an area of about 200,000 km². Lake sediments, especially minute particles of clay, sand, and silt, were deposited. When the glacial lake drained, much of the land was covered by peat, making it unsuitable for agriculture. In fact, less than 5 per cent of the Clay Belt contains arable land, only a portion of which has been cultivated. As a result, farmland in the Clay Belt is in scattered pockets, separated by large areas of forested country.

Forest Industry

Forests cover nearly half of Québec. In total, Québec accounts for nearly 760,000 km² of forest lands. This valuable natural resource consists of mixed forest in the St. Lawrence Lowland and the Appalachian Uplands (Figure 2.1). Here, commercial forests play a small role in logging operations but a key role in maple syrup production and in harvesting hardwoods for furniture-making. Most logging operations that produce softwood lumber and wood for pulp and paper mills are in the boreal forest found in the Canadian Shield.

VIGNETTE 6.9

Mapping the Road to Nunavik

The land now known as Nunavik became part of Canada in 1870 when Rupert's Land was transferred from Great Britain to Canada. In 1912, the Boundaries Extension Act assigned Nunavik to Québec on condition that outstanding Aboriginal rights be settled.

In 1971, the announcement of the James Bay Project triggered a court case over Aboriginal rights. In 1973, the court challenge led to Québec agreeing to fulfill its obligation and resulted in the signing of the James Bay and Northern Québec Agreement on 11 November 1975. One result was the creation of two corporations for the Inuit—Makivik Corporation and the Kativik Regional Government. Makivik administers the Inuit compensation funds from the JBNQA and represents the Inuit in political and economic matters, such as negotiating the Nunavik Inuit Offshore Land Claim (resolved in 2008). On the other hand, Kativik provides the various public services to residents of Nunavik, including education and security. These two organizations have allowed the Inuit to manage their own affairs and thus gain administrative experience for over 30 years. In the process, the dream of a regional government emerged. The breakthrough for this radical political goal came in 1983 when Premier René Lévesque stated unequivocally that an Inuit regional government within Québec was possible.

Like other Aboriginal peoples, the Québec Inuit are seeking a form of political autonomy within the existing structure that would respond to their needs, desires, and aspirations. To achieve that goal, three formidable challenges had to be resolved. First, how can a regional government function within a province? This calls into question the division of powers between the province of Québec and the soon-to-be-formed government of Nunavik. Second, how can Nunavik (a non-ethnic government) treat all its residents equally and still promote the Inuit culture? Lastly, how can 11,000 people living in 14 communities scattered over 500,000 km² govern themselves and also generate sufficient revenue to pay for their government?

The road to Nunavik is difficult but not impossible. Somehow the structure, operations, powers, and design of this new form of government within a Canadian province can be achieved. Since 2001, negotiations continued between Ottawa, Québec City, and Makivik Corporation. In 2003, a framework agreement was struck with the purpose of creating a new form of government in Nunavik. By 2007, an Agreement-in-Principle was signed by federal and Québec officials with the final agreement scheduled for approval in 2011. However, in an April 2011 referendum, the Inuit voted against proceeding with the formation of the Nunavik Regional Government, indicating that the Inuit leadership had not gained the confidence of the general population on this issue. The next step is to explain the benefit of the proposed Nunavik Regional Government to Quebec Inuit and, in doing so, gain their support in a future referendum.

While Québec has 22 per cent of Canada's productive forest lands, the natural vegetation zone known as the Tundra–Boreal Transition contains a non-commercial forest (Figure 2.8). In terms of productive forest, Québec ranks first among Canada's geographic regions, but it ranks second behind British Columbia in total volume of wood cut. However, Québec leads British Columbia in pulp and paper production and the output of newsprint. Québec's advantage is its close proximity to major US cities, especially New York, where, historically, the demand for paper and newsprint has been extremely high. For US buyers, Québec softwood is preferred over US softwood lumber because the cold climate in northern Québec results in slower growth, which increases the strength of the wood. Unfortunately, the demand for paper and newsprint has been declining as more and more readers are turning to the Internet for their news rather than buying "hard" copy newspapers and to e-readers for books. For that reason, the future for the pulp and paper industry appears gloomy.

While the forest industry contributes significantly to the province's economy in both value of production and employment, it has fallen on hard times due at first to the US duty on softwood lumber imposed in 2001 and later to the collapse of the US housing market since 2006. Consequently, the forest industry has contracted, causing many to lose their jobs. The Quebec forest industry generated more than 68,000 direct jobs as recently as 2009, but since then it has lost nearly 14,420 jobs.

Clearly, a rebound in the forest industry awaits the recovery of the US economy. In spite of the sluggish American economy, prices for softwood lumber increased in mid-2012, reaching US$310 by August 2012, up 24 per cent over the average price for January 2012 (Jang, 2012: B6). If this price trend continues, Québec forest industry may be on the rebound.

Mining Industry

Mining has always been important in northern Québec. As with the rest of the Canadian resource economy, globalization of ownership has taken a firm hold. Iron deposits in Québec represent world-class ore bodies that major mining companies and steel companies sought to place in their economic orbit. Two companies controlled iron mining in Québec and both are now controlled by global mining companies. In 2000, Rio Tinto became the principal shareholder of the Iron Ore Company of Canada, and in 2008 ArcelorMittal Mines Canada purchased the Québec Cartier Mining Company. In 2011, the value of Québec's metallic mineral production was close to $6.1 billion, with iron ore and gold making up close to two-thirds of the value, while non-metallic production was valued at $1.7 billion (Natural Resources Canada, 2012). Québec's mining industry, by value of production, placed third among Canada's six regions. In 2011, Québec accounted for 15.4 per cent or $7.8 billion of Canada's $50.3 billion total (metallic and non-metallic) mineral production (ibid.).

Table 5.6 , page 202, lists 2011 mineral production by province and territory. Table 7.3, page 294, shows 2011 value of Canadian mineral production by the leading minerals.

VIGNETTE 6.10

Nunavimmiut Benefit from Resource Profit-Sharing Agreement

The Raglan Agreement, signed in 1995 by Falconbridge Ltd, Makivik, and the communities of Salluit and Kangiqsujjuaq, has generated $65.4 million for Nunavik residents. Most benefits go to residents of Salluit and Kangiqsujjuaq, the two Inuit communities closest to the nickel mine at Raglan. Salluit receives the greatest share because it is closest to the mine and the port at Deception Bay. Last year, after $14 million was shared among Salluit's 1,100 beneficiaries, each adult got $15,000 and every child received $3,500. In Kangiqsujjuaq, beneficiaries, about 520 in all, received cheques for $4,700, with $4,000 coming from the Raglan profit-sharing agreement and the rest from other enterprises of Makivik, which are shared by all beneficiaries in Nunavik.

One reason for the importance of mining is the presence of the Canadian Shield, which contains many mineral deposits. In particular, its wealth is concentrated in the Labrador Trough where many rich ores, especially iron deposits, are found. As a consequence, Québec and Labrador provide most of the iron ore for North American steel mills and an increasing proportion for mills around the world.

The Labrador Trough is clearly one of the rich natural assets found in Québec and Labrador. This rare natural geological body extends for about 1,100 km southeast from Ungava Bay through both Québec and Labrador. Further south, it turns southwest past the Wabush and Mount Wright areas to within 300 km of the St Lawrence River. Deposits of iron were first reported in 1895 by A.P. Low of the Geological Survey of Canada, the first geologist to investigate the region's mineral potential. At that time, however, these deposits had no commercial value because more accessible mines could supply the needs of the iron and steel companies.

All that changed in the 1940s. American steel producers could no longer count on domestic supplies of iron ore, so they sought more reliable sources, including those in northern Québec and Labrador. Through a process of market integration initiated by US steel companies, Québec's resource hinterland became dependent on a particular group of steel companies in the industrial heartland of the United States for its economic well-being. Two mining companies, Québec Cartier Mining Company and the Iron Ore Company of Canada, developed iron mines in isolated areas of northern Québec and Labrador. By 1947, plans were laid for an open-pit mine in northern Québec near the border with Labrador. The Iron Ore Company built a town (Schefferville) for miners and their families; transmitted power from Churchill Falls to operate the mine and the town; and built a railway (the Québec North Shore and Labrador Railway) to deliver the iron ore to the port at Sept-Îles, from where the ore was transported to supply US steel mills in Ohio and Pennsylvania. The demand for iron ore rose in the

Photo 6.11
Some 1,000 employees work at the Mont-Wright Mining Complex owned and operated by ArcelorMittal. The mine site comprises an open-pit mine that extends some 20 km beyond the rise of the Canadian Shield shown in the right side of the background. Deep holes (15.8 metres) are drilled in the ore-bearing rock and an explosive mixture is poured and blasted to split the rock. Huge trucks haul the ore to the ore crusher facility after which the small rocks go to the concentrator to produce small pellets. These pellets, with an iron content around 70 per cent, are loaded onto ore cars and then shipped by rail to Port-Cartier where ore carriers take the product to steel plants around the world.

1960s, resulting in the establishment of three more mining towns—Wabush and Labrador City in Labrador and Fermont, Québec. At the same time, Québec Cartier Mining Company built a similar iron-mining operation by constructing the Cartier Railway from Port Cartier on the St Lawrence River to the resource town of Gagnon. But by the 1980s world steel production had surpassed the demand, causing a severe slump in the demand for iron ore. To add to this economic problem, US steel plants were now less efficient than the new steel mills in Brazil, Canada, Korea, and Japan, and lower-cost iron mines had opened in Australia and Brazil. As lower-priced steel from these countries undercut the price of US steel, American steel companies had to reduce their output, close plants, and sell their shares in the two mining companies. The mining companies were forced to restructure their operations to reduce costs. The repercussions for workers were severe. The mining workforce was reduced and two mines, at Schefferville and Gagnon, were closed. As the world business cycle improved in the twenty-first century, demand from iron and steel plants around the world, especially in China, increased. The two companies increased their production accordingly. In 2013, Rio Tinto's Iron Ore Company of Canada operates a pellet plant and mine at Labrador City and at Wabush Mines. The concentrated ore is shipped by rail to Sept-Îles for transshipment to world markets. ArcelorMittal Mines Canada, which purchased the Québec Cartier Mining Company in 2008, has two mines, Mont-Wright and Fire Lake, plus a pellet plant at Port-Cartier.

Mine (and town) closures are not unusual events. Several northern Québec communities are single-industry towns and rely on mining for their existence. Extraordinary measures are taken to survive. Residents of Malartic, for example, are relocating outside of town because one of the largest gold deposits in North America exists under the town. As in other resource towns, times can be tough. In 2008, half of Malartic's workers were either on unemployment insurance or on welfare. But this $1 billion project, located in the famous Abitibi gold belt, breathed new life into the community by creating nearly 500 permanent jobs and some 800 construction jobs (Séguin, 2009). Since 2011, Malartic gold mine outpaced other mines to become the largest gold mine in Canada (Osisko, 2013).

As noted earlier, the cyclical nature of the mining industry, due to its dependence on world markets, poses a problem for resource communities. Global companies operate to make profits and have no commitment to local communities, so community prosperity and employment are tied to global demand. Low demand means layoffs and even the closure of mines and ore-processing mills. This boom-and-bust cycle driven by fluctuations in world prices is particularly hurtful to the narrowly based economies of resource hinterlands. The world demand for mineral products reached a peak in the 1960s, declined sharply in the 1980s, recovered somewhat by the late 1990s, and then began to slide downward in 2002 and hit bottom in 2009. Since then, demand (and prices) for minerals has climbed significantly. Such price fluctuations have had a profound impact on Québec's iron-mining communities and their workers.

Québec's Urban Geography

Over 80 per cent of Québec's population live in urban centres. Two of the largest metropolitan cities in Canada, Montréal and Québec City, are located in the province. Other major population centres are Gatineau, Sherbrooke, Saguenay, and Trois-Rivières. In total, these six urban clusters have a population of 5.4 million (Statistics Canada, 2012e). Except for Sherbrooke, these large cities are situated in the St Lawrence Lowland physiographic region.

As in Ontario and most other regions of Canada, migration has played a major role in Québec's urbanization. Push and pull factors have attracted rural Québecers to cities. The key push factors in rural Québec have been limited job opportunities, a shrinking labour force in the primary sector, and an increasing number of young people entering the workforce.

From 2001 to 2011, the population of the major cities, with the exception of Saguenay, grew, led by Gatineau (formerly Hull), with 20.2 per cent increase, and followed by Sherbrooke, Québec City, and Montréal (Tables 6.8 and 6.9). The population growth in Gatineau, across the Ottawa River from the nation's capital, was largely due to the employment and business opportunities generated by the federal government. The population of Saguenay (formerly Chicoutimi–Jonquière) dropped by 2.1 per cent. Located in the resource hinterland of the Lac Saint-Jean region, this city has been losing population for several decades. The downward trend is related to its two main economic activities,

Table 6.8 Major Cities in the St Lawrence Lowland

City	Population 2001	Population 2011	% Change 2001–11
Montréal	3,451,027	3,824,221	10.8
Québec City	686,569	765,706	11.5
Gatineau	261,704	314,501	20.2
Trois-Rivières	137,507	151,773	10.4
Saint-Jean-sur-Richelieu	79,600	92,394	16.1
Saint-Hyacinthe	54,275	56,794	4.6
Shawinigan	56,412	55,009	–2.5
Sorel-Tracy	47,802	47,772	0.0
Joliette	39,720	46,932	18.2
Salaberry-de-Valleyfield	39,028	40,077	2.7
Total	4,853,644	5,395,269	11.2

Source: Statistics Canada (2007, 2012e).

THINK ABOUT IT
Does expanding the highway system improve automobile flow to suburbs and satellite commuter communities, or does it facilitate more urban sprawl?

the aluminum plants and the forestry mills. As in other resource hinterlands, employment prospects in these two industries have been declining, especially in the forest industry.

Montréal

Montréal, the metropolis of the province, is the industrial, commercial, and cultural focus of Québec. Montréal is the largest census metropolitan area in the province with a population in 2011 of 3.8 million. Nearly half of Québec's population lives in the Montréal CMA, which includes the cities of Laval and Longueuil as well as the municipalities of Beaconsfield, Baie-D'Urfé, Côte-Saint-Luc, Dollard-Des-Ormeaux, Dorval, Hampstead, Kirkland, L'Île-Dorval, Montréal-Est, Montréal-Ouest, Mont-Royal, Pointe-Claire, Sainte-Anne-de-Bellevue, Senneville, and Westmount. Like Toronto, Montréal has developed an effective combination of metro lines and surface bus transportation that provides quick and inexpensive city transportation. Still, Montréal's infrastructure, perhaps more than that of other Canadian cities, needs attention. For instance, in 2006, the collapse of the de la Concorde overpass in Laval killed five people, and six years earlier eight concrete beams toppled onto a Laval highway, killing one man.

Within Montréal, a dozen bridges and overpasses are well past their best years, as is the Honoré Mercier Bridge. Like other cities, the cost of improving its infrastructure is well beyond its means. City mayors have challenged Ottawa to provide more tax dollars to cities for infrastructure. Ottawa has responded by sharing a portion of the federal gas tax—in 2010–11, the gas tax transfers to cities and communities totalled $15.8 billion (Department of Finance, 2012).

Given its strategic location and economic size, Montréal serves as the transportation hub of Québec, making it the **regional core** of the province and the rest of Québec the periphery. At the national scale, Montréal is part of the national core because it is part of Canada's manufacturing belt.

Following the Free Trade Agreement, Montréal's manufacturing sector had to respond to strong foreign competition. Labour-intensive manufacturing firms experienced great difficulty in competing with foreign firms that had substantially lower labour costs. Montréal's manufacturing firms made two major changes: labour-intensive plants substituted machinery for workers to increase productivity, and high-technology firms expanded. By the late 1990s, Montréal's economy had become much more specialized in aerospace, computers, fibre optics, multilingual software, telecommunications, and other areas of industrial research and development. The provincial government has taken a leading role by providing subsidies for high-tech firms that relocate to Montréal and other Québec cities.

Montréal and Toronto

Beginning in the 1970s, especially with the election of the first PQ government in 1976, Toronto quickly replaced Montréal as the premier city in Canada. Montréal was no longer the largest city and the financial capital of the country. The principal reason for this shift in metropolitan power was the strong economic growth in Toronto and in southern Ontario powered by the Auto Pact and the resulting expansion of the automobile industry. A secondary factor was the economic and demographic fallout from the political unrest in Québec and the very real possibility of Québec separating from Canada. The unsettled political environment leading up to the 1980 Québec referendum worried the anglophone community. Many anglophones and some corporations moved to Toronto. While the francophone business community and the provincial government kept the Montréal economy growing, by 1981 Toronto had a population of 3.0 million compared to Montréal's 2.8 million (Table 6.9). Three

Table 6.9 **Population Change: Montréal and Toronto, 1951–2011 (000s)**

Year	Montréal	Toronto	Difference
1951	1,539	1,262	277
1961	2,216	1,919	297
1971	2,743	2,628	115
1981	2,828	2,999	−171
1991	3,127	3,893	−766
2001	3,426	4,683	−1,257
2006	3,636	5,113	−1,477
2011	3,824	5,583	−1,759

Source: Statistics Canada (2002, 2007, 2012e).

other economic factors explain Montréal's slow growth and Toronto's much faster growth:

1. Montréal's economy was much more dependent on labour-intensive manufacturing (which was in decline) such as that in the textile industry.
2. Montréal's industrialization had begun much earlier than Toronto's and, for that reason, its manufacturing firms tended to be older and less efficient than those in Toronto.
3. Toronto services a North American market while Montréal has more of a provincial market, and for that reason Toronto has a much larger "external" demand for its manufacturing, wholesale, tourist industries.

For those reasons, Toronto's growth rate exceeded that of Montréal. Even in the 10-year period from 1951 to 1961, when both cities grew at phenomenal rates—Montréal at 4.3 per cent per year and Toronto at 5.1 per cent—Toronto's rate was higher. From 1971 to 1981, the rate of population increase slowed in both cities, with that of Montréal close to zero. During that time, Montréal's annual growth rate was only 0.3 per cent, while Toronto's was 1.9 per cent. This gap continued in the early 1990s with the annual rates for Toronto and Montréal at 1.9 per cent and 1.3 per cent, respectively. By 2011, Toronto's population had reached 5.6 million while Montréal's was 3.8 million, leaving an ever widening population gap between the two cities.

Québec City

Québec City, the historic capital of French Canada and the capital of modern-day Québec, has the next largest urban concentration in the province, totalling nearly 770,000 people in 2011. Close to the Laurentides, Québec City has a magnificent physical setting on high banks along the St Lawrence River. As the only walled city in North America and with buildings over 300 years old, the city has a vibrant tourist industry with many international visitors. In 1985, Québec City was selected as a World Heritage Site by UNESCO.

The economic base of Québec City revolves around three functions. First, it is a government and university town. As the seat of government for the province, Québec City employs a large number of civil servants. Second, it has become a world-class tourist centre. The Old World charm of Québec City draws tourists from around the world, while special events such as its Winter Carnival are very popular. Third, Québec City is only minutes away from excellent seasonal recreation areas—from skiing in the winter to water sports in the summer. The economic base of Québec City remains heavily dependent on its political and cultural roles, but it also has a number of other economic functions: it is a port and rail centre, as well as a centre for resource processing, metal fabricating, manufacturing, and higher learning. Université Laval ranks highly among universities in North America. High-tech industries play an important role in Québec City's economy,

Table 6.10 Major Cities in the Western Appalachian Uplands, 2001–2011

Cities	Population 2001	Population 2011	% Change 2001–11
Sherbrooke	175,950	201,890	14.7
Drummondville	72,778	88,480	21.6
Granby	63,069	77,077	22.2
Victoriaville	38,841	46,354	19.3
Sainte-Georges	29,759	34,642	16.4
Thetford Mines	26,721	27,968	4.7
Total	407,118	476,411	17.0

Source: Statistics Canada (2007, 2012e).

including the highly specialized health and photonics/optics sectors.

Urban Development in the Western Appalachian Uplands

To the east of Montréal lies the subdued landscape of the Appalachian Uplands.[8] First settled by Loyalists following the American Revolutionary War, this physiographic region was named the **Eastern Townships**. These hilly lands lie between the St Lawrence Lowland and the border with the United States. The area is now known as **Estrie** (meaning east of Montréal).

Estrie is dominated by Sherbrooke, with a population of just over 200,000 in 2011. A number of small cities and towns are scattered across Estrie, but most communities west of Sherbrooke fall under the orbit of Montréal. The rate of population increase for Estrie cities from 2001 to 2011 outstripped those of the St Lawrence Lowland by nearly six percentage points—from 17.0 to 11.2 per cent (Tables 6.8 and 6.10). Two exceptions are the asbestos mining towns of Asbestos and Thetford Mines. With the closing of both asbestos mines, the small town of Asbestos, with a population just over 7,000, is on a steep downward spiral, unlike the much larger and more diversified Thetford Mines, with a population of near 28,000. The *Globe and Mail*'s Ingrid Peritz, writing in reference to Maxime Blake, a cook at Le Fou du Roi restaurant in Asbestos, explains: "Asbestos is becoming a ghost town and virtually his entire high school class has left town and he plans to do the same"(Peritz, 2012: A11).

Beyond the Urban Cores

Québec's urban geography takes on a distinct regional character. Southern Québec has experienced modest growth with the highest rate of increase from 2001 to 2011 taking place in the Western Appalachian area (Table 6.10). On the other hand, beyond this area and the St Lawrence Lowland, most urban centres saw very limited population increase—the main exceptions were Rimouski and Rivière-du-Loup (Table 6.11). Most of these cities and towns were affected by out-migration, just as those in Estrie and along the St Lawrence gained population by in-migration. Saguenay, the regional capital of the vast interior known as the Lac Saint-Jean region, experienced only a meagre population increase of 1.2 per cent over the past 10 years (2001–11). This demographic pattern is associated with weak regional economies found in the hinterlands, especially the depressed state of the forest industry. From 2001 to 2011, slow growth or population decline occurred in the urban centres of northern Québec, the North Shore, and the Gaspé Peninsula. Two forces were at play.

First, the resource-based economies in these outlying areas are contracting. Within the new North American marketplace, resource companies that export most of their products are under great pressure to reduce their costs, which translates into reducing the size of the labour force or, in the worst-case scenario, closing their operations. This global-induced trend was reinforced by the 2008–9 global crisis that saw demand for both forest products and minerals decline sharply. Urban centres are affected by the negative spinoff

Table 6.11 **Population of Cities of Northern Québec, the North Shore, and Gaspé Peninsula, 2001–2011**

City	Population 2001	Population 2011	% Change 2001–11
Northern Québec			
Saguenay	154,938	157,790	1.8
Rouyn–Noranda	39,621	41,798	5.0
Val-d'Or	32,423	33,265	2.5
Alma	32,930	33,018	0.0
North Shore			
Rimouski	46,012	50,912	10.7
Baie-Comeau	30,401	28,789	−5.3
Sept-Îles	27,623	28,487	3.1
Rivière-du-Loup	23,229	27,734	19.4
Gaspé Peninsula			
Gaspé	14,932	15,163	1.6
Percé	3,614	3,312	−8.3
Total	405,723	410,977	1.3

Source: Statistics Canada (2007, 2012e).

Photo 6.12

Founded in 1608, Québec City is one of the oldest cities in North America. With the ancient wall enclosing Old Québec in the foreground, the Château Frontenac dominates the background. On 3 July 2008, Québec City celebrated its 400th anniversary since its founding by French explorer Samuel de Champlain.

Luc-Antione Couturier Photographe/Andre Caron

from industrial restructuring, resulting in a smaller population, fewer well-paying jobs, and, therefore, fewer customers. Such a downward spiral reduces the size of city markets and forces some local businesses to close.

Second, the demographic outcome of a declining regional economy where few job opportunities exist leads to out-migration, especially by younger and more educated people. Such migration causes a contraction of the local population and a loss of potential community leaders, and signals a general economic malaise that results in a relentless downward spiral of both the regional economies and urban populations.

The French/English Faultline in Québec

After the Quiet Revolution, the Québec government took legislative measures to ensure that the French language, the basis of Québécois culture, prospered within the province. From the Québec government's perspective, such action was necessary because Québec was surrounded by a sea of English-speaking North Americans and because of the strong presence of the English business community in Montréal. Accordingly, the government passed a series of language laws, notably Bill 101 in 1977, that obliged businesses to use French and required allophone parents to send their children to French schools. After the 1995 referendum, the language issue gradually became less of a sore point within Québec because of a general acceptance of the primacy of the French language and the growing number of anglophones, allophones, and francophones who became bilingual.

Evidence of the primacy of the French language is revealed in the steady shift over the last decade—from 2001 to 2011, English-only speakers in Montréal declined from 11.1 per cent to 9.9 per cent while the number of bilingual Montréal residents increased from 8.8 per cent to 9.5 per cent (Statistics Canada, 2013: Table 5). Another sign of the well-being of the French language in Québec is that Lucien Bouchard and the other authors of the manifesto, *Pour un Québec lucide*, called on the government to encourage the learning of English to allow francophones to function in the larger English-speaking world: "The Government must also make far greater effort to ensure that all Québecers speak and write English, as well as a third language" (Bouchard et al., 2006: 8). With language as the central characteristic of Québécois identity, the election of a minority Parti Québécois government in September 2012, and its introduction of Bill 14 designed to promote the use of French, language tensions are once again heating up.

SUMMARY

Culture remains the fundamental distinguishing feature of this region of Canada. Québec's francophone culture and French language remain strong. As the heartland of francophones in Canada and North America, Québec has a special role to play. Cultural events like La Fête nationale du Québec evoke a sense of ethnic nationalism and a love of the land and its people, which is popularly expressed as "*J'ai le goût du Québec*." Within this cultural context, the French language serves as a linchpin.

Economics and demography are doing well but not as well as in most other regions. As one of the original British colonies to form Canada in 1867, Québec's position within Confederation is weakening as its share of Canada's population and economic output declines. The reason is simple: while Québec's economy and population have expanded, other regions of Canada—Ontario, British Columbia, and Western Canada—have expanded at a more rapid rate. If this trend continues, then Québec's place will be seriously eroded. Yet two bright lights suggest a turnaround. First, Québec has recognized the importance of the knowledge-based economy and is devoting more public funds for scientific research initiatives than any of the other five regions. Already, Québec has a few internationally established high-tech companies that can compete on the world stage; the challenge is to expand in this vital economic sector. Second, Hydro-Québec is flexing its muscle by building more hydroelectric dams, producing more electricity, and seeking to sell more power to the lucrative New England energy market.

CHALLENGE QUESTIONS

1. What are the political implications for Québec if its share of Canada's population and GDP continues to decline?

2. Why is Montréal the focal point of the language faultline in Québec?

3. Why was the Paix des Braves so fundamental to continued development of the James Bay Project and why did the Cree support this agreement?

4. When Nunavik gains its desired form of self-government, will this political development result in a semi-autonomous territory within Québec?

5. If it is logical for Lucien Bouchard, the former PQ Premier, to argue that the learning of English by francophones is beneficial to them and Québec because English allows them to function in the larger English-speaking global business world, what does that say about the English/French language faultline in Montréal?

FURTHER READING

De Courcy, Diane, Minister responsible for the Charter of the French language. 2012. Bill 14: *An Act to Amend the Charter of the French Language, the Charter of Human Rights and Freedoms and Other Legislative Provisions.* National Assembly, First Session, 40th Legislature. At: <www.assnat.qc.ca/en/travaux-parlementaires/projets-loi/projets-loi-40-1.html>.

Bill 14 is designed to strengthen the place of French in the workplace, to require that small businesses function in French, and to restrict access to English schools. The bill also proposes amendments to the Québec Charter of Human Rights and Freedoms that would emphasize that French plays an essential role in Québec's social cohesion. The main battleground for this linguistic fight is Montréal, where the majority of English-speaking Québecers live and work.

While the two main opposition parties are likely to oppose this legislation and thus defeat the bill, others warn that the province is not doing enough to promote the French language. The Conseil superieur de la langue française, the official advisory body for the Minister of Language, argues that the use of French is declining in Quebec and more legislation beyond Bill 14 is needed to ensure social cohesion (CTV News, 2013). On the other hand, English-speaking Montréalers see the bill as one more attempt to squeeze their language rights. Whether the bill passes or not, the French/English language debate has caught fire once again.

BRITISH COLUMBIA

Introduction

British Columbia remains a powerhouse within Canada. As the province continues to expand its economy, population and trade, its position as a leading region within Canada is reinforced. While its export-oriented economy expands and contracts in accordance with fluctuations in global demand, the growing number of its trading partners softens these fluctuations and, at the same time, diminishes its reliance of trade with the United States.

British Columbia's location on the Pacific coast binds it to the neighbouring US states, and to the Pacific Rim countries through global trade. BC benefits from trade with the US and, more and more, with the expanding economies of China and other Pacific Rim countries. While BC's economy becomes more dependent on its vital transportation role between Asia and North America, its traditional economic base, the forest industry, has lost some of its lustre because of the slump in US imports. Yet, BC's spectacular scenic beauty, especially its interface between sea and mountains, supports a vital tourist industry; its natural resources remain critical to this region's economic development and diversification. In fact, newly found natural gas reserves have sparked plans for pipelines to the Pacific and so have the huge bitumen reserves of the Alberta oil sands. In the long-run, however, BC's expanding knowledge-based industries hold the key to its future and are pushing other economic sectors into the information age. In the heart of these activities, Vancouver serves as a transportation hub between Canada, the United States, and the Pacific Rim countries.

British Columbia's place in Canada represents a paradox. On the one hand, its natural orientation is southward to the United States, following the grain of its physical geography; and westward across the Pacific Ocean to Asia. On the other hand, its political orientation is necessarily eastward. In this context, geography tends to either dull or inflame relations with Ottawa and Central Canada. This paradox shows its face in the centralist/decentralist faultline.

CHAPTER OVERVIEW

The "Key Topic" is the forest industry. Other topics in this chapter focus on:

- British Columbia's physical geography and history.
- BC's economy and population, especially as these factors relate to the region's physical setting and the drive to the Asian markets.
- The BC version of the centralist/decentralist faultline.
- Resources and their changing role in the BC economy.
- The importance of international trade, the Asia-Pacific Gateway, Prince Rupert's new role as a major port for container cargo, and Kitimat's rising star as an LNG terminal.
- Whether BC will remain on the outer edge of Canada's economy or turn into a powerhouse on the Pacific Rim.

Cross River Falls, Rockies, British Columbia. The Cross River is a tributary of the Kootenay River, which is part of the Columbia River Basin. The river takes its name from a cross that was erected in 1858 by a Jesuit missionary Pierre-Jean De Smet. Photo: John E Marriott/All Canada Photos.

British Columbia within Canada

British Columbia lies at the western edge of Canada's land mass, far from the economic/political heart of the country. This geographic fact underscores the basis for tensions between BC and Central Canada. For most British Columbians, however, the dramatic interface between the sea and the mountains defines the province. While BC consists of nearly 10 per cent of Canada's land mass and 13 per cent of its population, that tiny coastal zone with its spectacular rain forest holds the key to the province's identity. From a physiographic perspective, the province is divided into two parts: the Cordillera accounts for the bulk of the province and the Interior Plains occupies a small portion of its northeast corner. British Columbia has access to the Pacific Ocean except

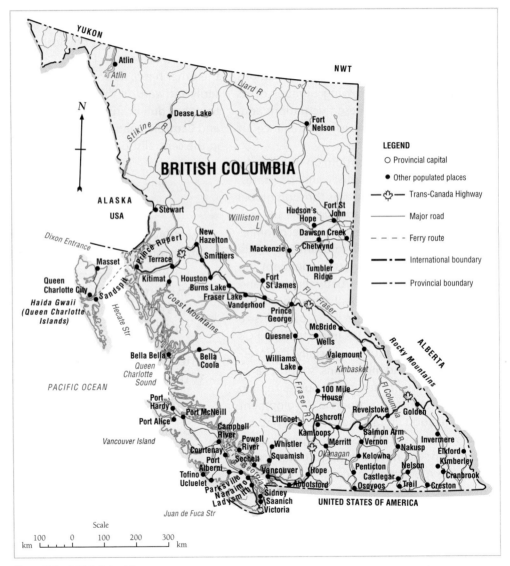

Figure 7.1 British Columbia

Vancouver's isolation from Central Canada and its close link to Washington, Oregon, and California are revealed in the following driving distances: Montréal, Ottawa, and Toronto are over 4,300 km east of Vancouver. Los Angles is less than half the distance, at 1,735 km, while Bellingham, Washington, is just 85 km away, making it the most popular US shopping destination for Vancouverites. Not surprisingly, then, a transborder region called Cascadia remains a popular talking point among academics. *Source: Atlas of Canada*, at: <atlas.nrcan.gc.ca/site/english/maps/reference/provincesterritories/british_columbia>, Natural Resources Canada, 2007. Reproduced with the permission of the Minister of Public Works and Government Services Canada, 2013.

Photo 7.1

The new seven-lane Pitt River Bridge just east of Vancouver was completed in late 2009. As part of the Asia-Pacific Gateway and Corridor Initiative, this bridge will serve to improve access to Vancouver along the north shore of the Fraser River, with easier and quicker flow for trucks carrying exports to ports in the Lower Mainland for shipment to Pacific Rim countries. Supported by both the provincial and federal governments, the Pitt River Bridge represents one phase of joint government efforts to create a superhighway corridor from Calgary to Vancouver.

---

for its far north, where the **Alaska Panhandle** blocks sea access.

British Columbia is an emerging giant within Canada's economic system. This west coast region's economy is heavily based on its natural resources and the export of those resources and those produced in Western Canada, namely coal, grain, and potash. Pipelines to Kitimat will soon add natural gas to its export list—and possibly bitumen from the Alberta oil sands. Already the Trans Mountain pipeline carries crude oil to Greater Vancouver, from where it is shipped by supertankers to US and Asian markets.

British Columbia's scenic beauty supports a vibrant tourist industry; and its expanding knowledge-based industries, including the film and high-technology industries, help drive economic growth in new directions. Trade is crucial. Most exports go to the United States, though exports to Pacific Rim countries continue to climb. Lumber, pulp, natural gas, and coal are the province's four main exports. Imports, especially from China, Japan, and South Korea, flow through Vancouver to markets across Canada. The expansion of CN and CP rail lines and twinning of sections of the Trans-Canada Highway are facilitating access to the Port of Vancouver and exports to Asian countries. While Prince Rupert remains in the shadow of Vancouver, new facilities at the Port of Prince Rupert signal its arrival as a potential major player in international shipping. Kitimat may soon follow with its bright prospects as an LNG port and, possibly, as a terminal for the Northern Gateway oil pipeline. The **British Columbia–Alberta–Saskatchewan Trade, Investment, and Labour Mobility Agreement** (TILMA) is another sign of BC's interest in promoting trade, in this case, with provinces in Western Canada.

BC's share of Canada's GDP and population reveals BC's important place within Confederation (Figure 7.2). Since 2001, British Columbia's rate of economic and population growth has outperformed the national figures even with a depressed forest industry. Yet, a divide exists between urban-core BC and rural-hinterland BC, which depends so heavily on the forest industry. Like other regions, BC was hit by the global economic crisis and its unemployment rate jumped from 4.2 per cent in 2007 to 7.6 per cent in 2009; by mid-2012 the rate had dropped to 6.6 per cent (Table 5.1). With ever-increasing immigration, especially from China and Hong Kong, BC's population has become more diverse. Richmond, located just 25 minutes from both downtown

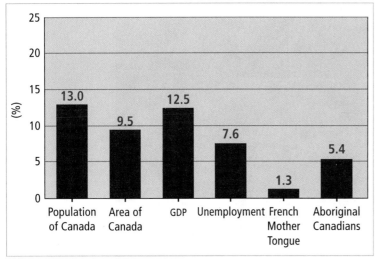

Figure 7.2 British Columbia basic statistics, 2011
BC's share of Canada's GDP and population reveals its solid place within Canada. The small number of those whose mother tongue is French indicates the relatively weak position of French-speaking Canadians in the province. The small percentage of Aboriginal people, however, does not reflect the extent of First Nations' potential ownership of land in BC through ongoing land claim negotiations.
Sources: Statistics Canada (2012a: Table 379–0025); Statistics Canada (2012b: Table 109–5324); Statistics Canada (2013: Table 2).

Vancouver and the US border, is a popular destination for Chinese immigrants. In the 2011 census, the percentage of French-speaking residents dropped from 1.4 per cent in 2006 to 1.3 per cent while the Aboriginal population, reached a new high of 5.4 per cent, up from the 2006 figure of 4.8 per cent.

 See Chapter 4, "Sense of Place: Now and Then," page 147, for background discussion pertinent to British Columbia.

Population is a measure of political power, i.e., the larger the population, the greater the political clout in Ottawa. Yet, political friction between BC and the federal government prevents this simple relationship from taking form. The struggle for political power and respect is ongoing and underscores the centralist/decentralist faultline discussed in this chapter. From time to time, various signs of disenchantment with Ottawa emerge. To those living in British Columbia, the province fits comfortably into the Pacific Northwest. In a sense, the Rocky Mountains are more than a physical divide. One expression of this regionalism that transcends the forty-ninth parallel is the concept of **Cascadia**—Oregon, Washington, and British Columbia. In part, Cascadia is a reaction to the negative feelings towards Ottawa. Within BC there is another faultline—the Aboriginal/non-Aboriginal faultline—whereby Aboriginal peoples are demanding more power through land claim agreements.

British Columbia's Physical Geography

The spectacular physical geography of British Columbia is perhaps the region's greatest natural asset. The variety of its natural features is unprecedented. Then, too, British Columbia is famous for its mild west coast climate. The combination of two contrasting climates (west coast and interior) with mountainous terrain has resulted in a wide variety of natural environments or ecosystems. Three examples of natural diversity are rain forests along the coast, desert-like conditions in the Interior Plateau, and alpine tundra found at high elevations in many BC mountains. As well, its fjorded coastline provides many deep harbours surrounded by the Coast Mountains. Burrard Inlet is one such fjord, although the prospect of daily calls by Suezmax supertankers challenges the depth of the inlet at the Second Narrows railway bridge (Figure 7.3).

The physical contrast between the wet BC coast and the dry Interior is largely due to the effect of the Coast Mountains on precipitation. Easterly flowing air masses laden with moisture from the Pacific are forced to rise sharply over this high mountain chain, and consequently most moisture falls as orographic precipitation on the western slopes while little precipitation reaches the eastern slopes (Vignette 7.1).

The climate of the west coast is unique in Canada. Winters are extremely mild and freezing temperatures are uncommon. Summer temperatures, while warm, are rarely as high as temperatures common in the more continental and dry climate of the Interior Plateau of British Columbia. Moderate temperatures, high rainfall, and mild but cloudy winters make the west coast of British Columbia an ideal place to live and a popular retirement centre for those Canadians wanting to escape long cold winters.

The Pacific Ocean has a powerful impact on BC's climate, resource base, and transportation system. Unlike in Atlantic Canada, the continental shelf in BC extends only a short distance from the coast. Within this narrow zone are many

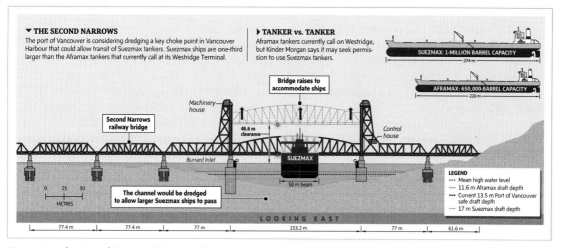

Figure 7.3 **The Second Narrows Bridge: Getting supertankers to port**
Source: "Sizing Up BC's Pipelines," *Globe and Mail*, 4 Aug. 2012, at: <www.theglobeandmail.com/report-on-business/industry-news/energy-and-resources/sizing-up-bcs-pipelines/article4461986/>. Illustration by John Sopinski and Michael Bird. *The Globe and Mail;* Sources: Port Metro Vancouver; Kinder Morgan; Enbridge; Google Maps; ESRI.

islands, the largest being Vancouver Island followed by Haida Gwaii (formerly known as the Queen Charlotte Islands).[1] The riches of the sea include salmon, which return to the rivers, such as the Fraser and the Skeena, to complete their life cycle. Most of BC's natural wealth, however, is not in the sea but in the province's diversified physical geography, which provides valuable resources, particularly forests, minerals, and rivers.

BC's Physiographic Regions

Most of British Columbia lies in the physiographic region known as the Cordillera—a combination of mountains, plateaus, and valleys. British Columbia's narrow and sometimes deep continental shelf is, in fact, a submerged mountain range. Vancouver Island and Haida Gwaii form the Pacific edge of the Cordillera and the boundary between the Pacific and North American tectonic plates extends underwater along the west coast of these islands. Not surprisingly, then, Canada's largest earthquake (magnitude 8.1) occurred in 1949 off the coast of the Haida Gwaii archipelago, and, in August 2012 another major quake (magnitude 7.7) struck the same area. A small portion of northeastern British Columbia that includes the Peace River country is part of the Interior Plains (Figure 2.1). Here,

THINK ABOUT IT
From studying Figure 7.4, would you conclude that the mountain chain found on Vancouver Island and on Haida Gwaii is geologically related and that a submerged mountain chain exists under the waters separating these two islands?

VIGNETTE 7.1

BC's Precipitation: Too Much or Too Little?

British Columbia receives the greatest amount of precipitation along its Pacific coast. Inland, the annual precipitation decreases sharply. In simple terms, there are two precipitation areas in British Columbia—one in the Pacific climatic zone, which receives from 800 to 2,000 millimetres of precipitation per year, the other in the Cordillera climatic zone, where less than 800 millimetres fall each year. These figures are average amounts of rain and snow. Fluctuations do occur. In August 2006, for instance, the west coast received very little precipitation. The small fishing/tourist town of Tofino, which is situated on the windward outer coast of Vancouver Island and is noted for its heavy rainfall, ran out of drinking water for the first time in its history. Two months later, in November, warm, moist subtropical air from Hawaii brought high temperatures and record amounts of precipitation to the BC coast. Known as the Pineapple Express, this subtropical air mass dumped heavy rains that flooded the low-lying Chilliwack area in the Fraser Valley, forcing an evacuation order for the residents of 200 houses.

the geological structure is part of the petroleum-rich Western Sedimentary Basin. Significantly, a massive natural gas deposit was discovered near the border with the Northwest Territories. If developed, this gas could reach Kitimat by pipeline and then would be shipped to Asian destinations by LNG tankers.

The Cordillera is a complex physiographic region. Extending from southern British Columbia to Yukon, this region encompasses over 16 per cent of Canada's territory. As explained in Chapter 2, the Cordillera was formed by severe folding and faulting of sedimentary rocks. This coastal zone is subject to earthquakes because of tectonic movement. The Cordillera has at least 10 mountain ranges, the most prominent of which are the Coast Mountains, which extend northward from Vancouver to the Alaskan panhandle, and the Rocky Mountains, which stretch from the US–Mexico border almost to Yukon (Figure 7.4). Other north–south mountain ranges include the Insular Mountain Range, which rises above the sea to form Haida Gwaii and Vancouver Island. The zone between the Insular Range and

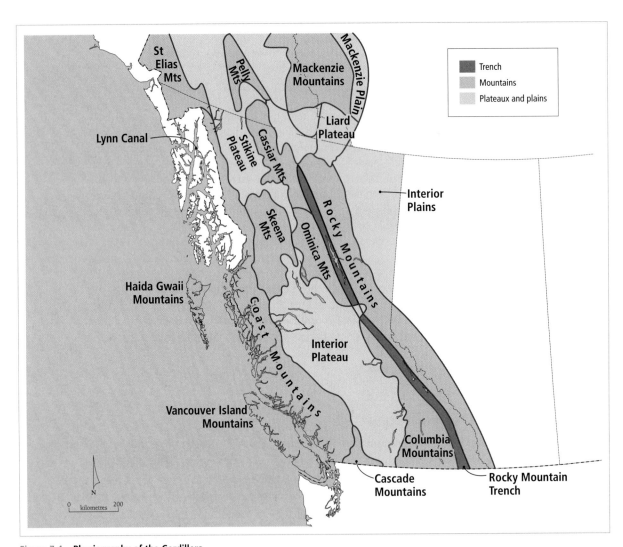

Figure 7.4 **Physiography of the Cordillera**
British Columbia's complex physiography is evident in the physiographic sub-regions of the Cordillera. The difficulty of constructing east–west transportation routes can be appreciated if one considers the number of north–south mountain ranges that must be traversed. For that reason, the importance of the Pacific Ocean for transportation prior to the completion of the CPR in 1885 becomes clear. As well, the grain of the land makes north–south land transportation construction into the Pacific Northwest of the United States relatively simple. Herein rest the natural factors behind the political concept of Cascadia and the feeling that BC is California North.

Photo 7.2
At Hell's Gate, located near Boston Bar, the canyon walls of the Coast Mountains rise 1,000 metres above the rapids. The Fraser River found a way through the Coast Mountains to the Pacific Ocean at Hell's Gate. Here, an Airtram carries tourists down to the visitors' centre by the river, and fishways enable salmon to head upstream to their spawning grounds. In 1914, the blasting of a route for the Canadian Pacific Railway caused a rockslide that blocked the salmon migration. The rail line occupies the lower reaches of the canyon while the Trans-Canada Highway is located in the upper reaches.

the Coast Mountains forms the Inland Passage. Sheltered by Vancouver Island and Haida Gwaii, the Inland Passage is the only body of water along the shores of British Columbia, Washington, Oregon, and California protected from the direct impact of the Pacific Ocean. Because of this protection, cruise ships, ferries, and private vessels travel along the Inland Passage in the waters of Georgia and Hecate straits (see Figure 7.1 for the location of these two straits, which combine to form the Inland Passage).

Two more mountain ranges, which lie in the Southern Interior of BC, are the Cascade Mountains, in south-central BC along the US border, and the Columbia Mountains. The Columbia Mountains consist of three parallel, north–south mountain ranges—the Purcell, Selkirk, and Monashee—and a fourth range, the Cariboo Mountains, forms the Columbia's northern extension. Further north are four mountain ranges—the Hazelton, Skeena, Ominica, and Cassiar mountains.

The Interior Plateau separates the Coast Mountains from the mountains of the Interior (Figure 7.4). North of the Interior Plateau is the Stikine Plateau. The topography of these plateaus consists of gently undulating land with occasional deeply trenched river valleys. The Fraser Canyon, one of BC's best-known landforms, drops about 300 to 600 m below the Interior Plateau and extends from just south of Quesnel to Hope. Just south of Lytton, the canyon walls rise about 1,000 m above the river. The Fraser Canyon is a major fault zone that separates the Cascades from the Coast Mountains. The Fraser River breached the Coast Mountains at the gorge known as Hell's Gate (photo 7.2).

Because of the rugged nature of the Cordillera, there is little arable land. Only about 2 per cent of the province's land is classified as arable. British Columbia's largest area of cropland lies outside of the Cordillera physiographic region in the Peace River country. Within the Cordillera, most arable land is in the Fraser Valley, while a smaller amount can be found in the Interior, especially the Okanagan and Thompson valleys. This shortage of arable land poses a serious problem for British Columbia. With urban developments spreading onto agricultural land, British Columbia lost some of its most productive farmland. From the end of World War II to the 1970s nearly 6,000 hectares of prime agricultural land were lost each year to urban and other uses. With only 5 per cent of BC's land mass classified as cropland, the provincial government responded to the serious erosion of its agricultural land base by introducing BC's Land Commission Act in 1973. This Act formed the **Provincial Agricultural Land Commission**, which is charged with preserving agricultural land from urban encroachment and encouraging farm businesses (Provincial Agricultural Land Commission, 2009). With continuing pressure from urban land developers, this legislation remains under fire from market economy-oriented groups such as the Vancouver-based Fraser Institute, the central argument being that more "value" can be derived from non-agricultural use of the land (Katz, 2010).

Climatic Zones

British Columbia has two climatic zones, the Pacific and the Cordillera. Because of the extremely high elevations in the Coast Mountains, few moist Pacific air masses reach the Interior Plateau. The spatial variation in precipitation is remarkable. Heavy orographic precipitation occurs along the western slopes of the Insular and Coast mountains where 2,000 mm of precipitation fall annually (Vignette 7.1). In sharp contrast, the Interior Plateau receives less than 400 mm per year. In the Thompson Valley and Okanagan Valley of the Interior Plateau, hot, dry conditions result in an arid climate with sagebrush in the valleys and ponderosa pines on the valley slopes. Most rain falls in the winter. For those along the Pacific coast, the so-called **Pineapple Express**, which originates over the warm waters around Hawaii, brings torrential rains but also relatively warm weather in the winter months, while those inland receive heavy snowfalls.

THINK ABOUT IT
Should the BC government allow development of agricultural land for greater individual profit, or does preservation of scarce farmland achieve more important social goals?

The three types of precipitation are described in Vignette 2.9, page 52.

The Pacific coast of British Columbia has the most temperate climate in Canada, dominated by the constant flow of moist Pacific air masses. The result is mild and often wet, cloudy weather. Because eastward-moving Pacific air masses must rise above the Insular Mountains and the Coast Mountains, orographic precipitation frequently occurs. This, combined with frontal precipitation, gives the west coast a high level of annual precipitation. Known locally as "liquid sunshine," the annual precipitation along the west coast of Vancouver Island can exceed 3,000 mm in some locations. Further east, the annual precipitation declines. At Vancouver, located near the Coast Mountains, the annual precipitation declines to about 1,000 mm. Victoria, however, lies in the partial rain shadow of the Insular Mountains of Vancouver Island and thus avoids the heavy orographic rainfall that affects Vancouver. Consequently, the capital city receives nearly 40 per cent less rainfall than Vancouver.

The Cordillera climatic zone consists of a number of microclimates. Changes in elevation (height above sea level), latitude (from 49° N to 60° N), variation in topography from mountain ranges to plateaus, and distance from the Pacific Ocean to the Rocky Mountains (600 km) control each microclimate, and in turn affect vegetation and soil conditions within these microclimates. In general, these areas become drier as distance from the Pacific Ocean increases and cooler as either elevation or latitude increases. The range of natural vegetation and soil types is enormous. In the Thompson Valley near Kamloops, an arid microclimate results in desert-like grassland vegetation with chernozemic soils. In sharp contrast, huge trees in the coastal rain forest grow in podzolic soils. Different still are the higher elevations of the Rocky Mountains extending beyond the treeline, where grasses, mosses, and lichens grow on cryosolic soils. For much of northern British Columbia, the northern coniferous forest (evergreen, needle-leaf trees) is the most common natural vegetation.

Because the Cordillera stretches from southern British Columbia to Yukon, the climate turns into a Subarctic mountain climate north of 58° N. Higher elevations in these latitudes have Arctic climatic conditions. South of 58° N, however, summers are longer and warmer, partly because the eastward-moving Pacific air masses are more dominant and partly because of greater annual solar energy.

Natural Vegetation Zones

The natural vegetation of British Columbia is far from homogeneous (Figure 2.8). Along the west coast, the mild, wet climate encourages a rain forest. The Insular Mountains of Vancouver Island and the Coast Mountains along the mainland both rise abruptly from the Pacific Ocean and therefore receive much rainfall throughout the year. The land responds with lush evergreen and deciduous trees. Characteristic tree species are western hemlock, Douglas fir, western red cedar, and Sitka spruce. Under this vegetation cover, podzolic soils are common. These soils are strongly acidic and low in plant nutrients, making it necessary for farmers to use fertilizers. This coastal region contrasts with the sagebrush and yellow grasses in the lower elevations of the southern Interior Plateau, where higher temperatures occur. Ponderosa pines are the predominant species in the southern Interior Plateau; at higher elevations and in more northerly areas, lodgepole pine replaces ponderosa pine along with other commercially valuable trees, including white and Engelmann spruce, Douglas fir, western hemlock, western red cedar, and two true firs—amabilis fir and subalpine fir. The series of forested mountain ranges, aligned north to south, include the Ominica Mountains located north of Prince George and the Columbia Mountains in the southern part of the province, centred on Revelstoke. Both mountain ranges are situated west of the Rocky Mountain Trench while the Rocky Mountains lie to its east (see Figure 7.4).

Environmental Challenges

British Columbia faces several environmental challenges, ranging from clear-cut logging and forest fires to the loss of agricultural land to urban and industrial sprawl. These challenges can be divided into those caused by human activities and those caused by nature. Human activities have subjected the seemingly limitless natural riches to mismanagement and wasteful practices that have led to resource loss, environmental degradation, and land-use conflicts. Without **sustainable resource use**, the future of BC's natural resources is not bright.

From the perspective of anthropogenic interference with nature, the greatest potential threat to the BC environment is the plan to ship bitumen by pipeline to the coast and then by supertankers to global markets. The danger is oil spills, especially in the coastal waters. The two proposed bitumen pipelines are the Northern Gateway Project and the second Trans Mountain pipeline (see Figures 8.12 and 7.5). The former terminates in Kitimat and the latter in Greater Vancouver. To reach Kitimat, ships must pass through Hecate Strait, the body of water between Haida Gwaii and the mainland of British Columbia, and then pass through Caamano Sound to Douglas Inlet. Weather can present problems for ships sailing into these waters. In winter, gale-force winds with gusts reaching 100 km/h or more generate huge waves that often exceed 10 metres high, but in extreme wind conditions waves over 20 metres have been recorded. In addition, ships at times must contend with thick fog conditions that make navigation more complicated. On the plus side, ships' pilots in British Columbia have gained many years of experience with ocean-going ships bringing bauxite to Kitimat and then leaving with smelted aluminum, and pilots in BC, it is said, must pass one of the hardest exams in the world (VanderKlippe, 2012b). Further south, winter weather is less of an issue. While the waters of the Strait of Georgia and Burrard Inlet are much more placid than those of Hecate Strait, the harbour is already a busy place and the addition of a super oil tanker running the length of the inlet each day does present potential traffic problems. The plan of Kinder Morgan, an American-based pipeline transport and energy storage company, is to twin its existing line from Edmonton to Vancouver, increase its storage capacity at its terminal in Burnaby, and bring in larger tankers to transport the added crude oil capacity that will come via pipeline from Alberta (see Figures 7.3 and 7.5 and photo 7.3).

Topping the list of natural challenges to the environment is the warming climate and the resulting increase in forest fires and the greater damage caused to forests by the pine beetle.

THINK ABOUT IT
Why do residents of White Rock near the US border receive much less annual precipitation than do residents of North Vancouver? Annual average precipitation for White Rock is 1,064 mm and that for North Vancouver is 1,770 mm. The locations of these communities are found in Figure 7.15.

THINK ABOUT IT
Does this sound like a good idea? At the Joint Review Panel for the Northern Gateway Pipeline project, Robin Allan, the former CEO of Insurance Corp. of BC, pointed out that Enbridge has limited its exposure to a catastrophic oil spill because the company formed a limited partnership for this project. Allan recommended that Enbridge purchase an insurance policy that would cover $1 billion in claims (O'Neil, 2012: C5).

Photo 7.3
The *Aqualegend*, a crude oil tanker, is escorted by tugboats as it arrives at Kinder Morgan's Westridge marine terminal in Burnaby, BC, in July 2012. While opposed by Vancouver and Burnaby city councils, Kinder Morgan plans to twin its Trans Mountain pipeline to transport Alberta crude oil and then load it on supertankers for foreign markets in Asia.

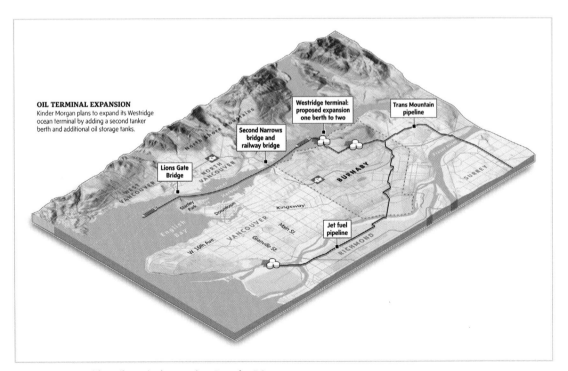

Figure 7.5 Westridge oil terminal expansion, Burnaby, BC
Kinder Morgan plans to expand its Westridge terminal by adding a second tanker berth and additional oil storage tanks. This expansion will accommodate the company's plan to twin its oil pipeline from Edmonton to Vancouver. *Source:* "Sizing Up BC's Pipelines," *Globe and Mail*, 4 Aug. 2012, at: <www.theglobeandmail.com/report-on-business/industry-news/energy-and-resources/sizing-up-bcs-pipelines/article4461986/>. Illustration by John Sopinski and Michael Bird/*The Globe and Mail*; Sources: Port Metro Vancouver; Kinder Morgan; Enbridge; Google Maps; ESRI.

CONTESTED TERRAIN 7.1

Cartographic simplification or purposely misleading?

Enbridge, the company wanting to build the Northern Gateway Pipeline, has prepared a video focusing on the shipping route and the measures to deal with marine challenges. An environmental group reacted to the Enbridge video as "cartographic lying." You can view both sites and then make your judgement—is it cartographic simplification or purposely misleading? The Enbridge video and the Sum of Us reaction are found at: <www.northerngateway.ca/economic-opportunity/northern-gateway-tanker-safety-video/> and <sumofus.org/campaigns/enbridge/?sub=fb>.

After viewing both video and the photograph below, do you feel that the Enbridge video is "balanced," that is, did the video show both the advantages and disadvantages of the natural conditions of the shipping route? In particular, did the company discuss the winter weather conditions described in the previous paragraph of your text? Also, why did the video refer to shipping "oil" rather than bitumen that would be converted into synthetic oil in Asian refineries?

Photo 7.4
One of the proposed LNG sites at Kitimat is taking shape on the shore of Douglas Channel. The Kitimat LNG project is financed by two US firms, Apache Corp and EOG Resources Inc., and a Canadian-based natural gas company, Encana Corp. Exports could begin as early as 2015—but maybe not. In 2011, the consortium received approval from the National Energy Board. The delay is securing contracts with foreign buyers, who are insisting on lower prices than those offered by the developers. For that reason, the pipeline construction is held up and the other LNG proposals remain in the planning phase (VanderKlippe, 2012b).

John Lehmann/The Globe and Mail

Exceptionally dry summers over the last decade, for example, have resulted in vast forest fires. Forests in the Interior become tinder dry, and the Okanagan Valley has three characteristics that can lead to fires spreading rapidly—strong winds, steep slopes, and a combination of dry underbrush, grass, and lodgepole pine trees. On 23 August 2003, such a fire reached Kelowna. As a forest fire raged out of control and swept into its suburbs, many thousands of residents had to

Photo 7.5
A raging wildfire forced 1,500 people from their homes in the community of Peachland. Across Skaha Lake, people in Kelowna watch smoke from the Peachland wildfire.

abandon their homes and more than 250 houses were destroyed (Boei and O'Brian, 2003). Six years later, in July 2009, a forest fire forced the evacuation of more than 10,000 Kelowna residents and destroyed a number of homes (CTV News, 2009). While the worst fire season took place in 2003, the summer of 2012 had several wildfires break out towards the end of a dry summer, including the Peachland wildfire. In September 2012, the Peachland fire broke out suddenly, and, encouraged by high winds, quickly destroyed buildings, a winery, and houses. More than 1,500 people were forced to flee Peachland, a community in the Okanagan.

Another natural impact on the environment has been the destruction caused by the pine beetle on the lodgepole and ponderosa pine trees in the Interior Forest and, more recently, in the boreal forest of Alberta and Saskatchewan. Vast areas have been affected and there is no way to stop the spread of the pine beetle. The lifespan of an individual mountain pine beetle is about one year. Pine beetle larvae spend the winter under bark, feeding on the tree, which is often a mature lodgepole pine. The adult pine beetle emerges from an infested tree and seeks another host. Logging activities are accelerating to salvage these trees because the beetles do not affect the wood's strength, its gluing characteristics, or its ability to be finished. While the accelerated logging means more timber for the sawmills in the short run, the loss of such a huge section of the Interior Forest means that mature timber will be in short supply in the future. For instance, a lodgepole pine matures in approximately 80 years. Why is the pine beetle so active? In the past, cold winters kept this pest under control. In recent years, however, pine beetles have spread and multiplied as a result of milder winters. Temperatures of at least $-38°C$ for four days or longer are required to kill off pine beetle infestation. If cold winters return, then the pine beetles will be controlled. On the other hand, if milder winters are a feature of climate change, then there will be no stopping the pine beetle from spreading across Canada's boreal forest.

British Columbia's Historical Geography

Indians lived along the Pacific coast of British Columbia for over 10,000 years before European explorers reached the northern Pacific coast in the mid-eighteenth century. The Spanish had already sailed northward from Mexico to California, but the Russians were the first to reach Alaska and establish fur-trading posts along its coast. In 1778, Captain James Cook established Britain's interest in this region by sailing into Vancouver Island's Nootka Sound, where he and his sailors found the Nootka village of Yuquot. The Nootka, now known as Nuu-chah-nulth, fished for salmon and hunted the sea otter. Upon landing, Cook engaged in trade for sea otter pelts, which opened up a profitable trade with China, although Cook, among the greatest of nautical explorers, did not live to see this trade flourish—he was killed in a skirmish with natives in the Hawaiian Islands on the return voyage. After the Royal Navy published Cook's record of his voyage, British and American traders came to the Pacific Northwest to seek the highly valued sea otters. Russian fur traders, based in Alaska, also harvested sea otters. Spain, which considered the lands Spanish territory, was disturbed by these interlopers and sent a fleet northward from Mexico in 1789. At Nootka Sound on the west coast of Vancouver Island, the Spanish seized several ships and built a fort to defend their claim. In 1792, Captain George Vancouver of Britain's Royal Navy sailed around Vancouver Island. In the following year, Alexander Mackenzie of the North West Company travelled overland from Fort Chipewyan to just south of Prince George and then to the Pacific coast near Bella Coola, which is just over 400 km north of Vancouver. Under the Nootka Convention (1794), the Spanish surrendered their claim to the Pacific coast north of 42° N, leaving the British and Russians in control.

In the early nineteenth century, the North West Company established a series of fur-trading posts along the Columbia River. From 1805 to 1808, Simon Fraser, a fur trader and explorer, explored the Interior of British Columbia on behalf of the North West Company. He travelled by canoe from the Peace River to the mouth of the Fraser River. As elsewhere, the strategy of the North West Company was to develop a working relationship with local Indian tribes based on bartering manufactured goods for furs. After 1821, when the North West Company merged with its rival, the Hudson's Bay Company, the HBC took charge of the Oregon Territory, which extended from the mouth of the Columbia River to Russia's Alaska. In the late nineteenth century, a bitter dispute arose over the boundary of the Alaska Panhandle and access to Dawson City and the Klondike gold rush. The boundary dispute was settled in 1903 by a six-man tribunal,

THINK ABOUT IT
If Canada had obtained the power to make international treaties in 1867, do you think its goal of access to Yukon from the Pacific Ocean along Lynn Canal (Figure 7.4) would have been more likely?

Photo 7.6
Captain James Cook's ships moored in Nootka Sound in 1778, as depicted in a watercolour by M.B. Messer. Four years earlier, the Spanish explorer Juan Hernandez sailed along the BC coast. However, there is the possibility that Francis Drake, while on a secret mission to find a western entrance to the Northwest Passage, reached these waters in 1579 (Hume, 2000: B1).

composed of American, Canadian, and British representatives. The settlement totally favoured the United States, suggesting that Canada's interests may have been sacrificed by Great Britain, which sought better relations with the United States.

In 1843, American settlers began to arrive on the coast from the eastern part of the United States. In the same year, the HBC relocated its main trading post from Fort Vancouver at the mouth of the Columbia River to Fort Victoria at the southern tip of Vancouver Island. The increasing number of American settlers who came west along the Oregon Trail represented a challenge to the authority of the Hudson's Bay Company. A few years later, the United States claimed the Pacific coast northward to Alaska, where Russian fur-trading posts existed. In 1846, Britain and the United States agreed to place the boundary between the two nations at 49° N and then to follow the channel that separates Vancouver Island from the mainland of the United States. While the loss of the Oregon Territory in present-day Washington and Oregon was substantial, Britain was fortunate to hold onto the remaining lands administered by the Hudson's Bay Company. Britain recognized that its hold on these lands through the HBC was tenuous and could not withstand the political weight of the growing number of American settlers. Indeed, without the presence of the HBC and Britain's negotiating skills, Canada might well have lost its entire Pacific coastline.

The gold rush of 1858 brought about 25,000 prospectors from California to the Fraser River. Prospectors walked upstream along the sandbanks and sandbars of the Fraser and its tributaries, panning for placer gold (small particles of gold in sand and gravel deposits). The major finds were made in the BC Interior, where the town of Barkerville was built near the town of Quesnel. By 1863, Barkerville had a population of about 10,000, making it the largest town in British Columbia. To ensure British sovereignty

VIGNETTE 7.2

Who Owns BC?

How much of BC do First Nations peoples own? The question still is only partially resolved, but it has terrific implications for BC and its 232,290 Aboriginal people, who form 198 First Nations. BC does not know who has rights to what lands—an uncertainty that stifles investment, cripples community development, and plagues the province with lawsuits. Other than the 14 pre-Confederation Douglas treaties signed between 1850 and 1854 with the Coast Salish on Vancouver Island and Treaty No. 8 of 1899, which extended into northeast British Columbia, the BC government for many years denied Aboriginal land claims.

In 1993, the provincial government accepted the concept of Aboriginal title and the process of negotiating land claim agreements under the BC Treaty Commission began. For BC, the challenge was to whittle down the total size of the land claims, which exceeded the size of BC, to around 10 per cent of the claimed "traditional" territory. In exchange for land surrender, cash and other benefits and rights are acquired by the First Nations. The stakes are very high, and the complex negotiations mean that progress is slow. By 2013, more than half of the 198 First Nations, comprising two-thirds of Aboriginal people in British Columbia, were participating in one of the 47 negotiations that ultimately could result in a final agreement, i.e., a treaty. Progress is slow and by 2013 only four final agreements had been reached under the BC Treaty Process. However, other agreements have been concluded, such as cut-off claims, land and resources agreements, and reconciliation agreements (see the website of the BC Ministry of Aboriginal Relations and Reconciliation at <www.gov.bc.ca/arr/treaty/default.html> for a full account of these other agreements). In the highly urban area of the **Lower Mainland**, the Tsawwassen First Nation signed the first Final Agreement under the BC Treaty Process in 2007. By 2013, five Nuu-chah-nulth First Nations on the west coast of Vancouver Island signed the Maa-nulth Treaty, followed by the Yale First Nation (located near the town of Yale in the Fraser Valley) and the Tla'amin First Nation (located just north of Powell River on the west coast. The Final Agreement with the Nisga'a signed in 2000 took place outside of the BC Treaty Process (Vignette 7.3).

over territory north of 49° N, the British government established the mainland colony of British Columbia in 1858 under the authority of Sir James Douglas, who was also governor of Vancouver Island. In 1866, the two colonies were united. During Douglas's time as governor, land claims were settled on Vancouver Island but not on the mainland (Vignette 7.2).

For further discussion of Aboriginal rights and the modern treaty process, see Chapter 3, "Modern Treaties," page 97, and Vignette 3.9, "From a Colonial Straitjacket to Aboriginal Power," page 102.

Confederation

By the 1860s, the British government was actively encouraging its colonies in North America to unite into one country. Once the first four colonies were united in 1867, Ottawa adopted the British strategy to create a transcontinental nation. An important part of that strategy was to lure British Columbia into the "national fabric." The Canadian Pacific Railway was the first expression of this national policy.[2] Ottawa promised to build a railway to the Pacific Ocean within 10 years after British Columbia joined Confederation. In Fort Victoria, however, some wanted to join the United States. By the middle of the nineteenth century, British Columbia had developed significant commercial ties with Americans along the Pacific coast. San Francisco was the closest metropolis and, with its railway to New York, offered the simplest and quickest route to London. In 1859, Oregon became a state and Washington was soon to follow. Commercial links with the United States were growing stronger. But the majority of people in Fort Victoria wanted to remain British. In 1871, British Columbia chose to become a province of Canada (Figure 3.5). British Columbians, however, had to wait 14 years for the Canadian Pacific

VIGNETTE 7.3

Facts about the Nisga'a Agreement

Significance:

The political significance of the Nisga'a Agreement goes back to 1973 with the Supreme Court of Canada ruling in *Calder*. Their split decision opened the door for land claims. According to Frank Cassidy (1992: 11) the Supreme Court decision represents "a centrepiece in the historical development of the province of British Columbia."

Location:

The Nass Valley extends from the Pacific Ocean across and beyond the Coast Mountains approximately 150 km inland. This valley is the homeland of the Nisga'a. Prince Rupert is the closest large city and it lies some 100 km southwest of the headwaters of the Nass Valley.

Population:

In 2000, the total population of the Nisga'a was almost 10,000, with approximately 6,000 living in British Columbia and the rest in other parts of Canada and the United States. Some 2,400 Nisga'a resided in the Nass Valley in the villages of New Alyansh (Gitlakdamiks), Canyon City (Gitwinksihikw), and Kincolith (Gingolx).

Agreement:

The Nisga'a and the federal and BC governments approved and signed the agreement in 1999. In exchange for surrendering their traditional lands, the Nisga'a received title to almost 2,000 km² in the Nass Valley of British Columbia; access to forest and fishery resources; self-government powers, including a Native judicial system and policing; and nearly $500 million in cash, grants, and program funds from Victoria and Ottawa. British Columbia supplies the land and 30 per cent of the cash settlement, while the federal government pays 70 per cent of the cash settlement. Under this agreement, the Nisga'a are no longer under the Indian Act and therefore cease to receive benefits as status Indians.

Railway to reach the Pacific coast at Port Moody on Burrard Inlet, near Vancouver. Later, the railway was extended 20 km westward to the small sawmill town of Vancouver, where there was a better harbour and terminal site for the railway (Figure 7.6).

When British Columbia joined Confederation in 1871, its official population was 36,247 (McVey and Kalbach, 1995: 35). This figure likely underestimated the number of Indians and prospectors who were living in the more remote areas of the province.[3] At the time of the first comprehensive census of British Columbia in 1881, there were approximately 4,200 Chinese, 19,000 white settlers (American, Canadian, and British), and about 30,000 Indians. Most British settlers lived around Fort Victoria. Beyond Fort Victoria, the vast majority of the inhabitants were Aboriginal people. The BC government denied that First Nations had a claim to land. In the previous decade, about 25,000 prospectors, mainly Americans, had been scattered in small camps along the Fraser River or at Barkerville. Most Americans had left at the end of the gold rush.

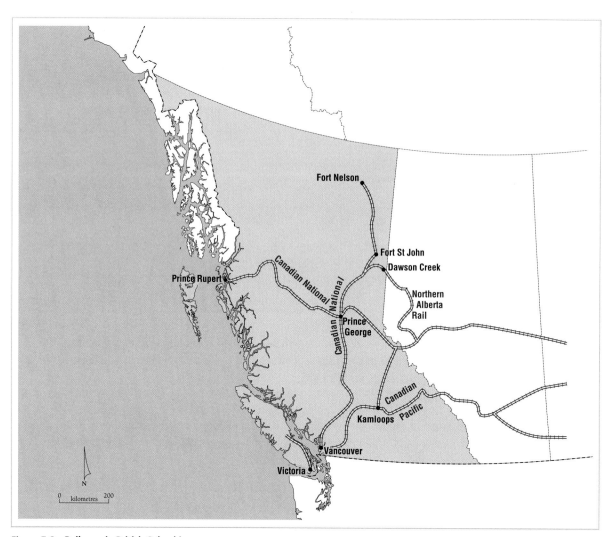

Figure 7.6 **Railways in British Columbia**
Railways have played a key role in BC's economic development. The first railway to cross the mountains of the Cordillera was the Canadian Pacific Railway in 1885. To open the Northern Interior of the province, Victoria built the Pacific Great Eastern Railway (later known as British Columbia Rail). After several extensions, BC Rail joined North Vancouver to Fort Nelson in 1971. An eastern link joins Dawson Creek to the Northern Alberta Railway. BC Rail was owned and operated by the province from 1918 to 2004, when it was sold to Canadian National. By 2008, Prince Rupert, the western terminus of the CN line, had the capacity to handle container traffic, thus relieving the congestion at the Port of Vancouver and providing another transportation route into the interior of North America.

Post-Confederation Growth

At first, Confederation had little effect on British Columbia. The province was isolated from the rest of Canada, Canada's fledgling factories, and even the halls of power in Ottawa. Goods still had to come by ship from San Francisco or London. When the Canadian Pacific Railway was completed in 1885, British Columbia truly became part of the Dominion, and BC's role as a gateway to the world began.

The main line of the Canadian Pacific Railway and its many branch lines were responsible for the formation of many of the province's towns and cities and for providing access to its forest and mineral wealth. The Esquimalt and Nanaimo Railway, the Canadian National Railways, and British Columbia Rail (BC Rail) added to the rail network in British Columbia. In 1886, the Esquimalt and Nanaimo Railway was built, connecting the coalfields of Nanaimo with the capital city of Victoria, and stimulated logging and sawmilling along its route. With the closure of the coal mines in the early 1950s, the railway lost its main function and now plays a minor role in the transportation system of Vancouver Island. By 1914, the Grand Trunk Pacific Railway (later the Canadian National Railways; today, CN) provided a trans-Canada rail route to Prince Rupert and an alternative rail service to Vancouver. The Pacific Great Eastern Railway (later BC Rail) was incorporated in 1912 but laid few rails until the early 1950s, when the provincial government made a commitment to complete the railway in order to facilitate resource development in the Interior of British Columbia. By 1956, this rail line extended from North Vancouver to Prince George. Two years later, BC Rail extended its rail system to Dawson Creek and Fort St John in the Peace River country, and by 1971 it reached Fort Nelson in BC's forested northeast. CN purchased BC Rail in 2004 and proceeded to expand its facilities at Prince Rupert (Figure 7.6).

After the completion in 1885 of the CPR line, Vancouver grew quickly and soon became the major centre on the west coast. By 1901, Vancouver had a population of 27,000 compared to Victoria's 24,000. As the terminus of the transcontinental railway, Vancouver became the transshipment point for goods produced in the Interior of BC and Western Canada. As coal, lumber, and grain were transported by rail from the Interior and the Canadian Prairies, the Port of Vancouver spearheaded economic growth in the southwest part of the province. It was then possible to tap the vast natural resources of BC and ship them to world markets. By the twentieth century, Vancouver had become one of Canada's major ports. Located on Burrard Inlet, Vancouver has an excellent harbour. Unlike Montréal, it is an ice-free harbour, thanks to the warm Pacific Ocean. As Canada's major Pacific port, Vancouver became the natural transportation link to Pacific nations. With the opening of the Panama Canal in 1914, British Columbia's resources were more accessible to the markets of the United Kingdom and Western Europe.

Between 1885 and the end of World War I in 1918, British Columbia underwent a demographic explosion. By 1921, BC had over half a million inhabitants. The province had been gradually transformed from a fragile political entity in 1871 into a self-confident political and social region within Canada. During this time, the populations of both European and Asian origin increased about tenfold. The European population had reached nearly half a million, while the Asian population was about 40,000. At the same time, a combination of disease and social dislocation led to a sharp decline in the number of Aboriginal people, perhaps from 40,000 to 20,000. This magnitude of demographic decline was matched in other regions of Canada. While a number of factors were at work, smallpox, tuberculosis, and other communicable diseases were responsible for the greatest losses.

While BC's economy and population continued to grow in the 1920s, the Great Depression of the 1930s caused the province's economy to stall and unemployment to rise sharply. Economic disaster struck British Columbia in 1929. Exports of Canadian products from BC, so necessary for its economic well-being, slowed and prices dropped. In the Prairies, the collapse in agricultural prices was accompanied by prolonged drought, turning the land into a dust bowl. Many prairie farmers abandoned their farms and fled to British Columbia, adding to the burden of unemployment in that province.

World War II and the Post-War Economic Boom

World War II called for full production in Canada, thereby pulling British Columbia's depressed

Photo 7.7
Vancouver's magnificent and busy harbour has facilitated trade with Pacific Rim countries.

resource economy out of the doldrums. Military production, including aircraft manufacturing, greatly expanded BC's industrial output. As well, resource industries based on forestry and mining (especially coal and copper) were producing at full capacity.

When the war ended in 1945, BC's resource boom continued. With world demand for forest and mineral products remaining high, the provincial government focused its efforts on developing the resources of its hinterland, the Central Interior of British Columbia. The first step was to create a transportation system from Vancouver to Prince George, the major city in the Central Interior. The highway system was improved and extended from Prince George to Dawson Creek in 1952. But the completion of the Pacific Great Eastern Railway to Prince George in 1956 and then to Dawson Creek in 1958 opened the country, allowed for exports to foreign markets, and triggered economic growth, especially in the forest industry. With rail access to Vancouver, forestry, as well as other resource industries, expanded rapidly, thereby leading to the integration of this hinterland into the BC and global industrial core.

Over the past two decades, BC's increasing economic strength, partly driven by the Asian economy, has outpaced that of all other regions in Canada. Trade is a dynamic force propelling British Columbia's economy, and as China and other Pacific Rim countries have led the world in economic growth, this growth has created new markets for Canadian products shipped through Vancouver. Trade opportunities are almost endless between British Columbia and the population of 2.5 billion people in the Pacific nations. Vancouver and, to a lesser degree, Prince Rupert and other Pacific ports serve as trade outlets for coal, lumber, potash, and grain from the BC Interior and the Prairie provinces to reach world markets. In BC's export-oriented economy, it is usually much less expensive to ship the raw material than the finished product. This has led to trade focused on resources. High labour costs, a relatively small local market, and distance from world markets have inhibited the development of manufacturing in BC.

British Columbia Today

British Columbia has evolved from a resource-dependent economy into a much more diversified one. The decline in exports to the United States has forced BC companies to turn to Asian markets. This trend of forced diversification seems unstoppable—at least until the US economy recovers and its demand for BC products, especially its forest products, returns to former levels. Even then, the rapidly growing Asian markets are bound to remain an important source of exports. One example lies in energy: Asia is hungry for energy imports while BC has an abundance of energy. In a few years, possibly by 2015, liquefied natural gas from the huge natural gas deposits in northeastern BC might be transported by super-size LNG tankers from Kitimat to Asian markets, thus cementing another trade link to Pacific Rim countries. The fly-in-the ointment is determining a long-term price that both the developers and the customers find acceptable. The developers hoped for a price similar to that in Japan ($15/million cubic feet) while the Asian customers are seeking a price closer to $5/million cubic feet (Angevine and Oviedo, 2012).

While the export of natural resources remains a key element in the BC economy, an emerging urban-based economy has taken hold where innovation, technology, and the export of non-resource products are the key elements. This economy is focused on global trade, tourism, high technology, innovative manufacturing, and even filmmaking. Shipbuilding has added another dimension to BC's economy. In 2011, Vancouver's Seaspan Shipyard was awarded a federal contract valued at $8 billion for building four Coast Guard science vessels, three joint support ships, and one polar-class icebreaker (Chase and Marotte, 2011: A1). The first seven ships will be built from 2013 to 2019, hopefully to be followed by up to two dozen more. The impact on Vancouver and the BC economy is huge. According to Jonathan Whitworth, CEO of Seaspan:

> This is going to be a boom. It's not like building a (liquefied natural gas) plant or a mine, where there are 1,500 jobs for about two years. This is work for decades. It's like winning the 2010 Olympics every two years. (Alldritt, 2012)

The sharp edge of the new economy lies in another arena and is based on innovative industries, such as fuel cell development by Ballard Power Systems; possible new technology applied to the shipbuilding contract, especially the polar class icebreaker; and Radarsat inventions by MacDonald, Dettwiler and Associates, Canada's principal space company. According to John MacDonald (2012), the co-founder of MacDonald, Dettwiler, "This cutting-edge project will create highly skilled jobs, and attract the world's best scientists, technicians and engineers to Canada's world-renowned space industry." The performing arts, filmmaking, and the rest of the entertainment industry centred in Vancouver add to BC magic, making Vancouver one of the three most "livable" cities in the world (Economist Intelligence Unit, 2012).

British Columbia's physical geography, especially the sea/mountain interface, provides a solid basis for a tourist industry and explains why BC is recognized as a world-class tourist destination. One tourist attraction involves the **Inland Passage**, a popular route for cruise ships heading for Alaska; internationally famous Whistler, the site of alpine events at the 2010 Vancouver Winter Olympics, draws skiers, snowboarders, and tourists from around the world. Filmmaking, too, takes advantage of BC's physical geography. Vancouver is known as "Hollywood North," and the Vancouver area has become a major film production centre. Filmmaking represents one aspect of the emerging highly diversified urban "knowledge economy" that has taken hold in Toronto and Montréal as well. The Vancouver film industry generates $1.2 billion annually and directly employs about 35,000 people (BC Facts, 2012).

BC's population continues to increase at a more rapid rate than most provinces. In turn, BC has gained more seats in the House of Commons. With more seats, the relationship between Ottawa and British Columbia is changing. In short, economic, demographic, and political power has shifted to the West. A test may arise if Ottawa approves of the two proposed oil pipelines—one to Kitimat and the other to Vancouver—while little support exists among BC voters. The consequences for the Conservative government may see a significant drop in its BC seats.

Economic Structure

Is the question, "Is British Columbia a core or a periphery?" relevant in a postmodern world? The

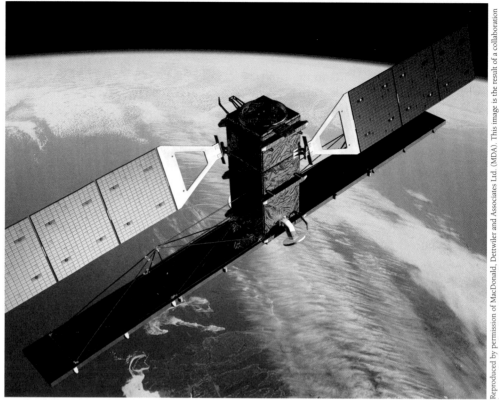

Photo 7.8
MacDonald, Dettwiler and Associates, a Vancouver firm, has developed and operates the Radarsat Constellation Mission, a set of three Earth observation satellites, which provide Canada with surveillance over the Arctic Ocean, including the Northwest Passage. See MacDonald (2012).

answer is no. In reality, developed nations are witnessing a shift from primary and secondary industries to tertiary ones, where most of the innovative activities take place, and this trend is also taking place within Canada's six regions, but especially BC (Tables 1.5 and 4.15). Economic advancement is now largely driven by knowledge-based activities that flourish mainly in the tertiary sector. In the post-industrial information society, these advances differ from past ones because they are more heavily based on scientific breakthrough and take the form of technical advances and digitized information. Some of BC's knowledge-based companies are local firms while others are associated with multinational corporations. Microsoft, for instance, with its home base in Seattle, opened its first Canadian Development Centre in Richmond, BC, employing 300 researchers from around the world. However, the knowledge-based economy is not limited to the computer world; it permeates all avenues of the economy. In the primary sector, for example, technical advances in oil and gas operations permit **horizontal drilling**, which, when combined with **hydraulic fracturing**, has unlocked natural gas deposits from the Horn River Basin in northeast BC. Plans for a pipeline and LNG terminal at Kitimat are well underway with most exports going to Asian countries.

For more on economic change in the twenty-first century, see Chapter 4, "Economic Structure," page 158.

An economic split exists within BC. The leading edge of economic change lies in the urban clusters where high-technology and tourism services are concentrated. Beyond those clusters, the resource economy remains strong. The trend, as observed in the economic structure, reveals a shift of workers from primary/secondary sectors to the tertiary sector (Table 7.1). This long-term trend of a growing tertiary sector is a mirror image of the new economy with the emphasis on technology and innovation, plus an expansion of service industries that support such an economy. Following Florida's concept of a creative class,

Table 7.1 **Employment by Economic Sector in British Columbia, 2005 and 2011**

Economic Sector	Workers, 2005 (000s)	Workers, 2005 (%)	Workers, 2011 (000s)	Workers, 2011 (%)	% Change
Primary	76.2	3.6	66.2	2.9	–0.7
Secondary	376.5	17.7	381.3	16.8	–0.9
Tertiary	1,677.8	78.7	1,827.2	80.3	1.6
Total	2,130.5	100.0	2,274.7	100.0	

Source: Statistics Canada (2006, 2012c).

these service industries include a vibrant entertainment sector coupled with a surprising growth in the performing arts. All in all, a west coast version of a laid-back, almost Bohemian lifestyle has taken root in these urban centres and in surrounding small towns like Tofino.

Manufacturing, a key element in the secondary sector, remains anchored on the processing of forest and other primary resources. Over the past five years, the value of manufacturing, the number of firms, and the number of workers have declined (Statistics Canada, 2012d: Table 301–0006). This decline is masked in Table 7.1 because another element in this economic sector, construction, has seen record levels of employment.

With such an abundance of natural resources, why does manufacturing play such a relatively weak role in the BC economy? One reason is that most resources are exported in a raw or semi-processed form. A second reason is the small size of BC's population, which translates into a small market. Exports to foreign markets are difficult because of the relatively high cost of manufactured products. High manufacturing costs are associated with two factors: (1) the absence of economies of scale, which would reduce per-unit costs of the manufactured product, and (2) the relatively high cost of labour. Technological breakthroughs, however, can create global demand for superior products. Such breakthroughs generally require long-term investment in research and development from the public sector.

In 2011, the tertiary sector accounted for just over 80 per cent of employment in BC (Table 7.1). Employment in transport and trade indicates the relative strength of BC's export sector. British Columbia's ever-expanding participation in world trade, a vibrant tourist industry, its growing domestic market, and its active **producer services** firms all add to its new economy.

The evidence is clear—British Columbia's economy is changing and diversifying, but growth in the manufacturing sector remains elusive. In the past two decades, however, BC has experienced growth in high technology. As the front edge of the structural shift in the manufacturing industry across Canada, high-tech firms are playing a greater role in BC's economy and are providing jobs for highly skilled workers. High technology involves cutting-edge research in the manufacture of electronics, telecommunications equipment, and pharmaceuticals.[4]

High-tech manufacturing employing highly skilled workers is the wave of the future for BC and Canada. As discussed in Chapters 5 and 6, Ontario and Quebec have lost many "low-end" manufacturing firms and jobs. The good news for BC, Ontario, and Québec is that most high-technology firms are located in these provinces; the bad news is the mismatch between the skill level of those losing their jobs in low-end manufacturing and the labour needs of high-tech firms.

British Columbia's Wealth

British Columbia's economic prosperity rests on three economic pillars: its geographic location as Canada's trade window to the Pacific; its natural resources; and its rapidly expanding high-technology sector. Exports are key to this prosperity. Diversification of foreign trade was a natural outcome of two events—the rising economic power of Asian countries led by China and the surprisingly sluggish US economy. Exports include much more than just BC products, as Western Canada's exports of coal, grain, potash, and petroleum move through BC and its ports. Imports consist of a variety of consumer goods, including automobiles and textiles, and they flow to all

THINK ABOUT IT
With large, high-quality timber reserves in BC, why is it so difficult for value-added forest manufacturing firms to increase their exports around the world? Should BC export logs to China rather than process them at home?

Table 7.2 **Exports through British Columbia, 2001, 2006, 2009, and 2011**

Country	2001 (%)	2006 (%)	2009 (%)	2011 (%)
US	69.8	61.3	51.3	42.8
Japan	12.8	14.1	13.7	14.2
China	2.3	4.4	10.2	14.8
South Korea	2.2	4.1	6.6	8.3
Other countries	12.9	16.1	18.2	19.9
Value in billions of dollars	31.7	33.5	25.2	32.7

Source: Adapted from BC Stats (2012a).

parts of Canada. In 2011, exports worth $33 billion and imports worth $39 billion passed through BC ports (Schrier, 2012a).

Four countries—the US, China, Japan, and South Korea—account for over 80 per cent of exports and imports moving through BC ports (Table 7.2). More recently, high technology has emerged as a serious component of BC's economy. Still, BC ranks a weak third behind Quebec and Ontario as these two provinces combine to account for 90 per cent of Canada's high-tech exports.

Pacific Rim Trade

Trade with Pacific Rim countries is growing rapidly. The prospect of huge amounts of natural gas and bitumen flowing to Asia marks a significant turning point in trade—with two provisos. First, the Asian customers must offer a fair price for BC natural gas in order for the companies to build the pipelines and LNG facilities. Second, the controversial Northern Gateway Pipeline must be approved and built along with marine facilities suitable for supertankers.

Much of BC's future economic well-being—as well as that of Canada—depends on access to these Asian markets, which provide two advantages. First, they represent new markets for BC and Western Canada commodities. Second, exports to Asia lessen dependency on the US market. While the US remains the most important trading partner, China, Japan, and South Korea occupy the next most significant positions. In terms of exports in 2001, the United States dominated trade by accounting for nearly 70 per cent of BC exports (Table 7.2). By 2011, the United States accounted for only 43 per cent, a drop of 27 percentage points. Over the same period, exports to China and South Korea jumped sharply while exports to Japan remained relatively stable. Another striking trend is the increase in exports to other countries, jumping from 13 per cent in 2001 to 20 per cent in 2011. Surprisingly, in comparing the value of these exports in 2001 and 2011, the figures changed little. On the other hand, the high point of $33.5 billion was reached in 2006 while the low point of $25.2 billion occurred following the global economic crisis of 2009.

Looking at Table 7.2, while trade with Japan is likely to remain relatively stable at around 14 per cent, trade with China has already jumped from 2.3 per cent in 2001 to nearly 15 per cent in 2011. Yet, the key to increased exports will depend on the recovery of the US economy, thus allowing US exports to regain lost ground (from a low of 43 per cent in 2011 back to the higher level of 70 per cent in 2001). To facilitate trade with Pacific Rim countries, a transportation corridor across Western Canada to Pacific ports is fundamental to Canada's and BC's future economic well-being. To take advantage of this potential expansion of trade, the BC and federal governments, in conjunction with governments in Western Canada and the two railway companies, CN and CP, are working together through the **Asia-Pacific Gateway and Corridor**: the highway and rail corridors from Western Canada to ports on BC's coast. One project that eliminated a critical bottleneck was the completion of the Pitt River Bridge in 2009 (photo 7.1). CN and CP are investing additional funds in rail line

THINK ABOUT IT
Does the federal commitment of $3 billion for the $6 billion Asia-Pacific Gateway and Corridor project counter the perception that Ottawa does little for BC?

Photo 7.9
CP's Port Metro in Vancouver, one of the busiest shipping hubs on the continent.

upgrading and port expansions (photo 7.9). Cost-sharing projects between the province and the federal government include expansion of container facilities at major ports, notably Prince Rupert, and highway improvements at critical bottlenecks, such as the Pitt River Bridge and South Fraser Perimeter Road. Vancouver alone benefited from these transportation developments with Port Metro Vancouver handling on average $75 billion worth of imports and exports a year (Port Metro Vancouver, 2012).

Container traffic between Asia and North America is growing rapidly. Over two decades ago, Peter Nemetz (1990) foresaw that BC and the rest of Canada could reap many economic benefits through trade with the growing industrial giants in Asia. Within the geopolitical world, BC has a decided advantage because its ports, especially Prince Rupert and Kitimat, provide the shortest North American sea links to Asia. However, since increasing amounts of this cargo come in the form of containers, port infrastructures to handle **container** shipments are critical. According to the BC government, container traffic to all west coast ports is forecast to rise a staggering 300 per cent by 2020 (BC Ministry of Transportation, 2007b). According to Port Metro Vancouver (2012), the Port of Vancouver handled 1.3 million TEU of container cargo (TEU refers to containers measuring 20 feet), an increase of 12 per cent over its 2009 figure.

The Asia-Pacific Gateway and Corridor strategy, a largely federal program, aims at expanding BC's transportation system, including its container port facilities. One goal of this strategy has been to increase container traffic imports from the Pacific Rim countries from 2 million TEU in 2005 to 9 million TEU in 2020 by expanding port facilities at Vancouver and by constructing a container port at Prince Rupert, which opened in 2007. This strategy is on track as increased volume of container traffic was recorded in 2011. For example, Prince Rupert in 2009 recorded its highest volume throughput since 1997, despite declining container traffic through other North American west coast ports (BC Ministry of Transportation, 2007b; Prince Rupert Port Authority, 2007, 2010). Figures for 2011 reveal this trend continued with an increase in container traffic of 12 per cent over its 2009 traffic (Port of Prince Rupert, 2012).

Natural Resources

Natural resources represent the basis of BC's wealth. Resource development, led by forest products and the export of forest and other resources, has provided the economic thrust for the province's growth. These resources include forests, fish, minerals, petroleum, water power, and physical beauty, and have driven growth in both the primary and secondary sectors. Moreover, a portion of the tertiary sector, the tourism industry, relies largely on the province's natural beauty and resources through ecotourism, hiking, skiing, sports fishing and hunting, and sailing, kayaking, and canoeing along BC's island-dotted coast of fjords and inlets. BC natural resources are found in its hinterland, far beyond BC's core. As in other hinterlands, the economic position of primary industries is declining and forms a small percentage of the total economy (Table 7.1).

Fishing Industry

British Columbia, like Atlantic Canada, has an important fishery. The BC fishery, as a renewable resource, ranks fourth in value of production among resource industries behind mining (including natural gas), forestry, and agriculture. Fish processing plants employ 25,000 full-time and part-time employees. More than 100 species of fish and marine animals are harvested from the Pacific Ocean, freshwater bodies, and aquaculture areas. Salmon is the most valuable species, followed by herring, shellfish, groundfish, and halibut. In 2010, the landed value from the sea reached $330 million (Figure 7.7). Most salmon today—ironically, Atlantic salmon—comes from fish farms. The province is the fourth largest producer of cultured salmon in the world, after Norway, Chile, and the United Kingdom. In 2010, the wholesale value of farmed salmon was $500 million, compared to $69 million for wild salmon (BC Ministry of Environment, Oceans and Marine Fisheries Branch, 2011). In recent years, excessive harvesting has reduced fish stocks and, in turn, the number of landings (fish caught). At the same time, production from fish farms has made fresh salmon available to the consumer throughout the year, thus popularizing the product. Environmentalists, however, have long expressed concern about fish farming—estuaries and inlets are subject to pollution from waste; fish meal used as feed in aquaculture results in the disruption of food chains and depletion of species in other parts of the globe; and predators break through the netting where farmed fish are kept, thus permitting their escape to the wild and the unknown effects of breeding with wild species of salmon. In May 2012, an outbreak of the highly infectious haematopoietic necrosis occurred at a fish farm in Dixon Bay near Tofino. All fish were destroyed and pens, cages, and equipment were cleaned and disinfected (CBC News, 2012).

Unfortunately, BC shares another similarity with Atlantic Canada—overexploitation of fish stocks, particularly the valuable salmon stocks. Pressure on

THINK ABOUT IT
As the CEO of one of the potash mines near Saskatoon, why would you select Prince Rupert over Vancouver to ship your produce to China?

VIGNETTE 7.4

Is the Manufacture of BC Natural Resources Important?

For BC, the key to manufacturing is related to making products from its natural resources. As David Emerson, the former federal Minister of Trade and Industry, observed, "There's almost unlimited scope to spin technology off our resource base" (Ebner, 2009: B2). Emerson suggests that BC should follow Finland's example with forestry, where massive public funding is directed to research and development within the Finnish Forest Research Institute of the Ministry of Agriculture and Forestry. To make such a commitment, the BC government has to recognize that a critical link exists between research and the region's most prominent industry. Put differently, does a long-term commitment of public funds for knowledge-based firms and universities to apply their research and development skills and techniques to the forest industry constitute a critical element in BC's future?

the fish stocks, especially salmon, comes from four sources—Canadian commercial fishers, American commercial fishers, the Aboriginal fishery, and the sports fishery. All want larger catches.

Management of the salmon fishery falls under the Pacific Salmon Treaty, which was signed by Canada and the United States in 1985 and sets long-term goals for the sustainability of this resource. The Pacific Salmon Commission, formed by the governments of Canada and the US to implement the Pacific Salmon Treaty, does not regulate the salmon fisheries but provides regulatory advice and recommendations to the two countries. In Canada, the Department of Fisheries and Oceans (DFO, now called Fisheries and Oceans Canada), in consultation with the Pacific Salmon Commission (www.psc.org/), has set annual quotas for salmon in Canadian waters.

Salmon are migratory fish, so regulating salmon fishing is particularly challenging. Like other fish, they are common property until caught. This principle is based on the "rule of capture." Fishers therefore try to maximize their share of a harvest so no one else will take "their" fish. The problem is complicated further because the Canadian government cannot regulate the "Canadian" salmon stocks, i.e., those that spawn in Canadian rivers, because they migrate to American waters, where the American fishing

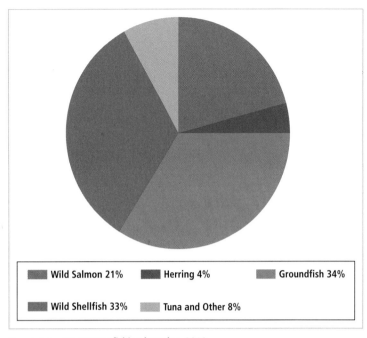

Figure 7.7 **BC capture fishing by value, 2010**
Source: BC Ministry of Environment, Oceans and Marine Fisheries Branch (2011).

fleet harvests them. The net result is that the salmon stocks are threatened. This problem is commonly referred to as the **tragedy of the commons** and is compounded by the economic cycle, whereby resource extraction accelerates

Photo 7.10
The Canadian Food Inspection Agency confirmed that a highly infectious and lethal virus infected a fish farm north of Tofino in May 2012.

during periods of high global demand for natural resources.[5]

Salmon catches by fishing fleets fluctuated, not because of conservation measures that restrict the allowable catch but simply because the number of salmon varied from year to year. Salmon (chinook, sockeye, coho, pink, and chum) spend several years in the Pacific Ocean before returning to spawn in the Fraser and other rivers. The principal BC salmon-spawning rivers are the Fraser and the Skeena. In 2010, the Fraser River sockeye salmon return was the largest since 1913, with an estimated run of 34 million. The previous year, the Fraser River experienced one of the biggest salmon disasters in recent history. Between 10.6 million and 13 million sockeye were expected to return to the Fraser but less than 2 million arrived (Hume, 2009).

 See Chapter 9, "Overfishing and the Cod Stocks," p. 387, for an account of the overexploitation of the northern cod in the waters of the Grand Banks.

Ottawa, which is responsible for managing salmon stocks within Canada and for negotiating international fishing agreements, must deal with four perplexing and interrelated issues:

1. the natural cycle (up to five years) of salmon to spawn in rivers, migrate to the sea, and then return to the rivers to spawn again;
2. the harmful effects of the forestry and hydroelectric industries on salmon spawning grounds and sea lice from salmon farms on juvenile salmon;
3. the division of the salmon catch among commercial, Native, and sports fishers;
4. the harvesting of "Canadian" salmon by American fishers in international waters.

Management of fish resources is based on the estimated size of the fish stock. The problem of determining the size of the stock, and therefore the quota to be set for each year's catch, is complicated by the natural population cycle of the fish. For salmon stocks, this management issue revolves around their migratory life cycle, from spawn in Canadian rivers, such as the Fraser and Skeena, then to most of their adult life in the Pacific Ocean, and finally a return to their original spawning ground to spawn and thereby begin the cycle for another generation. Other factors[6] are:

- pollution of fish habitat;
- warming ocean temperatures;
- overfishing;
- high fish quotas.

Vignette 9.4, "Mi'kmaq Become Lobster Fishers," page 370, discusses the outcome of the Supreme Court's Marshall decision.

Mining/Energy

The mineral wealth of British Columbia is found in both of the province's physiographic regions—the

VIGNETTE 7.5

Aboriginal Fisheries Strategy: A Temporary Arrangement?

First Nations people along the Pacific coast based their economy on the sea. After BC joined Confederation, fishing for subsistence was generally accepted, but commercial fishing was controlled by Ottawa and Aboriginal fishers could not sell their catch. The federal government's Aboriginal Fisheries Strategy (AFS) was established in 1992 following the 1990 Supreme Court of Canada decision in *Sparrow*, which recognized the "existing aboriginal right" to fish for food and ceremonial purposes. This new strategy rankled other commercial fishers, who argued that there should be but one commercial fishery. When DFO assigned commercial fishing licences and allocated fixed quotas of salmon to several First Nations along the Fraser River and then opened the annual salmon harvest to First Nations before it permitted fishing by the general commercial fleet, the reaction was swift and this soon became a political issue. The commercial fishers argued that the earlier opening date for First Nations fishing was a violation of the equality guarantees in the Charter of Rights and Freedoms. Only with the Supreme Court's 1999 decision in *Marshall*, a case originating in Nova Scotia, did it become clear in law that Aboriginal peoples whose ancestors traded fish have a right to fish for commercial profit.

Cordillera and the Interior Plains. Each has a distinct geological structure. The Cordillera contains a wide variety of minerals, while the Interior Plains is part of the Western Sedimentary Basin, which contains petroleum deposits. Minerals include copper, gold, silver, lead, zinc, molybdenum, coal, and industrial minerals. While coal is king among mineral production in BC, natural gas is soon expected to lead in terms of value of energy production. In 2011, BC produced $5.7 billion worth of coal (Natural Resources Canada, 2012a) while the value of natural gas was $4.4 billion (Canadian Association of Petroleum Producers, 2012)

In the coming years, natural gas is anticipated to increase in importance because of the huge Horn River shale gas reserves near Fort Nelson at the northerly latitude of 58°48′ and the more southerly Montney Basin around Fort St John and Dawson Creek. If an agreement with Asian customers over long-term price can be settled, production could come on stream as early as 2015 and be shipped by LNG supertankers from Kitimat. If no agreement is reached, these vast gas reserves would remain just that—vast reserves. Since the North American market is served by gas deposits much closer to major population centres, BC's gas deposits lie well beyond the economic sphere. With the US producing more and more natural gas, the answer must lie in Asian markets where natural gas prices are well above those in North America. Accordingly, two **liquefied natural gas** (LNG) terminals are planned for the west coast of British Columbia—one at Kitimat and the other at Prince Rupert. Specially designed ships with double hulls aim to protect the LNG cargo from damage or leaks. One concern is that future prices may drop because several countries are racing to build LNG terminals to export liquefied gas to the Asian market.

British Columbia provides the world with a host of minerals. As shown in Table 5.6, in 2011 British Columbia had the third-highest value in mineral production among Canada's provinces and territories, at $8.6 billion, just behind Ontario ($10.7 billion) and Saskatchewan ($9.2 billion). BC's principal mineral export was coal, which in 2011 accounted for $5.7 billion or 66 per cent of mineral production (Natural Resources Canada, 2012a).

For BC, location of its energy and mineral deposits, distance to foreign markets, and proximity to ocean ports are key factors. Unfortunately,

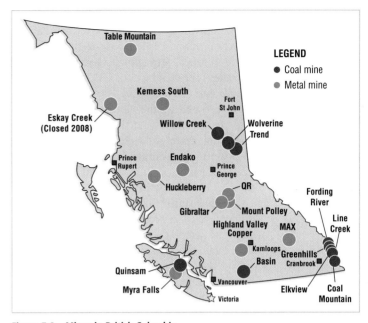

Figure 7.8 Mines in British Columbia
Source: Adapted from Mineral Resources Education Program of BC (2009).

these deposits are "locked in" because of the high cost of transporting them to distant markets. An expanded rail and pipeline system and rail-to-ship loading and LNG facilities provide part of the answer—though at a high construction cost. Risks are high as the failed Northeast Coal Project so clearly demonstrated back in the 1980s (Vignette 7.6).

THINK ABOUT IT
From a geopolitical perspective, there are two views on energy resources and markets: continental and global. First, the US considers energy resources such as the Horn River shale deposits as part of a continental system. Second, Canada, by selling this gas to Asian countries, is diversifying its global markets. Which view do you think will prevail?

Coal remains an important export for British Columbia, which is far and away Canada's largest producer (Figure 7.10 and Table 7.3). BC coal production depends heavily on world demand for metallurgical coal used in steel mills, which fluctuates according to the global economic cycle. For that reason, large-scale developments take on considerable risk. During resource booms, huge investments are made on resource developments. Such was the case in the late 1970s when private and public funds went into the development of the vast Peace River coalfields near the Alberta border. Known as the Northeast Coal Project, production began 1984 about the time demand for coal slumped. Struggling to hang on, the coal companies squeaked by until 2000 when the project collapsed (Vignette 7.6).

As illustrated by the Northeast Coal Project, world coal prices have a dramatic effect on BC coal mining. For instance, during the first decade of the twenty-first century, prices for thermal coal hit lows of nearly $20/metric tonne and highs of almost $140/metric tonne (Figure 7.9). Demand rose sharply starting around 2004 when large shipments to China began; these peaked in 2008. Then, coal prices collapsed in late 2008 with the global economic crisis, followed by a slow recovery over the next four years sparked mainly by China's voracious appetite for coal to fire its growing industrial economy.

Following the economic crisis of 2008–9, demand for BC coal slowly recovered and by 2011, demand was up due largely to long-term contracts with Chinese firms (Figure 7.10). In 2011, a record 27.4 million metric tonnes were produced from the Kootenay and Peace River coalfields.

Each coalfield is situated along the northwest-trending Rocky Mountain foothills in the northeast and southeast of the province. While some coal is shipped by CN rail from the Peace River mines to Prince Rupert, most goes to Westshore Terminals located at Roberts Bank, some 32 kilometres south of downtown Vancouver. Westshore is the leading single export coal terminal in North America, routinely shipping about 21 million tonnes of coal each year. Coal is shipped thousands of kilometres from mines in BC, Alberta, and the western US. Three major railways serve the terminal: CN, CP, and Burlington.

Transportation costs are critical and these have been minimized by unit trains and bulk-loading facilities. Unit trains consist of a large number of ore cars, sometimes over 100, pulled by one or more locomotives. The Roberts Bank terminal was designed as a large bulk-loading facility where coal in rail cars is dumped and then the coal is moved by conveyor belts to the ship's hold. These ships must be moored in deep water, which necessitated building a long causeway to reach the ships. Similar arrangements take place at Prince Rupert.

VIGNETTE 7.6

Northeast Coal Project: A Failed Megaproject

In the 1970s, the vast Peace River coal reserves in the northeast corner of British Columbia represented an enormous untapped energy resource. But distance from an ocean port made these rich coal deposits uneconomic. However, by the late 1970s, global prices for coal reached new highs, making the Peace River coal attractive. With sales secure with Japanese steel companies, the dream of developing the Peace River coal deposits became a reality as the Northeast Coal Project. The Northeast Coal Project comprised Quintette Coal Ltd at Tumbler Ridge, Teck Corporation at Bullmoose, and Gregg River Coal Ltd at Gregg River.

The goals were straightforward: a mega energy project for BC and a stable supply of coking coal for Japan's steel industry. Denison Mines and Teck Corporation were the major Canadian mining companies behind this $4.5 billion construction project (Bone, 2009: 174–5). With a 14-year contract to sell coal to Japanese steel companies, banks and other financial institutions were happy to supply much of the capital (about $2.5 billion) to Denison and Teck. The federal and provincial governments provided $1.5 billion of the $4.5 billion to upgrade the CNR line to Prince Rupert, to extend a BC Rail line to Tumbler Ridge, and to build a coal terminal to load ships that would carry the coal to Japan. The two mining companies contributed about $500 million. Both governments provided another $1 billion to build the coal town of Tumbler Ridge.

Like all megaprojects, the construction phase takes time and production began in 1984. By then, the world demand and price for coal had peaked and a slow but steady decline set in as the global economy slipped into a recession. At that point, the Japanese steel mills no longer needed so much coal and the dream of the Northeast Coal Project collapsed with the bankruptcy of Denison Mines. As well, the federal and the BC provincial governments had to write off their substantial public investment.

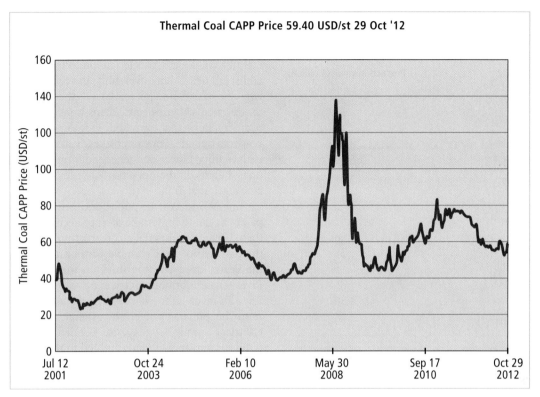

Figure 7.9 **Thermal coal prices, 2001–2012**
The price for metallurgical (coking) coal is traditionally around 15 per cent higher than for thermal coal. The acronym CAPP stands for Central Appalachian Coal Price, which sets the standard for thermal coal prices in North America. *Source:* InvestmentMine, "Historical Coal Prices and Price Chart," 29 Oct. 2012, at: <www.infomine.com/investment/metal-prices/coal/all/>.

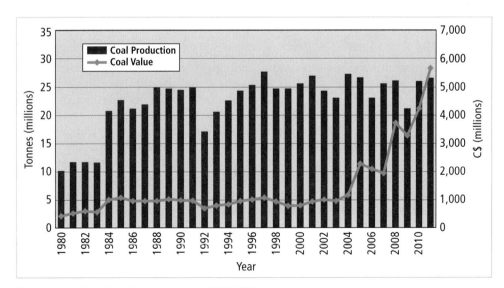

Figure 7.10 **BC coal production and value, 1980–2011**
While coal production has remained relatively stable over the 31 years shown in the graph, the price of coal has surged in the last seven years, causing the value of production to jump from $1 billion in 1990 to $5.7 billion in 2011. *Source:* BC Ministry of Energy, Mines, and Petroleum Resources, "Coal Production and Value 1980–2011," 14 May 2012, at: <www.empr.gov.bc.ca/Mining/MineralStatistics/MineralSectors/Coal/ProductionandValues/Pages/AnnualCoalProduction.aspx>.

Table 7.3 **Canada's Leading Minerals (Including Coal) by Value of Production, 2011**

Mineral	Production Value ($ billions)
Potash (1)	7.97
Coal	7.05
Iron ore	5.33
Gold	5.09
Copper	5.01
Nickel	4.74
Diamonds	2.52
Cement (2)	1.59
Sand and gravel (3)	1.54
Stone (3)	1.52
Zinc	1.30
Uranium (4)	1.09

Preliminary figures. (1) Excludes shipments to potassium sulphate plants. (2) Includes exported clinker. (3) Excludes shipments of sand, gravel, and stone to Canadian cement, lime, and clay plants. (4) Uranium value is calculated using spot market prices.

Source: Natural Resources Canada (2012b: Table 3). Reproduced with the permission of the Minister of Public Works and Government Services Canada, 2013.

THINK ABOUT IT
In examining Figure 7.10, why has the value of BC coal production risen so sharply since 2004?

Hydroelectric Power

Hydroelectric energy is a renewable energy source dependent on the hydrologic cycle of water, which involves evaporation, precipitation, and the flow of water due to gravity. British Columbia, with abundant water resources and a geography that provides many opportunities to produce low-cost energy, produces more hydroelectric power than all other provinces except Québec. Eighty per cent of BC's electrical power is generated by hydro generating stations on the Peace River and Columbia River. In 2011, BC Hydro produced 57 gigawatts of electric energy (BC Hydro, 2012: 6). Within the Cordillera, a combination of elevation, steep-sided valleys, and steady-flowing rivers provides ideal conditions for the construction of hydroelectric dams. The Columbia, Fraser, and Peace rivers offer many excellent hydroelectric sites. In addition, heavy precipitation and meltwater from the mountain snowpack ensure a regular and abundant supply of water. Major hydroelectric developments have taken place on the Columbia and Peace rivers as well as on the Nechako River, a tributary of the Fraser. Minor developments have occurred on Vancouver Island.

In the early days of hydroelectric development, small-scale hydroelectric projects were established near Vancouver and Victoria. These two cities are the major markets in the province for electricity. Hydroelectric dams were also built near smelters that required vast amounts of power to operate. For example, early in the twentieth century, power was generated from privately owned dams on the Kootenay River for the lead-zinc smelter at Trail.

After World War II, public and private companies began to undertake megaprojects. Among these giant industrial construction efforts, three—one on the Columbia River, another on the Peace River, and the third on the Nechako River—had enormous impacts on the economy. They all involved the harnessing of water power from the province's rivers to generate low-cost electrical power, but they also flooded valuable farmland and Indian lands and led to the loss of salmon spawning grounds. Such construction projects were considered major engineering feats. In 1951, Alcan completed the construction of the Kenney Dam across the Nechako River, impounding the water draining from an area more than twice the size of Prince Edward Island. Ten years later, Canada agreed to construct three dams on the Columbia River while the United States constructed a hydroelectric dam at Grand Coulee in Washington. The power was shared with BC. In 1968, BC Hydro built the W.A.C. Bennett hydroelectric dam on the Peace River, creating Williston Lake, the largest freshwater body in the province. From this giant reservoir, vast quantities of water flow through the turbines at the power station, generating much of British Columbia's electrical power and creating a surplus of power for sale in the US.

Aluminum production is one result of British Columbia's low-cost electrical power. For Alcan, obtaining the rights to the waters of the Nechako River for 50 years from the provincial government was key to their plans. In 1951, Alcan began construction of the first phase of this complex.[7] Three years later, Alcan built a new town called Kitimat, a power plant at Kemano, and a dam on the Nechako River. In 1989, Alcan began the second phase of its hydroelectric complex—the Kemano Completion Project. Estimated costs were $1.3 billion. The plan was to enlarge its hydroelectric facility by boring a second tunnel

Photo 7.11
One of BC's most important processing plants, Rio Tinto's Kitimat aluminum smelter, uses low-cost hydroelectric power from the Kemano generating station to process bauxite from foreign countries.

through Mount DuBose, thereby diverting more waters of the Nechako reservoir westward. This time, both the Carrier-Sekani Tribal Council (whose members include the Nechako River as part of their land claim) and environmental groups, spearheaded by the Rivers Defense Coalition, openly challenged the project, claiming that the social and environmental costs were too great. In 1995, the project was half-finished, but the provincial government cancelled its approval for the Kemano Completion Project. Alcan threatened to sue. Later, the British Columbia government agreed to compensate Alcan for the money spent on construction by selling the company electricity at a very low price until 2024 (BC was able to make use of BC Hydro surplus power). The next development came in 2003 when the global economy and the commodity cycle again began to swing upward. Alcan (now Rio Tinto Alcan) dusted off its plans for its BC operations. Modernization of Kitimat Works (formerly call the Kemano Completion Project) is underway at a cost of US$3.3 billion, with a completion date by 2014 (Bouw, 2011). By 2010, the aluminum price cycle had stalled and Rio Tinto began to shed its low-performing smelters, such as the aging Arvida and Shawinigan smelters in Québec. At the same time, the company increased its investments in new, more efficient plants such as the Kitimat Works and the $1.2 billion Saguenay smelter (Jordan, 2012: B10).

Tourism: A Growing Sector in BC's Economy

The natural beauty of BC's ocean, mountains, and forests provides the basis of tourism. While many tourists are attracted to the parks and wilderness areas in the province's forests, most are drawn to the larger urban centres, especially Vancouver and Victoria. Tourism is a central and growing element in the service sector. The 2010 Winter Olympics drew many thousands of tourists and, with world-class winter sports facilities, many more are expected to come year after year. Whistler ski resort, a primary site for the 2010 Games, benefited from the global exposure afforded by the Olympics, as will British Columbia generally. Tourism generated nearly $6.5 billion in 2010, making up over 4 per cent of the province's total real GDP. Employment in the tourism industry accounts for approximately one in every 15 jobs in the province, or 127,000 jobs (Lomas, 2012).

Given British Columbia's scenic landscapes and many parks, tourism has enormous potential for expansion. One example is BC communities tapping into the Alaska cruise ship business. Cruise ships to Alaska carry nearly a million tourists. Most cruises begin in Vancouver and sail north through BC's spectacular Inland Passage where whales and grizzly bears are often sighted. From Vancouver, the cruise ships next stop in Juneau, Alaska. A new and distinctive twist for the cruise ships is to stop along the BC coast to learn about local Indian tribes and their cultures. The world's first Indian-themed cruise ship terminal, the Wei Wai Kum Terminal, opened in 2007, is located near Campbell River on Vancouver Island. This recently constructed terminal is owned and operated by the Campbell River First Nation. It cost $24.5 million, with principal funding from the federal government and lesser amounts from the province and local governments, including the Campbell River First Nation (Indian and Northern Affairs Canada, 2007). Passengers visit a traditional village complete with totem poles and a Big House, both of which provide a window into Laichwiltach history and culture. In 2010, approximately 60,000 tourists came ashore (Turtle Island Native Network Culture, 2011).

While tourism has grown in importance, BC faces a dilemma. Tourist facilities do not always mesh with preserving the environment. For instance, the expansion of the **Sea-to-Sky Highway**

Photo 7.12
Two members of the Campbell River First Nation greet passengers on a cruise ship.

Photo 7.13
A hiking trail on Whistler Mountain, with Blackcomb Mountain and its ski runs in the background.

that winds along Howe Sound from North Vancouver to Squamish and then on to Whistler was opposed by environmentalists and residents of West Vancouver. Currently, the plans to increase the number of super tankers in Burrard Inlet has sparked opposition from both the Vancouver and Burnaby city councils, who fear an oil spill will devastate the waters and shoreline of Burrard Inlet. The basic question is: How does an industrial landscape affect BC's wilderness and harm the tourism trade? The challenge remains how to balance the need for more tourist development and the desire to maintain a natural landscape and an urban environment free of air and water pollution. The answer is complex—industry provides jobs but people want to live in a "clean" environment. Then, too, city governments are recognizing that an "urban/green environment" is the wave of the future where green spaces abound, bike paths are widespread, and dedicated downtown areas are designed for pedestrians rather than for automobiles.

Key Topic: Forestry

The forest is British Columbia's greatest natural asset. With just over 60 per cent of the province covered by forests, British Columbia contains about half of the nation's softwood timber and leads the nation in the export of forest products. With such a dominant position, this region is easily Canada's leading supplier of wood products for construction and finishing. With effective access to ports and rail, forest firms are ideally located to supply the needs of the US and other global markets. At the same time, forests play an important role in the tourist industry with the growing recreational use of forest and wilderness areas by campers, hikers, and whitewater adventurers.

The forest industry remains important, but it no longer dominates the province's economy. The immediate problem is the weak demand from the US, BC's major market (Table 7.4). While sales to Asia have increased, the BC forest industry remains anchored in sales to the US. For

THINK ABOUT IT
Is Vancouver's opposition to the Kinder Morgan plan to ship even greater quantities of crude oil by supertankers likely to stop the project?

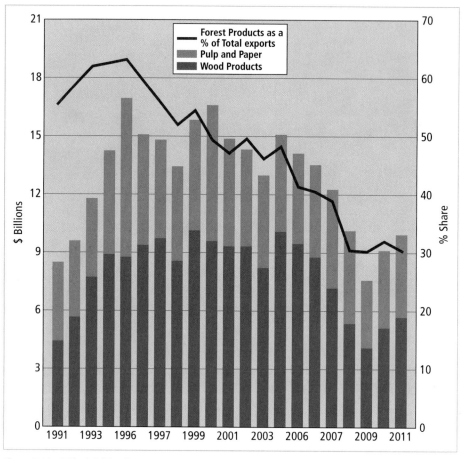

Figure 7.11 **BC's shrinking forestry exports, 1991–2011**
Since the turn of the century, the downward slide of the processing sector in the BC forest industry is readily apparent in this graph. With a recovery of the US economy, the export of wood products stands a better chance of rebounding than does pulp and paper due to the falling demand for paper products. *Source:* Schrier (2012b).

that reason, forestry has slipped from the economic pedestal. Fifty years ago forestry alone accounted for 50 per cent of the provincial economy and for most of the employment. Challenging markets coupled with US restrictions on BC lumber have played havoc with the entire industry. By the late 1990s, forestry comprised only 17 per cent of BC's economy and 14 per cent of employment (Barnes and Hayter, 1997: 5). This decline continued into the twenty-first century. By 2011, forest-sector jobs in BC numbered only 46,203, a decline of 52 per cent from 1991 (Schrier, 2012b: 1). A mid-2012 revival of the US housing market—hopefully a long-term trend—has given the softwood sector a boost. However, as Figure 7.11 indicates, the plight of wood processing and paper manufacturers remains particularly bleak and that situation partly accounts for the decline in BC's manufacturing sector (Figures 7.11 and 7.12).

Among the small coastal communities and those in the BC hinterland, forestry remains the number-one activity. The sharp difference in timber harvests between the Coast and Interior forests is revealed in Figure 7.13. These charts also reveal the much larger harvest in the Interior forest zone—in 2010, 47 million m^3 were taken from the Interior Forest while 16 million m^3 were harvested from the Coastal Forest.

British Columbia's forest consists almost entirely of coniferous **softwood forest**. Within Canada, British Columbia accounts for just over half of the total logging output. The main species harvested are lodgepole pine, spruce, hemlock, balsam, Douglas fir, and cedar (Figure 7.13). The processing of timber into lumber, pulp, **newsprint**, paper products, shingles, and shakes supports a major manufacturing industry in British Columbia, which leads in the production of such wood products as lumber and plywood, while Ontario and Québec produce a greater proportion of pulp and paper products. At first, the forest industry was concentrated along the coast where logs could be barged to sawmills and then the lumber exported by ship. The exploitation of the Interior Forest did not take place until the completion of the Pacific Great Eastern Railway (now a CN line) from North Vancouver to Prince George in 1956.

Fluctuations in total BC exports by value are affected by forest exports. For example, since 2004 softwood lumber exports have declined, mainly due to falling prices. As a result, the place of the forest industry within the BC economy has diminished. The major factor was the drop in exports, especially to the United States. Natural events added to the economic problems facing the industry—the invasion of the pine beetle and several years of massive forest fires have destroyed forest stocks. In spite of these difficulties, once the US economy recovers and demand for BC lumber and other wood products increases, the forest industry should regain its place in the BC economy.

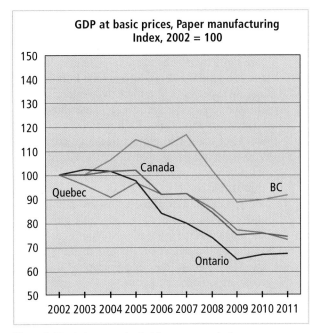

Figure 7.12 **A downward spiral for paper products**
The future for pulp and paper mills is bleak. While BC has bucked the downward price trend somewhat due to its high-quality paper, the consumption of paper continues to fall. One example is the crisis faced by newspapers across North America. *Source:* Wilson (2012).

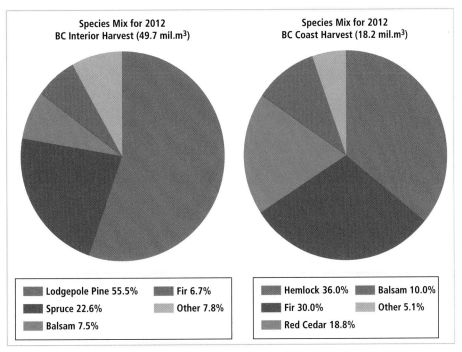

Figure 7.13 **Coast and Interior timber harvests**
Source: Ministry of Forests, Lands and Natural Resource Operations (2011: 3).

Forest Regions

Climate and topography have divided this vast forest into two distinct regions—the rain forest of the coast and the boreal forest of the Interior. Size, age, and species vary in these two regions. Within the rain forest, the mild, wet climate allows trees to grow to great heights for hundreds, even thousands, of years. Old-growth trees that are at least 250 years old are located along the Pacific coast, where fires are rare because of heavy precipitation throughout the year. The major species harvested are spruce and Douglas fir. In the Interior, the main species logged is the lodgepole pine. Since the climate in the Interior is much drier as well as colder in the winter and hotter in the summer than along coast, the lodgepole pine and other trees are smaller and forest stands less dense. Trees have a shorter lifespan (120 to 140 years). While forest fires occur in all parts of BC, most occur in the dry, hot Interior. Here, forest fires are a constant threat and, more recently, pine beetle infestations (discussed above) have become a significant environmental and commercial problem. Another terrain-related issue is clear-cut logging. BC forests often are found in mountainous areas. Clear-cut logging on steeply sloping land leads to erosion and sediment deposition in streams and rivers (photo 7.14). Beyond the environmental concern about clear-cut logging, this harvesting approach is antithetical to the sustainability of the BC forests. Along with this efficient but anti-sustainable logging system, economic cycles of high demand cause companies to accelerate logging in response to economic rather than environmental considerations (Clapp, 2008: 129).

British Columbia's forest is far from homogeneous, largely because of varying climatic zones and the varied relief in the Cordillera. BC's forest lands are divided into two major regions: the Coast Rainforest and the Interior Boreal Forest. Within the Interior Forest are four sub-regions: the Northern Forest, the Nechako Forest, the Fraser Plateau Forest, and the Columbia Forest

Photo 7.14
The practice of clear-cut logging keeps costs down but makes the goal of a sustainable industry problematic. Note the logs lying on the steep slope in this photograph. According to a 2005 report by the David Suzuki Foundation, clear-cutting remains the dominant system of harvesting the old-growth trees in British Columbia's rain forest. Two impacts of clear-cutting are blowouts (strong winds knock over isolated trees) and landslides (moderate to steep slopes are subject to mud and landslides during periods of heavy rain).

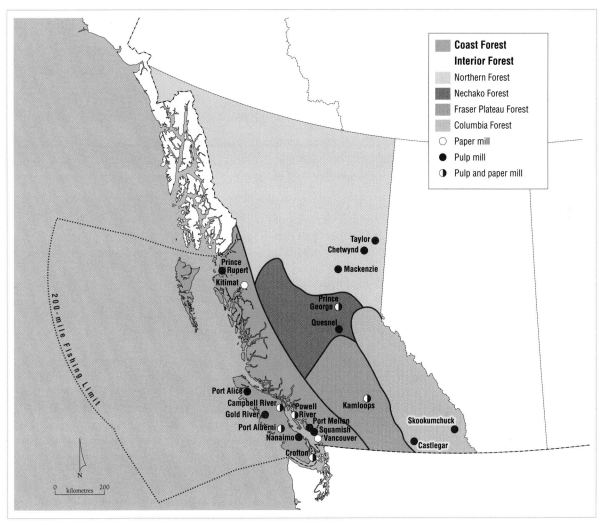

Figure 7.14 Forest regions in British Columbia
The two principal regions are the Coast Rainforest and the Interior Boreal Forest. The forest lands in the Coast Rainforest are almost entirely in mountainous terrain. The Interior Forest is found in many types of terrain. It is subdivided into four areas that reflect variations in growing conditions. The two main elements are lower temperature towards the north and, because of the warmer summer temperatures, drier conditions.

(Figure 7.14). Within each of these sub-regions there is great variation due to the differences in local growing conditions, which are affected by precipitation and length of growing season. Other factors are elevation, soil conditions, and topography.

The Coast Rainforest is the most luxuriant coniferous forest in Canada. With its wet and mild marine temperatures and abundant rainfall, this is one of the most densely forested areas of North America. The key species are Douglas fir, western red cedar, and western red hemlock. Under ideal conditions, mature cedar and hemlock trees reach 45 m in height and about 1 m in diameter. The Douglas fir is an even larger tree. Mature stands may average 60 m in height and over 1.5 m in diameter. Logs from the mature, old-growth stands are highly valued for lumber and plywood, while logs from immature stands (known as secondary growth) are used for pulp and paper.

Inland from the Coast Rainforest, the climate changes from a wet marine climate to a semi-arid one. At this point, the Interior Boreal Forest begins. However, differences exist within the Interior Forest. The Fraser Plateau sub-region (Figure 7.14) is an open woodland with much smaller trees than those in the Coast Rainforest. The controlling factor is the dry climate. Trees in the Fraser Plateau must be able to cope with drought conditions. Ponderosa pine and lodgepole pine are the most common species. They often attain heights of 25 m and a diameter of 1 m. These trees are usually converted into lumber.

The Nechako Forest sub-region, which lies to the north of the Fraser Plateau, receives more precipitation and experiences lower summer temperatures. The net result is more moisture for tree growth. Consequently, the forest cover is denser than that in the Fraser Plateau. A common species in the Nechako Forest is the Engelmann spruce, which can attain heights of 40 m and diameters of 60 cm. Timber from this sub-region is processed by pulp and paper mills at Prince George.

The Columbia Forest sub-region lies in the easternmost area of southern British Columbia. Again, a dry climate limits tree growth. Because of widely varying terrain, the forest cover is quite varied. Western red cedar and western hemlock are common species, though stands of ponderosa pine and even Douglas fir are also found in this region. Large timber is sent to sawmills, while the smaller logs are used for pulpwood.

The Northern Forest sub-region is the most remote of BC's forests. High transportation costs to ship wood to world markets hinder logging. Tree growth is hampered by cool growing conditions and poorly drained land. Large blocks of land that contain muskeg are either devoid of trees or have trees with little commercial value.

THINK ABOUT IT
If the high Canadian dollar inhibits exports of softwood lumber to the United States, should the Bank of Canada force the dollar lower?

Dependency on the US Market

The small domestic market compels the forest companies to seek foreign buyers, especially in the US. Forest exports, the real barometer of the industry, fell from $14.1 billion in 2005 to a low of $7.6 billion in 2009; recovering to $10.2 in 2012 (BC Stats, 2013).

In 2006, the US housing industry collapsed. Consequently, prices for Canadian softwood lumber dropped to half the 2005 prices and value of exports fell by 40 per cent. Plant closures and lay-offs followed. This sudden shift from high demand to low demand illustrates the vulnerability of the forest industry, and of single-industry forest communities. A strong Canadian dollar compounds the difficulty of regaining market share in the US.

Table 7.4 **The Declining Value of Softwood Lumber Exports to the US**

Year	Value (Billions)
2004	5.1
2005	4.8
2006	4.3
2007	3.4
2008	2.2
2009	1.5
2010	1.8
2012	2.0

Source: BC Stats (2013).

North American Free Trade Agreement

The US government has imposed a number of countervailing duties against Canadian softwood lumber—in 1982, 1986, 1991, and 2002. NAFTA has reduced trade barriers between the United States and Canada, but it has not prevented trade disputes involving lumber from erupting. Forest exports from British Columbia have been subjected to the heavy hand of Washington. In each case of countervailing duties, Washington's action has been driven by pressure from the American lumber lobby, the Coalition for Fair Lumber Imports. Most Canadians do not understand why NAFTA has not given Canadian products "free" access to US markets. After all, was not "free" access to each other's market the purpose of the original Free Trade Agreement and the subsequent NAFTA? The answer to that question is "freer" but not "free" trade. For example, when an American producer is adversely affected by imported products, the US company complains to the American government about the lower-priced products, claiming that such lower prices are a form of unfair trade. The company expects Washington to protect it by creating a trade barrier, which it has done on occasion. The trade agreement has a dispute-settlement mechanism, but resolving trade disputes takes time, during which the Canadian exporter loses sales and profits.[8]

From a Canadian perspective, the softwood lumber trade dispute has never been about subsidies but rather has served as a means by which

Washington protects the US lumber producers. In May 2002, as negotiations to extend the previous agreement reached an impasse, Washington announced a duty of 27 per cent on Canadian softwood lumber entering the United States. Four years later, a new agreement finally was signed in October 2006. During the time of the duty, larger firms were able to increase their efficiency through economies of scale but many small forest operations were hurt, causing closures and layoffs. These problems were exacerbated when the US housing industry collapsed in 2006. By mid-2012, the US housing market had recovered significantly, causing lumber prices to increase dramatically. In addition, the devastation along the American east coast resulting from Hurricane Sandy in late October 2012 will require major rebuilding, thus increasing the demand for softwood lumber and other wood products.

The latest softwood lumber agreement between Ottawa and Washington ended a bitter trade dispute—at least for the next seven years. The agreement provides protection to the US industry when lumber prices fall. Highlights of this agreement are:

- The seven-year agreement includes a possible two-year extension.
- Import duties of $4 billion the US charged Canadian companies since 2002 will be returned. The US keeps $1 billion.
- The US is banned from launching new trade actions against lumber imports from Canada.
- Canada is to impose an export tax if softwood lumber prices in the US fall below US $355 per thousand board feet. At that point, Canada must impose a tax on softwood lumber shipped to the United States of at least 5 per cent of the price per 1,000 board feet. The purpose of the export tax is to protect US lumber producers by increasing the price of Canadian lumber, thus giving American producers an advantage.
- Neutral trade arbitrators are to provide final and binding settlements of disputes.

The lumber industry in British Columbia is at a crossroads. The pessimistic interpretation is of an industry on the decline, especially among its pulp and paper firms. The optimistic interpretation, though, is of a kind of industrial renaissance, of a newly fashioned forest products industry that emphasizes high-value products, skilled labour, and leading-edge technology. Industry analysts say that the province has too many pulp mills facing a dwindling supply of wood fibre and a downward demand trend. The forest industry remains an important part of BC's economy, but a shift to high-value products is necessary, as is a recovery of BC's major market in the United States.

British Columbia's Urban Geography

The most striking aspect of the urban geography of British Columbia is the concentration of people in the southwest corner of the province popularly referred to as the Lower Mainland: over 60 per cent of BC's residents are concentrated in this area. Besides Vancouver, the major cities in the Lower Mainland include Abbotsford, Burnaby, Coquitlam, Langley, Maple Ridge, New Westminster, North Vancouver, Richmond, Surrey, and West Vancouver (Figure 7.15). Beyond this population core, a secondary population cluster is found on Vancouver Island, where Victoria forms the second-largest urban population and a series of towns and cities, including Nanaimo, stretch northward to Campbell River. In the Interior, Kelowna, Vernon, and Penticton constitute the major urban centres in the Okanagan Valley, while nearby Kamloops is in the Thompson Valley. As well, population centres are located along the north coast around Prince Rupert, in the Central Interior at Prince George, in the Peace River country where Dawson Creek and Fort St John are located, and along the US border, where Trail and Creston are located.

Canada's third-largest city, Greater Vancouver, has a population of over 2 million. The census metropolitan area of Vancouver dominates the urban geography of British Columbia (Table 7.5). Like the province as a whole, Vancouver's population has increased at a rate well above the national average. Until 1996 it was the fastest-growing large metropolitan area in Canada. Since then, Vancouver's spectacular growth rate has continued, but from 1996 to 2011 Calgary was the fastest-growing CMA in Canada.

Besides the rapid growth of Vancouver, BC's three other CMAs also grew quickly. From 2001 to 2011, Vancouver grew by 16.4 per cent, Kelowna by an astounding 21.7 per cent, Abbotsford 15.5 per cent, and Victoria by a modest 10.5 per cent. At

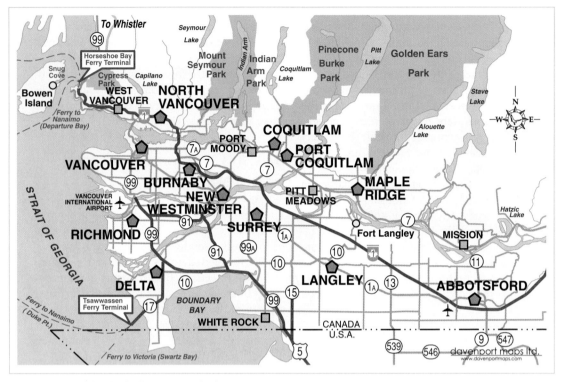

Figure 7.15 British Columbia's Lower Mainland
The Lower Mainland of British Columbia, with Vancouver as the focal point, dominates British Columbia's urban geography (British Columbia Adventure Network © 1996–2007 Interactive Broadcasting Corporation). *Source:* Davenport Maps Ltd, at: <www.davenportmaps.com>.

the other end of the scale, the greatest population losses (over 12 per cent) took place in Kitimat and Prince Rupert. The demographic decline of Prince Rupert may have halted since the last census now that its container port is in operation and traffic flows along the CN's much touted **North West Transportation Corridor**. Looking to the next decade, the prospects for a natural gas LNG facility look bright for Prince Rupert, and this would draw more workers and their families. Kitimat's future depends on the completion of the Kitimat Works, which will result in an expansion of aluminum production and, with it, employment. Also in the wind is the prospect for Kitimat to become a transshipment point for Fort McMurray oil and BC natural gas going to China. Even more exciting, David Black, a BC media owner, has plans for a game changer—processing the bitumen at a proposed heavy oil refinery at Kitimat rather than sending the raw bitumen directly to Asia for processing. Pipe dream or not, Black has the backing of Oppenheimer Investment Group, a Swiss-based firm, to raise funds to build the refinery at Kitimat, which would be the first in the world to use a new technology to reduce greenhouse gases by 50 per cent (CBC News, 2013).

BC can claim only the Vancouver area as a population cluster. The Vancouver CMA's 2011

Table 7.5 Census Metropolitan Areas in British Columbia, 2001–2011

Centre	Population 2001	Population 2011	Percentage Change
Abbotsford	147,370	170,191	15.5
Kelowna	147,739	179,839	21.7
Victoria	311,902	344,615	10.5
Vancouver	1,986,965	2,313,328	16.4

Source: Adapted from Statistics Canada (2007, 2011).

population was 2.3 million. Victoria, the second-largest urban centre, had 344,615 inhabitants (Table 7.5). As the capital of the province, Victoria is a "government" town, as well as an important tourist and service centre. Its mild climate has also attracted retired people, especially from the Prairie provinces. Kelowna and Abbotsford, with populations in 2011 of 179,839 and 170,191, respectively, are the only other CMAs in the province. Three cities—Kamloops, Nanaimo, and Chilliwack—have populations close to 100,000 and may, by the next census, qualify as CMAs. Prince George, the regional centre in northern BC, remained around 85,000 over this period. The remaining cities range in size from Kitimat at 8,335 to Vernon at 58,584 (Table 7.6). With a few exceptions, such as Fort St John, the population of urban centres in BC's northern hinterland declined from 2001 to 2011, with the major losses occurring in Terrace, Kitimat, and Prince Rupert.

Vancouver

With one of the most spectacular physical settings in the world, Vancouver, located on the shores of Burrard Inlet, lies across these waters from the snow-capped peaks of the North Shore Mountains. The Lions Gate Bridge passes over Burrard Inlet, thereby linking Vancouver with the North Shore and its main urban centres of West Vancouver and North Vancouver. From West Vancouver, the Sea-to-Sky Highway leads to the Whistler ski resort. To the west is the island-studded Strait of Georgia,

THINK ABOUT IT
By the next census in 2016, what economic developments might turn the population losses into gains for Terrace, Kitimat, and Prince Rupert?

Table 7.6 **Other Urban Centres in British Columbia, 2001–2011**

Centre	Population 2001	Population 2011	Percentage Change
Kitimat	10,285	8,335	−19.0
Dawson Creek	10,754	11,583	0.8
Prince Rupert	15,302	13,052	−14.7
Terrace	19,980	15,569	−22.1
Powell River	16,604	16,689	0.0
Squamish	14,435	17,479	21.1
Salmon Arm	15,388	17,683	14.9
Williams Lake	19,768	18,490	−6.4
Quesnel	24,426	22,096	−10.0
Cranbrook	24,275	25,037	−3.1
Port Alberni	25,299	25,465	0
Fort St John	23,007	26,380	9.3
Parksville	24,285	27,822	9.2
Campbell River	35,036	36,096	4.1
Penticton	41,564	42,361	4.2
Duncan	38,613	43,252	6.6
Courtenay	45,205	55,213	8.9
Vernon	51,530	58,584	7.5
Prince George	85,035	84,232	−2.1
Chilliwack	74,003	92,308	9.3
Nanaimo	85,664	98,021	7.8
Kamloops	86,951	98,754	4.4

Source: Adapted from Statistics Canada (2007, 2011).

Photo 7.15
Located between Lake Okanagan and Skaha Lake, Penticton lies in the Okanagan Valley of the Interior Plateau. Its hot summer weather and sandy beaches make Penticton a summer tourist hub. Irrigated fruit orchards and vineyards surround the city on the lower slopes of the Okanagan Valley. On the upper slopes, the natural vegetation becomes desert-like, with sagebrush, cactus, and other desert plants.

while the Fraser River and its deltaic islands (flat, low islands composed of silt and clay near the mouth of the river) mark Vancouver's southern edge. Vancouver has a mild, marine climate, though some find the frequent rain and overcast skies unappealing.

Like Montréal, much of Vancouver's commercial strength stems from its role as a trade centre. Vancouver is the largest port in Canada and one of the largest on the Pacific coast. With most of the world's population located along the Pacific Rim, Port Metro Vancouver handles over $65 billion worth of trade goods annually (Port Metro Vancouver, 2013). In fact, the Port of Vancouver is so busy that some shippers may opt for a longer route through the Panama Canal to Halifax.

Much of Vancouver's economic well-being is closely tied to the United States and Pacific Rim countries. The bulk of exports exit BC from the Port of Vancouver with most going to the United States, China, Japan, and South Korea. In 2011, 80 per cent of exports from British Columbia and 78 per cent of imports involved those four countries (BC Stats, 2012c). Energy comprised 30.7 per cent of 2011 exports with coal and natural gas playing a key role. Forest products came a close second at 30.5 per cent, led by lumber, which accounted for 17.4 per cent (BC Stats. 2012b).

Besides being a major global transshipment point, Vancouver is home to the head offices of various private companies (especially fishing, forestry, and mining companies), and federal and provincial public offices are located in Vancouver. While Vancouver (known in 1867 as Gastown) began as a sawmill centre, these air-polluting mills relocated to other centres or were dismantled. False Creek, in downtown Vancouver, had been the prime site for processing logs, but in 1986 False Creek became the site of Expo '86 and then the site of an upscale residential and farmers' market complex known as Granville Island (Vignette 7.7).

Photo 7.16
The Lower Mainland is a popular destination for Chinese immigrants and the urban landscape contains Chinese-Canadian "ethnoburbs" in Vancouver, Richmond, and other cities in the area. One example of the vibrant Chinese community is the Vancouver Chinatown Night Market, which is open on weekends throughout the summer.

Faultlines in BC

The centralist/decentralist faultline in BC originates in the province's perceived lack of power within Confederation. This perception poses an irritant in BC's relations with Ottawa and reinforces other grievances that tend to increase the faultline. Unexpectedly, a squabble between BC and Alberta over the proposed Northern Gateway Pipeline created a new political divide. The Premier of British Columbia, Christy Clark, has insisted that BC receive its fair share from this project, largely because most risk falls on BC lands, waters, and people. Accordingly, British Columbia should receive "a fair share of the fiscal and economic benefits of a proposed heavy oil project that reflects that level and nature of the risk borne by the province, the environment

VIGNETTE 7.7

Granville Island

Granville Island houses a public market in the heart of downtown Vancouver. Lying beneath Granville Bridge, the island has turned into a key gathering spot for both locals and tourists. Granville Island was a nondescript sandbar that was converted in 1915 into an industrial core for the growing forest industry. In the 1970s, it began to change into an upscale residential and specialized commercial area centred on the Granville Island Public Market. Other enterprises have widened its appeal—artists' studios and shops, a wide variety of restaurants, and features like the Kids Market, Maritime Market, and Coast Salish Houseposts, a joint endeavour between the Emily Carr College and First Nations. Granville Island is unique to Vancouver and has added another dimension to the wide-ranging tourist attractions in the Greater Vancouver area.

and taxpayers" (Clark, 2012). With Alberta Premier Alison Redford's rejection of the notion of sharing Alberta's oil revenues with BC, the dispute simmers on, fuelled by environmental groups and, especially, by numerous BC First Nations whose lands and waters the pipeline would cross. This one proposed project, then, which has been strongly supported by Stephen Harper's Alberta-friendly federal government, brings into sharp relief both the centralist/decentralist and Aboriginal/non-Aboriginal faultlines.

THINK ABOUT IT
Would BC have a "fair share" of the revenues from the proposed Northern Gateway Pipeline if a refinery was built at Kitimat?

SUMMARY

British Columbia, economically and demographically, is a growing power within Canada. As one of six regions, BC's population continues to increase well above the national rate while its export-oriented economy is slowly regaining ground lost from the 2008–9 global crisis, especially trade with the United States. After years of unprecedented economic growth, British Columbia's economy hit a rough patch. Its first rough patch was the sudden drop in forest exports to the United States when housing construction in the US collapsed. The second stumble was due to the broader global economic crisis. The 2010 Olympics may have signalled better times, but more important, in 2012 the US demand for BC softwood lumber suddenly surged and the shipbuilding industry obtained a huge contract from the federal government.

One geographic fact remains irrefutable: the **grooves of geography**, in this case the north–south mountain ranges of BC, tend to align this region with the adjacent US Northwest and, despite modern transportation methods, set it apart from the rest of Canada. Yet, the heavy investment in transportation infrastructure—the superhighway corridor to Vancouver and the expansion of port facilities at Vancouver and Prince Rupert, plus the promise of LNG terminals at Kitimat—creates an east–west groove and encourages links with Pacific Rim countries. The expansion of port facilities at Prince Rupert alone has provided another and shorter outlet for natural resources and agricultural products from Western Canada to reach markets in Asia. In addition, the prospect of a new energy corridor for both natural gas and bitumen across the Cordillera to Kitimat and then by ships to Asian countries would greatly strengthen trade. Once the global economy regains its steam, exports to the United States and Pacific Rim countries are anticipated to return to previous or even higher levels. Finally, BC's natural setting makes it a world-class tourist destination. The main challenge facing BC is the accelerating development of its high-technology industry and the need to apply some of that technology to the processing of its natural resources, especially its forestry products.

CHALLENGE QUESTIONS

1. Is the appeal of Cascadia rooted in the "grooves of geography"?
2. What new technology allowed natural gas to be extracted from the Horn River shale deposit in the Interior Plains of British Columbia?
3. Is the Aboriginal fishery a treaty right or a temporary political compromise?
4. Why is the proposed Northern Gateway Project a game changer for Alberta and Canada?
5. Why does the US want to limit BC's ability to ship lumber to US markets while the US places no such limits on BC's export of electricity to the Pacific Northwest?

FURTHER READING

Molloy, Tom. 2000. *The World Is Our Witness: The Historic Journey of the Nisga'a into Canada.* Calgary: Fifth House.

On 11 May 2000, the Nisga'a treaty passed into law, marking a historic agreement between this small group of Indians and the rest of Canadian society. In *The World Is Our Witness*, Molloy, who was the chief federal negotiator, describes how this agreement ends the centuries-old colonization by the British and, later, Canadians of the Nisga'a lands and people. Furthermore, this treaty begins the search for a place within Canadian society by the Nisga'a. For the other Indian tribes of British Columbia, this treaty has far-reaching implications. Molloy not only provides an insider's view of the struggle to achieve an agreement, but also explains its significance to Canadians. The Nisga'a treaty represents a compromise between the Nisga'a and other Canadians on how to share the lands and resources found in the traditionally occupied lands of the Nisga'a. Originally, there was no sharing. The early English settlers who established British colonies along the east coast of the United States nearly 400 years ago claimed the land by the right of "discovery" and acknowledged no Aboriginal right to land. Recognition of Aboriginal rights began with the Royal Proclamation of King George III in 1763, which declared that lands to the west of the Appalachian Mountains were Indian lands. While these lands were lost to the Americans in 1783, the legal precedent laid the foundation for treaty-making across British North America and, later, Canada. The Nisga'a treaty is not only one of the most recent agreements but it has expanded the scope of treaties into the area of resource-sharing.

WESTERN CANADA

Introduction

Western Canada, rich in natural resources, lies in the heart of North America. Global demand for its resources, especially its oil and potash, has caused Western Canada to experience record economic and population growth, thereby transforming the region into the economic engine for Canada. Farmers, too, are benefiting from an expanding global demand, with canola becoming the most profitable crop. Traditionally, exports of natural resources and agricultural products flowed mainly to the United States, but now significant exports are heading to Asian countries. One of the most challenging issues facing Western Canada, but especially Alberta and its Pacific neighbour, is the proposal to build two pipelines across British Columbia, thus connecting the oils sands to the Pacific coast and Asian markets. Another one, related to the proposed pipelines, is the heavy investment of state-owned Asian countries in the oil sands.

Three political jurisdictions, the provinces of Alberta, Saskatchewan, and Manitoba, comprise Western Canada. From a physiographic perspective, Western Canada divides into two distinct sub-regions:

- The Interior Plains, which is a relatively densely populated agricultural/industrial core.
- The Canadian Shield, where the majority of people are First Nations and Métis living in a northern resource region.

Geography presents unique challenges to Western Canada. First, a dry continental climate with hot, dry summers and long, cold winters limits the spatial dimension of agricultural lands

CHAPTER OVERVIEW

The main themes in this chapter are:
- Western Canada's physical and historical geography.
- The basic characteristics of the population and economy of the region.
- Western Canada's emergence as a global energy and potash powerhouse.
- The rise of canola as the chief cash crop and the importance of processing it within Western Canada.
- The threat of drought and its impact on agriculture activities.
- The proposals to ship bitumen by pipelines to the Pacific coast and then by supertankers to world markets.

while market conditions determine the mix of crops grown. Second, the region's continental position makes transportation a central cost factor in accessing global markets. The search for solutions to this transportation challenge is necessarily ongoing.

As an exporting region, Western Canada trades mainly with the United States, Pacific Rim countries, and the European Union. Prices for its resource products have followed a cyclical pattern. In the last decade, for example, low prices for grain and forest products have hampered these activities while high prices for energy, potash, and canola have had the opposite

Prairie slough and field of mature wheat, near Bruxelles, Manitoba. A slough is a glacial pond. The Prairie Pothole Region contains thousands of similar glacial ponds which were formed approximately 10,000 years ago at the end of the Wisconsin glaciation. Photo: Dave Reede/All Canada Photos.

effect. Canola now exceeds spring wheat in returns to farmers and in sown acres.

Western Canada, but particularly Alberta, has experienced a surge in economic and population growth. This growth is reflected in its major cities—Calgary, Edmonton, Winnipeg, Saskatoon, and Regina. In sharp contrast, rural areas have seen a huge population loss, the disappearance of rural villages, and fewer but larger farms. The loss of rural population is one element associated with the ongoing transformation of the region's agricultural economy. We examine this transformation more closely in this chapter's "Key Topic," agriculture.

Western Canada within Canada

Western Canada is composed of three provinces, Alberta, Manitoba, and Saskatchewan. Situated in the vast western interior of North America, its geographic location poses two formidable challenges. First, the region's dry continental climate (Vignette 8.1) makes farming much riskier than in other agricultural areas of Canada. Dry spells that lead to crop failure, while relatively rare, do strike with vengeance. The last dry spell took place from 1999 to 2002, when annual precipitation fell well below the long-term average. In fact, the variation from the average annual precipitation is a geographic fact of life, resulting in "wet and dry years." Second, Western Canada's continental position within North America makes distance to world markets extremely long and has proven to be a major stumbling block to the region's resource development. It remains an ongoing struggle to reduce the cost to access markets. For coal, grain, and potash, **unit trains** have cut transportation costs substantially, while pipelines have done the same for oil and natural gas. For that reason, sparks flew between the premiers of Alberta and British Columbia when BC's Christy Clark

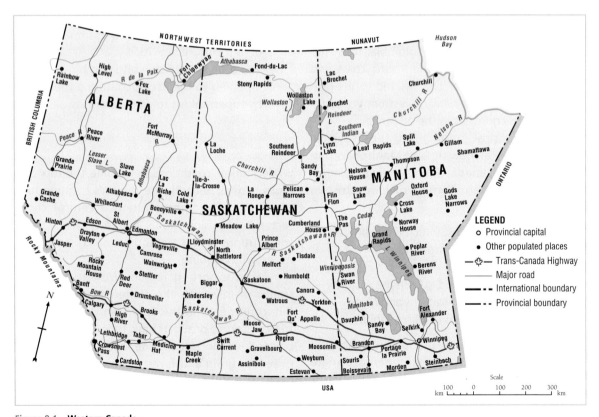

Figure 8.1 **Western Canada**

Source: Reference map of the Prairie Provinces, at: <atlas.nrcan.gc.ca/data/english/maps/reference/provincesterritories/prairie_provinces/map.pdf>, Natural Resources Canada, 2000. Reproduced with the permission of the Minister of Public Works and Government Services Canada, 2013.

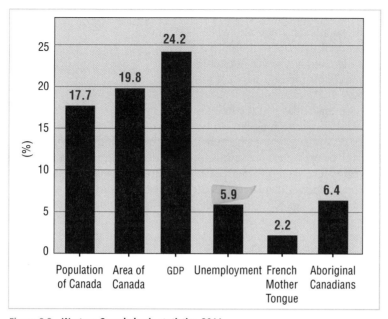

Figure 8.2 **Western Canada basic statistics, 2011**
Western Canada contains nearly 20 per cent of Canada's land mass and 18 per cent of its population. Its economic strength, as measured by its share of the national GDP, is just over 24 per cent. For comparison, Québec has 20 per cent of Canada's GDP but accounts for 24 per cent of its population. Significantly, Western Canada's population includes more Aboriginal Canadians than French-speaking Canadians, and the region—except for the Territorial North—has the highest percentage of Aboriginal Canadians. *Source:* Tables 1.1, 1.2, 5.1.

expressed opposition to the Northern Gateway Pipeline project unless her province was compensated for the potential environmental risks associated with a pipeline crossing its territory and oil tankers plying its coastal waters. Alberta's Alison Redford has refused to share royalties, citing a province's right to income from natural resources within its own borders.

Western Canada's renewable resources—the fertile soils of the Canadian Prairies, the boreal forest of its northern lands, and its rivers—offer the prospect of a sustainable economy, if managed properly. Added to these renewable assets, the huge deposits of oil sands and potash, while non-renewable, have over a hundred years of reserves and thus might be expected to provide support for the regional economy over that period of time. Western Canada's economic boom is based on the sale of oil and potash at prices well above their long-term average. These high prices are a result of the expanding demand from Asia.

VIGNETTE 8.1

Water Deficit and Evapotranspiration

The Canadian Prairies often has a "water deficit." This deficit is measured in terms of potential evapotranspiration, which is the amount of water vapour that can potentially be released from an area of the earth's surface through evaporation and transpiration (the loss of moisture through the leaves of plants). There is a water deficit if the evapotranspiration rate is greater than the average annual precipitation. For example, most of the grassland natural vegetation zone receives less than 400 mm annually, while its evapotranspiration rate exceeds 500 mm. The difference indicates a water deficit of over 100 mm. If a water deficit occurs in a dry year, plants and crops can use water in the soil. When this reserve is exhausted, crops suffer from a lack of water. The maps of precipitation and natural vegetation in Chapter 2 (Figures 2.7 and 2.8) indicate the geographic location of this water deficit area.

Western Canada leads the country in the rate of economic and population increase while Ontario and Québec lead the nation in absolute terms. But this gap is closing. Overall, Western Canada's economy accounts for 24.2 per cent of Canada's GDP (Figure 8.2). In addition, for the past decade Western Canada has had the lowest unemployment rate of the six regions. In 2011, the three provinces of Western Canada had unemployment rates well below the national average of 7.5 per cent: Saskatchewan, 5.0 per cent; Manitoba, 5.4 per cent; Alberta, 5.5 per cent. Led by Alberta, the rate of economic growth in Western Canada is outstripping those of Ontario and Québec, and record numbers of newcomers to Canada are choosing Western Canada because of job opportunities.

Table 5.1, "Before and After the Crash: Unemployment Rates by Province, 2007, 2009, and 2012," page 166, shows how well the three provinces in Western Canada have fared relative to the other provinces.

In 2011, the population of Western Canada totalled over 5.8 million, an increase of almost 9 per cent over its 2006 population. The vast majority live in the regional core, an area of major cities, towns, and villages located in the southern half of the region. Agricultural, industrial, and service activities are well established within this population core. Calgary, Edmonton, Winnipeg, Saskatoon, and Regina are the leading cities. In the northern half of Western Canada is the regional hinterland, with less than 5 per cent of the region's population. While most reside in resource towns, such as Fort McMurray and Thompson, nearly a hundred small Aboriginal settlements are scattered across this boreal landscape.

THINK ABOUT IT
Given the past boom-and-bust economic pattern associated with agriculture and resource development, is the author overly optimistic about Western Canada's economic prospects?

Western Canada's Physical Geography

Western Canada has two major physiographic regions—the Interior Plains and the Canadian Shield—as well as small portions of two others—the Hudson Bay Lowlands and the Cordillera (Figure 2.1). A thin portion of the Cordillera, the Rocky Mountains, forms a natural and political border between southern Alberta and British Columbia. Each physiographic region has a particular set of geological conditions, physical landscapes, and natural resources.

The four physiographic regions found in Western Canada are discussed in Chapter 2.

The tiny slice of the Cordillera, along the eastern flank of the Rocky Mountains in Alberta, provides logging and mining opportunities for Western Canada. Its spectacular mountain landscape attracts thousands of visitors each year. This region has two internationally acclaimed parks, Banff National Park (photo 8.1) and Jasper National Park, which attract visitors from around the world. Calgarians are especially fortunate in having easy access to the Kananaskis Country Provincial Park, where a number of mountain recreational activities—camping, hiking, and skiing—are available.

In the Interior Plains region the sedimentary rocks contain valuable deposits of fossil fuels. By value, the four leading mineral resources are oil, gas, coal, and potash. Most petroleum production occurs in a geological structure known as the **Western Sedimentary Basin**, which underlies most of Alberta and portions of British Columbia, Saskatchewan, and Manitoba. While potash and coal mining take place deep in the earth's crust, some minerals are close to the surface, thus permitting open-pit mining. In southeastern Saskatchewan, brown coal is extracted through open-pit mining and then burned to produce thermal electricity. In northeastern Alberta, the huge petroleum reserves in the **Athabasca tar sands** are exploited by both open-pit mining and sub-surface mining techniques that involve injecting steam deep underground to "liquefy" the **bitumen** and then pumping the slurry liquid to the surface.

The Canadian Shield, consisting mainly of bare rocks exposed at the surface, extends over one-third of northern Manitoba and Saskatchewan as well as a small part of Alberta. While logging takes place along the southern edge of the Canadian Shield, most economic activity is associated with mining and hydroelectricity. In northern Saskatchewan, uranium companies produce most of Canada's uranium from open-pit and underground mines. Manitoba's northern rivers, particularly the Nelson River, have huge dams and power stations.

Photo 8.1
Lake Louise, one of Banff National Park's most stunning natural features, lies within the alpine recreation zone of Calgary. The lake's famous turquoise colour is caused by fine rock particles, called "glacial flour," in the stream water from alpine glaciers. The steep-sided U-shaped valley behind the lake, known as a glacial trough, indicates the erosional effect of an alpine glacier.

A continental climate controls weather conditions. Far from moderating ocean influences, this climate is characterized by cold, dry winters and hot, dry summers. The resulting range of temperatures is extreme—from lows of −30°C in January to +30°C in July. During the winter, Arctic air masses often dominate weather conditions in the Prairies, placing the region in an Arctic "deep freeze" (Figure 2.5). The hot, dry summer weather, on the other hand, results from the northward migration of hot, dry air masses from the Southwest US (Figure 2.6).

Annual precipitation, whether in the form of snow or rain, is among the lowest in all regions except the Territorial North (Figure 2.7). The region is dry for two main reasons. First, distance from the Pacific Ocean reduces the opportunity of moist Pacific air masses to reach Western Canada. Second, **orographic uplift** of these Pacific air masses over the Rocky Mountains causes them to lose most of their moisture, leaving little precipitation for Western Canada. A combination of strong winds and sub-zero temperatures can produce blizzard-like weather. In southern Alberta, strong winds that become warm and dry as they flow down a mountain slope are known as **chinooks**. On the other hand, the **Alberta clippers** with their strong, frigid winds produce true blizzard conditions, due to severe blowing and drifting snow.

Photo 8.2
The Cypress Hills, located along the Alberta–Saskatchewan border, here tower above the wheat fields of Saskatchewan. At 600 metres above the surrounding Interior Plains, the Cypress Hills stand out as a physiographic, climatic, and vegetation anomaly. As an isolated island of lodgepole pine and white spruce surrounded by the semi-arid Prairies, few thought that the pine beetle could reach this remote area. But that assumption was incorrect and efforts to stem the attack of this beetle on the forest of the Cypress Hills began in the summer of 2012. See Vignette 2.4, "Cypress Hills," for more on this unique area of Western Canada.

In general, annual precipitation decreases from west to east in Alberta and Saskatchewan. On average, Calgary receives 413 mm annually while Saskatoon's figure falls to 350 mm. One exception is the Cypress Hills, which because of the higher elevation receives more precipitation, thus allowing a forest vegetation (photo 8.2). Further east, precipitation increases, with Winnipeg recording an average annual amount of 514 mm. While Saskatoon depends heavily on precipitation from Pacific air masses, southern Manitoba (including Winnipeg) receives heavy summer rainfall from the moist Gulf of Mexico air masses. As a result, southern Manitoba is rarely troubled by droughty conditions compared to the other prairie lands in southern Saskatchewan and Alberta. The driest lands are found in **Palliser's Triangle** where, on average, less than 400 mm fall each year (Vignette 8.2). **Evapotranspiration** provides one measure of water deficit/surplus by measuring transpiration and evaporation, thus providing an indication of dryness/drought (Vignette 8.1).

Crop agriculture in the semi-arid regions of southwestern Saskatchewan and southern Alberta depends on summer rainfall. Under normal conditions, most rainfall occurs in June when the growth of crops demands the most moisture. In dry years, little rainfall occurs in the summer. With regard to the irregularity of summer precipitation, Natural Resources Canada (2010) states that "All regions of Canada can experience seasonal [summer] dry spells, but only in the Prairie provinces can precipitation cease for a month, surface water disappear for entire seasons, and drought persist for a decade or more.

In Western Canada, the Prairies are the agricultural heartland. Lying within the Interior Plains physiographic region, two natural vegetation zones (parkland and grassland) and three chernozemic soil zones (black, dark brown, and brown) exist (Figures 2.9 and 8.3). The parkland is a transition zone between the boreal forest and the grassland natural vegetation zone (photo 8.3). Within the grassland, the dry climate becomes more prevalent

Photo 8.3
Within the Fertile Belt, black and dark brown chernozemic soils have formed under tall-grass natural vegetation and they now provide farmers with one of the most fertile soils in Canada. In this photograph taken near Biggar, Saskatchewan, the practice of crop rotation is illustrated by the three types of land use. The dark brown field represents summer fallow where chemicals are used to control weeds, known as chem fallow; the yellow field contains a spring wheat crop; and the green field consists of grassland used for pasture.

as the evaporation rate increases towards the American border. The result is a change in natural vegetation from tall grass to short grass. Beneath the two types of grasslands, chernozemic soils were formed; black chernozemic soils are associated with tall-grass natural vegetation and dark brown and brown soils with short-grass natural vegetation. Figure 8.3 illustrates the spatial expanse of these soils, while Figure 8.4 shows that the Fertile and Dry Belts closely parallel the location of these soil types.

VIGNETTE 8.2

Palliser's Triangle

In 1857, the Palliser Expedition set out from England to assess the potential of Western Canada for settlement. Well documented by Professor Spry (1963), this expedition spanned three years (1857–60). In his report to the British government, John Palliser identified two natural zones in the Canadian Prairies. The first zone was described as a sub-humid area of tall grasses, while the second, located further south, was described as a semi-arid area with short-grass vegetation. Palliser considered the area of tall grasses to be suitable for agricultural settlement. He named this area the Fertile Belt. In Manitoba this belt is south of the Canadian Shield and stretches to the border with the United States. Palliser believed that the semi-arid zone, located in southern Alberta and Saskatchewan, was a northern extension of the Great American Desert. His belief was reinforced by the Great Sand Hills (Vignette 8.7). According to Palliser, these semi-arid lands were unsuitable for agricultural settlement. The area described by Captain Palliser became known as Palliser's Triangle—its area overlaps with, but is slightly larger than, the agriculture zone known as the Dry Belt (see Figure 8.4). Homesteaders called these lands "heartbreak territory" and most abandoned their attempts at farming because of the frequency of drought-induced crop failures. David Jones's *Empire of Dust* (1987) captures the settling and abandonment of homesteads in the 1930s while Arthur Kroeger's *Hard Passage* (2007) is a personal recollection of the struggle of the author's family to homestead in Palliser's Triangle and their eventual defeat.

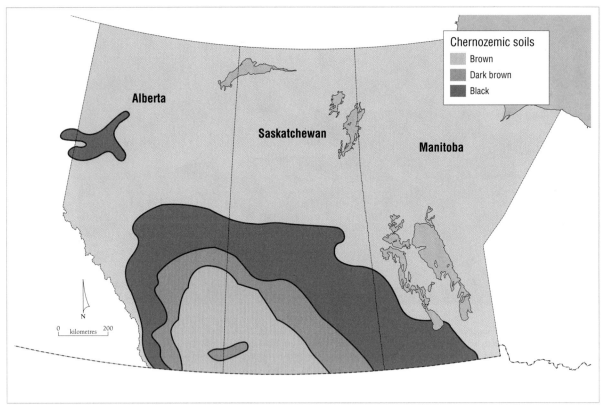

Figure 8.3 **Chernozemic soils in Western Canada**
Three types of chernozemic soils are found in the Canadian Prairies: black, dark brown, and brown. The differences in colour are due to the varying amount of humus in the soil, which, in turn, is a factor in the natural vegetation cover—short grass, tall grass, and parkland vegetation. The soil of the Peace River country, formed under an aspen forest, is "degraded" black soil.

Environmental Challenges

Nature provides Western Canada with its major environmental challenge—droughts. In a dry continental climate, dry spells and droughts are a common feature, but their occurrence is unpredictable. While normal precipitation is low, its variation from year to year is the critical factor. For instance, annual precipitation well below the normal amount leads to dry spells and, in extreme cases, to droughts. For that reason, adequate rainfall in a dry continental world is a highly valued commodity. Two factors, one natural and the other human, are placing more and more pressure on this water resource. First, climate warming results in slightly higher average monthly temperatures and these higher temperatures are increasing the evapotranspiration rate. Second, as the population and industry of Western Canada grow, so does the demand for its scarce water supplies.

Within Western Canada, the Dry Belt, the so-called Palliser's Triangle, represents the most vulnerable area for drought (Figure 8.4 and Vignette 8.2). While average annual precipitation in the Dry Belt is the lowest in Western Canada—from place to place, precipitation varies from 250 to 300 mm per year—below-average precipitation results in so-called "dry years." A series of dry years in the 1930s resulted in the disastrous Dust Bowl, with the longest-lasting drought conditions occurring in the Dry Belt, driving thousands of homesteaders off the land. The most recent dry spell extended from 1999 to 2002. By the summer of 2002, the cumulative effect of several dry years left reserves of soil moisture extremely low. With insufficient precipitation and low soil moisture, farmers in Saskatchewan, Manitoba, and Alberta saw their crops and hay fields fail, leaving livestock without feed or water. From 2003 to 2012, annual precipitation has rebounded, creating a decade of "wet years." When arid conditions will return is unknown, but in a dry continental climate, a spell of adequate precipitation for 10 years or so is often followed by several years of

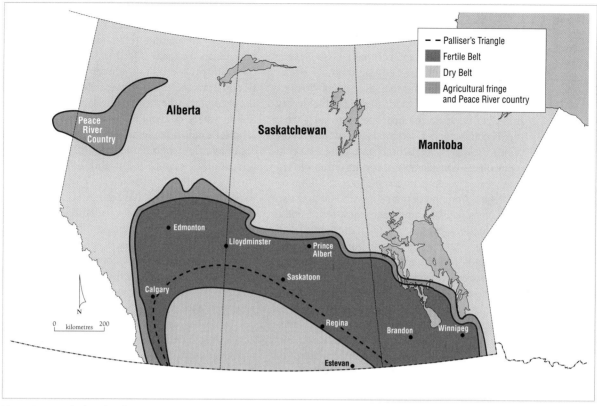

Figure 8.4 Agricultural regions in Western Canada
Farming in the Prairies can be divided into three areas: the Fertile Belt, the Dry Belt, and the agricultural fringe and Peace River country. Each region has a different type of agriculture because of variations in physical geography.

below-average precipitation. From a historical perspective, map-makers in the late nineteenth century still labelled the Great Plains of the United States as the "Great American Desert." American settlers following the Oregon Trail had no thoughts of stopping in the Great Plains but pushed across the Rocky Mountains to a more favourable climate for agriculture along the Pacific coast. The Palliser Expedition was charged with determining if the Canadian Prairies had the potential for agricultural settlement. Palliser's conclusion was mixed—short-grass lands were not suitable for farming but long-grass lands could sustain the tilling and support agriculture (Vignette 8.2).

One of the major threats to the environment is posed by the extraction of bitumen from the Alberta oil sands. With one of the world's largest deposits of oil sands, Alberta currently is benefiting from the oil sands. Yet, this mining operation poses three major environmental challenges to Alberta, Canada, and the world. First, the release of greenhouse gases to the atmosphere is among the largest in Canada. Second, open-pit mining has created a scarred industrial landscape and the reclamation process, if successful, will take time and money. Third, separating the oil from the bitumen requires large amounts of water and heat. The resulting toxic waste is drained into huge tailing ponds. While each challenge has serious consequences for Alberta and Canada, tailing ponds deserve special attention.

Alberta's oil sands industry produces 1.8 billion litres of toxic water each day. The problem facing industry is what to do with this vast quantity of non-renewable water. Industry's solution is tailing ponds, but these are not without problems. First, since the toxic waters cannot be released into the local rivers and lakes, they must be stored in large ponds for an indefinite time. Second, the amount of toxic water is increasing every day, thus forcing companies either to increase the size of existing tailing ponds or create new ones. Third, leakage from these ponds has a negative effect on the landscape, groundwater, and surface waters, including the Athabasca River. Native communities downstream from the tar sands development, notably at

THINK ABOUT IT
Oil from Alberta is a critical element in US geopolitical strategy, which calls for reducing oil imports from the Middle East. Yet, US environmentalists do not want imports of "dirty oil" from Alberta. Do you think the environmentalists will win the day?

Fort Chipewyan, have experienced health consequences, including unusually high rates of cancer. Since 1967 when tailing ponds were first established, no pond has been reclaimed. Over 40 years later, the geographic extent of these lake-sized ponds is enormous and they pose an obvious example of "dirty oil." Towards the end of the twenty-first century, when the first open-pit mines close, the oil companies are required to turn these tailing ponds back to their "natural" state. Since the companies have considerable scope in terms of "natural" state, two approaches are favoured. One option is to cover the depression with soil and then plant trees, bushes, and native grasses. Another less expensive option is to convert the tailing ponds into large "natural" lakes by burying the toxic sludge into deep pits and covering it with layers of earth, topped off with fresh water that will form a lake. These reclamation schemes for the toxic tailing ponds are experimental and pilot projects are pushing the oil companies into uncharted waters.

Western Canada faces other environmental challenges. One is the growing concern about the ever-increasing drawing of water from the South and North Saskatchewan rivers for agricultural, industrial, and urban uses. With the glaciers in the Rocky Mountains retreating, future supplies may be in jeopardy. Oddly, Manitoba faces the opposite problem—the annual threat of spring floods from the north-flowing Red River. Spring thaw comes earlier in North Dakota and, as the rising waters of the Red River reach Manitoba, river ice dams often form, causing the rising waters to overflow their banks into the extremely flat terrain of the Manitoba lowland. Efforts to divert the flood waters from reaching Winnipeg by the Red River Floodway have been successful, but other communities and farmland in the Red River Valley remain vulnerable. The extent of the April 2009 Red River flood is shown in photo 2.11. Two years later, both the Red and Assiniboine rivers breached their banks, suggesting that this catastrophic event may be occurring more frequently.

In Chapter 2, "Extreme Weather Events," page 52, the Red River Valley, spring floods, and the diversion of water around Winnipeg by the Red River Floodway are discussed.

Photo 8.4
Pincher Creek, located near the foothills of the Rockies in southwest Alberta, is famous for its livestock industry. More recently, Pincher Creek has become the site of wind energy production, helping to make Western Canada the leading region for wind-produced electricity in Canada. Powerful chinook winds from the Rocky Mountains make southwest Alberta a particularly attractive site for wind turbines.

Another challenge is the need to remove the radioactive wastes from abandoned uranium mines near Lake Athabasca before the radioactive waste seeps into the lake and eventually spreads throughout the Mackenzie River system. For years, these dangerous wastes were recognized as a serious hazard to the environment but a decommissioning agreement between the federal and Saskatchewan governments was stalled over the sharing of the cost of cleanup. In fact, estimating cleanup costs was difficult and actual costs may exceed the estimate. In 2007, the federal and provincial governments announced an agreement to share equally the estimated cost of $24.6 million. The Saskatchewan Research Council (SRC) is responsible for this multi-million dollar reclamation project, which involves assessing and reclaiming Gunnar Uranium Mine and Mill site, Lorado Uranium Mill site, and 36 satellite mine sites in northern Saskatchewan. By 2011, the assessment was completed and reclamation work began (SRC, 2012).

Western Canada's Historical Geography

Western Canada's history began long before the three Prairie provinces became part of Canada. In fact, Western Canada's recorded history goes back to the fur-trading days. Beginning in 1670, the Hudson's Bay Company (HBC) administered for 200 years much of Canada's western interior. This area was part of Rupert's Land (all the land draining into Hudson Bay). In 1821, when the company merged with its rival, the North West Company, the HBC acquired control over more land, known as the North-Western Territory (lands draining into the Arctic Ocean). Before 1870, when Canada was ceded these lands by the British government (Figure 3.4), the HBC used this vast territory exclusively for the fur trade.

The land in Western Canada began to be used for purposes other than fur trading at the beginning of the nineteenth century. In 1810, Lord Selkirk, a Scots nobleman who was concerned with the plight of poor Scottish crofters (tenants) evicted from their small holdings, acquired land in the Red River Valley from the Hudson's Bay Company. The first Scottish settlers arrived in 1812 to form an agricultural settlement near Fort Garry, the principal HBC trading post in the region. This became known as the Red River Settlement. Selkirk's settlers, however, faced an unfamiliar and harsh environment and had great trouble establishing an agricultural colony. Over the years, many gave up and left for Upper Canada and the United States.

At the same time, many former officers and servants of the Hudson's Bay Company, along with their Indian wives and children, settled at Fort Garry. In addition to these English-speaking people were the French-speaking Métis who had worked for the North West Company. Because the Métis were Catholic and spoke French, they formed a separate cultural group within the settlement. After the consolidation of the Hudson's Bay Company and the North West Company, many who worked for the North West Company were no longer employed by the new company. Many Métis, particularly those who were French-speaking, settled at Red River, where they turned their attention to subsistence farming, freighting, and buffalo hunting.

Desert or Arable Land?

By the middle of the nineteenth century, little was known about the suitability of the Canadian Prairies for settlement. What was known came from fur traders and explorers. South of the border, American explorers had dubbed Montana part of "the Great American Desert." Did that desert extend into the Interior Plains of British North America?

The British government was concerned about the political viability of these lands managed by the HBC. Without settlement, the British hold on these lands was tenuous, as the loss of the Oregon Territory in 1846 demonstrated. Was history to repeat itself? American settlers had begun to occupy land in the Dakota Territory (land west of the Mississippi River in what is now North Dakota). By 1854, a railway stretched across the United States from New York to St Paul on the Mississippi River and, after 1863, American settlers began to occupy the upper reaches of the Red River Valley. How long would it be before American settlers began to look northward to the unoccupied lands of British North America?

The events that led to the loss of the Oregon Territory in 1846 are considered in Chapter 3, "National Boundaries," page 82, and Chapter 7, "British Columbia's Historical Geography," page 277.

Action was needed. In 1857, the British government and the Royal Geographical Society sponsored an expedition into the Canadian West.

THINK ABOUT IT
Do you think that Palliser's original assessment of the semi-arid areas of Western Canada as being unsuitable for agricultural settlement is less valid today because of innovations in farming such as continuous cropping?

THINK ABOUT IT
If the Canadian Pacific Railway had existed in 1869, do you think negotiations between Riel's followers and the government of Canada would have taken place?

Their central task was to determine the suitability of the Canadian West for agricultural settlement. John Palliser, an explorer, led the British North American Exploring Expedition. After his party arrived from England, they quickly travelled by rail from New York to St Paul and then by steamboat to the Red River Colony. They travelled by horseback from Fort Garry across the Canadian Prairies to the Rocky Mountains. Palliser reported that fertile land in much of the western interior existed, but that the land in southern Alberta and Saskatchewan near the border with Montana was "treeless" and therefore far too dry for farming (Vignette 8.2). Palliser believed the **Great American Desert** label that American explorers had applied to the Great Plains extended into the grasslands of southern Alberta and Saskatchewan. Named after him, this semi-arid area is now known as Palliser's Triangle. Another expedition in 1858, organized in Canada West (Ontario) and led by Henry Hind, a geologist and naturalist, confirmed that the parkland (the natural vegetation zone between the grasslands and the boreal forest) offered the best land for agricultural settlement.

During the negotiations with Britain over Confederation, the subject of the annexation of Rupert's Land into the Dominion of Canada arose, and provision was made in the British North America Act for its admission into Canada. In 1869, the Hudson's Bay Company signed the deed of transfer, surrendering to Great Britain its chartered territory for £300,000—with the exception of the lands surrounding its posts and about 1,133,160 ha of farmland. In 1870, Great Britain transferred Rupert's Land to Canada.

In 1869, Canada's new government sent surveyors into the Red River Settlement to prepare a land registry system for the expected influx of settlers. In this **Dominion Land Survey**, the surveyors employed a rectangular grid system known as a township survey. A township consisted of 93 km^2, or 36 sections. Each section was subdivided into four quarter sections of 65 ha each. Each **homesteader** would receive a quarter section of land and would be required to till the land and build a house on that section. The opening of Western Canada for agricultural settlement officially began in 1870 and continued until 1914, when World War I halted the influx of European immigrants into Western Canada. Following World War I, veterans were encouraged to establish homesteads, especially in the Peace River country, where some arable land remained.

The key attraction for homesteaders in Western Canada was "free" land, but many settlers, especially those familiar with the mild climate and treed landscape of Western Europe, were ill-prepared for the harsh continental climate and the absence of forests. Unlike the agricultural lands of southern Ontario and Québec, arable land in Western Canada lies much further north. For example, Central Canada's arable lands fall within 42° N and 47° N while Western Canada's are between 49° N and 56° N. Hence, a shorter growing season in the Canadian Prairies affects selection of crops and chances of a successful harvest (Akinremi et al., 2001). Added to the disadvantage of a short growing season were grasshoppers, hail, and untimely frosts, and homesteaders quickly learned that a July bumper crop can turn into a September crop failure. From these experiences of farming in such a physical environment, farmers coined the term "Next Year Country" (Vignette 8.3).

VIGNETTE 8.3

"Next Year Country"

The dry continental climate in Western Canada makes farming a risky business. Prairie farmers often describe the land as "Next Year Country." The meaning behind these words is simple: Our crops did poorly this year, but we hope that they will do better next year. Often the reason for low yields is insufficient precipitation during the growing period. However, farmers face a whole range of natural hazards: summer frosts, which occur when a cold air mass slides unimpeded from the Canadian Arctic to Western Canada; hail and early snowfall; pests, such as grasshoppers; and diseases, such as stem rust. All of these hazards can have devastating effects—from delaying a harvest or lowering the grade of wheat, to reducing a yield or destroying an entire crop.

But What about the Original Inhabitants?

In the nineteenth century, Plains Indians and Métis formed the population of Western Canada. Both depended on the buffalo and the fur trade. In the last half of the century, commercial hunting of buffalo in the United States and Canada for the buffalo robe trade marked the end of the great buffalo herds. With the virtual extinction of the buffalo, the Plains Indians (Sarcee, Blood, Peigan, Stoney, Plains Cree, Nakota, Lakota, Blackfoot, and Saulteaux) could no longer support themselves. The transfer of Hudson's Bay lands to Canada, coupled with Ottawa's plan to settle the arable lands of Western Canada, meant that the only option for Plains Indians was to sign treaties and live on reserves (Figure 3.10).[1] The Métis were also confronted by the impact of these historic changes on their way of life. Yet, because they were not semi-nomadic like the Plains Indians and, instead, formed an organized settlement at Red River that was remote from Canada, the Métis were more able to resist Canada's desire to settle the Prairies with farmers.

The rebellions of 1869–70 and 1885 are examined in Chapter 3; see page 104 and page 106.

In 1867, the population at Red River was nearly 12,000, mostly Métis. The arrival of land surveyors and settlers led the Métis, under the leadership of Louis Riel, to mount the Red River Rebellion in 1869. The Métis wanted to negotiate the terms of entry into Canada from a position of strength—that is, as a government—and they obtained major concessions from Ottawa: guaranteed ownership of land, recognition of the French language, and permission to maintain Roman Catholic schools. In 1870, the fur-trading district of Assiniboia became the province of Manitoba. But the rebels' victory was hollow. First, Ottawa sent troops to exert Canada's control over the new province, forcing Riel and his followers to flee. Second, settlers began to pour into Manitoba, changing the demographic balance of power and overwhelming the Métis community.

Many Métis left the colony to search for a new place to settle in the Canadian West. Such a place was Batoche, just north of the site where Saskatoon now is situated. Within 15 years, however, settlers would again encroach on the Métis agricultural settlement. In 1885, as before, the Métis, led by Louis Riel, rebelled. This time, the Canadian militia defeated the Métis at the Battle of Batoche and Riel was captured, found guilty of treason, and hanged.

The experiences of Indian tribes during the early period of western settlement were somewhat different. Tribes such as the Blackfoot had roamed across the Canadian Prairies and the northern Great Plains of the United States long before the arrival of European explorers, fur traders, and settlers. The tribes were semi-nomadic and hunted buffalo. By the 1870s, the buffalo had virtually disappeared from the Prairies, leaving the Plains Indians destitute. They had little choice but to sign treaties with the federal government. Between 1873 and 1876, all the tribes (except for three Cree chiefs—Big Bear, Little Pine, and Lucky Man—and their followers) signed numbered treaties in exchange for reserves, cash gratuities, annual payments in perpetuity, the promise of educational and agricultural assistance, and the right to hunt and fish on Crown land until such land was required for other purposes. In 1882, impending starvation for his people also forced Big Bear to accept Treaty No. 6. Over the next few years, however, the Cree sought other concessions from the federal government. When these efforts failed, Cree warriors supported the doomed Métis rebellion in 1885 by attacking several settlements, including Fort Pitt, the Hudson's Bay post on the North Saskatchewan River near the present-day Alberta–Saskatchewan border.[2]

Treaties with Ottawa offered the Indians prospects for survival and time to find a place in a new economy, but the treaties also made them wards of the Crown. Living on reserves, Indians were isolated from the evolving Canadian society and became increasingly dependent on the federal government. Further north, the Woodland Cree and Dene (Chipewyan) tribes who lived in the boreal forest were not as affected by the encroachment of western settlers. Although they, too, signed treaties, these northern Indians continued their migratory hunting and trapping lifestyle well into the next century. In the 1950s, their dependency on Ottawa grew with the demise of the fur trade and their subsequent relocation to settlements.[3]

Canadian Pacific Railway

Once treaties ensured the peaceful availability of land for homesteading, the next steps were a land

Photo 8.5
The Métis search for a place within Confederation began with armed resistance in 1869–70 and 1885. Both rebellions were led by Louis Riel. After losing the Battle of Batoche, Riel surrendered to the Canadian forces. He was tried and convicted of high treason, and on 16 November 1885 Riel was hanged as a traitor. He remains a hero to the Métis to this day.

markets. For instance, in 1870, the new province of Manitoba was linked by steamboat to the rail centre of Fargo, North Dakota.

British and Canadian companies were not interested in a risky railway construction project across Canada unless they could obtain substantial financial assistance from Ottawa. Two reasons accounted for their lack of interest: the Canadian Shield and the Cordillera were two formidable (and therefore costly) barriers to overcome in building a railroad. Indeed, physical geography posed a much greater challenge to Canadian railway builders than to their American counterparts. In 1881, Ottawa announced generous terms: the Canadian Pacific Railway Company was awarded a charter, whereby the company received $25 million from the federal government, 1,000 km of existing railway lines in eastern Canada owned by the federal government, and over 10 million ha of prairie land in alternate square-mile sections on both sides of the railway to a maximum depth of 39 km. The terms were successful—the Canadian Pacific Railway was completed in 1885. As a result, the new Dominion achieved four important nation-forming goals:

- An east–west transportation link united Canada from coast to coast.
- The vast territory west of the Red River Valley was secured for Canada.
- The Canadian Prairies could be settled.
- A rail-based transportation system to eastern ports could be used to ship farm products to the world's major grain market in Great Britain and other European countries.

survey in the form of townships and a transcontinental railway. In fact, Prime Minister Macdonald's vision of Canada extending from the Atlantic to the Pacific hinged on a transcontinental railway. The third American transcontinental railway, the Northern Pacific Railway, was just south of the forty-ninth parallel. Without a Canadian counterpart, Ottawa feared that the West would be lost to the Americans. Even if Canada could retain its western territories in the absence of a Canadian transcontinental railway, the north–south transportation pull exerted by the American railways would prevent Ontario's fledgling industrial core from reaching the market in Western Canada, and western settlers would be unable to ship their products to eastern

Settlement of the Land

The settling of Western Canada by Europeans marks one of the world's great migrations and the transformation of the Prairies into an agricultural resource frontier. Under the Dominion Land Act that Ottawa had passed in 1872, homesteaders were promised "cheap" land in Manitoba—by building a house and cultivating some of the land, they could obtain 65 ha of land for only $10. Following 1872, an influx of prospective homesteaders began arriving, most coming from Ontario and, to a lesser degree, from the Maritimes, Québec, and the United States. However, settlement did not occur west of the Red River Valley until the Canadian Pacific Railway was completed. Then, homesteaders began to occupy lands in

Saskatchewan and Alberta. The first wave of homesteaders came from Ontario, Great Britain, and the United States. The second wave came from continental Europe.

By 1896, the federal government sought to increase immigration by promoting Western Canada in Great Britain and Europe as the last agricultural frontier in North America. The Canadian government initiated an aggressive campaign, administered by Clifford Sifton, Minister of the Interior, to lure more settlers to the Canadian West. Thousands of posters, pamphlets, and advertisements were sent to and distributed in Europe and the United States to promote free homesteads and assisted passages. Prior to 1896, most immigrants came from the British Isles or the United States—these were "desirable" immigrants. Sifton's campaign, however, cast a wider net to areas of Central and Eastern Europe that were not English-speaking and therefore provided "less desirable" immigrants. The strategy generated considerable controversy among some English-speaking Canadians who believed in the racial superiority of British people.

Nevertheless, Clifford Sifton's efforts paid off. At the end of the 1880s, the Canadian Prairies had few settlers beyond Manitoba, and most of them had taken land near the Canadian Pacific Railway line. Following the recruitment campaign, a flood of settlers arrived and the land was quickly occupied. Thus began the great migration to Western Canada. After 1896, the majority of settlers—about 2 million—were Central or Eastern Europeans from Germany, Russia, and Ukraine. This large influx of primarily non-English-speaking immigrants led to a quite different cultural makeup in Western Canada from that in Central Canada, where the French and English dominated. Within a remarkably short span of time, cultural and linguistic assimilation had forged a non-British but English-speaking society from the sons and daughters of these immigrants. Some, however, kept separate. For instance, Doukhobor settlers were Russian-speaking peasants whose adherence to communal living made adjustment and acceptance difficult if not impossible. By 1905, Alberta and Saskatchewan had sufficient populations to warrant provincial status. By the outbreak of World War I, the region of Western Canada was settled.

 See Chapter 3, "Sifton Widens the Net," page 108, for an examination of the plight of this pacifist religious sect.

The decision to build the CPR along a southern route (from Winnipeg to Regina to Calgary) meant that much of the land opened to homesteaders was in the driest part of Western Canada.

Life was not easy for homesteaders. Many were ill-prepared for farming, let alone farming in a dry continental environment. Securing supplies of wood and water often posed a problem. Those settlers who could not afford to import lumber were forced to live in sod houses and burn buffalo chips and cow dung for heat. While many members of ethnic and religious groups settled together in the same area, forming communities, isolation still posed a problem for many. The land survey system encouraged a dispersed rural population, with individual farmsteads rather than rural villages. As a result, farm families sometimes did not visit the nearest town or see their neighbours for weeks or even months. Such isolation was particularly hard on farm wives. In spite of these difficulties, the land was settled, towns sprang up, and institutions were created to meet the local and regional needs. In short, a new society was in the making.

By 1921, there were over 250,000 farms in Western Canada. Homesteaders now had to turn to the Peace River country for arable land. The Peace River country, part of the high Alberta Plain, is much further north and its short growing season makes agriculture risky. Even so, this area has a climate and soils that allow for mixed farming (a combination of grain and hay crops with livestock). The section of the Rocky Mountains located to the west of the Peace River country is considerably lower and thus allows more rainfall from Pacific air masses to reach the area. The final settlement of the Peace River country took place after World War I, when returning soldiers were encouraged to settle there.

Geographic Challenges Facing Homesteaders

Few had experienced growing crops at such high latitudes under a dry continental climate. Drought, hail, and frost threatened crops. Farmers situated in Palliser's Triangle—the semi-arid Dry Belt of southern Alberta and Saskatchewan—were most vulnerable to dry spells and crop failures. Innovations were sought that would lessen this natural threat. The strategy of dryland farming, where part of the land is left in **summer fallow**

> **THINK ABOUT IT**
> Which physiographic regions posed the greatest challenge to John A. Macdonald's dream of a railway stretching to the Pacific Ocean?

> **THINK ABOUT IT**
> Do you think that homesteaders from the steppes of the Russian Empire were better prepared for life on the Canadian Prairies than those from Great Britain?

each year, reduced the risk of crop failure. In this way, sufficient soil moisture is accumulated over the year, allowing for the seeding of the land every other year.[4] Another challenge is hail, which is associated with severe thunderstorms. Hailstones can reach golf-ball size (and even larger), and can flatten crops within minutes. While hail can occur anywhere, its frequency seems random, except for areas of relief such as the Alberta foothills. Farmers can purchase hail insurance. A third challenge is the short growing season, which leaves crops vulnerable to frost damage. In fact, freezing temperatures have occurred in every month of the summer! Fortunately, most years have frost-free summers. Spatially, the frequency of summer frost is greatest in the higher latitudes where the northern edge of the grain belt is situated. For farmers in the Fertile and Dry Belts, innovation came to the rescue. In 1910, a new strain, Marquis wheat, was developed. Its shorter maturation period overcame the threat of frost.[5] Marquis wheat also extended the growing area for wheat in the Prairies. By 1920, Marquis wheat was the most popular spring wheat in Western Canada. While the Peace River country, located between 54° and 56° N, benefits from long summer days, its northern location and the invasion of cold air masses from the Territorial North make it particularly subject to frost.

Geography placed the Prairies in the heart of North America. Farmers struggled with the problem of getting their crops to market. The high cost of shipping grain long distances by rail to reach ocean ports in Québec and Nova Scotia and then by ocean vessels left little profit for the farmer. An alternative and shorter route to Europe, the Hudson Bay Railway, was constructed with a rail line extending from The Pas, Manitoba, to Churchill, Manitoba (Vignette 8.4; see Figure 8.1). Unfortunately, this route proved ineffective because of the high cost of marine insurance.

"Have" and "have-not" provinces are considered in Chapter 3, "The Centralist/Decentralist Faultline," page 89.

Western Canada Today

The economy and population of Western Canada, led by Alberta and to a lesser degree by Saskatchewan, continue to grow. While agriculture remains a basic element, most wealth is created by resource industries, especially from oil sands, natural gas, potash, and uranium, and indirectly by the manufacturing and service industries that cater to resource industries. These resource industries require huge capital investments and more and more of this investment in energy projects is coming from Asian countries, led by China. In turn, an increasing volume of energy exports, such as coal, natural gas, and oil, is expected to reach these Asian countries once the necessary transportation infrastructure is completed.

The region is diversifying with a growing trend towards processing its agricultural products and widening its mineral production. However, only a small portion of the bitumen is processed in Alberta—and that remains a sore point with Albertans but not with the oil sands companies. Western Canada has a relatively high proportion of Aboriginal people, a reflection of the region's history, and this demographic reality has consequences

---

VIGNETTE 8.4

The Hudson Bay Railway

The search for a shorter route to European grain markets caused farmers to call for a railway to Hudson Bay where ocean ships could take their grain to British and other European markets. After all, why not follow the lead of the Hudson's Bay Company, which used this shorter route to send its furs to London, England? Building a railway, however, is a more onerous and major task in some ways than it is for individual trappers and traders to follow existing water routes by canoe. The Hudson Bay Railway had to cross difficult terrain, including the muskeg and permafrost found in the Hudson Bay Lowlands. Completed in 1929, the Hudson Bay Railway stretches from The Pas to Churchill. The railway was never a success and is now operated by an American company, Omni TRAX.

for the region and its future; as well, the region has the lowest unemployment rate in Canada, which underlies its robust economy. Added to these factors, innovations within the knowledge-based economy are accelerating economic change, especially in the field of biotechnology. Most research into plant breeding and seed development takes place in public institutions, though leading global companies such as Bayer and Monsanto now have research facilities in the region. Still, innovative activities extend to the more exotic areas, such as blimp-like transportation for mining and Aboriginal communities in remote areas of northern Canada (Vignette 8.5).

The role of research parks at universities is presented in Chapter 4, "The Information Society," page 160.

Alberta's economy, driven by the oil and gas industry, has become the most diversified in Western Canada. Saskatchewan and Manitoba have also diversified but, lacking Alberta's huge petroleum reserves, their economic growth is occurring at a much slower rate. The 1995 cancellation of the **Crow Benefit**, a transportation subsidy that allowed farmers to ship their grain by rail at a reduced cost, initiated massive changes in the agricultural sector, especially for the livestock industry. With the transportation subsidy gone, the trend is to process more agricultural products within Western Canada and to ship higher-valued processed products in containers by rail and then by ship to world markets. Canola provides such an example with crushing mills producing oil for global markets.

Alberta, the economic giant of the three provinces, has over half of the population in Western Canada and accounts for about 63 per cent of the region's GDP (Table 8.1). Each province has much natural wealth. Saskatchewan, for example, has

VIGNETTE 8.5

Reaching for the Sky

Resource development in remote areas is hampered by the high cost of air transportation. Winter roads, such as those to the diamond mines in the Territorial North and to Aboriginal communities, provide one solution but only a seasonal one. Global warming may reduce the winter period and thus time during which winter roads are viable. In spring 2010, ice roads connecting several First Nations settlements in northern Manitoba became impassable much earlier than normal, making it impossible for trucks to bring much-needed supplies. Could blimps provide a solution? The concept of blimps has long been touted as a possible solution to northern transportation and Calgary-based SkyHook International Inc. may have the answer with its heavy-lift rotorcraft—a blimp with four helicopter-like rotors underneath that has the capacity to lift a 40-tonne load. In 2008, Boeing was awarded a contract to design and build two prototypes by 2014 (photo 8.6). A competitor to Boeing is Hybrid Air Vehicles (HAV) based in the United Kingdom. Discovery Air based in Yellowknife has contracted this UK firm to develop an airship designed for northern conditions.

Photo 8.6
Artist's conception of the "SkyHook" being developed by SkyHook International of Calgary.

Table 8.1 Basic Statistics for Western Canada, 2011

Province	Population (000s)	GDP (%)	French (%)	Farm Area (acres)	Farm Area of Canada (%)
Alberta	3,645	16.8	1.9	50.5	31.5
Saskatchewan	1,033	4.2	1.6	61.6	38.5
Manitoba	1,208	3.2	3.5	18.0	11.2
Western Canada	5,886	24.2	2.2	130.1	81.2
Canada	33,477	100.0	100.0	160.2	100.0

Sources: Statistics Canada, 2013a; 2012a, 2012b, 2012c.

most of the cropland and is the leading producer of potash and uranium. In addition to having the richest agricultural land in the West, Manitoba produces vast amounts of hydroelectric power from the Nelson River. Even so, Alberta holds the trump resource card—oil and gas.[6] Indeed, by 2015 Alberta is forecast to rank as the fifth largest producer of oil in the world (Grauman, 2006).

Back in 1972, when a barrel of oil was worth $2 on the world spot market, Alberta oil was not valuable enough to dominate the western economy. Since then, prices, while fluctuating, have risen, generating great wealth for Alberta. The high point took place in July 2008 when a barrel of crude oil was worth nearly $150. Prices fell sharply during the 2008–9 global meltdown, and by December 2008 the price was $35/barrel. By December 2012, the price had recovered to over $85/barrel.

Though Western Canada exports both primary and processed products to other countries and has recently increased its exports to Pacific Rim countries, the petroleum industry is closely tied to the US market. However, two developments have placed access to the US market in jeopardy. First, in November 2011, President Obama delayed a decision on the Keystone XL Pipeline until 2013, and in early 2013 was under heavy pressure to move ahead on this major project and under at least equal pressure to quash it for good. Second, large discoveries of oil through the process of **fracking** may turn the US into an oil exporter. In response, Ottawa called on the oil industry to diversify its exports of Alberta oil. Two possibilities exist. One is to build the Northern Gateway Pipeline to Kitimat for export to Asian countries. The other is to twin the Trans Mountain pipeline that terminates in Burnaby and use that new line for transporting bitumen where it would be loaded onto supertankers.

World prices for commodities hold the key to economic growth in Western Canada. Unfortunately, global prices rise and fall with the world business cycle. Except for oil, **primary prices** (prices for primary resources) had not been robust until China became a major importer of a wide range of primary goods.

Economic Structure

Given the importance of agriculture and natural resources, it is not surprising that employment by industrial sector in Western Canada reveals the prominence of primary economic activities. While Table 8.2 presents only a generalized picture of the western economy, the importance of the primary sector is undeniable. For instance, the percentage of people employed in this sector (9.3 per cent) is over five times the figure for Canada's principal industrial core region—Ontario. Yet, the percentage of primary workers has decreased both absolutely and relatively from 2005 to 2011 (Table 8.2). This trend follows the same one found in the other regions, namely, that the total percentage of workers in and relative importance of the primary sector continues to decline. The tertiary sector showed the greatest absolute growth but its percentage of the labour force remained unchanged at 73.5 per cent. Only the secondary sector increased its share of the workforce over this period—from 16.5 per cent in 2005 to 17.2 per cent in 2011. The expansion of the oil sands and potash industries supports a robust mining-oriented manufacturing sector in Edmonton, Calgary, and Saskatoon as well as in industrial centres in Ontario and Québec. This expansion results in a strong construction industry and, together with the highly specialized manufacturing for that industry, accounts for the relatively high performance of the secondary sector.

Table 8.2 **Employment by Industrial Sector, Western Canada, 2005 and 2011**

Economic Sector	2005		2011		Difference, 2005 to 2011 (000s)
	Workers (000s)	Workers (%)	Workers (000s)	Workers (%)	
Primary	284.3	10.0	300.6	9.3	16.3
Secondary	468.6	16.5	557.9	17.2	89.3
Tertiary	2,095.3	73.5	2,386.0	73.5	290.7
Total	2,848.2	100.0	3,244.5	100.0	396.3

Source: Statistics Canada (2009b, 2012d).

Scientific breakthroughs and knowledge-based innovations help explain the continued growth in the non-renewable oil and gas industry. For example, oil and gas operations use horizontal drilling, which, when combined with hydraulic fracturing, has unlocked oil from the Bakken shale formation in the Interior Plains of southern Saskatchewan. Carbon capture and storage is another complex and innovative undertaking. Several companies and universities are engaged in perfecting this technology. Saskatchewan, with its experience in **carbon sequestration** at its Weyburn oil field, is a world leader in this fledgling technology, and the University of Regina is continuing its research in this area. Alberta has provided funds to large coal and oil sands companies to develop carbon capture and storage technology that will meet their specific needs (Vignette 8.6).

THINK ABOUT IT
Is it morally acceptable to consider trade-offs between jobs and the environment? For instance, would you favour exporting more bitumen to upgraders in the US, where jobs would be created and greenhouse emissions would occur, rather than employ people in Alberta and refine the bitumen there? Or is this the wrong question, and is there another alternative?

VIGNETTE 8.6

Technological Gamble: Carbon Capture and Storage

Climate change is arguably the most serious environmental issue society faces. The heavy reliance on coal and oil sands industries in Western Canada has placed these industries, especially companies developing the oil sands, under the spotlight. In response, these firms are under pressure to reduce their carbon footprints and to deal with the "dirty oil" label. One example is Shell's Quest Project. The Quest Project focuses on reducing greenhouse gas emissions from Shell's Scotford Upgrader at Fort Saskatchewan (just east of Edmonton). Once the project is completed in 2013, these emissions will be sent by pipeline to a site 80 kilometres north where they will be injected 2.3 kilometres below the surface. Over a 25-year period, the cost of capture and storage is estimated at $72 per tonne of carbon stored. However, without funding from Ottawa and Alberta and Saskatchewan, the project would not have been approved by Shell. At an estimated $1.35 billion cost, the company obtained 65 per cent in public subsidies to launch the project in 2012. Most ($745 million) came from Alberta's Carbon Capture and Storage Fund while Ottawa contributed $120 million from Ottawa's **Clean Energy Fund**. In addition, future payments from Alberta's carbon legislation, which assigns a $15 refund per tonne of sequestered carbon, may reduce the real cost to the company to less than 15 per cent (under $200 million) of the Quest Project (VanderKlippe, 2012).

Key Topic: Agriculture

Agriculture remains a key economic component of Western Canada's economy. While it was the driving force behind the settlement and development of Western Canada in the late nineteenth and early twentieth centuries, agriculture in the twenty-first century, while still important, is no longer the principal engine of economic growth. Natural resource extraction, led by the oil and gas industry, has taken over that role.

Climate, especially precipitation, plays a critical role in the success or failure of agricultural endeavours. Though the dry continental climate limits options for farmers, within the narrow range of crops farmers select those crops that offer the greatest profit. In short, prices determine the type of crops. For instance, prices for canola have risen sharply over the last decade while those for wheat have remained low, causing farmers to favour canola over spring wheat. Not surprisingly, then, farmers view canola as their "money-making" crop and spring wheat as their "rotation" crop. Unfortunately for farmers, canola must be rotated with another crop to maintain yields and to prevent attacks by diseases and insects.

In addition to the lure of profitability, the face of agriculture has changed because of advances in farm practices and technology. Notably, the now well-established practice in dry land farming of **zero-tillage** allows the farmer to cultivate and seed the land in one operation into a **stubble field**, thus minimizing effort and cost as well as conserving soil moisture. As well, the farmer, by avoiding the practice of **fallowing**, has more land in operation.

The critical trends in agriculture in Western Canada are:

- larger farms, fewer farmers, and older farmers;
- changing biotechnology;
- depopulation of rural Western Canada and the demise of rural towns;
- the marketing of grain by farmers rather than by the Canadian Wheat Board.

Larger Farms, Fewer Farmers, and Older Farmers

The irreversible trend of larger farms, fewer farmers, and older farmers continued into the twenty-first century. By 2011, Statistics Canada's "Census of Agriculture" reported fewer than 100,000 farms in Western Canada, with the average farm size over 1,000 acres (Tables 8.3 and 8.4). Within the three provinces, Saskatchewan's farm size was the largest, at nearly 1,700 acres. Over the last 30 years, the average size of farms has nearly doubled and the number of farms has declined by 45 per cent—from 175,000 farms to 96,000 farms. As the number of farms declined, so did the number of farmers. Another critical factor is the average age of farmers—over 50 years of age—which raises the question: "Who will farm the land 20 years from now?" The average age of Alberta and Saskatchewan farmers was 54.2 years, while Manitoba was not far behind at 53.1 years (Statistics Canada, 2012e).

Changing Biotechnology

Technology plays a critical role in agriculture, ensuring greater yields, improved farming practices, and more efficient farming implements. Innovations

Table 8.3 **Number of Farms in Western Canada, 1971–2011**

Year	Alberta	Saskatchewan	Manitoba	Western Canada
1971	62,702	76,970	34,981	174,653
1981	58,056	67,318	29,442	154,816
1991	57,245	60,840	25,706	143,791
2001	53,652	50,598	21,031	125,281
2011	43,234	36,952	15,877	96,063
Farm Loss 1971–2011	–19,468	–40,018	–19,104	–78,590

Source: Statistics Canada (1992, 2003, 2012f: Table 1).

in the form of advanced machinery and improved seeds have played a key role in agriculture. Spring wheat, for instance, has evolved over time thanks to the efforts of plant breeding research. Red Fife was brought from Ontario to Manitoba in the late nineteenth century. While these seeds produced high yields, they required a long growing season and so crop loss due to frost was a serious shortcoming. In 1910, Marquis wheat was developed by federal plant breeders and these seeds had a shorter maturation period. Canola represents a more recent innovation. It is an achievement of Canada's plant breeding community. The initial breakthrough took place at the University of Saskatchewan, where researchers were able to alter rapeseed into a superior product called canola. Today, Western Canada is the global centre for canola research with the major effort taking place in research facilities in Saskatoon, where experimental work is conducted by private seed developers, federal researchers, and university plant breeders. Their combined efforts have increased yields, reduced plant diseases, and improved the quality of the final product. Canola offers another advantage over wheat and other grain crops: farmers can greatly reduce their shipping costs by trucking canola to local processing plants. One of the private seed developers, the global giant Monsanto, developed and patented a canola plant that is resistant to their Roundup herbicide (Monsanto, 2012). With these seeds, farmers are able to control weed competition and obtain high yields, but, besides the cost of the seed, farmers must sign a formal agreement with Monsanto that specifies that new seed must be purchased from the company every year. Farmers commonly collect seeds from last year's crop to save the cost of purchasing new seed. Not so with the Monsanto product. The company has taken farmers to court who have not paid Monsanto for such seed, arguing that seeds with its modified plant cells belong to Monsanto. The most famous case in Saskatchewan, *Monsanto Canada Inc. v Schmeiser*, went to the Supreme Court of Canada. In 2004 the Court ruled in favour of Monsanto (photo 8.7).

Table 8.4 **Average Acreage of Farms in Western Canada, 1971–2011**

Year	Alberta	Saskatchewan	Manitoba
1971	790	845	543
1981	813	952	639
1991	898	1,091	743
2001	970	1,283	891
2011	1,168	1,668	1,135
Change 1971–2011	378	823	592

Source: Statistics Canada (1992, 2003, 2012f).

a homestead farm was 160 acres. By 1921, over 250,000 farms dotted the landscape of the Prairies, but by 2011 this number had dropped below 100,000. In 1921, over a million families lived on the farm or in rural villages. By 2011, most rural folks had left the land because the nature of farming had changed—mechanization of farm operations greatly reduced the need for farm labour and led to farm consolidation; larger farm units made more efficient use of the machinery. The net result was the depopulation of the rural countryside and the disappearance of many small villages. The introduction of machinery, such as self-propelled steam tractors and threshing machines, changed the way farms were run and reduced the need for farm labour. While those who pursue mixed livestock and grain farming have stayed on the land, grain farmers, known as suitcase farmers, live in cities and till their land in the summer. With fewer rural people, villages have seen their services—both private and public—drift away to larger regional centres and cities. Governments call this process "centralization of services" while private businesses react to the marketplace. The ripple effect is that many villages have been abandoned and eventually have disappeared from the landscape.

Depopulation of Rural Western Canada and the Demise of Rural Towns

At the beginning of the twentieth century, farmers worked the land with oxen and horses, planting wheat as their main cash crop. The average size of

Marketing Grain: The Demise of the Canadian Wheat Board

In the early years of agricultural settlement, grain farmers felt helpless because they had no control over the prices for their commodities. Private grain buyers and the railways, it seemed, made money,

Photo 8.7 Monsanto, the ownership of seed, and the courts
Farmer Percy Schmeiser leaves a press conference after the Supreme Court of Canada, in the 2004 case of *Monsanto Canada Inc. v. Schmeiser*, ruled that Monsanto has the right to collect fees for canola seeds Schmeiser collected. The Court determined that Monsanto's patented invention of genetically modified plant cells applies to all seeds with those cell characteristics. To many ecologists and environmentalists, global companies like Monsanto and the modified crops they produce pose a serious threat to the environment.

but not the farmers. This experience accounts for much of the bitterness that took root in Western Canada. In 1935, the answer to the plight of grain farmers came in the form of a state grain buyer, the Canadian Wheat Board. Prairie wheat farmers could only sell their crops to the Wheat Board. The twin goals of the CWB were to sell as much grain as possible at the best possible prices and to ensure that each producer got a "fair" share of the market. For some time, the Canadian Wheat Board was under attack from the United States, which claimed that it subsidized Canadian exports. For that reason, Washington placed duties on Canadian durum and spring wheat. Following a NAFTA decision in 2005, the US government was required to rescind those duties in the following year. In 2011, the Conservative government, with its free-market perspective, ended the CWB's monopolistic position. Starting with the 2012–13 crop year, farmers can still market their grain through the Board, but now it will be voluntary. The farm community is split over the ending of the Canadian Wheat Board's monopoly. Some farm organizations and large grain producers argue that individual farmers can get better prices on the open market than through the Canadian Wheat Board. Others believe that the farmers will be at the mercy of railways and international grain companies. One thing is certain: grain farmers in more remote areas are likely to lose while those close to the US border stand to win by trucking their product to US mills when prices are higher than in Canada.

Agricultural Regions

Western Canada is blessed with rich black, dark brown, and brown chernozemic soils that are well suited for growing cereal crops (Figure 8.3). Some crops, such as peas and lentils, require more moisture and are restricted to the black soil zone. Unlike canola, durum wheat thrives in hot, dry weather commonly occurring in the brown soil zone. It also requires a longer growing season than spring wheat and therefore is usually grown south of the fifty-first parallel. Consequently, the use of agricultural land varies, and three distinct agricultural regions can be identified: (1) the **Fertile Belt** (black soil associated with parkland

and long-grass natural vegetation; (2) the **Dry Belt** (brown soils with short-grass natural vegetation); and (3) the **agricultural fringe** (southern edge of the boreal forest) and the Peace River country (Figure 8.4). These sub-regions have very different growing conditions. The major factors controlling those conditions are the number of frost-free days and the soil moisture. The Fertile Belt provides the best environment for crop agriculture. The Dry Belt occupies semi-arid lands and has become a grain/livestock area. In the agricultural fringe, the short growing season encourages farmers to grow feed grains and raise livestock; in the Peace River country, farmers grow both grain and feed grain for livestock.

The Canadian Prairies are on the northern margin of crop agriculture where the length of the growing season is short. Slight variations in weather conditions have either a positive or negative impact on crops. Wheat, for example, is one of the more resilient crops and can grow in areas of low precipitation, but wheat harvests can vary widely from year to year in both size and quality due to adverse weather conditions. Because of its continental climate, temperature and precipitation vary from year to year so that the risk to the prairie farmer is high compared to that of the southern Ontario farmer. Other weather conditions affecting crop farming include late spring seeding due to cold or wet weather, summer frosts before the crop is mature, and wet weather in the fall that impedes harvesting. All of these weather conditions will reduce the quality of the harvest. Beyond weather conditions, pests, such as grasshoppers, have greatly reduced the size of the harvest in some years.

Canadian Climatic Zones are shown in Figure 2.4, page 47 and are described in Table 2.3, page 48; natural vegetation zones are shown in Figure 2.8, page 53.

The Fertile Belt

The Fertile Belt extends from southern Manitoba to the foothills of the Rocky Mountains west of Edmonton (Figure 8.4). The higher levels of soil moisture, an adequate frost-free period, and rich soils make this belt ideal for a variety of crops and livestock. The most popular crop since farmers first arrived in the West has been wheat. In recent years, however, the acreage in grain has declined, while the planting of canola and specialty crops, such as beans, field peas, and sunflowers, has increased. This change was fuelled by rising prices for these crops and low prices for wheat.

The predominance of wheat in the Fertile Belt is most evident in western Saskatchewan and eastern Alberta. In southern Manitoba and the adjacent parts of eastern Saskatchewan, there is more annual precipitation so farmers there can grow a wider variety of crops as well as keep livestock. As a result, mixed farming is common. Grain and specialty crops (canola, flax, sunflowers, and lentils) are combined with beef, pork, and poultry production. Near Winnipeg, for instance, a livestock industry has developed where cattle are fattened before shipment to meat-packing plants in Winnipeg and Ontario. Feedlots and nearby meat-packing plants are located in other parts of Western Canada, particularly in Brandon, Brooks, Calgary, Edmonton, Lethbridge, and Saskatoon. Since NAFTA, meat production has increased, especially pork, and most meat products are exported to markets in the United States.

As the urban markets grow, market gardens, dairy farms, and other specialized forms of intensive agriculture are developing near major cities in the Fertile Belt. The demand for specialty-crop products has spawned a number of smaller but very intense production units, including nurseries and greenhouses, to produce flowers and vegetables for local sale. For that reason, farm sizes are much smaller around the major cities.

THINK ABOUT IT
How would global warming affect farming? A longer, warmer growing season would be an advantage, but what about precipitation and evapotranspiration?

Photo 8.8
Foam Lake Terminal, between Yorkton and Saskatoon, is served by the Canadian Pacific Railway. The terminal is now owned by Viterra, which used to be called the Saskatchewan Wheat Pool. The export of grain to foreign markets requires an elaborate transportation and handling system. Inland terminals and elevators play a key role by serving as collection points along main railway lines. Here the grain is graded, cleaned, and weighed before being transported by railway cars to Canada's major ports.

The Dry Belt

The Dry Belt contains both cattle ranches and large grain farms. It extends from the Saskatchewan–Manitoba boundary to the southern foothills of the Rockies and north nearly to Saskatoon (Figure 8.4). However, the driest area, or heart of the Dry Belt, occupies a much smaller area, stretching southward from the South Saskatchewan River to the US border. The arid nature of the Dry Belt is due to a combination of low annual precipitation and to longer, hotter summers resulting in high evapotranspiration rates. Within the Dry Belt, feed grain and hay crops are grown to supply winter feed for the cattle ranching that dominates in this area. Cattle ranching began in this area in the 1880s. Today, ranches are large, often many times the size

VIGNETTE 8.7

The Great Sand Hills

The Great Sand Hills are situated in a semi-arid climatic zone where short-grass vegetation exists. Located in the centre of Palliser's Triangle, some sand dunes in the Great Sand Hills remain void of natural vegetation, making them subject to wind erosion, while others have been stabilized by a covering of native prairie grasses. Cacti, creeping juniper, and small shrubs like wild rose, saskatoon, chokecherry, and silver sagebrush grow in the Great Sand Hills. This area of southwestern Saskatchewan near the Alberta border covers about 1,900 km², and the hills were formed from wind action causing beach deposits of former glacial lakes to form desert-like sand dunes. Archaeologists and geologists believe that these dunes were active in the late eighteenth century due to wild grass fires and Indians using fire to drive buffalo, thus keeping the grass from encroaching into the sand dunes (Hugenholtz and Wolfe, 2005; Wolfe et al., 2007). With the settling of the land by homesteaders in the late nineteenth century, the natural vegetation slowly began to encroach on the edges of the sand dunes and thus started a stabilization process that continues today.

Photo 8.9
The Great Sand Hills of southwestern Saskatchewan.

of grain farms, because of the lower productivity of the dry land and the need for huge grazing areas to support a rotational grazing system.

Within the Dry Belt, grain farming is pursued, but the risk of crop failure is high. To conserve soil moisture and control weeds, summer fallowing was a widespread practice, but this has been largely replaced with **continuous cropping**. In Saskatchewan, for example, summer fallowing has declined from a high of 15 million acres in 1988 to 4 million in 2009 (Saskatchewan Ministry of Agriculture, 2009a). The purpose of summer fallowing is that two years of precipitation will accumulate sufficient soil moisture to germinate the seed and sustain the young wheat plant. The crop will still require summer rainfall to reach maturity. On average, one-fifth of the arable land in the Dry Belt is kept in summer fallow each year. The drawback to summer fallowing is that the ploughed (fallow) land is exposed to water and wind erosion. In a dry spring, windy weather can result in extensive loss of topsoil with huge clouds of dust stretching for many kilometres across the Prairies. Continuous cropping keeps short, stiff stalks of grain or hay remaining on a field after harvesting, thus protecting topsoil from water and wind erosion. Then, too, advances in technology have allowed for "one-pass" seeding, spraying, and fertilizing. The expense and time not spent on repeated tilling of a field reduce a farmer's costs and conserve soil moisture.

Irrigation on these semi-arid lands has provided another solution to dry conditions. The most extensive irrigation systems are in southern Alberta. In fact, nearly two-thirds of the 750,000 ha of irrigated land in Canada are located in Alberta. In the 1950s and 1960s, two major irrigation projects were developed in the dry lands of Alberta and Saskatchewan: the St Mary River Irrigation District is based on the internal storage reservoirs of the St Mary and Waterton dams in southern Alberta, and Lake Diefenbaker serves as a massive reservoir on the South Saskatchewan River. Since the two areas of irrigated land compete for the same potato market, Saskatchewan farmers are at a disadvantage because of the greater distance (shipping costs) to processing plants in southern Alberta. A second disadvantage facing Saskatchewan irrigators is that they experience a shorter growing season than farmers in southern Alberta. Consequently, their selection of crops is more limited. On the other hand,

Photo 8.10

In the dry lands of southern Alberta, water is king. Irrigation waters from the Oldman River reservoir supply valuable water to farmers who specialize in growing corn, sugar beets, potatoes, and other vegetables.

farmers in southern Alberta are able to grow corn, sugar beets, and other specialty crops that provide a high return per acre.

The Agricultural Fringe and Peace River Country

The agricultural fringe is a narrow transitional strip of forested land located just to the north of the Fertile Belt while the Peace River country lies in the Interior Plateau where the deeply entrenched Peace River flows from the Rocky Mountains and across the High Plains. Grain, livestock, and hay crops are the principal agricultural activities. Specialty crops include legumes and grass seeds. In these higher latitudes (55° to 57° N), crop agriculture is challenged by a short growing season and the threat of frost, hence hay crops are widespread.

As the last agricultural frontier, the Peace River country did not attract settlers until after World War I. By then, the best lands in the

Prairies had been settled. Two events laid the groundwork for settlement. First, the signing of Treaty No. 8 in 1899 opened large areas to agricultural settlement. Second, accessibility to export markets improved with the construction of a railway to Peace River with rail connections to Vancouver in 1915; then, in 1955, the Pacific Great Eastern Railway linked Peace River to Prince George, thus allowing wheat and other agriculture products to be shipped to world markets via Prince Rupert.

Global Demand and Higher Prices

Unlike the previous decades, global demand is outstripping supply in the early years of the twenty-first century. The result is higher prices for many crops. In the past, higher prices were a short-term phenomenon. So are these higher prices, like in the past, just a brief anomaly or do they represent an upward trend? Brian Oleson, an agribusiness professor at the University of Manitoba, believes that the higher prices are not anomalous but represent a dramatic shift to permanently higher prices for wheat and other agricultural products. Oleson states: "We are entering a new era. It's almost as if the platform [for wheat prices] has been raised and we're all dancing on a new dance floor" (Greenwood, 2007). His position is supported by the World Food Price Index, which remained relatively stable until 2003 when it rose sharply, but due to the global financial crisis the index declined for 2009 and 2010. The following year, however, the index reached an all-time high (Figure 8.5).

While agricultural prices fluctuate from year to year, the upward trend is supported by international circumstances, spelling good news for prairie farmers. High prices for grain, canola, and pork can be attributed to (1) strong and increasing demand for pork and pork feed from China, Japan, and South Korea; (2) the US biofuel industry use of corn, which has pushed grain prices up; (3) the price of canola, which remains high. Climate, too, has favoured prairie farmers. As noted earlier, precipitation on the Prairies has remained adequate for the last 10 years, resulting in higher than average yields. At the same time, weather conditions in other countries have been less favourable, and the 2012 drought in the Corn Belt of the United States greatly reduced American corn and wheat yields, pushing prices for those crops higher.

Exports for prairie agricultural products remain a critical element. For example, prices for pork depend heavily on the emerging middle class in Asian countries seeking a diet with pork as a staple. Imports of pork from Canada are increasing to China, South Korea, and Japan. And there is no turning back. As Keith Bradsher (2007) reported:

> Few things are as essential to the Chinese as their pigs. From pork spare ribs and mu shu pork to char siu bao—barbecued pork buns—pork is a staple of the Chinese diet. So in this Year of the Pig, an acute shortage of pork has been national news, as butchers raise prices almost daily and politicians scramble to respond.

Chinese officials offer several reasons for the high pig prices. The cost of animal feed has risen by one-quarter in the last year, partly because more corn is being made into ethanol and partly because more prosperous workers are eating more meat, which requires more animal feed.

The biofuel industry has become a major factor in the global agricultural world. For the European Union, the political rationale for subsidizing the biofuel industry is an attempt to reduce the emission of greenhouse gases to the atmosphere. In the case of the United States, the political case for federal financial support for the ethanol industry is to reduce dependency on foreign oil. In the US, ethanol is made from corn. But corn is also used to feed chickens, hogs, and cattle. The result is a rise in prices for meat, eggs, and dairy products and for other farm produce. The large subsidies provided by the EU and US governments to the biofuel industry may come under fire because both face immense pressure to reduce public spending and to tackle public debt. For that reason, the only dark cloud on the horizon for agricultural prices is the possible ending of subsidies for biofuel companies that would inevitably decrease their demand for corn.

Finally, the world's agricultural land is growing ever smaller as cities continue to sprawl outward—more so in China and India than in Canada. Often, this loss involves highly productive land rather than marginal lands. Of course, the counter-argument to a smaller arable land base is that technology will allow farmers to produce more on a smaller amount of land. Such an argument, however, is itself countered by the

THINK ABOUT IT
How do farmers reduce their transportation cost for canola while that option is not available to wheat?

realization that additional technological inputs have two outcomes—negative environmental consequences and higher costs for farmers.

Rural Land-Use Changes

Across the Canadian Prairies, the most dramatic land-use changes in the last 30 years were triggered by technological advances in machinery; innovations in plants and seeds; changes in cultivation practices; and higher rail transportation costs for grains due to the end of the Crow subsidy.

Since the early days of homesteaders, cultivation practices had followed an almost religious pattern of till the land one year and rest the land the next year. The practice of fallowing the land in an annual sequence was driven by the belief that soil moisture would accumulate over the fallow year and thus it would be available for the following crop year. As well, many farmers tilled the fallow land to kill weeds while others burned the stubble, making, in both cases, the bare soil susceptible to wind erosion. In fact, each spring when the westerly winds were whipped into a frenzy, the horizon would turn a brownish colour.

By the 1980s, some farmers adopted a continuous cropping system and used herbicides that keep weeds under control and inorganic fertilizers that maintained the fertility level of the soil. Eventually, a "one-pass" system was developed that involves tilling the soil, planting the seed, and applying fertilizers and herbicides in a single operation. While costs increased due to the necessary purchase of specialized farm equipment, herbicides, and fertilizers, these costs were more than offset by an increase in the amount of land under cultivation. With such evidence before their eyes, neighbouring farmers quickly adapted this new cultivation practice and, as a result, the amount of fallow land fell sharply, to an all-time low in 2012 (Figure 8.6).

In Saskatchewan, the increase in land available for cultivation is substantial, climbing from 21.1 million acres in 1981 to 35.1 million acres in 2012, a total gain of 14 million acres (Saskatchewan Ministry of Agriculture, 2012). Over the same period, the amount of land left in summer fallow dropped from 16.5 million acres to 2.6 million acres (ibid.) for a total of 13.9 million acres—which almost perfectly matches the amount of land added to production.

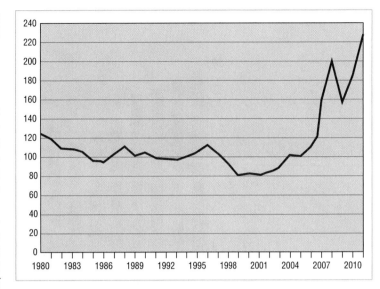

Figure 8.5 **World Food Price Index: Higher prices for short term or long term?**
Note: 2005 = 100
Source: FAO (2012).

THINK ABOUT IT
The Saskatchewan Farm Security Act restricts ownership by foreign individuals or companies of agricultural land to 10 acres. This legislation promotes family farms and rural communities rather than tenant farming controlled by absentee owners (Folk, 2011). However, the legislation has kept prices relatively low, thus tempting land speculators. Rumours that wealthy offshore investors from China are buying farmland in Saskatchewan as a speculative investment have prompted the province to investigate recent farmland sales (Coolican, 2012). Is the Saskatchewan legislation being realistic in a globalized world? As Minister of Agriculture, what would you do?

Canola: The Prairie Staple?

Grain production, particularly spring wheat, has been the prairie staple for over 100 years. Grains do well in dry conditions where other crops would fail. But two factors turned prairie farmers against spring wheat—low world prices and high rail transportation costs. Until 1995, the federal government subsidized grain exports. The cancellation of that subsidy, known as the Crow Rate, led grain farmers to seek alternative crops.[7] From 1971 to 1995, the acreage seeded in spring wheat in Saskatchewan never fell below 10 million acres and exceeded 14 million acres 18 times (Saskatchewan Ministry of Agriculture, 2009b: Tables 2.3, 2.9). Without a doubt, wheat was the prairie staple.

Since 1971, canola acreage jumped from 2.7 million acres to a record high of 11.1 million acres

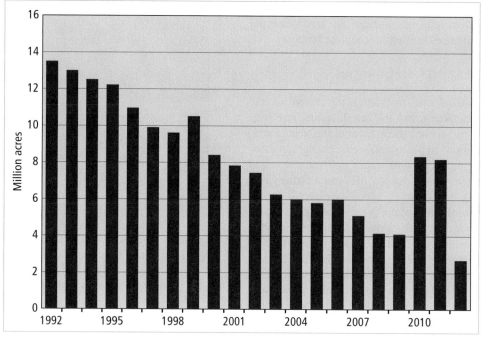

Figure 8.6 Saskatchewan summer fallow area, 1992–2012
In the past, prairie farmers left fields in fallow to conserve soil moisture. Now they use fallowing as part of their crop rotation system for canola, which must be rotated every three or four years to avoid diseases. In 2010 and 2011, the amount of land in fallow jumped largely because of crop rotation, but by 2012, much of this land was back in cultivation. *Source*: Saskatchewan Ministry of Agriculture (2012) and Statistics Canada.

in 2012 (Saskatchewan Ministry of Agriculture, 2012). A similar shift from wheat to canola took place in Alberta and Manitoba. As shown in Table 8.5, canola is the new prairie staple.

Canola's popularity with prairie farmers lies in its profitability. Not only does canola command a higher price than wheat, but farmers have the option to truck their crop to one of the many canola refineries. Yorkton has two canola refineries, but the largest canola crushing facility in North America is located just east of Saskatoon at the small town of Clavet. The Clavet facility was

Table 8.5 Seeded Area and Production for Western Canadian Canola

	Seeded area (000s of hectares)		Production (000s of tonnes)		Average production (000s of tonnes)
	2011	2010	2011	2010	2006–10
Manitoba	1,102.8	1,363.9	1,655.6	2,215.8	2,232.1
Saskatchewan	3,957.8	3,156.5	7,019.3	5,034.9	4,807.2
Alberta*	2,523.2	2,246.0	5,385.7	4,525.8	3,716.1
Western Canada	7,583.8	6,766.3	14,060.6	11,776.5	10,755.4

*Alberta includes the part of the Peace River area in British Columbia; 1 hectare = 2.471 acres.

Sources: Statistics Canada, Field Crop Reporting Series, No. 8, Vol. 90, 6 Dec. 2011; Statistics Canada, Field Crop Reporting Series, revised final estimates for 2006–10; Canadian Grain Commission, "Quality of Western Canadian Canola 2011," 6 Mar. 2012, at: <www.grainscanada.gc.ca/canola/harvest-recolte/2011/hqc11-qrc11-03-eng.htm>. Reproduced with permission of the Minister of Public Works and Government Services Canada, 2013.

Photo 8.11

Weather is a constant challenge facing prairie farmers. Just when a farmer thinks he/she has a bumper crop, nature throws a curveball. In this case, the swathed canola field near Crossfield, Alberta, looked great in August 2012 (*left*) but along came strong winds and the canola was tangled into a mess (*right*), resulting in a drop in yields and profits.

opened in 1996 and has undergone two expansions; with its current capacity the plant processes 1.5 million metric tonnes of canola each year. The Clavet plant and other canola crushing plants in Western Canada produce canola oil and specialty canola oils, as well as canola feed for livestock. The products are in high demand in North American and world markets. For Western Canada's economy, the processing of canola represents an important value-added industry.

A few decades ago, canola began its transformation from a non-descript plant known as rapeseed. Plant researchers from the University of Saskatchewan were able to create an oilseed fit for human consumption. The new plant's long descriptive name—Canada oil, low acid—was shortened to *can* (for Canada) and *ola* (for oil, low acid). In the 1990s, another breakthrough—this time by a private company, Monsanto—saw the emergence of a genetically modified seed that was tolerant to farm herbicides. With high prices for canola compared to spring wheat, farmers quickly inserted canola into their crop rotations, often with wheat, barley, and legume crops. Crop rotation is crucial for growing canola because the plant is susceptible to disease if grown in the same field for several years.

Livestock Industry

The livestock industry is also undergoing change—a combination of restructuring and consolidation of its processing plants. As a result of the Free Trade Agreement in 1989, competition within the North American market has become fierce, forcing Canadian operators to build larger processing plants, to specialize in a single product in each plant, and to demand lower wages from employees. As the cost of shipping grain by rail to Ontario and Québec jumped with the end of the Crow Benefit, prairie farmers searched for crops and livestock that can be sold and processed locally, thus reducing their transportation costs.[8] Canola was

Photo 8.12

Spring wheat has been replaced by canola as the principal crop in Western Canada because canola is more profitable. Still, spring wheat remains a popular crop for two reasons. First, spring wheat is able to survive droughty conditions. Second, it is used in most farmers' crop rotation scheme for canola. The farmer here is using an old model of swather to cut his wheat while a modern concrete grain terminal with a Saskatchewan Wheat Pool logo occupies the background. Viterra now owns all Saskatchewan Wheat Pool assets. In 2012, Glencore, a Swiss-based international company, purchased Viterra.

CONTESTED TERRAIN 8.1

Living with Pigs?

Hog barns, like cattle feedlots, produce a great deal of waste that could threaten the local water supply. Two perspectives on a proposed hog barn in a rural community are a predictable response to this often contested activity. (1) Local folks think that hog barns are going to ruin the land and the water supply, not to mention that they don't smell very nice. (2) Proponents argue that rural communities need economic development to survive.

one such crop that could be processed within the region. Livestock processing was a second means of dealing with the high transportation cost. Field grains, for example, used to be shipped to feedlots in Ontario and Québec, but without the Crow Benefit transportation costs were too high. As a result, livestock processing plants sprung up across Western Canada, especially in Alberta where the bulk of Canada's 15 million Canadian cattle are located. Two slaughterhouses, the XL Lakeside plant in Brooks and Cargill's plant in High River, account for more than 80 per cent of Canada's beef processing. Most processed meat is shipped to the United States. Unfortunately for the Brooks plant, its workers, and customers, an E. coli outbreak forced the plant to close for a month in late September 2012. While no one died from the tainted meat, 16 people in four provinces reported falling ill from eating XL meat products (D'Aliesio, 2012).

Like the beef industry, hog production and processing is undergoing restructuring and consolidation. For the last decade, older and less efficient processing plants have closed and a few larger plants have opened. Western Canada, led by Manitoba, accounts for nearly 40 per cent of hog production in Canada. Farmers who operate large-scale hog barns supply the hogs to the slaughterhouses. By 2012, the hog industry consisted of hundreds of hog barns and half a dozen modern hog slaughterhouses that supply carcasses to specialized pork processing plants across Western Canada. Outside of domestic sales, most pork products are exported to the US, Mexico, and Asian countries. Plant specialization is driven by cost of production—larger plants gain economies of scale, and specialized production lines achieve higher productivity. While it does cost to ship meat from the slaughterhouse to a number of processing plants, the saving gained from efficiencies in processing plants dedicated to a single product offsets the shipping costs. For example, Maple Leaf Foods' new hog-slaughtering operation at Brandon supplies carcasses to its Winnipeg "ham" plant and its North Battleford "bacon" plant. Michael McCain, president of Maple Leaf Foods, described competing for a place in the North American market for pork products as "akin to dancing with elephants. We have to be a little more nimble, a little more agile than our dancing partners" (Bell, 1999: 60). With thin margins, weaker hog plants are always struggling to survive—or must be a little more nimble than the competition, as McCain put it. In the fall of 2012, a sharp increase in feed caused by a severe drought in the Corn Belt of the United States triggered bankruptcies in two slaughter plants, Big Sky near Humboldt, Saskatchewan, and Puratone at Niverville, Manitoba, just south of Winnipeg. Both were badly hurt by these higher feed costs and were forced to sell their businesses to Olymel, Québec's giant pork processing company, thus continuing the trend to consolidation in this industry.[9]

Western Canada's Resource Base

Western Canada has a vast and rich resource base. Besides its agriculture, the region has a variety of natural assets. The leading one at present is Alberta's oil sands; another is the vast potash reserves of Saskatchewan. The development of the oil sands remains the driving force behind Canada's successful economy in a time of global recession, and its economic contribution is one reason why Canada has not been dragged into the economic quagmire the US has experienced for the past several years.

The oil sands are a powerful economic engine for Alberta and, to a lesser degree, for the rest of the country, as are other Western Canadian resources such as potash, natural gas, and uranium. Besides adding greatly to Canada's exports and to a positive balance of trade, the entire country benefits in varying degrees through the manufacture of specialized equipment for the oil sands companies, and air commuting allows workers from the far end of the country to access jobs in Fort McMurray. In addition, the development of technologies to ease the environmental impact of this industry may well serve other countries as they focus on extracting more unconventional hydrocarbon sources. Figure 8.7 shows the areal distribution of Alberta's oil resources, the exploitation of which began in earnest after the 1947 discovery of oil at Leduc in southern Alberta, near Edmonton. While the future of Alberta's oil production depends heavily on the oil sands, production of **conventional oil and gas** is further complemented by technological innovations that release vast quantities of light crude oil locked deep underground in impervious shale strata (Vignette 8.8). Over the last 10 years, oil from the **Bakken formation** in southern Saskatchewan and Manitoba has greatly increased those provinces' production.

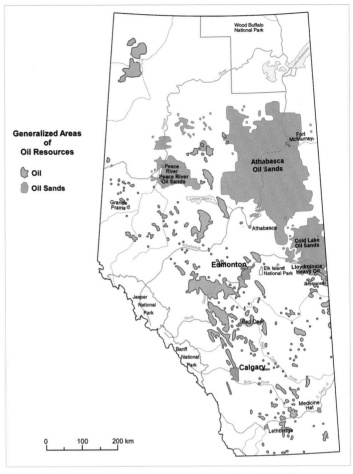

Figure 8.7   Alberta's hydrocarbon resources: Oil sands and oil fields
Source: Alberta (2007), from Energy Resources Conservation Board/Alberta Geological Survey.

Background

Western Canada's resource economy began to diversify in the 1970s, when oil and gas developments in Alberta benefited from rising prices, potash and uranium mining gained a foothold in Saskatchewan, and Manitoba expanded its hydroelectric facilities on the Nelson River. The dramatic rise in oil prices, resulting from the action of the Organization of Petroleum Exporting Countries in

VIGNETTE 8.8

Technological Breakthrough: Horizontal Drilling in Oil Shale

Horizontal drilling, often called directional drilling, has revolutionized the way that oil and gas wells are drilled. First, horizontal drilling will produce many times more oil and gas than a vertical well. A vertical well only penetrates a few feet of the oil or gas zone, but a well drilled horizontally may penetrate several thousand feet into this zone. Second, unproductive rock formations, such as the Bakken oil shale, have become productive due to horizontal drilling. But the story does not end here. Since the Bakken oil shale has poor porosity, oil could not flow along these horizontal drill holes. The solution is hydraulic fracturing, which involves pumping fluid and rounded beads to break and keep the fissures in the shale open and thus allow oil to flow. One downside to this technology is that the toxic fluid used in the process can enter the water table and underground aquifers, making the water unfit for drinking.

1973 to control supply, gave an extra boost to Alberta's economy in the 1970s.

The economic impact of petroleum on Western Canada, especially Alberta, has been remarkable. A combination of investments, demand for labour, and royalties has transformed Alberta into the fastest-growing province in Canada as well as the richest province. In terms of the core/periphery model, the export of oil reversed a century-old economic relationship—Central Canada was dependent on the western provinces. Ottawa interfered with this new relationship by introducing the National Energy Program in 1981 (which was later jettisoned when Brian Mulroney and the Progressive Conservatives took office in 1984). The federal government was seen by Alberta as intruding on provincial rights and attempting to transfer the wealth away from Alberta. The results of this federal program were to widen the political gap between Alberta and Ottawa and to deepen the sense of **western alienation** and mistrust towards Ottawa and Central Canada.

As the twenty-first century began, oil, natural gas, electricity, potash, and uranium all benefited from strong global demand and high prices. While these prices dropped sharply with the 2008–9 financial crisis, they have recovered to their previous price levels. By 2012, lagging agricultural prices show positive signs with growing demand and higher prices. Still, given the cyclical and uncertain nature of the global economy, demand for Western Canada commodities could soften and prices could fall.

While signs are positive for global prices for agricultural products to increase, reality tells Western Canada that fossil fuel, potash deposits, and water resources will generate the bulk of the region's wealth. Manitoba holds most of the rivers with high potential for hydroelectric power, and in 2012 the Wuskwatim generating project joined that province's electrical system.

Oil Sands, Pipelines, and Challenges

Energy obviously is the central element of Western Canada's resource base. Led by Alberta, Canada has the third largest oil reserves in the world (Figure 8.8). Without a doubt, these reserves place Western Canada in a dominant position within the globe oil economy now and in the future. For that reason, more and more foreign firms are investing in Alberta's oil sands. As oil production increases, Alberta faces a serious challenge of accessing world markets. The existing network of oil pipelines is stretched to the limit. Plans to build new pipelines to the refineries in Texas as well as those to Canada's Pacific coast have generated considerable resistance. Consequently, the price of Western Canadian oil is considerably less than the world price for oil.

With huge investments projected to increase production from the oil sands, the Canadian

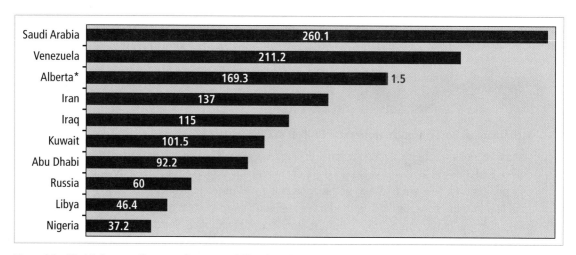

Figure 8.8 **World's largest oil reserves by country (billion barrels), 2010**
*Alberta's total oil reserves were 170.8 billion barrels, of which crude bitumen reserves accounted for 169.3 billion barrels and conventional crude oil reserves for 1.5 billion barrels. (*Source:* Alberta Energy (2012).)

Table 8.6 **Oil Production in Western Canada to 2030 (millions bbl/d)**

Production	2011	2015	2030	% 2011	% 2030
Conventional	1.1	1.3	1.1	37	18
Oil Sands	1.6	2.3	5.0	63	81
Canadian total	3.0	3.8	6.2	100	100

Sources: Canadian Association of Petroleum Producers (2012a).

Association of Petroleum Producers predicts that Alberta will account for a larger and larger share of Canadian oil production, reaching 81 per cent by 2030 (Table 8.6). In 2011, most output came from Alberta (74.1 per cent), Saskatchewan (14.3 per cent), and Newfoundland and Labrador (8.8 per cent) (Figure 8.9). By 2030, Alberta is forecast to produce around 90 per cent of Canadian oil.

As indicated in Figure 8.8, the Alberta oil sands are one of the world's largest deposits, with 169.3 billion barrels of proven reserves. Unlike conventional oil, the vast amounts of oil are trapped in the tar sands—a mixture of oil and sand known as bitumen. Extraction is costly. Approximately 20 per cent of the oil sands are recoverable by open-pit mining; the remainder can be extracted by in-situ methods, which means that the oil is drawn to the surface after being separated from the sand, which is left in place underground. One process is known as cyclic steam stimulation (Figure 8.10) and the other is steam-assisted gravity drainage. The general rule of thumb is that oil sands within 75 metres of the surface can be extracted by open-pit mining techniques. In 2010, Alberta's production of crude bitumen reached over 1.6 million bbl/d, with surface mining accounting for 53 per cent. Another critical fact—and a politically controversial one—is that only 58 per cent of crude bitumen production was processed within Alberta. The Canadian Association of Petroleum Producers estimates that this percentage will decline as oil sands production increases, meaning that more and more diluted bitumen, or **dilbit**, will be exported to foreign countries for processing.

The three oil sands fields are known as Athabasca (Fort McMurray), Peace River, and Cold Lake. With most oil sands deposits too deep to mine, an in-situ system is employed, which is similar to conventional oil production. Operations began with open-pit mining because of its low cost per unit of output; more recent operations have had to extract bitumen from much deeper deposits, necessitating the employment of the more expensive methods. The Cold Lake field, while the largest single oil sands deposit, lies 400 metres below the surface. In 1966, when Imperial Oil purchased the leases to these oil sands, the technology to extract the oil was not in place. Imperial Oil's research unit devised new recovery technologies known as cyclic steam stimulation and gravity drainage. In 2011, Cold Lake accounted for 35 per cent of the total output from the oil sands (Imperial Oil, 2012a).

Suncor Energy, the largest oil sands company, operates open-pit mines at Millennium and Steepbank and in-situ production at MacKay River and Firebag. These in-situ operations employ

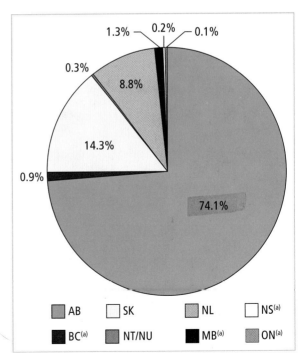

Figure 8.9 **Crude oil by province, 2011**
Note: (a) Estimates
Source: National Energy Board (2012). Reproduced with the permission of the Minister of Public Works and Government Services Canada, 2013.

344 | THE REGIONAL GEOGRAPHY OF CANADA

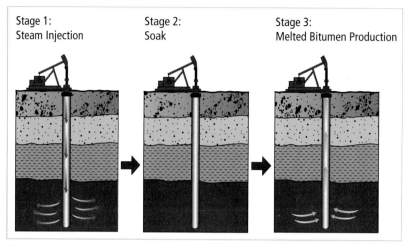

Figure 8.10 **Cyclic steam stimulation (CSS)**

steam-assisted gravity drainage, where parallel pairs of horizontal wells are drilled: one for steam injection and one for oil recovery. The bitumen is then sent by pipeline to Suncor's upgrading facility at Fort McMurray. The other oil sands company now in production is Syncrude Aurora, which operates open-pit mines. Other projects are at the proposal stage (Oilsands Developers Group, 2009).

The economic impact of petroleum on Alberta and Saskatchewan has been remarkable. A

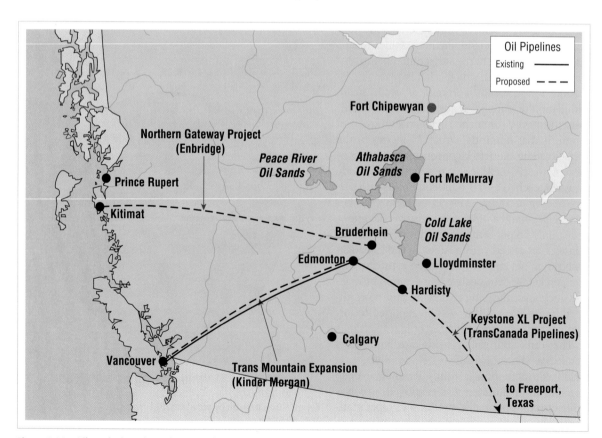

Figure 8.11 **Oil sands deposits and proposed pipelines**

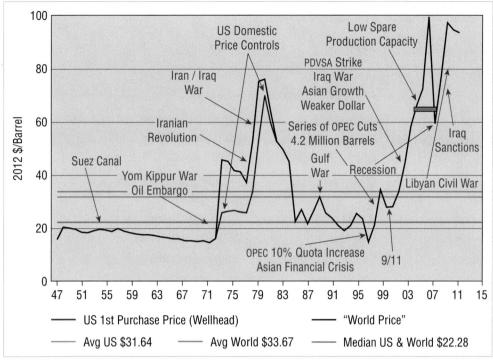

Figure 8.12 **Crude oil prices, 1947–March 2013**
Source: WTRG Economics (2013).

combination of rising oil prices (Figure 8.12), the influx of workers and immigrants, and huge investments by Canadian and foreign companies has helped to diversify the economy of Western Canada and generate revenue for the two provincial governments. For example, in 2010–11 Alberta received $5.9 billion in royalties from bitumen and conventional oil production ($3.7 billion and $2.2 billion, respectively), amounting to 17 per cent of all revenues in that year (Alberta, 2012: 51).

The manufacture of equipment for this industry has meant spinoff benefits in other parts of Canada in the form of producing manufactured equipment for this industry, and many thousands of Canadian workers, especially from Atlantic Canada, have found employment in the industry through relocation and air commuting. The capital and operating cost of oil sands development is very high. Originally, most capital came from American and Canadian sources. In the last decade, other companies, particularly Chinese state-run companies, have gained a foothold in the oil sands through purchases of smaller companies. Companies from China, France, Japan, Korea, Norway, and Saudi Arabia are investing in the oil sands with the goal of securing a supply of oil in the next decade. To further their ambitions, East Asian countries have invested just over $26 billion in acquiring assets in the oil sands (Table 8.7). While the oil sands required huge amounts of capital for development, the federal government was concerned about state-owned Chinese companies taking over existing companies. This concern was sparked by China's bid for Nexen through CNOOC (China National Offshore Oil Corporation), as well as the offer of the Malaysian state oil company, Petronas, to buy a natural gas company with holdings in northern BC. In December 2012, the federal government approved the takeover of these two Canadian-based companies to the Chinese and Malaysian state-owned companies. However, Ottawa declared that purchases by state-owned companies of oil sands companies was no longer possible and future investment would be limited to joint ventures, thus keeping the control of the company in Canadian hands but allowing the large capital investments to continue. A few days later, PetroChina's response was to spend $2.18 billion

Table 8.7 **Foreign Direct Investment in Oil Sands, 2007–2012**

Year	Buyer	Seller	Project	Value $MM
2012	CNOOC Limited	Nexen Inc.	Company – Long Lake	US$15,100
	Petro China	Athabasca Oil Sands Corp.	40% MacKay River	C$680
2011	CNOOC Limited	OPTI	OPTI interest in Nexen/OPTI Long Lake	$2,100
	PetroChina Company Limited	Dover Operating Company	Joint Venture Projects	$83
2010	Korea Investment Corp.	OSUM		$100
	Korea Investment Corp.	Laricina Energy Ltd.	Private Placement – 1.7 million common shares	$76
	PTT Expl. & Prod.	Statoil Canada	40% Kai Kos Dehseh	$2,280
	China Investment Corp.	Penn West	45% Seal Project JV	C$817
	Harvest Operations	Korea National Oil Corp.	BlackGold Oil Sands	$374
	Sinopec	ConocoPhillips	9% Syncrude	$4650
2009	Korean National Oil Co.	Harvest Energy	Incl. oil sands assets	
	Petro China	Athabasca Oil Sands Corp.	60% MacKay River & Dover	C$1900
	China Investment Corp.	Teck Resources	20% interest in Fort Hills	$1740
	Sinopec	Total Canada	10% interest in Northern Lights (now 50/50 Total & Sinopec)	
2008	Total	Synenco	Synenco	$381
2007	Statoil	NAOSC	Company	$2000

THINK ABOUT IT
Does the proposed heavy oil refinery at Kitimat refute the argument that market conditions discourage the construction of such refineries, and thus does this proposal break the "hewers of wood and drawers of water" myth?

to obtain 49.9 per cent of Encana's Duvernay natural gas holdings in west-central Alberta.

Alberta is faced with two challenges. First, should the pace of development be regulated rather than leaving it to the market? Some, such as former Alberta Premier Peter Lougheed, argue that, by regulating the number of new construction projects in the oil sands, Alberta would benefit from a more stable economy, reduce the boom construction conditions, and ensure royalties over a longer period of time.

Second, and related to the first challenge, others believe that the dependency on oil exports, especially unprocessed bitumen, is not in Alberta's interest and perpetuates the image of Canadians as **hewers of wood and drawers of water**. Yet, market conditions have discouraged companies to build more **upgraders** in Alberta and encouraged them to export more and more of the raw bitumen to foreign refineries. Currently, just over half of the bitumen is processed into synthetic oils in Alberta (Table 8.8), but by 2030 this figure could drop to less than 25 per cent.

Of the various environmental challenges posed by exploitation of the oil sands, a recent report (Kurek et al., 2013) reveals that oil sands production is polluting air and water at greater rates and over a larger geographic area than previously thought. The findings of these researchers, published in the highly regarded *Proceedings of the National Academy of Sciences*, clearly refute industry's argument that pollution is caused largely by natural seepage rather than by their industrial operations. Given the higher levels of pollution expected from the expanding oil sands industry, the authors call on industry to take immediate action to reduce the pollution as well as for governments to implement stronger regulations to ensure such reductions occur.

Open-pit mining represents the most visible impact on the landscape. Extraction by solution mining has relatively little impact on the surface but, like open-pit mining, requires huge amounts

CONTESTED TERRAIN 8.2

Who Pays for and Who Profits from the Oil Sands?

Development of the oil sands is not without environmental, economic, and political consequences. Two issues dominate. First is the argument that oil exports cause the Canadian dollar to increase relative to the US dollar. Known as Dutch disease, the higher Canadian dollar makes the export of manufactured goods to the US more difficult, and without that market manufacturing plants and their workers suffer. Second, many critics—notably NDP leader Thomas Mulcair—insist that the oil sands companies are not paying the real costs of environmental damage and greenhouse gas emissions.

Figure 8.13 Thomas Mulcair comes to Alberta
Thomas Mulcair, in light of the high unemployment figures for Ontario and Québec, sees the oil sands as a job killer for Central Canada's manufacturing sector, but on his visit to Alberta and the oil sands on 31 May 2012, Mulcair stressed the environmental impact and the need for oil sands companies to pay for the full cost of environmental damage.

---

of water to extract the tar-like bitumen and turn it into synthetic crude oil. Water is taken from the Athabasca River and the toxic sludge resulting from the procedure flows into tailing ponds. According to the Pembina Institute (2007), the ratio of water to bitumen is 2 to 4.5 cubic metres of water to 1 cubic metre of bitumen. During this process, the water taken from the Athabasca River becomes highly toxic and, thus, cannot be returned to the river but is stored in tailing ponds. Production is expected to double again in the next five years, which means that the demand for water will grow, as will the size of the tailing ponds, while water levels downstream and in Lake Athabasca will

Table 8.8 Upgraders Located in Alberta

Upgrader	Location	Capacity (bbl/d)
Athabasca Oil Sands Project (AOSP), Shell Scotford	Fort Saskatchewan	255,000
Suncor Base and Millennium	Fort McMurray	440,000
Syncrude, Mildred Lake	Fort McMurray	407,000
Canadian Natural Resources Ltd (CNRL) Horizon	Fort McMurray	135,000
Nexen Long Lake	Fort McMurray	72,000
Total		1,309,000

Source: Alberta Energy (2012).

continue to fall. According to Alberta Energy (2012), companies have reduced their needs for fresh water by reusing existing water:

> In open pit mining operations, 7.5 to 10 barrels of water is used for every barrel of SCO (synthetic crude oil) produced; however, with recycle rates of 40 to 70 per cent this means only 3 to 4.5 barrels of water . . . is required; and for in-situ operations about 2.5 to 4 barrels of water is used for every barrel of bitumen produced; however, with recycle rates of 70 to 90 per cent this means only 0.5 barrels of water . . . is required.

The oil sands are far from large markets. Pipelines to Canadian and US markets provided a solution for the initial oil production. However, the expansion of production in the twenty-first century has created a bottleneck at an oil hub in Cushing, Oklahoma. As a result, the price of oil at Cushing is depressed relative to the world price, and **oil prices** therefore vary. On average, oil at Cushing is sold at a $20 discount compared to the global price (known as the Brent price). The solution is another pipeline or two from Cushing to the refineries along the Texas Gulf coast. The Keystone XL Pipeline proposal would provide such a solution but a decision on this pipeline has been delayed, and by February 2013 the opponents of Keystone XL had ramped up their opposition, with large demonstrations in Washington putting immense pressure on President Obama to kill the proposal once and for all, especially after the emphasis in his 2013 State of the Union address on climate change and environmental stewardship. Two pipelines from Alberta to ports on the BC coast also are under consideration (Figures 8.12 and 11.3, and Table 8.9). Opposition from British Columbia is extremely strong, and even if the federal review board approves these projects, without the BC government, First Nations, and BC residents on side, construction may not occur—unless, of course, Ottawa declares the construction of these pipelines is in the "national interest."

Mining in Western Canada

The mining industry has helped to diversify the economy of Western Canada. Like the oil and gas industry, mining depends on exports to foreign markets. The variety and value of mineral production in Western Canada are significant, with coal, gold, nickel, potash, and uranium leading the way. In 2011, the value of mineral production, including coal, was $13.6 billion with Saskatchewan leading the way based on nearly $8 billion in potash production, followed by coal, gold, nickel, and uranium (Natural Resources Canada, 2012).

The geology of each province differs sufficiently to produce three distinct types of mining. Alberta contains rich bituminous coal reserves along the eastern slopes of the Rocky Mountains while Saskatchewan has large lignite deposits. Both provinces use coal for thermal-electric power generation. Manitoba, on the other hand, produces its electricity from hydro-power stations. Only Alberta exports coal, mainly to Japanese and South Korean steel plants. Alberta metallurgical coal is mined in the East Kootenay and Peace River coalfields; then shipped by rail to BC ports for export to Asia (ibid.).

Table 5.6, page 202, shows production values for minerals in 2011 by province and territory. Table 7.3, page 294, indicates Canada's leading minerals by value of production for 2011.

Potash, uranium, and diamonds are the major mineral deposits in Saskatchewan, though only

Table 8.9 Proposed Oil Pipelines* to the West Coast

Pipeline	Point of Origin	Destination	Status	Capacity (thousand bbl/d)
Kinder Morgan Trans Mountain Expansion	Edmonton, Alberta	Burnaby, BC	Considering 2017	590
Enbridge Northern Gateway	Bruderheim, Alberta	Kitimat, BC	Proposed 2017	525

*Kinder Morgan proposes to twin its existing pipeline and use the new pipeline to bring more bitumen to its ocean vessel loading facility on Burrard Inlet. Enbridge plans two pipelines, one described above designed to bring bitumen to Kitimat and the other one to bring **diluent** to the oil sands.

Source: Canadian Association of Petroleum Producers (2012a: 27).

potash and uranium have producing mines. The **potash** deposit lies approximately one kilometre below the surface of the earth, reaching its thickest extent around Saskatoon, where six of the nine potash mines are located. Royalties from potash make up a surprisingly small portion of the revenues of the Saskatchewan government. In 2011–12, potash contributed just under $440,000 while oil contributed $1.5 billion (Saskatchewan, 2012: 31). Such low potash royalties result from the complex regulations designed by the previous NDP government that allow for a considerable reduction in royalties for companies expanding their operations.[10] Consequently, while the province is benefiting from a very high level of capital investment and employment resulting from an expansion of existing potash mines, the trade-off is negligible annual potash royalty payments to the province for the foreseeable future (ibid.).

Canada (which, in effect, means Saskatchewan) is the world's largest producer and exporter of potash. Canadian Potash Exporters (Canpotex), based in Saskatoon, manages the entire Saskatchewan potash exports outside of North America. It is the world's largest exporter of potash, with its main customers being China, India, and Japan, as well as the US. The province has the largest and highest-quality deposit in the world, and except for a potash mine in New Brunswick, all of Canada's potash production—which is used to produce potassium-based fertilizers—comes from Saskatchewan (Vignette 8.9). World demand varies, but Canadian potash producers adjust production volume to maintain price—at the expense of their workers, who are laid off, and farmers, who must pay high prices.

THINK ABOUT IT
Do you agree with Jack Mintz of the University of Calgary, who gave the Saskatchewan government a failing grade for its potash royalty regulations that result in low payments to the province, or with the province in wanting to expand the potash industry? Mintz's article can be found at: <opinion.financialpost.com/2010/10/14/jack-mintz-the-potash-royalty-mess/>.

Uranium mining also takes place in Saskatchewan (Table 8.10). Production began in 1953 on the northern shores of Lake Athabasca near Uranium City. Since the late 1970s uranium mining has shifted south to the geological area of the Canadian Shield known as the Athabasca Basin. Currently, three mines are operating—McArthur River, McClean Lake, and Rabbit Lake—and one, Cigar Lake, is under construction. Three mills process the uranium ore into yellow cake at Key Lake, McClean Lake, and Rabbit Lake before shipment by truck to Ontario refineries at Blind River and Port Hope and to undisclosed refineries in the United States.

Manitoba has two major mineral deposits—copper-zinc and nickel—both located in the Canadian Shield. Mining for copper and nickel in Manitoba is relatively expensive. As well, the high cost of shipping the processed product to distant markets adds to their economic disadvantage. Copper and zinc ore bodies are found near

VIGNETTE 8.9

Potash: Saskatchewan's Underground Wealth

Potash is a general term for potassium salts. Potassium (K), a nutrient essential for plant growth, is derived from these salts. Roughly 95 per cent of world potash production goes into fertilizer, while the remainder is used in a wide variety of commercial and industrial products, ranging from soap to explosives.

In Saskatchewan, potassium salts are found in the Prairie Evaporite, which extends over much of southern Saskatchewan at varying depths. The three potash mines near Saskatoon operate at just over 1,000 metres below the surface while Belle Plain mine near Regina operates at the 1,600 m level. The deposit at each mine has a maximum thickness of 210 metres. Since the Prairie Evaporite slopes downward towards the border with the United States, this potash formation reaches its greatest depth in Montana and North Dakota—depths that are not economical to mine. The more accessible potash deposits are found near Saskatoon, where six of the nine mines are located. In fact, Saskatoon claims to be the "Potash Capital of the World."

Table 8.10 **Uranium Mines and Mills in Northern Saskatchewan**

Facility	Licensee	Licence	Type
Cigar Lake Mine Project	Cameco Corporation	Construction	Under construction for underground mining activities
Key Lake Mill	Cameco Corporation	Operation	Licensed to mill up to an average of 7,200,000 kg of uranium per year
McArthur River Mine	Cameco Corporation	Operation	Licensed to mine up to an average of 7,200,000 kg of uranium per year
McClean Lake Mill	AREVA Resources Canada Inc.	Operation	Licensed to mine and mill up to 3,629,300 kg of uranium per year
Rabbit Lake Mine and Mill	Cameco Corporation	Operation	Licensed to mine and mill up to 4,250,000 kg of uranium per year

Source: Canadian Nuclear Safety Commission (2012).

Flin Flon, a mining and smelter town in northern Manitoba that began production in 1930, shortly after a rail link to The Pas was completed in 1928. At one time, Flin Flon produced most of Canada's copper and zinc, but today it is an aging resource town with a declining population. Thompson, located some 740 km north of Winnipeg, is a nickel-mining town. In 1957, after a rail link to the Hudson Bay Railway was completed, the mine facility, smelter, and town were constructed. Unlike Flin Flon, Thompson was a specially designed resource town with a complete array of urban amenities. The first nickel was produced in 1961. During the 1960s, Thompson's population soared to 20,000. Since then, the population of Thompson and of many other resource towns has dwindled: greater mechanization in the mining process results in a smaller labour force, and this, in turn, leads to a contracting service sector. In 1991, Thompson had a population of 14,977, but by 2011 it had fallen to 12,839.

Forest Industry: Is a Revival in Sight?

The boreal forest stretches across the northern part of Western Canada and along the foothills on the eastern slopes of the Rocky Mountains. The bulk of the forested area lies in Alberta[11] (Table 8.11). The forest industry depends on exports to the United States. As with other regions, Western Canada's forest industry had hit the skids with the US housing crisis. Plants had closed and little logging was taking place. Aboriginal workers and businesses have been particularly hard hit because they depend heavily on this industry for their livelihood. In an attempt to support pulp and paper mills, the federal government, under its Green

Table 8.11 **Forested Areas by Province, Western Canada**

Province	Area (ha)	% of Western Canada Forest
Manitoba	13,648	16.9
Saskatchewan	13,000	15.9
Alberta	54,981	67.2
Western Canada	81,629	100.0
Percentage Canada		11.1

Source: National Resources Canada, 2013.

Transformation Program, has provided financial assistance for investments in capital projects that improve environmental performance and economic efficiency. But hope may be around the corner as the 2012 housing starts in the Unites States picked up.

For discussion of forestry exports to the US, see Chapter 7, "Dependency on the US Market," page 302, and "North American Free Trade Agreement," page 302.

The federal government's Green Transformation Program is discussed in Chapter 9, "Forest Industry: A Weak Sister," page 391.

Western Canada's Population

Western Canada's population has undergone significant changes since European settlers first came to this region, and these changes have affected and been affected by many aspects of prairie life. In 1921, Western Canada had a population of nearly 2 million. At that time, it comprised 22 per cent of Canada's population. By 2011, its population had increased to 5.9 million but accounted for only 18 per cent of the national population.

Another demographic feature of Western Canada is the size of the Aboriginal population. By 2011, just over half a million Aboriginal people lived in Western Canada, forming nearly 10 per cent of Western Canada's population. The high natural rate of increase of Aboriginal people accounts for the fact that the populations of Indian reserves and Métis communities are increasing, as is the urban Aboriginal population. The attraction of cities is due to the pull of urban amenities and opportunities and the push of insufficient employment opportunities and social problems on reserves and in Native communities. While an Aboriginal middle class has emerged in cities, poverty-driven social problems take the form of street gangs, drug and alcohol abuse, and prostitution. Winnipeg's North End is a troubled neighbourhood where the Indian Posse, formed in 1988, holds sway. The 2004 film, ***Stryker***, by Noam Gonick, captures the brutal turf war between two street gangs in Winnipeg's North End.

Photo 8.13
Saskatoon, situated on the South Saskatchewan River, is Saskatchewan's largest city with a population of 260,600 (2011). Known as the "Bridge City," Saskatoon has witnessed rapid population growth over the last 10 years, outpacing its southern rival, Regina. During that period, many Indians and Métis have settled in Saskatoon. By 2011, the Aboriginal population comprised over 10 per cent of Saskatoon's residents.

Western Canada's Urban Geography

The process of urbanization in Western Canada has lagged behind that of Ontario, British Columbia, and Québec. Even so, nearly three-quarters of Western Canada's residents live in urban centres. Most (3.6 million) reside in the five major cities—Regina, Saskatoon, Winnipeg (Vignette 8.10), Edmonton, and Calgary. A second order of urban centres includes Airdrie, Lethbridge, Red Deer, Grande Prairie, Medicine Hat, Wood Buffalo (Fort McMurray), Brandon, Prince Albert, Moose Jaw, and Lloydminster. In 2011, these 15 cities had a combined population of 4.1 million (Table 8.12). From 2001 to 2011, the rate of urban growth varied considerably for the towns and cities of Western Canada. This variation reflects differences in local economic growth and in the pace of consolidating populations into regional centres. From 2001 to 2011, the fastest-growing cities in Western Canada were in Alberta: Wood Buffalo

THINK ABOUT IT
The Indian Posse is only one street gang. The larger question is, why has the gritty world of street gangs, which cuts across ethnic lines, become a common feature of Canadian cities?

Table 8.12 **Major Urban Centres in Western Canada, 2001–2011**

Centre	Population 2001	Population 2011	% Change
Lloydminister	20,988	30,798	46.7
Moose Jaw	33,519	34,421	2.7
Prince Albert	41,460	42,673	2.9
Brandon	46,273	53,229	15.0
Grande Prairie	36,983	55,032	48.8
Wood Buffalo	42,581	66,896	57.1
Medicine Hat	61,735	72,807	17.9
Red Deer	67,829	90,564	33.5
Lethbridge	87,388	105,999	21.3
Regina	192,800	210,556	9.2
Saskatoon	225,927	260,600	15.3
Winnipeg	676,594	730,018	7.9
Edmonton	937,845	1,159,869	23.7
Calgary	951,494	1,214,869	27.7
Total	3,423,416	4,128,331	20.6

Sources: Adapted from Statistics Canada (2007b, 2012g).

VIGNETTE 8.10

Winnipeg

Winnipeg, the capital and the largest city in Manitoba, is located at the confluence of the Red River and the Assiniboine River—the location of Lord Selkirk's early nineteenth-century settlement for Scottish crofters that became the predominantly Métis Red River Settlement. With the building of the Canadian Pacific Railway, Winnipeg became known as the "Gateway to the West." By the turn of the twentieth century, Winnipeg was the principal city in Western Canada, controlling the grain trade and also acting as the administrative, financial, and wholesale hub for Western Canada. Today, Winnipeg dominates the economy of Manitoba, but its role within Western Canada has been considerably diminished by the growth of Calgary and Edmonton and, to a lesser degree, by Saskatoon and Regina.

After the completion of the Panama Canal in 1914, Alberta farmers and some Saskatchewan farmers began to ship their wheat through Vancouver instead of through Winnipeg. Next, service industries began to emerge in Edmonton, Calgary, Saskatoon, and Regina. Each city captured some of the trade previously held by Winnipeg merchants. Winnipeg's stranglehold over the sale and distribution of agricultural machinery and products to the farmers of Western Canada was broken.

Calgary and Edmonton grew rapidly in the years following World War II, partly because of developments in the oil and gas industries, but by 1951 Winnipeg remained the largest city in Western Canada, with a population of 357,000 compared to 177,000 for Edmonton and 142,000 for Calgary. By 1981, both of these cities had surpassed Winnipeg in population. This demographic trend continued. By 2011, the populations of Calgary and Edmonton had jumped to over one million. Winnipeg saw its population expand more slowly, to 730,000. Winnipeg, however, is attracting many Aboriginal people. These urban Aboriginals make up 10 per cent of the city's population.

Table 8.13 Aboriginal Population* by Province, Western Canada

Province	Population 2001	Population 2011	Change (%)	Regional (%)**
Alberta	156,225	220,695	41.3	6.2
Manitoba	150,045	195,900	30.6	16.7
Saskatchewan	130,185	157,740	17.5	15.6
Western Canada	436,455	378,435	31.6	41.0
Canada	976,305	1,400,685	41.0	100.0

*Population based on responses to the identity (not ancestry) question in the 2011 National Household Survey.
**Percentage of total Aboriginal population to total provincial population.

Source: Statistics Canada (2013b).

at 57 per cent, Grande Prairie at 49 per cent, and Red Deer at 34 per cent. Lloydminster, which straddles the Alberta/Saskatchewan border, came in at 47 per cent. The fastest-growing cities in Saskatchewan and Manitoba were Saskatoon (15.3 per cent) and Brandon (15.0 per cent) (Table 8.12).

Calgary–Edmonton Corridor

The Calgary–Edmonton Corridor has emerged as the most urbanized region in the Western Canada region and one of the densest in Canada. With major universities and colleges in its cities, it is a hub for the knowledge-based activities where high-tech industries thrive and cutting-edge research takes place. Over the past decade, this urban corridor has had a high rate of population growth, exceeding 25 per cent.

Anchored by Calgary and Edmonton, the 400-kilometre corridor includes the cities of Airdrie and Red Deer and a host of smaller centres. Calgary has become one of Canada's key corporation headquarters, especially for the oil and gas industry. British Petroleum, Encana, Imperial Oil, Suncor Energy, Shell Canada, and TransCanada are headquartered in Calgary, as is Canadian Pacific Railway. With its proximity to Banff, the Rocky Mountains, and Kananaskis Country, Calgary offers easy access to mountain recreation. Edmonton, besides being the provincial capital, is a petrochemical industrial node. Known as the "Gateway to the North," Edmonton serves as a staging centre for the oil sands and for diamond mining in the Northwest Territories.

Faultline: Aboriginal/Non-Aboriginal Populations

Western Canada has the highest proportion of Aboriginal population of the five regions in southern Canada. Not surprisingly, many Aboriginal/non-Aboriginal achievements, challenges, and problems are found in this region. These issues range from employment to housing and health care. Demography helps explain this focus. First, most Aboriginal people reside in Western Canada (Statistics Canada, 2013c). Second, cities and towns in Western Canada contain the highest percentages of Aboriginal people (Statistics Canada, 2009d). Winnipeg leads the way with 78,810 Aboriginal people, representing 11.9 per cent of that city's total population. Edmonton and Calgary, with 71,930 and 44,765, have the next largest urban Aboriginal populations, comprising 8.9 per cent and 4.1 per cent of their respective populations. Saskatoon (23,870) and Regina (19,430) also have large Aboriginal populations, making up 14.8 and 10.1 per cent of these cities' respective populations. The highest percentages are found in smaller cities near large First Nation reserves. Thirty-one per cent of the population of the northern mining town of Thompson, Manitoba, is Aboriginal, while Prince Albert, Saskatchewan, the gateway to northern Saskatchewan, has 39.5 per cent.

Further discussion of Aboriginal demography is found in Chapter 4, "The Expanding Aboriginal Population", page 150. Table 1.2, page 13, explains the Statistics Canada distinction between Aboriginal identity and Aboriginal ancestry.

THINK ABOUT IT
Do urban places offer a solution to the economic and social issues facing First Nations peoples?

SUMMARY

Geography has dealt Western Canada a full hand of resources, but also two deuces—distance and climate. Locked in the heart of North America, the long haul to ocean ports makes its export products more expensive while its same interior location generates a dry, continental climate with low precipitation and therefore uncertainty for crop agriculture. Price plays a critical role. For example, low prices for forest products have stalled this industry while high prices for energy, potash, and canola have had the opposite effect. In the last decade, canola has exceeded spring wheat in returns to farmers and has surpassed spring wheat in sown acres.

Exports are the lifeblood of this region, and the emergence of Asian industrial powerhouses and their burgeoning middle classes have created a strong demand for many of Western Canada's resources, including foodstuffs, potash, and uranium. Added to this growing export market, the United States remains the principal importer of oil, gas, and hydroelectricity. Alberta's oil sands, which contain the third-largest oil reserves in the world, supply the US with over 10 per cent of its oil needs and, with further expansion, could supply more than 20 per cent. From a geopolitical perspective, Alberta's oil sands play a key role in a US strategy of "North American energy security" and this strategic fact means that the construction of additional oil sands plants, upgraders, and pipelines likely will continue and thus anchor Western Canada's economic boom for the foreseeable future. More recently, however, the US rhetoric on energy has emphasized "American energy self-sufficiency."

The transformation of Western Canada's economy is clear—more processing of agricultural products, a larger service sector, and growing urban centres where knowledge-based research clusters focus on technological innovations in agriculture, oil sands extraction, and mining. In spite of a few dips, if the upward-moving trend line for global prices continues for the region's underground wealth and agricultural products, a robust economy is secure for the short and middle terms. These positive signs raise the question: Has Western Canada reached a turning point where the promise of Next Year Country finally arrived? The only dark clouds on the horizon are:

- lower commodity prices;
- a return of dry weather and drought;
- failure to resolve the environmental issues facing oil sands companies;
- possible political decisions in the United States to rely increasingly on alternative energy sources and shale oil deposits;
- global warming and climate change, which could result in a shortage of water for agriculture, industry, and cities and create havoc for trucking goods along winter roads.

CHALLENGE QUESTIONS

1. Nature plays a critical role in prairie agriculture, especially in the area known as Palliser's Triangle. Besides a dry summer, what other natural factors affect the size and quality of the harvest?
2. For Western Canada, long distance to ports and then global markets translates into high transportation costs. In the 1920s, farmers called for an "on to the Bay" solution based on the old Hudson's Bay Company route to its major market in England. Nearly a hundred years later, oil sands companies are calling for access to the Chinese market via the Northern Gateway Project. Yet, opposition to this project is strong. Who are the main opponents and what are their reasons for opposing the project?
3. Wheat is no longer king. Why are more and more farmers switching from wheat to canola?
4. Why doesn't Alberta insist that all bitumen produced in its oil sands be refined in the province, thus reversing the image of Canada as a nation of "hewers of wood and drawers of water"?
5. Is the current level of foreign ownership of the oil sands by Chinese state corporations in Canada's best interest?

FURTHER READING

Casséus, Luc. 2009. "Canola: A Canadian Success Story," *Canadian Agriculture at a Glance*. Ottawa: Statistics Canada Catalogue no. 96–325–X.

In the 1970s, canola (an abbreviation for "Canadian oil, low acid") was developed by plant breeders in Saskatchewan and Manitoba to yield food-grade oil. Canola is often grown in rotation with Canada's traditional cereal crops of wheat, oats, and barley. The area seeded with canola varies according to prevailing economics, among other factors, but the trend is to increasing area. Wheat, on the other hand, has been showing the opposite trend in planted area. Increasing yields, improved marketing, higher-quality crops, and increases in both canola prices and quantity sold have helped boost Canada's cash receipts for canola. By 2005 the crop had surpassed wheat to become the most valuable field crop in Canada. In 1976, canola accounted for only 4.9 per cent of total crop receipts; by 2006, this percentage had climbed to 17.2 per cent. Over the same period, wheat receipts dropped from 35.5 per cent to 15.0 per cent of total crop receipts. Farm cash receipts for wheat totalled $2.2 billion in 2006, while those for canola totalled $2.5 billion.

9

ATLANTIC CANADA

Introduction

Atlantic Canada, located on the eastern margins of the country, is far from the centres of economic and political power in Central Canada. Yet, its proximity to Europe allowed French settlers to reach its shores as early as the seventeenth century; English and Scottish settlers would follow later. Fish and timber were among the first exports, but harvesting extended beyond sustainable levels and gradually Atlantic Canada became an "old" resource hinterland.

For some time, Atlantic Canada has lagged far behind the other southern Canadian regions in terms of economic and population growth. Heavily dependent on equalization payments, Atlantic Canada is classified as a "have-not" region. Not surprisingly, then, the region suffers from out-migration of its younger members, thus accelerating the greying of its population and creating a shortage of skilled workers. For rural Atlantic Canada, the painfully slow recovery of the northern cod stocks remains a hope for the future. Over the last two decades the coastal fishing economy collapsed, resulting in the relocation of fisher families to larger centres but especially St John's. For many, the cultural loss—the end of a way of life—has been at least as devastating as the economic loss of the cod.

Today, however, Atlantic Canada shows signs of economic rejuvenation, driven largely by offshore oil and gas developments, its role as an energy supplier to New England, and its position as a gateway for container traffic into Central Canada and the US Midwest. Under these positive circumstances, the region's economic pace shows promise, but virtually all has taken place in the major urban centres, leaving rural Atlantic Canada in a seemingly relentless decline. Still, the task of revitalizing Atlantic Canada's economy remains. Until then, the region's dependency on Ottawa and the outflow of its younger population to the faster-growing areas in Canada will undercut its rejuvenation. In this chapter, the "Key Topic," the fishing industry, examines the demise of cod fishing and processing in coastal communities as well as the growing importance of the shellfish industry.

CHAPTER OVERVIEW

Important issues and topics examined in this chapter include:

- Atlantic Canada's physical and historical geography.
- The basic characteristics of Atlantic Canada's population and economy as seen from two perspectives: an "old" resource hinterland and a rejuvenating one.
- The population and economy within the context of Atlantic Canada's fragmented geography.
- The cod fishery as an example of the "tragedy of the commons."
- The future: offshore oil and hydro energy plus military shipbuilding.

Fishing boats tied up at the Malpeque Harbour wharf in Malpeque Bay, Prince Edward Island. Malpeque Bay is a wetland of international importance and is home to nesting colonies of herons and cormorants. The bay is also well known for its oysters. Photo: Barrett & MacKay/All Canada Photos.

Atlantic Canada within Canada

Stretching along the country's eastern coast, the region of Atlantic Canada consists of two parts: the Maritimes (Nova Scotia, Prince Edward Island, and New Brunswick) and Newfoundland and Labrador[1] (Figure 9.1). Separated by Cabot Strait, the island of Newfoundland stands alone in the Atlantic Ocean while its Labrador territory abuts Québec. Still, in spite of its fractured geography, Atlantic Canada remains a key region of Canada, bonded by its rich and enduring sense of place that has grown out of the region's history, its original British and French settlements, and its long-standing and occasionally tragic relationship with the Atlantic Ocean. Through all of this history and geography, the Atlantic Ocean has dominated this region from the early days of the fishery to the "Golden Age of Wooden Ships and Iron Men," to the dream of an Atlantic Gateway. Offshore petroleum developments, the energy corridor to New England, and the wealth returning to Atlantic Canada from its commuters to Alberta's oil sands have added new dimensions to the regional economy. But most important of all, the fading image of outports like Diamond Cove (photo 9.1) is now superseded by visions of vibrant urban centres led by Halifax (photo 9.2) and St John's.

Photo 9.1
The sharp interface between the land and the sea demonstrates why fishing was a way of life. Outports like Diamond Cove, situated along the isolated rocky and indented coastline of the southwest tip of Newfoundland, are a dying breed. Since the closure of the cod fishery in 2003, outports no longer have an economic base and many families abandoned these tiny fishing communities. Diamond Cove is hanging on, but its demographics tell the sad tale: a tiny and dwindling population (66 in 2006, down to 53 in 2011); an older population with an average age of 56 and no one under the age of 20.

Dale Wilson/All Canada Photos

ATLANTIC CANADA

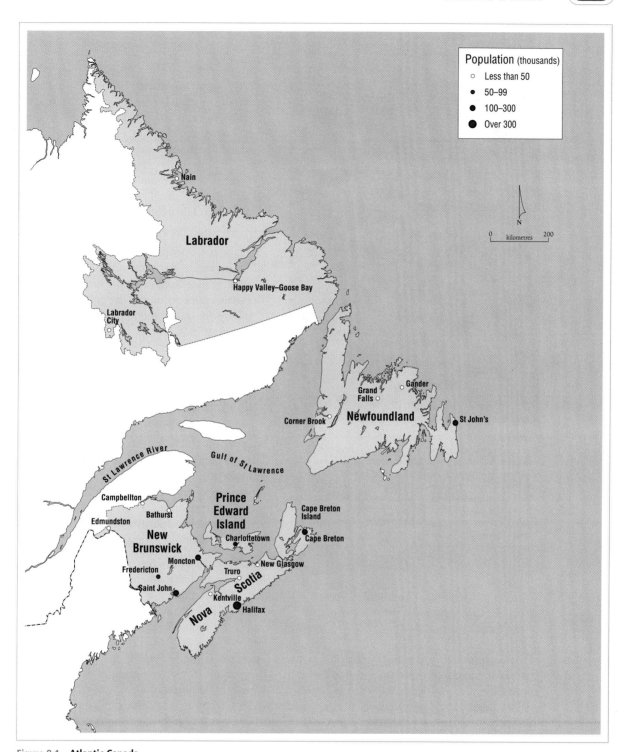

Figure 9.1 **Atlantic Canada**
Atlantic Canada contains four provinces, Nova Scotia, New Brunswick, Newfoundland and Labrador, and Prince Edward Island. Atlantic Canada has the smallest population and the weakest economy—except for the Territorial North—of Canada's six regions, although its offshore oil resources recently have made Newfoundland and Labrador a "have" province. The largest city is Halifax, with a population of just under 400,000, followed by St John's at just under 200,000.

Photo 9.2

When the British founded Halifax in 1749, they were attracted by its magnificent harbour. The high hill overlooking the harbour offered a perfect location for a fortress to defend the new town and its naval base. Named the Halifax Citadel (upper left), this fortress is an impressive star-shaped masonry structure complete with defensive ditch, earthen ramparts, musketry gallery, powder magazine, garrison cells, guard room, barracks, and school room. The Citadel is now a National Historic Site.

 In Chapter 1, "Canada's Geographic Regions," page 5, the rationale for Canada's six geographic regions, including Atlantic Canada, is elaborated.

As Canada's oldest hinterland, Atlantic Canada has experienced both growth and decline over the last 200 years. Atlantic Canada has become, in Friedmann's regional version of the core/periphery model, a downward transitional region. The region was troubled by past exploitation of its renewable resources and the exhaustion of its most accessible and richest non-renewable resources. Atlantic Canada's troubles are epitomized by the subpar performance of its economy relative to the rest of Canada and the seemingly unstoppable out-migration to faster-growing regions of Canada.

Table 1.3, "National Version of the Core/Periphery Model," page 18, describes the four regions of Friedmann's model.

One measure of Atlantic Canada's overall economic performance is reflected in the region's per capita gross domestic product figures and its level of unemployment. In 2011, except for the Territorial North, Atlantic Canada's GDP per capita was the lowest in Canada, while the region's 2011 unemployment rate was the highest (Figure 9.2 and Table 9.1). The primary reasons for Atlantic Canada's weak economic performance include the following:

- The political division of Atlantic Canada into four provinces discourages the emergence of an integrated economy in which economies of scale might occur and the cost of a single government might reduce public expenditures.
- Atlantic Canada has been exploiting its resources for a long time and some of these, such as coal and iron, have been exhausted, while its renewable resources, such as the northern cod, have been overexploited.
- Atlantic Canada's population is widely dispersed and, with the exception of Halifax and St John's, consists of small markets. From 2006 to 2011, the region's population only increased by 1.9 per cent, or 42,859 people.
- Distance from national and global markets stifled its manufacturing base.

All these factors have made it extremely difficult for economic development to flourish in the region. Furthermore, over the years Atlantic Canada has become heavily dependent on Ottawa for economic support through equalization payments and social programs. Yet, Atlantic Canada has received a second chance with the discovery of offshore oil and gas deposits and a huge shipbuilding contract from Ottawa, and if a trade agreement with the European Union is achieved in the future the region's potential as an Atlantic gateway could become a reality. In addition, the prospect of hydro power from the Lower Churchill reaching Maritimes and New England markets is not out of the question. By taking full advantage of these new opportunities, Atlantic Canada may well become a more resilient and prosperous region and thus shed its moniker as a "have-not" region.

Atlantic Canada's Physical Geography

Atlantic Canada contains two of Canada's physiographic regions: the Appalachian Uplands and the Canadian Shield. The Appalachian Uplands are found in the Maritimes and the island of Newfoundland while Labrador is part of the Canadian Shield. In terms of geologic time, the Appalachian Uplands represents the worn-down remnants of an ancient mountain chain. Formed in the Paleozoic Era, the Appalachian Mountains have been subjected to erosional forces for some 500 million years. As photo 2.8 illustrates, streams have cut deeply into the Cape Breton Highlands of the Appalachian Uplands, resulting in rugged, hilly terrain. In Labrador, the most prominent feature of this portion of the Canadian Shield is the uplifted and glaciated Torngat Mountains (photo 9.3). Unlike the rest of the Canadian Shield, the Labrador portion was subjected to the mountain-building process (**orogeny**) in which the rocks were folded and faulted some 750 million years ago. More recently in geologic time, these mountains were covered with glaciers, which, as the glaciers slowly moved down slope, carved the mountain features and eventually reached the sea, where they created a fjorded coastline.

See photo 2.2, page 35, which illustrates Labrador's fjorded coastline.

The weather of Atlantic Canada is quite varied because of the frequent meeting of continental air masses with marine air masses. The flow of continental air masses from the northwest brings warm weather in the summer and cold weather in the winter. Yet, with no part of Atlantic Canada more than 200 km from the Atlantic Ocean, moderate, marine-type weather predominates. The result is generally unsettled weather. Still, Atlantic Canada, especially Labrador, has a strongly continental aspect to its climate and, coupled with the cold Labrador Current, takes on a more Arctic-like climate. Storms are not uncommon, especially in the fall when hurricanes reach the Maritimes. Usually tropical storms lose

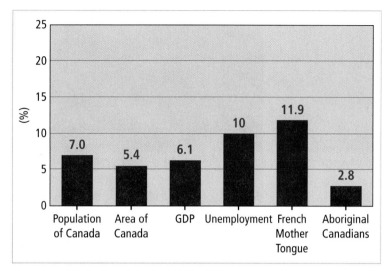

Figure 9.2 Atlantic Canada basic statistics, 2011
Atlantic Canada has a slow-growing economy. Until 2009, all four provinces were considered "have-not" provinces and received equalization payments. Since 2009, Newfoundland and Labrador has not received equalization payments because of its oil revenues. Atlantic Canada has 7.0 per cent of Canada's population but produces only 6.1 per cent of the country's GDP. The high rate of unemployment is another indication of its poor economic performance. Acadians account for the high number of French-speaking Canadians in Atlantic Canada, which has the second-highest percentage by region in Canada. In 2011, Aboriginal people formed only 2.8 per cent of the region's population.
Source: Tables 1.1, 1.2, 5.1.

> **THINK ABOUT IT**
> Given its offshore oil and gas resources and Lower Churchill hydro potential, might Atlantic Canada soon be reclassified as a "have" region?

Table 9.1 Basic Statistics for Atlantic Canada by Province, 2011

Province	Population (000s)	Population Density	% of National GDP	Unemployment Rate (%)
Prince Edward Island	140	24.7	0.3	11.3
Newfoundland and Labrador	514	1.4	1.9	12.7
New Brunswick	751	10.5	1.8	9.5
Nova Scotia	922	17.4	2.1	8.8
Atlantic Canada	2,327	4.7	6.1	10.0
Canada	33,477	3.7	100.0	7.5

Source: Statistics Canada (2012a, 2012b, 2012c).

Photo 9.3

The Torngat Mountains, a national park reserve since 2005, stretch along the fjorded coast of northern Labrador. One such fjord is Ramah Bay (with an iceberg floating in its waters). The mountains were recently, in geologic time, subjected to alpine glaciation, resulting in extremely sharp features, including arêtes, cirques, and horns. These mountains, including Mount Caubvik (also known as Mont D'Iberville), straddle the Québec/Labrador boundary. They attain heights of 1,652 m (5,420 ft) above sea level and are located near the sixtieth parallel. For both reasons—high elevation and high latitude—the Torngat Mountains lie beyond the tree line.

their punch by the time they reach Nova Scotia, but sometimes this is not the case. Hurricane Juan, for example, made landfall on 29 September 2003, bringing its full fury.

Winter storms are caused by the meeting of cold Arctic air with warmer, humid air from the south. In summer, occasional incursions of hot, humid air from the Gulf of Mexico occur, but the

VIGNETTE 9.1

Consequences of the Geographic Fragmentation of Atlantic Canada

According to Macpherson (1972: xi), Atlantic Canada is not a homogeneous region but "a region of geographical fragmentation" consisting of three sub-regions: the Maritimes, Newfoundland, and Labrador. Macpherson argues that these physical divisions have affected all aspects of life in Atlantic Canada, hindering the emergence of a common political will, a common regional consciousness, and an economic union among the four Atlantic provinces. Whether or not an economic union is a desirable political arrangement for Atlantic Canada is uncertain, though the topic of union has long been debated. As early as 1949, Joey Smallwood, Premier of Newfoundland, proposed a union of the four eastern provinces to enable them to increase their bargaining power in Ottawa. In the 1970s, the Deutsch Commission on Maritime Union recommended a "full political union as a definite goal." However, there was no support within Atlantic Canada, and before the 1970s had ended Nova Scotia Premier Gerald Regan pronounced the proposal for a Maritime Union "as dead as a doornail." For the residents of Atlantic Canada, the advantages of a union are not readily apparent, and Atlantic premiers have not shown much enthusiasm for this proposal.

dominant weather is cool, cloudy, and rainy. In the winter, influxes of moist Atlantic air produce relatively mild snowy weather except in Labrador (and to a lesser extent in the New Brunswick interior), where it can become extremely cold for extended periods.

Annual precipitation is abundant throughout Atlantic Canada, averaging around 100 cm in the Maritimes and 140 cm in Newfoundland, but this gradually diminishes further north in Labrador. Much precipitation comes from **nor'easters**—strong winds off the ocean from the northeast—that draw their moisture from the Atlantic Ocean. Atlantic Canada, especially the Maritimes and the island of Newfoundland, has foggy weather. Thick, cool fog forms in the chilled air above the Labrador Current when it mixes with warm, moisture-laden air from the Gulf of Mexico. With onshore winds these banks of fog move far inland, but the coastal communities experience the greatest number of foggy days. Both St John's and Halifax, for instance, experience considerable foggy and misty weather.

With such varied weather conditions, Atlantic Canada has three climatic zones—Atlantic, Subarctic, and Arctic zones. The great north–south extent of this region is one reason. For example, the distance from the southern tip of Nova Scotia (44° N) to the northern extremity of Newfoundland and Labrador (60° N) is over 2,000 km. Then, too, Atlantic Canada is the meeting place of Arctic and tropical air masses and ocean currents, resulting in wet, cool, and foggy weather. In addition, close proximity to the Atlantic Ocean exerts a moderating effect on the region's climate.

North of 55 degrees latitude, the Arctic zone of Labrador appears in northern Labrador, in the Torngat Mountains and then along its coast. An Arctic storm track funnels extremely cold and stormy weather along the Labrador coast while the **Labrador Current** (Figure 9.3) brings icebergs from Greenland to the Labrador and Newfoundland coastlines and its cold waters contribute to the formation of land-fast ice along the Labrador and northern coastlines of Newfoundland. Beyond the land-fast ice in the open sea, ice floes and icebergs are carried by the Labrador Current as far south as the Grand Bank. By July, Labrador waters are ice-free.

The Arctic zone is associated with tundra vegetation as the summers are too cool for tree growth. The Subarctic climate zone exists over the interior of Labrador. The interior of Labrador experiences much warmer summer temperatures than its coastal areas and this area is associated with the boreal forest. The Atlantic zone includes the Maritimes and the island of Newfoundland. For most of the year, this more southerly area is influenced by warm, moist air masses that originate in the tropical waters of the Caribbean Sea and the Gulf of Mexico. Only in the winter months does the Arctic storm track dominate weather conditions. The coastal areas of the Maritimes and Newfoundland can be affected by tropical storms in the late summer and fall.

The main air masses affecting the region originate in the interior of North America and from the Gulf of Mexico and the North Atlantic Ocean. Consequently, summers are usually cool and wet, while winters are short and mild but often associated with heavy snow and rainfall. Most precipitation falls in the winter, and temperature differences between inland and coastal locations are striking. Temperatures are usually several degrees warmer in the winter near the coast than at inland locations. During the summer, the reverse is true, with coastal areas usually several degrees cooler.

Along the narrow coastal zone of Atlantic Canada, the climate is strongly influenced by the Atlantic Ocean. The summer temperatures of coastal settlements along the shores of Newfoundland are markedly cooled by the cold water of the Labrador Current (a cold ocean current in the North Atlantic Ocean), which brings with it icebergs as far south as Newfoundland. Besides filling St John's harbour with pack ice, icebergs are often sighted offshore as late as June. Another effect of the sea occurs in the spring and summer—fog and mist result when the warm waters of the northeastward-flowing **Gulf Stream**, which originates in the Gulf of Mexico, mix with the cold, southerly flowing Labrador Current. In the winter, the clash of warm and cold air masses sometimes results in severe winter storms characterized by heavy snowfall. St John's, for example, had a record 648.2 cm of snow (almost 22 feet) in the winter of 2000–1, and a vicious winter storm hit Nova Scotia and Prince Edward Island on 19–20 February 2004, leaving behind snowfall accumulations of 50–70 cm (Conrad, 2009: 36–7).

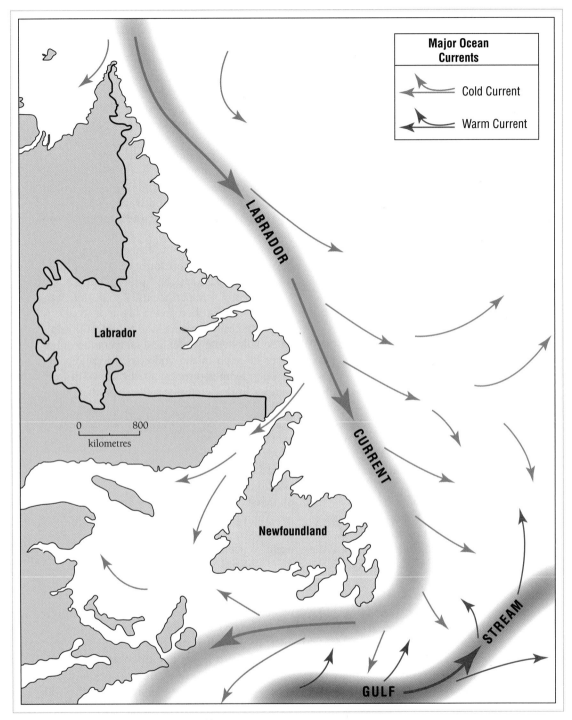

Figure 9.3 **The Labrador Current and the Gulf Stream**
Source: Macpherson (1997). Modified by Duleepa Wijayawardhana with permission, 1998. Reproduced by permission of Gary E. McManus and Clifford H. Wood, *Atlas of Newfoundland and Labrador* (St. John's: Breakwater, 1991), Plate 5.2.

VIGNETTE 9.2

Weather in St John's

While occupying a more southerly location than Victoria, British Columbia, St John's has a cooler climate, partly because of its very cold offshore waters. Even so, St John's, like Victoria, is an ice-free port. However, unlike Victoria, a winter ice pack lies offshore, and in the late spring icebergs are not uncommon off the coast (photo 9.4). Melting of sea ice begins in spring, retreating northward along the land-fast ice attached to the Labrador coast.

Generally speaking, weather is characterized by foggy days, overcast skies, and stormy, rainy weather. Fog is common from April to September. Throughout the year, a salty smell is in the air. While most of the year has mild temperatures, strong northwest winds result in heavy winter snowfalls. Also, St John's does not escape the wrath of tropical storms—such as Hurricanes Igor (2010) and Leslie (2012)—that leave behind a path of destruction that includes toppled trees, torn power lines, and roofs ripped apart from buildings. Jutting into the Atlantic Ocean, St John's feels the full brunt of strong winds from the North Atlantic Ocean. As David Phillips (1993: 155), a climatologist and the spokesperson for Environment Canada's Meteorological Service, explains, "Of all major Canadian cities, St John's is the foggiest, snowiest, wettest, windiest, and cloudiest."

Photo 9.4
Icebergs, calved from the Greenland Ice Sheet, frequently float near shore and into harbour entrances, as here in the Narrows off St John's harbour. While tourist viewing of the icebergs in late spring has become a popular attraction, the icebergs pose a threat to vessels and shipping.

Environmental Challenges

Without a doubt, the major environmental disaster to strike Atlantic Canada has been the collapse of the cod fishery. But this disaster was due more to technological advances that enable much larger catches, coupled with federal mismanagement, than to natural factors such as the warming of the Atlantic waters or the expanded seal population consuming vast quantities of cod. Clearly a case of a "tragedy of the commons," this ecological collapse of one of the world's greatest fish stocks is documented later in this chapter. Here, our attention is focused on an unwanted remnant of Cape Breton's glory days as a major iron and steel centre in Canada—the so-called tar ponds.

By the end of the twentieth century, Sydney, Nova Scotia, not only had lost its industrial sector

---

VIGNETTE 9.3

The Annapolis Valley

The Annapolis Valley is a low-lying area in Nova Scotia. At its western and eastern edges the land is at sea level, but it rises to about 35 m in the centre. The area is surrounded by a rugged, rocky upland that reaches heights of 200 m and more. The fertile sandy soils of the Annapolis Lowlands originate from marine deposits that settled there about 13,000 years ago. After glacial ice retreated from the area, sea waters flooded the land, depositing marine sediments that consisted of minute sand and clay particles. Isostatic rebound then caused the land to lift and slowly these lowlands emerged from the sea. In the seventeenth century, the favourable soil of the Annapolis Valley attracted early French settlers, the Acadians, who built dikes to protect parts of this low-lying farmland from the high tides of the Bay of Fundy and Minas Basin. Today, the Annapolis Valley's stone-free, well-drained soils and its gently rolling landforms provide the best agricultural lands in Nova Scotia. In photo 9.5, at high tide, waters from the Minas Basin (seen in the background) extend into the low, wet land in the foreground. Land use is changing with vineyards replacing apple orchards.

Photo 9.5

Nova Scotia's Annapolis Valley, just north of Wolfville near Cape Blomidon. In the foreground is a small apple orchard, for which the Annapolis Valley is famous; in the middle is a tidal stream; in the background are the waters of the Minas Basin. Like other apple-growing areas in Canada, vineyards are replacing fruit trees.

Barrett & MacKay/All Canada Photos

but the community was saddled with an environmental disaster—the Sydney Tar Ponds, which are composed of tar and a wide variety of chemicals that are dangerous to human beings. Arsenic, lead, and other chemicals are found in the tar ponds. These toxic wastes came from the operation of the Sydney Steel Co., or Sysco, especially its coke ovens, where coal was heated at extremely high temperatures. Waste products from coke ovens include tar and toxic gases. These toxic wastes included benzene, kerosene, and naphthalene.

In the case of Sysco, the wastes were discharged into a nearby stream and gradually seeped into Muggah Creek. Since the 1980s, residents living near the tar ponds complained of orange goo seeping into their basements, while others found that dust from the tar ponds was blowing into their yards and houses. When Health Canada stated that those living closest to the tar ponds were in greater danger than those living further away, Ottawa and Halifax began to pay attention. Action began after a cancer specialist concluded that there was a higher risk of dying from cancer in Whitney Pier and Ashby, communities closest to the tar ponds, than the national average. At that point, the federal and provincial governments joined forces with the local government to fund a cleanup of the tar ponds.

In 1998, Ottawa and Halifax began one of the biggest environmental cleanup projects in Canadian history at a cost of $400 million, with the federal government contributing 70 per cent. Since 2001, the tar ponds, coke ovens site, and other areas within the Muggah Creek watershed have been the scene of intense activity to repair the environmental damage. The community has seen improvements on a number of fronts: the construction of a sewer interceptor that will divert the tonnes of raw sewage that used to flow into the tar ponds; the demolition and removal of derelict structures on the coke ovens site; and the closure and capping of the old Sydney landfill. By 2012, the reclamation project was completed and, instead of an old industrial eyesore, the area will soon consist of a recreation area with sports field, outdoor stage, and walking trails with bridges (CBC News, 2012).

Atlantic Canada's Historical Geography

Atlantic Canada was the first part of North America to be discovered by Europeans. In 1497, John Cabot reached the rocky shores of Atlantic Canada (the exact location, Cape Bonavista, Newfoundland, or Cape Breton Island, Nova Scotia, is in dispute). Yet, Newfoundland and Labrador, the first stretch of North America's Atlantic coastline explored by Europeans, was one of the last to be settled and formally colonized. In sharp contrast, French colonies found root in the Maritimes, where the land and climate were more favourable for agriculture, early in the seventeenth century.

In England, Cabot's report of the abundance of **groundfish**—cod, grey sole, flounder, redfish, and turbot—in the waters off Newfoundland lured European fishers—chiefly English, French, and Basque—to make the perilous voyage across the Atlantic to these rich fishing grounds, though some, especially the Basques, possibly had been fishing these waters at an earlier date. In any event, the Newfoundland coast quickly became a popular area for European fishers and though landings on shore took place—for drying the fish and establishing temporary habitation during the fishing season—permanent settlements were slow to take hold in this part of North America. This pattern of migratory fishing dominated the Newfoundland fishery for some 300 years until political circumstances changed in Europe and North America. During this time, the French presence in the Newfoundland fishery was particularly strong, and Plaisance on the Avalon Peninsula's southwest coast was its largest settlement. With the defeat of the French in 1760, two events shaped the settlement of Newfoundland. First, French access was limited to what was called the French Shore, after 1783 stretching from Cape St John on the north coast around the Northern Peninsula and along the west coast of the island to Cape Ray in the far southwest, and permanent settlement was restricted to the islands of Saint-Pierre and Miquelon. Second, the emergence of a strong resident fishery marked the foundation of a Newfoundland society. English and Irish families settled along the Newfoundland coast, each locating in distinct places.

While French settlers had established themselves in the Maritimes by the early seventeenth century, English settlers began arriving in 1610 in another part of Atlantic Canada—at Cupids in Conception Bay on Newfoundland's Avalon Peninsula. By the 1750s, there were over 7,000 permanent residents, mostly English, living in hundreds of small fishing communities along the Newfoundland coast. Along the French Shore, French fishers enjoyed treaty rights to fish. These rights were first granted by the British in the 1713

Figure 9.4 **Atlantic Canada in 1750**
European settlement in Atlantic Canada was concentrated in the Maritimes. In 1605, a handful of French settlers, at Port Royal on the Bay of Fundy coast of present-day Nova Scotia, established the first permanent European settlement in North America north of Florida. During the seventeenth and part of the eighteenth centuries, French settlers spread into the Annapolis Valley and other lowlands in the Maritimes. By 1750, French-speaking Acadians numbered over 12,000. These French settlements, united by culture, language, and a common economy, became known as Acadia. In the coming decade, Acadians were deported by the British to various English colonies in North America. *Source:* Emanuel Bowen, A new & accurate map of the islands of Newfoundland, Cape Breton, St John and Anticosta. London, William Innys et al., 1747 BAnQ, G 3400 1750 B6.

Treaty of Utrecht, when the French surrendered any territorial interest in Newfoundland, and though the Treaty of Versailles in 1783 cut back on the northern and eastern extent of the French Shore, it replaced what was taken away by extending the French Shore to the southwest corner of the island at Cape Ray. These French rights did not end until 1904. Today, several communities along Newfoundland's southwest coast are populated by descendants of early French settlers.

At the dawn of the eighteenth century, British possessions in Atlantic Canada were little more than names on a map. On the ground, the French colony of l'Acadie and their allies, the Mi'kmaq, were the most numerous inhabitants of the Maritimes, while French and English settlers occupied coastal settlements in Newfoundland with the **Beothuk** still occupying the interior of Newfoundland.

Over the first half of the eighteenth century, war between the two European colonial powers in North America—England and France—was almost continuous. During that time, the French forged an alliance with the Mi'kmaq and Maliseet, drawing them into the conflict with the English and their Iroquois allies.[2] Under the terms of the 1713 Treaty of Utrecht, France surrendered Acadia to the British. However, many French-speaking settlers, the Acadians, remained in this newly won British territory, which was renamed Nova Scotia. During the previous century, the Acadians had established a strong presence in the Maritimes with settlements and forts. Most Acadians lived in the Annapolis Valley, near the Bay of Fundy coast, tilling the soil; others farmed on Île Saint-Jean (Prince Edward Island). Until the mid-1700s, Britain made little effort to colonize these lands, leaving the Acadians to till the land in this British-held territory. The Mi'kmaq remained close to the Acadians, but, as the British power took hold, they became strangers in their own land. Between 1725 and 1779, the Mi'kmaq signed a series of peace and friendship treaties with Great Britain. However, the Mi'kmaq remained close to the Acadian settlers. Events turned against the Mi'kmaq and Acadians with the founding of Halifax in 1749. The expulsion of the Acadians in 1755 from Nova Scotia, and in 1758 from Île Royale (Cape Breton Island) and Île Saint-Jean, eliminated the Mi'kmaq's ally and relations between the British and the Mi'kmaq deteriorated. British rangers were unleashed to harass the Natives, to destroy their villages, and to drive them far beyond the British settlements. After the British defeated the French, the Treaty of Paris in 1763 ceded all French territories in North America to the British except for the islands of Saint-Pierre and Miquelon near the southern coast of Newfoundland.

The next event to influence the evolution of the Maritimes was the American Revolution. Following victory by the American colonies, approximately 40,000 Loyalists made their way to Nova Scotia and New Brunswick where they occupied the fertile lands of the recently departed Acadians and the prime hunting lands and fishing areas of the Mi'kmaq. With its superb harbour for ships of the British navy, Halifax became known as the "Warden of the North." Over the next 100 years, more and more British settlers came to the Maritimes. Nova Scotia alone received 55,000 Scots, Irish, English, and Welsh. Most Scots went to Cape Breton and the Northumberland shore. The driving forces pushing them from the British Isles were the **Scottish Highland clearances** and the **Irish famine**, which resulted in large influxes of migrants with Celtic cultural roots. These immigrants helped to define the dominantly Scottish character of Cape Breton and the Irish character of Saint John. The cultural impact of these Celtic peoples still resonates, and people of Scottish descent are still the largest ethnic group in Nova Scotia ("New Scotland").

For more details on Loyalist migrations to Canada, see "The Loyalists" in Chapter 3, page 112.

By the time impoverished settlers from Britain arrived in the Maritimes, the old world of the Mi'kmaq had collapsed. Denied a place in the new economy, the Mi'kmaq were forced into a state of subsistence in which they soon became outcasts in their own land. It would not be until some 200 years later that the Supreme Court of Canada ruled that the Mi'kmaq have a right to fish commercially in waters where they once ruled (Vignette 9.4).

See Vignette 5.7, "Historical Timeline of the Caledonia Dispute," page 186, for an outline of another modern-day confrontation between Aboriginal Canadians and non-Aboriginal Canadian law and society.

Head Start, Slow Start

In the early nineteenth century, the harvesting of Atlantic Canada's natural wealth increased. This frontier hinterland of the British Empire exploited its rich natural resources—the cod off the Newfoundland coast and the virgin forests in the Maritimes—and became heavily involved in transatlantic trade of these resources. Furthermore, the availability of timber and the region's favourable seaside location provided the ideal conditions for shipbuilding. By 1840, Nova Scotia and New Brunswick entered the "Golden Age of Sail," becoming the leading shipbuilding centres in the British Empire. While the Maritimes enjoyed an economic "head start" coupled with an expanding population, Newfoundland remained focused on the cod fishery and its coastal settlements. This slow start to Newfoundland's economy meant a much slower rate of population increase than in the Maritimes.

Exports from Atlantic Canada were primarily cod and timber, while imports were manufactured goods from England and sugar and rum from the British West Indies. Several world events in the mid-nineteenth century added strength to this economic boom in the Maritimes. Demand for its exports rose, especially for timber, which was used to build British merchant ships and warships. Overseas trade with other British colonies in the Caribbean, as well as with Britain, formed the cornerstone of Atlantic Canada's trade and prosperous economy. In the first half of the nineteenth century, trade with New England was relatively limited because the Maritimes and New England had almost identical resource-oriented economies. However, after the American Civil War, New England industrialized, leading to greater trade between the two regions. In

THINK ABOUT IT
Did the Six Nations of the Iroquois Confederacy, who were allies of the British during the American Revolution, fare better than the Mi'kmaq, who earlier aligned themselves with the French?

addition, Britain's move to free trade in 1849 meant the loss of Atlantic Canada's protected markets for its primary products, resulting in even greater interest in the American market, especially in the Maritimes. Just before Confederation, however, external events dampened the Maritimes' resource-based economy. Iron was replacing wood in shipbuilding; the three-way trade with Britain and its Caribbean possessions collapsed, largely because a world glut of sugar caused prices to drop; and the end of the Reciprocity Treaty in 1866 cut off access to the Maritimes' natural trading partner, New England. These events resulted in the deterioration of the Maritimes' economic position among the Atlantic trading nations and colonies.

Confederation

The provinces of Atlantic Canada joined Canada at different times and for different reasons. Nova Scotia and New Brunswick entered at the time of Confederation; Prince Edward Island followed in 1873; Newfoundland rejected the proposal.

The Maritimes Join Confederation—Reluctantly

Maritimers saw little advantage to joining Canada. Each colony had to be pushed into Confederation. In spite of their historic head start and their excellent access to world markets, they were on the margins of the proposed country.

VIGNETTE 9.4

Mi'kmaq Become Lobster Fishers

In 1999, the Supreme Court of Canada, in its landmark *Marshall* decision, acquitted Donald Marshall Jr of a federal Fisheries Act violation for catching fish (eels) for sale outside of the regulated season. In effect, the ruling meant that the Mi'kmaq should have a share of the commercial fishery. This led, in 2000, to confrontations between non-Aboriginal lobster fishers and Mi'kmaq at Burnt Church, New Brunswick, when the Mi'kmaq—as their Aboriginal and treaty right—began to set lobster traps in Miramichi Bay at a time when the non-Aboriginal fishery was closed. The non-Native fishers took it upon themselves to destroy the Native traps. Fisheries and Oceans officials, in their attempt to impose Fisheries Act regulations, rammed and capsized a Mi'kmaq fishing boat, and as the trouble escalated, the Supreme Court took the unprecedented step of seeking to qualify its earlier decision. Former Ontario Premier Bob Rae was brought in by the federal government to negotiate a solution among the involved parties, and in October 2000 the Mi'kmaq pulled their traps from the water earlier than they had planned.

Photo 9.6
A Mi'kmaq lobsterman.

In the typical Canadian fashion of resolving political confrontation, compromise was reached, though not without some residual bitterness on the part of the involved parties. Five years later, the Mi'kmaq had gained access to about 3 per cent of the federal lobster allocation by receiving about 320 commercial fishing licences. When operating at full capacity, these licences should allow the Mi'kmaq fishers to generate around $25 million worth of lobster each year. "Further Reading," at the end of this chapter, offers a summary of Ken Coates's book on the *Marshall* decision.

New England remained their closest and largest market. So strong was the opposition in Nova Scotia that the colonial government argued that Confederation was no more than the annexation of the province to the pre-existing province of Canada. So strong were feelings that, a year after joining Confederation, the Nova Scotia House of Assembly passed a motion refusing to recognize the legitimacy of Confederation.

See "The Territorial Evolution of Canada" in Chapter 3, page 79.

The federal government's National Policy of 1879 worked against manufacturing in the Maritimes and, at the same time, reinforced Central Canada's population advantage. In doing so, the National Policy placed the Maritimes' fledgling manufacturing base in danger.

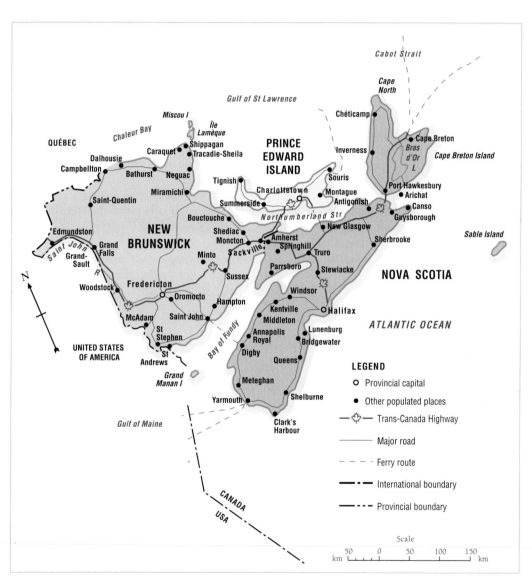

Figure 9.5 The three Maritime provinces: First to enter Confederation
New Brunswick and Nova Scotia joined the Province of Canada (Québec and Ontario) to form the new Dominion of Canada in 1867; Prince Edward Island entered Confederation four years later. The region's manufacturing sector, especially its iron and steel complex on Cape Breton Island, was disadvantaged by the political union because of the relatively long distance to the major market in Central Canada and the expanding market in the West and the curtailment of trade with New England. *Source: Atlas of Canada* reference map–Maritime Provinces at: <//atlas.nrcan.gc.ca/data/english/maps/reference/provincesterritories/maritimes/map.pdf>. Natural Resources Canada, 2000. Reproduced with the permission of the Minister of Public Works and Government Services Canada, 2013.

To offset the disadvantage of geography, Ottawa's answer was the Intercolonial Railway (completed in 1876) that linked the Maritimes with Central Canada. The Intercolonial was operated and subsidized by the federal government: freight rates were kept low to promote trade, and substantial annual deficits were paid by Ottawa. With low-cost transportation to the national market and the possibility of firms achieving economies of scale, a general economic surge occurred in the Maritimes. Cotton mills, sugar refineries, rope works, and iron and steel manufacturing plants were established or expanded to serve the much larger Central Canadian market and to grab a share of the expanding western market. The railway boom in the Canadian Prairies had a positive impact on Nova Scotia in the form of heavy investment in the province's steel mills, where steel production was directed to manufacturing steel rails and locomotives. By taking advantage of Cape Breton's coalfields and iron ore from Bell Island, Newfoundland, the steel industry in Sydney prospered, accelerating Nova Scotia's economic growth well above the national average. However, in 1919, the Maritime economy suffered a deadly blow when federal subsidies for freight rates were eliminated. Immediate access to the national market became more difficult and, with the loss of sales, many firms had to lay off workers, while others were forced to shut down. Even before these troubled times, the Maritimes economy was unable to absorb its entire workforce, leading many to migrate to the industrial towns of New England and Central Canada. From then on, out-migration was a fact of life in the Maritimes and, after 1949, in Newfoundland. Since then, the Maritimes' economy has languished because of its scattered and relatively small population, its narrow resource base, the high cost of transporting goods to the national market, and barriers to trade with New England. Even during periods of national affluence—for instance, the 1950s and 1960s—the Maritimes and Newfoundland and Labrador experienced limited economic growth; numerous federal loans and grants to firms in Atlantic Canada made little difference; and the forlorn struggles of its iron and steel industry typified the problem of manufacturing in the Maritimes and the cod fishery in Newfoundland and Labrador, while maintaining the coastal communities, could not take its economy to a higher level. In 1968, Professor David Erskine at the University of Ottawa drew a rather dismal picture of Atlantic Canada as a hinterland:

> The region is, in the Canadian context, one of "effort" rather than of "increment." Small scale resources once encouraged

VIGNETTE 9.5

Steel, Iron, and Coal—The Rise and Fall of Nova Scotia's Industrial Base

For over 100 years, coal and iron mining provided the basis for the Nova Scotia iron and steel industry. The iron and steel industry in Cape Breton Island near Sydney, Nova Scotia, was for a long time the heavy industrial heartland of Atlantic Canada. During that time, Sydney was the principal city on Cape Breton Island and the second-largest city in Nova Scotia.[3] Sydney's fate was closely tied to its major industrial firm, the Sydney steel mill, and to local coal mines. The mill and mines prospered in the early part of the twentieth century, and both the town of Sydney and steel production expanded, with much steel exported as rails for the construction of railways in Western Canada. Two dark sides to this iron and steel complex existed. One was the loss of life in the coal mines. While the pay was good, danger went with the job and, for some, the cost was their lives. In 1956 and again in 1958, the Springhill mine near the head of the Bay of Fundy was the site of two underground explosions that took the lives of nearly 100 miners. In 1992, the Westray mine in Pictou County, Nova Scotia, was the site of another mine disaster that took 26 lives. The second dark side was the environmental degradation caused by seepage of toxic fluids from the steel mill, which resulted in the Sydney Tar Ponds discussed above.

A major turning point took place after World War II when demand for steel dropped and the size of the labour force was reduced. This process of deindustrialization, while delayed by federal and provincial subsidies, eventually saw the steel mill closed. By 2001, the dream of a heavy industrial base in Nova Scotia was gone. Without the steel and coal industry, Sydney and Cape Breton fell on hard times.

small scale development, but only large scale resources encourage modernization. The small scale and lack of concentration of its resources makes the region one in which government investment is easily dispersed without bringing about growth. Low levels of professional services and low levels of education result from the high taxes from low incomes; thus, still further retardation of economic growth occurs. (Erskine, 1968: 233)

Newfoundland Rejects Confederation

The political process of forming Canada took place in two meetings in 1864—in Charlottetown in September and then at Québec in October. While Newfoundland had observers at the Charlottetown session, its government sent two delegates, Frederic Carter, Speaker of the House of Assembly, and Ambrose Shea, leader of the Liberal opposition, to the Québec Conference. They signed the resolution that formed the basis of the British North America Act but this resolution required the approval of the Newfoundland House of Assembly. Opposition was strong, especially from those of Irish descent who were suspicious of the anti-Catholic sentiment in Ontario as espoused by the Orange Lodge. On the other hand, the merchants were concerned about their business situation and could not see any advantages of a union with Canada. Furthermore, the merchants were strong advocates of a North Atlantic community led by Britain rather than a North American one led by Canadians. The rally call of the anti-confederationists was:

> Men, hurrah for our own native isle, Newfoundland, Not a stranger shall hold one inch of her strand; Her face turns to Britain, her back to the Gulf, Come near at your peril, Canadian Wolf. (Hiller, 1997)

Newfoundland Votes for Confederation—Narrowly

For Newfoundland, the question remained: join Canada or remain an independent political entity. Independence required a sound economy. In 1934, Newfoundland was forced to suspend responsible government and a colonial version of government known as the Commission of Government, run by a governor and six commissioners appointed by the Crown, was instituted. During World War II, the American military presence in Newfoundland made a huge imprint on the people and economy. In Ottawa circles, concerns were expressed about Newfoundland slipping under the US influence and joining the southern colossus. Newfoundlanders, on the other hand, preferred their independence but were attracted to Confederation because they stood to gain from social programs available to Canadians. Two were particularly popular. First, mothers received monthly family allowance payments for each child and that payment provided a reliable source of cash. Second, unemployment insurance payments to seasonal workers like fishers and loggers greatly stabilized family income and, through spending, supported rural business. The option for independence was not encouraged when Britain made it clear that future subsidies to an independent but financially troubled Newfoundland were not in the cards. Britain, as it had with its Maritime colonies nearly a century earlier, favoured a union with Canada. Newfoundlanders faced a difficult choice. In the first referendum of 1948, Newfoundlanders faced three choices—continuance of the Commission of Government for five years, joining Canada, and a return to responsible government (i.e., quasi-independence within a fast-fading British Empire). An independent Newfoundland took 44.5 per cent of the vote, followed by joining Canada at 41.1 per cent. In the second referendum, Commission of Government was dropped off the ballot and voters faced two choices with a slim majority, 52.3 per cent, voting for joining Canada over an independent Newfoundland. From a geopolitical perspective, the vote was split, with solid support for Confederation outside of St John's and the Avalon Peninsula and little support within that more urbanized and Roman Catholic area.

Atlantic Canada Today

Atlantic Canada contains a natural beauty and captivating cultural roots that continue to foster a quality of life for Atlantic Canadians. Yet, young people of the region continue to seek economic opportunities in other parts of the country. The prospect for strong economic growth remains elusive and the challenge facing Atlantic Canada is to translate its fossil fuel resources, minerals, forests, and marine and freshwater resources into a strong sustainable

Figure 9.6 **Newfoundland and Labrador**
When Newfoundland entered Confederation in 1949, Canada gained a territory, population, and the remaining part of Britain's North American Empire. While an integral part of Atlantic Canada, Newfoundland and Labrador occupies a different space, economy, and culture from the Maritimes. In many ways, each has gone its own way, but the proposed Lower Churchill hydroelectric system might well bind them closer together. Source: *Atlas of Canada* reference map–Newfoundland and Labrador, at: <atlas.nrcan.gc.ca/data/english/maps/reference/provincesterritories/newfoundland/map.pdf>. Natural Resources Canada, 2002. Reproduced with the permission of the Minister of Public Works and Government Services Canada, 2013.

economy. Although tourism remains an important part of the regional economy, the high Canadian dollar in recent years—ranging around par with the US dollar—has negatively affected this industry.

Barriers to a sustainable economy are formidable. The overriding barrier is the handicap of being an old resource hinterland. As such, many of the region's most valuable natural resources have already been harvested. Second, Atlantic

Canada's population continues to decline and this trend is predicted to continue according to a 2012 report from the Atlantic Provinces Economic Council (Chaundry, 2012: 95):

> According to the Conference Board, the declines in the labour force between 2011 and 2031 range from 13 per cent in Newfoundland and Labrador to 6 per cent in Nova Scotia and 3 per cent in New Brunswick, compared with a gain of 2.5 per cent in Prince Edward Island and 16 per cent nationally. The cumulative decline in the region's labour force by 2031 is 73,600 people.

Consequently, the region remains troubled by a small market that divides its population into a series of small provincial markets, and the four separate provincial governments are very costly, with the highest per capita cost of government in Canada. Rural Atlantic Canada is confronted with a mismatch of resources and population, resulting in a depopulation of rural Atlantic Canada. The principal factor driving this retreat from rural areas is the loss of jobs in the primary sector, especially those associated with the cod fishery and the forest industry. Rural Atlantic Canada's predicament is summed up by its subpar economic performance, high unemployment rates, and the seemingly unstoppable out-migration to faster-growing towns and cities within Atlantic Canada and to other regions of Canada. Finally, the out-migration of Atlantic Canada's workforce to more prosperous parts of Canada drains away many of the more ambitious, younger, and skilled people.

But opportunities do exist. Atlantic Canada has a second chance. The first opportunity was the discovery and exploitation of offshore oil and gas deposits, which began the process of rejuvenating Atlantic Canada's economy. Geography dictated that, because large oil deposits are located 200–300 km east of Newfoundland beneath the seafloor of the Grand Banks, Newfoundland and Labrador received the economic stimulus and royalties. Since 2008, offshore royalties have exceeded $2 billion annually and form over one-third of that province's revenues (Figure 9.7).

Nova Scotia also has promising offshore petroleum-bearing geological strata. So far only small deposits of natural gas, located about 225 km offshore near Sable Island, have been developed. Unlike Newfoundland and Labrador, Nova Scotia's natural gas production is relatively small and is declining. New gas production from the Deep Panuke Project has encountered numerous delays and production is at least a year away. Gas royalties accruing to Nova Scotia amounted to only $117.9 million in 2011–12 and are projected to fall to $27.7 million in the following fiscal year (Taylor, 2012). Two factors are at play: (1) the relatively low price of natural gas compared to oil; and (2) declining gas production. But the dream situation for Nova Scotia

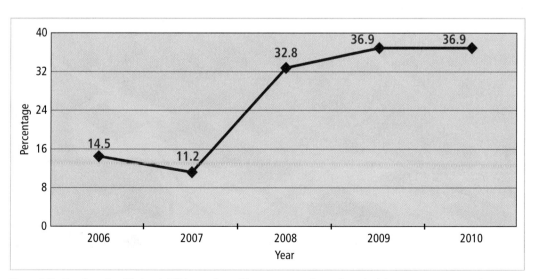

Figure 9.7 **Newfoundland jackpot: Offshore oil royalties as a percentage of provincial revenues**
Source: Department of Natural Resources, Newfoundland and Labrador (2011: 3).

is that its offshore geology contains vast quantities of oil so that, like its neighbour to the north, it will hit the royalty jackpot.

The second opportunity—again, centred on Newfoundland and Labrador—stems from the mining of the huge nickel deposit at Voisey's Bay, Labrador, and the **hydrometallurgy** processing of nickel ore at the Long Harbour facility, scheduled for completion in late 2013, near St John's on the island of Newfoundland. The combination of these two projects maximizes the long-term economic gain for the province.

Location plays a role, too. One example is that New Brunswick, though not a producer of oil or natural gas, is well situated to receive foreign LNG shipments and regasify the liquid natural gas (LNG) at its facilities at Saint John. Afterwards, the natural gas is shipped by pipeline to New England. A second example provides a template for Saint John's role as an energy hub. Five years ago, Maine and New Brunswick signed the **Northeast Energy Corridor** agreement designed to facilitate the flow of electricity from Québec and natural gas from the Saint John regasification plant to markets in New England.

As in the other regions of Canada, economic growth is found in urban centres. This fact remains true for Atlantic Canada. In these centres, not only do some economies of scale prevail but also an atmosphere of innovation can be nurtured. Halifax, St John's, Moncton, and Saint John are benefiting from an ever-widening tertiary base to their urban economies. Most knowledge-based businesses, however, are located in Halifax, the de facto regional capital of the Maritimes. In addition, with its deep harbour, Halifax is ideally suited to participate in global container trade.

Finally, even a positive spin can be put on out-migration. The movement of workers to Alberta's oil sands has had one positive impact on Atlantic Canada, especially its smaller communities—the transfer of wealth back home. Many of these workers commute to the oil sands in a cycle of 20 days at work followed by eight days back home,

Photo 9.7
Artist's sketch of the Long Harbour nickel processing and docking facility on Placentia Bay, southwest of St John's.

and thus bring home much of their wages. Others send money back to their parents and close relatives. Some have made their fortunes and have returned to Atlantic Canada. Unfortunately, the size and impact of this transfer of wealth from Alberta to Atlantic Canada are difficult to estimate. Anecdotal stories abound, such as that Newfoundland and Labrador's second-largest city is Fort McMurray. Yet, does economic opportunity in distant places trump sense of place?

Equalization Payments

Atlantic Canada has long benefited from equalization payments. Only recently has Newfoundland and Labrador broken from the pattern of dependency as a consequence of its offshore oil royalties. Atlantic Canada currently receives just over $3 billion in equalization payments each year (Table 9.2). Before equalization payments to Newfoundland and Labrador ceased in the fiscal year 2008–9, the previous two annual payments were $632 million and $477 million. Since 2008–9, equalization payments to the three Maritime provinces have remained relatively unchanged, though New Brunswick's payment decreased from $1.6 billion in 2008–9 to $1.5 billion in 2012–3. Those provinces receiving equalization payments are labelled "have-not" provinces by the press, but that label is not well received in Atlantic Canada. Unless some spectacular event takes place of the magnitude of the discovery of vast offshore oil deposits, the Maritimes is destined to remain dependent on equalization payments.

Is that spectacular event the $25 billion federal contract to build 21 combat vessels over 30 years by the Irving shipyard based in Halifax? More importantly, could this shipbuilding contract trigger a series of complementary economic events to create an ocean industry cluster in Halifax? A maintenance contract alone would be worth billions and extend over another 30–50-year period. Yet, "cutting steel" is still some years away as the design phase is just starting (MacDonald, 2012a). In 2012, Irving Shipbuilding Inc. of Halifax received $9.3 million to review the design and specification of patrol ships. In March 2013, a federal contract for $288 million to actually design new Arctic offshore patrol ships was awarded to Irving. However, because the company lacks the design expertise, Irving has subcontracted the design phase to Odense Maritime Technology, a Danish engineering and naval architectural firm (Doucette, 2013).

For more on equalization payments, see Chapter 3, "Major Federal–Provincial Programs," page 91.

Can the Lower Churchill Hydro Project Mend Atlantic Canada's Fractured Geography?

The physical setting of Atlantic Canada has prevented this region of Canada from achieving a cohesive economic and political unit. The island of Newfoundland stands alone, separated from the rest of Atlantic Canada by the Cabot Strait and from Labrador by the Strait of Belle Isle. These water barriers hamper Newfoundland from having close economic links with the Maritimes and Labrador. At the same time, Labrador's land connection to Québec has had profound economic implications and has drawn Labrador into Québec's economic orbit. Examples include: (1) the transmission of Labrador hydroelectric energy across

Table 9.2 **Equalization Payments to Atlantic Canada Provinces ($ millions)**

Year	NL	PEI	NS	NB	Atlantic Canada	% of Total Federal Payments
2006–7	632	291	1,386	1,451	3,760	33.3
2007–8	477	294	1,465	1,477	3,713	28.7
2008–9	0	310	1,465	1,584	3,359	25.0
2009–10	0	340	1,391	1,689	3,420	24.1
2010–11	0	330	1,110	1,581	3,021	21.0
2011–12	0	329	1,167	1,483	2,979	20.3
2012–13	0	337	1,268	1,495	3,100	20.1

Source: Canada, Department of Finance (2011). Reproduced with permission of the Minister of Public Works and Government Services, 2011.

Québec; (2) the shipment of Labrador's iron ore to the port of Sept-Îles, Québec; and (3) the Labrador–Québec highway that runs from Labrador's largest town, Happy Valley–Goose Bay, to Baie-Comeau, thereby connecting to Québec's provincial highway system. Connections also exist along the border between New Brunswick and Québec, a shared culture and language prominent among them.

At a cost of at least $6.5 billion and more than 10 years of construction, the Lower Churchill hydroelectric project either represents the megaproject of the twenty-first century for Atlantic Canada or a white elephant for **Nalcor Energy Corporation** (Figure 9.8). The proposal involves two power stations, at Gull Island and Muskrat Falls on the Churchill River. Together, these

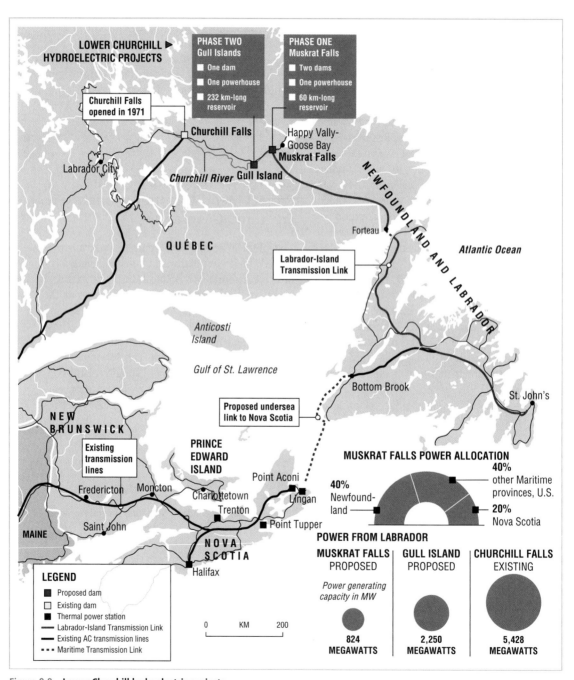

Figure 9.8 **Lower Churchill hydroelectric projects**
Source: Adapted from McCarthy (2011).

CONTESTED TERRAIN 9.1

Whose Energy? Whose Revenue?

From Hydro-Québec's perspective, a contract is a contract. The terms reached in the late 1960s between the Newfoundland government and Québec for the sale and transmission of hydroelectric power from the Churchill Falls Project were negotiated when energy prices were at a rock bottom and the Churchill Falls Project was on the verge of failure. Hard bargaining with a desperate opponent resulted in very favourable terms for Hydro-Québec—purchase of virtually all the power for 40 years at a price of under 30 cents per 1,000 kWh and the option to renew for another 25 years at only 20 cents per 1,000 kWh.

From the perspective of Nalcor Energy, Newfoundland and Labrador's recently formed Crown corporation responsible for energy production and transmission, the contract represents an enormous windfall to Québec because electrical prices have risen sharply over the decades, well above power, maintenance, and transmissions costs. For example, retail customers in Montréal paid $6.76 per 1,000 kWh as of April 2012 while those in St John's paid $11.80, in Halifax $15.01, and in Boston $15.91 (Hydro-Québec, 2012).

Sadly, those who lost the most were the Innu, whose lands, traplines, hunting grounds, and burial sites were altered and flooded without consultation or compensation, as well as other Labradorians, such as the mixed-race Settlers, whose lives revolved around a relationship with the land. The New Dawn Agreement, reached between the Innu and the province in 2008 and ratified in 2011, finally provides compensation for the Upper Churchill Project as well as for the proposed Lower Churchill development.

stations would produce 3,074 megawatts (MW) of electricity (Churchill Falls power station, which achieved full production in 1971, produces 5,425 MW). Rather than transmit the electrical energy through Québec, an all-Atlantic Canada route was selected. The attraction of this route lies in the unification of the electrical power system for Atlantic Canada and exports to New England. The transmission route involves undersea cables, one from Labrador to Newfoundland and the other from Newfoundland to Nova Scotia. From Nova Scotia, the power would move through high-voltage transmission lines to New Brunswick and then to New England. The high cost of the transmitting the energy, especially by untested undersea cables, represents a disadvantage.

 See Chapter 1, "Sense of Place," page 8, for discussion of the powerful cultural bond to a "home" region

Industrial Structure

Atlantic Canada's primary resources are fish, forests, minerals, and petroleum deposits, but the region's economic future lies in its tertiary sector, especially high-technology industries that focus on ocean technology. The rapid growth of ocean technologies in Atlantic Canada is fuelled in part by the strong growth of the offshore oil and gas and the defence/marine security industries. Atlantic Canada's highly developed skills in cold-water engineering are in full view with the $1 billion Confederation Bridge that spans the Northumberland Strait between New Brunswick and Prince Edward Island (photo 9.8).

Employment statistics provide a picture of the basic economic structure of Atlantic Canada. In 2011, employment in primary activities accounted for 5.1 per cent of the labour force; secondary employment constituted 15.8 per cent; and tertiary employment was 79.1 per cent (Table 9.3). Like other regions, Atlantic Canada has experienced percentage decreases in the primary and secondary sectors and an increase in the tertiary sector. But unlike other regions, the total number of workers has remained relatively constant, indicating a stagnant economy (Table 9.3). Another indicator of the weak economy is the high rate of unemployment (Table 9.4).

Atlantic Canada has exceedingly high rates of unemployment, especially in rural and coastal communities. In 2011, unemployment figures for Atlantic Canada remained well above the national average. The good news is that from 1996 to 2011, unemployment figures decreased (Table 9.4), but the gap with western provinces, particularly Alberta, helps to explain the out-migration from the region. Of the four provinces, Newfoundland

Photo 9.8
The Confederation Bridge connects Prince Edward Island with New Brunswick. As an integral part of the Trans-Canada Highway network, the 12.9-km Confederation Bridge is the longest bridge over ice-covered waters in the world. Opened in 1997, the economic impact on Prince Edward Island has been significant in four areas: increased tourism; a real estate boom; expanded potato production and potato-based processed foods; and greater export of time-sensitive and high-priced seafood. These economic gains help to account for the province's increased population.

and Labrador made a remarkable improvement in its unemployment rate over this period, dropping from 25.1 per cent in 1996 to 12.7 per cent in 2011. The offshore oil and gas developments account for much of the change but so do the iron and nickel mining in Labrador as well as the Big Commute to and from the Alberta oil sands.

While urban centres hold their own, out-migration from rural Atlantic Canada remains strong and will continue as the region adjusts to twenty-first-century realities. Many rural towns and villages, for example, have their roots in a fishing industry that can no longer support many of them. Great Harbour Deep is one such isolated community that has disappeared from the map of Newfoundland (Vignette 9.6). On the other hand, economic growth and high rates of employment are concentrated in the major urban

Table 9.3 Employment by Economic Sectors in Atlantic Canada, 2005, 2008, and 2011

Economic Sector	2005 Workers (%)	2008 Workers (%)	2011 Workers (%)
Primary	5.8	5.4	5.1
Secondary	16.0	16.4	15.8
Tertiary	78.2	78.2	79.1
Total Workers	1,075,800	1,109,900	1,102,200

Source: Statistics Canada (2007b, 2009a, 2012c).

Table 9.4 Unemployment Rates in Alberta and Atlantic Provinces, 1996, 2006, and 2011

Province	1996	2006	2011
Newfoundland and Labrador	25.1	14.8	12.7
Prince Edward Island	13.8	11.0	11.3
New Brunswick	15.5	8.8	9.5
Nova Scotia	13.3	7.9	8.8
Alberta	7.2	3.4	5.5
Canada	10.1	6.1	7.5

Source: Statistics Canada (2007b, 2009a, 2012a).

centres, especially Halifax and St John's. Ironically, Atlantic Canada, and especially Newfoundland and Labrador, has a mismatch between the demand for skilled workers and the education/training level of its labour force. The shortage of skilled workers is partly due to the exodus of these workers to such locations as Alberta's oil sands, where wage levels are much higher and overtime is readily available. The Irving shipyard is concerned about the shortage of skilled trade workers, and this issue may become a drag on its plans to begin work on the large federal shipbuilding contract in 2013.

Key Topic: The Fishing Industry

Nature has given Atlantic Canada a vast continental shelf that provides an excellent physical environment for fish: the warm ocean currents from the Gulf of Mexico and the cold Labrador Current create ideal conditions for fish reproduction and growth. The continental shelf extends almost 400 km offshore (Figure 9.9). In some places, where the continental shelf is raised, the water is relatively shallow. Such areas are known as banks. The largest banks are the Grand Banks off Newfoundland's east coast and Georges Bank off the south coast of Nova Scotia (Vignette 9.7).

VIGNETTE 9.6

The End of Great Harbour Deep

For 400 years, people fished cod in the waters off Great Harbour Deep. Located on the Northern Peninsula, Great Harbour Deep was one of the most isolated communities in Newfoundland. Access to the outside world was by a three-hour boat ride in the summer and by a ski-equipped aircraft in the winter. While residents cherished their independence, the need for outside amenities, including health services, had become more and more pressing. In the 1990s, the loss of the inshore fishery and the closure of the town's fish plant led to the economic collapse of this outport. Some left, while others worked on the large trawlers for months at a time. By 2003, the population had dropped to 180 residents. After a series of public meetings, community residents decided to accept the government's $5 million package and leave their community for good. In the summer of 2003, the 50 families of Great Harbour Deep left their community, leaving behind memories and empty homes but hoping to start a new life in a new community. The census tells the story: population in 2001 was 312; 119 in 2006; and 77 in 2011 (Statistics Canada, 2012d).

Photo 9.9
The sheltered bay of Great Harbour Deep in winter.

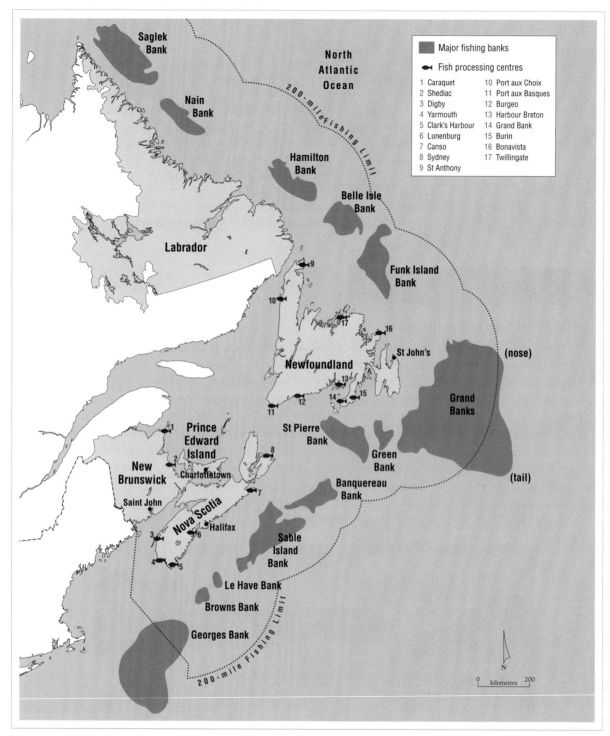

Figure 9.9 **Major fishing banks in Atlantic Canada**
The Atlantic coast fishery operates within a vast continental shelf that extends some 400 km eastward into the Atlantic Ocean, southward to Georges Bank, and north to Saglek Bank. Within these waters are at least a dozen areas of shallow water known as "banks." The Grand Banks of Newfoundland is the most famous fishing ground while Georges Bank contains the widest variety of fish stocks. Scallops, for instance, are harvested in beds on Georges Bank and Browns Bank.

The diversity that characterizes Atlantic Canada extends to the fishery as well. Although each province relies on the fishery, there are striking differences (particularly between Newfoundland and Labrador and the Maritime provinces) in the type of catch and the location. Before the collapse of the cod stocks due to overfishing, which led to federal restrictions on cod fishing in 1992, Newfoundland fishers relied heavily on cod, while Maritimers have harvested a variety of sea life, including cod, lobsters, and scallops. Fishers from the Maritimes ply the waters of Georges Bank and smaller banks just offshore of Prince Edward Island and Nova Scotia. In these waters, Maritimers find cod, grey sole, flounder, redfish, and shellfish. Newfoundlanders, on the other hand, historically fished in the waters of the Grand Banks and, especially, in the inshore fishery around the island of Newfoundland and along the shore of Labrador, where cod and other groundfish congregate in the shallow areas of the continental shelf. Since the collapse of the cod fishery, the Newfoundland fishing industry has diversified and shellfish have become the most important and valuable species.

The Commercial Value of the Atlantic Fisheries

The value of the Atlantic fisheries reached $1.5 billion in 2011. Shellfish, led by lobster, crab, and shrimp, made up 85 per cent of the total value (Table 9.5). The Newfoundland and Labrador fishery lies beyond the ideal habitat for lobster. Here, the value of lobster landings is less than 3 per cent of total landings of $650 million. On the other hand, Newfoundland and Labrador leads in the harvesting of queen crab and shrimp, which account for $470 million or 72 per cent of the province's catch.

Lobster Fishery

In the Maritimes, lobster is the most valued species, making up 62 per cent or $718 million of the total value of the Maritimes fishery. Interestingly, lobster was not always considered a valuable species, and in the eighteenth century was only eaten by poor families—and as a last resort—and was used to fertilize potato mounds. The disdain for lobster slowly waned, and by the nineteenth century the poor man's chicken became highly desired—and high-priced.

While lobster fishing grounds are found in the shallow waters surrounding the Maritimes, the most productive grounds are near Yarmouth on the southern tip of Nova Scotia. Traps are set on the seabed either individually or in groups of up to eight on a line. The traps are hauled to the fishing vessel by powered winches, emptied, re-baited, and lowered again. The catch is transported live

VIGNETTE 9.7

Georges Bank

As part of the Atlantic continental shelf, Georges Bank is a large, shallow-water area that extends over nearly 4,000 km². Water depth usually ranges from 50 to 80 m, but in some areas the water is 10 m or less. Georges Bank is one of the most biologically productive regions in the world's oceans because of the tidal mixing that occurs in its shallow waters. This brings to the surface a continuous supply of regenerated nutrients from the ocean sediments. These nutrients support vast quantities of minute sea life called plankton. In turn, large stocks of fish (such as herring and haddock), as well as scallops, feed on plankton. Cod also feed in these waters, but their numbers have dwindled due to overfishing, especially in international waters. Properly managed, Georges Bank can sustain annual fishery yields of about 400,000 tonnes.

Fishers from both the Maritimes and New England harvest the fish stocks on Georges Bank, which is located at the edge of the Atlantic continental shelf between Cape Cod and Nova Scotia. A boundary decision made by the International Court at The Hague in October 1984 allocated five-sixths of Georges Bank to the United States. However, the easternmost one-sixth that was awarded to Canada is rich in groundfish and scallops. The Nova Scotia fishers have greatly benefited from this decision. With the scallops now under Canadian management, sustainable harvesting of the scallops and groundfish has brought a measure of stability to this sector of the fishing industry.

Table 9.5 Value of Atlantic Coast Commercial Landings, by Region, 2011 ($000s)

Species	NS	NB	PEI	QC	NL	Atlantic Total
Atlantic Cod	5,142	43	42	790	11,226	17,243
Other Groundfish	76,000	718	547	11,393	64,086	152,744
Pelagic and Other Finfish	40,949	14,298	8,840	3,377	19,913	87,377
Lobster	373,653	106,730	84,515	37,947	16,895	619,739
Queen Crab	108,735	33,854	7,122	58,458	250,978	459,147
Shrimp	54,366	8,502	0	29,899	219,154	311,921
Other Shellfish	97,050	10,578	8,327	7,403	49,542	172,900
Other	251	896	390	3	12,103	13,643
Totals	750,146	175,619	109,783	149,269	643,896	1,828,714

Totals may not add up due to rounding.
Source: Adapted from Fisheries and Oceans Canada (2012a). Reproduced with the permission of the Minister of Public Works and Government Services Canada, 2013.

to harbour where it is often kept in seawater-permeated wooden crates for later sale. Since most are exported to New England, the US demand for lobster sets the price. With the US economy in trouble, US lobster prices have dropped below those in the Maritimes, thus greatly reducing the income for fishers.

Cod Fishery

The cod fishery collapsed because of overfishing. Both Canadian and foreign fishers contributed to the demise of the cod stock. At its peak, the cod fishing industry consisted of an inshore fishery, an offshore fishery, and fish processing plants. Traditionally, Newfoundland and Maritime fishers used a flat-bottomed skiff called a dory and stayed close to the shore. Only larger boats ventured to the Grand Banks and other offshore banks. Today, inshore fishers still use smaller boats to fish close to the shore. During the winter, rough waters and ice prevent the smaller craft from leaving port. Offshore fishers travel in steel stern trawlers to the Grand Banks and other offshore fishing banks where they fish for weeks. Upon returning to port, they unload hundreds of tonnes of iced groundfish at a fish-processing plant. Offshore vessels, often much larger than 25 tonnes, are equipped with radar, radios, sonar detectors, and other sophisticated equipment that enable them to operate in winter as well as summer.

The modernization of the Canadian fishing industry, with its larger trawlers and more efficient nets and gear, was accompanied by huge investments by fish companies and individuals and an expansion of the fishing fleet in number and size. Atlantic Canada's fishery, particularly in Newfoundland and Nova Scotia, was able to increase its catch and its income. Now, fishing companies usually own the ships, so fishers no longer share in the value of the catch but receive a salary. However, modernization had many downsides:

- Fewer fishers were required by this more efficient fishing system.
- The new fishing system, based on technological advances and huge dragger nets, created the circumstances for overfishing.
- Overfishing and the collapse of the northern cod stocks forced inshore fishers to turn to government assistance to survive.

Fishers from both the Maritimes and New England harvest the fish stocks on Georges Bank, which is located at the edge of the Atlantic continental shelf between Cape Cod and Nova Scotia. A boundary decision made by the International Court at The Hague in October 1984 allocated five-sixths of Georges Bank to the United States. However, the easternmost one-sixth that was awarded to Canada is rich in groundfish and scallops. The Nova Scotia fishers have greatly benefited from this decision. With the scallops now under Canadian management, sustainable harvesting of the scallops and groundfish has brought a measure of stability to this sector of the fishing industry.

Table 9.6 Cod Landings for Newfoundland/Labrador and Atlantic Canada, 1990–2011 (metric tonnes live weight)*

Year	NL	Atlantic Canada	% NL
1990	245,896	395,266	62.2
1991	178,687	309,031	57.8
1992	75,138	187,804	40.0
1993	37,068	76,644	48.4
1994	2,292	22,719	10.1
1995	863	12,438	6.9
1996	1,147	15,541	7.4
1997	12,317	29,899	41.2
1998	22,764	37,894	60.0
1999	38,663	55,527	69.6
2000	30,216	46,177	65.4
2001	23,774	40,440	58.8
2002	21,083	35,741	59.0
2003	14,290	22,768	62.8
2004	14,534	24,729	58.8
2005	16,257	26,156	62.2
2006	17,050	27,307	62.4
2007	17,784	26,593	66.9
2008	17,594	26,783	65.7
2009	14,473	19,948	72.6
2010	12,228	17,257	70.9
2011	9,744	13,013	74.9

*Fisheries and Oceans Canada has two operational regions for Atlantic Canada, the Maritime region and the Newfoundland and Labrador region. The department reports its statistics by fisheries that correspond to its operational regions. However, it includes the Maritime and Newfoundland fisheries within the larger, non-operational region of Atlantic fisheries.
Since 1992, a modest cod catch was allocated to inland fishers.
Source: Adapted from Fisheries and Oceans Canada (2012b).

However, overexploitation of international fish stocks remains a worldwide problem. The root of the problem is that common, i.e., public, resources are decimated by the selfish actions of individuals who have no regard for the well-being of the resource. Such mismanagement is known as the "tragedy of the commons." Such ecological disasters occur when no one is responsible for ensuring the proper management of resources—or when those entrusted with management do a poor job. Northern cod stocks suffered huge losses from overfishing and mismanagement (Vignette 9.8). Mismanagement was based on three factors: (1) the estimates of cod stocks by the federal Department of Fisheries and Oceans (DFO) were overly optimistic and too often disregarded the anecdotal local ecological knowledge of fishers, and consequently DFO set the quotas too high; (2) strong pressure for high cod quotas came from Newfoundland politicians and from the major fish processors whose trawlers worked the Grand Banks, as well as from some outport communities; and (3) Canada had no control over foreign fishing along the nose and tail of the Grand Banks. The circumstances and consequences of the demise of cod stocks in

Photo 9.10
Neil's Harbour is a small fishing village on the northern tip of Cape Breton. The protected harbour is ideal for mooring small fishing boats.

VIGNETTE 9.8

Ecological Crises and the Resource Cycle

The decline in groundfish stocks in Atlantic Canada is an ecological crisis. But history tells us that many natural resources have suffered such a crisis. Professor Robert Clapp (1998: 129) examined this issue and concluded that "the long record of failure to sustain the yield of biological resources suggests that such restriction is possible only in theory." Clapp offers the resource cycle as an explanation for ecological crises; in other words, what begins as a rich resource leads to overexploitation and the collapse of the resource.

The cause of overexploitation of the northern cod is well known. Much is due to the use of draggers by the Canadian and foreign fishing fleets. The attraction of this form of fishing is its cost-efficiency. The Canadian Atlantic Fisheries Scientific Advisory Committee revealed in mid-1992 that stocks had declined sharply since the 1960s, when the **biomass** (the volume of fish aged three years or older) was estimated at over 3 million tonnes. By 1991, the biomass had dropped to between 530,000 and 700,000 tonnes, the lowest level ever calculated. **Spawning biomass** (generally seven years or older) represents the reproductive part of the fish stock. In 1991 it amounted to between 50,000 and 110,000 tonnes. Thirty years before, the spawning biomass had been as high as 1.6 million tonnes, and even in 1987 it was estimated at 400,000 tonnes. Faced with such a serious crisis, the Canadian government announced in 1992 a moratorium on cod fishing in the waters of Atlantic Canada. Twenty years later, signs of a recovery were observed but federal officials declared that cod stocks had not yet recovered sufficiently to allow the reopening of the cod fishery. Fisheries and Oceans Canada reported in 2012 that the northern cod stock remains 90 per cent below levels measured in the 1980s. "There's a poor prognosis for the stock and recovery," said Don Power, the head of federal groundfish research in St John's. "We're still a good ways down into this critical zone" (MacDonald, 2012b).

As shown in Table 9.6, the last large catch occurred in 1991. Since then, cod fishing has been restricted, on a limited basis, to inshore fishers. Inshore fishers claim the cod stocks are recovering but Fisheries and Oceans Canada (2012c) maintains that cod stocks remain well below levels measured in the 1980s.

Atlantic Canada can lead to a greater appreciation of the need for sustainable harvesting in the fishing industry.

Overfishing and the Cod Stocks

The Atlantic cod (*Gadus morhua*) was the major natural resource in Atlantic Canada. While the Atlantic cod (sometimes called the northern cod) is found as far north as the southern tip of Baffin Island and as far south as Cape Hatteras, cod stocks are concentrated off Newfoundland and along the Labrador coast. The largest single concentration of northern cod, historically, was on the Grand Banks.

The Atlantic cod has been one of the world's leading food fishes for centuries. Cod is sold fresh, frozen, salted, and smoked. Salted cod was very popular in the past, prior to quick freezing and refrigeration, but now most of the cod sold is frozen. During the sixteenth century, the catch was small, probably averaging less than 25,000 metric tonnes a year. By the end of the seventeenth century, however, the annual catch may have reached 100,000 metric tonnes. Annual cod landings during the nineteenth century were about 150,000 to 400,000 metric tonnes. As fishing technology advanced, catches of cod rose to nearly 1 million metric tonnes in the 1950s. By the 1960s, the annual catch reached a peak of almost 2 million metric tonnes. European trawlers accounted for most of this catch.

With the emergence of a modern fishing fleet, fish stocks around the world were overfished. Several fish stocks, including the California sardine, Peruvian anchovy, and Namibian pilchard, collapsed in the 1980s from overfishing—not so much from local fishing operations but from the technology-enhanced fishing fleets that scan the world's oceans for fish. The simple hook-and-line fishing system did not have the capacity to overfish, but foreign and Canadian fishing fleets that employ sophisticated fishing equipment and that drag huge, weighted nets across the ocean floor indiscriminately trapping all groundfish regardless of age, species, and value do have that capability. Known as draggers, these fleets create enormous waste because "non-commercial" fish—the bycatch—are simply discarded. In addition, the draggers destroy fragile ocean-floor ecosystems, including coral reefs and breeding habitat. In spite of the many excuses for the collapse of the northern cod stocks, including changing ocean temperatures and an increase in seal population, the primary cause has been the multinational seafood companies that ran the technologically advanced fishing fleets with factory ships that could stay at sea for extended periods of time. These fleets exploited the cod and other fish of the sea, to the point where a study by 14 international scholars published in late 2006, which analyzed all existing datasets, forecast that all of the world's seafood fisheries will collapse by the year 2050 (Physorg.com, 2006).

The fishing industry in Newfoundland and Labrador is now focused on the shell fishery, especially crab and shrimp. While returns per fisher are high, the downside is that far fewer fishers are engaged in this industry because of the limited number of licences issued by Fisheries and Oceans Canada. Only about 1,200 licences to catch crab are issued to Newfoundland fishers each year. On the other hand, before the cod moratorium, approximately 7,000 licences were issued for groundfish. As a result, the fishery is concentrated in fewer and fewer hands.

Atlantic Canada's Resource Wealth

Atlantic Canada is depending more and more on energy and mineral development than on its fisheries. Since 1997, offshore oil deposits have sparked a boom focused on St John's (Figure 9.10), while Labrador has enjoyed expanding nickel and iron ore output. Agriculture and forestry, two renewable resource activities, trail far behind the non-renewable sector in value of output and number of workers.

Petroleum Industry: The Leading Edge

The petroleum deposits off Canada's east coast were identified in the late twentieth century, but the high cost and technological challenges delayed full-scale exploitation. The oil deposits consist of a **light sweet crude**. These oil fields are situated in sedimentary basins off the east coast of Newfoundland on the Grand Banks. Within the Jeanne d'Arc Basin, oil and natural gas deposits have been discovered and three oil projects—Hibernia, Terra Nova, and White Rose—are now operating. In 2011, production totalled

THINK ABOUT IT
In the late twentieth century, did politicians, both federal and provincial, support the high quotas for the northern cod because they believed that full exploitation of the cod fishery would lead to a more prosperous Atlantic Canada?

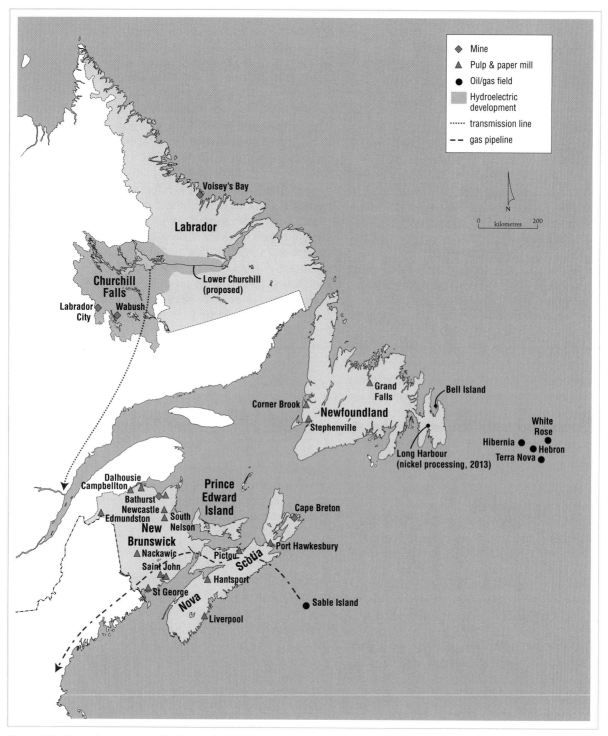

Figure 9.10 **Natural resources in Atlantic Canada**

Besides fish, Atlantic Canada contains important natural resources. Enormous iron and nickel deposits are located in Labrador, huge oil and gas reserves are found below the continental shelf, and one of Canada's largest hydroelectric facilities is situated on the Churchill River. Especially important for Newfoundland and Labrador is that, in late August 2007, Premier Danny Williams negotiated a stake in the Hebron–Ben Nevis offshore project, including a 4.9 per cent equity share at a cost of $110 million. Depending on the price of oil, this share could amount to $16 billion over the 25-year life of the project (McCarthy, 2007). In January 2013, ExxonMobil finally accepted the "Williams terms" and announced that the $14 billion Hebron oil production is moving ahead with production expected to flow by 2017 (Tait, 2013).

15.5 million m³, which amounted to 21 per cent of Canada's output (Canadian Association of Petroleum Producers, 2012a). Newfoundland and Labrador are enjoying record levels of royalties from this oil production, transforming this province from a "have-not" to a "have" province.

These megaprojects have added a new dimension to Newfoundland and Labrador's economy. In 1997, the Hibernia oil project began producing oil; Terra Nova followed in 2002, and White Rose in 2005. All three oil fields are located some 300 km east of St John's, in water ranging from 80 to 120 m deep. The oil deposits extend another 2,500 to 4,000 m below the seabed. These projects required huge capital investments. In turn, they generated construction booms by creating a high demand for workers, especially skilled tradesmen, and for a variety of products and services. Once these developments were operational, employment in the construction industry dropped sharply, and the number of permanent workers required for the oil production is relatively small. Fortunately, Exxon is moving forward with development of its Hebron oil field and the company agreed to build most of the platform locally and compensate the province $150 million for that portion built outside the province (Tait, 2013). Hibernia (Vignette 9.9) uses a fixed platform while the Terra Nova and White Rose oil fields employ floating production storage and offloading (FPSO) vessels. The FPSO facility is a ship-shaped vessel with integrated oil storage from which oil is offloaded onto a shuttle tanker.

The Hibernia drilling site has an annual output of about 30 million barrels and adds greatly to Newfoundland's mineral output. Based on an average price of $80 per barrel, the annual value of production is $2.4 billion. The Brent oil price hit a low of US$41/barrel in February 2009, but more recently the daily Brent price has increased and, since 2011, has frequently been over US$110/barrel; on 11 March 2013, it was at US$109/barrel (Index Mundi, 2013). Over the next 10 years, with a Brent price of just over a $100 per barrel, the value of production is estimated at $3 billion. Oil is exported to American and other foreign refineries. According to the owners, the Hibernia Consortium, production should continue until 2030.

Despite the lure of megaprojects, they often come at a cost in human life. In February 1982, the *Ocean Ranger*, a semi-submersible mobile drilling rig involved in exploratory drilling on the Grand Banks east of St John's, sank in a violent storm. Its crew of 84 died. More recently, in March 2009, a helicopter crashed in bad weather while transporting workers to oil platforms off Newfoundland's east coast, killing 17.

Megaprojects also present problems for regional development. First, they are capital-intensive undertakings. During the construction phase a large labour force is required, but in the operational phase relatively few employees are needed. Second, megaprojects in resource hinterlands lose much of their spinoff effects to industrial areas. As a consequence, economic benefits related to the manufacture of the essential parts for building a megaproject go outside the hinterland, as does the processing of the resource once the project is up and running. Interventions

Table 9.7 **Agriculture, Fishing, Forestry, Mineral, and Petroleum Production in Atlantic Canada, 2011 ($ millions)**

Product	NL	PEI	NS*	NB	Atlantic Canada**
Minerals	5,190	3	247	1,308	6,748
Petroleum	10,772	0	343	0	11,115
Farm Cash Receipts	125	475	539	539	1,678
Fishing	643.9	109.8	750.1	175.6	1,679.4
Forest	263	3	856	2,574	3,696
Total	17,000	586	2,448	4,571	24,605

*Gas production has dropped sharply in the offshore fields of Nova Scotia. Value of production in 2008 was $1.3 billion.
**The figure for the Atlantic Canada fishery excludes Québec's North Shore (see Table 9.5).
Sources: Natural Resources Canada (2012); Canadian Association of Petroleum Producers (2012b); Canada Forest Service (2012); Fisheries and Oceans Canada (2012a); Prince Edward Island, Department of Agriculture and Forestry (2012: Table 2).

by the government of Newfoundland and Labrador to address this classic problem have had mixed results. The biggest success story comes not from the petroleum industry but from the agreement with the developers of the Voisey's Bay nickel mine whereby the government of Newfoundland and Labrador obtained a commitment from the company, Vale,[4] to process the ore at Long Harbour in Placentia Bay. Production is expected to begin in 2013.

The Mining Sector: Solid Performer

Atlantic Canada is endowed with world-class mineral deposits. The Canadian Shield in Labrador has rich deposits of iron ore and nickel. Historically, coal mining on Cape Breton Island was extremely important and was the basis of Nova Scotia's iron and steel industry. But those days are gone and the last coal mine was closed in 2001.[5] By the twenty-first century, iron and nickel mines, owned and operated by global mining companies, dominated the mining sector and these mines are located in Labrador. Following the worldwide trend, these Canadian deposits have been purchased by global companies. In the case of Labrador Iron Mines, it is a subsidiary of India's Tata Steel company (Canadian Press, 2013); Brazil's Vale controls nickel mining in Labrador and its processing plant at Long Harbour.

VIGNETTE 9.9

The Hibernia Platform

The Hibernia oil project is located 315 km east of St John's, Newfoundland, on the Grand Banks. To tap the estimated 615 million barrels of oil from the Hibernia deposit, an innovative offshore stationary platform was needed. About 4,000 workers built a specially designed offshore oil platform that can withstand the pounding storms of the North Atlantic and crushing blows from huge icebergs. The massive concrete and steel construction sits on the ocean floor, with 16 "teeth" in its exterior wall designed to absorb the impact of icebergs.

The 111-metre-high Hibernia platform, which includes oil-storage units, weighs over 650,000 tonnes. In the summer of 1997 the platform was placed on the ocean floor just above the oil deposits. The depth of the water at this point is about 80 m, leaving the oil platform approximately 30 metres above the ocean surface. This structure is designed as a platform for the oil derricks, and houses pumping equipment and living quarters for the workers, as well as a storage facility for the crude oil. The rig extracts oil from the Avalon reservoir (2.4 km under the seabed) and from the Hibernia reservoir (3.7 km deep). The crude oil is then pumped from the Hibernia storage tanks to an underwater pumping station and then through loading hoses to three 900,000-barrel supertankers for shipment to foreign refineries.

Adapted from Cox (1994).

Photo 9.11
The Hibernia platform has a massive concrete base that supports its drilling and production facilities as well as the workers' accommodations. Since the platform was positioned on the ocean floor in 1997, the province has joined the ranks of other oil-producing provinces. The government of Newfoundland and Labrador at first received half of the royalties generated by these offshore developments. With a new agreement with Ottawa, in 2006 the province received all of the oil revenues. By 2009, oil revenues had made Newfoundland and Labrador a "have" province.

Photo 9.12
Discovered in 1993, the Voisey's Bay nickel deposit lies along the coast of Labrador approximately 350 km north of Happy Valley–Goose Bay. At that time, the Labrador Innu and Inuit had not settled their land claims (Vignette 9.10). In 1996, Inco Ltd acquired the rights to the Voisey's Bay property and then sold it nearly a decade later to Vale, the giant Brazilian-owned mining company. Vale began its production in 2005 and employs approximately 450 workers. The nickel ore body is estimated at 150 million tonnes, giving the mine a lifespan of 25 years. At its Ovoid site, mining operations are employing an open-pit system. The more expensive underground mining will take place at two other deposits—the Western Extension and Eastern Deeps. The initial processing of the ore begins at the mine site with the reduction of the ore into a more concentrated form of nickel. From Voisey's Bay, the ore is shipped to Sudbury and Thompson for further processing until the hydromet processing plant at Long Harbour is up and running in 2013.

In 2011, the annual value of mineral production in Atlantic Canada was $6.8 billion (Table 9.7), with nickel valued at $1.7 billion and iron ore at $2.7 billion (Natural Resources Canada, 2012). Much smaller production by value comes from the lead-zinc mine operated by Brunswick Mining and Smelting Corp. near Bathurst, New Brunswick. In 2011, its production came to $557 million.

The exciting mineral development taking place at the huge nickel deposit, the Ovoid deposit, at Voisey's Bay, Labrador, has powerful implications for both Labrador and the island of Newfoundland, where the processing will take place. Employment will reach 1,000 workers when the hydromet processing plant swings into play, hiring 500 workers. Since the Ovoid deposit lies close to the surface and is only a short distance from open water, the Voisey's Bay mine is one of the lowest-cost nickel mines in the world. It consists of 32 million tonnes of relatively rich ore bodies—2.8 per cent nickel and 1.7 per cent copper. Another deposit, Eastern Deeps, though larger (about 70 million tonnes), has a lower grade of nickel and copper (similar grade to that found at the Sudbury mines in Ontario). The Voisey's Bay deposit has three attractive features: high-grade nickel, a surface deposit with the potential for open-pit mining, and proximity to ocean shipping. Processing of the ore will add to the value of this development for the province. In April 2009, construction began on the $2.8 billion nickel processing plant at Long Harbour, a short distance north of Argentia. This plant is expected to be fully operational by 2013 (Roberts, 2009; Vale, 2012).

Forest Industry: A Weak Sister

Beyond the sea, Atlantic Canada's most important renewable resource, historically, has been its forests. The rugged Appalachian Uplands in Atlantic Canada encompass 22.9 million ha of forest land. The forest industry, both logging and pulp and paper processing, is concentrated in

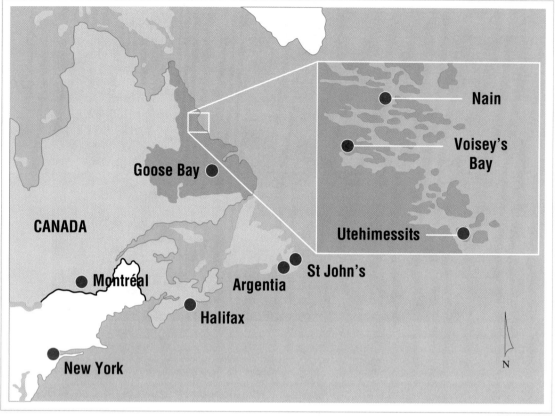

Figure 9.11 **Voisey's Bay mine site**
Source: North of 56 (2012).

New Brunswick. Logging is both an important employer and an income generator for woodlot owners. Unlike in the rest of Canada, where forest land is usually Crown land (public), the proportion of private timberlands to Crown lands in the Maritimes is extremely high. Private timberlands make up 92 per cent of the commercial forest in Prince Edward Island, 70 per cent in Nova Scotia, and 50 per cent in New Brunswick. In Newfoundland and Labrador, like the rest of

VIGNETTE 9.10

Land Claims and the Labrador Inuit and Innu

The Voisey's Bay mining development sparked a renewed interest in comprehensive land claim settlements with the Inuit and with the Labrador Inuu.[6] At the time the proposed development was taking form in 2003, neither Aboriginal group had reached such an agreement. Consequently, Inco (now Vale) negotiated Impact and Benefits Agreements (IBAs) with the Labrador Inuit and the Labrador Innu, thus allowing the project to proceed before comprehensive agreements were reached. For the Aboriginal peoples, the IBAs included employment opportunities, provided the Aboriginal workers had a basic command of English necessary for the workplace and at least a high school level of education. Unfortunately, relatively few qualified. By 2005, the Labrador Inuit and the federal government signed the comprehensive land claim agreement that created the Inuit government of Nunatsiavut. By 2013, the Innu had not yet reached a similar agreement with the federal government, although the New Dawn Agreement between the Innu Nation of Labrador and the province, which was achieved in 2008 and ratified by the Innu three years later, has established the parameters—in regard to claim area and compensation—for an eventual agreement with Ottawa.

Canada, private ownership makes up only 2 per cent of the forested area. On average, the rate of logging on private lands is very high, sometimes exceeding the annual allowable cut estimated by the province. The high rate of logging often takes place on farms where timber sales are an important source of income.

Yet, the best days for this industry are long gone. In fact, the forest industry in Atlantic Canada (and across Canada's boreal forest) is contracting due to diminishing demand and low prices in the US. Other factors contributing to the weak state of the forest industry are the appreciating Canadian dollar against the US dollar, making forest products more expensive for American buyers; declining demand for newsprint due to a slumping newspaper industry; and rising electricity costs in Atlantic Canada, which constitute a large portion of operating costs in pulp and paper. The weak nature of the forest industry is revealed by a slight drop in the value of forest products from $4 billion in 2001 to $3.7 billion in 2011 (Table 9.7). The downward cycle, as elsewhere in Canada, is linked to the collapse of the US housing market. All forest operations have suffered: logging, sawmilling, and pulp and paper plants. Pulp and paper mills have been hit hard. In 2002, nine pulp plants operated in New Brunswick, three in Nova Scotia, and three in Newfoundland. Since then, each province has seen plant closures. The most recent occurred at Liverpool, Nova Scotia, with the closure of the mill putting 320 employees out of work. As well, logging operations ceased (Ware, 2012).

A new issue in the forest industry in Atlantic Canada relates to Aboriginal people's right to timberlands. Aboriginal Canadians are demanding greater access to timber. Most Crown land in Atlantic Canada has been leased to large companies, mainly to the Irving Forest Corporation. In 1998, a dispute erupted over cutting rights on Crown land. The dispute went to court. In a provincial court ruling in 2003, the right of the Mi'kmaq to harvest timber on Crown lands claimed by these Indians was confirmed. However, in 2005, the Supreme Court of Canada, in a decision described

THINK ABOUT IT
In spite of great natural wealth, Atlantic Canada, unlike Western Canada, remains unable to shift from a resource economy to a more diversified one. Does this mean that Innis's staples thesis, described in Vignette 1.4, is flawed?

Photo 9.13
The AV Nackawic mill is a joint venture of India's Aditya Birla Group and Québec's Tembec, and produces rayon from wood fibre for the Asian market. Like many other mills, it is on the financial edge. The mill closed in 2004, but the present ownership group took over and reopened the mill in 2006 with millions of dollars of provincial loans and loan guarantees. In April 2008 the New Brunswick government provided the owners with another $10 million loan, and then added another $10 million loan in July 2009 to keep the mill running and preserve 300 jobs. Public monies were still sought. By November 2012 the mill was still operating, though the company planned to reduce its labour force over the next three years.

as a setback for Native rights, ruled that the Mi'kmaq of Nova Scotia and New Brunswick need permits if they want to cut down trees on Crown land for commercial gain. The Mi'kmaq are still seeking a compromise with the provincial governments whereby some Crown land currently leased to large companies would be reallocated to them. Just how this ruling of the Supreme Court will translate into a negotiated settlement remains unclear at this time.

Agriculture: Limited in Size

Agriculture is limited by the physical geography in Atlantic Canada. Arable land constitutes less than 5 per cent of the Maritimes. Arable land is even rarer in Newfoundland and Labrador, making up less than 0.1 per cent of its territory. Though limited in size, agricultural production significantly contributes to the economy of Atlantic Canada. In 2011, the value of agricultural production in the region was about $1.7 billion (Table 9.7). New Brunswick, Nova Scotia, and Prince Edward Island provide 98 per cent of this figure. Specialty crops, especially potatoes and apples, contributed heavily to the value of this production.

Atlantic Canada has nearly 400,000 ha in cropland and pasture. Almost all of this farmland is concentrated in three main agricultural areas—Prince Edward Island, the Saint John River Valley in New Brunswick, and the Annapolis Valley in Nova Scotia. Potatoes and tree fruit are important cash crops, though vineyards are gaining ground in the Annapolis Valley. In all three agricultural areas, dairy cattle graze on pasture land. The dairy industry in Atlantic Canada has benefited from the orderly marketing of fluid milk products through marketing boards.

Prince Edward Island is the leading agricultural area in Atlantic Canada. It has almost half of the arable land in the region. Most of Prince Edward Island's 155,000 ha of farmland are devoted to potatoes, hay, and pasture, with the principal cash crop being potatoes. Since the 1980s, most potato growers have had contracts with the island's major potato-processing plants—Irving's processing plant near Summerside and McCain's plant at Borden–Carleton now dominate the potato industry on the island. The second major agricultural area, the Saint John River Valley, is in New Brunswick. Its 120,000 ha of arable land make up about one-third of Atlantic Canada's farmland. The Saint John River Valley has the best farmland in New Brunswick. Nova Scotia has nearly one-quarter of Atlantic Canada's farmland, with 105,000 ha. Nova Scotia's famous Annapolis Valley, the region's third agricultural area, is the site of fruit orchards and market gardens. The valley's close proximity to Halifax, the major urban market in Atlantic

Photo 9.14

The rich, red soils of Prince Edward Island are famous for growing potatoes, which are the primary cash crop in the province. Prince Edward Island is Canada's leading potato province, responsible for almost one-third of Canadian production. Its potatoes are grown for three specific markets: seed, table potatoes, and processing. Seed potatoes are sold to commercial potato growers and home gardeners to produce next year's crop; table potatoes go to the retail and food service sectors; and processing potatoes are manufactured into french fries, potato chips, and other processed potato products.

Canada, has encouraged vegetable gardening. In both New Brunswick and Nova Scotia, potatoes are a major cash crop. Almost all potato farmers in these two provinces seed their potatoes under contract to McCain Foods, a multinational food-processing corporation based in New Brunswick. The company has benefited from NAFTA after the removal of tariffs on its food products, especially french fries and potato chips, for export to the United States. Newfoundland has the least amount of farmland—just over 6,000 ha.

The Politics of Hydro Development: The Lower Churchill Project

Labrador is rich in hydroelectric sites, but far from markets and limited to one land route—through Québec—for transmission of the power produced. The geopolitical dynamics of pursuing the risky and hugely expensive Lower Churchill Project, including undersea transmission lines spanning the Strait of Belle Isle and the Cabot Strait, are attractive to Newfoundland and Labrador because this project could challenge Hydro-Québec's monopoly on exports to the United States. The project would also please Newfoundlanders who still feel cheated by the 1969 Churchill Falls agreement.[7]

Geography had a hand in the conditions of the Churchill Falls agreement because the immediate market for Churchill Falls power was just across the border at the iron mines near Schefferville, Québec. Other factors were the huge cost of the project and its long construction period before returns would be realized. The long-term contract signed in 1969 allowed Churchill Falls a secure market for its power but at a low and "fixed" price. In this way, Hydro-Québec made the deal of a lifetime by obtaining vast quantities of low-cost power over 65 years at 1960 rates. While the initial agreement was made in 1969, its starting date began with the production of power in 1976. However, a "controversial" 25-year extension was added, pushing the expiry date to 2041. The problem with the agreement—at least for Newfoundland and Labrador—is obvious. Since the price of Churchill Falls power is fixed at 1960 levels (Feehan and Baker, 2005), the returns to Newfoundland and Labrador were always far below current market prices. In hindsight, fixing the price was a mistake for the province, but then, who would have predicted the jump in energy prices in the 1970s due to OPEC? Added to that unexpected boost in energy prices, the desire for "clean" energy towards the end of the twentieth century forced utilities in New England to reduce their power production from local coal-burning generating stations.

It remains to be seen if the Lower Churchill Project will alter the playing field between Québec and Newfoundland and Labrador over the old Churchill Falls contract. First, let us assume that Hydro-Québec expects to negotiate a new agreement in 2041 with Newfoundland and Labrador. However, if the Lower Churchill Project proceeds, then another option is opened, whereby, in 2041, the power from Churchill Falls could flow through the Lower Churchill transmission system to the Maritimes and New England. For Hydro-Québec, the question becomes: "Should the old Churchill Falls contract be renegotiated now to discourage the construction of the very expensive Lower Churchill transmission line with its two undersea cables—one to Newfoundland and the other to Nova Scotia?" From Hydro-Québec's point of view, keeping the power from Churchill Falls plus gaining the electricity generated from the Lower Churchill Project would secure Québec's dominant position not only in Québec and Atlantic Canada but also in the very profitable New England energy market for the twenty-first century and beyond. But then, the bad feelings among Newfoundlanders and Labradorians might only be appeased by the Lower Churchill Project, regardless of its cost.

Atlantic Canada's Population

Since Confederation, Atlantic Canada's population has increased, but at a rate well below the national average. Over the last 15 years, Atlantic Canada's population has actually declined and this decline is predicted to continue into the future (Chaundry, 2012: 95). From 1996 to 2011, the region's population declined by just over 6,000, from 2,333,764 to 2,327,638 (Table 9.8). Over this span of 15 years, the Maritimes saw a modest increase, but Newfoundland and Labrador lost just over 37,000 residents. For Newfoundland and Labrador, the loss took place in the smaller communities, especially the outports. Many have taken advantage of high-paying jobs in Alberta and moved permanently to that province. Others have joined in the **Big Commute**, working in Alberta but keeping family and home in

THINK ABOUT IT
Thinking of regional faultlines, would you rank BC's current opposition to the Northern Gateway Project higher or lower than Newfoundland and Labrador's feelings towards the 1969 Churchill Falls agreement?

Table 9.8 **Population Change in Atlantic Canada, 1996–2011**

Province	1996	2006	2011	Change 1996–2011	% Change 2006–11
PEI	134,557	135,851	140,204	5,647	4.2
NL	551,792	505,469	514,536	–37,256	–6.8
NB	738,133	729,997	751,171	13,038	1.8
NS	909,282	913,462	921,727	12,445	1.4
Atlantic Canada	2,333,764	2,284,779	2,327,638	–6,126	–0.3

Source: Statistics Canada (2002a, 2007a, 2012a).

THINK ABOUT IT
Which has the higher social costs for a family, unemployment or air commuting with 14 days at the work site and 7 days at home (minus air travel time)?

Newfoundland (Vignette 9.11). Professor Keith Storey (2009) notes that commuting became so popular that some 5,000 workers in St John's lined up for jobs in Alberta, and that by 2008 about 7 per cent of the Newfoundland and Labrador workforce was employed outside of the province but living in the province. The demographic split, with the Maritimes gaining population and Newfoundland and Labrador losing population, is likely to continue for the next decade.

The main demographic factors accounting for this decline have been a falling birth rate and a steady death rate, which resulted in a very low rate of natural increase; little in-migration; and a massive out-migration, especially from Newfoundland. In Newfoundland and Labrador, the unemployment rate was the highest of the 10 provinces, thus providing a push to seek work in other parts of Canada.

Atlantic Canada's Urban Geography

The urban geography of Atlantic Canada is characterized by few large cities and, as we have seen, many small towns. Statistics Canada's definition of an urban place requires a minimum population concentration of 1,000 persons and a population density of at least 400 persons per square kilometre (Statistics Canada, 2009c). It is not surprising, then, that Atlantic Canada in 2011 was the least urbanized region of Canada with just over half of its population living in urban centres. In comparison with other regions, the difference is both striking and an indicator of how much more urban growth (or rural decline) is likely. For instance, in Ontario and British

VIGNETTE 9.11

The Big Commute

Although an untold number of Atlantic Canadians have moved to Alberta for work in the oil fields and, more generally, in construction, many others commute from their homes on the east coast on a cycle of about two weeks of work in the oil patch and one week back home. Estimates suggest that as many as 10,000 Newfoundland trades workers regularly commute to the Alberta oil sands to work for salaries that start above $100,000 a year, not including overtime. Their air travel from St John's to Fort McMurray is paid by the company and once there they are fed and housed at company expense (CBC News, 2007, 2009; Storey, 2009). The earnings they bring back are believed to amount to hundreds of millions of dollars—for Newfoundland and Labrador, a hidden economic boost for a recently anointed "have" province and a monetary infusion that is keeping some communities viable, and at the same time dependent on the fortunes of an industry practically at the other end of the country. The hidden costs—to families, to social structure, to individual lives and values—are perhaps even more difficult to discern. But as Kirk Penney, a 30-year-old electrician who earns $150,000 a year at Suncor Energy's Firebag site, said, "it provides a good lifestyle for my kids. That's the only reason I'm doing it [working 14 days on site, flying for 16 hours, and spending six days at home]. Living in Newfoundland is mostly hand-to-mouth." He stays in touch by morning phone calls to his kids and night calls to his wife (Quinn, 2012).

Photo 9.15
With the city of Saint John in the background, the strategic location of Canaport LNG facility, in this artist's conception, is ideal for access to the huge New England energy market. Liquefied natural gas arrives from a variety of locales, including the Caribbean and Middle East. Here, the liquefied gas is stored in containers at −162°C, then is regasified and sent by pipeline to New England markets. The first shipment of Qatar natural gas arrived in December 2009 in the *Q-Flex*, the world's second-largest LNG tanker.

Canaport LNG

Columbia the urban population accounts for over 85 per cent of total population. As well, Atlantic Canada has none of Canada's largest cities: Halifax ranks thirteenth in population and St John's is twentieth (see Table 9.9 and Table 4.3). Equally significant, Atlantic Canada's fractured geography results in these two cities being the dominant metropolitan centres in the region. Halifax serves as the urban focal point for the Maritimes, while St John's fills a similar role for Newfoundland and Labrador. Halifax's advantage is its deep, ice-free harbour, its role as a naval base, and its relatively large population/market. Halifax serves as a major container port and has the potential of becoming the hub of the Atlantic Gateway. On the other hand, St John's today is focused on offshore oil, the fishing industry, and government services, and is a centre for Arctic marine research and resupply. The only other CMAs in the Atlantic region are Moncton and Saint John, New Brunswick. Within the Maritimes, Saint John is ideally situated as an energy hub to New England. The natural gas shipped to the United States comes from foreign sources and is shipped to Saint John in a liquefied state (see photo 9.15). Moncton, geographically, is positioned as a "gateway" to both Nova Scotia and PEI, and with a large francophone population is a "gateway," too, to the Acadian French area of northern New Brunswick.

The larger urban centres, especially those with a population of 50,000 or more, had the highest rates of increase from 2001 to 2011. Moncton led with a 16.8 per cent increase over its 2001 population, followed by Fredericton at 15.9 per cent and St John's at 13.9 per cent. Surprisingly, Halifax only increased over that period by 8.7 per cent. The exception to the large-city rule was Labrador City (19.2 per cent increase), which was undergoing an expansion of its iron mining operation, and Gander (6.0 per cent), which serves as an international airport and refuelling station between Europe and North America. On the other hand, urban centres losing population over this 10-year period were led by Cape Breton,

THINK ABOUT IT
Would national unity be served if Alberta crude supplied the oil refineries in Montréal and Saint John? Keep in mind that both eastern refineries pay the Brent price for crude oil while the Alberta oil, because it is geographically distant from ocean ports, receives a much lower price, the so-called Cushing price (see "oil prices" in Glossary).

Table 9.9 Urban Centres in Atlantic Canada, 2001 and 2011

Urban Centre	Population 2001	Population 2011	% Change
Labrador City, NL	7,744	9,228	19.2
Gander, NL	9,651	10,234	6.0
Bay Roberts, NL	10,531	10,871	3.2
Grand Falls–Windsor, NL	13,340	13,725	2.9
Campbellton, NB	18,820	17,842	−5.2
Edmundston, NB	22,173	21,903	−1.2
Kentville, NS	25,172	26,359	4.7
Corner Brook, NL	26,153	26,623	1.5
Bathurst, NB	32,523	33,484	3.0
New Glasgow, NS	36,735	35,809	−2.5
Truro, NS	44,276	45,888	3.6
Charlottetown, PEI	57,234	64,487	12.7
Fredericton, NB	81,346	94,268	15.9
Cape Breton, NS*	109,330	101,619	−7.1
Saint John, NB	122,678	127,761	4.1
Moncton, NB	118,678	138,644	16.8
St John's, NL	172,918	196,966	13.9
Halifax, NS	359,183	390,328	8.7

*Cape Breton includes Sydney, North Sydney, Glace Bay, and other Cape Breton Island municipalities and, therefore, does not have the population density to be considered a census metropolitan area.

Source: Adapted from Statistics Canada (2002b, 2012a).

a former coal mining centre, which saw its population decline by 7.1 per cent, followed by Campbellton (–5.2 per cent); New Glasgow (–2.5 per cent) and Edmundston (–1.2 per cent).

See Chapter 4, "Urban Population," page 129, including Table 4.4, "Percentage of Urban Population by Region, 1901–2011," for further understanding of Atlantic Canada's lagging urbanization.

VIGNETTE 9.12

Halifax

Halifax, the capital of Nova Scotia and the largest city in Atlantic Canada, was founded in 1749. By 2011, Halifax had a population of 390,000. As in the past, its strategic location allows Halifax to play a major role on the Atlantic coast as a naval centre, an international port, and a key element in the Atlantic Gateway concept. Along the east coast of North America, its deep, ice-free harbour is ideally suited for huge post-Panamax ships. Yet, because of its relative distance from the major markets in North America and its reliance on transferring goods between ships, trains, and trucks, Halifax cannot provide lower transportation costs than New York. The economic strength of Halifax rests on its defence and port functions, its service function for smaller cities and towns in Nova Scotia, and its role as a provincial administrative centre. Halifax also has a small manufacturing base and a growing service sector, as well as a small but growing high-technology industry. In 2013, the 30-year military shipbuilding contract with the federal government has begun at the design stage. Overall, the total contract is expected to stimulate the economy of Halifax and result in a surge in population.

SUMMARY

Atlantic Canada lies on the eastern rim of Canada. Atlantic Canada remains Canada's economically troubled region with high unemployment figures and record out-migration of workers. Yet, while unemployment and out-migration continue, Newfoundland and Labrador has a second chance due to its offshore petroleum resources, its rich nickel deposit at Voisey's Bay, and the potential of hydroelectric power from the proposed Lower Churchill Project. Newfoundland and Labrador has already shaken off the mantle of a "have-not" province but its population continues to decline. Even the booming economy of Alberta has helped through the Big Commute, which transfers to Atlantic Canada much-needed wages. Even though Atlantic Canada's cities are growing ever so slowly, they play a critical role in the region's second chance.

Thus, the future for Atlantic Canada in the twenty-first century looks promising, but the prospects for sustainable economic growth remain problematic. On the one hand, mega-resource developments, especially offshore oil, have the potential to transform the regional economy, and the possibility of an Atlantic Gateway—a trade corridor connecting Atlantic Canada with North America and the rest of the world—if realized, would be a tremendous boost. For Newfoundland and Labrador and Nova Scotia, the expansion of offshore oil and gas extraction has stimulated their economies, as will the start of military shipbuilding in Halifax for Nova Scotia, while New Brunswick is banking on serving as an energy hub for New England and Nova Scotia. On the other hand, two nagging questions persist. First, what future does rural Atlantic Canada have since the new economy is taking place in the major cities? Second, does energy production hold the key to economic rejuvenation and sustainability? The answer to the latter question seems more hopeful because oil-producing Newfoundland and Labrador receives billions in royalties each year. As the global economy recovers, higher prices for its commodities, including softwood lumber, could propel Atlantic Canada's economy into a more prosperous state—one in which unemployment rates decline, business opportunities increase, and out-migration slows and even reverses. Such an economic state may create that elusive balance between economic growth and a "down East" way of life.

CHALLENGE QUESTIONS

1. Offshore oil and iron/nickel mining have provided Newfoundland and Labrador with an opportunity to break its downward economic spiral, but why did these developments fail to slow the province's population decline?
2. Given the difference between Brent (world) pricing and Cushing (US) pricing, why does it make economic sense for both Alberta and New Brunswick to refine Alberta crude at the Irving refinery in Saint John?
3. Fishing technology was critical to the destruction of the cod stocks. But why were draggers the worst culprit in destroying the cod stocks and other marine life?
4. Why do you think that Québec might offer to renegotiate the 1969 Churchill Falls contract?
5. While the forest industry is in trouble, why are the pulp and paper mills in Atlantic Canada the most vulnerable to closure?

FURTHER READING

Coates, Ken S. 2000. *The Marshall Decision and Native Rights*. Montréal and Kingston: McGill-Queen's University Press.

On 7 September 1999 the Supreme Court of Canada ruled in a case involving Donald Marshall Jr that the Mi'kmaq could earn a "modest income" from the fishery in the Maritimes. In one swoop, Atlantic Canada woke up to the Aboriginal desire and right to participate in the fishery, especially the lucrative lobster fishery. This is easier said than done because sharing of natural resources, such as lobsters, means taking away from those who already have the right to harvest such resources. The lobster solution involved the federal government obtaining lobster fishing licences from non-Aboriginal fishers and allocating them to the Mi'kmaq fishers.

Coates places the *Marshall* decision, which changes the relationship between the Mi'kmaq and other Maritimers, within the larger Canadian context where a search for a new relationship between Aboriginal Canadians and other Canadians is underway.

10

THE TERRITORIAL NORTH

Introduction

The Territorial North is Canada's last frontier, but it is also a homeland for Aboriginal peoples, who form the majority of its population (Table 1.2). Located in the highest latitudes of Canada, the Territorial North, remote and permafrost-affected, remains a paradox—rich in natural resources but slow to develop. Two other challenges unique to Canada face this most northerly region: finding a more secure place for Aboriginal peoples within the unfolding modern version of territorial society and its resourced-based economy; and dealing with Arctic sovereignty issues. With these challenges, plus global warming, land claim settlements, and increasing world demand for its resources, what does the future hold for the Territorial North?

In Friedmann's regional scheme of the core/periphery model, the Territorial North would be described as a resource frontier (Table 1.3). In a developing region in a remote part of the country, economic and social development involves huge capital investments and substantial public funding through transfer payments from Ottawa. Like developing frontiers around the world, the Territorial North's economic performance is limited to non-renewable resources, and reliance on these resources makes the region vulnerable to sharp fluctuations in global demand for its exports. For these reasons, the "Key Topic" in this chapter explores megaprojects.

CHAPTER OVERVIEW

Topics and issues examined in this chapter include the following:

- The Territorial North's physical and historical geography, including the search for the Northwest Passage, the birth of Nunavut, and the impact of modern land claims.
- The dualistic nature of the region's population and economy.
- The environment, Arctic sovereignty, and control of the Northwest Passage, as seen in the context of global warming.
- The region's changing economic and political position within Canada, the circumpolar world, and the emerging global economy.
- The role of megaprojects in the North's development.

A tupqujaq near Cape Dorest, Baffin Island, Nunavut. A tupqujaq is a man-made rock formation shaped like a doorway. In Inuit tradition, the tupqujaq represents a portal through which shaman enter the spirit world. Photo: Radius Images/Alamy.

The Territorial North within Canada

The Territorial North has the largest geographic area of the six regions, but the smallest population and economy (Figure 10.2). Its population of just over 100,000 persons is spread over nearly 4 million km², making the Territorial North one of the world's most sparsely populated areas. Its narrowly based mining economy depends on global markets and prices, making it extremely vulnerable to **boom-and-bust cycles**.

The Territorial North's demographic features have been shaped by three factors. The first, its small population, is due to the limited capacity of the land to support people. The second is the Aboriginal population—especially the Inuit, whose very high birth rate and extremely low death rate account for the population growth in the Territorial North. Third, the education and job experience levels found in the Aboriginal

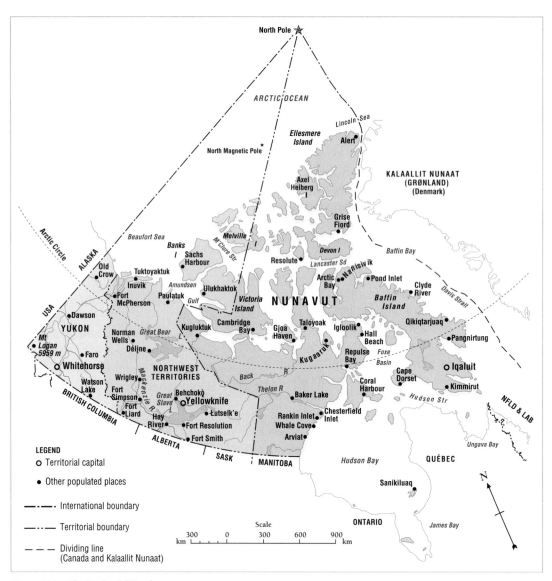

Figure 10.1 **The Territorial North**
The Territorial North consists of three territories. Its boundaries are fixed, although two areas where international boundaries are unclear are the sea border between Yukon and Alaska and the unclaimed international waters of the Arctic Ocean. *Source: Atlas of Canada*, 2006, "The Territories," at: <atlas.nrcan.gc.ca/site/english/maps/reference/provincesterritories/northern_territories>. Natural Resources Canada, 2006. Reproduced with the permission of the Minister of Public Works and Government Services Canada, 2013.

labour force fall short of the non-Aboriginal counterpart. Fourth, the non-Aboriginal population moves to job opportunities in other regions when the North's economy stalls. The Northwest Territories is currently experiencing this type of out-migration

 The discussion in Chapter 4, "Population Density," page 123, provides insight into the issue of the limited capacity of the land to support people.

As a **resource frontier**, the Territorial North has an economy based on the exploitation of its energy and mineral resources. Such industrial activities require huge capital investments and run the risk of failure. The Mackenzie Gas Project is an example. Megaprojects also disturb the natural environment and wildlife, and they provide few jobs for local workers, who often lack the education and job experience necessary to gain employment. Exploitation of non-renewable resources has spurred economic growth in the region, but this type of economic development lacks stability because it is entirely dependent on exports to national and global markets. Variation in demand makes the Territorial North's economy vulnerable to boom-and-bust cycles. Such cycles are common to commodity production, which fluctuates with world demand. In addition, the finite nature of non-renewable resources adds to the unsteady nature of a resource-based economy.

Two Visions

The people of the Territorial North have two powerful and seemingly contradictory visions—one is of a **northern frontier**, while the other is of a **homeland**. The traditional image of the northern frontier is one of great wealth just waiting to be discovered. For example, during the Klondike gold rush (1897–8), prospectors flooded the Yukon to pan for gold along the Klondike River and its tributaries. A more contemporary version of this image comprises large multinational corporations with their vast capital and advanced technology undertaking megaprojects—mining for gold, diamonds, lead, and zinc, and drilling for oil and gas. **Megaprojects** are large-scale resource developments financed and managed by multinational corporations designed to meet global needs for primary products. Such projects create an economic boom during the construction

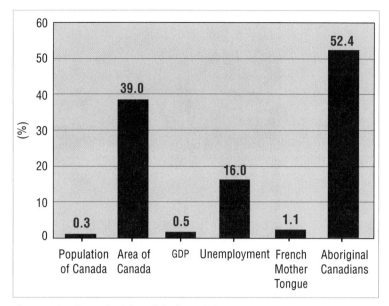

Figure 10.2 The Territorial North basic statistics, 2011
Though the region is the largest in Canada, its population and economy are smallest. The Aboriginal population continues to increase, jumping from 51.7 per cent of the total population in 2001 to 52.4 per cent in 2011. The Territorial North suffers from high **unemployment** as well as **underemployment**, i.e., when no jobs are available, individuals do not seek employment and hence are not classified as unemployed. *Sources:* Tables 1.1, 1.2, and 10.3.

period, but in their operational phase fewer employment opportunities are available and economic spinoffs for local businesses are limited. Because of the risks associated with developing resources in a frontier—from overcoming physical barriers unique to the Territorial North to coping with downturns in world prices for resources—such projects usually are undertaken by large corporations. In return, these corporations reap large profits and supply the industrial cores of the world with raw materials and energy.

Northerners, particularly Aboriginal peoples, see the North as a homeland where their culture and language can flourish. This perception is based on a special, deep commitment to the North, which cultural geographers often attribute to a sense of place. Local people have a strong appreciation for natural features, cultural traits, and the political and economic issues affecting their homeland. A sense of place evokes a feeling of belonging as well as a commitment to a particular place.

Sense of place is discussed further in Chapter 1, page 8.

The concepts of homeland and resource frontier are not always compatible. For instance, the interests of multinational companies relate to the

THINK ABOUT IT
If population density were to be measured by physiological density or the carrying capacity of the land to sustain life, how would the Territorial North compare with the other five geographic regions?

Photo 10.1

The South Nahanni is one of the world's great wild rivers. Located in the boreal wilderness of Nahanni National Park Reserve in the southwestern part of the Northwest Territories, this untamed river is seen surging through the steep-walled First Canyon. Downstream, its waters rush past hot springs, plunge over a waterfall twice the height of Niagara, and cut through canyons more than one kilometre deep.

profitability of the northern frontier's untapped natural wealth. The interests of northerners, however, especially Aboriginal northerners, are best served when the long-term environmental and social well-being of their communities is considered. "Frontier" economic activities often bring jobs and investment, but the Aboriginal community has had minimal involvement, partly because of a mismatch between the needs of the companies and the skill sets of Aboriginal workers. Nevertheless, on balance, positive change is taking place with these two visions interacting to place a "duality" stamp on the regional character of the Territorial North. The very interplay between actual events representing visions indicates a capacity to change over time, largely due to acculturation and accommodation processes. This dynamic produces a powerful social current described in this chapter, especially in comprehensive land claim agreements and the formation of Nunavut.

In the Territorial North, the concepts of homeland and regional consciousness have resulted in the devolution of political power from the federal government to the territorial governments. Elected governments exist in the three territories and six land claim agreements with First Nations have been concluded, the most recent being the Tlicho Final Agreement (2003). Significantly, the Tlicho Final Agreement included provisions for self-government for the Dogrib who live between Great Bear and Great Slave lakes in the Northwest Territories. Future comprehensive land claim agreements are likely to include a section on self-government. Aboriginal self-government received an enormous boost in 1992 when the provision for the territory of Nunavut was placed in the Nunavut Land Claims Agreement under the Nunavut Political Accord. Unlike Yukon and the Northwest Territories, Nunavut is an expression of "ethnic" regional consciousness and yet Nunavut is a public government and therefore is different from First Nations' "ethnic" self-governments. The next challenge facing the Territorial North, but especially Nunavut, is to generate sufficient economic growth and to create a labour force that can take advantage of such growth. The challenge is to overcome the mismatch between the education/job experience of the Aboriginal labour force and the employment needs of the companies and governments. While a tall order, if achieved, then the Territorial North's economic dependency on Ottawa would diminish. At the same time, the Territorial North, while blending Western and Aboriginal ways, would create a homeland accepted by all.

Physical Geography of the Territorial North

The Territorial North extends over four of Canada's physiographic regions: the Canadian Shield, the Interior Plains, the Cordillera, and the Arctic Lands (including the Arctic Archipelago) (Figure 2.1). As a result, the region encompasses a more varied topography than the other five regions. Its topography ranges from mountainous terrain and forest stands in the west, to barren plains, ice-covered islands, and Arctic seas further east and north.

One special feature is the aurora borealis or **northern lights**. While not unique to the Territorial North, these illuminations occur regularly in the long winter nights, causing spectacular displays of shifting or streaming coloured light in northern skies. These northern lights are caused by the interaction of charged solar particles with gases in the atmosphere under the influence of the earth's magnetic field, thus they coincide with geomagnetic disturbances that can disrupt communication and electrical systems.

The vegetation in this region is quite varied and includes a small portion of the boreal forest in the southwest, the tundra with its mosses and lichens further north, and a polar desert in the highest reaches (Figure 2.8). The region is known for the several rivers that wind through it, the many lakes that dot its landscape, and the Arctic Ocean that supports a range of aquatic wildlife.

The physical geography of the Territorial North is governed not so much by physiography as by a cold environment. Cold persists throughout most of the year and in many ways affects human activities. The cold environment includes permafrost (Figure 2.10) and long winters with sub-zero temperatures. Except for the northern half of Yukon, the Territorial North was subjected to glaciation. For that reason the rich Klondike placer deposits were not affected by the scouring effects of the Cordillera ice sheets that affected southern Yukon. The region's main climate zones, the Arctic and the Subarctic (Figure 2.4), are characterized by very short summers.[1] In the Arctic climate, summer is limited to a few warm days interspersed with colder weather, including freezing temperatures and snow flurries. The Subarctic climate has a longer summer that lasts at least one month. During the short but warm summer, the daily maximum temperature often exceeds 20°C and sometimes reaches 30°C.

Arctic air masses dominate the weather patterns in the Territorial North. They are characterized by dry, cold weather and originate over the ice-covered Arctic Ocean, moving southward in the winter. The Arctic zone has an extremely cold and dry climate. Distinguished by long winters and a brief summer, the Arctic climate is normally associated with high latitudes and lower levels of solar energy. The ice-covered Arctic Ocean and continuous permafrost keep summer temperatures cool even though the sun remains above the horizon for most of the summer. These cool summer temperatures, which Köppen defined as an average mean of less than 10°C in the warmest month, prevent normal tree growth. For that reason, the Arctic climate region has tundra vegetation, which includes lichens, mosses, grasses, and low shrubs. In the very cold Arctic Archipelago, much of the ground is bare, exposing the surface material. As there is little precipitation in the Arctic Archipelago (often less than 20 cm per year), this area is sometimes described as a "polar desert."

Beyond 70° N, growing conditions for the hardy tundra vegetation reaches its polar margins. With lower temperatures and less precipitation than in the lower latitudes of this climatic zone, tundra vegetation cannot survive, giving the land a "Mars"-like landscape. The Arctic climate, however, does extend into lower latitudes in two areas: along the coasts of Hudson Bay and the Labrador Sea. These cold bodies of water chill the summer air along the adjacent land mass. In this way, the Arctic climate extends along the coasts of Ontario, Québec, and Labrador well below 60° N, sometimes extending as far south as 55° N.

The Arctic Ocean was called the "Frozen Sea" by early explorers. This extensive ice cover, known as the **Arctic ice pack** or polar pack ice, drifts in a clockwise motion in the Beaufort Sea. Because of the extent and thickness of the Arctic ice pack, few ships can navigate these waters without the

THINK ABOUT IT
Does Nunavut represent a merging of the two visions of the Territorial North?

THE REGIONAL GEOGRAPHY OF CANADA

Photo 10.2
The Yukon River Valley at Dawson City. For Aboriginal peoples as well as for fur traders and prospectors, this long and winding river has been an important transportation route in the history of the Territorial North.

assistance of icebreakers. Two former mining operations in the High Arctic—the Nanisivik mine on northern Baffin Island and the Polaris mine on Little Cornwallis Island—stored their ore until the summer navigation season, when ships reinforced against ice, sometimes aided by icebreakers, transported the ore to European and other world markets. However, if the impact of global warming continues at its present pace, more open water will appear in the short summer, making the Northwest Passage a reality.

The geology of the Territorial North provides much of its wealth. For example, the sedimentary basins of the Interior Plains and Arctic Lands contain large deposits of oil and natural gas (Vignette 10.1). Even the sedimentary strata beneath the Arctic Ocean just beyond Canada's current jurisdiction hold vast energy deposits. The Cordillera and Canadian Shield of the Territorial North have already yielded some of their mineral wealth to prospectors and geologists. These minerals include diamonds, gold, lead, uranium, and zinc. Since its discovery in 1962, an extremely high-grade iron deposit on Baffin Island has attracted mining companies. The Mary River Project,[2] proposed by Baffinland Iron Mines

VIGNETTE 10.1

Sedimentary Basins

The Territorial North has many sedimentary basins, some of which contain petroleum deposits. Those containing petroleum are the Western Sedimentary Basin, the Mackenzie Basin, and the Canadian Arctic Basin. The Canadian Arctic Basin contains several smaller basins, including the Sverdrup Basin. Offshore reserves in the seabed of the Arctic Ocean are vast, but for the most part these remain undiscovered and, ultimately, are potential deposits. While drilling has taken place in discovered resources and are thus "proven" deposits, such is not the case with undiscovered and ultimate potential resources. When Arctic navigation becomes a reality, these offshore petroleum reserves could become a commercial reality.

> **CONTESTED TERRAIN 10.1**
>
> **Less Ice, More Whales**
>
> Global warming is threatening the Inuit way of life, which is so dependent on shore ice for hunting seals and other sea mammals. For example, thinner ice than normal endangers hunters travelling on snowmobiles, while less ice than normal reduces the time available for fishing and hunting marine mammals. Yet, global warming of Arctic waters is causing a variety of whales found in the North Atlantic Ocean to spend more time in the Arctic Ocean, possibly providing a secure marine food source for the Inuit. But is this a realistic expectation?

Corporation, calls for mining the ore, shipping to port by rail and then to markets in Europe by ship (Figure 10.7).

Environmental Challenge: Global Warming

The Territorial North is faced with a major environmental challenge—the warming of its lands and waters. Temperature increase in the Arctic is much higher than in the provinces because of the **albedo effect**, whereby greater solar warming of the land and water will occur because of the reduction of ice and snow cover. The Arctic Ocean is particularly vulnerable to global warming. In the late summer of 2012, Arctic Ocean sea ice reached its lowest extent in the twenty-first century with only 3.41 million km^2 of sea ice (Environment Canada, 2012). The previous low of 4.2 million km^2 occurred in 2007.

As discussed in Chapter 2, global warming is the increase over time of the earth's average surface temperature. Several factors are involved in greater temperature increases occurring in the Arctic, such as heat transfer from lower latitudes to higher ones, but the primary factor is the albedo effect. Light from the sun takes the form of short-wave radiation while energy emitted from the earth's surface takes the form of long-wave radiation. Long-wave radiation is more readily absorbed by the atmosphere and thus warms the atmosphere whereas short-wave radiation escapes into outer space without warming the atmosphere. Currently, the Arctic has a high albedo because of its cover of snow and ice, which means most solar energy is reflected back into outer space without warming the atmosphere.

However, as ice and snow cover decreases in the Arctic, its albedo will shift from high to low, meaning that the solar energy reaching the Arctic will be more effective in warming the atmosphere and thus raising temperatures well above their long-term averages.

The impact of global warming in the Arctic is expected to have both positive and negative impacts on the wildlife and northern peoples. By the end of the twenty-first century, global warming may result in an ice-free Arctic Ocean each summer, thus allowing for ocean transportation across the Northwest Passage, including shipments of petroleum and mineral deposits in the Arctic. The vast copper and zinc deposits located inland at Izok and High Lake are a case in point. Land transport, with the melting of permafrost, will be another matter. Wildlife will be affected. Already scientists have noted negative impacts on polar bears but a positive effect on seal populations. The huge migrating caribou herds that have their calving grounds in the Arctic could be affected. The reduction of the size of calving grounds would have a negative impact on preferred space for reproduction. The Dene and Inuit communities that rely on these herds for much of their country food may have to purchase more of their food from local stores. On the other hand, more open and warmer seas would permit a return of large numbers of bowhead whales, which used to sustain the Thule. Perhaps their descendants, the Inuit, could again harvest these huge mammals (although restrictions under the International Convention for the Regulation of Whaling and environmental groups certainly could make such an activity more difficult, in one respect, than it was for the Thule). The cultural and nutritional implications of a return of a large whale population or of significantly diminished caribou herds are large. Caribou meat forms an important part of the culture of Dene and Inuit while the superior nutritional value of all country food compared to store-bought frozen meat and poultry is well recognized. As the Arctic Ocean warms and opens up, northern cod and shrimp stocks may appear, too.

Historical Geography of the Territorial North

European Contact

At the times of initial contact with Europeans, seven Inuit groups and seven Indian groups belonging to the Athapaskan language family (also known as Dene) occupied the Territorial North. The Inuit stretched across the Arctic: the Mackenzie Delta Inuit lived in the west; further east were the Copper Inuit, Netsilik Inuit, Iglulik Inuit, Baffinland Inuit, Caribou Inuit, and Sadlermiut Inuit. Inuit also lived in northern Québec and Labrador. By the early twentieth century, two groups (most of the Mackenzie Delta Inuit and all of the Sadlermiut Inuit) would succumb to diseases that European whalers brought to the Arctic. The Indian tribes that resided in what is today the territorial Subarctic were the Kutchin, Hare, Tutchone, Dogrib, Tahltan, Slavey, and Chipewyan. More recently, these tribes are known in the aggregate as the Dene.

These Aboriginal peoples had developed hunting techniques well adapted to two cold but different environments. The Inuit employed the kayak and harpoon to hunt seals, whales, and other marine mammals, which enabled them to occupy the Arctic coast from Yukon to Labrador. As a result, the Inuit depended extensively on marine mammals and fish. The Indians hunted and fished in the northern coniferous forest, where the birchbark canoe, the bow and arrow, and snowshoes enabled them to hunt in summer and winter. The Dene tribes relied heavily on big game like the caribou, the Chipewyans often following the caribou to their calving grounds in the northern barrens of the Arctic. Both the Inuit and the Dene moved across the land in a seasonal rhythm, following the migratory patterns of animals. Operating in small and highly mobile groups, these hunting societies depended on game for their survival. Cultural traits, such as the ethic of sharing, developed from this dependency on the land and sea for food.[3]

Early European Exploration

Though the Vikings were the first to make contact with northern Aboriginal peoples around 1000, little is known of those encounters.[4] At that time, the Arctic Ocean had much less ice cover because of a warmer climate. Five centuries later, the Arctic had become a much colder place. In 1576, Martin Frobisher, in searching for a Northwest Passage to the Far East, reached Baffin Island. Unfortunately for Frobisher, his expedition took place at the height of the "Little Ice Age," and his ships met with heavy ice conditions in Davis Strait (which separates Baffin Island from Greenland).

VIGNETTE 10.2

Toxic Time Bombs: The Hidden Cost of Mining

Mining brings jobs and wealth to the North but it also leaves behind toxic wastes. The short lifespan of most mines—less than 20 years—results in a geography of toxic time bombs. But why don't companies accept the social responsibility for cleaning up their mess? The answer is that they wish to skirt the high cost of cleanup. New and more stringent regulations have corrected this situation—except in cases of bankruptcy, which are not uncommon in the mining industry. For example, gold mining near Yellowknife has ended but hidden costs remain. The refining of gold at the Giant gold mine at Yellowknife left residues of arsenic. Now closed, this mining operation has left behind 237,000 tonnes of arsenic trioxide, a by-product of gold refining. Since the last mining company (Royal Oak Mines) declared bankruptcy, the cost of the cleanup, estimated at a quarter of a billion dollars, is left to the federal government—and that means the Canadian taxpayer (Bone, 2012: 217; Danylchuk, 2007).

The Little Ice Age is discussed in Vignette 2.12, "Fluctuations in World Temperatures," page 63.

Frobisher reached Baffin Island where he encountered a group of Inuit, some on land, others in their kayaks. Relations quickly soured. During a skirmish between Frobisher's men and the Inuit, five of his men were lost, three of the Inuit were captured, and Frobisher was hit by an arrow. The Inuit and one kayak were taken back to England, as often was done in the early years of European exploration, as proof of Frobisher's discovery. All three of the captives soon succumbed to European illness. Over the next three centuries, the search for a Northwest Passage through Arctic waters led to misadventure for various European explorers, including John Franklin, whose famous last expedition ended in disaster with all hands lost (Vignette 10.3). However, cultural exchange between Europeans and the original inhabitants of these lands remained limited until the nineteenth century, when the trade in fur pelts and whaling peaked in North America.

Whaling and the Fur Trade

Whaling, the first commercial venture in the Arctic, began in the late sixteenth century in the waters off Baffin Island. During those early years of whaling, whalers had little opportunity or desire to make contact with the Inuit living along the Arctic coast. The Inuit probably felt the same, particularly those who had heard stories of the nasty encounter with Frobisher's men. During early summer, whaling ships set sail from British, Dutch, and German ports for Baffin Bay, where they hunted whales for several months. By September, all ships would return home. In the early nineteenth century, the expeditions of John Ross (1817) and William Parry (1819) sailed farther north and west into Lancaster Sound. Their search for the Northwest Passage had limited success but opened virgin whaling grounds for whalers. These new grounds were of great interest as improved whaling technology had reduced the whale population in the eastern Arctic. In fact, the period from 1820 to 1840 is regarded as the peak of whaling activity in this area. At that time, up to 100 vessels were whaling in Davis Strait and Baffin Bay.

As whaling ships went further to find better whaling grounds, it became impossible to return to their home ports within one season. By the 1850s, the practice of "wintering over" (that is, allowing ships to freeze in sea ice along the coast) was adopted by English, Scottish, and American

THINK ABOUT IT
Photo 3.3, page 76, illustrates a skirmish between Frobisher's men and the Baffin Island Inuit. Such hostilities were not unusual. In the late nineteenth century, for example, the Labrador Inuit and Labradorians still had bloody encounters. Do such encounters still occur in Canada along the Aboriginal/non-Aboriginal faultline?

VIGNETTE 10.3

The Northwest Passage and the Franklin Search

In 1845, Sir John Franklin headed a British naval expedition to search for the elusive Northwest Passage through the Arctic waters of North America. This British naval expedition took place at the end of the Little Ice Age, meaning that ice conditions would have been much more challenging than those occurring today. He and his crew never returned. Their disappearance in the Canadian Arctic set off one of the world's greatest rescue operations, which was conducted on land and by sea and stretched over a decade. The British Admiralty organized the first search party in 1848. Lady Franklin sent the last expedition to look for her husband in 1857. These expeditions accomplished three things: (1) they found evidence confirming the loss of Franklin's ships (the *Erebus* and *Terror*) and the death of their crews; (2) one rescue ship under the command of Robert McClure almost completed the Northwest Passage; and (3) the massive rescue effort resulted in a greater knowledge and mapping of the numerous islands and various routes to the north and west of Baffin Island in the Arctic Ocean. The exact sequence of events that led to the Franklin disaster is not known. However, archaeological work, conducted in the early 1980s on the remains of members of the expedition, revealed that lead poisoning, possibly caused by the tin cans in the ships' food supplies, may have contributed to the tragic demise of the Franklin expedition. In 2010, Parks Canada began a serious underwater search for the two ships. At the end of the summer of 2012, Parks Canada had not yet found any of the ships, but the investigation is expected to continue.

whalers. This allowed whalers to get an early start in the spring, providing for a long whaling season before the return trip home at the onset of the next winter. Wintering over took place along the indented coastlines of Baffin Island, Hudson Bay, and the northern shores of Québec and Yukon. Permanent shore stations were established at Kekerton and Blacklead Island in Cumberland Sound, at Cape Fullerton in Hudson Bay, and at Herschel Island in the Beaufort Sea. Life aboard whaling ships was dirty, rough, and dangerous, and many sailors died when their ships were caught in the ice and crushed.

Despite the early unfortunate encounters with European explorers, the Inuit welcomed the whaling ships because of the opportunity for trade. The Inuit were attracted to shore stations and often worked for the whalers by securing game, sewing clothes, and piloting the whaling ships through difficult waters to promising sites for whale hunting. Some Inuit men signed on as boat crew and harpooners. In exchange for this work, the Inuit obtained useful goods, including knives, needles, and rifles, which made domestic life and hunting easier. While this relationship brought many advantages for the Inuit, there were also negative social and health aspects, including the rise in alcoholism and the spread of European diseases (Vignette 10.4). Perhaps the most devastating result of this trade relationship for the Inuit was the unexpected end of commercial whaling, which represented the loss of access to highly valued trade goods. Just as the twentieth century began, demand for products made from whales—whalebone corsets, lamp oil—decreased sharply, halting the flow of whalers, and thus trade goods, that were sailing into Arctic waters. By this time, however, the Inuit depended on trade goods for their hunting activities. Somehow, they had to find other means of obtaining these useful goods.

Fortunately for the Inuit, European fashion had taken a liking to Arctic fox pelts, which caused the Hudson's Bay Company to establish trading posts in the Arctic. This fortuitous event provided a replacement for whaling and associated trade. The fur trade had already been successfully operating in the Subarctic for some time—a relationship between European traders and the Subarctic Indian tribes was established through the trade of fur pelts, especially beaver. Soon the Inuit were deeply involved in the fur trade. The working relationship between the Hudson's Bay Company and the Inuit was based on barter: white fox pelts could be traded for goods.

Until the 1950s, the fur trade dominated the post-contact Aboriginal land-based economy. It lasted for less than 100 years in the Arctic and for over three centuries in the Subarctic. Did the fur trade, as well as Arctic whaling, create a form of

VIGNETTE 10.4

European Diseases

Whalers, fur traders, and missionaries introduced new diseases to the Arctic. As the Inuit had little immunity to measles, smallpox, and other communicable diseases such as tuberculosis, many of them died. In the late nineteenth century, the Sadlermiut and the Mackenzie Delta Inuit were exposed to these diseases. According to Dickason (2002: 363), in 1902 the last group of Sadlermiut, numbering 68, died of disease and starvation on Southampton Island, "a consequence of dislocations that ultimately derived from whaling activities." The Mackenzie Delta Inuit, whose numbers were as high as 2,000, almost suffered the same fate but managed to survive.

The Mackenzie Delta Inuit occupied the northwestern Arctic coast, in present-day Yukon, the Northwest Territories, and part of Alaska. Herschel Island, lying just off the Yukon coast, was an important wintering station for American whaling ships. Whalers often traded their manufactured goods with the local Mackenzie Delta Inuit, who became involved with the commercial whaling operations. Through contact with the whalers, the Inuit were infected by European diseases. By 1910, only about 100 Mackenzie Delta Inuit were left. Gradually, Inupiat Inuit from nearby Alaska and white trappers who settled in the Mackenzie Delta area intermarried with the local Mackenzie Delta Inuit, which secured the survival of these people. Today, their descendants are called Inuvialuit.

dependency whereby Indians and Inuit could not survive without trade goods? The answer is a qualified "yes." At first, Aboriginal people had a form of partnership with European traders and whalers. Each side had power—for instance, the European traders needed the Indians to trap beaver and, often, to show them how to survive in the harsh climate, and the Indians needed the traders to obtain European goods and technology. Gradually, however, the power relationship shifted in favour of the European traders. By the nineteenth century, the fur companies controlled the fur economy. Fur-trading posts dotted the northern landscape. Indians, who had long ago integrated trade goods into their traditional way of life—including their hunting techniques and their migration patterns—were therefore heavily dependent on trade. In fact, when game was scarce, tribes relied on the fur trader for food. Ironically, by securing game for the traders, Indians reduced the number of animals that would be available for their own sustenance. In the Territorial North, game became scarce around fur-trading posts from overexploitation.

The problems of a growing dependency on European goods and a changing way of life for northern Aboriginal peoples were compounded after the arrival of Anglican and Catholic missionaries in the 1860s and the North West Mounted Police (NWMP) in the 1890s. The Indians and Métis were confronted with the full force of Western culture in the late nineteenth century, as were the Inuit in the early twentieth century. Western ideas and rules introduced by the missionaries and police who now lived near the trading posts had a profound impact on Aboriginal culture. The NWMP (which added "Royal" to its name in 1904 and, in 1920, was renamed the Royal Canadian Mounted Police) imposed Canada's system of law and order on Aboriginal people, while the missionaries challenged their spiritual values and encouraged the Inuit, Indians, and Métis to remain in the settlements. Also, on behalf of the Canadian government, both Anglican and Catholic missionaries placed young Aboriginal children in church-run residential schools, where they were taught in either English or French. In this attempted assimilation, most children learned to read and write in English or French, but they were inadequately prepared for northern life. As they lost the opportunity to learn from their parents about how to live on the land, they became trapped between the two very different worlds of their Aboriginal communities and the Euro-Canadians. Under these circumstances, many lost their indigenous language, animistic beliefs, and cultural customs. Fur traders opposed many of these induced Western cultural adaptations because they needed the Aboriginal people on the land to trap. Nevertheless, the influence of the churches, the power of the state, and the number of non-Aboriginal residents in the North increased in the twentieth century, placing Aboriginal cultures under siege and crippling their land-based economy. However, political and social changes were occurring at this time that would lead to territorial governments, then to relocation of Aboriginal people to settlements, and most recently to land claim agreements and self-government.

From the Land to Native Settlements

Perhaps the greatest impacts on Aboriginal people in the twentieth century were (1) residential schools and (2) relocation to settlements. In the 1950s and 1960s, both resulted in an enormous cultural shock. In the case of resettlement, the freedom and risks of living on the land were exchanged for a regulated and more secure life in settlements. The relocation from the land to these tiny settlements marks the beginning of a new way of life with some good aspects and some bad ones. Advantages included food security, access to medical services, and public education. Food security eliminated hunger and starvation but it also meant more store foods and less country food in their diet. Unfortunately, relocation had many negative impacts, including destroying the traditional social hunting/trapping unit from a family-based one to a male one.[5] Children were required to attend schools so mothers stayed in the settlements. In fact, the much-needed family allowance payments to the mothers, introduced in 1945, only took place if the children remained in school. Finally, being based in settlements, it became virtually impossible for the people to follow the seasonal cycle of wildlife movements.

Relocation continues to be a controversial subject in northern history. Williamson (1974), Elias (1995), Marcus (1995), and Rowley (1996) provide different perspectives. However, leaving the Aboriginal people in what was seen to be a failing hunting/trapping economy was not an option. Living off the land was, from time to time, a challenge but the shortage of cash/credit from trapping to purchase goods was critical. One option might have been to subsidize the hunting/trapping economy by providing the necessary cash to Aboriginal families as well as more time to adjust to relocation, as is now done in northern Québec where Cree hunters and trappers are paid for living on the land and acquiring country food. When relocation to settlements became the order of the day, few of the people had a full command of English (or French in Québec), which would become so necessary for participating in the affairs of settlement life. Nevertheless, the apparent political urgency of the day caused Ottawa to push for relocation; but these settlements and the newcomers were both ill-prepared. Few communities had "extra" housing, schools, or nursing stations and none had a sufficiently vigorous economic base to offer much employment. The former hunters and their families had their lives turned upside down.

Federal officials saw relocation from two perspectives. First, it was seen as a necessary step in protecting northern peoples from the hardships of living on the land, such as life-threatening food shortages. Second, concentrating Aboriginal people in settlements allowed Ottawa to provide a variety of services, including schooling for their children. These two perspectives were part of the overall strategy of "modernization" of northern Aboriginal peoples. One could argue that, since few Aboriginal families abandoned settlement life to return to the land, the attractions of settlement life outweighed the disadvantages.

Yet, how serious was the hardship of living on the land? The search for game (food) might not always be successful and so hardship and hunger were characteristic of their ancient culture. Most of the time, however, their hunting, fishing, and trapping lifestyle was entirely satisfying. When game was scarce, hardships and hunger could be severe, but these troubles were not an accepted or acceptable part of the culture of modern Canadian society. Some areas were more risky for hunting peoples. The **Barren Grounds** of the central Arctic were a particularly challenging place to live off the land because of the heavy dependence for sustenance on the migrating caribou herds. Failure to find the caribou translated into hunger and even starvation. Reports of deprivation and even cases of death by starvation among the Caribou Inuit had reached Ottawa before, but no action was taken. In the early 1950s, the Canadian media reported that about 60 Caribou Inuit starved to death. How could people living in a modern country like Canada starve to death? This sad event pushed the government into action, leading to a relocation policy. By 1958, Ottawa made the decision to relocate the Caribou Inuit to settlements, such as Baker Lake and Eskimo Point, but by then starvation had taken its toll—the Caribou Inuit population had dropped from "about one hundred and twenty in 1950 to about sixty in 1959" (Williamson, 1974: 90). At the same time, Ottawa extended this relocation program to coastal Inuit and to Indians and Métis in the Subarctic who also lived off the land. However, Ottawa was unprepared for the economic, psychological, and social consequences of settlement life for hunting peoples.

After 60 years of settlement life, what has happened? On the one hand, access to store food and medical services resulted in a population boom. Today, the Aboriginal population forms a clear majority in the Territorial North but especially in Nunavut. On the other hand, the increased population has not matched the availability of public housing and jobs, resulting in overcrowding and chronic underemployment. As well, Aboriginal communities face deep-rooted social dysfunctions, including extremely high suicide rates among their young people. The causes are various, and include the cultural dislocation and devaluation resulting from the residential schools. When Native people speak of the need to consider the impact of current actions and inactions on those seven generations from now, this is not merely a poetic manner of expression. Past actions—whether social engineering or the environmental impacts of megaprojects—can reverberate for generations to come. Another, more tangible factor contributing to social dysfunction is the fact that Native communities have no solid economic base, resulting in heavy dependency on government for the impoverished and few opportunities for young people. The two principal sources of income are wages and various forms of government payments, including social assistance. The major employer is the government. As well, lower-income households, especially the elderly, have limited access to country food, which comes mainly through sharing, and insufficient income to purchase store food. Another alarming trend reported by Chan (2006) is that the increase in the consumption of store food rich in carbohydrates, particularly by younger generations, has already caused obesity and diabetes. Perhaps the most positive outcome of settlement life is the emergence of a more educated population. From these ranks, Aboriginal leaders have had a hand in transforming the Aboriginal

> **THINK ABOUT IT**
> If you were the Minister of Northern Affairs and Natural Resources in the 1950s, what policy would you propose for northern peoples?

Photo 10.3

Pangnirtung is a small but fast-growing hamlet on the coast of Baffin Island. Like most Inuit communities, its natural increase far outstrips other Canadian urban centres. From 2006 to 2011, Pangnirtung's population grew from 1,325 to 1,425, an increase of 7.5 per cent. Most significant, 35 per cent of its population was under the age of 15. Such an age structure is common in developing countries but not in developed nations. At a cost of $17 million, Public Works and Government Services Canada expected to complete harbour dredging and a small-boat wharf in September 2012, and that development should spark renewed activities in the region's turbot fishing industry.

society and economy in new directions through successful negotiations for comprehensive land claim agreements, for the first effectively Aboriginal territory within Canada (Nunavut), and for their international involvement in the Arctic Council.

 Comprehensive land claim agreements are discussed in Chapter 3, "Modern Treaties," page 97.

Territorial Expansion: Rupert's Land, the Arctic Islands, and the Arctic Seabed

The Territorial North fell under Canadian jurisdiction in three stages. First, the transfer of Rupert's Land to Canada by Britain took place in 1870. Second, Great Britain transferred the Arctic Islands to Canada in 1880. Third, in 1985, Canada declared a 200-mile economic zone that extended its control over the Arctic Ocean. In addition, in the same year Canada announced its Arctic Waters Pollution Prevention Act. Still, the last remaining territory that may become part of Canada consists of a portion of the seabed of the Arctic Ocean, which now lies in international waters. Canada's claim must be submitted to the United Nations by 2013.

When Canada was formed in 1867, much of what is now Canada remained under British control. Britain had claimed British North America on the basis of settlement, trade, and exploration. In the Territorial North, the British declared ownership of Rupert's Land as a result of early discovery and exploration, and in 1670 this wide territory had been granted by Charles II to the newly formed Hudson's Bay Company, which had maintained a presence in the region from that time. The claim to the Arctic islands was based on the British Navy's efforts to find the Northwest Passage, including the search for the missing Franklin expedition. In the rest of the Territorial North, beyond Rupert's Land, the Hudson's Bay Company, after its 1821 amalgamation with North West Company, had established and maintained a number of fur-trading posts in the forested lands of the Mackenzie Basin and the Yukon. By extending its fur-trading economy over this area, the British government claimed these Subarctic lands.

Forgotten Frontier: Confederation to 1939

Canada never paid much attention to the Territorial North. In fact, the region was a forgotten part of Canada for several reasons. First, it had little value for agricultural settlement or, because of its remote location, for resource development. Second, Ottawa had its hands full with the provinces where almost all Canadians lived. Third, the fur trade in the North depended on the Aboriginal peoples' living on the land. In short, the Territorial North was not a "priority" region and thus received minimum attention from Ottawa. With the exception of the Klondike gold rush in the Yukon, the North's economy was left in the hands of the nomadic Dene and Inuit who hunted and trapped, moving seasonally with the wild animals, such as the caribou. Ottawa had adopted a laissez-faire policy to minimize federal expenditures, leaving the fur traders and missionaries to deal with the food and health needs of a hunting society.

Strategic Frontier

With the outbreak of World War II, the Territorial North became a strategic frontier. Military investments and activities included military bases, highways, landing fields, and radar stations. While the nature of its strategic role changed over time, the Territorial North served as a buffer zone between North America and the Soviet Union for over 50 years. This role ceased with the collapse of the Soviet Union and the end of the Cold War in 1991.

For the Americans, Canada's North provided a secure transportation link to the European theatre of war and, in 1942, to Alaska, where the threat of Japanese attack was real. The air routes consisted of the Northwest Staging Route and Project Crimson. Each consisted of a series of northern landing strips that would enable American and Canadian warplanes to refuel and then continue their journey to either Europe or Alaska. In the Northeast, Project Crimson involved constructing landing fields at strategic intervals to allow Canadian and American airplanes to fly from Montréal to Frobisher Bay (now Iqaluit) and then to Greenland, Iceland, and, finally, England. In Canada's Northwest, American aircraft came to Edmonton and then flew along the Northwest Staging Route to Fairbanks, Alaska, where their major military base was located. The Alaska Highway, built at the same time, provided road access to the various landing fields and to Alaska. The US Army command had decided that the oil needed by the American armed forces in Alaska must be made secure by increasing the oil production at Norman Wells in the NWT and sending it by pipeline across several mountain ranges to Whitehorse and then northward to the military facilities at Fairbanks. Known as the Canol Project, the oil pipeline was completed in 1944, but with the disappearance of the Japanese threat it was closed within a year.

After World War II, the geopolitical importance of northern Canada changed. The North's new strategic role was to warn of a surprise Soviet air attack. The defence against such an attack was a series of radar stations that would detect Soviet bombers and allow sufficient response time for American fighter planes and (later) American missiles to destroy the Soviet bombers. In the 1950s, 22 radar stations, called the Distant Early Warning line, were constructed in the Territorial North along 70° N. Before the end of the Cold War, these radar stations were abandoned and replaced with more sophisticated methods of detecting incoming Soviet planes or missiles. With the collapse of the Soviet Union, Ottawa withdrew its military establishment at Inuvik, did not proceed with its plans for a military base at Nanisivik, and downsized its operation at Alert.

American military investment did expand the Norman Wells oil field, along with a pipeline to Whitehorse, but most of their investment went into improving the transportation system, such as the construction of the Alaska Highway. Private resource development moved along much more slowly as it responded to world demand.

Arctic Sovereignty: The Arctic Ocean and the Northwest Passage

In the twenty-first century, Arctic sovereignty has taken on a fresh urgency (Bone, 2012). This urgency has two elements. First, global warming has meant more open water in the Arctic Ocean, making transarctic shipping a reality. Second, circumpolar nations are in the process of claiming the Arctic seabed. How does Canada maintain surveillance of these remote waters? One method is aircraft sent from bases in southern Canada to patrol this vast region. Another method is Radarsat satellite surveillance. A possible third method could involve polar drones (Figure 10.3). Then, too,

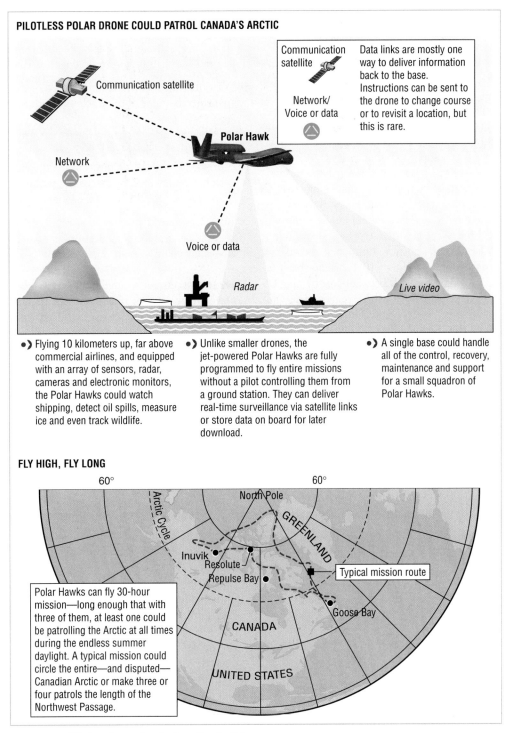

Figure 10.3 **Pilotless drones for the Canadian Arctic?**
The proposal of the giant American arms manufacturer. Northrup Grumman, to build Canada three Polar Hawk drones to patrol the Canadian Arctic is intriguing, but the cost of such a project is unknown. Rough estimates suggest a price tag of $1.6 billion, but military procurement contracts often far exceed the initial estimated costs. And if one of the three drones went down, the plan for constant surveillance would be compromised. *Source:* Adapted from Koring (2012).

plans for an Arctic naval base and patrol ships are expected to become a reality in the future, perhaps as soon as 2025.

In 2007, Prime Minister Harper firmly declared "the first principle of Arctic sovereignty is use it or lose it" (BBC News, 2007). What prompted this sense of urgency is pressure from Russia and other **circumpolar countries** that are actively staking their claims to the Arctic seabed and from shipping nations that want the Northwest Passage to be defined as lying in international waters. National boundaries have yet to be set for the Arctic Basin and global warming appears to have opened a summer ice-free shipping route through the Northwest Passage. In addition, the Yukon/Alaska offshore boundary, shown as a blue dashed line in Figure 10.4, has not been resolved. Consequently, does Canada or the US control management of fish resources, deal with environmental issues, and authorize deep-sea drilling? This disputed area will likely be resolved by negotiations under the United Nations Convention on the Law of the Sea (UNCLOS).

For Canada, the Arctic Basin possibly is its last territorial acquisition. Ottawa recognizes that:

- Vast quantities of petroleum deposits lie beneath the floor of the unclaimed zone of the Arctic Ocean (Table 10.1).
- Global warming may turn the frozen Arctic Ocean into a commercial ocean route.
- Canada's international position within the Arctic Council and, by extension, within the Circumpolar World is at stake (Vignette 10.5).

While Canada did propose the Arctic Council, Ken Coates and others question if Canada is ready to lay claim to seabed deposits and to control the Northwest Passage, or if Ottawa will continue to treat Arctic sovereignty as a "zombie" of its foreign policy. Prominent Canadian scholars question Canada's readiness. In a provocative book, *Arctic Front: Defending Canada in the Far North*, Coates et al. (2008: 1) probe Ottawa's past, present, and hopefully more vigorous future efforts to claim Arctic waters and the resources below:

> Arctic sovereignty seems to be the zombie—the dead issue that refuses to stay dead—of Canadian public affairs. You think it's settled, killed and buried, and then every decade or so it rises from the grave and totters into view again. In one decade the issue is the DEW Line, then it's the American oil tanker *Manhattan*, steaming brazenly through the Northwest Passage, then the *Polar Sea* doing the same thing. In August 2007, a Russian submarine planted a flag at the North Pole. Or perhaps it was under the North Pole, as the UK *Daily Telegraph* reported, raising an image of a striped pole floating in the ocean, with the devious Russians diving underneath it. Perhaps the flag did land on the pole, though good luck with that, since the pole is a point with no size at all, so the Russians likely missed it. However it was, they are up there, and the zombie has come to life once more.

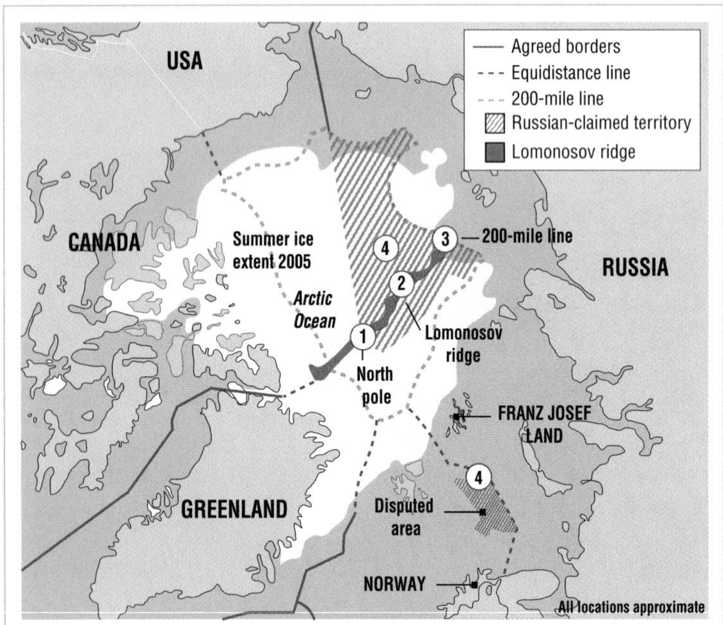

With the exception of Hans Island, a 1.3-square-kilometre rock that Denmark and Canada both claim, and which they are about to agree to divide the island into two parts, one Canadian and the other Danish, the international

THINK ABOUT IT
Should only those nations bordering the Arctic Ocean be able to lay claims to the remaining international waters or should other nations, such as other members of the Arctic Council, be allowed to make a claim?

Figure 10.4 The Arctic Basin and national boundaries
The "land rush" to claim the Arctic seabed directly involves five countries: Canada, the United States, Russia, Norway, and Denmark (Greenland). The four designated areas shown on the map are: (1) North Pole: in 2007 Russia planted its flag on the seabed, 4,000 m (13,100 ft) beneath the surface, as part of its claim for oil and gas reserves. (2) **Lomonosov Ridge**: Russia argues that this underwater feature is an extension of its continental territory and is looking for evidence. (3) 200-nautical-mile (370-km) line indicates how far countries' agreed economic areas extend beyond their coastlines; this often is set from outlying islands. (4) Russian-claimed territory: the bid to claim a vast area is being closely watched by other countries. Some could follow suit.
Source: BBC News (2007). Based on "Maritime jurisdiction and boundaries in the Arctic region" by International Boundaries Research Unit, Durham University.

Table 10.1 Petroleum Resources in the Territorial North

Oil Resources

Region	Discovered Resources		Undiscovered Resources		Ultimate Potential	
	$10^6 m^3$	Million bbl	$10^6 m^3$	Million bbl	$10^6 m^3$	Million bbl
Northwest Territories and Arctic offshore	187.9	1,182.5	799.7	5,032.6	987.6	6,215.0
Nunavut and Arctic offshore	51.3	322.9	371.8	2,339.4	423.1	2,662.3
Arctic offshore Yukon	62.5	393.8	412.7	2,596.8	475.2	2,990.6
Total	301.7	1,899.1	1,584.1	9,968.8	1,885.9	11,867.9

Gas Resources

Region	Discovered Resources		Undiscovered Resources		Ultimate Potential	
	$10^9 m^3$	Trillion cubic feet	$10^9 m^3$	Trillion cubic feet	$10^9 m^3$	Trillion cubic feet
Northwest Territories and Arctic offshore	457.6	16.2	1,542.2	54.8	1,999.8	71.0
Nunavut and Arctic offshore	449.7	16.0	1,191.9	42.3	1,641.6	58.3
Arctic offshore Yukon	4.5	0.2	486.6	17.3	491.1	17.4
Total	911.8	32.4	3220.7	114.3	4132.6	146.7

Sources: Compiled and integrated from several published sources that may underestimate or overestimate actual field resources. Volumes and distribution should be regarded as approximate. Numbers may not add due to rounding.
Aboriginal Affairs and Northern Development Canada, Northern Oil and Gas Branch (2012: Table 2). Reproduced with the permission of the Minister of Public Works and Government Services Canada, 2013

VIGNETTE 10.5

The Arctic Council and the Circumpolar World

The Circumpolar World is an enormous area, sprawling over one-sixth of the earth's landmass and spanning 24 time zones. The Arctic Council focuses its attention on this massive land area, its environment, and its peoples. In 1996, Canada played a key role in the establishment of the Arctic Council, which is designed to serve as a high-level intergovernmental forum to provide a means for promoting co-operation, co-ordination, and interaction among the Arctic states and peoples, with the involvement of the Arctic indigenous communities and other Arctic inhabitants on common Arctic issues—issues of sustainable development and environmental protection in the Arctic. The member states are Canada, Denmark (including Greenland and the Faeroe Islands), Finland, Iceland, Norway, the Russian Federation, Sweden, and the United States of America. The various national organizations of indigenous peoples are represented as permanent participants, and some European nations, including the UK and Germany, have observer status. Canada has a number of permanent participants on the Arctic Council, including the Athabaskan Council, the Gwich'in Council, and through its participation in the International Inuit Circumpolar Council. The Arctic Council provides information to its member states, especially in six areas—Arctic contaminants; Arctic monitoring and assessments; conservation of Arctic flora and fauna; prevention and response to environmental problems; protection of the Arctic marine environment; and sustainable development. In 2013, Canada is once again chair of this council for two years, thus giving Ottawa an opportunity to shape its agenda. Prime Minister Stephen Harper has named his Minister of Health, Leona Aglukkaq, as the new chairperson.

community recognizes Canada's ownership of the islands in the Arctic Ocean. However, the ownership of Arctic waters between the islands in Canada's archipelago remains unclear. The United States considers the Northwest Passage as an international sea route and claims ownership of a narrow strip of the Beaufort Sea—some 21,436 km². While this route is not used for commercial purposes because of the difficulty of navigating through ice, a number of "unknown" countries have sent their nuclear submarines under the ice cover, and the Americans have made several trips through the Northwest Passage on the surface. In 1969, an American tanker, the SS *Manhattan*, made a voyage through the Northwest Passage without asking Canada's permission. It was an attempt to prove the passage was a viable route for shipping oil from Alaska's Prudhoe Bay oil fields. Ottawa did provide a Canadian icebreaker to escort the *Manhattan*. In 1970, the *Manhattan* made another trip through the passage. In 1985, the US Coast Guard icebreaker *Polar Sea* transited the passage—once again, without asking the Canadian government for permission. The political fallout over what was considered the most direct challenge to Canada's sovereignty in the Arctic led to the signing of the Arctic Co-operation Agreement in 1988 by Prime Minister Brian Mulroney and US President Ronald Reagan. The document states that the US is to refrain from sending icebreakers through the Northwest Passage without Canada's consent; in turn, Canada will always give consent. However, the issue of whether the waters are international or internal was left unresolved.

Over the years, Canada has sought to legalize its sovereignty over the Arctic. In 1907, Canada first announced the "Sector Principle," which divided the Arctic Ocean among those countries with territory adjacent to the Arctic Ocean. Accordingly, Canada could claim title over a wedge of the Arctic Ocean north of its territory between 60° W and 141° W longitudes and extending to the North Pole. While the Soviet Union supported this principle, the United States and other countries have not.

More recently, Canada has looked to environmental legislation as a means of exercising its sovereignty over Arctic waters. In the age of supertankers and container ships, the threat of toxic spills is more likely than ever before. Ottawa has two concerns. First, how would it ensure sovereignty over Arctic waters in order to manage such ocean traffic? Second, how would Canada protect its waters and adjacent islands from toxic wastes discharged from foreign ships passing through the Northwest Passage? In 1985, Ottawa's response took the form of the Arctic Waters Pollution Prevention Act, which gives Canada the right to control navigation in its sector of the Arctic Ocean and to manage its ocean environment. These complicated geopolitical issues are discussed more fully in Robert Bone's *The Canadian North: Issues and Challenges* (Bone, 2012: 262–86).

In 2003, Canada ratified the UN Convention on the Law of the Sea, which specifies that coastal countries have the right to control access to their coasts. This access zone is 12 nautical miles (22.2 kilometres) wide. While the Convention on the Law of the Sea supports Canada's claim to Arctic waters, it leaves some grey areas, including the fact that some of the islands in the Arctic Archipelago are separated by more than 90 kilometres of water. For that reason, the greatest threat to Canada's sovereignty claims remains the presence of foreign ships, including tour ships, passing through the Northwest Passage without obtaining permission from the federal government. As well, Canada's response capacity to a ship in distress or to an oil spill is extremely limited. Plans for a naval base in the Arctic exist, but the cost of such a project keeps delaying approval by Ottawa.

The Territorial North Today

The Territorial North remains a resource frontier far from the Canadian ecumene. Surprisingly, almost everyone lives in a settlement, town, or city. Rural communities and farms, as found in southern Canada, do not exist. Another surprising fact is that mining sites, rather than being associated with resource towns, often involve workers being flown to and from the mine. The three diamond mines provide examples. Each is located in an isolated area but the companies house their workers in camps and fly them to and from Yellowknife and other centres rather than build a permanent community where the workers' families would live.

A difference from the rest of Canada is that most Arctic urban centres are very small and isolated from one another (Figure 10.5). From Statistics Canada's perspective, these settlements

THINK ABOUT IT
Can you think of any reasons why fly-in, fly-out air commuting is more attractive to mining companies than building permanent communities?

Table 10.2 **Capital Cities Dominate the Demographics of the Territorial North**

Capital	Population 2006	Population 2011	Percentage Change	Percentage of Territory Population, 2006	Percentage of Territory Population, 2011
Iqaluit	4,796	6,254	30.4	16.3	19.6
Yellowknife	18,700	19,234	2.9	45.1	46.4
Whitehorse	22,898	26,888	13.7	75.4	76.8

Source: Statistics Canada (2012a).

do not qualify as **urban areas** because their populations are less than 1,000. In 2011, for example, approximately three-quarters of these centres had populations under 1,000, and more than 40 per cent of the Territorial North's population lived in three cities: Whitehorse (26,898), Yellowknife (19,234), and Iqaluit (6,254).

As measured by function and reflecting the Aboriginal/non-Aboriginal spatial duality, centres in this region fall into two principal categories: Native settlements and regional service centres. Most Aboriginal people reside in Native settlements, where they form more than half the population but where there are few job opportunities.

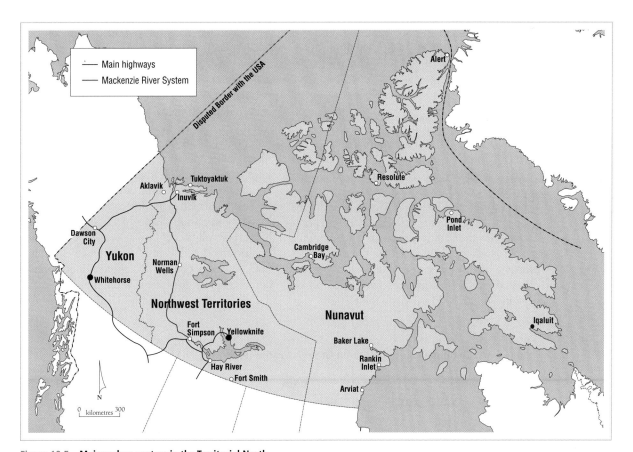

Figure 10.5 **Major urban centres in the Territorial North**
The major cities are the territorial capitals, Whitehorse, Yellowknife, and Iqaluit. With most government jobs in these cities, underemployment is much less a problem than in many smaller centres, especially Native settlements. Alert, located at the northern tip of Ellesmere Island, remains a military base from the Cold War. Resolute, on the other hand, soon will become a naval base as part of Canada's effort to exert more control over its Arctic waters, including the Northwest Passage.

Table 10.3 **Population and Aboriginal Population, Territorial North**

Territory	Population 2001	Population 2011	% Change, 2001–11	Aboriginal Population, 2011	% Aboriginal of Total 2011 Population
Yukon	28,674	33,897	18.2	7,705	23.1
Northwest Territories	37,300	41,462	11.0	21,160	51.9
Nunavut	26,745	31,906	19.3	27,360	86.3
Territorial North	92,719	107,265	15.6	56,225	52.4

Source: Statistics Canada (2012a and 2013).

This mismatch between Native settlements and jobs has its roots in the relocation of Aboriginal people to settlements in the 1950s. On the other hand, Native communities are more closely linked to traditional ways of life and therefore provide a cultural link with the land and the past. A growing number, however, have become regional centres, particularly the three capital cities where employment opportunities in the public sector are greatest. As well, a small but growing number of Aboriginal families have relocated to cities in southern Canada.

Population

From 2001 to 2011, the population of the Territorial North increased by nearly 16 per cent, reaching a figure of 107,265 (Table 10.3). This population growth is due exclusively to a high rate of natural increase (Table 10.4). Overall, Aboriginal people make up 52 per cent of the northern population. However, the percentage of Aboriginal people varies widely between the three territories. Nunavut has the highest percentage at 86 per cent, followed by the Northwest Territories at 52 per cent and Yukon at 23 per cent (Table 10.3).

In 2011–12, the Territorial North had a natural increase of nearly 2,000 persons, or 1.8 per cent (Statistics Canada, 2012a). Within the three territories, the highest rate (2.1 per cent) occurred in Nunavut. In comparison, the national figure was only 0.4 per cent. The reason for this significant difference is the much higher birth rate in the Territorial North (23.5 births per 1,000 compared to the national figure of 11.4 births per 1,000) and a lower death rate (5.5 deaths per 1,000 compared to the national figure of 7.5 deaths per 1,000), the result of a younger population. While birth and death rates are not collected by ethnicity, given the proportion of Aboriginal people in each territory, the association of high birth rates with a high percentage of Aboriginal population indicates the highest birth rates are among Aboriginal people, especially among the Inuit in Nunavut (Table 10.4).

The population of the Territorial North also is affected by migration (Table 10.4). Migration to the North normally occurs when economic expansion creates jobs, thus drawing workers and their families from southern Canada. When economic contraction takes place, these same workers and their families often return to southern Canada. As Table 10.4

Table 10.4 **Components of Population Growth for the Territories, 2011–2012**

Demographic Event	Canada	Yukon	NWT	Nunavut	Territorial North
Births/1,000 persons	11.4	11.3	17.0	26.2	23.5
Deaths/1,000 persons	7.5	6.4	4.8	5.5	5.5
Natural rate of increase (%)	0.4	0.5	1.2	2.1	1.8
Net interprovincial migrants		265.0	−1,491.0	−492.0	−1,718.0

Source: Statistics Canada (2012b).

indicates, the outflow from the Northwest Territories and Nunavut reflects their depressed economic situation while the inflow in Yukon marks the upswing in its economy. Another feature is the increase in mobility of the Aboriginal population. While the numbers remain small, more and more Aboriginal migrants have moved to southern cities in search of jobs and urban amenities, such as post-secondary education and training and specialized medical care, and to escape from a depressed social environment. While precise figures are not available, the number of Aboriginal Canadians living in large urban centres has increased dramatically over the last 20 years and these numbers are expected to continue to increase because of the economic and social state of many Aboriginal communities.

Economic Structure

As a northern frontier, the Territorial North's economy depends heavily on private investment to develop its natural resources and on transfer payments for its public sector. Added to these purely economic matters, the region is caught up in the race for control over the Northwest Passage, the unclaimed zone of the Arctic Ocean, and the ongoing adjustment of its Aboriginal workforce to the needs of the market economy. All require strong financial support from Ottawa. In sum, the Territorial North is a high-cost area for economic development, social programs, and geopolitical challenges. With a limited tax base and without the right to collect royalties from resource companies, the three governments of the Territorial North must depend on Ottawa. All of this translates into the simple fact: Canadians in other parts of the country will be called on to invest in the country's last frontier for decades to come. Canadians should take solace from the territorial version of equalization payments called Territorial Formula Financing—transfer payments and the cost of sovereignty are essential to nation-building. In a broader sense, perhaps, this public financing is the true meaning of the Prime Minister's phrase, "use it or lose it."

See Chapter 3, "Major Federal–Provincial Programs," page 91, for more on how the federal government aims to level the playing field for the various political jurisdictions within Canada.

Resource extraction and transfer payments are the principal commercial elements of the northern economy. The economic structure of the Territorial North mirrors these two forces, so that the tertiary and primary sectors are of greatest significance. The secondary sector plays a small role, primarily in construction and in the processing of ore. Manufacturing is almost non-existent in the Territorial North as most products are imported from southern Canada and foreign countries (Table 10.5).

In terms of employment, the primary sector in the Territorial North accounts for a much larger proportion of the workforce than it does in other geographic regions. For example, in 2011, approximately 15 per cent of the workers were in the primary sector compared to less than 2 per cent in Ontario. The Territorial North has approximately 83 per cent of its workforce allocated to the service sector, making it the dominant sector. With a small tax base, large transfer payments from Ottawa to the three territorial governments play a controlling role in their budgets. Given the late start in the process of economic and social development, the Territorial North has a great deal of catching up in terms of infrastructure and in terms of equipping its Aboriginal population with the skills and tools to complete in the market economy. From this perspective, transfer payments, as large as they are, are insufficient to narrow the gap between the North and the South.

Most workers are employed by one of three governments: federal, territorial, or local. Local governments are settlement councils, band councils, and other organizations funded by a higher level of government. The reason the tertiary sector is so large in the Territorial North is that geography demands such an investment of people and capital to ensure the delivery of public services. Territorial governments must spend more

Table 10.5 **Estimated Employment by Economic Sector, Territorial North, 2011**

Economic Sector	North Workers (%)	Ontario Workers (%)	Difference (percentage points)
Primary	15	1.9	13.1
Secondary	2	19.2	−17.2
Tertiary	83	78.9	4.1
Total	100.0	100.0	

Source: Table 5.3; author's estimate.

money per resident than do provincial governments to provide basic services. Much of the cost differential is attributed to overcoming distance in the Territorial North and hiring staff for small communities. The drawback is that economies of scale are difficult to achieve in small communities where teachers and nurses often have a relatively small number of students and patients compared to those working in larger centres. However, the social importance of the public service sector goes beyond the number of employees. The wide geographic distribution of public jobs across the North is a major social benefit to those in small communities. As a result, employment opportunities, while concentrated in the three capital cities, exist in every community. In small, remote communities, where few jobs exist and where potential workers have given up searching for jobs, underemployment rates are often as high as 60 per cent, and virtually all jobs are associated with one or more levels of government. The territorial governments' ability to implement social policies in their hiring practices can make a difference. For example, Nunavut's government seeks to employ mainly Inuit in its civil service, but the success of this policy so far has been held back by the low levels of educational attainment among the Inuit population.

Government Structure

The Territorial North consists of three territorial governments: Yukon, the Northwest Territories, and Nunavut. The capital cities of these three territories are Whitehorse, Yellowknife, and Iqaluit. Territorial governments have fewer powers than provincial governments, and in this sense they are political hinterlands. For example, the federal government retains power over natural resources and collects a substantial amount of tax revenue from companies extracting natural resources, so this important revenue source is not available to territorial governments. Instead, they depend on Ottawa for transfer payments. Without these transfer payments, the territorial governments could not offer the basic services available in southern Canada. Nunavut receives 90 per cent of its revenue from Ottawa. The level of fiscal dependency is somewhat lower for Yukon and the Northwest Territories, and currently the Northwest Territories is challenging Ottawa over the issue of resource royalties.

The Territory of Nunavut

The new territory of Nunavut was made possible through a land settlement agreement between Canada and the Inuit of the eastern Arctic in 1993. The terms of the agreement included the use of Crown lands for the Inuit to hunt, fish, and trap, and the transfer of part of the land to the Inuit, with a portion of this area involving rights to subsurface minerals. The same year the land agreement was reached, the federal government made the commitment to create a new territory by passing the Nunavut Act. This Act, which provided the legal basis for the creation of a distinct territory and territorial government, also allowed for a six-year transition period, giving the Inuit time to form the government, recruit civil servants, and select a capital city. By means of a plebiscite, Iqaluit was selected as the capital of the new territory. Following an election in February 1999, the 19 members of the Nunavut Assembly took office on 1 April 1999. Unlike First Nations, the Inuit created a public form of government, meaning that every resident—Aboriginal and non-Aboriginal—has the same political rights.

The creation of a separate territory for the Inuit brought hopes for a brighter future. Through an Inuit government, a sustainable economy was thought to be achievable within 20 years. The Bathurst Mandate (Nunavut, 1999) gave voice to that hope:

> As Nunavummiut we look to support ourselves and contribute to Canada through the potential of our land, the responsible development of our resources and the contributions of our peoples and our cultures.

The word "Nunavut" means "our land" in Inuktitut. The vast majority of the approximately 34,000 people who reside in Nunavut are Inuit. A primary goal of the Nunavut government is to reflect the people's aims and aspirations. In addition to expanding economic development and increasing the number of jobs in the new public service for residents of the highly decentralized administration of Nunavut, the government strives to promote Inuit culture and the Inuktitut language. For many Inuit, Nunavut is a dream come true where their cultural survival within Canadian society is assured. The dream also brings with it the challenge of building a northern economy that can provide jobs for its growing labour force and thereby reduce its financial dependency on Ottawa. However, the challenges facing the Inuit government are formidable. As Légaré (2008a: 367) writes:

> For now, though, the urgent socio-economic plight of Nunavut does not bode well for the future. The vision of a viable Nunavut society by the year 2020, as expressed through the Bathurst Mandate, seems to be, at least for now, an illusion.

Culture is a critical issue among Inuit, who are spread across the Arctic in 53 communities and who have four regional governments (Figure 10.6). In Nunavut, for instance, the struggle to maintain an Inuit culture depends heavily on the combination of education and language. Transfer payments from the federal government fall short of the needs of the schooling system and therefore for the maintenance of Inuit culture and language. The 2007 report of Nunavut Tunngavik, *Saqqiqput: Kindergarten to Grade 12 Education in Nunavut*, stated the problem as:

> The fact that most Inuit children in Nunavut drop out of school before graduating is a serious societal problem. Without renewed attention and investment to improve kindergarten to Grade 12 (K–12) education outcomes, Inuit will not be able to fully access the government's obligations under the *Nunavut Land Claims Agreement* (NLCA), access benefits of economic development, and fulfill desires to build a fully functioning Inuit society through a public government model. The

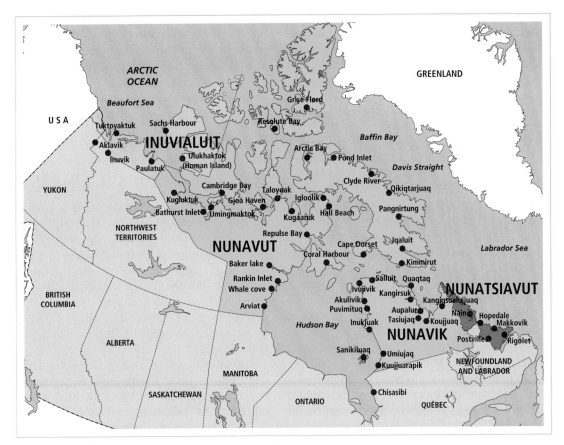

Figure 10.6 **Inuit Nunangat**

This map shows the four Inuit regions of Canada. Three regions—Inuvialuit in the NWT, Nunavik in Québec, and Nunatsiavut in Newfoundland and Labrador—have attained some degree of autonomy within long-established political jurisdictions, but only the territory of Nunavut is a stand-alone political entity within Canada. *Source:* Aboriginal Affairs and Northern Development, Inuit Relations Secretariat, Inuit Nunangat Map, 15 Oct. 2010, at: <www.aadnc-aandc.gc.ca/eng/1100100014250/1100100014254>.

THINK ABOUT IT

Margaret Wente, a *Globe and Mail* columnist, asks the question: "Who's in trouble, polar bears or people?" In Wente's opinion, the polar bears are likely to survive for another 100,000 years but the Inuit are in deep trouble. Wente writes: "Nunavut has the highest rate of lone-parent families and the lowest education levels in the country. Most kids don't finish high school" (Wente, 2012). Does the Nunavut Tunngavik report partly answer this question?

education system does not currently fully entrench Inuit language, values, culture and society into its administration and delivery, thereby denying Inuit from fully utilizing one of the most powerful formal resources for empowerment.

While 72.4 per cent of the Inuit population in Nunavut stated that Inuktitut is their first language and 79.2 per cent of Inuit stated that Inuktitut was the only or main language spoken at home, only two schools offer Inuktitut instruction beyond Grade 3 and then only to Grade 6. (Nunavut Tunngavik, 2007)

Aboriginal Economy

Aboriginal people participate in both the land-based and wage economies. While the "mix" varies among Aboriginal peoples, harvesting wildlife from the land and sea tends to be highest in Native settlements where traditional practices are strong and job opportunities in the formal economy are limited. Trapping, hunting, and other land-based activities persist in the North not so much because of their commercial value but largely because of their cultural importance. For instance, hunting produces food for the family and country food remains a core cultural feature among northern Aboriginal families. While trapping and hunting sometimes go hand in hand, interest in trapping has diminished because of low prices for furs and because of the effective lobbying of the European Union by animal rights groups. Since peaking in the late 1980s at a value of just over $6 million for the Northwest Territories, fur production has declined. The value of fur production in the first decade of the twenty-first century ranged from a low of $830,000 in 2010–11 to $1.3 million in 2006–7. Yet, the number of pelts remained in the 20,000–30,000 range (NWT Bureau of Statistics, 2011). With low prices and rising costs, few trappers can make a profit and most supplement their trapping activities from other income.

Efforts of animal rights groups have focused on the seal hunt on the ice floes near the island of Newfoundland where their case is based on "inhumane" killing practices. However, with the ban on importing seal pelts, the Inuit have been affected as well. In the past, the seal provided both food and cash for Inuit families but without a use for the pelt the incentive to hunt seal has diminished, pushing Inuit to rely more on store food.

The Aboriginal economy took a new turn with comprehensive land claim agreements. These agreements provided the structure and capital to participate in the marketplace. The Inuvialuit, for example, are major shareholders in the Aboriginal Pipeline Group.

Comprehensive Land Claim Agreements

Aboriginal land claims are based on the peoples' long-time use of the land for hunting, fishing, and trapping. In Canada, Aboriginal land claims are settled by treaty: the Aboriginal tribe surrenders its claim to all the land in exchange for title to a smaller amount of land, a cash settlement, and, in most instances, usufructuary right to a larger territory owned by the Crown for hunting and fishing. Other issues involved in land claim negotiations are self-government, management of wildlife resources, and preservation of language and culture. In simple terms, land claims are an attempt by Aboriginal peoples and the federal government to resolve the issue of Aboriginal rights.

For further discussion of land claims, see Chapter 3, "Modern Treaties," page 97.

During the latter part of the twentieth century, Indian, Inuit, and Métis organizations in the Territorial North made land claims on behalf of their members. In the 1970s, three major land claims were made in the region: Yukon First Nations (Yukon), Dene/Métis (Denendeh), and Inuit (Arctic). Over the next two decades, the Inuit land claim was split into the western Arctic (Inuvialuit) and the eastern Arctic (Inuit/Nunavut), and the Dene/Métis land claim for Denendeh was divided into five separate land claims (Gwich'in, Sahtu/Métis, North Slavey—also referred to as Dogrib or Tlicho, South Slavey, and Deh Cho).[6]

Until the late 1970s, Ottawa did not recognize traditional claims to Crown land. The evolution of claims policy and objectives has been greatly influenced by jurisprudence. In the *Calder* case in 1973, the Supreme Court of Canada acknowledged the possible existence of Aboriginal title in Canadian law, that is, some claims to land-ownership might be valid but such claims had to prove earlier occupancy. The Supreme Court ruling caused Ottawa to reconsider its position.

Between 1973 and 1985, the rethinking of the federal government's position on land claims resulted in the comprehensive and specific land claims policy, which was formally announced in 1986. Prior to this time, negotiations with the Inuvialuit followed the guidelines of the new policy, and in 1984 the Inuvialuit reached a final agreement with Ottawa. They were followed by the Gwich'in in 1992, and, in the next year, the Sahtu/Métis, Inuit, and Yukon First Nations. In 2003, the Dogrib or Tlicho Agreement was completed, leaving only two outstanding claims—Deh Cho and South Slavey.

Figure 3.11, "Modern treaties," page 99, shows the geographic extent of modern land claim agreements in the Territorial North.

As with all comprehensive land claim agreements, the Inuvialuit Final Agreement (IFA) was based on the Inuvialuit's traditional use and occupancy of lands in the western Arctic, that is, the land that they and their ancestors used for hunting and trapping. The Inuvialuit land claim was originally part of the Inuit land claim to the entire Arctic. However, the Inuvialuit broke away from the Inuit and, as a result, there are two large **settlement areas** in the Arctic, the Inuvialuit and the Nunavut settlement areas.[7] The goals of the Inuvialuit were to preserve their cultural identity and values within a changing northern society; to enable themselves to be equal and meaningful participants in the northern and national economy and society; and to protect and preserve the Arctic wildlife, environment, and biological productivity (Canada, 1985: 1). Canada's goal, on the other hand, was to extinguish the Inuvialuit claim to the western Arctic.

Though the JBNQA (1975) represents the first modern treaty, Ottawa, as discussed in Chapter 3, instituted a new negotiating system composed of specific and comprehensive agreements. As the first comprehensive land claim agreement achieved under this system, the IFA served as a model for subsequent ones. Since 1984, as shown in Table 10.6, five more comprehensive land claim agreements have been reached. Each agreement is very similar to the IFA. One common feature is that each one has created economic and environmental administrative sectors. In the case of the IFA, the economic sector is the Inuvialuit Regional Corporation, which manages and invests the cash settlement received as part of the agreement.[8] The second sector, the Inuvialuit Game Council, is responsible for environmental issues that affect their hunting economy. The creation of these two sectors was the Inuvialuit's attempt to straddle two worlds; their old world was based on harvesting game from the land, while their new world is part of the global industrial economy. However, the IFA was silent on one important element so necessary for Aboriginal peoples—self-government. Over time, the issue of self-government became part of comprehensive land claim agreements. The Nunavut and Tlicho final agreements do spell out the specific nature and unique structures of self-government for the Inuit and the Dogrib.

The Yukon First Nation agreement, signed in 1993, was different in that it only established the basic elements of final agreements for each of the 14 Yukon First Nations, leaving the negotiations for a final agreement to each Yukon First Nation. Known as the Umbrella Final Agreement, this 1993 arrangement provided the basic framework within which each of the 14 Yukon First Nations (Carcross/Tagish; Champagne and Aishihik; Tr'ondek Hwech'in; Kluane; Kwanlin Dun; Liard; Little Salmon/Carmacks; Nacho Nyak Dun; Ross River Dena; Selkirk; Ta'an Kwäch'än Council; Teslin Tlingit Council; Vuntut Gwitchin; and White River) can conclude a final claim settlement agreement. White River First Nation, Ross River Dena Council, and Liard First Nation are the remaining Yukon First Nations that have not concluded agreements (AANDC, 2010). The 11 First Nations that have achieved a final agreement

Table 10.6 **Comprehensive Land Claim Agreements in the Territorial North**

Aboriginal People	Date of Agreement	Cash Value	Land (km²)
Inuvialuit	1984	$45 million (1977 $)	90,650
Gwich'in	1992	$75 million (1990 $)	22,378
Sahtu/Métis	1993	$75 million (1990 $)	41,000
Inuit (Nunavut)	1993	$580 million (1989 $)	350,000
Yukon First Nations	1993	$243 million (1989 $)	41,440
Dogrib (Tlicho)	2003	$152 million (1997 $)	39,000

Sources: Canada (1985: 6, 31; 1991: 3; 1993a: 3; 1993b: 81, 215; 2004); *Globe and Mail* (1993).

are: Teslin Tlingit Council (1995), Vuntut Gwitchin First Nation (1995), Champagne and Aishihik First Nations (1995), First Nation of Nacho Nyak Dun (1995), Selkirk First Nation (1997), Little Salmon/Carmacks (1997), Tr'ondek Hwech'in (1998), Ta'an Kwäch'än Council (2002), the Kluane First Nation (2004), Kwanlin Dun First Nation (2005), and Carcross/Tagish First Nation (2006).

For the Inuvialuit, the agreement was an important step towards defining their place within Canadian society. In exchange for cash and land, the Inuvialuit gave up their Aboriginal rights, including their claim to the vast lands of the western Arctic. In exchange, they received legal title to about 90,650 km² of land, which is slightly less than 20 per cent of the settlement area (Canada, 1985: 6). The land to which the Inuvialuit gained title lies within the Inuvialuit settlement area, which extends from the Arctic mainland into the Arctic Ocean (Figure 3.11). A number of islands, including Banks Island, Prince Patrick Island, and the western parts of Melville Island and Victoria Island, are situated here. Of the total settlement land, 12,950 km² include surface and subsurface mining rights for the Inuvialuit. The Inuvialuit enjoy exclusive rights to hunting, trapping, and fishing over the remaining 77,700 km² and the offshore waters (as noted in Figure 10.4, the sea boundary with Alaska is not yet resolved).

Country Food

Country food remains a key element of Aboriginal culture and practices. While Aboriginal cultural identities vary across the Territorial North, certain core elements maintain their distinctness. These core elements include a strong attachment to the land, to country food, and to the ethic of sharing. **Country food** is food obtained from the land, a preferred source of meat and fish. As equivalent store-bought foods are expensive, most Aboriginal northerners keep their food costs low by consuming country food. While it is true that there are substantial costs expended in harvesting country food, to some degree this cost is offset by the pleasure and spiritual rewards of being on the land and participating in hunting and fishing. Sharing also remains an important component in the harvesting and distribution of country food among family members, relatives, and close friends. Respect for cultural diversity is a hallmark of Canadian identity. In May 2009, the issue of respect took on a particularly nasty face with the negative reaction in some quarters to Governor General Michaëlle Jean eating a slice of seal heart at an Inuit festival in Rankin Inlet, Nunavut. By participating in this traditional Inuit practice, the Governor General showed respect and implied solidarity with their hunting of seals and the selling of seal pelts (photo 10.4). The negative reaction from animal rights groups and the European Union reflects a form of cultural relativism. Who sets the standards for social conduct? Mary Simon, president of Inuit Tapiriit Kanatami, Canada's national Inuit organization, applauded Jean for her public support of traditional Inuit culture. "Hunting and eating a seal is not a political act, nor is it 'bizarre' or 'disgusting' as the anti-sealing lobby have commented," she said. "To us, this kind gesture is an acknowledgement by the Governor General of our culture and our dependence upon our wildlife as an important resource for our communities today" (O'Neil, 2009). In March 2010, Canadian parliamentarians followed the example of the Governor General when the parliamentary restaurant, for the first time, featured various seal dishes.

The European Union, formerly a major importer of seal pelts, has banned their import, with economic consequences for the Inuit that are hurtful. The issue is complicated because of the connection between seal as a food and seal pelts as a commercial product. For centuries,

Photo 10.4
Governor General Michaëlle Jean, centre, helps an Inuit elder skin two seals during a community feast in Rankin Inlet, Nunavut, 26 May 2009

European nations have imported furs and pelts from Canada. Ironically, Britain and France originated the fur trade, which has entailed the killing of seals.

Respect for the diverse cultures in Canada is discussed in Chapter 4, "Multiculturalism," page 144.

Figures on the cost, size, and value of harvesting country food are not available. However, estimates suggest that the "substitute value" of country food (the value of equivalent store-bought food) was about $40–$50 million per year in the mid-1980s (Usher and Wenzel, 1989). A visit to an Aboriginal community quickly reveals the importance of wildlife harvesting to the daily diet. While some store-bought foods can be found on the tables of Aboriginal families, the preference is for game and fish from the land and waters.

Country food is so important to Aboriginal northerners that they have made wildlife an essential issue in land claim negotiations. The agreements establish co-management committees that administer the environment and wildlife. For example, these committees approve (or reject) proposals for new industrial projects and determine the total allowable harvest of wildlife based on biological principles and **traditional ecological knowledge**. In the case of the Inuvialuit Final Agreement, the members of the co-management committees are named by the Inuvialuit Game Council and the three governments (Canada, Northwest Territories, and Yukon). The members of the Inuvialuit Game Council are selected by the Hunters and Trappers Associations found in each of the six Inuvialuit communities. Under the direction of these committees, professional scientists record and monitor the state of the environment and wildlife. As a result of comprehensive land claim agreements, power to control the use of the environment and wildlife now rests, in part, with most Aboriginal peoples in the Territorial North. Still, resistance to traditional knowledge remains alive and well in a few corners of academe (Widdowson and Howard, 2008).

Early contacts and European hopes of discovering wealth in the North are discussed in Chapter 3, "Initial Contacts," page 71, and in Vignette 3.1, "Contact between Cartier and Donnacona," page 72.

Resource Development in the Territorial North

Since the sixteenth century, explorers—now replaced by multinational companies—sought to exploit the North's natural wealth, particularly its minerals. The search for precious metals began soon after explorers reached North America. In Mexico and South America, the Spanish and Portuguese were particularly successful in obtaining vast quantities of gold and silver. In North America, English and French explorers searched for similar wealth. Martin Frobisher, on three separate expeditions in the 1570s, returned to England with hundreds of tons of "sandstone flecked with mica" in the hope he had discovered gold—his shiploads of Arctic rock ended up being used in Elizabethan-era road-building; years earlier, Jacques Cartier had returned to France with iron pyrites (fool's gold) (Bumsted, 2010: 56, 54). Northern Indians trading at the Hudson's Bay posts often wore copper ornaments and had copper tools. Richard Norton, the factor or chief trader at Fort Prince of Wales, ordered one of his men, Samuel Hearne, to find the source of this copper and assess its commercial worth to the company. In one of the great overland journeys of all time, Hearne travelled with Chipewyan leader Matonabbee's tribe from Fort Prince of Wales on the west coast of Hudson Bay (present-day Churchill, Manitoba) to the mouth of the Coppermine River at Coronation Gulf on the Arctic Ocean. In 1771, Hearne found the source of copper, but this natural copper occurred only at the surface in small amounts. Others seeking mineral wealth in the North were more fortunate. In 1896, George Carmack, an American prospector, and his Indian brothers-in-law, Skookum Jim and Tagish Charley, found gold nuggets in a small tributary (later renamed Bonanza Creek) of the Yukon River. When word of their find reached the outside world, the Klondike gold rush sparked the greatest gold rush on record. Dawson City was the centre of this frenzy. Thousands of prospectors flooded the area to scour the sandbars of the rivers and pan for gold. Within two summers, the more accessible placer gold was gone and by 1900 many left for a new gold find in Alaska. The only remaining gold was deep in the frozen terraces found along the sides of the Klondike River. Permafrost is widespread at latitudes above 60° N. Companies with both capital and technology took over gold mining using hydraulic and steam-mining techniques to thaw the ground and force the sand, gravel, and fine gold particles into separating devices commonly known as sluice boxes. The commercial, large-scale gold mining signalled the birth of megaprojects and the depredation of the northern environment—the erosion and tailings from hydraulic techniques and dredging are still visible in the area around Dawson City, Yukon.

Resource Development and Transportation

Until Prime Minister Diefenbaker's "Northern Vision" in the late 1950s, Ottawa paid little attention to the Territorial North. With the Development Road Program, Ottawa began to invest funds into highway construction. Still, the North is so vast and so sparsely populated, the case for spending large sums on road-building is difficult to make—the cost of highway construction is extremely expensive, which accounts for the paucity of highways in the Territorial North. One reason for the high cost is the presence of permafrost; another is the distance between places, which translates into extra expenses to assemble road-building equipment and materials. The federal government plays a major role in financing highway construction. Mining companies lobby Ottawa for roads to their mining sites because the costs of air transportation are either impractical or too expensive, or both. For the three diamond mines in the Northwest Territories, winter roads provide relatively low-cost trucking routes.

The Territorial North has few highways. Nunavut has none that connect to the national highway system. The most prominent highways in the North are Yukon's Alaska Highway and the Dempster Highway and several in the Northwest Territories, including the Mackenzie Highway, Yellowknife Highway, Fort Smith Highway, and Liard Trail. Winter roads greatly extend the road system, making the trucking of heavy equipment,

Photo 10.5

Dominion Diamond Corporation's Ekati Diamond Mine is located approximately 310 km northeast of Yellowknife and 200 km south of the Arctic Circle. It is Canada's first surface and underground diamond mine. Operations at Ekati Mine officially began in October 1998. Mining and processing occur continuously 24 hours per day, 365 days per year. The mine has produced on average over three million carats per year of rough diamonds annually with production transitioning from predominantly open-cut to a mix of open-cut and underground mining. Ekati has development options for future open-cut and underground production. Annual sales from Ekati Diamond Mine represent approximately 2 per cent of current world rough diamond supply by weight and approximately 6 per cent by value in 2012.

Photo 10.6
A drawing of the Deh Cho Bridge that crosses the Mackenzie River near Fort Providence. The bridge replaced the ferry system in the summer and the ice road in the winter.

building supplies, and other goods possible in the winter. In late 2012, the completion of the Deh Cho Bridge (photo 10.6) greatly reduced the cost of trucking goods to Yellowknife and other centres along the Mackenzie River served by road. CN operates on a single track to Hay River, Northwest Territories, where it connects with the river barge system on the Mackenzie River.

The pattern of mining developments in the North clearly indicates the importance of water transportation and winter roads. In 1935, the first major gold discovery in the Northwest Territories occurred near Yellowknife on Great Slave Lake. Water transportation was critical to the opening of the Con-Rycon gold mine, which was near the shore of Great Slave Lake and therefore had excellent access to the Mackenzie River transportation system. Mines located far from water transportation, such as the now-closed Lupin gold mine (about 250 km northeast of Yellowknife), required the construction of **winter roads**. The need for roads in mining is not so much for transporting minerals to external markets as for bringing equipment and supplies to the mine site. Mines along the Arctic coast can take advantage of ocean transportation. Three examples are the former mine at Rankin Inlet on the shores of Hudson Bay, the abandoned lead-zinc mine at Nanisivik near the northern tip of Baffin Island, and the now-closed Polaris lead-zinc mine on Little Cornwallis Island in the Arctic Archipelago (see Figure 10.7).

In assessing the cost of transportation, the value of the mineral is taken into account. For example, copper, lead, nickel, and zinc are low-grade ores (much of the ore has no commercial value). Even after the first-stage separation of some of the waste material from the valuable mineral, the enriched ore still remains a bulky product, with significant waste material remaining that can be removed only through smelting. Shipping such a low-value commodity is very expensive. Ideally, such ore is transported to a smelter by ship. Railways are the second-most effective transportation carrier for low-grade ore. The critical nature of transportation for such mines is clear in the example of the lead-zinc deposit at Pine Point in the Northwest Territories. Discovered in 1898 by prospectors heading overland to the Klondike gold rush, the Pine Point deposit was not developed until 1965, when a railway was extended to the mine site. With a means of transporting the ore to a smelter, the company, Cominco, could send massive amounts

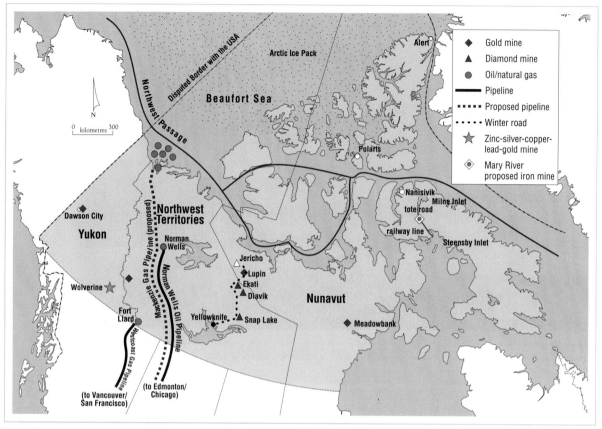

Figure 10.7 **Resource development in the Territorial North**
The mineral wealth of the Territorial North lies mainly in the Northwest Territories. Diamonds, gold, natural gas, and oil drive this territory's resource economy. In contrast, the resource economies of Yukon and Nunavut are much smaller. Petroleum exploration in the Beaufort Sea is complicated by the border dispute between Canada and the United States. The US claims a narrow strip of the Canadian section of the Beaufort Sea—an area of 21,436 km². Mines often have a short lifespan. Jericho, Polaris, and Nanisivik are abandoned mines. The proposed Mary River Project is an exciting iron mining development for Nunavut, including a railway and all-season port.

VIGNETTE 10.6

Sea Transportation on the Arctic Ocean

Arctic transportation takes advantage of nature. In late summer when the shore ice melts along the western Arctic, a narrow stretch of water opens between the shore and the polar pack ice. Small ships take advantage of this open water to bring supplies to communities located along the coast of the Beaufort Sea. Most of these supplies are transported by barge northward along the Mackenzie River. In the eastern Arctic, after the shore ice has retreated, ocean-going ships bring fresh supplies to Arctic communities. In other places, most of the surface of the Arctic Ocean is permanently covered by floating sea ice known as the Arctic pack ice. Here, only specially reinforced ships and icebreakers can traverse the ice-covered waters. How thick is the ice? One-year ice is about 1 m thick while older ice can reach 5 m thick. Such voyages by icebreakers in older ice are not common. In August 1994, two icebreakers, the American *Polar Sea* and the Canadian *Louis S. St Laurent*, ploughed through thick ice on a scientific voyage to the North Pole. Canada has commissioned a new icebreaker, CCGS *John G. Diefenbaker*, costing just over $1 billion. The construction job for this icebreaker was awarded to the Vancouver shipyard with a delivery date of 2017.

of ore by rail to its smelter at Trail in British Columbia. The mine closed in 1983, leaving Pine Point a ghost town.

In the past, companies built resource towns near their mining sites—Pine Point, for example. However, the life expectancy of resource towns tends to be short because when the mine closes, no economic support remains for the community. Some companies have avoided this problem by turning to **air commuting**, which may be an effective way to obtain skilled southern workers for remote resource projects but does have drawbacks for the North. Southern-based air commuting systems hurt the North's economy because: (1) workers spend their wages in their home communities in southern Canada, thereby stimulating provincial, not territorial, economies; and (2) workers who reside in a province but work in the territories pay personal income tax to provincial rather than to territorial governments, thereby depriving territorial governments of valuable personal income tax. For Aboriginal communities, air commuting is advantageous because it provides access to high-paying jobs in the mining industry and, at the same time, allows workers' families to remain in their Aboriginal communities. Known as "cultural commuting," the fly-in and fly-out work schedule has another attraction by allowing workers time to hunt and fish on their week(s) off.

Key Topic: Megaprojects

The Territorial North has entered a new phase of resource development characterized by megaprojects controlled by multinational companies. Megaprojects usually cost more than $1 billion and require several years to complete the construction stage. Megaprojects have integrated the Territorial North's economy into the global economy, thereby firmly locking the North into a resource hinterland role in the world economic system. The most recent megaprojects in the Territorial North have involved diamond mining but two enormous megaprojects planned for the Arctic are on the drawing board, including the Izok Corridor proposal of the Chinese state-owned MMG Ltd (formerly Minmetals Corporation), which calls for five underground and open-pit mines producing lead, zinc, and copper (Weber, 2012: B1), and Baffin Island's massive Mary River iron ore project proposed by Luxembourg-based ArcelorMittal and Baffinland Iron Mines. In early 2013, however, Baffinland announced that, at least initially, the Mary River Project was being scaled back significantly (Jordan, 2013). The two projects, in their original conception, would involve port construction and year-round shipping of the ore to markets in Asia and Europe, respectively.

In 1998, diamond mining rejuvenated the economy of the Northwest Territories. The first diamond mine to come into production was the Ekati mine. As part of the NWT Diamonds Project (MacLachlan, 1996), preliminary construction of buildings, roads, and mines began in the early 1990s, but the Ekati mine was not completed until 1998. Two other mines, Diavik and Snap Lake, are now in production, while the Gahcho Kué deposit, at Kennedy Lake 85 km southeast of Snap Lake, has reached the environmental impact review state. Supplies for these remote mines are trucked along a winter road (Figure 10.7). All are located in the Slave geological region that straddles the border between the Northwest Territories and Nunavut. The zinc and copper deposits at Izok and High Lake, also located in the Slave geological region, are far from reaching a commercial stage because of the high cost of transporting low-value ore. In the early 1990s a German firm, Metall, considered developing the Izok lead/zinc deposit, but without the promise of federal government support for a 350-km highway to the Arctic coast, Metall turned its back on this very expensive project. Fast forward to 2012: the same deposit is under consideration by the Chinese firm, MMG Ltd. MMG has submitted its Izok Corridor Project proposal to the Nunavut Impact Review Board (Jordan, 2012).

Proponents of resource development describe megaprojects as the economic engine of northern development, though others challenge this assumption, claiming that they offer few benefits to the region and, more particularly, to the Aboriginal communities (Bone, 2012: 162–4). These large-scale ventures are designed for the export market. By injecting massive capital investment into the construction of giant engineering projects, megaprojects create a short-term economic boom. However, most construction expenses are incurred outside hinterlands because the manufactured equipment and supplies are produced not in hinterlands but in core industrial areas. This reduces the benefits of megaprojects to the hinterland economy and virtually eliminates any opportunity for economic diversification. As well, since all megaprojects in the Territorial North are based on non-renewable resources, these developments have a limited lifespan. At the end of these projects, the local economy suffers a collapse. Three examples are the closure of mines at Faro, Yukon; Pine Point, Northwest Territories; and Rankin Inlet, Nunavut.

However, megaprojects can inject much-needed capital and create development in the region. Megaprojects in resource hinterlands are high-risk ventures. Multinational companies can reduce their risks in three ways. First, they can create a consortium of companies and thereby spread the investment risk among several firms. Second, they can arrange for long-term sales of the product at a fixed price before proceeding with construction. Third, they can obtain government assistance, which often takes the form of low-interest loans, cash subsidies, and tax concessions. Four megaprojects are discussed in the following sections: the Mackenzie Valley Pipeline Project; the Mackenzie Gas Project; the Norman Wells Oil Expansion and Pipeline Project; and the NWT Diamonds Project.

The Mackenzie Valley Pipeline Project: 1970s

By 1970, the plans formulated by oil and pipeline companies to bring natural gas from Prudhoe Bay, Alaska, to major markets in the United States, primarily in the Midwest, were submitted to Ottawa.⁹ One plan was to build a pipeline, known as the Mackenzie Valley Pipeline Project, which would transport natural gas from the Arctic coast of Alaska to the Mackenzie Delta and then south along the Mackenzie Valley, eventually reaching Alberta and then the United States.

In Canada, this proposal was subjected to the Mackenzie Valley Pipeline Inquiry (also known as the Berger Inquiry). The Berger Inquiry began in 1974 and reported in 1977. Led by British Columbia Justice Thomas Berger, who a few years earlier had been chief counsel for the Nisga'a in their landmark land claim case (*Calder*), this federal investigation examined the potential environmental, social, and economic impacts of the proposed gigantic construction project. The Berger Inquiry established a new precedent by holding community hearings in numerous remote Native settlements, and its final report, *Northern Frontier, Northern Homeland* (1977), recommended that a natural gas pipeline from the Mackenzie Delta to Alberta was feasible but should only proceed after further study and after settlement of the Dene land claims. To complete these two tasks, Berger recommended a 10-year delay in the construction of such a pipeline. At the same time, he rejected the construction of a pipeline from Prudhoe Bay to the Mackenzie Delta because of the danger of harming the delicate Arctic environment and wildlife found along the North Slope of Alaska and Yukon. By the end of the Berger Inquiry, Prudhoe Bay natural gas was no longer in demand because of the discovery of natural gas deposits in Alberta and Mexico that offered North American markets a lower-cost source of natural gas.

For discussion of the *Calder* case, see Chapter 3, "Modern Treaties", page 97.

The Mackenzie Gas Project: 2000

Near the mouth of the Mackenzie River, large deposits of natural gas exist. The three natural gas fields in the Mackenzie Delta are Taglu, Parsons Lake, and Niglintgak. In 2000, Imperial Oil proposed the Mackenzie Gas Project, which would see a 1,220-km pipeline system along the Mackenzie Valley, linking northern natural gas sources to southern US markets. In 2003 the pipeline was estimated to cost $5 billion, but by 2007 Imperial Oil reassessed the cost of the pipeline, the gas fields, and the gas-gathering system at $16.2 billion (CBC News, 2007). Could such a jump in estimated costs be a ploy to obtain concessions from Ottawa? With the Northwest Territories government in tow, Imperial Oil sought federal support, arguing that the project was in the "national interest" because, like the CPR, it will open up the last major frontier in Canada's North. Needless to say, with more and more gas deposits discovered due to the fracturing technique that releases gas from shale deposits, North America is awash in natural gas, driving its price to new lows. The Mackenzie Gas Project, like the Mackenzie Valley Pipeline Project, failed the commercial test and Imperial Oil was forced to shelve it.

The Norman Wells Oil Field, 1920–2012

The Norman Wells oil field was discovered in 1920. Until the pipeline to southern Canada was built, production was limited to providing for local communities and mining sites along the Mackenzie River and briefly, during World War II, as the central component of the Canol Project. In 1982, Esso Resources Canada (Imperial Oil) obtained federal permission to build a pipeline and ship the oil to Canadian and US markets. Prior to the pipeline, annual output was less than 180,000 m³. With the completion of the pipeline in 1985 and the expansion of oil production by a factor of 10, the Norman Wells oil field became a major player in the Territorial North. The Norman Wells Project, which cost almost $1 billion in 1996 dollars, was hailed by industry as a model megaproject for the North. Both the federal government and Esso Resources Canada believed that the project would not harm the natural and social environments in the construction impact zone, which stretched from Norman Wells to Fort Simpson and then to the Alberta border. Aboriginal organizations, such as the Dene Nation, and environmental groups, such as the Canadian Arctic Resources Committee, did not agree. Nevertheless, the Norman Wells Project was successful and continues to provide profits to Esso Resources Canada. Since 2002, however, production from the Norman Wells field has decreased from nearly 1.4 million m³ to just over 600,000 m³ in 2011 (Figure 10.8). Over the same time span, rising oil prices have offset the production decline—in 2010, its value was $456.6 million; with higher production in 2004, the value was $390 million (NWT Bureau of Statistics, 2012: 37).

The NWT Diamonds Project

Canada is now the third-largest producer of diamonds in the world, behind Botswana and Russia, and accounts for 15 per cent of the world supply, thanks in large part to the three diamond mines in the Northwest Territories—Ekati, Diavik, and Snap Lake. De Beers has submitted its environment impact statement to the federal government for another mine at Gahcho Kué. In 2011, the Northwest Territories accounted for $2.1 billion of diamonds (Table 10.7).

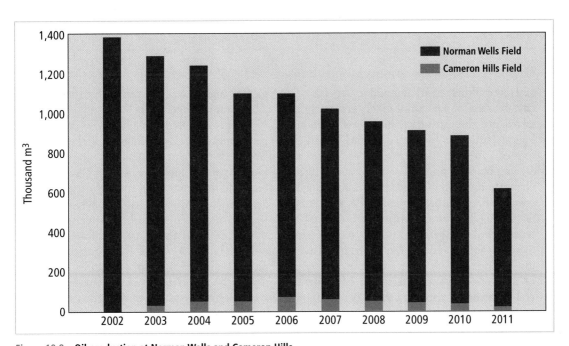

Figure 10.8 **Oil production at Norman Wells and Cameron Hills**
Norman Wells is in the Central Mackenzie Valley; the much smaller Cameron Hills deposit is in southern NWT, near the Alberta border.
Source: Adapted from AANDC, Northern Oil and Gas Branch (2012: Figure 6). Reproduced with the permission of the Minister of Public Works and Government Services Canada, 2013.

Photo 10.7

A drilling rig near Norman Wells (second island from the left). The Norman Wells Project allowed for a significant increase in oil production and supply to southern markets. The Norman Wells Project represents a successful and profitable operation.

How did this remarkable and unexpected discovery come about? Until 1991, geologists believed that the Canadian Shield was not a geological structure where diamond could be formed. Two prospectors, Charles Fipke and Stewart Blusson, proved them wrong and turned conventional thinking on its head when they discovered diamond-bearing kimberlite near Lac de Gras in the Northwest Territories. Following extensive drilling, their find proved to have commercial viability, and BHP Diamonds Inc. decided to develop a mine. The company applied to Ottawa to secure approval, and in 1996 the federal government approved the NWT Diamonds Project, which entails the operation of the Ekati diamond mine in the Lac de Gras area of the Northwest Territories for about 25 years. Five diamond-bearing **kimberlite pipes** are being mined, four of them located within a few kilometres of each other. All five kimberlite pipes are located under lakes that had to be drained to facilitate mining operations.

Diamond mining has quickly become the backbone of the mining industry in the Territorial North. By value of production, number of employees, and spinoff effects, BHP Diamonds has been the leading mining company in the North. Starting in 2006, the two mines, Ekati and Diavik, produced $1.6 billion worth of diamonds; by 2011 the figure had jumped to $2.1 billion (Table 10.7). The mining operation employs about 800 workers with an estimated annual wage bill of $40 million. BHP has used an air-commuting system to bring workers from Yellowknife to its mine site on a two-weeks-in and two-weeks-out rotation. The effect on the economy of the Northwest Territories has been profound. One spinoff was the designation of Yellowknife as the pickup point for miners who commute by air on a rotational scheme to the mine site. This reverses the usual trend of cash and tax flowing out of the territories. An attempt at another spinoff was three diamond-cutting and -polishing businesses in Yellowknife, but they failed. By 2012, mining costs had risen for two reasons. First, the companies have had to shift from open-pit mining to tunnel mining. Second, processing/sorting costs have risen since the higher-quality ore has been extracted, so more ore is required to produce the same amount of raw diamonds. In November 2012, BHP Billiton—a multi-billion dollar global mining enterprise headquartered in Australia—sold its stake in Ekati to Harry Winston Diamond Corporation, a Toronto-based firm that also has a 40 per cent stake in the Diavik mine. The sale to Harry Winston was for the surprisingly low price of $500 million.

How important is the diamond industry to the Northwest Territories? The figures are dazzling—diamond production has gone from zero in 1997 to $2.1 billion in 2011. Furthermore, diamond production came on stream just as gold mining was ending. With the closing of the Giant gold mine at Yellowknife, the last gold was minted in 2004 but closing-down operations lasted until 2005. In the next decade, several diamond mines may exhaust their reserves, raising the question, what non-renewable resource will replace diamonds?

Megaprojects: Achilles Heel?

Megaprojects in the Territorial North are based on non-renewable resources, that is, petroleum and mineral deposits. Forests, water, and wildlife are classified as renewable resources. The value of non-renewable resources in the North far outweighs the economic value derived from renewable resources. In 2011, energy and mineral production in the Territorial North was valued at $2.5 billion (Table 10.7). The key benefits from resource development are the spinoff effects. Unfortunately for the Territorial North, most spinoff effects leak to southern Canada because the economic structure in the Territorial North is so small. Accordingly, diversification has not occurred and, in the long run, the Territorial North might slip into the staples trap.

Non-renewable resource development, because of the limited lifespan of such projects, subjects the Territorial North to a boom-and-bust economic cycle. Since mineral and petroleum

Table 10.7 **Mineral and Petroleum Production in the Territorial North, 2011 ($ millions)**

Mineral Product	Yukon	NWT	Nunavut	Territorial North
Metals	395 (37.7)	65 (55.6)	44 (0)	504 (93.3)
Non-metals	7 (55.7)	2,080 (1,573.3)	0 (29.2)	2,087 (1,658.2)
Petroleum	0 (28)	400 (501)	0 (0)	400 (529)
Total	402 (121.4)	2,545 (2,129.9)	44 (29.2)	2,991 2,280.5

Note: 2006 figures are in parentheses.

Source: Adapted from Natural Resources Canada (2007, 2012); Indian and Northern Affairs Canada (2007); Northwest Territories Bureau of Statistics (2012).

deposits are in finite quantities, they do not offer long-term economic stability for the region. This economic cycle is the Achilles heel of northern development. In addition, because resource development is based on world demand, production fluctuates with this demand, creating great instability in resource communities. Fluctuations in mineral production provide a measure of this instability, and although production has been on the upswing in recent years, especially with the development of diamond deposits, it fell precipitously in 2009 with the global recession—Canadian diamond production by value fell nearly 30 per cent from 2008 to 2009 (Natural Resources Canada, 2010: Table 2).

Table 5.6, "Mineral Production by Province and Territory, 2011," page 56, includes statistics for the three northern territories. Table 7.3, page 294, shows "Canada's Leading Minerals by Value of Production, 2011."

Megaprojects have made important contributions to the northern economy. Often, they are touted in the press as "the engines of economic growth." Certainly, megaprojects give the regional economy a boost, but is that boost enough to lead to regional diversification in the Territorial North? True, megaprojects generate the bulk of the North's GDP, expand the northern transportation infrastructure, and, during the construction phase, provide high wages for large numbers of employees. The diamond industry holds out some hope of even greater spinoffs, such as the processing of diamond gems in Yellowknife, but as with other fledgling businesses, the global economic downturn saw a loss of business and the closure of Arslanian Cutting Works in June 2009 (CBC News, 2009). Megaprojects, however, have so far failed to diversify the Territorial North's economy in a significant manner. While generating profits for the corporations, megaprojects have been unable to broaden the northern economy or solve the massive underemployment problem in Native settlements, and they have increased the region's economic vulnerability to sudden changes in world demand for its primary products. Worse, all northern megaprojects are based on non-renewable resources, which, having a fixed lifespan, cannot provide the basis for a stable economy over the long term. From this perspective, megaprojects are not the engine, but the Achilles heel, of the northern economy.

The potential downside of megaprojects relates to Innis's staples thesis. See Chapter 1, Vignette 1.4.

SUMMARY

The geography of the Territorial North consists of four dominating factors—the Territorial North is huge, sparsely populated, Canada's last resource frontier, and the homeland for numerous Aboriginal peoples. These facts have dominated and will continue to dominate the course of events in the region. At present, Arctic sovereignty once again looms large on the northern political landscape because circumpolar nations are laying claim to the seabed of the Arctic Ocean that lies beyond traditional national boundaries.

As Canada's last frontier, it serves as a resource hinterland with a tiny local market and a great distance to world markets, which means that resource companies must depend on external demand for their products. These characteristics contrive to make the Territorial North's economy extremely sensitive to fluctuations in world prices for its resources. While three levels of government—federal, territorial, and local—seek sustainable and robust economies, the geographic realities make those goals virtually impossible to achieve without substantial financial support. Added to these structural characteristics are weaknesses in the northern labour force and economy. For instance, few skilled and professional workers are in the northern labour force, forcing companies and governments to seek such workers from other regions of Canada. Then, too, the weak northern economy results in **economic leakage**. This raises the question: Are these circumstances an indication of a staples trap?

In the next decade, megaprojects, such as the Mary River Project on Baffin Island, will continue to draw the Territorial North more closely into the global economy as a resource frontier. While diamond production is now the leading industry, if the Mary River Project proceeds, iron ore will become the leading resource by value. The implication of such resource development, based on its non-renewable nature, is that stable economic growth and economic diversification remain elusive goals. Without a more diversified economy, the Territorial North will continue to suffer from boom-and-bust economic cycles and the economic instability thus created. Even if ownership of natural resources is transferred to territorial governments, these governments will remain financially dependent on Ottawa because of the "have-not" character of the region's geography. One question remains: Do Canadians and their governments recognize that this financial dependency represents the cost of incorporating the Territorial North and its Aboriginal peoples into the rest of Canada?

CHALLENGE QUESTIONS

1. How have comprehensive land claim agreements equipped Aboriginal peoples to chart a new and brighter future?

2. Why would the British naval expedition led by Sir John Franklin stand a better chance of navigating the Northwest Passage today than in 1845?

3. Are there any weaknesses in Canada's case for claiming ownership of the Northwest Passage?
4. Megaprojects are the driving force behind the economy of the Territorial North. What is it about these projects that prohibits sustainable growth?
5. Should Imperial Oil rethink its Mackenzie Gas Project by considering shipping the natural gas by LNG supertankers via the Arctic Ocean to Asian markets?

FURTHER READING

Bone, Robert M. 2012. *The Canadian North: Issues and Challenges*, 4th edn. Toronto: Oxford University Press.

In this book the author discusses, among other subjects, the geopolitics of the Territorial North (Chapter 8). Bone identifies and analyzes the impact of climate warming on sea ice, and, with the loss of summer sea ice, the possibility of the Northwest Passage becoming a reality, thus connecting Asia and Europe. The author expresses his concern for the preparedness of the federal government to monitor and control shipping across the Northwest Passage; for the huge investment to build ice-worthy patrol ships in Halifax and a Class 1 icebreaker in Vancouver. Both take both time and money, with the first ships not entering Arctic waters for some 20 years. In the meantime, the last large land rush—for the unclaimed seabed of the Arctic Ocean—is underway. Here, perhaps 30 per cent of the world's untapped oil and gas deposits are found. Canada is expected in 2013 to submit its well-documented claim to the United Nations and it is hoped will gain a share in the potentially great storehouse of mineral wealth and petroleum locked in the sediments under the Arctic Ocean.

CANADA: A COUNTRY OF REGIONS WITHIN A GLOBAL ECONOMY

Introduction

"Canada is big—preposterously so," wrote geographer Kenneth Hare (1968: 31). The sheer size of Canada has meant that the country spans a diverse set of physical and cultural geographies, thus transforming the nation into a country of regions. From another perspective, now that Canada seeks to participate actively within the global economy, the six regions have felt the impact of world trade—sometimes positively and other times negatively—on their regional economies. As we discussed earlier, manufacturing-oriented regions like Ontario and Québec have lost their economic momentum while resource-rich regions, but particularly Western Canada, have seized the global economic opportunities.

The heterogeneous nature of Canada, in extreme cases, can pit one region against another. Currently, regional stresses have arisen because of Canada's economic and social diversity, and no doubt new strains will emerge in the future. Fortunately, Canadians have recognized that regional economic differences cannot be obliterated by decree but must involve a negotiated settlement. What is important—beyond the heated words—is the existence of a political process that addresses these regional differences even though the agreements may be temporary and incomplete. Final solutions are not the basis of democracies, but negotiation, compromise, and resolution keep society flexible and open to ongoing solutions to problems as they arise or alter. Canada's future, while full of challenges and uncertainty, contains one truth: Canada's strength lies in its ability to reconcile cultural and regional disputes. In John Ralston Saul's view (1997: 8–9), Canada is a "soft" country "because the classic nation-state is hard—hard in the force of its creation and its maintenance." In this concluding chapter, four critical topics are addressed: Canada's regional character and structure, its faultlines, the impact of the global economy on Canada's regions, and the future.

A portion of the Trans-Canada Highway at Lake Superior Provincial Park, Ontario. The Trans-Canada highway extends from St. John's, Newfoundland to Victoria, BC. At 8,030 km, it is one of the world's longest national highways. Photo: Don Johnston/All Canada Photos.

Regional Character and Structure

Canada's strength—you might even say what makes it interesting—is its regional diversity and ensuing political struggle to balance regional interests with national ones. As a country of regions, the character of Canada flows from its six geographic regions—Ontario, Québec, British Columbia, Western Canada, Atlantic Canada, and the Territorial North. These geographical regions have been shaped by history. Canada evolved from a small country consisting of four former British colonies to the second-largest country in the world—with a strong possibility of further territorial gains in the unclaimed Arctic Ocean.

In the twenty-first century, the dynamic element of regional geography has been strongly influenced by global trade. While each region has a special regional character with strengths and weaknesses, global trade is playing a more powerful role in shaping the economies of Canada's six regions.

Ontario has the most favourable natural conditions and location for agriculture, economic growth, and trade with the United States. Complementing these natural advantages were federal trade policies, such as the Auto Pact in 1965, the Free Trade Agreement in 1989, and the North American Free Trade Agreement in 1994, that greatly benefited Ontario's manufacturing sector, especially its automobile industry, by providing access to the lucrative US market. On the other hand, such dependency on one market has its risks and Ontario has felt the harsh hand of the US recession on its manufacturing section. By late 2012, signs of a resurgence in automobile sales in the United States have appeared, although increased sales are more a result of pent-up demand than due to a broader economic recovery. Yet, if the US demand continues to strengthen, then, with 80 per cent of Canadian

Photo 11.1
Researchers aboard the Canadian Coast Guard icebreaker *Louis S. St Laurent* and the US Coast Guard cutter *Healy* engaged in a joint Arctic survey mission in 2009. Their objective was to explore largely unknown parts of the Canada Basin, north of the Beaufort Sea. By combining their efforts in this joint mission, the results will provide a much more convincing database for Canada and a better-documented submission to the United Nations in 2013 for its share of the unclaimed waters of the Arctic Ocean. At stake are areas rich in resources and, if polar ice melts, international ocean routes.

Jessica K. Robertson, US Geological Survey

automobile output exported to the United States, Ontario's automobile industry may well be on the rebound to better times. Such a rebound would reverberate throughout Ontario's manufacturing sector, suggesting that the fear of a hollowing-out of Ontario's manufacturing base was overstated.

Québec remains Canada's second core region. Its culture, shaped by its past relations with France, then Britain, has meant that it has a unique place within Confederation. Remarkably, its French-speaking inhabitants have maintained their culture and language and, consequently, represent a resilient cultural anomaly in North America. But Québec's journey within the confines of Confederation continues as La Belle Province continues its struggle to define its place within Canada.

Geography has presented Québec with great natural assets such as the St Lawrence River, which serves as a transportation gateway to Europe and the rest of the world as well as to the Great Lakes and the heart of North America. Nature also has blessed Québec with extensive hydroelectric resources, and this formidable source of "green" energy supplies consumers and businesses within Québec with low-cost electricity while surrounding provinces and New England pay somewhat higher rates. Like Ontario, much of Québec's economic strength lies in its manufacturing sector; and like Ontario, Québec is suffering from the drop in demand from American customers and from fierce competition from Asian manufacturers. On the other hand, major Québec manufacturers have a strong global footprint. With its announcement on 27 November 2012 of a blockbuster sale of 56 business jets worth $3.1 billion, Bombardier—which began as a local snowmobile manufacturer in 1942—has secured both its global presence and its manufacturing base in Québec for years to come (Marotte, 2012). In a surprising political development that could strengthen national bonds, Québec and Alberta are exploring the prospect of a national oil pipeline that would fuel Montréal refineries. Such an arrangement would secure a supply of oil for Montréal's two refineries and ensure jobs for Québecers. In addition, since the price of Alberta oil is lower than the global price, Québec may also benefit from lower gasoline and fuel oil prices. The prospect of shipping Alberta oil sands bitumen to Montréal refineries (one of which is owned and operated by Suncor Energy), once unthinkable because of environmental and political concerns, is driven by potential economic benefits that now rule the day.

In spite of isolated examples of "good news," Québec remains the recipient of the largest equalization payments, suggesting that, overall, its economy is not healthy.

British Columbia, cut off from the rest of Canada by the Cordillera, has a distinctly Pacific coast frame of mind—one that Garreau (1981) calls **Ecotopia**. British Columbia continues to expand its population and economy. With its trade future tied to trade with the Pacific Rim countries, its traditional leading industry, forestry, remains stalled because its major market is the depressed US housing market. Though exports to China and Japan have increased, a rebound to past export levels must wait for an upturn in the US construction sector. Faint signs of a US housing recovery appeared in late 2012 but whether or not this recovery will be sustained in the years to come is a question mark. On more secure economic grounds, the huge natural gas deposits in northern British Columbia have the green light for a pipeline and LNG facilities at Kitimat, but the project has hit a stumbling block in trying to obtain long-term contracts at an attractive price from Asian customers. Bitumen is another matter. Price is not a problem but the environment remains a sticking point. Whether BC likes it or not, the construction of two oil pipelines transporting bitumen from the Alberta oil sands remains a troubling question. Cast as projects in the national interest, the Northern Gateway Project and the twinning of the Trans Mountain pipeline are pitted against the threat to BC's land and sea environment, plus opposition from First Nations.

Western Canada, with its agricultural output and vast underground resources, benefits from trade with Pacific Rim countries. Then, too, China, Japan, and South Korea are investing heavily in those resources needed to keep their economies going: one target is Alberta's oil sands while another is Saskatchewan's potash. Western Canada's agricultural activities have diversified to meet the demands of world markets. For instance, the chickpea, a relatively new crop, is well suited to the dry prairie soils, and Western Canada has captured a large share of the chickpea market in India. The value of farm exports continues to increase, led by canola, wheat, canola oil, soybeans, and pork. Farmland purchases are on the rise, too. Locked in the heart of the continent, Western Canada must factor the cost of moving its products to ocean ports where low-cost ocean shipping completes the journey to markets. Potash, for example, is delivered by unit trains to Vancouver and Portland, Oregon, where specially designed ships take the product to Asian markets. Oil, on the other hand, is transported to American and Canadian markets by pipelines. While a small amount of diluted bitumen is shipped along the Trans Mountain pipeline to Vancouver, plans to twin the Trans Mountain route and to build the Northern Gateway Project to Kitimat are under discussion.

Atlantic Canada remains tied to the sea. Fishing continues to play a major role in its regional economy and the character of its coastal communities, but the real driving force in the economy of Atlantic Canada is found in energy and mineral resources. Fortunately for Newfoundland and Labrador, most energy and mineral resources lie within its boundaries. Royalties from offshore oil production fuel the provincial coffers, pushing Newfoundland and Labrador into the "have" category of provinces. Future megaprojects could well add to Newfoundland and Labrador's newly found economic well-being. The proposed hydroelectric dams on the Lower Churchill River are one aspect of a huge megaproject; another is the construction of a high-voltage transmission system—part of which would lie on the sea bottom—connecting the power-generating stations in Labrador with the island of Newfoundland and then Nova Scotia and beyond to Prince Edward Island, New Brunswick, and possibly New England. Other potential megaprojects are an offshore oil development for Nova Scotia and, if the proposed oil pipeline from Alberta reaches Saint John, the transformation of Saint John's oil refinery into the largest one in Canada.

The *Territorial North* has three prominent features. First, it is the largest region. Second, its cold environment is a dominating geographical feature—though global warming may affect the Territorial North more than any of the other five regions. Third, its Aboriginal population forms a majority, which has political and social implications for the region. Unlike the other five regions, a decision at the United Nations may greatly increase its geographic size and offshore energy reserves. The economy of the Territorial North is heavily oriented to the non-renewable resource sector. Energy and mining dominate its economy, though the public sector provides the bulk of employment opportunities. The non-renewable resource cycle is at three different stages for Yukon, Northwest Territories, and Nunavut. Yukon is enjoying a surge in mining activities that has given its economy a temporary boost; the Northwest Territories remains based on diamond mining but, since several deposits are close to exhaustion, this industry is nearing the end of its productive cycle; and Nunavut stands on the cusp of a possible huge iron mining operation on Baffin Island. Yet, the Inuit workforce is ill-equipped to take advantage of the job opportunities because most lack the necessary education, skills, and experience for employment at the Mary River Project.

What is the connection between the economic performance of Canada's regions and equalization payments? The economic performance of Canada's six regions varies. Some regions have grown swiftly while others have lagged behind. One measure of economic performance comes from the federal equalization payments to less well-off provinces. Ottawa's intervention through equalization payments has played a critical role in reducing the economic gap between regions, but these payments have not solved the problem of regional dependency (Table 3.7). "Have-not" regions use the payments to ensure a degree of parity in terms of education and social services, but such payments may also prevent provincial governments from dealing with core economic problems.

Global trade has introduced a new variable in the regional equation. In the last decade, a seismic shift took place among Canada's six regions, moving Ontario from its lead position to a secondary one. For decades, Ontario led Canada's regions in economic growth, but suddenly it was cast into a slow-growing province in need of equalization payments. At the same time, Newfoundland and Labrador joined the elite club of "have" provinces because of its offshore oil production. For Ontario, the main causes of its economic slump are (1) the global impact of imported low-cost manufactured goods that have disrupted Ontario's manufacturing sector; (2) the restructuring of the automobile industry and the shift to lower wages for its employees; and (3) the sharp decline in exports to the United States. The last two factors may turn around soon, thus returning Ontario to its former dominant economic position within Canada.

Canada's Faultlines

As a country of regions, tensions naturally arise. The most recent eruption along the regional faultline took place in 2012 when the premiers of

THINK ABOUT IT
If Ottawa were to discontinue its equalization program and use those funds to pay down the national debt, which provinces would suffer the most?

Alberta and British Columbia took strikingly different positions on the proposed Northern Gateway Project. This new controversy revolves around three issues:

1. the threat of a major oil spill;
2. opposition from First Nations and concerned BC citizens;
3. BC, which would take the bulk of the environmental risks and costs, wanting a "fair" share of the revenue.

Canada and the Global Economy

The global economy was turned on its head by the entry of China into the World Trade Organization in 2001. Within 10 years, China blossomed into the world's greatest exporter and second-biggest importer (Figure 11.2). The marriage of foreign know-how, Chinese labour, and the open, global market has succeeded beyond anyone's predictions. Canada, like other industrialized countries, has seen its manufacturing sectors eroded and replaced by low-cost Chinese goods, while its resource industries have expanded to meet the unprecedented demand from China.

Much of Canada's export trade depends on its resource economy. Alberta oil sands production plays a key role in Canadian exports to the United States and the completion of the Keystone XL Pipeline would not only allow for a substantial increase in the volume of oil shipments but also permit Canadian oil firms to receive the higher global price known as Brent. But the long-run view is less promising because the *World Energy Outlook* is calling for a "sea change" in global energy flows by 2035 when the United States, by exploiting its shale oil deposits, will not only become self-sufficient but a net exporter (International Energy Agency, 2012). During the same time frame, the *World Energy Outlook* predicts that most oil and natural gas imports will flow to Asian countries,

Figure 11.1 Fighting over oil revenues
British Columbia and Alberta have had difficulty resolving their differences over whether BC should be compensated with a significant piece of the oil revenue pie if the Northern Gateway Pipeline is built across British Columbia to port facilities at Kitimat.

THINK ABOUT IT
Does the core/periphery spatial structure, even if bent or broken, illustrate the value of a theoretical framework in regional geography?

VIGNETTE 11.1

Canadian Identities

Saul's concept of a "soft" Canada is based on its heterogeneous nature, which fits nicely with the concept of Canada as a country of regions. An identity quandary, based in the regional nature of the country, is due to the absence of a unified, powerful Canadian identity. Pluralism, which is hardly a national identity, prevails. Each region has its own character, though "Canadian" values thread their way through regional identities. Québec may be the exception to this rule because its values centre on Québécois culture. Unlike some other parts of Canada, Québec sees multiculturalism as a threat because its people do have a clear and relatively unified identity, though the issue of accommodation to Canadian values, as we have seen, remains to be resolved. The identity quandary across the country, however, has helped make Canada freer of prejudice than is the case in some other countries, and, at the same time, more of a place where immigrants can feel comfortable.

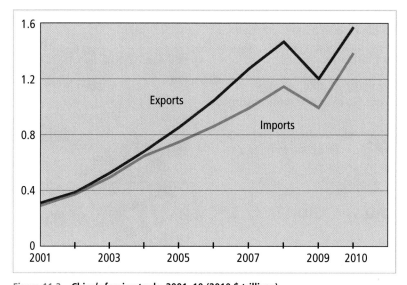

Figure 11.2 China's foreign trade, 2001–10 (2010 $ trillions)
The combination of China turning into a trade powerhouse and the slump in the US economy has severe consequences for Canada and its regions. In response to this turn of events, the federal government has lurched from tepidly seeking to diversify its foreign trade to pulling out all stops to sign trade deals with a host of countries. Already trade with China and other Asian countries is on the rise, while a trade agreement with the European Union may be close. *Source: The Economist* (2011).

THINK ABOUT IT
If President Obama does not approve the Keystone XL Pipeline, would that "force" the current Canadian government to declare the construction of the Northern Gateway Project "in the national interest"?

especially China. Such a scenario places more pressure on Canada to diversify its energy exports, thus making the debate over the Northern Gateway Project (Figure 11.3) even more critical to Canada's long-term interests.

The Future

Change is inevitable. The recalibrated global economy is revising the nature and structure of Canada's economy and its regional geography. By

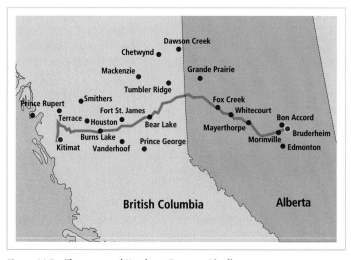

Figure 11.3 The proposed Northern Gateway Pipeline route
The controversial Northern Gateway Project calls for the construction of two pipelines and the construction and operation of the Kitimat Marine Terminal. Each pipeline extends approximately 1,170 km from Bruderheim, Alberta, to Kitimat, BC. The larger pipeline (914 mm or 36 inches in diameter) would carry 525,000 barrels/day of diluted bitumen to Kitimat. The smaller line (508 mm or 20 inches in diameter) would ship 193,000 barrels of condensate per day to Bruderheim, which would be used to thin the bitumen, making it able to pass through the larger pipeline. Plans for the Kitimat Marine Terminal entail two ship berths and storage for three condensate tanks and 11 petroleum tanks. The terminal would be equipped with a radar monitoring station and, in the case of an oil spill, first-response capabilities. *Source*: Enbridge, "Route Map," 2012, at: <www.northerngateway.ca/project-details/route-map/>.

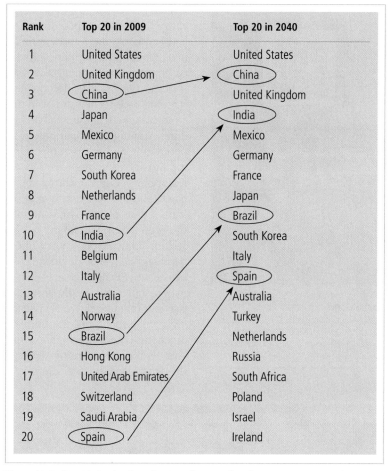

Figure 11.4 **Top 20 destinations for Canadian merchandise exports, 2009 and 2040**
The key to Canada's future lies in global trade, and trade agreements empower greater trade. Ottawa is aggressively pursuing such trade agreements. *Source:* Foreign Affairs and International Trade Canada (2012). Reproduced with the permission of the Minister of Public Works and Government Services Canada, 2013.

2040, the top five export destinations for Canadian goods are predicted to be the United States, China, United Kingdom, India, and Mexico (Figure 11.4).

Within Canada, the resource-rich regions are destined to expand and increase their populations at a rate well above the national average, because this century belongs to the fast-growing Asian countries, led by China, and these countries have an insatiable demand for energy and resources. At the same time, the manufacturing regions of Ontario and Québec will undergo a resurgence based on increased trade with the United States and the European Union.

Already, this global tidal wave has overturned the basic principle of the core/periphery model by benefiting Western Canada and British Columbia while hurting Ontario and Québec. Added to their economic strength, population growth has resulted in more political power, with Alberta receiving five new seats and British Columbia seven new seats in the House of Commons. However, Ontario, according to legislation proposed in April 2010, would receive the most new seats—18—thus enhancing its place as the political centre but weakening Québec and Atlantic Canada (Ibbitson, 2010).

Submerged within this resurgence are two significant trends:

- Major cities drive the knowledge-based economy.
- Urban Canada is moving to a softer, gentler, and greener place where the **creative class**, as proposed by Richard Florida (2002, 2005, 2008, 2010, 2012), flourishes, where downtowns become more pedestrian- and cycle-friendly, and where Aboriginal residents and newcomers find their footing in Canada's economy and society.

Figure 11.5 Selling Canada's energy to the world
In seeking new trade agreements, Prime Minister Harper uses his ace in-the-hole argument that Canada is a "global energy superpower" (Cattaneo, 2012). Ottawa has signed five free trade agreements in the last decade, and 12 such agreements are currently under negotiation, including one with the European Union.

Related to these trends, Canada's increasingly pluralistic society can expect to see tensions subside between newcomers and old-timers. At the same time, greater diversification of trade exports should be able to take advantage of fast-growing economies in Asia without decoupling from our long-standing trade relationship with the United States.

In sum, geographic regions have and will endure. Although regional power, as measured by population and economic performance, is shifting in two directions—westward and inward to major cities, Canada will remain a country of regions, still led by Ontario and Québec. Another truth is that tensions will continue to flare up and the well-tested Canadian "ad hoc" process of seeking adjustments leading to compromises, so necessary in a "soft" democratic country, will still be needed. Finally, Canada is not a homogeneous nation-state but a pluralistic country where regional differences are cherished. Canada's survival will depend on the ability of its citizens to accept the country's geographic structure together with its regional identities (Frye, 1971: i–iii). In the next decades, the country needs to build

Photo 11.2
One dream of resurgence sees Toronto as a leading global financial centre and as Canada's leading centre for knowledge-based activities (Greenwood, 2009). As the headquarters for Canadian banks—seen on the world stage as reliable, stable, and profitable—plus several important universities, a creative class, and **ethnoburbs** (Li, 2009), Toronto is well positioned to take the next step as a world financial hub and centre for knowledge-based firms. But will it?

cultural as well as economic bridges between its regions; to provide more powers and revenue to its cities; and, in so doing, to strengthen national identity and unity. Such a national identity, however frail or fractured it sometimes appears, is the linchpin holding Canada and its six regions together; if strengthened, it can empower economic and cultural bonds that will secure the future of Canada and its regions.

The future, while full of uncertainty, contains two certainties: Canada will remain a large, diverse, and highly urbanized country and, as such, will continue to experience regional tensions. By engaging in open and vigorous debate, compromises—not solutions— will follow. Compromise works at the heart of the exercise of democratic power in a diverse society by demanding a balance between individual and collective rights, between regional and national interests, and between those with power and those in need of it. The ability to strike this balance is critical to Canada's well-being and regional harmony.

CHALLENGE QUESTIONS

1. "Canada is big—preposterously so," wrote geographer Kenneth Hare in 1968. But how, when, and where can its territory enlarge?

2. If push comes to shove, does the national interest trump the opponents to the Northern Gateway Project?

3. With increasing immigration from non-Christian countries, it seems inevitable that the demand for public funding for all religious schools will become a major political issue. How would you resolve this religious/political issue and thus prevent rumblings from becoming an earthquake on the faultline between newcomers and old-timers?

4. The original core/periphery model was based on exploitation of poor regions by rich regions. Do federal programs, such as equalization and transfer payments, repudiate the notion of regional "exploitation" and replace it with regional "dependency"?

5. Has John Ralston Saul fallen either into postmodern relativism (a recent trap for social thinkers) or into **environmental determinism** (an old trap for geographers) when he writes that "in spite of intellectual claims to the contrary, not religion, not language, not race but place is the dominant feature of civilizations. It decides what people can do and how they will live"?

FURTHER READING

Saul, John Ralston. 2009. *The Collapse of Globalism and the Reinvention of the World*, 2nd edn. Toronto: Penguin Canada.

John Ralston Saul's *The Collapse of Globalism* brings a fresh argument to the question: Has globalization spread wealth to all nations through international trade? No one doubts the world is a "richer" place, but the bumps along an unregulated road have displaced lesser economies. His argument has implications for Canada's regions. Essentially, Saul calls for more regulations of capitalism: "Are political decisions meant to be made in deference to the economy and markets, or are political institutions meant to shield us from harsh realities of globalization?" As a case in point, should Alberta allow the marketplace to determine where its bitumen is upgraded, even if this means that the processing and the jobs involved are lost to Alberta and Canada?

GLOSSARY

Aboriginal ancestry Statistics Canada defines the Aboriginal ancestry population as those people reporting an Aboriginal identity as well as those who report being Aboriginal to the ethnic origin question, which focuses on the ethnic or cultural origins of a person's ancestors.

Aboriginal identity According to Statistics Canada, those persons identifying with at least one Aboriginal group, that is, North American Indian, Métis, or Inuit, and/or those who report being a treaty Indian or a registered Indian, as defined by the Indian Act, and/or those who report they were members of an Indian band or First Nation.

Aboriginal peoples All Canadians whose ancestors lived in Canada before the arrival of Europeans; includes status and non-status Indians, Métis, and Inuit.

Aboriginal rights The practices, customs, and traditions that Aboriginal peoples practised prior to contact with or large-scale settlement by Europeans. One Aboriginal right, the right to hunt and fish on Crown lands, has legal status and is protected by the Canadian Constitution. Given the diversity among Aboriginal peoples, Aboriginal rights vary from group to group. Indians whose chiefs signed treaty agreements on behalf of their tribe also have treaty rights.

Aboriginal title A legal term that recognizes an Aboriginal interest in traditionally occupied land.

acculturation Adaptation by new arrivals to a society with different cultural and social values than those of their country of origin. While the rate of acculturation varies, the long-term or generational trend for the vast majority of immigrants is to adopt most of the cultural and social values of the host society, such as language and customs, in the second generation but to retain certain core values such as religion.

age dependency ratio The ratio of the economically dependent sector of the population to the productive sector, arbitrarily defined as the ratio of the elderly (those 65 years and over) plus the young (those under 15 years) to the population of working age (those 15 to 64 years); the old-age dependency ratio is similar to the age dependency ratio except that it focuses only on those over 64.

agricultural fringe Agriculture at its physical limits. Along the southern edge of the boreal forest and in the Peace River country, farmers have cleared the land, but the short growing season prevents most crops from maturing, and consequently many farmers turned instead to cattle.

air commuting Travel to a work site in a remote area, such as a mine, by aircraft owned or hired by the company. Until the 1970s, companies built resource towns to house workers and their families. Employees remain at the work site for a week or longer, working long shifts (often 12 hours per day), then have a week or two at home. The company pays for the air transportation and for food and lodging at the work site.

air drainage The movement of colder, heavier air to lower elevations, leaving warmer, lighter air in the higher elevations.

air pollution Any chemical, physical, or biological agent that modifies the natural characteristics of the atmosphere.

Alaska Panhandle A strip of the Pacific coast north of 54°40′ N latitude that was awarded to the United States in 1903 following what is known as the Alaska boundary dispute.

albedo effect Proportion of solar radiation reflected from the earth's surface back into the atmosphere.

Alberta clipper A low pressure system that begins when warm, moist winds from the Pacific Ocean come into contact with the Rocky Mountains and then the winds form a chinook in southern Alberta; winter storms occur over the Canadian Prairies when it becomes entangled with the cold Arctic air masses. Eventually, the storm reaches Ontario and Québec. Also, "Alberta Clipper" is the name of a natural gas pipeline that runs from Alberta to the US Midwest.

allophones A term used by Statistics Canada to identify those whose mother tongue is not English, French, or one of the Aboriginal languages.

alpine permafrost Permanently frozen ground found at high elevations.

anglophones Those whose mother tongue is English.

Arctic Circle An imaginary line that signifies the northward limit of the sun's rays at the time of the winter solstice (21 December). At a latitude of 66°32′ N, the sun does not rise above the horizon for one day of the year (the winter solstice). Except for a short period of twilight, darkness prevails for 24 hours. At the summer solstice (21 June), the sun's rays do not fall below the horizon, providing constant daylight for 24 hours.

Arctic ice pack Floating sea ice in the Arctic Ocean that has consolidated into an ice pack, with an extent of over 10 million km^2. New sea ice (less than one year in age) is often about 1 m thick; old sea ice can reach 5 m in thickness. Ice ridges are formed, reaching 20 m in thickness. Scientists have found that higher temperatures are reducing the geographic extent and thickness of the Arctic ice pack.

arêtes Narrow serrated ridges found in glaciated mountains. Arêtes form when two opposing cirques erode a mountain ridge.

Asia-Pacific Gateway and Corridor A system of transportation infrastructure, including British Columbia's Lower Mainland and Prince Rupert ports, road, and rail connections, that reaches across Western Canada and into the economic heartlands of North America, including major airports and border crossings.

Athabasca tar sands The largest reservoir of crude bitumen in the world and the largest of three major oil sands deposits in Alberta; also known as the oil sands.

backward linkages Investments in production of infrastructure, such as railways, to move products from source to market.

Bakken formation A geological structure containing large quantities of oil trapped in shale. It lies in Williston Basin in southern Saskatchewan and Manitoba and extends into North Dakota, Montana, and South Dakota.

Barren Grounds The area of tundra stretching from the west coast of Hudson Bay to Great Slave and Great Bear lakes and northward to the Arctic Ocean. The barren ground caribou use this region for calving each summer before migrating to the boreal forest.

basins Structural depressions in sedimentary rock caused by a bending of sedimentary strata into huge bowl-like shapes. Petroleum may accumulate in sedimentary basins.

Beothuk Before the arrival of fishing boats from Europe, the Beothuk Indians, who probably spoke an Algonkian language, hunted and fished on the island of Newfoundland. Relations with fishers and settlers often resulted in conflicts, which confined the Beothuk to the inland. With access to coastal resources cut off and under attack by settlers, the Beothuk struggled to survive in the resource-poor interior. In 1829, the last of the Beothuk died.

Big Bear The last of the great chiefs, who had a vision to unite the Plains Cree to stand together against the impending wave of settlers and to find a way to sustain their culture. Dissatisfied with Treaty No. 6 and the other numbered treaties, he wanted a new treaty with one huge reserve for all Plains Indians. His people facing starvation, Big Bear, like Poundmaker, could not control his warriors at the time of the 1885 Northwest Rebellion. Big Bear was tried for treason and sentenced to prison for three years; released from prison early, he died within months of his release.

Big Commute Air travel by Newfoundland trades workers to and from the Alberta oil sands, on a cycle such as 20 days in Alberta and eight days back home in Newfoundland.

biomass The total quantity or weight of an organism (e.g., codfish) in a given area.

bitumen A tar-like mixture of sand and oil.

boom-and-bust cycles A rapid increase in economic activities in a resource-oriented economy, often based on the value of a single commodity, quickly followed by a downturn when the commodity price(s) falls due to a drop in world demand, which is usually associated with a contraction in the business cycle.

BP (before present) A time scale with a base year of 1950. Past time is measured from this base year, which is considered "the present" or the beginning of the present era. For example, 18,000 BP means 18,000 years before 1950, or 16,050 BC (before Christ).

British Columbia–Alberta–Saskatchewan Trade, Investment, and Labour Mobility Agreement An agreement giving businesses and workers in all three provinces seamless access to a larger range of opportunities across all sectors, including energy, transportation, labour mobility, business registration, and government procurement. In 2009, Saskatchewan signed this agreement.

business cycle The world market (capitalist) economy follows a series of irregular fluctuations in the pace of economic activity. These fluctuations consist of four phases: "contraction" (a slowdown in the pace of economic activity); "trough" (the lowest level of economic activity); "expansion" (a sharp increase in the pace of economic activity); and "peak" (the maximum level of economic activity).

Calder In 1973, the Supreme Court of Canada ruled that "Indian title" (now called "Aboriginal title") has some undefined collective rights, based on historic occupation, possession, and use of traditional territories. These rights are not property rights, which involve the full weight of ownership.

capitalist world-systems theory Wallerstein's theoretical Marxian framework on the workings of modern capitalism. His division of the world economy into three spatial units and their economic relationships forms a key aspect of the capitalist world system. Industrial nations represent the core; developing nations form the periphery; and semi-peripheries are those countries that are partly industrialized.

carbon sequestration Carbon capture and storage technology involving the capturing of CO_2 and other greenhouse gas (GHG) emissions from fossil-fuel power stations or other large carbon emitters and the storing of the CO_2 in deep, stable geological formations.

Cascadia The name proposed for an independent sovereign state advocated by a grassroots movement in the Pacific Northwest, which would include British Columbia, Washington, and Oregon; see *Ecotopia*.

census metropolitan area An urban area with a population of at least 100,000, together with adjacent smaller urban centres and even rural areas that have a high degree of economic and social integration with the larger urban area.

Château Clique The political elite of Lower Canada, composed of an alliance of officials and merchants who had considerable political influence with the British-appointed governor; similar to the Family Compact in Upper Canada.

Chernozemic A soil order identified by a well-drained soil that is often dark brown to black in colour; associated with the grassland and parkland natural vegetation types and located in the Prairies climatic zone.

"China Syndrome" With the virtual elimination of trade barriers by 2002, imported goods from China and other countries with low wages have displaced manufacturing companies in developed countries where wages are much higher. The net outcome is the loss of manufacturing jobs and companies in developed countries like Canada. In the long term, when rising wages in China and other developing countries are more in balance with wages in developed countries, a revival of manufacturing in developed countries is anticipated. Such a shift is described as "reshoring."

chinook A dry, warm, downslope wind in the lee of a mountain range. Also called a rain shadow wind because it has dropped most of its moisture on windward slopes.

circumpolar countries The eight nations associated with the circumpolar area are Canada, Denmark (including Greenland and the Faeroe Islands), Finland, Iceland, Norway, Russian Federation, Sweden, and the United States of America. Five of these—Canada, Denmark, Norway, Russia, and the US—have a territorial claim to portions of the Arctic seabed.

cirques Large, shallow depressions found in mountains at the head of glacial valleys that are caused by the plucking action of alpine glaciers.

Clean Energy Fund Federal $1 billion fund for research, development, and demonstration projects to advance carbon capture and storage technology and other clean energy research.

climate An average condition of weather in a particular area over a very long period of time.

climatic zone A geographic area where similar types of weather occur.

commercial forests Forest lands able to grow commercial coniferous (softwoods), deciduous (hardwoods), and mixed woods timber within an acceptable time frame.

comprehensive land claim agreement Agreement based on territory claimed by an Aboriginal group that was never ceded or surrendered by treaty. Such agreements extinguish the Aboriginal land claim to vast areas in exchange for a relatively small amount of land, capital, and the organizational structure to manage their lands and capital.

container A sealed steel "box" of standardized dimensions (measured in 20-foot equivalent units or "TEU") for transporting cargo.

continental air masses Homogeneous bodies of air that have taken on moisture and temperature characteristics of the land mass of their origin. Continental air masses are normally dry and cold in the winter and dry and hot in the summer.

continentalism Policies, like the Free Trade Agreement, that promote Canadian trade and economic ties with the United States. Washington views continentalism along the same lines as Canada. However, earlier in its history, some in the US saw that country spreading across all of North America and this ideology was expressed by the concept of Manifest Destiny.

continuous cropping A popular farming practice in grain-growing areas where the stubble left after harvest is not tilled; stubble serves to control weeds and reduce soil erosion by wind.

continuous permafrost Extensive areas of permanently frozen ground in the Arctic, where at least 80 per cent of the ground is permanently frozen.

convectional precipitation An upward movement of moist air that causes the air to cool, resulting in condensation and then precipitation.

conventional oil and gas Deposits that can be recovered through natural flow or pumping to the surface and that, therefore, have a higher ratio of extracted energy to the energy needed in extracting and refining. In contrast, "unconventional" oil and gas must be "unlocked" from the deposit through special means such as fracking (as with shale deposits) or must undergo a complicated extraction/refining/separation process (as with in-situ extraction and upgrading of oil sands). "Tight" oil and gas refers to oil trapped in impermeable rock.

core An abstract area or real place where economic power, population, and wealth are concentrated; sometimes described as an industrial core, heartland, or metropolitan centre.

core/periphery model A theoretical concept based on a dual spatial structure of the capitalist world and a mutually beneficial relationship between its two parts, which are known as the core and the periphery. While both parts are dependent on each other, the core (industrial heartland) dominates the economic relationship with its periphery (resource hinterland) and thereby benefits more from this relationship. The core/periphery model can be applied at several geographical levels: international, national, and regional.

country food Food, primarily game, such as caribou, fish, and sea mammals, obtained by Aboriginal people from the land and sea. Although Indians, Inuit, and Métis now live in settlements, they fish and hunt for cultural and economic reasons.

creative class Culture workers, from artists to computer programmers, who, Richard Florida argues, are the key to a flourishing and progressive city and who are attracted to urban centres rich in diversity and culture.

Crow Benefit Signed in 1897, the Crow's Nest Pass Agreement between the Canadian Pacific Railway and the federal government ensured that the rail rates for grain were low, and in this way helped to overcome the disadvantage of a long rail distance to ports. The rail subsidy ended in 1995, placing the full burden on farmers.

crude birth rate The number of births per 1,000 people in a given year.

crude death rate The number of deaths per 1,000 people in a given year.

Cryosolic A soil order associated with permafrost and poorly drained land; soil is either lacking or extremely thin; associated with the tundra and polar desert vegetation types and located in the Arctic climatic zone.

culture The sum of attitudes, habits, knowledge, and values shared by members of a society and passed on to their children.

culture areas Regions within which the population has a common set of attitudes, economic and social practices, and values.

Delgamuukw Supreme Court of Canada case in 1997 that defined how the Aboriginal title may be proved in a court of law. The Court left the issue of compensation for infringements, that is, the private use of part of these lands, to negotiations between the two parties, the affected First Nation and the developer.

demographic transition theory The historical shift of birth and death rates from high to low levels in a population. The decline in mortality precedes the decline in fertility, resulting in a rapid population growth during the transition period.

demography The scientific study of human populations, including their size, composition, distribution, density, growth, and related socio-economic characteristics.

denudation The process of breaking down and removing loose material found at the surface of the earth. In this way, erosion and weathering lead to a reduction of elevation and relief in landforms.

deposition The deposit of material on the earth's surface by various processes such as ice, water, and wind.

dilbit Bitumen diluted with a diluent.

diluent A hydrocarbon substance used to dilute crude bitumen so that it can be transported by pipeline.

discontinuous permafrost Permanently frozen ground mixed with unfrozen ground in the Subarctic. At its northern boundary about 80 per cent of the ground is permanently frozen, while at its southern boundary about 30 per cent of the ground is permanently frozen.

dispute settlement provisions Binding arbitration to resolve trade disputes, as built into the FTA and NAFTA. Each country has a panel of five members. After deliberations, the panel makes a recommendation to the Free Trade Commission, which comprises cabinet-level representatives or their designates, as to what action should be taken. The Commission is then expected to take the appropriate political action.

doctrine of Manifest Destiny The belief and subsequent political actions in nineteenth-century America that the United States, by divine right, should expand to the Pacific coast; in the view of some, this expansion was to include all of North America, thus incorporating Canada. The term was coined by journalist John L. O'Sullivan in 1845, in the context of the annexation of Texas and the Oregon Territory.

Dominion Land Survey This survey method divided Western Canada into one-square-mile sections to allow ownership of specific land units by homesteaders and others.

drainage basin Land sloping towards the sea; an area drained by rivers and their tributaries into a large body of water.

drumlins Low, elliptical hills created by the deposit of glacial till, believed to be from subglacial megafloods, and shaped by the movement of the ice sheet; also called whalebacks or hogbacks.

Dry Belt An agricultural area in the semi-arid parts of Alberta and Saskatchewan, primarily devoted to grain farms and cattle ranches, where crop failures due to drought are more common.

Dutch disease A theory describing the apparent relationship within a country between its expanding energy resource sector and a subsequent decline in the manufacturing sector. In 1977, the term first appeared in *The Economist* to describe the phenomenon of a declining manufacturing sector in the Netherlands, which at the same time was enjoying increased revenues from the export of its natural gas but also seeing its exchange rate with other countries increasing.

Eastern Townships An area of Québec in the Appalachian Uplands lying south and east of the St Lawrence Lowland and near the US border. Originally occupied by less than 2,000 Loyalists, most English-speaking settlers arrived after 1791 from neighbouring New England and the British Isles. By the 1870s, French Canadians formed the majority. The region is now referred to as Estrie.

economic leakage The loss of economic benefits from one jurisdiction to another, which dulls the potential impact of large-scale projects in the local area. In the case of the Territorial North, construction of megaprojects sees the demand for goods and labour satisfied largely by businesses and labour pools outside the region and, similarly, the spending of wages taking place outside the region.

economic structure The sectors of a national, regional, or local economy—primary (e.g., resource extraction and harvesting), secondary (e.g., manufacturing, construction), and tertiary (services)—and the extent to which the whole economy is driven by each of these sectors.

economies of scale A reduction in unit costs of production resulting from an increase in output.

Ecotopia A narrow band along the Pacific coast from northern California to Alaska, defined as a vernacular region with distinctive economic and cultural features by Joel Garreau in his 1981 book, *The Nine Nations of North America*. Ecotopia—an ecological utopia—focuses on a "green" world with an emphasis on "quality of life" and a sustainable economy.

ecumene The portion of the land that is settled.

Eeyou Istchee Created in 2007 as a political unit equivalent to a Québec regional county municipality and administered by the Cree Regional Authority; it consists of Jamésie (Cree lands defined following the James Bay and Northern Québec Agreement) plus Whapmagoostui and Cree Village, which are north of the fifty-fifth parallel and therefore within Nunavik.

environmental determinism The assumption that human activities are controlled by the physical environment. Now considered far too deterministic, it was a popular philosophical position of geographers in the late nineteenth and early twentieth centuries.

erosion The displacement of loose material by geomorphic processes such as wind, water, and ice by downward movement in response to gravity.

eskers Long, sinuous mounds of sand and gravel deposited on the bottom of a stream flowing under a glacier; eskers appear on the land surface after the glacier has retreated.

Estrie An administrative region that overlaps most of the area formerly called the Eastern Townships.

ethnic group People who have a shared awareness of a common identity and who identify themselves with a particular culture.

ethnic origin A Statistics Canada definition, which refers to the ethnic or cultural origins of the respondent's ancestors. An ancestor is someone from whom a person is descended and is usually more distant than a grandparent.

Ethnoburbs Suburban residential and business areas with a significant ethnic character composed of new Canadians. These cultural enclaves in urban areas reflect the pluralistic nature of Canadian society and the very malleable Canadian identity, which allows Canadians to accept and relish such ethnic footprints in their urban landscapes.

ethnocentricity   The viewpoint that one's ethnic group is central and superior, providing a standard against which all other groups are judged.

evapotranspiration   Part of the water cycle: the sum of evaporation of water from the soil to the air and the transpiration of water from plants and its subsequent loss as vapour.

fallowing   The practice of leaving land untilled for one year, thus allowing the soil to accumulate moisture for the following year.

Family Compact   A group of officials who dominated senior bureaucratic positions, the executive and legislative councils, and the judiciary in Upper Canada

fast card   A document that allows the paper work concerning contents of the truck to be reviewed by US border officials before the truck arrives.

faulting   The breaking of the earth's crust as a result of its differential movement; often associated with earthquakes.

fault line   A crack or break in the earth's crust. A complex fault line is known as a fault zone; major fault lines exist between two tectonic plates.

faultlines   Application of a geological phenomenon to the economic, social, and political cracks that divide regions and people.

Fertile Belt   Area of long-grass and parkland natural vegetation in Western Canada associated with black and dark-brown chernozemic soils. It supports a mixed farming area where crop failures due to drought are less common.

fertility rate   The number of live births per 1,000 women aged 15 to 44 in a given year; also known as the general fertility rate. The fertility rate is much more indicative of changes in fertility behaviour than is the crude birth rate because it is based on those women of child-bearing age rather than the general population.

final demand linkages   Expenditures of income generated in the production and export of staples.

First Nations people   By Statistics Canada definition, those Aboriginal persons who report a single response of "North American Indian" to the Aboriginal identity census question.

folding   The bending of the earth's crust.

forward linkages   Investments in staples production and processing, such as a crushing mill for processing canola.

fracking   A technique involving the injection of water and unidentified chemicals underground at very high pressure to create fractures in the underlying shale rock formations, thus releasing the oil or gas for extraction; hydraulic fracturing.

francophones   Those whose mother tongue is French.

Free Trade Agreement (FTA)   Trade agreement between Canada and the United States enacted in 1989.

French Canadians   Canadians with roots to Québec and who likely still speak French.

friction of distance   The effect of distance on spatial interaction; that is, as distance increases, the number of spatial interactions (such as telephone calls or trade in goods) diminishes.

frontal precipitation   When a warm air mass is forced to rise over a colder air mass, condensation and then precipitation occur.

glacial erosion   The scraping and plucking action of moving ice on the surface of the land.

glacial spillways   Deep and wide valleys formed by the flow of massive amounts of water originating from a melting ice sheet or from water escaping from glacial lakes.

glacial striations   Scratches or grooves in the bedrock caused by rocks embedded in the bottom of a moving ice sheet or glacier.

glacial troughs   U-shaped valleys carved by alpine glaciers.

global circulation system   The movement of ocean currents and wind systems that redistribute energy around the world.

globalization   An economic/political/social process driven by international trade and investment that leads to a single world market and wide-ranging impacts on the environment, cultures, political systems, and economic development. While most large businesses position globalization as a positive force for world economic growth, reducing economic and social disparities between nations, and encouraging the spread of democracy, others consider it a negative force causing environmental degradation, exploitation of the developing world, and domination of world politics and culture by a few powerful countries led by the US.

Great American Desert   The treeless Great Plains as described by American explorers in the nineteenth century; in fact, this region has a semi-arid climate and a grasslands vegetation cover.

Green Energy Act   Ontario legislation, passed in 2009, that makes it easier to bring renewable energy projects into production.

greenhouse effect   The absorption of long-wave radiation from the earth's surface by the atmosphere.

greenhouse gases   Water vapour, carbon dioxide, and other gases that make up less than 1 per cent of the earth's atmosphere but are essential to maintaining the temperature of the earth.

grooves of geography   Physiographic structure that facilitates exchanges between adjacent regions, as with Canada and the United States, where the physiography has a north–south alignment.

gross domestic product (GDP)   An estimate of the total value of all materials, foodstuffs, goods, and services produced by a country or province in a particular year.

groundfish   Fish that live on or near the bottom of the sea. The most valuable groundfish are cod, halibut, and sole.

Gulf Stream   A warm ocean current paralleling the North American coast that flows from the Gulf of Mexico towards Newfoundland.

habitants   French peasants who settled the land in New France under a form of feudal agriculture known as the seigneurial system. After the British Conquest, the seigneurial system continued until the mid-nineteenth century, marking a significant difference between Upper and Lower Canada.

GLOSSARY

"hard" and "soft" countries Saul portrays countries as "hard" and "soft" in terms of their relationships with minority groups. Hard nations with homogeneous populations tend to treat minorities harshly. Soft nations have more diverse populations and as a result of a history of interaction among different cultural groups, the value of harmonious relations has taken root.

heartland A geographic area in which a nation's industry, population, and political power are concentrated; also known as a core.

"hewers of wood and drawers of water" Biblical phrase applied by sociologists and others to the labouring classes of capitalism doing the most menial, low-paid work necessary for the operation of capitalist society. Within the context of the core/periphery model, this term refers to periphery regions where primary production prevails; core areas, on the other hand, focus on the processing of those raw products. Its application to a country's economy refers to the export of raw materials rather than of finished goods.

hinterland A geographic area based on resource development that supplies the heartland with many of its primary products; also known as a periphery.

Hobson's choice A choice of taking what is offered or nothing.

hollowing-out The relocation of manufacturing plants in one country to another, which leaves the economy of the original country much weakened.

Holocene epoch The current geological division of the Geological Time Chart. It began some 11,000 years ago and is associated with the warm climate following the last ice age.

homeland A land or region where a relatively homogeneous people and their ancestors have been born and raised, and thus have developed a strong attachment to that place; a sense of place.

homesteader A settler who obtained land. In Western Canada, quarter sections were available as homesteads under the federal government's plan known as the Dominion Lands Act where a settler paid a $10 fee for a quarter section.

horizontal drilling Recently developed technology used in drilling for oil and gas, as opposed to vertical oil and gas drilling, which has existed for a long time.

hydraulic fracturing A method used to fracture rock formations in order to allow oil or natural gas to flow from impervious geological strata.

hydrometallurgy A process that produces nickel, copper, and cobalt directly from ore, thus avoiding the smelting process and eliminating environmentally unfriendly sulphur dioxide and dust emissions.

ice age A geological period of severe cold accompanied by the formation of continental ice sheets. The most recent ice age, the Pleistocene Ice Age, began some 2 million years ago and ended with the beginning of the Holocene Epoch some 11,000 years ago.

igneous rocks Rock formed when the earth's surface first cooled or when magma or lava that has reached the earth's surface cools.

Inland Passage The protected waterway of the Pacific Ocean lying between the BC mainland and Vancouver Island and Haida Gwaii.

Inuit People descended from the Thule, who migrated into Canada's Arctic from Alaska about 1,000 years ago. The Inuit do not fall under the Indian Act, but are identified as an Aboriginal people under the Constitution Act, 1982.

Irish famine The great famine in Ireland that took place between 1845 and 1852 when the principal crop and source of food, the potato, was devastated by blight, causing widespread crop failures. Many Irish immigrated to Atlantic Canada, especially to Saint John, New Brunswick.

isostatic rebound The gradual uplifting of the earth's crust following the retreat of an ice sheet that, because of its weight, depressed the earth's crust. Also known as post-glacial uplift.

James Bay and Northern Québec Agreement The 1971 announcement of the James Bay Project triggered a series of events that quickly led to a negotiated settlement and, in 1975, an agreement between the Cree and Inuit of northern Québec and the federal and Québec governments. In this modern treaty, Aboriginal title was surrendered by the Inuit and Cree in exchange for specific rights, including self-government and benefits (cash and financial support for the hunting economy).

just-in-time principle A system of manufacturing in which component parts are delivered from suppliers at the time required by the manufacturer, so that manufacturers do not bear the cost burden of building and maintaining large inventories; air pollution from heavier vehicular traffic is an ancillary consequence.

Kativik Regional Government Administrative organization for Inuit in Nunavik. Formed in 1978 after the James Bay and Northern Québec Agreement.

kimberlite pipes Intrusions of igneous rocks in the earth's crust that take a funnel-like shape. Diamonds are sometimes found in these rocks.

knowledge-based economy Sector of post-industrial economy based on the use of inventions and scientific knowledge to produce new products and/or services, often in engineering, management, and computer technology fields. Some consider it as part of the information society.

Labrador Current Cold ocean current flowing south in the North Atlantic from Greenland and Labrador.

Lake Agassiz Largest glacial lake in North America that covered much of Manitoba, northwestern Ontario, and eastern Saskatchewan.

latitude A measure of distance north or south, in degrees and minutes, along imaginary lines that encircle the globe parallel to the equator.

light sweet crude The most highly valued crude oil, which because of its low level of sulphur has a pleasant smell and, more importantly, requires little processing to become gasoline, kerosene, and diesel fuel.

liquefied natural gas (LNG) A liquid form of natural gas chilled to −162°C. The cooling process, called liquefaction, reduces the volume to one six-hundredth of its original volume. As a liquid, it can be loaded onto special tankers and transported by sea around the world, making a once regional commodity a global one. At regasification terminals, the LNG is warmed until it returns to a gaseous state.

Lomonosov Ridge An 1,800-km-long underwater ridge with a height of 3,500 m above the seabed stretching from the New Siberian Islands of Russia over the central part of the Arctic Ocean to Ellesmere Island of Canada. Its width varies from 60 to 200 km. The Lomonosov Ridge was discovered in 1948 by a Soviet scientific expedition.

Longitude The distance east or west from the prime meridian at Greenwich, England, an imaginary line that runs through both the North and South poles, as measured in degrees and minutes.

Lower Mainland A local term describing Vancouver and the surrounding area extending from the North Shore Mountains to the border with the United States and eastward to include the Fraser Valley and the town of Hope.

Loyalists Colonists who supported the British during the American Revolution. About 40,000 American colonists who were loyal to Britain resettled in Canada, especially in Nova Scotia and Québec.

Luvisolic A soil order identified by a well-drained soil that is often grey-brown in colour; associated with the broadleaf and mixed forest natural vegetation types in the Great Lakes–St Lawrence climatic zone.

Makivik Corporation A non-profit organization owned by the Inuit of Nunavik and created in 1978 pursuant to the JBNQA. Its central mandate is to protect the integrity of the JBNQA, and Makivik focuses on the political, social, and economic development of the Nunavik region.

Manitoba Act of 1870 This Act created Canada's fifth province. Also known as the "postage stamp" province, its territory only encompassed that of the Red River Colony. The remainder of the former Hudson's Bay lands became the North-West Territories. This Act provided substantial land grants to the Métis as well as guarantees for the French language and Roman Catholic schools.

marine air masses Large homogeneous bodies of air with moisture and temperature characteristics similar to the ocean where they originated. Marine air masses normally are moist and relatively mild in both winter and summer.

megaprojects Large-scale construction projects, often related to resource extraction, that exceed $1 billion and take more than two years to complete.

metamorphic rocks Rocks formed from igneous and sedimentary rocks by means of heat and pressure.

Métis People of mixed biological and cultural heritage, usually either French–Indian or English– or Scottish–Indian. The joining of blood lines between Indians and Europeans took place during the fur trade and continues today. Originally, the term was more narrowly applied to French–Indian people who settled in the Red River area and who developed a distinct hunting economy and society based on the French language and the Roman Catholic religion.

mortality rate The number of deaths per 1,000 people in a given year; also called crude death rate.

muskeg A wet, marshy area found in areas of poor drainage, such as the Hudson Bay Lowlands. Muskeg contains peat deposits.

Nalcor Energy Corporation An energy Crown corporation created by the Newfoundland and Labrador government in 2007. Nalcor is responsible for Newfoundland and Labrador Hydro, the Churchill Falls Generating Station, the Lower Churchill Project, oil and gas, and the Bull Arm Fabrication Site.

nation A territory that is politically independent; a group of people with similar cultural characteristics and a shared historical experience that make them self-consciously aware of their uniqueness as a group.

National Energy Program A bold policy of the federal Liberal government that took the form of the National Energy Program in 1980. Its purpose was to keep Canadian oil prices lower than the rapidly rising world oil prices, provide manufacturers in Ontario and Québec with low-priced western oil, foster oil exploration in the Arctic, and increase federal government revenues from oil sales. In effect, the National Energy Policy kept domestic oil prices lower than global oil prices, thus forcing oil-rich Alberta to forgo market prices and, in doing so, subsidize oil consumers in the rest of Canada.

National Policy A policy of high tariffs instituted in 1879 by the federal government of John A. Macdonald to insulate Canada's infant manufacturing industries from foreign competition and thus create a national industrial base.

National Research Council of Canada The government of Canada's leading agency for research, development, and technology-based innovation. The NRC conducts research and provides research funding to universities and research parks.

Native settlements Small Aboriginal centres, often found on reserves or in remote, northern locations.

net migration The net effect of immigration and emigration on a country's population in a given period.

newsprint A general term used to describe very thin paper used primarily in the publication of newspapers.

non-status Indians Those of Amerindian ancestry who are not registered as Indians under the Indian Act.

NORAD North American Air Defence Command, created in September 1957 by Canada and the US and headquartered in Colorado Springs, Colorado, as a binational command, centralizing operational control of continental air defences against the threat of Soviet bombers. In March 1981, the name was changed to North American Aerospace Defence Command. Since 11 September 2001, NORAD is responsible for protecting North America from domestic as well as from foreign air attacks.

nor'easters Strong winds off the North Atlantic from the northeast that bring stormy weather.

North American Free Trade Agreement (NAFTA) Trade agreement between Canada, the United States, and Mexico that came into effect in January 1994, forming the world's largest free trade area. NAFTA has increased trade among the three countries and rearranged the location of labour-intensive manufacturing firms to Mexico, where wages are much lower than in either the United States or Canada. The agreement was not extended to include state, provincial, and local purchasing markets because Canadian provinces did not wish to participate.

North American Security Perimeter An array of strategies intended to protect America from terrorist attacks. As the US continues to search for a balance between security and trade, cross-border commerce and tourism have been hampered. For example, Americans and Canadians must now have a valid passport to cross the Canada–US border.

Northeast Energy Corridor A proposed electric transmission system from Saint John, New Brunswick, to Maine in the US. The plan includes transmission lines capable of carrying between 1,200 and 1,500 megawatts of electricity from wind generators, as well as a natural gas co-generator that would supply base-load power for the line.

northern frontier View of Canada's North as a place of resource wealth to be exploited.

northern lights (aurora borealis) The visible portions of the dissipation of solar energy carried to the earth's magnetosphere by solar winds. The energy is visible, most commonly in higher latitudes, as rapidly moving light that appears as white or green or red flashes or "curtains" of light across the sky.

Northwest Passage Sea route(s) through the Arctic Ocean connecting the Atlantic and Pacific Oceans that can be traversed only in the summer.

North West Transportation Corridor Stretching from Prince Rupert across northern BC and into Western Canada, the North West Transportation Corridor centres on the CN rail route and the Yellowhead Highway.

Nunavik Homeland of the Inuit of northern Québec and a semi-autonomous political region within that province.

oil prices Two pricing systems exist for oil in North America. The Brent crude oil price is the standard benchmark for global oil prices, and gets its name from the Brent oil field in the North Sea. Oil from the interior of the US and Canada is priced in Cushing, Oklahoma, the largest crude oil hub in North America, and is known as the West Texas Intermediate (WTI) price. Over the past five years, the WTI prices have traded $15 to $30 a barrel below Brent prices as oil production has overwhelmed pipeline capacity in the US.

Oregon Territory Territory in the Pacific Northwest stretching from 42° N to 54°40′ N, the possession of which was disputed between the US and Great Britain (now composed of British Columbia, Washington, and Oregon). The Treaty of 1846 between the United States and Great Britain determined the boundary at the forty-ninth parallel but with Vancouver Island remaining within British North America.

orogeny Mountain-building, a geologic process that takes place as a result of plate tectonics (movement of huge pieces of the earth's crust). The result is distinctive structural change to the earth's crust where mountains often are formed.

orographic precipitation Rain or snow created when air is forced up the side of a mountain, thereby cooling the air and causing condensation followed by precipitation.

orographic uplift Air forced to rise and cool over mountains. If the cooling is sufficient, water vapour condenses into clouds and rain or snow occurs.

outsourcing Arrangement by a firm to obtain some parts or services from other firms.

Paleo-Indians Considered by archaeologists the first people of North America because they shared a common hunting culture, which was characterized by its uniquely designed fluted-point stone spearhead.

Palliser's Triangle Area of short-grass natural vegetation in southern Alberta and Saskatchewan determined by Captain John Palliser, who led an expedition organized by the British Colonial Office and the Royal Geographical Society to survey the Canadian West in 1857–60, to be a northern extension of the Great American Desert and therefore unsuitable for agricultural settlement.

patterned ground The natural arrangement of stones and pebbles in polygonal shapes found in the Arctic where continuous permafrost is subjected to frost-shattering as the principal erosion process.

peneplain A more or less level land surface caused by the wearing down of ancient mountains; represents an advanced stage of erosion.

periphery The weakly developed area surrounding an industrial core; also known as a hinterland.

permafrost Permanently frozen ground.

physiographic region A large geographic area characterized by a single landform; for example, the Interior Plains.

physiography A study of landforms, their underlying geology, and the processes that shape these landforms; geomorphology.

Pineapple Express A strong and persistent flow of warm air associated with heavy rainfall that originates in the waters adjacent to the Hawaiian Islands.

pingos Hills or mounds that maintain an ice core and that are found in areas of permafrost.

placelessness The reverse of "sense of place." Placeless landscapes have no distinguishing features and could be found in very many places. Examples are strip malls, cookie-cutter theme parks, and service stations.

plate tectonics The study of movement in the seven large pieces or "plates" of the Earth's outermost layer, the lithosphere. These plates are floating on the molten material comprising the rest of the interior of the Earth. When the plates collide, earthquakes occur. The seven plates are the African, North American, South American, Eurasian, Australian, Antarctic, and Pacific plates.

Pleistocene epoch A minor division of the Geological Time Chart beginning nearly 2 million years ago. The glaciation of the Pleistocene was not continuous but consisted of several (likely, four) glacial advances interrupted by interglacial stages during which the ice retreated and a comparatively mild climate prevailed. The last advance, called the Wisconsin, ended about 11,000 years ago.

Pluralism The social acceptance of a diversity of views in preference to a single approach or method of interpretation. Under pluralism, conflicting values may be considered equally important.

pluralistic society A society where small groups within the larger society are permitted to maintain their unique cultural identities; multiculturalism.

podzolic A soil order, often grey in colour, identified by poor drainage; associated with the boreal forest and the coastal rain forest and with climates that have large amounts of precipitation, such as the Pacific, Atlantic, and Subarctic climatic zones.

population density The total number of people in a geographic area divided by the land area; population per unit of land area.

population distribution The dispersal of a population within a geographic area.

population growth The rate at which a population is increasing or decreasing in a given period due to natural increase and net migration; often expressed as a percentage of the original or base population.

population increase The total population increase resulting from the interaction of births, deaths, and migration in a population in a given period of time.

population strength The equating of population size with economic and political power.

Port-Royal The settlement founded in the summer of 1605 on the north shore of the Annapolis Basin near the mouth of the Annapolis River by a French colonizing expedition led by Pierre du Gua de Monts and Samuel de Champlain.

postglacial uplift The gradual rising of the earth's crust following the retreat of an ice sheet that, because of its weight, depressed the earth's crust. Also known as isostatic rebound.

potash A general term for potassium salts. The most important potassium salt is sylvite (potassium chloride). Potassium (K) is a nutrient essential for plant growth.

Poundmaker An outstanding political leader of the Plains Cree. During the tumultuous years surrounding treaty-making, he played a key role in setting the terms for Treaty No. 6. Later, Poundmaker sought a peaceful solution for the desperate plight of his people. Unable to control his warriors, he took a conciliatory position in the Cree uprising and arranged for a surrender of his people in 1885. Poundmaker was charged with treason and sentenced to three years in prison. He died shortly after his early release, a man broken by the losses his people had endured and by his time in prison.

primary prices The prices for commodities such as foodstuffs, raw materials, and other primary products.

primary products Goods derived from agriculture, fishing, logging, mining, and trapping; products of nature with no or little processing.

primary sector Economic sector involving the direct extraction/production of natural resources that includes agriculture, fishing, logging, mining, and trapping.

producer services Services that have enabled firms and regions to maintain their specialized roles in marketing, advertising, administration, finance, and insurance industries. Producer services are one of several parts of the growing service sector of the economy.

Provincial Agricultural Land Commission An independent British Columbia agency responsible for administering the province's land-use zone in favour of agriculture. This agency also manages the Agricultural Land Reserve (ALR), which extends over 4.7 million hectares.

Provisional Government Government formed in 1869 by the Métis, under the leadership of Louis Riel, in order to negotiate the terms to permit the Red River Colony to join Canada as a province. Ottawa accepted the terms and they were incorporated into the Manitoba Act of 1870.

push-pull model One of the laws of migration, as devised by Ravenstein, whereby certain negative factors (e.g., lack of employment opportunities or human security) in the migrant's present location push him/her to migrate, just as certain positive factors (e.g., economic opportunities, more amenable climate, or greater human security) in the location of destination pull the migrant to relocate.

Quaternary Period The geological period consisting of the Pleistocene and Holocene epochs.

Québécois A term that has evolved from referring to French-speaking residents of Québec to meaning all residents of Québec.

Quiet Revolution A period in Québec during the Liberal government of Jean Lesage (1960–6) characterized by social, economic, and educational reforms and by the rebirth of pride and self-confidence among the French-speaking members of Québec society, which led to a resurgence of francophone ethnic nationalism. During this time, the secular nationalist movement gained strength.

rain shadow effect A dry area on the lee side of mountains where air masses descend, causing them to become warmer and drier.

rate of natural increase The surplus (or deficit) of births over deaths in a population per 1,000 people in a given time period.

recent immigrants Statistics Canada term that refers to landed immigrants who arrived in Canada within five years prior to a given census.

Red River migration British-instituted migration organized by the Hudson's Bay Company whereby perhaps as many as 1,000 settlers from Fort Garry travelled by horse-drawn wagons to Fort Vancouver in 1841 to shore up the British claim to the Oregon Territory.

region An area of the earth's surface defined by its distinctive human and/or natural characteristics. Boundaries between regions often are transition zones where the main characteristics of one region merge into those of a neighbouring region. Geographers use the concept of regions to study parts of the world.

regional consciousness Identification with a place or region, including the strong feeling of belonging to that space and the willingness to advocate for regional interests.

regional core Within the core/periphery model, cores can occur at different geographic levels. A regional core is an area (often a large city) that dominates trade and stimulates economic growth in the region.

regional geography The study of the geography of regions and the interplay between physical and human geography, which results in an understanding of human society, its physical geographical underpinnings, and a sense of place.

regional identity Persons' association with a place or region and their sense of belonging to a collectivity.

regionalism The division of countries or areas of the earth into different natural/political/cultural parts.

regional self-interest The aspirations, concerns, and interests of people living in a region and acted on by local politicians. Sometimes such efforts are designed to improve the prospects of their region at the expense of other regions or of the federal government.

regional service centres Urban places where economic functions are provided to residents living within the surrounding area.

relief A measure of elevation of the land relative to sea level, which is designated as zero; a relief map indicates elevation and/or topographic features, such as a mountain range, by different colours.

reserve Under the Indian Act, lands "held by her Majesty for the use and benefit of the bands for which they were set apart; and subject to this Act and to the terms of any treaty or surrender."

residual uplift The final stages of isostatic rebound.

resource frontier The perception of the Territorial North as a place of great mineral wealth that awaits development by outsiders.

resource town An urban place where a single economic activity focused on resource extraction (e.g., mining, logging, oil drilling) dominates the local economy; single-industry town. Also, a company town built near an isolated mine site to house the mine workers and their families.

restrained rebound The first stage of isostatic rebound.

restructuring Economic adjustments made necessary by fierce competition, whereby companies are driven to reduce costs by reducing the number of workers at their plants.

Robinson treaties Two 1850 treaties signed between the Crown (Canada West) and the Ojibwa Indians of Lake Superior and the Ojibwa Indians of Lake Huron. Under the terms of both agreements, the Crown secured an area of 52,400 square miles mostly in central and northern Ontario. For the first time, reserves were part of the agreement. The Crown paid the Ojibwa Indians a lump sum as well as an annual payment in perpetuity.

Scottish Highland clearances Forced displacements of poor tenant farmers in the Scottish Highlands during the eighteenth and nineteenth centuries. Migration ensued and, in 1812, Scottish settlers arrived at Fort Garry to found Lord Selkirk's experimental colony. Most, perhaps 100,000, settled in Nova Scotia. The clearances were part of a process of change of estate land use from small farms to large-scale sheep herding.

scrip Under the Manitoba Act of 1870, certificates issued by Ottawa to the Métis to settle their land claims and to allow them to obtain land. This scrip was issued to individuals and was redeemable in Dominion lands in Manitoba.

sea ice Ice formed from ocean water. Types of sea ice include: (1) "fast ice" frozen along coasts and that extends out from land; (2) "pack ice," which is floating, consolidated sea ice detached from land; (3) the "ice floe," a floating chunk of sea ice that is less than 10 km (six miles) in diameter; and (4) the "ice field," a chunk of sea ice more than 10 km (six miles) in diameter.

Sea-to-Sky Highway Highway that winds through the spectacular Coast Mountains, linking communities from West Vancouver to Whistler. Since rock slides occur frequently, the highway was widened and straightened to reduce the chances of rock slides and to improve safety and reliability for the 2010 Winter Olympic Games.

secondary sector The sector of the economy involved in processing and transforming raw materials into finished goods; the manufacturing sector of an economy.

sedimentary rocks Rocks formed from the layered accumulation in sequence of sediment deposited in the bottom of an ocean.

seigneurs Members of the French elite—high-ranking officials, military officers, the nascent aristocracy—who were awarded land in New France by the French king. A seigneur was an estate owner who had peasants (habitants) to work his land.

sense of place The special and often intense feelings that people have for the area where they live. Considered a social product, sense of place can be applied to different geographic levels, i.e., local, regional, and even national levels. These feelings are derived from a combination of experiences: some from natural factors such as climate, others from cultural factors such as language. A sense of place is a powerful bond between people and their region.

settlement area The geographic extent of a comprehensive land claim. While less than 25 per cent of this land is allocated to the Aboriginal beneficiaries as a collective (not individual) landholding, the entire area is subject to the environmental and wildlife regulations exercised by the settlement area's co-management boards.

sex ratio The ratio of males to females in a given population; usually expressed as the number of males for every 100 females.

shariah law Islamic religious law based on the Koran, the Muslim holy book; under most interpretations, Islamic law gives men more rights than women in matters of inheritance, divorce, and child custody.

softwood forest The predominant forest in Canada. Softwood forests consist mainly of coniferous trees, characterized by needle-like foliage.

soil orders Classes of soil based on observable soil properties and soil-forming processes. In Canada there are nine soil orders, including chernozemic, cryosolic, and podzolic.

sovereignty-association A concept designed by the Parti Québécois under the Lévesque government and employed in the 1980 referendum. This was based on the vision of Canada as consisting of two "equal" peoples. Sovereignty-association called for Québec sovereignty within a partnership with Canada based on an economic association.

spawning biomass The total quantity or weight of a species at sexual maturity in a given area that can reproduce.

Cod, for example, reach sexual maturity around the age of seven.

specific land claims Claims made by treaty Indians to rectify shortcomings in the original treaty agreement with a band or that seek to redress failure on the part of the federal government to meet the terms of the treaty. Many of these have involved the unilateral alienation of reserve land by the government.

sporadic permafrost Pockets of permanently frozen ground mixed with large areas of unfrozen ground. Sporadic permafrost ranges from a trace of permanently frozen ground to an area having up to 30 per cent of its ground permanently frozen.

staples thesis Harold Innis's idea that the history of Canada, especially its regional economic and institutional development, was linked to the discovery, utilization, and export of particular staple resources in Canada's vast frontier. It was expected that economic diversification would take place, making the region less reliant on primary resources. Innis proposed this thesis in the early 1930s, and his ideas continue to influence Canadian scholars.

staples trap The economic and social consequences on a region and its population following the exhaustion of its resources; the opposite outcome of the positive outcome of economic diversification anticipated in the staples thesis.

status (registered) Indians Aboriginal peoples who are registered as Indians under the Indian Act.

strata Layers of sedimentary rock.

stubble field A field in which the crop, such as wheat, hay, or another grain crop, has been harvested, leaving behind short stalks.

Stryker Slang term for a prospective gang member and the title of a 2004 film by Winnipeg filmmaker Noam Gonick.

subsidence A downward movement of the ground. Subsidence occurs in areas of permafrost when large blocks of ice within the ground melt, causing the material above to sink or collapse.

summer fallow The farming practice of leaving land idle for a year or more to accumulate sufficient soil moisture to produce a crop or to restore soil fertility; summer fallowing is being replaced by continuous cropping.

super cycle theory Theory based on two premises: (1) that demand will tend to outstrip supply and thus keep prices high; and (2) that in a global economic downturn, demand from industrializing countries will keep price declines to a minimum.

Sustainable Development Technology Canada A foundation created by the Canadian government to support the development and demonstration of clean technologies—solutions that address issues of clean air, greenhouse gases, clean water, and clean soil to deliver environmental, economic, and health benefits to Canadians.

sustainable resource use Use of renewable resources (e.g., forests, fish stocks) when the rate of consumption equals (or is less than) the resource's natural rate of replenishment.

terraces Old sea beaches left after the sea has receded; old flood plains created when streams or rivers cut downward to form new and lower flood plains. The old flood plain (now a terrace) is found along the sides of the stream or river.

terra nullius The doctrine according to which European countries claimed legal right to ownership of the land occupied by Indians and Inuit because the land was not cultivated and lacked permanent settlements.

tertiary/quaternary sector The economic sector engaged in services such as retailing, wholesaling, education, and financial and professional services; the quaternary sector, for which at present statistical data are not compiled, involves the collection, processing, and manipulation of information.

till Unsorted glacial deposits.

topography The shape of the surface of the land; contour maps, using isolines (contour lines), are one representation of topography and/or topographic features, with each line on the map representing the same elevation above sea level.

traditional ecological knowledge Familiarity with and knowledge of the natural surroundings. Such knowledge of the environment, climate, and wildlife in a particular area or ecosystem accumulates over centuries among a group of people who live close to the land/sea; also called local ecological knowledge.

tragedy of the commons The destruction of renewable resources that are not privately owned, such as fisheries and forests. Historically, common pasture was available for the livestock of all people within a community, but in the absence of some form of collective control some individuals may maximize the use of the resource for personal gain; such use, in total, overwhelms the capacity of the resource to maintain and regenerate itself. The result, for those who seek to maximize profit, is short-term gain; the result for everyone is middle- and long-term loss.

transmission technology High-voltage transmission lines that reduce power loss and thus make shipping electricity over long distances viable.

Trans-Pacific Partnership Eleven countries (Canada, United States, Mexico, Australia, New Zealand, Peru, Chile, Singapore, Vietnam, Malaysia, and Brunei) that are negotiating a free trade agreement.

treaty Indians Aboriginal peoples who are descendants of Indians who signed a numbered treaty and who benefit from the rights described in the treaty. All treaty Indians are status Indians, but not all status Indians are treaty Indians.

treaty rights Specific rights that apply only to the First Nation(s) that signed the treaty in question. While no two treaties are identical, the list of rights always included land (reserves). These rights are protected in the Constitution Act.

Tyrrell Sea Prehistoric Hudson Bay as the Laurentide Ice Sheet receded. Its extent was considerably greater than that of present-day Hudson Bay because the land had been depressed by the weight of the ice sheet.

underemployment Several definitions may be used; a classical one refers to workers who are employed, but not in the desired capacity, whether in terms of compensation, hours, or level of education, skill, and experience. In this text, "underemployment" refers to persons in small communities where the very few jobs are already filled and, because these potential workers are aware that no job opportunities exist, they do not seek jobs elsewhere.

unemployment Lack of paid work, but this term and statistics based on it measure only those who are seeking paid employment.

unit trains A set of 150 or so specialized coal or potash rail cars pulled by one or two diesel engines.

upgrader A processing plant that breaks large hydrocarbon molecules (such as bitumen) into smaller ones by increasing the hydrogen-to-carbon ratio. The product is supplied to refineries, which will process it into gasoline, jet fuel, diesel, propane, and butane.

urban areas Communities with economic and social functions that differentiate them from rural places; the common practice of defining urban population is by a specified size that assumes the presence of urban economic and social functions. Statistics Canada considers all places with a combination of a population of 1,000 or more and a population density of at least 400 per km^2 to be urban areas. People living in urban areas make up the urban population. People living outside of urban areas are considered rural residents and, by definition, constitute the rural population.

value-added production Manufacturing that increases the value of primary (staple) goods.

weathering The decomposition of rock and particles in situ.

western alienation Feeling on the part of those in Western Canada and BC—derived from past government actions and a natural periphery response to the core—that they have little influence on federal policy and that Central Canada controls the government in Ottawa.

Western Sedimentary Basin Within the geological structure of the Interior Plains, normally flat sedimentary strata that are bent into a basin-like shape. These basins often contain petroleum deposits.

winter roads Temporary ice roads over muskeg, lakes, and rivers built during the winter to provide ground transportation for freight and travel to remote communities.

zebra mussel A small freshwater mollusc (*Dreissena polymorpha*). In 1988 the zebra mussel gained a foothold in the Great Lakes, likely from having attached itself to ships coming from Europe. It has since spread and only the higher salinity in the lower St Lawrence River has halted its expansion downstream.

zero-tillage A farming system in which the seeds are directly planted into an untilled field, in most cases, a stubble field.

WEBSITES

CHAPTER 1

atlas.nrcan.gc.ca/site/english/maps/reference/provincesterritories
The *Atlas of Canada* has prepared a set of provincial, territorial, and regional maps, which appear in Chapters 5–10 of this book.

geography.about.com/od/politicalgeography/a/coreperiphery.htm
A summary of the core/periphery theory is presented here.

www.statcan.gc.ca/kits-trousses/projet-cyber-project/manufact2-eng.htm
Teachers' tool box: Using selected data from the *Canada Year Book*, students will test the accuracy of the premise: Ontario and Quebec were the Canadian leaders in manufacturing at the time of Confederation and they maintained their dominance well into the twentieth century.

www.allacademic.com//meta/p_mla_apa_research_citation/0/7/3/8/7/pages73870/p73870-1.php
The North American Security Perimeter and its implications for North American community are discussed.

www.international.gc.ca/trade-agreements-accords-commerciaux/agr-acc/eu-ue/can-eu.aspx?view=d
The EU/Canada trade agreement would be a major agreement. As of January 2013, negotiations continue.

CHAPTER 2

www.ec.gc.ca/cc/default.asp?Lang=En
Environment Canada's website for climate change.

atlas.nrcan.gc.ca
The Atlas of Canada site contains a variety of maps describing the physical nature of Canada.

nrcan-rncan.gc.ca
Natural Resources Canada provides information on many resource topics.

www.nasa.gov/topics/earth/features/arctic_thinice.html
NASA's website has satellite images of the Arctic Ocean, including late summer when the Northwest Passage is ice-free.

climate.nasa.gov
NASA presents a time-series of climate change, including annual temperatures, sea levels, and global ice.

CHAPTER 3

www.american-indians.net
North American Aboriginal culture regions are found at this site.

http://geogratis.gc.ca/api/en/nrcan-rncan/ess-sst/c9bc8461-8893-11e0-baff-6cf049291510.html
The Hudson's Bay Company undertook a census of Aboriginal peoples in 1822. The Atlas of Canada has produced a map and accompanying text describing this very detailed census.

http://geogratis.gc.ca/api/en/nrcan-rncan/ess-sst/?q=territorial+evolution

atlas.nrcan.gc.ca/sitefrancais/english/maps/historical/territorialevolution/territorial_animation.gif/image_view
Maps that illustrate the territorial evolution of Canada and an animated version of Canada's territorial development are found here.

faculty.marianopolis.edu/c.belanger/quebechistory/events/quiet.htm
An account of the Quiet Revolution in Québec.

archives.cbc.ca/politics/federal_politics/topics/1891/
CBC Digital Archives provides a close look at "separation anxiety" in Canada associated with the 1995 referendum.

www.mhs.mb.ca/docs/pageant/13/selkirksettlement1.shtml
The Manitoba Historical Society has documented Lord Selkirk's land grant of 1811.

http://en.wikipedia.org/wiki/Haldimand_Proclamation
The Haldimand Proclamation.

CHAPTER 4

www.idlenomore.ca
Official website of the grassroots Aboriginal movement that emerged in late 2012.

www.statcan.gc.ca/daily-quotidien/121218/dq121218f-eng.htm?HPA
Statistics Canada's population estimates are provided quarterly.

www12.statcan.ca/census-recensement/2006/as-sa/97-558/index-eng.cfm?CFID=2953573&CFTOKEN=43245102
Statistics Canada provides a detailed account of the demography and social characteristics of Aboriginal peoples from the 2006 census. Figures for 2011 are scheduled for publication in May 2013.

www.statcan.gc.ca/pub/11-008-x/2009001/article/10864-eng.htm
Linda Gionet's article, "First Nations People: Selected Findings of the 2006 Census," *Canadian Social Trends*, provides another perspective.

CHAPTER 5

www.statcan.gc.ca/pub/75-001-x/2009102/pdf/10788-eng.pdf
Statistics Canada article (2009) on the factors behind the decline of manufacturing in Ontario and in Canada and other industrial countries.

www.ic.gc.ca/eic/site/auto-auto.nsf/eng/h_am01302.html
Industry Canada's website for statistics, analysis, and industrial profiles of the automobile industry.

www.nationalpost.com/m/story.html?id=1100168
Good public policy? Ottawa and Ontario dole out $4 billion to Chrysler and General Motors.

www.greenenergyact.ca/
Ontario's Green Energy Act.

www.ainc-inac.gc.ca/ai/mr/is/eac-eng.asp
A chronology of the Six Nations land dispute is maintained by Indian and Northern Affairs Canada.

www.sdtc.ca/en/about/index.htm
Sustainable Development Technology Canada provides start-up funding for "clean" technologies.

CHAPTER 6

www.knowyourmeme.com/memes/events/2012-quebec-student-protests
An account in English of the Quebec student protests in 2012.

www.greatlakes-seaway.com/en/
Official site of the Great Lakes–St Lawrence Seaway system.

www.canadiangeographic.ca
History of the St Lawrence Seaway plus archival photographs and a tour of the seaway.

www.qc.ec.gc.ca/csl/pub/pub004_e.html
A scientific account of the environmental impact of the zebra mussel in the St Lawrence River by the St Lawrence Centre of Environment Canada.

www.mrnf.gouv.qc.ca/mines/strategie/index.jsp
Details of the Québec government's mineral strategy are presented at this site.

www.mrnf.gouv.qc.ca/presse/communiques-detail.jsp?id=7699
Québec's wind-generated electric program is described here.

www.mrnf.gouv.qc.ca/energie/hydroelectricite/developpement.jsp
Québec government announcement of its hydroelectric projects to 2010.

CHAPTER 7

earthquakescanada.nrcan.gc.ca/histor/20th-eme/1949-eng.php
Details on the 1949 earthquake off Haida Gwaii.

www.northerngateway.ca
Enbridge, the company proposing to build this bitumen pipeline, present its case.

www.livingoceans.org/initiatives/tankers/issues/enbridge-northern-gateway-project
An environment organization, Living Oceans, presents its view of the Northern Gateway Project.

www.th.gov.bc.ca/PacificGateway/documents/PGS_Action_Plan_043006.pdf
Pacific Gateway Strategy Action Plan.

forum.skyscraperpage.com/showthread.php?t=151987&page=2
Maps and diagrams of highway construction projects designed to create a super east–west highway corridor.

www.llbc.leg.bc.ca/public/PubDocs/bcdocs/322107/north_east_coal_facts.pdf
In 1982, the BC government described its high hopes for the Northeast Coal Project.

CHAPTER 8

www.saskmining.ca/commodity_info/Commodities/1/potash.html
The history of potash mining in Saskatchewan.

www.oilsands.alberta.ca
Alberta's oil sands: facts, statistics, and video produced by the Alberta government.

http://www.energy.alberta.ca/Initiatives/3223.asp
Alberta government report: *Responsible Actions: A Plan for Alberta's Oil Sands*.

www.cbc.ca/edmonton/features/dirtyoil/
CBC Edmonton: "Alberta Oil Sands: Black Gold or Black Eye?"

pubs.pembina.org/reports/climate-leadership-report-en.pdf
Pembina Institute and David Suzuki Foundation report on greenhouse-gas emissions and carbon tax.

www.defenseindustrydaily.com/Boeings-Skyhook-Shot-Redefining-the-Aerial-Heavy-Lifting-Market-04970/
Information on Skyhook.

CHAPTER 9

http://www.parl.gc.ca/content/lop/researchpublications/prb0820-e.pdf
Canada's new equalization formula.

www.eleanorbeaton.com/userfiles/file/Chatelaine%20-%20Oil%20Patch%20Widows.pdf
"No Man's Land"—story of Newfoundland and commuting.

www.geography.ryerson.ca/jmaurer/702art/702Clapp1999.pdf
The resource cycle in forestry and fishing.

nbwoodlotowners.ca///uploads//Website_Assets/APEC%27s_Atlantic_Lumber.pdf
Atlantic Canada's forest industry.

www.offshore-technology.com/projects/hibernia/
Hibernia, Jeanne d'Arc Basin, and its technology.

www.sysco.ns.ca/history.htm
The history of the steel plant at Cape Breton.

www.nalcorenergy.com/Lower-Churchill-Project.asp
The Lower Churchill hydroelectric project has the potential to create an Atlantic Canada electric power network.

CHAPTER 10

www.mmg.com/en/Our-Operations/Development-projects/Izok-Corridor.aspx
A Chinese company, MMG, proposes to develop the vast lead/zinc deposit in the Izok Corridor and ship the ore to China through the Northwest Passage.

apecs.arcticportal.org/index.php?option=com_jreviews&Itemid=131
The Association of Polar Early Career Scientists (APECS) has launched a discussion forum called "Polar Literature Discussion Webpage" where recent literature is cited.

www.arctic-council.org/index.php/en/
Formation and mandate of the Arctic Council, as well as Canada's role in the Arctic Council.

www.scientificamerican.com/article.cfm?id=drawing-lines-in-the-sea&page=2
In this issue of Scientific American, circumpolar countries are described as drawing lines in the Arctic Ocean to define their territorial limits. These lines represent internationally accepted boundaries but also those that are in dispute.

www.uphere.ca/node/175
In December 2007, the northern magazine, *Up Here*, featured an article by Jack Danylchuk on the unwanted legacy of arsenic.

http://www.statsnwt.ca/
The Bureau of Statistics of the Northwest Territories provides not only a wide variety of statistical data but also links to Yukon and Nunavut statistical data.

www.ipy-api.gc.ca
Canada's website for International Polar Year scientific activities.

CHAPTER 11

www.canadiangeographic.ca/blog/posting.asp?ID=30
What's at stake for Canada in the Arctic?

www.capp.ca/GetDoc.aspx?DocID=141879
Oil sands economic impacts across Canada.

NOTES

CHAPTER 1

1 In the late nineteenth century, geographers believed that the physical environment determined human affairs. That position was rejected, although geographers recognize that the environment does exert a strong influence on the nature of human activities in various regions of the world. Students can find a more complete discussion of environmental determinism and other philosophical options in geography, including possibilism, positivism, humanism, and Marxism, in William Norton, *Human Geography*, 7th edn (Toronto: Oxford University Press, 2010). The writings of most geographers reflect either one of these philosophical positions or some combination. The core/periphery model, for example, sprang from Marxist scholarship. Because of the model's powerful spatial implications, other scholars holding different philosophical positions have modified this theory by removing its economically deterministic character. Instead, they accept that external forces (such as global institutions like the World Trade Organization [WTO]) and internal forces (such as federal–provincial agreements like equalization payments) can modify the impact of the physical environment on regional development. Hinterlands, therefore, are not locked into a single outcome because of their physical geography.

2 While Canadians and Americans occupy the same continent, historic events and geographic differences laid the foundation for the emergence of two different societies within one continent. These differences are found in many aspects of the two societies. Each country has a different approach to gun control legislation, multiculturalism, and a national health-care system. As a result of these and other differences, some scholars believe that Canadians are more trusting of their governments and more tolerant of social diversity than are Americans (Hartz, 1995; Lipset, 1990; Lemon, 1996; Saul, 1997).

3 A group of Muslim women visited the town council to present a fuller picture of Islam and its customs. "There was a real exchange," Najat Boughaba told Radio-Canada. "There were people who reached out to us. I really think our visit to Hérouxville benefited both sides" (CBC News, 2007). Najat Boughaba was struck by how little the townspeople knew about Islam and its customs. Following this exchange, the town council in Hérouxville amended its immigrant code of conduct to remove references to "no stoning of women in public" and "no female circumcision." On 8 February 2007, Québec Premier Jean Charest announced the establishment of the Bouchard-Taylor Commission, headed by sociologist Gérard Bouchard and philosopher Charles Taylor, formally known as the Consultation Commission on Accommodation Practices Related to Cultural Differences, in response to public discontent concerning reasonable accommodation.

4 Yet, by their very nature, theories must simplify the real world. In doing so, when applied to actual regions, they often lose touch with the complexity and variety of economic, political, and social forces at play. Added to this shortcoming, economic theories are based on past events, which means that, as the world changes, they must either be adjusted to fit the new economic circumstances or be replaced by more robust hypotheses.

5 Trade disputes remain troublesome. In April 2009, for example, the United States rekindled the softwood lumber dispute by imposing a 10 per cent tariff on lumber from four provinces that, according to Washington, exported more to the US than the 2006 lumber agreement specified. The Free Trade Agreement, which was replaced by the North American Free Trade Agreement, really means freer trade rather than free trade. As we have seen, three trade agreements—the Auto Pact (1965; nullified in 2001), the FTA (1989), and NAFTA (1994)—led to a realignment of the Canadian economy so that it was more thoroughly integrated with the North American economy. These trade agreements saw more and more manufactured goods exported to the United States, thus breaking the old pattern of exporting primarily low-value unprocessed or semi-processed resource products, and the volume of exports to the United States grew dramatically. In spite of these trade agreements, however, Washington is prepared to defend US business interests by imposing trade barriers, as has been the case with duties on softwood lumber and grain, to restrict the natural flow of certain Canadian goods into the United States. The purpose of these duties is twofold: (1) to protect US farmers and forest companies in the short run by imposing duties on targeted Canadian exports, and (2) to force Canada to accept a long-term agreement that will limit its grain and lumber exports.

CHAPTER 2

1 Canadians have various visions of themselves, their region, and their country. For the most part, these visions are rooted in the physical nature and historical experiences that have affected Canada and its regions. For example, people see Canada as a northern country because of its location in North America and because of its climates, which are often noted for long, cold winters. Professor Louis-Edmond Hamelin's concept of nordicity exemplifies the impact of "northernness" from a geographer's perspective. Hamelin (1979) provides a measure of "northernness" according to five geographic zones—Extreme North, Far North, Middle North, Near North, and Ecumene (southern Canada). Songwriters, too, have been intrigued by Canada's northern nature, and well-known writers, from Jack London and Robert Service to Margaret Atwood, Pierre Berton, and Farley Mowat, have written about the North and, in doing so, have etched out another parameter of Canadian identity.

2 The Champlain Sea covered Anticosti Island and the northern tip of the island of Newfoundland. For the purposes of

this text, the eastern extent of this physiographic region ends just east of Québec City.

3 The distance between latitudes is almost constant at 111 kilometres. On the other hand, the distance between longitudes varies from 111 kilometres to nearly zero because of the spherical shape of the earth, which causes longitude lines to become closer and closer towards the North Pole and South Pole. The distance between the Equator (0°) and 1° N is approximately 111 kilometres and between 89° N and the North Pole (90° N) is nearly zero.

4 The mean annual temperature of a location on the earth's surface is a measure of the energy balance at that point. Solar energy is the source of heat for the earth, and this energy is returned to the atmosphere in a variety of ways. Therefore, a global energy balance exists. However, there are regional energy surpluses and deficits in different parts of the world. For example, the Arctic has an energy deficit, while the tropics have a surplus. These energy differences drive the global atmospheric and oceanic circulation systems. When the mean annual temperature is below zero Celsius, it indicates that an energy deficit exists.

CHAPTER 3

1 Under the terms of the British North America Act, the Dominion of Canada was composed of four provinces (Ontario, Québec, New Brunswick, and Nova Scotia). Modelled after the British parliamentary and monarchical system of government, the newly formed country had a Parliament made up of three elements: the head of government (a governor general who represented the monarch), an upper house (the Senate), and a lower house (the House of Commons). This Act was modified several times to accommodate Canada's evolving political needs and its gradual movement to independent nationhood. The patriation of Canada's Constitution in 1982 removed the last vestige of Canada's political dependence on the United Kingdom, although Canada still recognizes the British monarch as its symbolic head.

The British North America Act was based on the highly centralized government of the United Kingdom in the 1860s. However, this Act assigned specific powers to the provinces in order to satisfy Québec's demand for control over its culture. The Canadian political system that emerged, therefore, allowed for regionalized politics. For example, political parties in the House of Commons sometimes serve regional interests. In the 1920s, the Progressive Party represented the concerns of farmers in Western Canada, while the pro-independence Bloc Québécois, which was formed in 1990, not only serves the interests of Québec but is also active in the separatist movement. Furthermore, while the House of Commons is based on the principle of representation by population, Senate membership is based on the principle of equal regional representation. However, because senators are appointed by the Prime Minister and not elected by the people in the different regions of the country, the Senate fails to provide a regional counterweight to the House of Commons.

2 A group of Irish Americans, known as Fenians, was struggling for Irish independence. They believed that attacking British possessions in North America would advance the cause of a free Ireland. Between 1866 and 1870, the Fenians launched several raids across the border into Canada. The United States did not encourage these raids and eventually forced the Fenians to disband. By the end of the American Civil War, Anglo-American relations again were strained because of Britain's tacit support for the Confederacy in the American Civil War. For that reason, the United States withdrew from the Reciprocity Treaty in 1866. This treaty, a free trade agreement between British North America and the United States, began in 1854; the subsequent years were prosperous ones for British North America, and the end of this agreement was a factor in the Province of Canada seeking an alternative economic union with the other British colonies in North America.

3 After the defeat of the French by the British in 1763, Pontiac, the Odawa chief in the Ohio Valley, led a successful uprising against the British. By capturing the forts in the Ohio Territory, he exposed Britain's precarious hold on this region, which the British had just obtained from the French. However, Pontiac and his followers could not hold these forts against the British because, with French forces driven back to France, Pontiac had no source of ammunition and muskets. Pontiac concluded that his best move would be to make peace with Britain. The British came to the same conclusion, though for other reasons. Without the help of Pontiac and the other chiefs in this region, Britain would lose control over these lands. Britain therefore had to form an alliance with them. With that objective in mind, George III announced an important concession to these Indians in the Royal Proclamation of 1763, namely, that the King recognized them as valued allies and that the land they used to hunt and trap was "Indian land" within the British Empire. In 1783, the American revolt against Great Britain ended with an American victory, which opened the lands west of the Appalachian Mountains to settlement by New Englanders and, at the same time, ended the dream of a vast Indian land within the British Empire.

4 The Manitoba Act of 1870 recognized the legal status of farms and other lands occupied by the Métis as "fee simple" private property. As well, the Act provided that 1.4 million acres (566,580 ha) be reserved for the children of Métis. The land was allocated to these Métis in 240-acre (97-ha) parcels, plus 160 acres (65 ha) in "scrip" for each adult head of a family. These lands were distributed after 1875, but much of the scrip land was sold and then occupied by non-Métis. For more on this subject, see Tough (1996: ch. 6). The 2013 Supreme Court of Canada ruling stated that the federal government, over 140 years earlier, "acted with persistent inattention and failed to act diligently" in regard to the land grant provision of the Manitoba Act of 1870, and emphasized that "repeated mistakes and inaction . . . persisted for more than a decade" (CBC News, 2013b).

5 The original Magna Carta goes far back in history to 1215, when King John of England was forced to sign a charter

known as the Magna Carta. In this charter, he promised to consult regularly with the country's nobles before collecting new taxes, and to stop interfering in affairs of the Church.

6 While the French language was not recognized by the Québec Act, 1774, the Governor made use of the French language in conducting his business with local officials. For example, the judges appointed by the Governor had to know both languages in order to facilitate the business of the court. In short, while English was the official language of British North America, the British colony of Québec functioned in both the French and English languages.

7 Louis Riel was the Métis political and spiritual leader in the late nineteenth century. This controversial figure is considered both a Father of Confederation and a traitor to the country. Riel, who was born in the Red River Colony in 1844, studied for the priesthood at the Collège de Montréal. The founder of Manitoba and the central figure in both the Red River Rebellion (1869–70) and the Northwest Rebellion (1885), he was captured shortly after the Battle of Batoche, where the Métis forces were defeated. After a trial in Regina, the jury found Riel guilty of treason but recommended clemency. Appeals were made to Manitoba's Court of Queen's Bench and to the Judicial Committee of the Privy Council. Both appeals were dismissed. A final appeal went to the federal cabinet, but the government of John A. Macdonald wanted Riel executed. Riel was hanged in Regina on 16 November 1885. His body was interred in the cemetery at the Cathedral of St Boniface in Manitoba.

Riel's execution has had a lasting effect on Canada. In Québec, French Canadians felt betrayed by the Conservative government and federalism. Riel's execution was proof for Canada's French-speaking population that they could not count on the federal government to look after French-Canadian interests. It was also a blow against a francophone presence in the West. In Ontario the hanging of Louis Riel satisfied the anti-Catholic and anti-French majority. For the Orange Order (the Protestant fraternal society that blamed Riel for the death of one of their members, Thomas Scott, who was executed by a Métis firing squad during the Red River Rebellion), Riel's execution was long overdue. In the West, Riel's hanging resulted in the marginalization of both the Métis and Indian tribes, especially those who participated in the uprising. Today, historians and many Canadians accept the view that these rebellions were, from the point of view of the Métis, resistances to a threat to their way of life.

8 An example is the political fallout from the Québec referendum in 1995. Prime Minister Chrétien sought to fulfill his verbal promises made in the closing days before the referendum vote. In a House of Commons resolution, the federal government proposed three concessions to Québec: (1) a veto over constitutional changes; (2) recognition of Québec's distinct society status; and (3) devolution of federal powers to Québec. In the case of the veto, Ottawa was prepared to "lend" its constitutional veto to Québec, Ontario, Atlantic Canada, and the four western provinces. Not only was the federal government committing itself to seeking permission from these four regions before putting its stamp of approval on any constitutional change, it was also recognizing that Canada consisted of four major regions. The premiers of Alberta and British Columbia reacted negatively to that concept of regionalism. British Columbians in particular saw this arrangement as another example of Ottawa's failure to recognize the west coast as a "distinct and powerful" part of Canada. The federal government retreated from this issue and quickly amended its resolution to extend the veto to British Columbia. In December 1995, this resolution passed in both the House of Commons and the Senate. It then became the law of the land that Canada consists of five major regions!

9 Separatists argue that Québec does not have enough powers, that is, Québec is subordinate to Ottawa. For separatists, the solution lies in independence, whether achieved through Parizeau's "chicken-plucking" strategy or Bouchard's "winning" referendum strategy. A federalist counter-argument is that, as a member of a federation, Québec has many "exclusive" powers, such as power over language and education. However, circumstances may force Ottawa to make a decision that adversely affects some provinces while favouring others. In 1982, the patriation of the British North America Act, renamed the Constitution Act, 1867, was such a decision. The Constitution Act, 1982, which was entrenched at the same time, added to the British North America Act in several ways, but without a doubt the most important addition has been the Charter of Rights and Freedoms. These rights and freedoms strengthen the rights of individuals and weaken collective rights. Prime Minister Trudeau, who conceived of society as an agglomeration of individuals (not collectivities), whose rights accrued to them as individuals, saw the Charter as protecting individuals from governments that try to suppress individual rights. Then, too, there is the Supreme Court of Canada's changed role, which has become more proactive with the adjudication of Charter cases.

CHAPTER 4

1 "Race," unlike ethnicity, is based on physical characteristics. Racial types are frequently assigned a set of social characteristics, which is known as "stereotyping." Sociologists define "race" as the socially constructed classification of persons into categories on the basis of real or imagined physical characteristics such as skin colour. Others consider "race" a means of creating major divisions of humankind on the basis of distinct physical characteristics.

2 French and English, as the official languages, represent the traditional duality of Canadian society. With the establishment of the Province of Canada in 1841, the relationship flowered into a partnership with English-speaking Canada West and French-speaking Canada East sharing political power. In 1867, the British North America Act (section 133)

established that both French and English "may be used by any Person in the Debates of the Houses of Parliament of Canada and of the Houses of the Legislature of Quebec," as well as in federal courts, but it was not until the Official Languages Act of 1969 that French truly began to be entrenched in the institutional fabric of Canadian society. The assignment of education to the provinces in the BNA Act (section 93) and the stipulation that Roman Catholic (i.e., francophone) schools had equal standing with Protestant (i.e., anglophone) schools ensured that French would retain a dominant position in Québec, at least for the time being.

3. In 1974, the Québec Liberal government passed Bill 22 (Loi sur la langue officielle), which made French the language of government and the workplace. In 1977, the Parti Québécois government introduced a much stronger language measure in the form of Bill 101 (Charte de la langue française). This legislation eliminated English as one of the official languages of Québec and required the children of all newcomers to Québec to be educated in French. Four years later, Bill 178 required all commercial signs to use only French. The French language has made modest gains outside of Québec. In 1969, New Brunswick passed an Official Languages Act, which gave equal status, rights, and privileges to English and French, and the federal Parliament passed the Official Languages Act, which declared the equal status of English and French in Parliament and in the Canadian public service.

CHAPTER 5

1. By December 2012, the Ontario government decision to cancel the construction of two gas-fired plants, in Oakville and Mississauga, and relocate the projects elsewhere was estimated to cost the Ontario taxpayers between $800 million and $1.3 billion (Leslie, 2012).
2. According to the Ontario Electricity Financial Corporation, as of 31 March 2011 the debt stood at $13.4 billion. For more on this subject, see: <www.oefc.on.ca/debtmanage.html>.
3. Both the federal and Ontario governments supported the automobile industry by supporting research into advanced technologies and by providing training in automotive manufacturing technologies at Ontario universities. For example, McMaster University and the University of Waterloo have established an Automotive Manufacturing Innovation initiative with 35 industry partners and support from the Ontario government. Another example is the launching of Auto21 at the University of Windsor with funding from the federal government.
4. Until 2001, foreign motor vehicle manufacturers had to pay a 6.1 per cent import duty. Based on the fair trade stipulations that grew out the final round of the General Agreement on Tariffs and Trade and led to the establishment of the WTO, Japan and the European Union could argue that the Auto Pact discriminated against their imported vehicles. In July 2001, as a consequence of a WTO ruling in favour of Japan and the EU that it did indeed amount to an unfair trade practice, the Auto Pact, which had helped to build Canada's automobile assembly and parts industry, ceased to exist.
5. In an effort to reduce costs, the three US automobile companies transferred their responsibility for their retired workers to the UAW. Since 2007, retired auto workers receive their health benefits through the Volunteer Employees' Beneficiary Association (VEBA).
6. At the beginning of the twentieth century, the Ontario government sought to develop the northeastern section of the province, especially agricultural lands in the Clay Belt, forest stands, and mineral deposits in the Canadian Shield. Between 1903 and 1909, the government-financed Temiskaming and Northern Ontario Railway was built from the Canadian Pacific Railway line at North Bay northward to the small town of Cochrane on the National Transcontinental Railway (now the CN). Over the next decade, branch lines were extended to Cobalt, Timmins, and Iroquois Falls. Overall, the Ontario government was pleased with the railway's impact on resource development. Plans were made to build the railway farther north—into the Hudson Bay Lowlands. By 1932, the Temiskaming and Northern Ontario Railway stretched from Cochrane to Moosonee at the southern tip of James Bay, but further developments did not occur in the resource-scarce Hudson Bay Lowlands.
7. In 1970, mercury was discovered in the fish near the Grassy Narrows Reserve, which is about 500 km downstream from the pulp mill. Levels of methyl mercury in the aquatic food chain were 10 to 50 times higher than those in the surrounding waterways (Shkilnyk, 1985: 189). These levels were similar to those found in the fish of Minamata Bay, Japan. Over 100 residents of this Japanese village died from mercury poisoning in the 1960s, and over 1,000 people suffered irreversible neurological damage. Because they depend on fish and game, the Ojibwa at the Grassy Narrows Reserve ate fish on a daily basis and many complained of mercury-related illnesses. While, unlike the Minamata incident, no one died, the economic and social impact on the Ojibwa was nevertheless an industrial tragedy of immense proportions.
8. Crossworks Manufacturing Ltd is the pre-eminent manufacturer of branded Canadian diamonds. Based in Vancouver, the company has three factories, in Sudbury, Vancouver, and Yellowknife. Crossworks sells its "Mine of Origin" diamonds around the world.
9. Toronto has had various forms of regional government. About 40 years ago, the Ontario government established a regional government by combining the City of Toronto with the surrounding centres of Etobicoke, North York, Scarborough, York, and East York into the municipality of Metropolitan Toronto (Metro Toronto). These six jurisdictions became one large urban area for regional planning purposes, but each retained its city government. As Metro Toronto continued to grow, it spread into adjacent jurisdictions. In 1988, the Ontario government created the Greater Toronto Area (GTA). Often called the Metro Toronto Region, it consists of Metro Toronto and the regional municipalities of Halton, Peel, York, and Durham. In 1997, the provincial government passed a bill to change the municipal

government structure to ensure more efficient, cost-effective services. The legislation came into effect in 1998, amalgamating six former municipalities (Etobicoke, York, Toronto, East York, North York, and Scarborough) and the regional municipalities into the new City of Toronto.

10 Toronto might solve its traffic problem by imposing a tax on automobiles entering the city's downtown. The first global city to experiment with a road toll was London, England. In 2002, London began charging drivers of automobiles £5 a day (about $12) to enter and leave the centre of London between 7 a.m. and 6:30 p.m. on weekdays, and this has resulted in a sharp decline in the volume of traffic, a great improvement in urban mobility, and a modest reduction in air pollution. By 2012, the London congestion fee had doubled to £10.

CHAPTER 6

1 Stephen Jarislowsky, the CEO of Jarislowsky Fraser Ltd in Montreal, an investment management firm, is sometimes referred to as the Warren Buffet of Canada.

2 By responsible government, Lord Durham meant a "political system in which the Executive is directly and immediately responsible to the Legislature, in which the ministers are members of the Legislature, chosen from the party which includes the majority of the elected representatives of the people" (Lucas, 1912: I, 138).

3 In 1912, Québec gained northern territories inhabited by the Inuit and Cree. Ottawa ceded these lands to Québec with the understanding that the Québec government would be responsible for settling land claims with the Aboriginal peoples in these territories. At the time of the 1995 referendum, the Cree in northern Québec, in response to the separatist claim to territorial independence, declared that they have the right to secede from Québec. They argued that if Québec has the right to secede from Canada, then the Cree have the right to secede from Québec. From a geopolitical perspective, the partitioning of Canada or Québec makes sense only to those supporting ethnic nationalism.

4 For the 2011 federal election, the House of Commons had 308 seats. Reapportionment takes place every 10 years based on population figures from the census. The last reapportionment, based on the 2011 census, added 30 new seats to the Commons: Ontario (15), BC (6), Alberta (6), and Québec (3). For the next election, expected to be in 2015, the total number of seats will be 338.

5 Camille Laurin, the father of Bill 101, declared that French was the province's only official language. Bill 101 required the children of immigrants to go to French schools and made the presence of French compulsory in the workplace and on commercial signs.

6 The case of Churchill Falls is an interesting one. The divide between the Atlantic Ocean and Hudson Bay drainage basins marks the Labrador–Québec boundary. For historical reasons, the Québec government does not formally recognize this boundary, but it does treat the area as part of Newfoundland and Labrador. Newfoundland and Labrador owns the large Churchill Falls hydroelectric project, but virtually all this power is purchased by Hydro-Québec. Hydro-Québec then transmits it across Québec to markets in the Great Lakes–St Lawrence Lowlands and the United States. In 2009, for the first time, Newfoundland and Labrador is selling its small share of electricity directly to the United States through Hydro-Québec transmission lines. Under a five-year arrangement, Hydro-Québec will receive $20 million a year to transport approximately 130 megawatts of electricity from the Churchill Falls power site.

7 While the same low electricity rates were provided to aluminum producers in Québec, US aluminum firms did not challenge the rates because they own facilities in both countries. The 13 companies with risk-sharing contracts are: Norsk Hydro Canada Inc., Aluminerie Alouette Inc., Québec-Cartier Mining Co., Cafco Industries Ltd, Timminco Ltd, QIT-Fer et Titane Inc., PPB Canada Inc., Reynolds Metals Co., Argonal, Hydrogenal, SKW Canada Inc., ABI Inc., and Aluminerie Lauralco Inc.

8 Because of the Loyalists, the land survey and naming of towns took on a British flavour. Unlike the land grants in the St Lawrence Valley, land was surveyed into townships similar to the method used in New England. However, farming in the Appalachian Uplands was difficult and by the middle of the nineteenth century English-speaking residents began to leave for Montréal and other cities in Canada. At the same time, land shortages in the St Lawrence Lowland caused French Canadians to move into the Eastern Townships.

CHAPTER 7

1 In December 2009 the British Columbia government, as part of a reconciliation agreement with the Haida, officially renamed the Queen Charlotte Islands as Haida Gwaii, which, in Haida means "islands of the people." Since the 1980s the archipelago of more than 150 islands had commonly been referred to both as Haida Gwaii and as the Queen Charlottes (CBC News, 2009).

2 Besides the issue of luring British Columbia into Confederation, there were other political reasons for constructing a transcontinental railway. First, there was the urgent need to exert political control over the newly acquired but sparsely settled lands in Western Canada. As in British Columbia, the perceived threat to this territory was from the United States. Second, there was the need to create a larger market for manufactured goods produced by the firms in southern Ontario and Québec.

3 The exact number of people in British Columbia in 1871 is not known. How many Aboriginal peoples is a guess because their numbers declined sharply as they came into contact with European diseases. Similarly, the number of Americans and people from other countries who remained in the country after the gold rush is unknown. Certainly most moved on to the next gold rush, but some stayed. The gold rush of 1858 may have attracted about 25,000 Americans who sailed from San Francisco to New Westminster at the mouth of the Fraser River.

4 The provincial government created the Technical University of British Columbia in 1997 and this virtual university began to provide online courses in 2000. The objective is to prepare students for the high-tech industry. The university offers certificate programs in electronic commerce and software development. High-technology companies are encouraged to locate offices and research laboratories near its Surrey campus. Students undertake internships and co-operative work sessions with high-technology firms. Its theoretical basis lies in the concept of high-tech clusters around a university. In 2002, the BC government placed the Technical University under the aegis of Simon Fraser University and it was renamed SFU Surrey.

5 In 2000, Ottawa tried to alleviate the pressure on salmon stocks by reducing the fleet of 4,500 fishing vessels by about one-third. However, this announcement sparked a strong reaction and little was accomplished. At the same time, Ottawa allowed Indian fishers, who have treaty rights to harvest fish for subsistence purposes, a share of the commercial stock. In 1999 Ottawa successfully negotiated a new Pacific Salmon Treaty with the United States, which extends to 2018 (Fisheries and Oceans Canada, 2009). Ottawa is able to exert some management of fish stocks in the Pacific Ocean because of its 200-mile fishing zone and because of its role in the Pacific Salmon Commission. Salmon fishing on the Pacific coast is regulated, based on the Pacific Salmon Treaty, which determines the size of the catch taken by each nation.

6 *Pollution of fish habitat*. Forestry companies have affected salmon habitat through their logging practices, which have blocked streams and thereby interfered with migration routes to spawning areas. Pulp and paper plants also discharged toxic wastes into rivers and the ocean. Fish farms, too, play a role because parasitic sea lice from these farms may be killing the juvenile salmon before they mature and return to the Fraser River to spawn (Morton et al., 2004).

Warming ocean temperatures. With global temperatures increasing, the North Pacific Ocean temperatures may have risen above levels suitable for salmon. Also, the El Niño effect provides a natural explanation for the declining salmon stocks: salmon tend to survive poorly in a warmer ocean and better in a cooler environment due to changes in ocean plankton and fish community composition (McKinnell et al., 2009).

Overfishing. The most likely explanation for the variation in salmon stocks is overfishing. With a combination of a larger fishing fleet and the application of new technology (radar and sonar equipment), the capacity to track and catch salmon and other fish has improved greatly. Since larger fish are normally caught in the fish nets, too few adult fish are left to reproduce and thus replenish the fish stocks.

High fish quotas. Within Canadian waters, DFO is responsible for estimating the size of the fish stocks and, based on this information, sets fish harvest quotas. Since estimating the size of the fish stocks is a chancy business, based as it is on historic records and insufficient sampling of fish stocks, DFO estimates tend to be on the high side. Quotas are set from these estimates, though pressure from fishers and the Americans has resulted in some past quotas set too high.

The Aboriginal fishery. First Nations peoples along the BC coast have always harvested fish for subsistence and cultural purposes. More recently, Indian fishers have sold some fish, mainly salmon, as a means of generating cash income. They are seeking a share of the fish quotas in order to participate more openly in the market economy. But how large a share should they receive? This is a difficult question, yet an Aboriginal commercial fishery was created in 1992 (see Vignette 7.5). As modern treaties are concluded with BC's First Nations, it is hoped that the Aboriginal fishery as it now exists will disappear and be replaced by individual harvesting agreements with each First Nation. As part of the Nisga'a Agreement, for example, their share of the fishery harvest for commercial and subsistence purposes was defined.

7 In 1950, Alcan secured the rights to the water in the Nechako River system until 1999. Further hydroelectric development would extend this right to the Nechako waters. Hence, Alcan commenced construction of the Kemano Completion Project. When the British Columbia government signed this long-term agreement, it saw the industrial development as the key to opening up the province's northwest. At that time, Victoria imagined that Kitimat would become an industrial complex with a population quickly reaching 20,000.

In the late 1940s, the Aluminum Company of Canada (Alcan) proposed building an aluminum complex in northwest British Columbia. The Alcan hydroelectric development in British Columbia illustrates the advantages and pitfalls of megaprojects. Alcan is one of the giants in the world aluminum industry. To keep its production costs low, Alcan is attracted to sites where low-cost hydroelectric power can be developed and then used by its smelting plants to reduce bauxite (a clay-like mineral) into alumina (the chief source of aluminum) and then aluminum. In Canada, water resources are under provincial jurisdiction, so Alcan seeks long-term arrangements with provincial governments to develop the water power for its smelting plants. In turn, provincial governments are attracted to megaprojects because they help to develop a region and employ large numbers of people in the construction phase.

8 Problems over free and fair trade have been particularly frequent in the forest industry. American lumber production, which operates mainly in the Pacific Northwest and Georgia, can produce a maximum of 15 billion board feet per year. Canadian lumber production nearly doubles the American figure. In fact, lumber production from British Columbia is roughly equal to that of the entire United States. The main difference is that US-produced lumber serves its domestic market, while most Canadian-produced lumber is exported to the United States, Japan, and other foreign customers. Trade disputes over lumber exports can therefore significantly affect BC's forest industry.

When American lumber producers lose market share to their Canadian counterparts, they turn to their lobby organization. The US political system is extremely sensitive to lobby efforts because individual members of the House of Representatives and the Senate have the power to intercede on trade matters. In addition, these elected officials rely on lobby organizations for support at election time. Since 1982, the US Coalition for Fair Lumber Imports has managed to convince the American government to reduce Canadian forestry imports into the United States. In 1986, for example, the US launched a trade action, claiming that Canadian softwood lumber production was subsidized by low provincial stumpage fees. In the following year, the US Department of Commerce ruled that Canadian stumpage rates constituted a countervailing subsidy. Ottawa agreed to impose a 15 per cent export tax on lumber exported to the US. British Columbia and several other provinces raised their stumpage rates to counter the American claim of low stumpage rates and to replace the federal export tax. In 1991, Canada terminated the 15 per cent export tax. In response, the US Coalition for Fair Lumber Imports called for another trade action to restrict the flow of Canadian lumber into the US. In 1992, the US government imposed a countervailing duty on Canadian lumber. Over a two-year period, the US government collected more than $800 million from Canadian exporters. In 1994, a bilateral trade panel (the dispute mechanism created by NAFTA) declared the tax invalid under NAFTA and ordered the US government to reimburse the Canadian companies. Not to be outdone, the US Coalition for Fair Lumber Imports called for new restrictions.

In 1996, the US government insisted on a ceiling for Canadian lumber shipments to the US, and Ottawa and Washington reached a new agreement: a limit of 14.7 billion board feet would be allowed into the US in exchange for five years of non-interference with the lumber trade by the US government. The agreed-to figure was an average of the volume of Canadian lumber exported to the United States over the previous three years. The first 650 million board feet in excess of the figure faced a US tax of $50 per 1,000 board feet. Even with this additional cost, Canadian lumber was still competitive in the US market. Once the 650 million board feet figure was exceeded, the tax jumped to $100 per 1,000 board feet. Even at that tax level, Canadian lumber remained competitive in the US market. All of these efforts by the US lumber lobby and the US government have been attempts to bypass NAFTA in order to protect US lumber interests. This US–Canada agreement ended on 31 March 2001. A new agreement finally was pulled together in 2006, though some observers and industry insiders considered it a Canadian sellout, and in 2012 this agreement was extended until 2015.

CHAPTER 8

1 Two novels illustrate the powerful impact of settlement on the Indians and Métis. Rudy Wiebe's *The Temptations of Big Bear* (1973) focuses on the Cree; Guy Vanderhaegh's *The Englishman's Boy* (1996) looks at the Cypress Hills Massacre.

2 The Cree attacked the outpost at Frog Lake, laid siege to Fort Battleford, and defeated the North West Mounted Police at Fort Pitt and Cut Knife Hill. With the arrival of the Canadian militia from eastern Canada, the Cree were defeated.

3 It was only in 1960 that the federal government extended the rights and privileges accorded to Canadian citizens to its Aboriginal peoples. Until then, Indians could not vote in provincial and federal elections without losing their status. The right to vote marked the start of a long journey to find a place in Canadian society. This journey is far from over.

4 Summer fallowing accomplished two goals: it conserved moisture and controlled weeds. In the 1990s, farmers began to abandon this technique because advances in technology enabled them to accomplish these goals without having to resort to summer fallowing.

5 The search for a quicker-maturing wheat began in 1892 when a cross was made between Red Fife, a popular wheat grown on the Prairies, and an earlier maturing wheat. After a decade of trials, Marquis wheat was tested at the Dominion Experimental Farms at Indian Head, Saskatchewan. By 1910, this variety of wheat made Canada famous for producing an exceptionally high-quality, hardy spring wheat.

6 Alberta, Manitoba, and Saskatchewan did not gain full control over Crown lands, and therefore over natural resources, until 1930. At that time, Ottawa transferred federal Crown lands to these provinces, giving them access to lucrative sources of taxation associated with natural resource development. Until then, taxes from these lands went to Ottawa, supposedly to pay for railway-building in the West. At first, the revenue from natural resources was small, but with the discovery of oil at Leduc, Alberta, in 1947, royalties from the production of petroleum provided most of the revenue for Alberta. Equally important, the exploitation of the oil and gas deposits in Alberta grew rapidly and eventually resulted in a variety of petroleum-processing plants and pipeline-construction firms in Alberta. Later, more modest energy and mineral developments were found in Saskatchewan, British Columbia, and Manitoba.

7 Farmers' decisions are affected by a complex set of variables, but ultimately the cost of doing business is the determining factor. For instance, the increased production of canola may affect the price of livestock feed. A protein by-product derived from the crushing of canola seeds is suitable for livestock. Instead of importing a similar protein supplement and paying for the built-in freight costs, using a local canola-based product to feed livestock may reduce costs.

8 Farmers are becoming less sheltered from economic forces. Through contract farming, agribusiness—the sector of the economy that provides inputs to farms, such as chemical fertilizers, and procures agricultural products from farms for processing and distribution to consumers—is increasingly affecting the daily lives of farmers. Some fear that in the next severe economic downturn or drought, family farming operations will be replaced by corporate farms. In fact, the trend towards contract farming by agribusiness

may be the first sign of such an ownership shift. While not yet apparent in the grain industry, agribusiness has established itself in specialty crops, poultry, and hog farming.

9 For a candid discussion of the factors behind the collapse of the Manitoba-owned pork processing business, see Wilkins (2013).

10 The potash royalty is composed of two payments: the base payment and profit tax. The base payment is a 35 per cent resource profits tax subject to a minimum $11 per tonne and maximum $12.33 per tonne of production reduced by Crown and Freehold Royalties and certain credits. The profit tax is progressive, with a 15 per cent rate applied to profits up to $59.95 per tonne and 35 per cent for profits in excess of that level. Under the profits tax, the government provides accelerated depreciation equal to 120 per cent of capital costs for new capital expenditures in excess of 90 per cent of 2002 capital expenditures. Losses are carried forward for five years. An adjusted base payment is credited against the profit tax.

11 Since the early 1990s, sawmills in the interior of British Columbia have exhausted local supplies of timber. They have had to purchase logs from Alberta ranchers who own timber lands in the foothills around Calgary.

CHAPTER 9

1 On 30 April 1999, the Newfoundland House of Assembly gave unanimous consent to a constitutional amendment that would officially change the name of the province to Newfoundland and Labrador.

2 The Maritime region was a focal point for the struggle between France and Britain. Indian allies were critical for both European powers. "Involvement with the French brought benefits—and immediate consequences. Before they realized the full scope of the French–British rivalry, the Mi'kmaq and the Maliseet discovered they had already chosen sides" (Coates, 2000: 31).

3 As with Fort McMurray (Wood Buffalo), Alberta, Statistics Canada has created a geographically sprawling census area—Cape Breton—out of Sydney and other Cape Breton Island municipalities.

4 In 2006 the Brazilian mining company CVRD (now named Vale) purchased the Canadian nickel-mining company Inco, including its Sudbury operations and the Voisey's Bay mine site, for approximately $17 billion. Its Canadian-based subsidiary, Vale Inco, is now named Vale Canada Limited or simply Vale.

5 Coal, a combustible sedimentary rock formed from the remains of plant life during the Carboniferous Age (a geological period in the Paleozoic era), is classified into four types: anthracite, bituminous, sub-bituminous, and lignite. Anthracite is the highest grade of coal, while bituminous is used in the iron and steel industry and for generating thermoelectric power. Most coal mined in Nova Scotia has been bituminous coal.

6 On 26 May 2004 the Labrador Inuit voted 76 per cent in support of the agreement, with an 86.5 per cent voter turnout. The provincial and federal governments passed legislation in 2004 and 2005, respectively, giving legal effect to the Labrador Inuit Land Claims Agreement Act. This agreement contains the provision that the Labrador Inuit will receive 25 per cent of the revenue from mining and petroleum production on their settlement land, as well as 5 per cent of provincial royalties from the Voisey's Bay project. The federal environmental review panel examining the possible environmental and social impacts of the Voisey's Bay Nickel Company's (now Vale Newfoundland and Labrador Limited) mine proposal expressed concern about the disposal of the 15,000 tonnes of mine tailings to be produced each day. The mining company proposed to deposit the toxic tailings in a pond and prevent this from draining into surrounding streams and rivers by building two dams. While federal and provincial environmental officials are satisfied with the company's solution to the tailings problem, local people, especially the Inuit and Innu, are skeptical and worried about the effects of these toxins on the wildlife they depend on for food. In 2002, the Labrador Innu signed a Memorandum of Agreement in regard to Voisey's Bay that gives them 5 per cent of provincial revenues from the project and the assurance that a final land claim agreement will include a chapter on Voisey's Bay that would detail any further compensation and environmental protections. In 2011 the Innu Nation ratified the New Dawn Agreement with the province, an agreement that is expected to form the basis of a final agreement that also involves the federal government.

7 Of course, Québec still feels cheated over the 1927 Québec–Labrador boundary decision of the British Privy Council, a decision it continues not to recognize as final. In that year the colony of Newfoundland won the argument over the meaning of the word "coast." In previous historical descriptions of the Labrador boundary, the vaguely worded geographic zone of "coast" was not defined, but in the 1927 decision Newfoundland won the day when Britain's Privy Council accepted that the inland extent of "coast" actually meant watershed (Budgel and Stavely, 1987).

CHAPTER 10

1 There is a third climatic zone in the Territorial North—the Cordillera. However, the Cordillera climate, often described as a mountain climate, is affected by elevation (as elevation increases, temperature drops). North of 60° N, the Cordillera climate is also affected by latitude so that boreal natural vegetation is found at lower elevations and tundra natural vegetation at high elevations.

2 The Mary River Project involves an iron ore mine at Mary River on Baffin Island; a landing strip at the mine site that will serve the air-commuting system for workers and another one at Steensby Inlet seaport; a railway line from the mine site to the Steensby Inlet all-season port; a landing site at Milne Inlet and a tote road to the mine site. Supplies and materials for the mine site will use the Milne Inlet port and tote road while the iron ore will be shipped by rail to Steensby Inlet and then by ship to markets in Europe; iron

ore ships will have ice-breaking capacity to allow for year-round operations.

3 Sharing food was essential for small hunting groups to be able to live together harmoniously. By sharing in all hunters' successes, they could adjust for the vagaries of individual luck and reduce the threat of starvation. Today, the sharing of food remains a pivotal component of Aboriginal culture and the Native economy.

4 Sometime around the year 1000 the Vikings made contact with the ancestors of the Inuit, the Thule (c. AD 1000 to 1600). The Thule originated in Alaska where they hunted bowhead whales and other large sea mammals. They quickly spread their whaling technology across the Arctic, travelling in skin boats and dogsleds. With the onset of the Little Ice Age in the fifteenth century, climate conditions affected the distribution of animals and the Thule who were dependent on them. An increased amount of sea ice blocked the large whales from their former feeding grounds, resulting in the collapse of the Thule whale hunt. With the loss of their main source of food, the Thule had to rely more and more on locally available foods, usually some combination of seal, caribou, and fish. By the eighteenth century, the Thule culture had disappeared and had been replaced by the Inuit hunting culture.

5 Today, capital and operating costs are still serious problems for hunters. Transportation from a settlement to hunting/trapping areas is a major expenditure. Snowmobiles, for example, may cost as much as $20,000 in a northern retail store (in the 1990s, they cost $5,000 to $10,000), and the price of gasoline to fuel them is significantly higher in the North. With a continuing rise in the prices of manufactured goods, the cost of a snowmobile 10 years from now may double again. Since a snowmobile used for long-distance travel has a short lifespan, the hunter/trapper must replace it about every three or four years.

6 The first comprehensive land claim submission came from the Dene/Métis in 1974. The land claimed by the Dene/Métis extended over most of the Mackenzie Basin north of 60° N. This land was called Denendeh. In 1988, an Agreement-in-Principle was signed by the two negotiating parties (the federal government and Dene Nation). This agreement called for a cash payment of $500 million over 15 years plus title to 181,230 km² of land. Nearly 6 per cent of this land (10,000 km²) was to include subsurface rights for the Dene/Métis. As well, the Dene/Métis would obtain a share of federal resource royalties, including those generated by the Norman Wells oil field. Chiefs and elders from the Great Slave Lake area refused to approve this agreement for two main reasons—it contained no reference to self-government and it called for the surrender of Aboriginal rights. With this rejection, the Gwich'in and Sahtu/Métis subsequently negotiated their own comprehensive land claim agreements with Ottawa, as did the Tlicho (Dogrib). The Deh Cho and South Slavey still have not reached final agreements.

7 The Inuvialuit broke away from the other Inuit who were seeking a common land claim settlement that would have stretched from the Arctic Coast of Yukon to Baffin Island. The Inuvialuit wanted to advance their own land claim before the huge oil and gas deposits in the Beaufort Sea were developed. The Inuvialuit, with their separate agreement, hoped to obtain land with oil and gas deposits and achieve a taxation arrangement similar to that of the Arctic Slope Regional Corporation of the Inupiat in Alaska near Prudhoe Bay. This Alaskan municipality received substantial tax revenues from the oil companies.

8 Under the IFA, Canada agreed to transfer $45 million in 1977 dollars. By 1984, these funds were valued at $152 million. The payments to the Inuvialuit Regional Corporation, the business corporation of the IFA, began on 31 December 1984 with $12 million. By 1997, there had been three annual payments of $1 million beginning 31 December 1985; five annual payments of $5 million beginning 31 December 1988; four annual payments of $20 million beginning 31 December 1993; and a final payment on 31 December 1997 of $32 million (Canada, 1985: 107).

9 This pipeline would be constructed by Canadian Arctic Gas Pipeline Ltd, which represented 27 Canadian and American oil producers, such as Exxon, Gulf, and Shell, and Trans-Canada Pipelines. Later, a rival bid was made by Foothills Pipeline Ltd, which consisted of Alberta Gas Trunk Line and Westcoast Transmission. Prudhoe Bay is located along the North Slope of Alaska. Alaskan oil accounts for 20 per cent of US oil production. An oil pipeline, known as the Trans-Alaska Pipeline System (TAPS), transports Alaskan oil from Prudhoe Bay to Valdez and then by tanker to markets along the west coast of the United States.

BIBLIOGRAPHY

CHAPTER 1

Adams, Michael. 2003. *Fire and Ice: The United States, Canada and the Myth of Converging Values*. Toronto: Penguin.

Agnew, John. 2002. *American Space/American Place: Geographies of the Contemporary United States*. New York: Routledge.

Atlas of Canada. 2009. "Official Languages, 1996," 25 Feb. At: <atlas.nrcan.gc.ca/site/english/maps/peopleandsociety/lang/officiallanguages/1>.

Azzi, Stephen. 2006. "Debating Free Trade," *National Post*, 14 Jan., A18.

Bailey, Ian, and Bill Curry. 2011. "Health Proposal Divides the Provinces," *Globe and Mail*, 20 Dec., A1, A12.

Barnes, Trevor, ed. 1993. "Focus: A Geographical Appreciation of Harold A. Innis," *Canadian Geographer* 37, 4: 352–64.

Blackwell, Tom. 2006. "Be 'Less Religious' about Sovereignty, Manley Urges," *National Post*, 7 Feb., A7.

Blanchfield, Mike. 2011. "Canada Won't Be 'Captive Supplier' of U.S. Energy, PM Says," *Globe and Mail*, 2 Dec., B1.

Burney, Derek. 2007. "From the FTA, Lessons in Leadership: Free Trade Turns 20," *Globe and Mail*, 9 Oct., A19.

Burney, Derek, and Fen Osler Hampson. 2012. "The Last Thing We Need Is Another Foreign Policy Review," *Globe and Mail*, 20 Jan., A13.

CBC News. 2007. "Hérouxville Drops Some Rules from Controversial Code," 13 Feb. At: <www.cbc.ca/canada/montreal/story/2007/02/13/qc-herouxville20070213.html>.

Chase, Steven, and Greg Keenan. 2012. "Iron, Steel for New Windsor Bridge Must Come from Canada or U.S.," *Globe and Mail*, 15 June, B1, B8.

Cresswell, Tim. 2005. *Place: A Short Introduction*. London: Blackwell.

Columbo, John Robert, ed. 1987. *New Canadian Quotations*. Edmonton: Hurtig.

De Blij, H.J., and Alexander B. Murphy. 2006. *Human Geography: Culture, Society, and Space*, 8th edn. Toronto: John Wiley.

Dicken, Peter. 2007. *Global Shift: Mapping the Changing Contours of the World Economy*, 5th edn. New York: Guilford.

Frank, A.G. 1969. *Capitalism and Underdevelopment in Latin America*. New York: Monthly Review Press.

Friedmann, John. 1966. *Regional Development Policy: A Case Study of Venezuela*. Cambridge, Mass.: MIT Press.

Garreau, Joel. 1981. *The Nine Nations of North America*. Boston: Houghton Mifflin.

Hamilton, Graeme. 2007. "Welcome! Leave Your Customs at the Door," *National Post*, 30 Jan., A1, A7.

Hare, Kenneth F. 1968. "Canada," in John Warkinton, ed., *Canada: A Geographical Interpretation*. Toronto: Methuen, 3–12.

Hart, Michael. 1991. "A Lower Temperature: The Dispute Settlement Experience under the Canada–United States Free Trade Agreement," *American Review of Canadian Studies* (Summer/Autumn): 193–205.

Hartz, Louis. 1995. *The Liberal Tradition in America*. New York: Harcourt Brace Jovanovich.

Hayter, Roger, and Trevor J. Barnes. 2001. "Canada's Resource Economy," *Canadian Geographer* 45, 1: 36–41.

Hazell, Elspeth, Har-Fai Gee, and Andrew Sharpe. 2012. *The Human Development Index in Canada: Estimates for the Canadian Provinces and Territories, 2000–2011*. Research Report of the Centre for the Study of Living Standards. At: <www.csls.ca/reports/csls2012-02.pdf>.

Hiller, Harry. 2000. "Region as a Social Construction," in Keith Archer and Lisa Young, eds, *Regionalism and Party Politics in Canada*. Toronto: Oxford University Press.

Howlett, Karen, Dawn Walton, and Shawn McCarthy. 2012. "McGuinty–Redford War of Words Keeps Simmering," *Globe and Mail*, 29 Feb., A1.

Hutchison, Bruce. 1942. *The Unknown Country: Canada and Her People*. Toronto: Longmans, Green and Company.

Innis, Harold. 1930. *The Fur Trade in Canada: An Introduction to Canadian Economic History*. New Haven: Yale University Press.

Khondaker, Jafar. 2007. "Canada's Trade with China: 1997 to 2006," *Canada Trade Highlight Series*, No. 1 Statistics Canada Catalogue no. 65–508–XWE. At: <www.statcan.gc.ca/pub/65-508-x/65-508-x2007001-eng.htm>.

Konrad, Victor, and Heather Nicol. 2008. *Beyond Walls: Reinventing the Canada–US Borderlands*. Aldershot: Ashgate.

Lemon, James T. 1996. *Liberal Dreams and Nature's Limits: Great Cities of North America Since 1600*. Toronto: Oxford University Press.

Lipset, Seymour M. 1990. *Continental Divide: The Values and Institutions of the United States and Canada*. New York: Routledge.

Macdonald, Neil. 2009. "Interview with US Homeland Security Secretary Janet Napolitano," CBC News, 20 Apr. At: <www.cbc.ca/canada/story/2009/04/20/f-transcript-napolitano-macdonald-interview.html>.

McParland, Kelly. 2012. "Dalton McGuinty Blames Dog for Eating Province," *National Post*, 28 Feb. At: <fullcomment.nationalpost.com/2012/02/28/dalton-mcguinty-blames-the-dog-for-eating-his-province/>.

Norcliffe, Glen. 2001. "Canada in a Global Economy," *Canadian Geographer* 45: 14–30.

Northwest Territories, Bureau of Statistics. 2006. *Northwest Territories—2005…By the Numbers*. At: <www.stats.gov.nt.ca/Statinfo/Generalstats/bythenumbers/BTNhome(dvo).html>.

Northwest Territories. 2011. *Statistics Quarterly* 33, 2. At: <www.stats.gov.nt.ca/publications/statistics-quarterly/sqsep2011.pdf>.

Norton, William. 2010. *Human Geography*, 7th edn. Toronto: Oxford University Press.

Paasi, Anssi. 2003. "Region and Place: Regional Identity in Question," *Progress in Human Geography* 27, 4: 475–85.

Parkinson, David. 2009. "Despite Some Price Woes, Commodities Bull Market Is Thriving," *Globe and Mail*, 17 July, B8.

Peet, Richard, and Elaine Hartwick. 2009. *Theories of Development: Contentions, Arguments, Alternatives*, 2nd edn. New York: Guilford.

Potter, Mitch. 2010. "Secretive Family and a Vital U.S. Link Create International Span of Mystery," *Toronto Star*, 16 Feb., A1, A17.

Relph, E.C. 1976. *Place and Placelessness*. London: Pion.

Resnick, Philip. 2000. *The Politics of Resentment: British Columbia Regionalism and Canadian Unity*. Vancouver: University of British Columbia Press.

Rumney, Thomas A. 2010. *Canadian Geography: A Scholarly Bibliography*. Lanham, Md: Scarecrow Press.

Samuelson, Paul. 1976. *Economics*. New York: McGraw-Hill.

Saul, John Ralston. 1997. *Reflections of a Siamese Twin: Canada at the End of the Twentieth Century*. Toronto: Viking.

Simco, Luke. 2009. "Wallin Addresses SARM Convention," *StarPhoenix* (Saskatoon), 13 Mar., A5.

Simpson, Jeffrey. 1993. *Faultlines: Struggling for a Canadian Vision*. Toronto: HarperCollins.

———. 2011. "An Eagle with Clipped Talons," *Globe and Mail*, 12 Oct., A15.

Statistics Canada. 1997. *A National Overview: Population and Dwelling Counts*, 1996 Census of Canada (Catalogue no. 93–357–XPB). Ottawa: Industry Canada.

———. 2006a. "Land and Freshwater Area, by Province and Territory, 2006." At: <www40.statcan.ca/l01/cst01/phys01.htm>.

———. 2006b. "Trade." At: <www41.statcan.ca/1130/ceb1130_000_e.htm>.

———. 2007a. "Population and Dwelling Counts for Canada, Provinces and Territories, 2006 and 2001 Censuses—100% data." At: <www12.statcan.ca/english/census06/data/popdwell/Table.cfm?T=101>.

———. 2007b. "Gross Domestic Product, Expenditure-based, by Province and Territory." At: <www40.statcan.ca/l01/cst01/econ15.htm>.

———. 2007c. *The Evolving Linguistic Portrait, 2006 Census: The Proportion of Francophones and of French Continue to Decline*. 2006 Census: Analytic Series. At: <www12.statcan.ca/english/census06/analysis/language/continue_decline.cfm>.

———. 2007d. "Canadian Statistics: Distribution of Employed People, by Industry, by Province." At: <www40.statcan.ca/l01/cst01/labor21c.htm>.

———. 2007e. "Canada's International Merchandise Trade," *The Daily*, 13 Feb. At: <www.statcan.ca/Daily/English/070213/td070213.htm>.

———. 2007f. "Export of Goods on a Balance-of-Payments Basis, by Product." At: <www40.statcan.ca/l01/cst01/gblec04.htm>.

———. 2008. "Aboriginal Identity Population by Age Groups, Median Age and Sex, 2006 Counts, for Canada, Provinces and Territories—20% Sample Data." At: <www12.statcan.ca/english/census06/data/highlights/Aboriginal/pages/Page.cfm?Lang=E&Geo=PR&Code=01&Table=1&Data=Count&Sex=1&Age=1&StartRec=1&Sort=2&Display=Page>.

———. 2011. "Canada's International Merchandise Trade: Annual Review," *The Daily*, 7 Apr., Table 1. At: <www.statcan.gc.ca/daily-quotidien/110407/t110407b1-eng.htm>.

———. 2012a. "Population and Dwelling Counts, for Canada, Provinces and Territories, 2011 and 2006 Censuses." 2011 Census, Jan. At: <www12.statcan.gc.ca/census-recensement/2011/dp-pd/hlt-fst/pd-pl/Table-Tableau.cfm?LANG=Eng&T=101&S=50&O=A>.

———. 2012b. "Gross Domestic Product (GDP) at Basic Prices, by North American Industry Classification System (NAICS) and Provinces," 27 May, Table 379–0025. At: <www5.statcan.gc.ca/cansim/a26?lang=eng&retrLang=eng&id=3790025&paSer=&pattern=&stByVal=2&p1=-1&p2=31&tabMode=dataTable&csid=>.

———. 2012c. "Distribution of Employed People, by Industry, by Province." At: <www.statcan.gc.ca/tables-tableaux/sum-som/l01/cst01/labor21a-eng.htm>.

———. 2012d. "Population by Mother Tongue and Age Groups (Total), Percentage Distribution (2011) for Canada, Provinces and Territories," 24 Oct. At: <www12.statcan.gc.ca/census-recensement/2011/dp-pd/hlt-fst/lang/Pages/highlight.cfm?TabID=1&Lang=E&Asc=1&PRCode=01&OrderBy=999&View=2&tableID=401&queryID=1&Ag>.

———. "Number and distribution of the population reporting an Aboriginal identity and percentage of Aboriginal people in the population, Canada, provinces and territories", *National Household Survey* (NHS). Table 2. 13 May at: http://www12.statcan.gc.ca/nhs-enm/2011/as-sa/99-011-x/99-011-x2011001-eng.cfm.

Tuan, Yi-Fu. 1977. *Space and Place: The Perspective of Experience*. Minneapolis: University of Minnesota Press.

———. 1980. *Landscapes of Fear*. Oxford: Blackwell.

VanderKlippe, Nathan. 2009. "T. Boone Pickens Bets on Natural Gas," *Globe and Mail*, 19 June. At: <www.theglobeandmail.com/globe-investor/t-boone-pickens-bets-on-natural-gas/article1186003/>.

Wallerstein, Immanuel. 1974. *The Modern World System: Capitalist Agriculture and the Origins of the European World Economy in the Sixteenth Century*. New York: Academic Press.

———. 1979. *The Capitalist World Economy*. Cambridge: Cambridge University Press.

Watkins, M.H. 1963. "A Staple Theory of Economic Growth," *Canadian Journal of Economics and Political Science* 29, 2: 141–58.

———. 1977. "The Staple Theory Revisited," *Journal of Canadian Studies* 12, 5: 83–95.

Yukon. 2006. *Yukon Labour Force Survey Review, 2005*. At: <www.eco.gov.yk.ca/stats/onetime/lforcerev05.pdf>.

———. 2011. *Yukon Economic Outlook, 2011*. 17 May. At: <economics.gov.yk.ca/Files/Economic%20Outlook/Outlook2011.pdf>.

CHAPTER 2

BBC News. 2010. "Q & A: Professor Phil Jones," 13 Feb. At: <news.bbc.co.uk/2/hi/8511670.stm>.

Bone, Robert M., Shane Long, and Peter McPherson. 1997. "Settlements in the Mackenzie Basin: Now and in the Future 2050," in Cohen (1997: 265–74).

Bouchard, Mireillle. 2001. "Un défi environnemental complexe du XXIe siècle au Canada: L'identification et la compréhension de la réponse des environnements face aux changements climatiques globaux," *Canadian Geographer* 45, 1: 54–70.

Broecker, Wally. 1975. "Are We on the Brink of a Pronounced Global Warming?" *Science* 189: 460–3.

Bryden, Joan. 2012. "Mulcair Stands Behind 'Dutch Disease' Diagnosis," *StarPhoenix* (Saskatoon), 17 May, B6.

CBC News. 2004. "Inside Walkerton: Canada's Worst-ever E. Coli Contamination," 20 Dec. At: <canadaonline.about.com/gi/dynamic/offsite.htm?site=http://www.cbc.ca/news/background/walkerton/>.

Christopherson, Robert W. 1998. *Geosystems: An Introduction to Physical Geography*, 3rd edn. Upper Saddle River, NJ: Prentice-Hall.

Cohen, Stewart J., ed. 1997. *The Final Report of the Mackenzie Basin Impact Study*. Downsview, Ont.: Environment Canada.

Conrad, Cathy T. 2009. *Severe and Hazardous Weather in Canada: The Geography of Extreme Events*. Toronto: Oxford University Press.

Curry, Bill, and Shawn McCarthy. 2011. "Canada Formally Abandons Kyoto Protocol on Climate Change," *Globe and Mail*, 12 Dec., A4.

Dearden, Philip, and Bruce Mitchell. 2012. *Environmental Change and Challenge: A Canadian Perspective*, 4th edn. Toronto: Oxford University Press.

de Loë, R. 2000. "Floodplain Management in Canada: Overview and Prospects," *Canadian Geographer* 44, 4: 354–68.

Environment Canada. 2012. "Facility Greenhouse Gas Emissions by Province/Territory," Canada's Greenhouse Gas Emissions, Table 2, 11 Apr. At: <ec.gc.ca/ges-ghg/default.asp?lang=En&n=12365A33-1>.

Fisheries and Oceans Canada. 2003. "Fast Facts." At: <www.dfo-mpo.gc.ca/communic/facts-info/facts-info_c.htm>.

French, H.M., and O. Slaymaker, eds. 1993. *Canada's Cold Environments*. Montréal and Kingston: McGill-Queen's University Press.

Hamelin, Louis-Edmond. 1979. *Canadian Nordicity: It's Your North, Too*, trans. William Barr. Montreal: Harvest House.

Hare, F. Kenneth, and Morley K. Thomas. 1974. *Climate Canada*. Toronto: Wiley.

Intergovernmental Panel on Climate Change (IPCC). 2010. *The IPCC Assessment Reports*. At: <www.ipcc.ch/>.

Laycock, A.H. 1987. "The Amount of Canadian Water and Its Distribution," in M.C. Healey and R.R. Wallace, eds, *Canadian Arctic Resources*. Ottawa: Department of Fisheries and Oceans, 13–42.

Leach, Andrew. 2011. "The Nuts and Bolts of Kyoto Withdrawal," *Globe and Mail*, 6 Dec. At: <www.theglobeandmail.com/report-on-business/economy/economy-lab/the-economists/the-nuts-and-bolts-of-kyoto-withdrawal/article2263236/>.

McCarthy, Shawn. 2011. "Kent Rejects 'Guilt Payment' on Climate," *Globe and Mail*, 30 Nov., A4.

McKibben, Bill. 2010. *Earth: Making a Life on a Tough Planet*. New York: Times Books.

Miller, G.H., et al. 2012. "Abrupt Onset of the Little Ice Age Triggered by Volcanism and Sustained by Sea-Ice/Ocean Feedbacks," *Geophysics Research Letters* 39 (Jan.). At: <www.agu.org/pubs/crossref/2012/2011gl050168.shtml>.

Muller, Richard A. 2012. "The Conversion of a Climate-Change Skeptic," *New York Times*, 28 July. At: <www.nytimes.com/2012/07/30/opinion/the-conversion-of-a-climate-change-skeptic.html?pagewanted=all>.

NASA. 2008. "Arctic Sea Ice Reaches Lowest Coverage for 2008." At: <www.nasa.gov/topics/earth/sea_ice_nsidc.html>.

Norton, William. 2010. *Human Geography*, 7th edn. Toronto: Oxford University Press.

Pachauri, R.K., and A. Reisinger, eds. 2007. *Climate Change 2007: Synthesis Report*. Geneva: IPCC Secretariat.

Rasid, Harun, Wolfgang Haider, and Len Hunt. 2000. "Post-flood Assessment of Emergency Evacuation Policies in the Red River Basin, Southern Manitoba," *Canadian Geographer* 44, 4: 369–86.

Shabbar, Amir, Barrie Bonsal, and Madhav Khandekar. 1997. "Canadian Precipitation Patterns Associated with Southern Oscillation," *Journal of Climate* 10: 3016–27.

Slocombe, D. Scott, and Phillip Dearden. 2009. "Protected Areas and Ecosystem-Based Management", in Dearden and Rick Rollins, eds, *Parks and Protected Areas in Canada: Planning and Management*, 3rd edn. Toronto: Oxford University Press, 342–70.

Statistics Canada. 2013. "Aboriginal identity population, Canada, 2011", *National Household Survey* (NHS). Table 1. 13 May at: <http://www12.statcan.gc.ca/nhs-enm/2011/as-sa/99-011-x/99-011-x2011001-eng.cfm.>

Vaughan, Scott. 2012. *2012 Spring Report of the Commissioner of the Environment and Sustainable Development*, 8 May. At: <www.oag-bvg.gc.ca/internet/English/parl_cesd_201205_00_e_36772.html>.

Weart, Spencer R. 2008. *The Discovery of Global Warming*. Cambridge, Mass.: Harvard University Press.

York, Geoffrey. 2011. "China, India Are Winners in the Deal to Delay," *Globe and Mail*, 12 Dec., A1.

CHAPTER 3

Aboriginal Affairs and Northern Development Canada. 2013. *Registered Indian Population by Sex and Residence, 2011*, 6 Feb. At: <www.aadnc-aandc.gc.ca/eng/1351001356714/1351001514619#chp9_1>.

Bélanger, Claude. 2007. "North-West Rebellion—Canadian History," *The Quebec History Encyclopedia*. At: <faculty.marianopolis.edu/c.belanger/quebechistory/encyclopedia/North-WestRebellion-CanadianHistory.htm>.

Brownlie, Robin Jarvis. 2003. *A Fatherly Eye: Indian Agents, Government Power, and Aboriginal Resistance in Ontario, 1918–1939*. Toronto: Oxford University Press.

Bumsted, J.M. 2007. *A History of the Canadian Peoples*, 3rd edn. Toronto: Oxford University Press.

Canada. 1882. *Census of Canada 1880–81*, vol. 1. Ottawa: MacLean, Rogers & Company.

———. 1892. *Census of Canada 1890–91*, vol. 1. Ottawa.

Canada, Department of Finance. 2009a. "Equalization Program." At: <www.fin.gc.ca/fedprov/eqp-eng.asp>.

———. 2009b. "Federal Transfers to Provinces and Territories." At: <www.fin.gc.ca/access/fedprov-eng.asp#Major>.

———. 2011. "Equalization Program," 19 Dec. At: <www.fin.gc.ca/fedprov/eqp-eng.asp>.

Cardinal, Harold. 1969. *The Unjust Society: The Tragedy of Canada's Indians*. Edmonton: Hurtig.

Carter, Sarah. 2004. "'We Must Farm To Enable Us To Live': The Plains Cree and Agriculture to 1900," in R. Bruce Morrison and C. Roderick Wilson, eds, *Native Peoples: The Canadian Experience*. Toronto: Oxford University Press, 320–40.

CBC News. 2013a. "At Least 3,000 Died in Residential Schools, Research Shows," 18 Feb. At: <www.cbc.ca/news/canada/story/2013/02/18/residential-schools-student-deaths.html>.

———. 2013b. "Métis Celebrate Historic Supreme Court Land Ruling," 8 Mar. At: <www.cbc.ca/news/politics.story/2013/03/08/pol-metis-supreme-court-land-dispute.html>.

Cook, Ramsay. 1993. *The Voyages of Jacques Cartier*. Toronto: University of Toronto Press.

Cuthand, Doug. 2012. "Time to Move On for Sake of Future Generations," *StarPhoenix* (Saskatoon), 22 June, A11.

Delacourt, Susan. 2010. "Ontario to Get 18 New MPs," *Toronto Star*, 2 Apr., A1, A19.

Dickason, Olive Patricia, with David T. McNab. 2009. *Canada's First Nations: A History of Founding Peoples from Earliest Times*, 4th edn. Toronto: Oxford University Press.

Flanagan, Thomas, 1991. *Métis Lands in Manitoba*. Calgary: University of Calgary Press.

Garreau, Joel. 1981. *The Nine Nations of North America*. Boston: Houghton Mifflin.

Globe and Mail. 1995. "No—by a Whisker", 31 Oct., A1.

Harris, R. Cole. 1997. *The Resettlement of British Columbia: Essays on Colonialism and Geographical Change*. Vancouver: University of British Columbia Press.

———. 2003. *Making Native Space: Colonialism, Resistance, and Reserves in British Columbia*. Vancouver: University of British Columbia Press.

———. 2009. *The Reluctant Land: Society, Space, and Environment in Canada before Confederation*. Vancouver: University of British Columbia Press.

——— and John Warkentin. 1991. *Canada before Confederation: A Study in Historical Geography*. Ottawa: Carleton University Press.

Indian and Northern Affairs Canada (INAC). 2009. "Registered Indian Population by Sex and Residence 2007." At: <www.ainc-inac.gc.ca/ai/rs/pubs/sts/ni/rip/rip07/rip07-eng.asp#Sum>.

Kerr, Donald, and Deryck W. Holdsworth, eds. 1990. *Historical Atlas of Canada, Volume III: Addressing the Twentieth Century 1891–1961*. Toronto: University of Toronto Press.

Laxer, James. 2012. *Tecumseh & Brock: The War of 1812*. Toronto: Anansi.

Library and Archives of Canada. 2012. "Métis Scrip Records," 1 Mar. At: <www.collectionscanada.gc.ca/metis-scrip/005005-3100-e.html>.

Lower, J. Arthur. 1983. *Western Canada: An Outline History*. Vancouver: Douglas & McIntyre.

McVey, Wayne W., and W.E. Kalbach. 1995. *Canadian Population*. Toronto: Nelson Canada.

Miller, J.R. 2000. *Skyscrapers Hide the Heavens: A History of Indian–White Relations in Canada*, 3rd edn. Toronto: University of Toronto Press.

Milloy, John S. 1999. *A National Crime: The Canadian Government and the Residential School System*. Winnipeg: University of Manitoba Press.

Milne, Brad. 1995. "The Historiography of Métis Land Dispersal," *Manitoba History* 30: 30–41.

Moffat, Ben. 2002. "Geographic Antecedents of Discontent: Power and Western Canadian Regions 1870 to 1935," *Prairie Perspectives* 5: 202–28.

Nemni, Max. 1994. "The Case against Quebec Nationalism," *American Review of Canadian Studies* 24, 2: 171–96.

Romaniuc, Anatole. 2000. "Aboriginal Population of Canada: Growth Dynamics under Conditions of Encounter of Civilisations," *Canadian Journal of Native Studies* 20: 95–137.

Saul, John Ralston. 1997. *Reflections of a Siamese Twin: Canada at the End of the Twentieth Century*. Toronto: Viking.

Sifton, Clifford. 1922. "The Immigrants Canada Wants," *Maclean's* 35, 7: 16.

Simpson, Jeffrey. 1993. *Faultlines: Struggling for a Canadian Vision*. Toronto: HarperCollins.

Smith, P.J. 1982. "Alberta Since 1945: The Maturing Settlement System," in L.D. McCann, ed., *Heartland and Hinterland: A Regional Geography of Canada*. Scarborough, Ont.: Prentice-Hall.

Sprague, D.N. 1988. *Canada and Métis, 1869–1885*. Waterloo, Ont.: Wilfrid Laurier University Press.

Statistics Canada. 2003. *Historical Statistics of Canada*. Ottawa: Statistics Canada Catalogue no. 11–516–XIE. At: <www.statcan.ca/english/freepub/11-516-XIE/sectiona/sectiona.htm>.

———. 2008. "Aboriginal Identity Population by Age Groups, Median Age and Sex, 2006 Counts, for Canada, Provinces and Territories." At: <www12.statcan.gc.ca/english/census06/data/highlights/Aboriginal/pages/Page.cfm?Lang=E&Geo=PR&Code=01&Table=1&Data=Count&Sex=1&Age=1&StartRec=1&Sort=2&Display=Page>.

Taylor, Charles. 1993. *Reconciling the Solitudes*. Montréal and Kingston: McGill-Queen's University Press.

Thomas, David Hurst. 1999. *Exploring Ancient Native America: An Archaeological Guide*. New York: Routledge.

Tough, Frank. 1996. *As Their Natural Resources Fail: Native Peoples and the Economic History of Northern Manitoba, 1870–1930*. Vancouver: University of British Columbia Press.

Tracie, Carl J. 1996. *Toil and Peaceful Life: Doukhobor Village Settlement in Saskatchewan, 1899–1918*. Regina: Canadian Plains Research Centre, University of Regina.

Villeneuve, Paul. 1993. "Allocution présidentielle: L'invention de l'avenir au nord de l'amérique," *Le géographe canadien* 37, 2: 98–104.

Williams, Glyndwr. 1983. "The Hudson's Bay Company and the Fur Trade, 1670–1870," *The Beaver* (Autumn): 4–81.

CHAPTER 4

Agrell, Siri. 2011. "Calgary Mayor Spreads a Gospel of Revenue Sharing," *Globe and Mail*, 22 Sept., A6.

Association of University Research Parks. 2009. "AURP Names 2009 Awards of Excellence Recipients." At: <www.aurp.net/more/awardrecipients2009.cfm>.

Bank of Canada. 2010. *Monetary Policy Report Summary*, Jan. At: <www.bank-banque-canada.ca/en/mpr/pdf/2010/mprsumjan10.pdf>.

Beaujot, Roderic. 1991. *Population Change in Canada: The Challenges of Policy Adaptation*. Toronto: McClelland & Stewart.

Bélanger, Alain, and Stéphane Gilbert. 2006. "The Fertility of Immigrant Women and Their Canadian-born Daughters," *Report on the Demographic Situation in Canada 2002*, Sept. Catalogue no. 91–209–XPE. At: <www.statcan.gc.ca/pub/91-209-x/91-209-x2002000-eng.pdf>.

Bell, Daniel. 1976. *The Coming of the Post Industrial Society*. New York: Basic Books.

Bernard, André. 2009. "Trends in Manufacturing Employment," *Perspective*. Statistics Canada Catalogue no. 75–001xs. At: <www.statcan.gc.ca/pub/75-001-x/2009102/pdf/10788-eng.pdf>.

Bernier, Jacques. 1991. "Social Cohesion and Conflicts in Quebec," in Guy M. Robinson, ed., *A Social Geography of Canada*. Toronto: Dundurn Press.

Bone, Robert M. 2012. *The Canadian North: Issues and Challenges*, 4th edn. Toronto: Oxford University Press.

Bourne, Larry S., Tom Hutton, Richard Shearmur, and Jim Simmons. 2011. *Canadian Urban Regions: Trajectories of Growth and Change*. Toronto: Oxford University Press.

——— and Damaris Rose. 2001. "The Changing Face of Canada: The Uneven Geographies of Population and Social Change," *Canadian Geographer* 45, 1: 105–19.

Brean, Joseph. 2012. "The Changing Meaning of Citizenship in Canada," *National Post*, 16 Mar. At: <news.nationalpost.com/2012/03/16/the-changing-meaning-of-citizenship-in-canada/>.

Britton, John N.H., ed. 1996. *Canada and the Global Economy: The Geography of Structural and Technological Change*. Montréal and Kingston: McGill-Queen's University Press.

Canada. 1884. *Census of Canada, 1880–1881*. Ottawa: Department of Agriculture.

Canadian Press. 2012. "CAQ Legault: Kids in Quebec Should Work Harder, Like Asians," *Globe and Mail*, 14 Aug. At: <www.theglobeandmail.com/news/politics/elections/caqs-legault-kids-in-quebec-should-work-harder-like-asians/article4480487/>.

Cardinal, Harold. 1969. *The Unjust Society: The Tragedy of Canada's Indians*. Edmonton: Hurtig.

Carney, Mark. 2009. "Canada's Economic Outlook and Framework for Unconventional Monetary Policy," opening statement to the Standing Senate Committee on Banking, Trade and Commerce, Ottawa, 6 May. At: <www.bis.org/review/r090507d.pdf>.

Carrick, Rob. 2012a. "2012 vs. 1984: Yes, Young Adults Do Have It Harder Today," *Globe and Mail*, 8 May, B12.

———. 2012b. "The Steep Climb," *Globe and Mail*, 30 Aug., B1.

Chui, Tina, Kelly Tran, and Hélène Maheux. 2008. "Immigration in Canada: A Portrait of the Foreign-born Population, 2006 Census: Immigration: Driver of Population Growth." At: <www12.statcan.ca/census-recensement/2006/as-sa/97-557/figures/c1-eng.cfm>.

CIA. 2007. *The World FactBook: Canada*. Washington. At: <www.cia.gov/library/publications/the-world-factbook/index.html>.

Citizenship and Immigration Canada. 2011a. *Annual Report to Parliament on Immigration 2011*, 27 Oct. At: <www.cic.gc.ca/english/resources/publications/annual-report-2011/section2.asp#part2_1>.

———. 2011b. "Canada—Permanent Residents by Source Country," *Facts and Figures 2010 Overview*. 30 Sept. At: <www.cic.gc.ca/english/resources/statistics/facts2010/permanent/10.asp#countries>.

Curry, Bill, and Gloria Galloway. 2012. "PM Sees Jobs as Key to First Nations' Future," *Globe and Mail*, 25 Jan., A6.

Denevan, William M. 1992. "The Pristine Myth: The Landscape of the Americas in 1492," *Annals, Association of American Geographers* 82, 3: 369–85.

Dickason, Olive Patricia, with David T. McNab. 2009. *Canada's First Nations: A History of Founding Peoples from Earliest Times*, 4th edn. Toronto: Oxford University Press.

Drucker, Peter. 1969. *The Age of Discontinuity*. London: Heinemann.

Fife, Robert. 2002. "Migrants Must Spread Out: Ottawa," *National Post*, 22 June, A1, A9.

Florida, Richard. 2002a. "The Economic Geography of Talent," *Annals, Association of American Geographers* 92: 743–55.

———. 2002b. *The Rise of the Creative Class: And How It's Transforming Work, Leisure, Community and Everyday Life*. New York: Basic Books.

———. 2005. *Cities and the Creative Class*. London: Routledge.

———. 2008. *Who's Your City? How the Creative Economy Is Making Where to Live the Most Important Decision of Your Life*. New York: Basic Books.

———. 2012. *The Rise of the Creative Class—Revisited*. New York: Basic Books.

Foot, David, with D. Stoffman. 1996. *Boom, Bust and Echo: How to Profit from the Coming Demographic Shift*. Toronto: Macfarlane, Walter & Ross.

Gee, Ellen, and Gloria Gutman, eds. 2000. *The Overselling of Population Aging: Apocalyptic Demography, Intergenerational Challenges, and Social Policy*. Toronto: Oxford University Press.

Gilbert, Anne. 2001. "Le français au Canada, entre droits et géographie," *Canadian Geographer* 45, 1: 175–9.

Gionet, Linda. 2009. "First Nations People: Selected Findings of the 2006 Census," *Canadian Social Trends* no. 87. At: <www.statcan.gc.ca/pub/11-008-x/2009001/article/10864-eng.htm>.

Hampson, Sarah. 2012. "A Farewell to Violence," *Globe and Mail*, 29 May, L3.

Hare, Kenneth. 1968. "Canada," in John Warkinton, ed., *Canada: A Geographical Interpretation*. Toronto: Methuen, 3–12.

Harrison, Brian, and Louise Marmen. 1994. *Focus on Canada: Languages in Canada*. Catalogue no. 96–313E. Ottawa: Minister of Industry, Science and Technology.

Harvey, David. 1989. *The Urban Experience*. Baltimore: Johns Hopkins University Press.

Herberg, Edward N. 1989. "Identity, Cultural Production and the Vitality of Francophone Communities outside Quebec," in Leen d'Haenens, ed., *Images of Canadianness: Visions on Canada's Politics, Culture, Economics*. Ottawa: University of Ottawa Press.

Hiebert, Daniel. 2000. "Immigration and the Changing Canadian City," *Canadian Geographer* 44, 1: 25–43.

——— and David Ley. 2003. "Assimilation, Cultural Pluralism and Social Exclusion among Ethnocultural Groups in Vancouver," *Urban Geography* 24, 1: 16–44.

Ibbitson, John. 2012. "Tories Prepare New Native Land Plan," *Globe and Mail*, 25 Aug., A1, A4.

Javed, Noor. 2010. "'Visible Minority' Will Mean 'White' by 2031," *Toronto Star*, 10 Mar., A3.

Jiminez, Marina, and Kim Lunman. 2004. "Canada's Biggest Cities See Influx of New Immigrants," *Globe and Mail*, 19 Aug.

Kyle, Cassandra. 2009. "Solido Receives $1.5M Tax Rebate," *StarPhoenix* (Saskatoon), 17 July, C9. At: <www.thestarphoenix.com/Technology/Solido+receives+rebate/1800053/story.html>.

Leo, Geoff. 2012. "Blind Spot: What Happened to Canada's Aboriginal Fathers?" CBC TV, 12 Jan. At: <www.cbc.ca/sask/community/mt/2012/01/blind-spot---what-happened-to-canadas-aboriginal-fathers.html>.

Lewington, Jennifer. 2007. "Immigrants and Integration—Is the City Ready to Listen?" *Globe and Mail*, 29 Jan. At: <www.theglobeandmail.com/servlet/story/RTGAM.20070116.wimmig16home/BNStory/National/home>.

Ley, David. 1999. "Myths and Meanings of Immigration and the Metropolis," *Canadian Geographer* 43, 1: 2–18.

——— and Daniel Hiebert. 2001. "Immigration Policy as Population Policy," *Canadian Geographer* 45, 1: 120–5.

Li, Peter S. 2003. *Destination Canada: Immigration Debates and Issues*. Toronto: Oxford University Press.

McVey, Wayne W., and W.E. Kalbach. 1995. *Canadian Population*. Toronto: Nelson Canada.

Martin, Don. 2005. "The Paycheque Exiles," *National Post*, 6 Sept., A18.

Mooney, James. 1928. *The Aboriginal Population of America North of Mexico*. Smithsonian Miscellaneous Collections. Washington: Smithsonian Institution.

Newhouse, David R. 2000. "From the Tribal to the Modern: The Development of Modern Aboriginal Societies," in Ron F. Laliberte et al., eds, *Expressions in Canadian Native Studies*. Saskatoon: University of Saskatchewan Extension Press, 395–409.

———. 2007. "Aboriginal Languages in Canada: Emerging Trends and Perspectives on Second Language Acquisition," *Canadian Social Trends* no. 83, Statistics Canada, Catalogue no. 11–008.

Ostry, Bernard. 2005. "Canada's Ethnic Makeup Is Branching Out in New Directions," *Globe and Mail*, 15 Nov., A25.

Peters, Evelyn J. 2010. "Aboriginal People in Canadian Cities," in Trudi Bunting, Pierre Filion, and Ryan Walker, eds, *Canadian Cities in Transition: New Directions in the Twenty-First Century*, 4th edn. Toronto: Oxford University Press, ch. 22.

PriceWaterhouseCoopers Canada. 2009. *SR&ED Tax Clips: 2009 Provincial and Territorial R&D Tax Credits (May 22, 2009)*. At: <www.pwc.com/ca/en/sred/tax-clips/2009-provincial-territorial-credits.jhtml>.

Ravenstein, Ernest George. 1885. "The Laws of Migration," *Journal of the Statistical Society of London* 48, 2: 167–235.

———. 1889. "The Laws of Migration," *Journal of the Royal Statistical Society* 52, 2: 241–305.

Royal Commission on Bilingualism and Biculturalism. 1970. *Report. Book IV: Cultural Contributions of the Other Ethnic Groups*. Ottawa: Queen's Printer.

Saul, John Ralston. 1997. *Reflections of a Siamese Twin: Canada at the End of the Twentieth Century*. Toronto: Viking.

Saunders, Doug. 2012. *The Myth of the Muslim Tide: Do Immigrants Threaten the West?* Toronto: Knopf Canada.

Scoffield, Heather. 2009a. "Recession Toll Harsh, but Easing," *Globe and Mail*, 1 June. At: <www.theglobeandmail.com/report-on-business/recession-toll-harsh-but-easing/article4275150/>.

———. 2009b. "Canada's Recovery in Sight, Says Report," *Globe and Mail*, 30 July. At: <www.theglobeandmail.com/report-on-business/canadas-recovery-in-sight-report-says/article1236130/>.

Statistics Canada. 1997a. "1996 Census: Mother Tongue, Home Language and Knowledge of Languages," *The Daily*, 2 Dec. At: <www.statcan.ca/Daily/English/>.

———. 1997b. *A National Overview: Population and Dwelling Counts*. Catalogue no. 93–357–XPB. Ottawa: Minister of Industry.

———. 1997c. *Mortality—Summary List of Causes, 1995*. Catalogue no. 84–209–XPB. Ottawa: Minister of Industry.

———. 2002a. *Census of Canada 2001—Census Geography. Highlights and Analysis: Canada's 2001 Population*. Ottawa. At: <www12.statcan.ca/English/census01/>.

———. 2002b. "Census of Population: Language, Mobility and Migration," *The Daily*, 10 Dec. At: <www.statcan.ca/Daily/English/02120/d021210a.htm>.

———. 2002c. *A Profile of the Canadian Population: Where We Live, 2001 Census*. At: <www.statcan.ca/bsolc/english/bsolc?catno=96F0030XIE2001001>.

———. 2003a. *Census of Canada 2001—Canada's Ethnocultural Portrait: The Changing Mosaic*. Analytical Series 96F0030XIE2001008. Ottawa, 21 Jan. At: <www12.statcan.ca/english/census01/products/analytic/companion/etoimm/canada.cfm>.

———. 2003b. "Components of Population Growth", 28 Feb. Ottawa. At: <www.statcan.ca/English/Pgdb/demo33a.htm>.

———. 2003c. *Census of Canada 2001—Aboriginal Peoples of Canada: A Demographic Profile*. Analysis series

96F0030XIE2001007. Ottawa, 21 Jan. At: <www12.statcan.ca/english/census01/products/analytic/companion/abor/contents.cfm>.

———. 2003d. *Census of Canada 2001—Religions in Canada*. Analysis series 96F0030XIE2001015. At: <www12.statcan.ca/english/census01/products/analytic/companion/rel/contents.cfm>.

———. 2006a. "Components of Population Growth, by Province and Territory, July 1, 2005 to June 30, 2006." At: <www40.statcan.ca/l01/cst01/demo33c.htm>.

———. 2006b. *Annual Demographic Statistics 2005*. Catalogue no. 91–213–XIB. At: <www.statcan.ca/english/freepub/91-213-XIB/0000591-213-XIB.pdf>.

———. 2006c. "Labour Force Survey," *The Daily*, 10 Feb. At: <www.statcan.ca/Daily/English/060210/d060210a.htm>.

———. 2006d. "Canada's Population by Age and Sex," *The Daily*, 26 Oct. At: <www.statcan.ca/Daily/English/061026/d061026b.htm>.

———. 2007a. "Population by Year, by Province and Territory." At: <www40.statcan.ca/l01/cst01/demo02a.htm?sdi=population%20change%20provinces>.

———. 2007b. "Community Profiles." At: <www12.statcan.ca/english/census06/data/profiles/community/Search/SearchForm_Results.cfm?Lang=E>.

———. 2007c. "Portrait of the Canadian Population in 2006: Subprovincial Population Dynamics: Canada's Population Becoming More Urban." At: <www12.statcan.ca/english/census06/analysis/popdwell/Subprov1.cfm>.

———. 2007d. "Portrait of the Canadian Population in 2006: Subprovincial Population Dynamics: Vast Majority of Canada's Population Growth Is Concentrated in Large Metropolitan Areas." At: <www12.statcan.ca/english/census06/analysis/popdwell/Subprov3.cfm>.

———. 2007e. *2006 Census Dictionary*. At: <www12.statcan.ca/english/census06/reference/dictionary/index.cfm>.

———. 2007f. "Dwelling Counts, for Canada, Provinces and Territories, and Census Subdivisions (Municipalities), 2006 and 2001 Censuses—100% Data." At: <www12.statcan.ca/english/census06/data/popdwell/Filter.cfm?T=308&S=1&O=A>.

———. 2007g. "Population and Growth Components (1851–2001 Censuses)." At: <www40.statcan.ca/l01/cst01/demo03.htm>.

———. 2007h. "Births and Birth Rate, by Province and Territory." At: <www40.statcan.ca/l01/cst01/demo04b.htm>.

———. 2007i. "Deaths and Death Rate, by Province and Territory." At: <www40.statcan.ca/l01/cst01/demo07b.htm>.

———. 2007j. "Population and Dwelling Counts, for Canada, Provinces and Territories, 2006 and 2001 Censuses—100% Data." At: <www.12.statcan.ca/english.census06/data/popdwell/Table.cfm?T=101>.

———. 2008a. "Aboriginal Peoples in Canada in 2006: Inuit, Métis and First Nations, 2006 Census," *The Daily*, 15 Jan. At: <www.statcan.gc.ca/daily-quotidien/080115/tdq080115-eng.htm>.

———. 2008b. "Language (including Language of Work)," Releases no. 4 and 6. At: <www12.statcan.gc.ca/census-recensement/2006/rt-td/lng-eng.cfm>.

———. 2008c. "Aboriginal Identity Population, 2006 Counts, Percentage Distribution, Percentage Change for Both Sexes, for Canada, Provinces and Territories—20% Sample Data," *Aboriginal Peoples Highlight Tables*. 2006 Census. Statistics Canada Catalogue no. 97–558–XWE2006002. At: <www12.statcan.ca/english/census06/data/highlights/aboriginal/index.cfm?Lang=E>.

———. 2008d. "The Evolving Linguistic Portrait, 2006 Census: The Proportion of Francophones and of French Continue to Decline." At: <www12.statcan.ca/census-recensement/2006/as-sa/97-555/p6-eng.cfm>.

———. 2008e. "Mandate, Responsibilities and Objectives." At: <www.statcan.gc.ca/about-apercu/mandate-mandat-eng.htm>.

———. 2009a. "Components of Population Growth, by Province and Territory, 2007–2008." At: http://www.statcan.gc.ca/tables-tableaux/sum-som/l01/cst01/demo33a-eng.htm.

———. 2009b. "Ethnocultural Portrait of Canada: Ethnic Origins, 2006 Counts, for Canada, Provinces and Territories." At: <www12.statcan.ca/english/census06/data/highlights/ethnic/pages/Page.cfm?Lang=E&Geo=PR&Code=01&Data=Count&Table=2&StartRec=1&Sort=3&Display=All&CSDFilter=5000>.

———. 2009c. "Table 2. Quarterly Demographic Estimates." At: <www.statcan.gc.ca/daily-quotidien/090326/t090326a2-eng.htm>.

———. 2009d. "Annual Demographic Estimates: Canada, Provinces and Territories." At: <www.statcan.gc.ca/pub/91-215-x/2008000/5200700-eng.htm>.

———. 2009e. "Labour Force, Employed and Unemployed, Numbers and Rates, by Province." At: www.statcan.gc.ca/tables-tableaux/sum-som/l01/cst01/labor07c-eng.htm.

———. 2009f. "Gross Domestic Product, Expenditure-based, by Province and Territory." At: www.statcan.gc.ca/tables-tableaux/sum-som/l01/cst01/econ15-eng.htm.

———. 2009g. "Population, Age Distribution and Median Age by Province and Territory, as of July 1, 2008." At: <www.statcan.gc.ca/daily-quotidien/090115/t090115c1-eng.htm>.

———. 2009h. *Aboriginal Peoples in Canada in 2006: Inuit, Métis and First Nations, 2006 Census*. 2006 Analysis Series. At: <www12.statcan.gc.ca/census-recensement/2006/as-sa/97-558/p2-eng.cfm#02>.

———. 2009i. "Births," *The Daily*, 22 Sept. At: <www.statcan.gc.ca/daily-quotidien/090922/dq090922b-eng.htm>.

———. 2009j. *Immigration in Canada: A Portrait of the Foreign-born Population, 2006 Census: Immigrants Came from Many Countries*. At: <www12.statcan.ca/census-recensement/2006/as-sa/97-557/figures/c2-eng.cfm>.

———. 2009k. *Canadian Demographics at a Glance: Population Growth in Canada*. At: <www.statcan.gc.ca/pub/91-003-x/2007001/figures/4129879-eng.htm>.

———. 2010a. "Ethnocultural Portrait of Canada," Highlight Tables, 2006 Census. At: <www12.statcan.ca/census-recensement/2006/dp-pd/hlt/97-562/sel_geo.cfm?Lang=E&Geo=PR&Table=2>.

———. 2010b. "Projections of the Diversity of the Canadian Population: Analysis of Results," 9 Mar. At: <www.statcan.gc.ca/pub/91-551-x/2010001/ana-eng.htm>.

———. 2011a. "Birth and Total Fertility Rate, by Province and Territory," 20 Dec. At: <www.statcan.gc.ca/tables-tableaux/sum-som/l01/cst01/hlth85b-eng.htm>.

———. 2011b. "Crude Birth Rate, Age-Specific and Total Fertility Rates (Live Births), Canada, Provinces and Territories," 19 Dec. At: <www5.statcan.gc.ca/cansim/pick-choisir?lang=eng&p2=33&id=1024505>.

———. 2011c. "Components of Population Growth, by Province and Territory," 28 Sept. At: <www.statcan.gc.ca/tables-tableaux/sum-som/l01/cst01/demo33a-eng.htm>.

———. 2011d. *Ethnic Origin Reference Guide, 2006 Census*. Catalogue no. 97–562–GWE200625, 5 May. At: <www12.statcan.gc.ca/census-recensement/2006/ref/rp-guides/ethnic-ethnique-eng.cfm>.

———. 2012a. "Population and Dwelling Counts for Canada, Provinces and Territories, 2011 and 2006 censuses," 11 Apr. At: <www12.statcan.gc.ca/census-recensement/2011/dp-pd/hlt-fst/pd-pl/Table-Tableau.cfm?LANG=Eng&T=101&S=50&O=A>.

———. 2012b. "Population and Dwelling Counts, for Population Centres, 2011 and 2006 Censuses," 11 Apr. At: <www12.statcan.gc.ca/census-recensement/2011/dp-pd/hlt-fst/pd-pl/Table-Tableau.cfm?LANG=Eng&T=801&SR=1&S=51&O=A&RPP=25&PR=0&CMA=0>.

———. 2012c. "Population of Census Metropolitan Areas," 3 July. At: <www.statcan.gc.ca/tables-tableaux/sum-som/l01/cst01/demo05a-eng.htm>.

———. 2012d. "The Canadian Population in 2011: Population Counts and Growth," 30 May. At: <www12.statcan.gc.ca/census-recensement/2011/as-sa/98-310-x/2011001/fig/fig2-eng.cfm>.

———. 2012e. "Population by Broad Age Groups and Sex, 2011 Counts for Both Sexes, for Canada, Provinces and Territories," 17 July. At: <www12.statcan.gc.ca/census-recensement/2011/dp-pd/hlt-fst/as-sa/Pages/highlight.cfm?TabID=1&Lang=E&Asc=1&PRCode=01&OrderBy=999&Sex=1&View=1&tableID=21&queryID=1>.

———. 2012f. "Figure 2.4, "Number of Children Age 14 and Under and Persons Aged 65 and Over, Canada, 1921 to 2011," Visual Census—Age and Sex, Canada, 5 July. At: <www12.statcan.gc.ca/census-recensement/2011/dp-pd/vc-rv/index.cfm?Lang=eng>.

———. 2012g. "Population by Broad Age Groups and Sex, Counts, Including Median Age, 1921 to 2011 for Both Sexes—Canada," 25 July. At: <www12.statcan.gc.ca/census-recensement/2011/dp-pd/hlt-fst/as-sa/Pages/highlight.cfm?TabID=1&Lang=E&PRCode=01&Asc=0&OrderBy=1&Sex=1&View=1&tableID=22>.

———. 2012h. "Labour Force Survey Estimates (LFS), by North American Industrial Classification System (NAICS), Sex and Age Group, Unadjusted for Seasonability, May 1977 to June 2012," 6 July. CANSIM Table 282–007. At: <www5.statcan.gc.ca/cansim/a26?lang=eng&retrLang=eng&id=2820007&pattern=282-0001..282-0042&tabMode=dataTable&srchLan=-1&p1=-1&p2=-1>.

———. 2012i. "Canada Economic Accounts First Quarter 2012 and March 2012," *The Daily*, 1 June, Chart 1. At: <www.statcan.gc.ca/daily-quotidien/120601/dq120601a-eng.htm>.

———. 2012j. *Immigrant Languages in Canada*, Oct. At: <www12.statcan.gc.ca/census-recensement/2011/as-sa/98-314-x/98-314-X2011003_2-eng.pdf>.

———. 2012k. "Population by Mother Tongue and Age Groups (Total), 2011 Counts, for Canada, Provinces and Territories," 24 Oct. At: <www12.statcan.gc.ca/census-recensement/2011/dp-pd/hlt-fst/lang/Pages/highlight.cfm?TabID=1&Lang=E&Asc=1&PRCode=01&OrderBy=999&View=1&tableID=401&queryID=1&Age=1>.

———. 2013a. "Population and Growth Rate of Metropolitan and Non-Metropolitan Canada, 2006 and 2011," 24 Jan. At: <www12.statcan.gc.ca/census-recensement/2011/as-sa/98-310-x/2011001/tbl/tbl2-eng.cfm>.

———. 2013b. "Immigration", Immigration and Ethnocultural Diversity in Canada, NHS document 99-010-x. 10 May. At: <http://www12.statcan.gc.ca/nhs-enm/2011/as-sa/99-010-x/99-010-x2011001-eng.cfm>.

———. 2013c. "Most common ethnic origins", Immigration and Ethnocultural Diversity in Canada. NHS document 99-010-x. 10 May. At: <http://www12.statcan.gc.ca/nhs-enm/2011/as-sa/99-010-x/99-010-x2011001-eng.cfm>.

———. 2013d. *Focus on Geography Series, 2011 Census: Province of Quebec*, 15 Jan. At: <www12.statcan.gc.ca/census-recensement/2011/as-sa/fogs-spg/Facts-pr-eng.cfm?Lang=Eng&GK=PR&GC=24>.

———. 2013e "Number and distribution of population reporting an Aboriginal identity and percentage of Aboriginal people in the population, Canada, provinces and territories, 2011", *Aboriginal Peoples in Canada: First Nations People, Métis and Inuit.* National Household Survey document 99-011-x. Table 2. 7 May. At: <http://www12.statcan.gc.ca/nhs-enm/2011/as-sa/99-011-x/2011001/tbl/tbl02-eng.cfm>.

Steyn, Mark, 2006. *America Alone: The End of the World As We Know It*. Washington: Regnery Publishing.

Taylor, Charles. 1994. *Multiculturalism: Examining the Politics of Recognition*. Princeton, NJ: Princeton University Press.

Trewartha, G.T., A.H. Robinson, and E.H. Hammond. 1967. *Elements of Geography*, 5th edn. New York: McGraw-Hill.

Walks, R. Alan, and Larry S. Bourne. 2006. "Ghettos in Canada's Cities? Racial Segregation, Ethnic Enclaves and Poverty Concentration in Canadian Urban Areas," *Canadian Geographer* 50, 3: 273–97.

Woods, Alan. 2006. "Dual Citizenship Faces Review," *National Post*, 21 Sept. At: <www.canada.com/nationalpost/news/story.html?id=fb2d75ab-8880-4945-8537-1508186a4964&k=61921>.

CHAPTER 5

Anderson, Fiona. 2006. "Dropping Lumber Demand Threatens Canadian Exporters," *StarPhoenix* (Saskatoon), 23 Sept., D7.

Arcand, Alan, Mario Lefebvre, Jane McIntyre, Greg Sutherland, and Robin Wiebe. 2012. *Metropolitan Outlook 1: Economic Insights into 13 Canadian Metropolitan Economies: Winter 2012*. Conference Board of Canada, Jan. At: <www.conferenceboard.ca/e-library/abstract.aspx?did=4606>.

Atkin, David. 2000. "In a Hot-Wired World, Everything's Personal," *National Post*, 15 Nov., C1, C6–7.

Avery, Simon. 2006. "In Ottawa, It's a Low-Budget Tech Rebound," *Globe and Mail*, 8 Nov. At: <www.theglobeandmail.com/servlet/story/RTGAM.20061108.wottawaa1108/BNStory/Technology/home>.

Bagnall, James. 2003. "Ottawa Slipping Off the Tech Map," *National Post*, 22 Aug., FP5.

Beatty, Stephen. 2009. "Speech: Toyota Canada's Managing Director at UofG," 10 Mar. At: <news.guelphmercury.com/printArticle/450846>.

Bernard, André. 2009. "Trends in Manufacturing Employment," *Perspectives*, Feb. Statistics Canada Catalogue no. 75-001-X. At: <www.statcan.gc.ca/pub/75-001-x/2009102/pdf/10788-eng.pdf>.

Bertin, Oliver, and Greg Keenan. 2003. "Algoma to Axe 600, Cut Costs in Latest Bid to Stay Profitable," *Globe and Mail*, 15 May, B1.

Bone, Robert M. 2012. *The Canadian North: Issues and Challenges*, 4th edn. Toronto: Oxford University Press.

Brean, Joseph. 2003. "Outbreak Predicted to Cost City $1-billion," *National Post*, 3 May, A6.

Brent, Paul. 2002. "Car Output Still Higher in Canada," *National Post*, 15 Aug., FP5.

———. 2003. "Daimler Calls Off Windsor Plant Plans," *National Post*, 23 May, FP5.

Britton, John N.H. 1996. "High-Tech Canada," in Britton, ed., *Canada and the Global Economy: The Geography of Structural and Technological Change*. Montréal and Kingston: McGill-Queen's University Press.

Campion-Smith, Bruce. 2006. "Mayor Fumes at Kashechewan Move," *Toronto Star*, 9 Nov. At: <www.thestar.com/NASApp/cs/ContentServer?pagename=thestar/Layout/Article_Type1&c=Article&cid=1163026213121&call_pageid=968332188774&col=968350116467>.

Canadian Auto Workers Union. 2012. *Re-thinking Canada's Auto Industry: A Policy Vision to Escape the Race to the Bottom*, Apr. At: <https://d3n8a8pro7vhmx.cloudfront.net/caw/pages/29/attachments/original/1335189435/554AutoPolicyDocumentweb.pdf?1335189435>.

CBC News. 2006a. "Caledonia Land Claim: Historical Timeline," 1 Nov. At: <www.cbc.ca/news/background/caledonia-landclaim/historical-timeline.html>.

———. 2006b. "Ford to Close Ontario Plant, Cut 10,000 More Salaried Jobs," 15 Sept. At: <www.cbc.ca/money/story/2006/09/15/ford-cuts.html>.

———. 2009. "Ontario Pushes Electric Cars as Auto-Sector Boost," 15 July. At: <www.cbc.ca/canada/story/2009/07/15/ont-electric-cars511.html>.

———. 2010. "New Canada–U.S. Border Route Dispute Ends," 9 Apr. At: <www.cbc.ca/canada/windsor/story/2010/04/09/wdr-border-access-route-100409.html?ref=rss>.

———. 2012. "$1 Billion Windsor–Detroit Bridge Deal Struck," 15 June. At: <www.cbc.ca/news/business/story/2012/06/14/wdr-prime-minister-bridge-windsor-detroit.html>.

Chartrand, Terre. 2012. "Ontario Premier Celebrates Waterloo Region Tech Sector Growth at the Communitech Hub," *Communitech*, 2 Feb. At: <www.communitech.ca/ontario-premier-celebrates-waterloo-region-tech-sector-growth-at-the-communitech-hub/>.

Chevralier, Patrick. 2009. "Gold," *Canadian Mineral Yearbook 2005*. At: <www.nrcan-rncan.gc.ca/mms-smm/busi-indu/cmy-amc/2005revu/gol-eng.htm>.

Church, Elizabeth. 2012. "MIT to Honour Two Professors Transforming the World from Toronto," *Globe and Mail*, 21 Aug., A5.

Cleroux, N., and M. Woods. 2012. "The Plasco Energy Group's Plasma Gasification System," 22 May. At: <wiki.telfer.uottawa.ca/ci-wiki/index.php/The_Plasco_Energy_Group's_Plasma_Gasification_System>.

Cole, Trevor. 2009. "Hamilton's Dead. Or Is It?" *Globe and Mail*, 28 Aug. At: <www.theglobeandmail.com/report-on-business/rob-magazine/hamiltons-dead-or-is-it/article1264739/>.

Courchene, Thomas J., with Colin R. Telmer. 1998. *From Heartland to North American Region State: The Social, Fiscal and Federal Evolution of Ontario*. Monograph Series on Public Policy, Centre for Public Management. Toronto: Faculty of Management, University of Toronto.

Cross, Philip. 2009. "2008 in Review," *Canadian Economic Observer*. At: <www.statcan.gc.ca/pub/11-010-x/2009004/part-partie3-eng.htm>.

CTV News. 2006. "Canada, U.S. Agree on Softwood Framework," 26 Apr. At: <www.ctv.ca/servlet/ArticleNews/story/CTVNews/20060426/softwood_deal_060426/20060426?hub=Canada>.

Darling, Graham. 2007. "Land Claims and the Six Nations in Caledonia Ontario," Centre for Constitutional Studies. At: <www.law.ualberta.ca/centres/ccs/Current-Constitutional-Issues/Land-Claims-and-the-Six-Nations-in-Caledonia-Ontario.php>.

Detroit River International Crossing. 2009. *Forum for the Private Sector—April 23, 2009*. At: <www.partnershipborderstudy.com/pdf/2009-04-23DRICForumBooklet.pdf>.

Dickason, Olive Patricia, with David T. McNab. 2009. *Canada's First Nations: A History of Founding Peoples from Earliest Times*, 4th edn. Toronto: Oxford University Press.

Dicken, Peter. 1992. *Global Shift: The Internationalization of Economic Activity*, 2nd edn. New York: Guilford Press.

Drummond, Don, and Derek Burleton. 2008. "Time for a Vision of Ontario's Economy." At: http://www.td.com/document/PDF/economics/special/td-economics-special-db0908-ont.pdf.

Environment Canada. 2006. "Information on Greenhouse Gas Sources and Sinks." At: <www.ec.gc.ca/pdb/ghg/onlineData/dataAndReports_e.cfm>.

Evans, Mark. 2006. "The 'Quiet Boom': High-Tech Turnaround Drives Ottawa, Waterloo Growth," *National Post*, 31 Jan. At: <www.canada.com/nationalpost/financialpost/story.html?id=2a0b745b-eded-4269-af0f-f454d25a8ca6>.

Export Development Canada. 2007. "Commodity Price Update," 28 May. At: <www.edc.ca/english/docs/commpric_e.pdf>.

Ferley, Paul, and Robert Hogue. 2009. "Has the Storm Passed?" *Provincial Outlook*, June. Royal Bank of Canada. At: <www.rbc.com/economics/market/pdf/provfcst.pdf>.

Flavelle, Dana. 2012. "Canadian Auto Workers at Ford Accept Contract," *The Star.com*, 23 Sept. At: <www.thestar.com/business/2012/09/23/canadian_auto_workers_at_ford_accept_contract.html>.

Florida, Richard. 2002. *The Rise of the Creative Class: And How It's Transforming Work, Leisure, Community and Everyday Life*. New York: Basic Books.

———. 2012. *The Rise of the Creative Class—Revisited*. New York: Basic Books.

——— and S. Jackson. 2010. "Sonic City: The Evolving Economic Geography of the Music Industry," *Journal of Planning, Education and Research* 29, 3: 310–21.

———, C. Mellander, and K. Stolarick. 2010. "Talent, Technology, and Tolerance in Canadian Regional Development," *Canadian Geographer* 54, 3: 277–304.

French, Cameron. 2007. "U.S. Security-related Delays at Border Cost Billions, Canadian Producers Say," *International Herald Tribune*. At: www.nytimes.com/2007/03/05/business/worldbusiness/05iht-canada.4799532.html?_r=0.

Gomes, Carlos. 2009. "Auto Lending Conditions Improve in North America," *Global Auto Report*, 25 June. Scotia Economics. At: <www.scotiacapital.com/English/bns_econ/bns_auto.pdf>.

———. 2010. "Demographic Boom Turning To Bust for North American Auto Sales," *Global Auto Report*, 26 Feb. Scotiabank Group. At: <www.scotiacapital.com/English/bns_econ/bns_auto.pdf>.

Hoffman, Andy. 2009. "At a Secret Facility in Sudbury, 27 Highly Skilled Workers Cut and Polish Diamonds as Part of Ontario's Ambition to Become a Key Line in the Global Diamond Trade: All Are from Vietnam," *Globe and Mail*, 15 Oct., B1, B6.

Hracs, Brian J., Jill L. Grant, Jeffry Haggett, and Jesse Morton. 2011. "Tale of Two Scenes: Civic Capital and Retaining Musical Talent in Toronto and Halifax," *Canadian Geographer* 55, 3: 365–82.

Indian and Northern Affairs Canada (INAC) 2009."Chronology of Events at Caledonia." At: <www.ainc-inac.gc.ca/ai/mr/is/eac-eng.asp>.

Industry Canada. 2012. *North American Production Quarterly*, 24 Feb. At: <www.ic.gc.ca/eic/site/auto-auto.nsf/eng/am02360.html>.

International Great Lakes Study. 2012. *Final Report to the International Joint Commission: Lake Superior Regulation: Addressing Uncertainty in Upper Great Lakes Water Levels*, Mar. At: <www.ijc.org/iuglsreport/wp-content/report-pdfs/Lake_Superior_Regulation_Full_Report.pdf>.

JAMA Canada. 2010. "Monthly and Annual Statistics for JAMA Canada Member Companies," 18 Mar. At: <www.jama.ca/jamastats/>.

Keenan, Greg. 2003. "Layoffs Mount among Auto Makers," *Globe and Mail*, 26 Mar., B4.

———. 2007. "Chrysler to Axe 20% of Canadian Work Force," *Globe and Mail*, 8 Feb. At: <www.theglobeandmail.com/servlet/story/RTGAM.20070208.wchrysler08/BNStory/Business/home>.

———. 2009. "Windsor Gets Bailout Dividend: New Chrysler Minivan," *Globe and Mail*, 20 Aug. At: <www.theglobeandmail.com/report-on-business/windsor-gets-bailout-dividend-new-chrysler-minivan/article1257692/>.

———. 2012. "In Tough Times, Cut Auto Workers' Wages, Report Urges," *Globe and Mail*, 21 Mar., B3.

——— and Steven Chase. 2013. "Auto Makers Get $250-Million as Ottawa Renews Innovation Fund," *Globe and Mail*, 4 Jan. At: <www.theglobeandmail.com/report-on-business/economy/auto-makers-get-250-million-as-ottawa-renews-innovation-fund/article6934817/>.

Kowaluk, Russell, and Rob Larmour. 2009. "Manufacturing: The Year 2008 in Review," Statistics Canada. At: <www.statcan.gc.ca/bsolc/olc-cel/olc-cel?lang=eng&catno=11-621-M2009077>.

Leeder, Jessica. 2012. "Chefs Cook Up Plan to Save Farmland," *Globe and Mail*, 15 Oct. A5.

Leslie, Keith. 2012. "TransCanada Inks Deal to Operate Controversial Ontario Power Plant," *Globe and Mail*, 17 Dec. At: <www.theglobeandmail.com/news/national/transcanada-inks-deal-to-operate-controversial-ontario-power-plant/article6480885/>.

McCarthy, Shawn. 1997. "Ottawa Trades Bureaucrats for Bytes," *Globe and Mail*, 30 Aug., B1, B4.

———. 2013. "As Cost Pressures Mount, Ontario's Wholesale Power Prices to Soar," *Globe and Mail*, 30 Jan. At: <www.theglobeandmail.com/report-on-business/industry-news/energy-and-resources/as-cost-pressures-mount-ontarios-wholesale-power-prices-set-to-soar/article8025566/>.

Marr, Lisa Grace. 2010. "Hamilton Losing Siemens Turbine Jobs," *Toronto Star*, 12 Mar. At: <www.thestar.com/business/article/778745-hamilton-losing-siemens-turbine-jobs>.

Matteo, Livio Di. 2006. "Breakaway Country," *National Post*, 6 Sept., FP19.

National Post. 2007. "Ipperwash Report," 1 June, A14.

Natural Resources Canada. 2006. *Mill Closures and Mill Investments in the Canadian Forest Sector: The State of Canada's Forests, 2005–2006*. At: <www.nrcan.gc.ca/cfs-scf/national/what-quoi/sof/sof06/mergers_e.html>.

———. 2008. "Mineral Production of Canada, by Province and Territory." At: <mmsd.mms.nrcan.gc.ca/stat-stat/prod-prod/PDF/2008p.pdf>.

———. 2010a. *Price Monitor*, 17 Mar. At: <canadaforests.nrcan.gc.ca/article/pricemonitor>.

———. 2010b. "The Global Recession Reduced Canada's Mineral Production in 2009," *Information Bulletin*, Mar. At: <mmsd.mms.nrcan.gc.ca/stat-stat/prod-prod/PDF/ib2010_e.pdf>.

———. 2012a. "Weekly Softwood Lumber Prices in North America," 15 Aug. At: <cfs.nrcan.gc.ca/pages/249#softwood>.

———. 2012b. "Mining Statistics," 23 Aug. At: <mmsd.mms.nrcan.gc.ca/stat-stat/prod-prod/2011-eng.aspx>.

Norcliffe, Glen. 1996. "Mapping Deindustrialization: Brian Kipping's Landscape of Toronto," *Canadian Geographer* 40, 3: 266–73.

OICA (International Organization of Motor Vehicle Manufacturers). 2009. "Production Statistics." At: <oica.net/category/production-statistics/>.

Ontario. 2012. "Ontario Trade Factsheet," 12 Mar. At: <www.sse.gov.on.ca/medt/ontarioexports/en/Pages/tradefactsheet_ontario.aspx>.

Ontario. Ministry of Natural Resources. 2012. "The Value of Ontario's Forest Sector." At: <www.mnr.gov.on.ca/en/Business/Forests/2ColumnSubPage/STEL02_167493.h>.

Ontario Medical Association (OMA). 2005. *The Illness Costs of Air Pollution in Ontario: A Summary of Findings*. Toronto. At: www.oma.org/Resources/Documents/e2005HealthAndEconomicDamageEstimates.pdf.

Potter, Mitch. 2010. "Secretive Family and a Vital U.S. Link Create International Span of Mystery," *Toronto Star*, 16 Feb, A1, A17.

Preston, Valerie, and Lucia Lo. 2000. "Canadian Urban Landscape Examples—21: 'Asian Theme' Malls in Suburban Toronto: Land Use Conflicts in Richmond Hill," *Canadian Geographer* 44, 2: 182–90.

Preville, Philip. 2012. "Secrets to Being a Global Leader," *Globe and Mail*, 6 Sept., B8.

Royal Bank of Canada. 2012. *Provincial Outlook: Ontario*, Sept. At: <www.rbc.com/newsroom/pdf/provfcst-09-2012.pdf>.

Scotiabank. 2013. *Global Forecast Update*, 28 Feb. At: <www.gbm.scotiabank.com/English/bns_econ/forecast.pdf>.

Shiell, Leslie, and Robin Somerville. 2012. *Bailouts and Subsidies: The Economics of Assisting the Automotive Sector in Canada*, IRPP Study No. 28, Mar. At: <www.irpp.org/pubs/IRPPstudy/IRPP_Study_no28.pdf>.

Shkilnyk, A.M. 1985. *A Poison Stronger Than Love: The Destruction of an Ojibwa Community*. New Haven: Yale University Press.

Shufelt, Tim. 2012. "Ontario Moves to Open Up Far North with $5.1-billion Chromite Deal," *National Post*, 9 May. At: <business.financialpost.com/2012/05/09/ontario-movesto-open-up-far-north-with-5-1-billion-chromite-deal/>.

Spears, John. 2013. "Ontario Coal-burning Power Plants to Close This Year," *The Star.com*, 28 Feb. At: <www.thestar.com/business/2013/01/10/ontario_coalburning_power_plants_to_close_this_year.html>.

Statistics Canada. 2002. "2001 Census: Population and Dwelling Counts, for Census Metropolitan Areas and Census Agglomerations, 2001 and 1996 Censuses," 16 July. At: <www.statcan.ca/English/IPS/Data/93F0050XCB2001013.htm>.

———. 2003. "Exports of Goods on a Balance-of-Payments Basis," *Canadian Statistics, International Trade*. At: <www.statcan.ca/english/Pgdb/gblec04.htm>.

———. 2006a. "Exports of Goods on a Balance-of-Payments Basis," *Canadian Statistics, International Trade*. At: <www40.statcan.ca/l01/cst01/gblec04.htm>.

———. 2006b. "Canadian Statistics: Distribution of Employed People, by Industry, by Province." At: <www40.statcan.ca/l01/cst01/labor21c.htm>.

———. 2007a. "Population Urban and Rural, by Province and Territory: 1851 to 2001." At: <www40.statcan.ca/l01/cst01/demo62g.htm>.

———. 2007b. "Census of Agriculture Counts 57,211 Farms in Ontario," *2006 Census of Agriculture*. At: <www.statcan.ca/english/agcensus2006/media_release/on.htm>.

———. 2007c. "Population and Dwelling Counts, for Census Metropolitan Areas and Census Agglomerations, 2006 and 2001 Censuses—100% Data." At: <www12.statcan.ca/english/census06/data/popdwell/Table.cfm?T=201&S=3&O=D&RPP=150>.

———. 2007d. "Study: Canada's Changing Auto Industry," *The Daily*, 17 May. At: <www.statcan.ca/Daily/English/070517/d070517c.htm>.

———. 2009a. "Provincial and Territorial Economic Accounts," *The Daily*, 27 Apr. At: <www.statcan.gc.ca/daily-quotidien/090427/dq090427a-eng.htm>.

———. 2009b. "Canada's Population Estimates: Quarterly Population Estimate," Table 2, 23 Dec. At: <www.statcan.gc.ca/daily-quotidien/091223/t091223b2-eng.htm>.

———. 2009c. "Chart 21.1 (data) Unemployment Rate," 16 Jan. At: <www41.statcan.gc.ca/2008/2621/grafx/htm/ceb2621_000_1-eng.htm#table>.

———. 2009d. "Distribution of Employed People, by Industry, by Province." At: <www40.statcan.ca/l01/cst01/labor21c-eng.htm>.

———. 2009e. "Net Interprovincial Migration," *Quarterly Demographic Estimates*, Table 2-12. At: <www.statcan.gc.ca/pub/91-002-x/2009001/t032-eng.htm>.

———. 2009f. "Toronto: Place of Birth for the Immigrant Population," *Immigration and Citizenship Highlight Tables, 2006 Census*. At: <census2006.ca/census-recensement/2006/dp-pd/hlt/97-557/T404-eng.cfm?SR=1>.

———. 2009g. "Population and Demography." At: <www41.statcan.gc.ca/2008/3867/ceb3867_000-eng.htm>.

———. 2010. "Labour Force, Employed and Unemployed, Numbers and Rates, by Province," 29 Jan. At: http://www.statcan.gc.ca/tables-tableaux/sum-som/l01/cst01/labor07b-eng.htm.

———. 2011. "Province Gives Shine to Ontario Diamonds." At: <news.ontario.ca/mndmf/en/2011/12/province-gives-shine-to-ontario-diamonds.html>.

———. 2012a. "Labour Force Survey, July 2012," *The Daily*, 10 Aug. At: <www.statcan.gc.ca/daily-quotidien/120810/dq120810a-eng.htm>.

———. 2012b. "Unemployment Rate, Canada, Provinces, Health Regions (2011 Boundaries) and Peer Groups," 24 May. At: <www5.statcan.gc.ca/cansim/pick-choisir?lang=eng&p2=33&id=1095324>.

———. 2012c. "Population by Broad Age Groups and Sex, 2011 Counts for Both Sexes, for Canada, Provinces and Territories, and Census Divisions," 25 July. At: <www12.statcan.gc.ca/census-recensement/2011/dp-pd/hlt-fst/as-sa/Pages/highlight.cfm?TabID=1&Lang=E&Asc=1&Orde

rBy=1&Sex=1&View=1&tableID=21&queryID=5&PRCode=35>.

———. 2012d. "Census of Agriculture," *The Daily*. 10 May. At: <www.statcan.gc.ca/daily-quotidien/120510/dq120510a-eng.htm>.

———. 2012e. "Distribution of Employed People, by Industry, by Province," 1 June. At: <www.statcan.gc.ca/tables-tableaux/sum-som/l01/cst01/labor21a-eng.htm>.

———. 2012f. "Export of Goods on a Balance-of-Payments Basis, by Product," 10 May. At: <www.statcan.gc.ca/tables-tableaux/sum-som/l01/cst01/gblec04-eng.htm>.

———. 2012g. "Population and Dwelling Counts, for Census Metropolitan Areas and Census Agglomerations, 2011 and 2006 Censuses," 11 Apr. At: <www12.statcan.gc.ca/census-recensement/2011/dp-pd/hlt-fst/pd-pl/Table-Tableau.cfm?LANG=Eng&T=201&S=3&O=D&RPP=150>.

———. 2013. "Number and distribution of population reporting an Aboriginal identity and percentage of Aboriginal people in the population, Canada, provinces and territories, 2011," *Aboriginal Peoples in Canada: First Nations People, Métis and Inuit*. National Household Survey document 99-011-x. Table 2. 7 May. At:<http://www12.statcan.gc.ca/nhs-enm/2011/as-sa/99-011-x/2011001/tbl/tbl02-eng.cfm>.

Transport Canada. 2009. "Detroit River International Crossing Project Receives Environmental Approval," 3 Dec. At: <www.tc.gc.ca/eng/mediaroom/releases-2009-h166e-5754.htm>.

CHAPTER 6

Alcoa. 2013. "Alcoa's Project in Baie-Comeau Looking Ahead with Confidence." At: <www.alcoaprojets.com/en>.

Aubin, Henry. 2013. "Henry Aubin: Taxes, Bill 101 Drive People Away," *The Gazette* (Montréal), 19 Feb. At: <www.montrealgazette.com/news/Henry+Aubin+Taxes+Bill+drive+people+away/7981947/story.html#ixzz2Mn1IOTOD>.

Blackwell, Richard. 2008. "Quebec Snags New Silicon Plant," *Globe and Mail*, 25 Aug., B1.

Blatchford, Andy. 2012. "Quebec Fracking Ban? PQ Eyes Banning Shale Gas, Shutting Nuclear Reactor, Ending Asbestos Industry," Canadian Press, 20 Sept. At: <www.huffingtonpost.ca/2012/09/20/quebec-fracking-ban_n_1900807.html>.

Bombardier. 2010. "Media Centre." At: <www.bombardier.com/en/corporate/media-centre/press-releases>.

Bone, Robert M. 2012. *The Canadian North: Issues and Challenges*, 4th edn. Toronto: Oxford University Press.

Bordeleau, Stéphane. 2011. "How It Works: Where Canada's Surplus Energy Goes," CBC News, 31 Mar. At: <www.cbc.ca/news/canada/story/2011/03/17/f-power-2020-provincial-energy-export.html>.

Bouchard, Gérard, and Charles Taylor. 2008. *Building the Future: A Time for Reconciliation (Abridged Report)*. Québec: Commission de consultation sur les pratiquesd'accommodementreliées aux differences culturelles. At: <www.accommodements.qc.ca/index-en.html>.

Bouchard, Lucien, et al. 2006. Manifest—Pour un Québec lucide. At: <www.pourunquebeclucide.com/cgi-cs/cs.waframe.content?topic=28226&lang=1>.

Bumsted, J.M. 2007. *A History of the Canadian Peoples*, 3rd edn. Toronto: Oxford University Press.

Canada. 2006. "Net Merchantable Volume of Roundwood Harvested by Category and Province/Territory, 1940–2005," National Forestry Database Program, Table 5.1. At: <www.nfdp.ccfm.org/compendium/products/tables_index_e.php>.

Canada's Innovation Leaders. 2006. "Canada's Top 20 Research Communities 2006," *National Post*, Research infosource supplement, 3 Nov., 7.

Canadian Dairy Information Centre. 2012. "Dairy Cash Receipts by Province – 2011," 28 June. At: <www.dairyinfo.gc.ca/index_e.php?s1=dff-fcil&s2=farm-ferme&s3=rev-dep>.

Canadian Federation of Independent Businesses. 2012. "Public Sector Debt Clock," CFIB Economics, 12 Sept. At: <www.cfib-fcei.ca/debt-clock-en.html>.

Canadian Hydropower Association. 2009a. *Hydropower in Canada: Past, Present and Future*. At: <www.canhydropower.org/hydro_e/pdf/hydropower_past_present_future_en.pdf>.

———. 2009b. "The Canadian Hydropower Association Applauds the Launch of the Romaine Hydroelectric Project." At: <www.canhydropower.org/hydro_e/pdf/2009-05-14_La_Romaine_news_release.pdf>.

Canadian Press. 2012a. "2 Former SNC-Lavalin Employees Charged with Corruption," CBC News, 22 June. At: <www.cbc.ca/news/business/story/2012/06/22/toronto-snc-lavalin-corruption.html>.

———. 2012b. "Cree Nation, Quebec Sign Land Management Deal for Province's North," *Globe and Mail*, 24 July. At: <www.theglobeandmail.com/news/politics/cree-nation-quebec-sign-land-management-deal-for-provinces-north/article4438763/>.

CBC News. 2010. "Québec, Vermont Sign Hydro Deal," 12 Aug. At: <www.cbc.ca/news/business/story/2010/08/12/qc-vermont-power-deal-hydro.html>.

Cousineau, Sophie. 2012a. "Quebec Premier Floats Idea of Tax Credits to Attract Plan Nord Mining Investment," *Globe and Mail*, 18 Oct., B5.

———. 2012b. "Decision to Close Nuclear Plant Expected to Cost Quebec $1.3-Billion," *Globe and Mail*, 6 Oct., A15.

Cross, Philip. 2009. "2008 in Review," *Canadian Economic Observer*. At: <www.statcan.gc.ca/pub/11-010-x/2009004/part-partie3-eng.htm>.

CTV News. 2013. "Bill 14 Not Strong Enough, Say Language Advisers," 6 Mar. At: <montreal.ctvnews.ca/bill-14-not-strong-enough-say-language-advisors-1.1184163>.

Department of Finance. 2012. *Your Tax Dollar: 2010–2011 Fiscal Year*, 18 Jan. At: <www.fin.gc.ca/tax-impot/2011/html-eng.asp>.

Deveau, Scott. 2012. "Bombardier Sales Should Improve on Chinese Rail Contract Changes," *National Post*, 5 Sept. At: <business.financialpost.com/2012/09/05/bombardier-sales-should-improve-on-china-rail-contract-changes/>.

Dougherty, Kevin. 2007. "Quebec the First to Announce Carbon Tax," *National Post*, 7 June, A1, A10.

Duhaime, Gérard, Nick Bernard, and Robert Comtois. 2005. "An Inventory of Abandoned Mining Exploration Sites in Nunavik, Canada," *Canadian Geographer* 49, 3: 260–71.

Ferrao, Vincent. 2006. "Recent Changes in Employment by Industry," *Perspectives on Labour and Income* (Statistics Canada) 7, 1. At: <www.statcan.ca/english/freepub/75-001-XIE/10106/art-1.htm>.

Fischer, David Hackett. 2008. *Champlain's Dream*. New York: Simon & Schuster.

Fisheries and Oceans Canada. 2012. "Aquatic Species at Risk—Beluga Whale (St. Lawrence Estuary)," 17 Apr. At: <www.dfo-mpo.gc.ca/species-especes/species-especes/belugaStLa-eng.htm>.

George, Jane. 2009. "Makivik Nickel Share Dwindles to $6.8 Million," *Nunatsiaq News*, 19 June. At: <www.nunatsiaqnews.com/archives/2009/906/90619/news/nunavik/90619_2260.html>.

Gray, Jeff. 2012. "SNC-Lavalin Sees Profits Dip, Warns of Lower Year-End Results," *Globe and Mail*, 3 Aug. At: <www.theglobeandmail.com/globe-investor/snc-lavalin-sees-profit-dip-warns-of-lower-year-end-results/article4459714/>.

Hodgins, Bruce W., Richard P. Bowles, James L. Hanley, and George A. Rawlyk. 1974. *Canadiens, Canadians and Québécois*. Scarborough, Ont.: Prentice-Hall.

Hodgson, Glen. 2012. "Quebec's Economic Future: A Hard Road Ahead", Conference Board of Canada. 6 Sept. At: <www.conferenceboard.ca/economics/hot_eco_topics/default/12-09-06/quebec_s_economic_future_a_hard_road_ahead.aspx>.

Hornig, James F., ed. 1999. *Social and Environmental Impacts of the James Bay Hydroelectric Project*. Montréal and Kingston: McGill-Queen's University Press.

Human Resources and Skills Development Canada. 2012. "The Manufacturing Sector in Canada and Québec," *Labour Market Bulletin: Quebec*, 8 Aug. At: <www.rhdcc-hrsdc.gc.ca/eng/workplaceskills/labour_market_information/bulletins/qc/qc-lmb-2012summer.shtml>.

Hydro-Québec. 2006a. "Eastmain-1 Hydroelectric Development." At: <www.hydroquebec.com/eastmain1/en/batir/fiche_7.html>.

———. 2006b. "Eastmain-1 Hydroelectric Development." At: <www.hydroquebec.com/eastmain1/en/batir/etapes_photos.html?list=4>.

———. 2009a. "Eastmain-1-A Powerhouse." At: <www.hydroquebec.com/rupert/en/batir/etapes_photos.html?list=7>.

———. 2009b. "Projet de la Romaine." At: <www.hydroquebec.com/romaine/projet/galerie.html>.

———. 2009c. *Romaine Complex: Project Information*, Feb. At: <www.hydroquebec.com/romaine/pdf/doc_info_2009_en.pdf>.

———. 2013. *Eastman-1-A/Sarcelle/Rupert Project*. At: <www.hydroquebec.com/rupert/en/index.html>.

Jang, Brent. 2012. "U.S. Orders Stack Up for B.C. Lumber," *Globe and Mail*, 18 Sept., B1, B6.

Jarislowsky, Stephen. 2012. "The French Myth Isolates Quebec," *Globe and Mail*, 28 Sept., A19.

Jenish, D'Arcy. 2009. "Inland Superhighway," *Canadian Geographic* 129, 4: 34–48.

Kativik Regional Government. 2009. *Proposed Timeline for the Creation of the Nunavik Regional Government*. At: <www.nunavikgovernment.ca/en/documents/NRG_Timeline_En_Oct_09.pdf>.

Keenan, Greg. 2013. "Bombardier Targets Big Rivals as C Series Makes Its Debut," *Globe and Mail*, 8 Mar., B1, B4.

Koplow, Douglas. 1994. "Federal Energy Subsidies and Recycling: A Case Study," *Resource Recycling*, Nov. At: <www.p2pays.org/ref/06/05940.pdf>.

Lagacé, Patrick. 2012. "The Case For," *Globe and Mail*, 5 May, A8.

La Presse. "Élections Québec 2012," 7 Sept. At: <www.lapresse.ca/actualites/elections-quebec-2012/>.

Legault, Josée. 2009. "Sabia Meeting Shows How Much Quebec Inc. Has Changed," *The Gazette* (Montréal), Apr. At: <www.vigile.net/Sabia-meeting-shows-how-much>.

Leger, Kathryn. 2000. "Biochem Sold for $5.9-Billion," *National Post*, 12 Dec., C1, C6.

Lemay, Martin. 2010. "Francophones Have Reason to Be Paranoid," *The Gazette* (Montreal), 30 Apr. At: <www.vigile.net/Francophones-have-reason-to-be>.

Lucas, C.P. 1912. *Lord Durham's Report of the Affairs of British North America*, vol. 1. Oxford: Clarendon Press.

McArthur, Greg, and Graeme Smith. 2012. "SNC-Lavalin's Gadhafi Disaster: The Inside Story," *Globe and Mail*, 27 Sept. At: <www.theglobeandmail.com/report-on-business/rob-magazine/snc-lavalins-gadhafi-disaster-the-inside-story/article4570115/?page=all>.

McCutcheon, Sean. 1991. *Electric Rivers: The Story of the James Bay Project*. Montréal: Black Rose Books.

McKenna, Barrie. 2012. "U.S. Cities Boosting Transport Makers," *Globe and Mail*, 19 Sept., B5.

Marotte, Bertrand. 2009a. "With Glowing Hearts, an Algerian City Rises," *Globe and Mail*, 9 July. At: <www.theglobeandmail.com/globe-investor/with-glowing-hearts-an-algerian-city-rises/article1212917/>.

———. 2009b. "Bastion of Quebec Inc. Cuts Its Final Deal," *Globe and Mail*, 17 July, B1.

———. 2012. "Canada's Shrinking Aerospace Horizon," *Globe and Mail*, 3 Oct., B10.

———. 2013. "Ericsson plans $1.3-billion Quebec R&D centre", *Globe and Mail*, 4 June, B3.

Marowits, Ross. 2009. "Bombardier Cancels Learjet Order," *Globe and Mail*, 20 Aug. At: <www.theglobeandmail.com/globe-investor/bombardier-cancels-learjet-order/article1258306/>.

———. 2010. "SNC-Lavalin Focuses on Global Growth," *Globe and Mail*, 17 Mar., B3.

Martin, Thibault, and Steven M. Hoffman, eds. 2008. *Power Struggles: Hydro Development and First Nations in Manitoba and Quebec*. Winnipeg: University of Manitoba Press.

Mills, David. 1988. "Durham Report," in James H. Marsh, ed., *The Canadian Encyclopedia*, 2nd edn. Edmonton: Hurtig, 637–8.

Moreault, Éric. 2009a "Le Saint-Laurent encore 'vulnérable' à la pollution," *Le Soleil*, 30 juin. At: <www.cyberpresse.ca/le-soleil/actualites/environnement/200906/30/01-880288-le-saint-laurent-encore-vulnerable-a-la-pollution.php>.

———. 2009b. "Une autre 'erreur boréale'?" *Le Soleil*, 6 julliet. At:<www.cyberpresse.ca/le-soleil/actualites/environnement/200907/05/01-881433-une-autre-erreur-boreale.php>.

Murphy, Michael. 2007. *Canada: The State of the Federation, 2005. Quebec and Canada in the New Century: New Dynamics, New Opportunities*. Montréal and Kingston: McGill-Queen's University Press.

Natural Resources Canada. 2009. "Key Facts: Quebec," 17 July. At: <canadaforests.nrcan.gc.ca/keyfacts/qc>.

———. 2010. "The Global Recession Reduced Canada's Mineral Production in 2009," *Information Bulletin*, Mar. At: <mmsd.mms.nrcan.gc.ca/stat-stat/prod-prod/PDF/ib2010_e.pdf>.

———. 2012. "Mining Statistics, 2011," 18 Sept. At: <mmsd.mms.nrcan.gc.ca/stat-stat/prod-prod/2011-eng.aspx>.

Osisko. 2013. "Canadian Malartic at a Glance." At: <www.osisko.com/mines-and-projects/canadian-malartic/canadian-malartic-in-brief/>.

Padova, Allison. 2004. *Trends in Containerization at Canadian Ports*. Ottawa: Library of Parliament. At: <www.parl.gc.ca/information/library/PRBpubs/prb0575-e.htm>.

Paquin, Stéphane. 1999. *Invention d'un mythe: Le pacte entre deuxpeuplesfondateurs*. Montréal: VLB.

Peritz, Ingrid. 2012. "An Industry Collapses, a Town's Fibre Is Torn," *Globe and Mail*, 22 Sept., A10, A11.

Port of Montréal. 2012. "Statistics: Traffic Summary." At: <www.port-montreal.com/PMStats/html/frontend/statistics.jsp?lang=en&context=business>.

Québec. 2012. *Rapport annuel de gestion, 2010–2011*, 5 Sept. At: <www.tourisme.gouv.qc.ca/publications/media/document/publications-administratives/RAG-2010-2011.pdf>.

Québec and Canada. 2012. *The St. Lawrence Action Plan: 2011–2026*. At: <planstlaurent.qc.ca/en/home/about_us.html>.

Reuters. 2009. "Lavalin Buys Stake in Russian Firm," *Globe and Mail*, 16 Aug. At: <www.theglobeandmail.com/report-on-business/snc-lavalin-buys-stake-in-russian-firm/article1252431/>.

Richardson, Boyce. 1975. *Strangers Devour the Land: The Cree Hunters of the James Bay Area versus Premier Bourassa and the James Bay Development Corporation*. Toronto: Macmillan.

Robitaille, Antoine. 2008. "Charest mise sur le Nord," *Le Devoir.Com*, 29 Apr. At: <translate.google.com/translate?hl=en&sl=fr&u=http://www.ledevoir.com/2008/09/29/208131.html&ei=b6JnSpWOHYfQlAfSpLnACQ&sa=X&oi=translate&resnum=6&ct=result&prev=/search%3Fq%3Dplan%2Bnord%2BPremier%2BCharest%26hl%3Den>.

Rodrigue, Jean-Paul, Claude Comtois, and Brian Slack. 2009. *The Geography of Transportation Systems*, 2nd edn. At: <www.people.hofstra.edu/geotrans/eng/ch4en/appl4en/ch4a3en.html>.

Roslin, Alex. 2001. "Cree Deal a Model or Betrayal?" *National Post*, 10 Nov., FP7.

Royal Bank of Canada. 2010. *Provincial Outlook: March 2010*. At: <www.rbc.com/economics/market/pdf/provfcst.pdf>.

———. 2012. *Provincial Outlook: Quebec*. Sept. At: http://www.rbc.com/newsroom/pdf/provfcst-09-2012.pdf.

St Lawrence Seaway Management Corporation. 2012. "Delivering Economic Value," *Annual Report 2011-2012*. At: <www.greatlakes-seaway.com/en/pdf/slsmc_ar2012_nar_en.pdf>.

St-Pierre, Karine, and Sylvain Carrier. 2012. "Québec Exports Up in 2011," Institut de la statistique Québec, 29 Feb. At: <www.stat.gouv.qc.ca/salle-presse/communiq/2012/fevrier/fev1229_an.htm>.

Salisbury, Richard Frank. 1986. *A Homeland for the Cree: Regional Development in James Bay, 1971–1981*. Montréal and Kingston: McGill-Queen's University Press.

Schonbek, Amelia. 2012. "The Long March," *The Walrus* (Sept.): 15–16.

Scoffield, Heather. 2009. "Few Bumps in La Belle Province's Recession Ride," *Globe and Mail*, 4 Aug., B3.

Séguin, Rhéal. 2009. "Tiny Quebec Town Is Sitting on a Gold Mine," *Globe and Mail*, 14 July. At: <www.theglobeandmail.com/news/national/tiny-quebec-town-is-sitting-on-a-gold-mine/article1217078>.

———. 2013. "PQ Cancels Program to Teach English," *Globe and Mail*, 8 Mar., A6.

——— and Daniel Leblanc. 2012. "Toughening Language Laws, Halting Tuition Hike on Agenda," *Globe and Mail*, 6 Sept., A1, A9.

Service Canada. 2006. "Sectorial Outlook: Province of Québec." At: <www150.hrdc-drhc.gc.ca/imt/regional/english/sec_out/2006-2008/introduction/index.html>.

Simard, Jean-Jacques, et al. 1996. *Tendances Nordiques: Les changements sociaux 1970–1990 chez les Criset les Inuit du Québec: Un enquête statistique exploratoire*. Québec: GETIC, Université Laval.

Simpson, Jeffrey. 2009. "Champlain's Dream Lives on in North America," *Globe and Mail*, 23 Oct. At: <www.theglobeandmail.com/news/opinions/champlains-dream-lives-on-in-north-america/article1336484/>.

SouthAfrica Info. 2008. "Major Milestone for Gautrain," 11 July. At: <www.southafrica.info/business/economy/infrastructure/gautrain-trains.htm>.

Statistics Canada. 2002. "2001 Census: Population and Dwelling Counts, for Census Metropolitan Areas and Census Agglomerations, 2001 and 1996 Censuses," 16 July. At: <www.statcan.ca/english/IPS/Data/93F0050XCB2001013.htm>.

———. 2006a. "Distribution of Employed People, by Industry, by Province." At: <www40.statcan.ca/101/cst01/labor21c.htm>.

———. 2006b. "Canadian Statistics—Net Merchantable Volume of Roundwood Harvested." At: <www.statcan.ca/english/Pgdb/prim41.htm>.

———. 2007. "Population and Dwelling Counts, for Census Metropolitan Areas and Census Agglomerations, 2006 and 2001 Censuses—100% Data." At: <www12.statcan.ca/english/census06/data/popdwell/Table.cfm?T=201&S=3&O=D&RPP=150>.

———. 2008. "Total Farm Area, Land Tenure, and Land in Crops, by Provinces, 1986 to 2006," *Census of Agriculture*. At: http://www.statcan.gc.ca/tables-tableaux/sum-som/l01/cst01/agrc25f-eng.htm.

———. 2009a. "Individuals by Total Income Level by Province and Territory." At: http://www.statcan.gc.ca/tables-tableaux/sum-som/l01/cst01/famil105a-eng.htm.

———. 2009b. "Gross Domestic Product, Expenditure-based, by Province and Territory." At: <www40.statcan.gc.ca/cbin/fl/cstprintflag.cgi>.

———. 2009c. "Canada's Population Estimates," *The Daily*, 23 June. At: <www.statcan.gc.ca/daily-quotidien/090623/dq090623a-eng.htm>.

———. 2009d. "Distribution of Employed People, by Industry, by Province." At: http://www.statcan.gc.ca/tables-tableaux/sum-som/l01/cst01/labor21c-eng.htm.

———. 2009e. *Using Languages at Work in Canada, 2006 Census: Provinces and Territories*, 2006 Analysis Series. At: <www12.statcan.gc.ca/census-recensement/2006/as-sa/97-555x/p4-eng.cfm>.

———. 2009f. "Birth Rate, by Province and Territory." At: <www40.statcan.gc.ca/l01/cst01/demo04b-eng.htm>.

———. 2011. "Births and Total Fertility Rate, by Province and Territory," 20 Dec. At: <www.statcan.gc.ca/tables-tableaux/sum-som/l01/cst01/hlth85b-eng.htm>.

———. 2012a. "Unemployment Rates, Canada, Provinces, Health Regions (2011 Boundaries) and Peer Groups," 24 May. At: <www5.statcan.gc.ca/cansim/pick-choisir?lang=eng&p2=33&id=1095324>.

———. 2012b. "Gross Domestic Product by Industry: Provinces and Territories, 2011 (Preliminary Data)," *The Daily*, 27 Apr. At: <www.statcan.gc.ca/daily-quotidien/120427/dq120427a-eng.htm>.

———. 2012c. *Canada at a Glance 2012*. At: <www.statcan.gc.ca/pub/12-581-x/12-581-x2012000-eng.htm>.

———. 2012d. "Distribution of Employed People, by Industry, by Province," 1 June. At: <www.statcan.gc.ca/tables-tableaux/sum-som/l01/cst01/labor21a-eng.htm>.

———. 2012e. "Population and Dwelling Counts, for Quebec, by Census Metropolitan Areas and Census Agglomerations, 2011 and 2006 Census," 11 May. At: <www12.statcan.gc.ca/census-recensement/2011/dp-pd/hlt-fst/pd-pl/Table-Tableau.cfm?LANG=Eng&T=202&PR=24&S=0&O=D&RPP=50>.

———. 2013. *Linguistic Characteristics of Canadians*. At: <www12.statcan.gc.ca/census-recensement/2011/as-sa/98-314-x/98-314-x2011001-eng.cfm>.

Toronto Transit Commission. 2013. "New Subway Train: The Toronto Rocket." At: <www.ttc.ca/About_the_TTC/Projects_and_initiatives/New_Subway_Train/index.jsp>.

Tower, Courtney. 2010. "Port Montreal Sees a Gradual Recovery in 2010," *Journal of Commerce*, 4 Jan. At: <www.joc.com/maritime/port-montreal-sees-gradual-recovery-2010>.

Turgeon, Pierre. 1992. *La Radissonie: le pays de la Baie James*. Montréal: Libre Expressions.

VanderKlippe, Nathan. 2010. "Shell's 76-year-old Gasoline Facility Becomes Second Major Canadian Plant to Close in Recent Years," *Globe and Mail*, 8 Jan., B1.

Wyman, Diana. 2006. "Trade Liberalization and the Canadian Clothing Market," *Canadian Economic Observer*. Catalogue no. 11–010. Ottawa. At: <www.statcan.ca/english/ads/11-010-XPB/pdf/dec06.pdf>.

Yakabuski, Konrad. 2008. "Environment? Economics? Hydro-Québec Project Fails Grade," *Globe and Mail*, 14 Aug., B1.

CHAPTER 7

Alldritt, Benjamin. 2012. "This Is Going to Be a Boom: Seaspan CEO," 25 Apr. At: <www.nsnews.com/story.html?id=6514682>.

Angevine, Gerry, and Vanadis Oviedo. 2012. *Laying the Groundwork for BC LNG Exports to Asia*. Vancouver: Fraser Institute, Oct. At: <www.fraserinstitute.org/publicationdisplay.aspx?id=18902&terms=laying+the+groundwork+for+BC+lng+exports+to+asia>.

Barnes, Trevor, and Roger Hayter. 1997. *Trouble in the Rainforest: British Columbia's Forest Economy in Transition*. Canadian Western Geographical Series, vol. 33. Victoria: Western Geographical Press.

BC Facts. 2009. "Hollywood North." At: <www.gov.bc.ca/bcfacts/>.

BC Hydro. 2009. "Bioenergy Call for Power," 18 Nov. At: <www.bchydro.com/planning_regulatory/acquiring_power/bioenergy_call_for_power.html>.

———. 2012. *Annual Report 2012*. At: <www.bchydro.com/etc/medialib/internet/documents/annual_report/2012_BCH_AnnualReport.Par.0001.File.2012-BCH-Annual-Report.pdf>.

BC Ministry of Environment, Oceans and Marine Fisheries Branch. 2011. "Capture Fisheries Statistics." At: <www.env.gov.bc.ca/omfd/fishstats/capture/index.html>.

BC Ministry of Transportation. 2007a. *BC Transportation Financing Authority—Statement of Earnings*. At: <www.bcbudget.gov.bc.ca/2007/sp/trans/default.aspx?hash=8>.

———. 2007b. "Pacific Gateway." At: <www.th.gov.bc.ca/PacificGateway/index.htm>.

BC Stats. 2006. *Quick Facts about British Columbia 2006 Edition*. Victoria. At: http://www.bcstats.gov.bc.ca/StatisticsBySubject/KeyIndicators/QuickFactsAboutBC.aspx.

———. 2012a. "Annual data for BC Exports with Selected Destinations and Commodity Details," Aug. At: <www.bcstats.gov.bc.ca/StatisticsBySubject/ExportsImports/Data.aspx>.

———. 2012b. "Softwood Lumber Exports." At: <www.bcstats.gov.bc.ca/StatisticsBySubject/BusinessIndustry/Forestry.aspx>.

———. 2012c. "Exports and Imports—Data." At: <www.bcstats.gov.bc.ca/StatisticsBySubject/ExportsImports/Data.aspx>.

———. 2013. "BC Origin Exports to the United States—Selected Commodities", Exports and Imports—Data. June. At: <http://www.bcstats.gov.bc.ca/StatisticsBySubject/ExportsImports/Data.aspx>.

BC Treaty Commission. 2007. "Lheidli T'enneh First Nation Vote to Reject the Treaty," Apr. At: <www.bctreaty.net/files/pdf_documents/April07update.pdf>.

Boei, Bill, and Amy O'Brian. 2003. "10,000 Flee Kelowna, B.C., Blaze," *National Post*, 23 Aug., A1.

Bone, Robert M. 2009. *The Canadian North: Issues and Challenges*, 3rd edn. Toronto: Oxford University Press.

Bouw, Brenda. 2011. "Rio Digs Deep for Kitimat with Eye on Future Payoff," *Globe and Mail*, 2 Dec., B4.

Canadian Association of Petroleum Producers. 2012. "Value of Producers' Sales: 1988–2011," *Statistical Handbook*. At: <membernet.capp.ca/SHB/Sheet.asp?SectionID=4&SheetID=311>.

Cassidy, Frank. 1992. "Aboriginal Land Claims in British Columbia: A Regional Perspective", in K. Coates, ed., *Aboriginal Land Claims*. Toronto: Copp Clark, 10–43.

CBC News. 2009. "Queen Charlotte Islands Renamed Haida Gwaii in Historic Deal," 11 Dec. At: <www.cbc.ca/canada/british-columbia/story/2009/12/11/bc-queen-charlotte-islands-renamed-haida=gwaii.html>.

———. 2012. "Tests Confirm Virus at B.C. Salmon Farm," 30 May. At: <www.cbc.ca/news/canada/british-columbia/story/2012/05/30/bc-salmon-virus-dixon-bay.html>

———. 2013. "Media Mogul Close to $24B Kitimat, BC Refinery Deal," 6 Mar. At: <www.cbc.ca/news/canada/edmonton/story/2013/03/06/bc-kitimat-refinery-black.html?autoplay=true>.

Chase, Steven, and Bertrand Marotte. 2011. "Halifax, Vancouver Win $33-Billion in Shipbuilding Sweepstakes," *Globe and Mail*. 20 Oct., A1.

Christensen, Bev. 1995. *Too Good to Be True: Alcan's Kemano Completion Project*. Vancouver: Talonbooks.

Clapp, R.A. 2008. "The Resource Cycle in Forestry and Fishing," *Canadian Geographer* 42, 2: 129–44.

Clark, Christy. 2012. "Premier Christy Clark's Letter to Alison Redford," 26 Sept. At: <www.newsroom.gov.bc.ca/2012/09/premier-christy-clarks-letter-to-alberta-premier-alison-redford.html>.

CN. 2007. "Port of Prince Rupert Container Terminal." At: <www.cn.ca/specialized/ports_docks/prince_rupert/fairview/en_KFPortsPrinceRupert_fairview.shtml>.

CTV News. 2009. "Huge Fire Impacts 10,000 in Kelowna, BC," 18 July. At: http://bc.ctvnews.ca/huge-fire-impacts-10-000-in-kelowna-b-c-1.417884.

David Suzuki Foundation. 2005. *Clearcutting Canada's Rainforest, Status Report 2005: Canada's Rainforests under Threat*. At: <www.davidsuzuki.org/files/Forests/DSF-rainforests-2005-5.pdf>.

———. 2009. "Sea Lice." At: <www.davidsuzuki.org/Oceans/Aquaculture/Salmon/Sea_Lice.asp>.

Ebner, David. 2009. "Waiting for His Ships to Come In," *Globe and Mail*, 24 Sept., B3.

Economist Intelligence Unit. 2012. "Melbourne, Vienna and Vancouver Named World's Most Liveable Cities," 15 Aug. At:<www.citymayors.com/environment/eiu_bestcities.html>.

Export Development Canada. 2009. "British Columbia's Exports to Drop Significantly in 2009 amid Forest and Energy Woes, Says EDC," 6 May. At: <www.edc.ca/english/docs/news/2009/mediaroom_16357.htm>.

Fisheries and Oceans Canada. 2009. "Pacific Salmon Treaty Renewal," 5 Jan. At: <www.dfo-mpo.gc.ca/media/backfiche/2009/pr01-eng.htm>.

Harris, Cole. 1997. *The Resettlement of British Columbia: Essays on Colonialism and Geographical Change*. Vancouver: University of British Columbia Press.

Hayter, Roger. 1992. "The Little Town That Did: Flexible Accumulation and Community Response in Chemainus, British Columbia," *Regional Studies* 26: 647–63.

Hume, Mark. 2009. "Millions of Missing Fish Signal Crisis on the Fraser River," *Globe and Mail*, 12 Aug., B1.

Hume, Stephen. 2000. "Did Francis Drake Discover B.C.?" *National Post*, 5 Aug., B1–B2.

Hutchison, Bruce. 1942. *The Unknown Country: Canada and Her People*. Toronto: Longmans, Green and Company.

Indian and Northern Affairs Canada. 2007. "Backgrounder—Wei Wai Kum Cruise Ship Port." At: <www.ainc-inac.gc.ca/nr/prs/m-a2007/2-2892-bk-eng.asp>.

Industry Advisory Group. 2006. *Pacific Gateway Strategy Action Plan 2006–2020*. Vancouver. At: <www.th.gov.bc.ca/PacificGateway/documents/PGS_Action_Plan_043006.pdf>.

Jordan, Pav. 2012. "Rio Tinto Wants to Reopen Union Deal in Quebec," *Globe and Mail*, 13 Oct., B10.

Katz, Diane. 2010. "The Agricultural Land Reserve Doesn't Work—So Let's Get Rid of It," *BCBusiness Online*, 6 Jan. At: <www.bcbusinessonline.ca/bcb/business-sense/2010/01/06/alr-tear-down-wall>.

Lomas, Peter. 2012. "Tourism Sector Recovers in 2010," BC Stats. At: <www.bcstats.gov.bc.ca/StatisticsBySubject/BusinessIndustry/Tourism.aspx>.

MacDonald, John. 2012. "Cuts Threaten Canada's Satellite Eye on the Arctic," *Vancouver Sun*, 24 Apr. At: <www2.canada.com/vancouversun/news/archives/story.html?id=af7f961f-8bfd-45fd-a7f2-5765220e7a93>.

McCullough, Michael. 1998. *Granville Island: An Urban Oasis*. Vancouver: CMHC.

McKinnell, Skip, Robert Emmett, and Joseph Orsi. 2009. "2009 Salmon Forecasting Forum," North Pacific Marine Science Organization, *PICES Press* 17, 2: 32–3.

McVey, Wayne W., and W.E. Kalbach. 1995. *Canadian Population*. Toronto: Nelson Canada.

Mineral Resources Education Program of BC. 2009. At: <www.bcminerals.ca/files/bc_mine_information.php>.

Ministry of Forests, Lands and Natural Resource Operations. 2011. *The Forest Industry Snapshot: A Selection of Monthly Economic Statistics*, Mar. At: <www.for.gov.bc.ca/ftp/het/external/!publish/web/snapshot/201103.pdf>.

Mittelstaedt, Martin. 2004. "Farmed Salmon Laced with Toxins, Study Finds," *Globe and Mail*, 9 Jan., A1.

Morton, Alexandra, Richard Toutledge, Corey Peet, and Aleria Ladwig. 2004. "Sea Lice (*Lepeophtheirussalmonis*) Infection Rates on Juvenile Pink (*Oncorhynchusgorbuscha*) and Chum (*Oncorhynchusketa*) Salmon in the Nearshore Marine Environment of British Columbia, Canada," *Canadian Journal of Fisheries and Aquatic Sciences* 61, 2: 147–57.

Natural Resources Canada. 2006. "About Hydroelectric Energy." At: <www.canren.gc.ca/tech_appl/index.asp?CaId=4&PgId=26>.

———. 2012a. "Mining Statistics," 8 Aug. At: <mmsd.mms.nrcan.gc.ca/stat-stat/prod-prod/2011-eng.aspx>.

———. 2012b. "Canada's Mineral Production Reaches a Record $50 Billion," *Information Bulletin*, Mar. At: <www.nrcan.gc.ca/minerals-metals/publications-reports/3575>.

———. 2013. *The State of Canada's Forest, 2012*. At:<http://cfs.nrcan.gc.ca/pubwarehouse/pdfs/34055.pdf>.

Nemetz, Peter N. 1990. *The Pacific Rim Investment, Development and Trade*, 2nd rev. edn. Vancouver: University of British Columbia Press.

Nisga'a. n.d. "Nisga'a History." At: <www.schoolnet.ca/aboriginal/nisga1/hist-e.html>.

O'Neil, Peter. 2012. "Former Exec Calls for Oil Spill Insurance," *StarPhoenix* (Saskatoon), 6 July, C5.

Padova, Allison. 2004. *Trends in Containerization at Canadian Ports*. Ottawa: Library of Parliament. At: <www.parl.gc.ca/information/library/PRBpubs/prb0575-e.htm>.

Port Metro Vancouver. 2012. *Statistical Overview 2011*. At: <www.portmetrovancouver.com/Libraries/ABOUT_Facts_Stats/PMV_2011_Stats_Overview.sflb.ashx>.

———. 2013. "Facts & Stats." At: <www.portmetrovancouver.com/en/about/factsandstats.aspx>.

Prince Rupert Port Authority. 2007. "The New World Port." At: <www.rupertport.com/about.htm>.

———. 2010. "Prince Rupert Records 12-Year-High Cargo Volumes in 2009," 25 Jan. At: <www.rupertport.com/pdf/newsreleases/ppr_records_12yearhigh2009cargovolumes.pdf>.

———. 2012. "Port of Prince Rupert Celebrates Unprecedented Growth and Economic Expansion," 13 June. At: <www.rupertport.com/news/releases/annual-public-meeting-2012>.

Provincial Agricultural Land Commission. 2009. "The Commission." At: <www.alc.gov.bc.ca/commission/alc_main.htm>.

Rio Tinto Alcan. 2010. "Kitimat Works Modernization." At: <www.kitimatworksmodernization.com/pages/modernization-project/about-kitimat-modernization.php>.

Schrier, Dan. 2012a. "Exports by Province, 2002–2011," Aug. At: <www.bcstats.gov.bc.ca/StatisticsBySubject/ExportsImports/Data.aspx>.

———. 2012b. "BC's Exports Moving Out of the Wood," BC Stats, 11 May. At: <www.bcstats.gov.bc.ca/Publications/AnalyticalReports.aspx>.

Statistics Canada. 2006. "Distribution of Employed People by Industry, by Province, 2005." At: <www40.statcan.ca/l01/cst01/labor21c.htm>.

———. 2007. "Population and Dwelling Counts, for Census Metropolitan Areas and Census Agglomerations, 2006 and 2001 Censuses—100% Data." At: <www12.statcan.ca/english/census06/data/popdwell/Table.cfm?T=201&S=3&O=D&RPP=150>.

———. 2008. *Oil and Gas Extractions—2007*, Table 5-1, "Commodity Data, Oil and Gas Extractions—Net Production Withdrawals." Catalogue no. 26–213–X. At: <www.statcan.gc.ca/pub/26-213-x/2007000/t018-eng.pdf>.

———. 2009. "Distribution of Employed People by Industry, by Province, 2008." At: http://www.statcan.gc.ca/tables-tableaux/sum-som/l01/cst01/labor21c-eng.htm.

———. 2011. "Population and Dwelling Counts, for Canada, Provinces and Territories, Census Metropolitan Areas and Census Agglomerations, 2011 and 2006 Censuses." At: <www12.statcan.gc.ca/census-recensement/2011/dp-pd/hlt-fst/pd-pl/Table-Tableau.cfm?LANG=Eng&T=202&PR=59&S=0&O=D&RPP=50>.

———. 2012a. "Gross Domestic Product at Basic Prices, by North American Industry Classification System and Provinces," Table 379–0025.

———. 2012b. "Unemployment rate, Canada, Provinces, Health Regions and Peer Groups," Table 109–5324.

———. 2012c. "Distribution of Employed People, by Industry, by Province," 6 Jan. At: <www.statcan.gc.ca/tables-tableaux/sum-som/l01/cst01/labor21a-eng.htm>.

———. 2012d. "Principal Statistics for Manufacturing Industries," CANSIM Table 301–0006, 2 Mar. At: <www5.statcan.gc.ca/cansim/a26?lang=eng&retrLang=eng&id=3010006>.

———. 2013. "Number and distribution of population reporting an Aboriginal identity and percentage of Aboriginal people in the population, Canada, provinces and territories, 2011," *Aboriginal Peoples in Canada: First Nations People, Métis and Inuit*. National Household Survey document 99-011-x. Table 2. 7 May. At:<http://www12.statcan.gc.ca/nhs-enm/2011/as-sa/99-011-x/2011001/tbl/tbl02-eng.cfm>.

Turtle Island Native Network Culture. 2011. "Wei Wai Kum Cruise Ship Terminal." At: <www.turtleisland.org/business/cruise.htm>.

VanderKlippe, Nathan. 2007a. "Lumber Waste a Hot Commodity," *National Post*, 6 June, FP1, FP6.

———. 2007b. "A Town's Ship Comes In," *National Post*, 9 June, FP4.

———. 2012a. "Trans Mountain: The Other Pacific Pipeline," *Globe and Mail*, 1 Aug., B1.

———. 2012b. "West Coast LNG Advocates Face Tough Road Ahead," *Globe and Mail*, 5 Oct, B10.

White, P., M. Michalowski, and P. Cross. 2006. "The West Coast Boom," *Canadian Economic Observer*, May. Ottawa: Statistics Canada Catalogue no. 11–010.

Wilson, Miriam. 2012. *B.C.'s Economy in 2011*, BC Stats (June). At: <www.bcstats.gov.bc.ca/Publications/AnalyticalReports.aspx?subject=forestry>.

Wynn, Graeme, and Timothy Oke, eds. 1992. *Vancouver and Its Region*. Vancouver: University of British Columbia Press.

CHAPTER 8

Agriculture and Agri-Food Canada. 2009. *Overview of the Canadian Agriculture and Agri-Food System 2008*. At: <www4.agr.gc.ca/AAFC-AAC/display-afficher.do?id=1228246364385&lang=eng#a2>.

Akinremi, O.O., S.M. McGinn, and H.W. Cutforth. 2001. "Seasonal and Spatial Patterns of Rainfall on the Canadian Prairies," *Journal of Climate* 14, 9: 2177–82.

Alberta. 2007. "Oil Reserves." At: <www.energy.gov.ab.ca/docs/oil/pdfs/AB_OilReserves.pdf>.

———. 2012. *Budget 2012: Investing in People*. At: <www.finance.alberta.ca/publications/budget/budget2012/fiscal-plan-revenue.pdf>.

Alberta Energy. 2012. "Facts and Statistics," At: <www.energy.alberta.ca/oilsands/791.asp>.

Bell, Ian. 1999. "Dancing with Elephants," *Western Producer*, 2 Sept., 60–1.

Bonsal, B.R., X. Zhang, and W.D. Hogg. 1999. "Canadian Prairie Growing Season Precipitation Variability and Associated Atmospheric Circulation," *Climate Research* 11: 191–208.

Bradsher, Keith. 2007. "Rise in China's Pork Prices Signals End to Cheap Output," *New York Times*, 8 June. At: <www.nytimes.com/2007/06/08/business/worldbusiness/08prices.html>.

Bramley, Matthew, Pierre Sadik, and Dale Marshall. 2008. *Climate Leadership, Economic Prosperity: Final Report on an Economic Study of Greenhouse Gas Targets and Policies for Canada*. Pembina Institute and David Suzuki Foundation. At: <pubs.pembina.org/reports/climate-leadership-report-en.pdf>.

Buchanan, Peter. 2013. "Changing Global Realities Buffet Canada's Oil Patch," *Economic Insights* (CIBC World Markets), 3 Apr. At: <research.cibcwm.com/economic_public/download/feature2.pdf>.

Canadian Association of Petroleum Producers. 2012a. *Crude Oil: Forecasts, Markets & Pipelines*, June. At: <www.capp.ca/getdoc.aspx?DocId=209546&DT=NTV>.

———. 2012b. "Safeguarding the Public." At: <www.capp.ca/ENVIRONMENTCOMMUNITY/HEALTHSAFETY/Pages/Public.aspx>.

Canadian Grain Commission. 2012. "Quality of Western Canadian Canola 2011," 6 Mar. At: <www.grainscanada.gc.ca/canola/harvest-recolte/2011/hqc11-qrc11-03-eng.htm>.

Canadian Nuclear Safety Commission. 2012. "Northern Saskatchewan," 25 Oct. At: <nuclearsafety.gc.ca/eng/my-community/facilities/northsask/index.cfm#Sectionb>.

Coolican, Lori. 2012. "Farmland Deals Probed," *StarPhoenix* (Saskatoon), 19 Dec., A1, A2.

D'Aliesio, Renata. 2012. "Brooks, Alberta on the Brink as It Awaits Fate of XL Plant," *Globe and Mail*, 22 Oct. At: <www.theglobeandmail.com/news/national/brooks-alberta-on-the-brink-as-it-awaits-fate-of-xl-plant/article4627920/>.

Folk, Mark. 2011. "Farm Ownership," Farmland Security Board. At: <www.farmland.gov.sk.ca/ownership/overview.shtml>.

Food and Agriculture Organization of the United Nations (FAO). 2012. "FAO Food Price Index," 4 Nov. At: <www.fao.org/worldfoodsituation/wfs-home/foodpricesindex/en/>.

Gonick, Noam. 2004. *Stryker*. Winnipeg: Telefilm Canada. At: <www.strykerthemovie.com>.

Grauman, Meny. 2006. "Alberta Profile—Economic View," *Provincial Pulse* (Scotiabank), 3 Feb., 1.

Greenwood, John. 2007. "Surging Prices Separate Wheat from the Chaff," *National Post*, 15 June, FP1.

Hall, Angela. 2003. "Hogs Touchy Issue in Rural Sask.," *StarPhoenix* (Saskatoon), 25 Feb., C8.

Hugenholtz, D.H., and S.A. Wolfe. 2005. "Biogeomorphic Model of Dune Activation and Stabilization on the Northern Great Plains," *Geomorphology* 70: 53–70.

Hursh, Kevin. 1999. "Subsidy Complaints Don't Hold Water," *StarPhoenix* (Saskatoon), 28 July, C9.

Imperial Oil. 2012a. "Cold Lake." At: <www.imperialoil.ca/Canada-English/operations_sands_cold.aspx>.

———. 2012b. "Oil Sands 101." At: <www.imperialoil.ca/Canada-English/operations_sands_glance_101.aspx>.

Jones, David C. 1987. *Empire of Dust: Settling and Abandoning the Prairie Dry Belt*. Edmonton: University of Alberta Press.

Jubinville, Mike. 2009. "Market Focus—CWB PRO Review," *Farm Credit Canada*, 2 Oct. At: <www.fcc-fac.ca/newsletters/en/express/articles/20091002_e.asp>.

Kroeger, Arthur. 2007. *Hard Passage: A Mennonite Family's Long Journey from Russia to Canada*. Edmonton: University of Alberta Press.

———. 2009. *Retiring the Crow Rate: A Narrative of Political Management*. Edmonton: University of Alberta Press.

Kurek, Joshua, Jane L. Kirk, Derek C.G. Muir, Xiaowa Wang, Marlene S. Evans, and John P. Smol. 2013. "Legacy of a Half Century of Athabasca Oil Sands Development Recorded by Lake Ecosystems," *Proceedings of the National Academy of Sciences* 110, 5: 1761–6.

MacLachlan, Ian. 2001. *Kill and Chill: Restructuring Canada's Beef Commodity Chain*. Toronto: University of Toronto Press.

Mintz, Jack M. 2010. "Jack Mintz: The Potash Royalty Mess," *Financial Post*, 14 Oct. At: <opinion.financialpost.com/2010/10/14/jack-mintz-the-potash-royalty-mess/>.

Monsanto Canada. 2012. "Monsanto Canada Unveils DEKALB® Canola Seed Processing Plant in Lethbridge, Alberta," *Monsanto in the News*, 7 Oct. At: <www.monsanto.ca/newsviews/Pages/NR-2012-07-10.aspx>.

National Energy Board. 2006. *Canada's Oilsands: Opportunities and Challenges to 2015—An Update*. Calgary: Publication Office, National Energy Board. At: http://www.neb-one.gc.ca/clf-nsi/rnrgynfmtn/nrgyrprt/lsnd/pprtntsndchllngs20152006/qap-prtntsndchllngs20152006-eng.html.

———. 2012. "Canadian Energy Overview 2011," 26 July. At: <www.neb-one.gc.ca/clf-nsi/rnrgynfmtn/nrgyrprt/nrgyvrvw/cndnnrgyvrvw2011/cndnnrgyvrvw2011-eng.html>.

Natural Resources Canada. 2009. *The State of Canada's Forests 2009*. At: <canadaforests.nrcan.gc.ca/rpt>.

———. 2010. "The Global Recession Reduced Canada's Mineral Production in 2009," *Information Bulletin*, Mar. At: <mmsd.mms.nrcan.gc.ca/stat-stat/prod-prod/PDF/ib2010_e.pdf>.

———. 2010. *Canada in a Changing Climate*, "Prairies," 20 Aug. At: <www.nrcan.gc.ca/earth-sciences/climate-change/community-adaptation/642>.

———. 2012. Annual Statistics: Table 2, Revised Statistics of the Mineral Production of Canada, by Provinces, 2011. At: <http://sead.nrcan.gc.ca/prod-prod/2011-eng.aspx>.

Oilsands Developers Group. 2009. "Oilsands Projects," 30 June. At: <www.oilsandsdevelopers.ca/index.php/test-project-table/>.

Paul, Alec H. 1997. "Shortlines, Mainlines, Branchlines, Dead Lines: Rural Railways in Southwestern Saskatchewan in the 1990s," in John Welsted and John Everitt, eds, *The Yorkton Papers: Research by Prairie Geographers*. Brandon Geographical Studies No. 2. Brandon, Man.: University of Brandon.

Pembina Institute. 2007. "Athabasca River Expedition: Connecting the Drops" At: <http://www.connectingthedrops.ca/river/stresses>."

Price, Jacqueline D. 2003. "Are Factory Farms Fouling Our Water?" *Alberta Views* (May–June): 34–9.

Richards, J. Howard. 1968. "The Prairie Region," in John Warkentin, ed., *Canada: A Geographical Interpretation*. Toronto: Methuen, ch. 12.

Rodrigue, Jean-Paul, Claude Comtois, and Brian Slack. 2009. *The Geography of Transport Systems*. Toronto: Routledge.

Saskatchewan Ministry of Agriculture.2009a. "September Estimate of 2009 Crop Production." At: <www.agriculture.gov.sk.ca/Default.aspx?DN=d8fe5165-80e6-46b4-ba5f-217e2a0184a2>.

———. 2009b. "Crop Statistics." At: <www.agriculture.gov.sk.ca/agriculture_statistics/HBv5_P2.asp>.

———. 2009c. *Agricultural Statistics Fact Sheet*. At: <www.agriculture.gov.sk.ca/Saskatchewan_Agricultural_Statistics_Fact_Sheet>.

———. 2012. "Factsheet," June. At: <www.agriculture.gov.sk.ca/Default.aspx?DN=274b2ec5-4d55-4a74-9a9d-6bc1105105da>.

Saskatchewan Ministry of Energy and Resources. 2012. *Annual Report, 2011/2012*, July. At: <www.finance.gov.sk.ca/PlanningAndReporting/2011-12/201112ERAnnualReport.pdf>.

Saskatchewan Research Council. 2012. "Project CLEANS (Cleanup of Abandoned Northern Sites)." At: <www.src.sk.ca/About/Featured-Projects/Pages/Project-CLEANS.aspx>.

Schwartz, Jennifer. 2009. "US Oil Imports from Canada Hit New High in July," 7 Oct. At: <www.heatingoil.com/blog/us-oil-imports-from-canada-hit-new-high-in-july-1007/>.

Spry, Irene M. 1963. *The Palliser Expedition: An Account of John Palliser's British North American Expedition 1857–1860*. Toronto: Macmillan.

Statistics Canada. 1992. *Census Overview of Canadian Agriculture 1971–1991*. Catalogue no. 93–348. Ottawa: Ministry of Industry, Science and Technology.

———. 2003. "Agriculture 2001 Census Farm Operations: Provincial/Regional Trends." At: <www.statcan.ca/english/agcensus2001/first/regions/contents.htm>.

———. 2006. "Exports of Goods on a Balance-of-Payments Basis, by Product." At: <www40.statcan.ca/l01/cst01/gblec05.htm>.

———. 2007a. "Total Farm Area, Land Tenure and Land in Crops, by Province," *2006 Census of Agriculture*. At: <www40.statcan.ca/l01/cst01/agrc25j.htm>.

———. 2007b. "Population and Dwelling Counts, for Canada, Provinces and Territories, 2006 and 2001 Censuses—100% Data." At: <www12.statcan.ca/english/census06/data/popdwell/Table.cfm?T=101>.

———. 2008. "Aboriginal Identity Population, 2006 Counts, Percentage Distribution, Percentage Change for Both Sexes, for Canada, Provinces and Territories—20% Sample Data." At: <www12.statcan.gc.ca/english/census06/data/highlights/Aboriginal/pages/Page.cfm?Lang=E&Geo=PR&Code=48&Table=>.

———. 2009a. *Energy Statistics Handbook*. Catalogue no. 57–601–X. At: <www.statcan.gc.ca/pub/57-601-x/57-601-x2009002-eng.pdf>.

———. 2009b. "Distribution of Employed People by Industry, by Province, 2008." At: http://www.statcan.gc.ca/tables-tableaux/sum-som/l01/cst01/labor21c-eng.htm.

———. 2009c. "Population, Age Distribution and Median Age by Province and Territory, as of July 1, 2008," Table 1. At: <www.statcan.gc.ca/daily-quotidien/090115/t090115c1-eng.htm>.

———. 2009d. *Aboriginal Peoples in Canada in 2006: Inuit, Métis and First Nations*, 2006 Census. At: <www12.statcan.gc.ca/census-recensement/2006/as-sa/97-558/p3-eng.cfm#01>.

———. 2010. "Exports of Goods on a Balance-of-Payments Basis, by Product," *Merchandise Exports*, 11 Mar. At: http://www.statcan.gc.ca/tables-tableaux/sum-som/l01/cst01/gblec04-eng.htm.

———. 2012a. "Gross Domestic Product (GDP) at Basic Prices, by North American Industry Classification System (NAICS) and Provinces," 27 May, Table 379–0025. At: <www5.statcan.gc.ca/cansim/a26?lang=eng&retrLang=eng&id=3790025&paSer=&pattern=&stByVal=2&p1=-1&p2=31&tabMode=dataTable&csid=>.

———. 2012b. "Population by Mother Tongue and Age Groups (Total), Percentage Distribution (2011) for Canada, Provinces and Territories," 24 Oct. At: <www12.statcan.gc.ca/census-recensement/2011/dp-pd/hlt-fst/lang/Pages/highlight.cfm?TabID=1&Lang=E&Asc=1&PRCode=01&OrderBy=999&View=2&tableID=401&queryID=1&Ag>.

———. 2012c. "2011 Census of Agriculture," *The Daily*, 10 May. At: <www.statcan.gc.ca/daily-quotidien/120510/t120510a001-eng.htm>.

———. 2012d. "Distribution of Employed People, by Industry, by Province." At: <www.statcan.gc.ca/tables-tableaux/sum-som/l01/cst01/labor21c-eng.htm>.

———. 2012e. "Farm and Farm Operator Data," *2011 Census of Agriculture*, 5 June. At: <www.statcan.gc.ca/pub/95-640-x/2012002-eng.htm>.

———. 2012f. "2011 Census of Agriculture," *The Daily*, 10 May. At: <www.statcan.gc.ca/daily-quotidien/120510/t120510a001-eng.htm>.

———. 2012g. "Population and Dwelling Counts, 2011 Census." At: <www5.statcan.gc.ca/bsolc/olc-cel/olc-cel?catno=98-310-XWE2011002&lang=eng>.

———. 2013a. "Population and Dwelling Counts, for Canada, Provinces and Territories, 2011 and 2006 Censuses." 2011 Census, Jan. At: <http://www12.statcan.gc.ca/census-recensement/2011/as-sa/98-310-x/98-310-x2011001-eng.cfm>.

———. 2013b. "Number and distribution of population reporting an Aboriginal identity and percentage of Aboriginal people in the population, Canada, provinces and territories, 2011", *Aboriginal Peoples in Canada: First Nations People, Métis and Inuit*. National Household Survey document 99-011-x. Table 2. 7 May. At: <http://www12.statcan.gc.ca/nhs-enm/2011/as-sa/99-011-x/2011001/tbl/tbl02-eng.cfm>.

———. 2013c. "Aboriginal Ancestry (6) by Canada, Province, Territory, CMA and Census Agglomeration", 2011 National Household Survey: Data Tables. 99-011-X2011029. At: <http://www12.statcan.gc.ca/nhs-enm/2011/dp-pd/dt-td/Ap-eng.cfm?LANG=E&APATH=3&DETAIL=0&DIM=0&FL=A&FREE=0&GC=0&GID=0&GK=0&GRP=1&PID=105402&PRID=0&PTYPE=105277&S=0&SHOWALL=0&SUB=0&Temporal=2013&THEME=94&VID=0&VNAMEE=&VNAMEF=>.

Vanderhaeghe, Guy. 1996. *The Englishman's Boy*. Toronto: McClelland & Stewart.

VanderKlippe, Nathan. 2012. "Shell Launches First Canadian Oil Sands Carbon-Capture Project," *Globe and Mail*, 5 Sept. At: <www.theglobeandmail.com/globe-investor/shell-launches-first-canadian-oil-sands-carbon-capture-project/article4520968/>.

Wiebe, Rudy. 1973. *The Temptations of Big Bear*. Toronto: McClelland & Stewart.

Wilkins, Charles. 2013. "This Little Piggy Went to Market . . . and the Farmer Lost Money," *Report on Business, The Globe and Mail* (Mar.): 32–41.

Wolfe, S.A., C.H. Hugenholtz, C.P. Evans, D.J. Huntley, and J. Ollerhead. 2007. "Potential Aboriginal-Occupation-Induced Dune Activity, Elbow Sand Hills, Northern Great Plains, Canada", *Great Plains Research* 17: 173–92.

WTRG Economics. 2012. "Oil Price History and Analysis." At: <www.wtrg.com/prices.htm>.

CHAPTER 9

Bennett, Margaret. 1989. *The Last Stronghold: Scottish Gaelic Traditions in Newfoundland*. Edinburgh: Canongate.

Bissett, Kevin. 2010. "N.B. Premier Pulls Plug on Power Sale to Quebec," *Toronto Star*, 25 Mar., A8.

Blades, Kent. 1995. *Net Destruction: The Death of Atlantic Canada's Fishery*. Halifax: Nimbus.

Bradfield, Michael. 1991. *Maritime Economic Union: Sounding Brass and Tinkling Symbolism*. Halifax: Canadian Centre for Policy Alternatives.

Budgel, Richard, and Michael Stavely. 1987. *The Labrador Boundary*. Labrador Institute of Northern Studies, Memorial University of Newfoundland. At: <www.mun.ca/labradorinstitute/projects/labrador_boundary2.pdf>.

Canada. Department of Finance. 2009. "Equalization Program." At: <www.fin.gc.ca/fedprov/eqp-eng.asp>.

———. 2011. "Federal Transfers to Provinces and Territories." At: <www.fin.gc.ca/fedprov/eqp-eng.asp>.

Canada Forest Service. 2012. "Overview," Statistical Data, 27 Oct. At: <cfs.nrcan.gc.ca/statsprofile/overview/pe>.

Canadian Association of Petroleum Producers. 2012a. "Canada Crude Oil Production, 1971–2011. At: <membernet.capp.ca/SHB/Sheet.asp?SectionID=3&SheetID=76>.

———. 2012b. *Statistical Handbook for Canada's Upstream Petroleum Industry*, Sept. At: <www.capp.ca/getdoc.aspx?Docid=204304&DT=NTV>.

Canadian Human Rights Commission. 2003. *Report to the Canadian Human Rights Commission on the Treatment of the Innu of Canada by the Government of Canada*, by Professors Constance Backhouse and Donald M. McRae. At: <www.chrc-ccdp.ca/publications/Rapport_Innu_Report/RapportInnuReport_Page3.asp?l=e>.

Canadian Press. 2013. "Labrador Iron to Get $30-Million Cash in Pact with Tata Steel," *Globe and Mail*, 12 Mar. At: <www.theglobeandmail.com/report-on-business/industry-news/energy-and-resources/labrador-iron-to-get-30-million-cash-in-pact-with-tata-steel/article9656180/>.

Cashin, Richard. 1993. *Charting a New Course: Towards the Fishery of the Future*. Ottawa: Department of Fisheries and Oceans.

CBC News. 2007. "Long Commute, Huge Rewards," 29 Oct. At: <www.cbc.ca/canada/newfoundland-labrador/story/2007/10/29/big-commute.html>.

———. 2009. "The Big Commute," 29 Oct. At: <www.cbc.ca/nl/features/bigcommute/>.

———. 2010. "NB Power Sale Cancelled," 24 Mar. At: <www.cbc.ca/canada/new-brunswick/story/2010/03/24/nb-nb-power-graham-1027.html>.

———. 2012. "Final Plan for Sidney Tar Ponds Clean-up Announced," 28 Oct. At: <www.cbc.ca/news/canada/nova-scotia/story/2012/10/28/ns-tarpond-announcement.html>.

Chaundry, David. 2012. *Meeting the Skills Challenge: Five Key Labour Market Issues Facing Atlantic Canada*. Report of Atlantic Provinces Economic Council, Oct. At: <www.apec-econ.ca/files/pubs/%7BBA615AD5-336A-4448-980F-A121DD036733%7D.pdf?title=Meeting%20the%20Skills%20Challenge%3A%20Five%20Key%20Labour%20Market%20Issues%20Facing%20Atlantic%20Canada&publicationtype=Research%20Reports>.

Clapp, R.A. 1998. "The Resource Cycle in Forestry and Fishing," *Canadian Geographer* 42, 2: 129–44.

Coates, Ken S. 2000. *The Marshall Decision and Native Rights*. Montréal and Kingston: McGill-Queen's University Press.

Conrad, Cathy T. 2009. *Severe and Hazardous Weather in Canada: The Geography of Extreme Events*. Toronto: Oxford University Press.

Cox, Kevin. 1994. "How Hibernia Will Cast Off," *Globe and Mail*, 12 Nov., D8.

Department of Labrador and Aboriginal Affairs. 2009. "Land Claims: Innu Nation of Labrador." At: <www.laa.gov.nl.ca/laa/land_claims/index.html>.

Department of Natural Resources, Newfoundland and Labrador. 2011. *Strategic Plan 2011–14*. At: <www.nr.gov.nl.ca/nr/publications/2011-14_StrategicPlan.pdf>.

Doucette, Keith. 2013. "$288-Million Deal will Kick-start Design of Arctic Patrol Ships, Ottawa Announces," *National Post*, 7 Mar. At: <news.nationalpost.com/2013/03/07/288-million-deal-will-kick-start-design-of-arctic-patrol-ships-ottawa-announces/>.

Erskine, David. 1968. "The Atlantic Region," in John Warkentin, ed., *Canada: A Geographical Interpretation*. Toronto: Methuen, 231–80.

Faragher, John Mack. 2005. *A Great and Noble Scheme: The Tragic Story of the Expulsion of the French Acadians from Their American Homeland*. New York: Norton.

Feehan, James P., and Melvin Baker. 2005. *The Renewal Clause in the Churchill Falls Contract: The Origins of a Coming Crisis*, Papers in Political Economy No. 96. London, Ont.: University of Western Ontario, Political Economy Research Group.

Fisheries and Oceans Canada. 2012a. "2011 Value of Atlantic Coast Commercial Landings, by Region," 16 Oct. At: <www.dfo-mpo.gc.ca/stats/commercial/sea-maritimes-eng.htm>.

———. 2012b. "Sea Fisheries: Atlantic Region (Quantities)," 25 Aug. At: <www.dfo-mpo.gc.ca/stats/commercial/sea-maritimes-eng.htm>.

———. 2012c. *Stock Assessment Update of Northern (2J3KL) Cod*, Apr. At: <www.dfo-mpo.gc.ca/csas-sccs/Publications/ScR-RS/2012/2012_009-eng.pdf>.

Griffiths, N.E.S. 2005. *From Migrant to Acadian: A North American Border People 1604–1755*. Montréal and Kingston: McGill-Queen's University Press.

Hardin, Garrett. 1968. "The Tragedy of the Commons," *Science* 162: 1243–8.

Hiller, J.K. 1997. "The Debate: Confederation Rejected, 1864–1869," *Newfoundland and Canada: 1864–1949*. At: <www.heritage.nf.ca/law/debate.html>.

Holden, Michael. 2008. *Canada's New Equalization Formula.* Ottawa: Library of Parliament, PRB 08–20E.

Hydro-Québec. 2012. *Comparison of Electricity Prices in Major North American Cities: Rates in effect April 1, 2012.* At: <www.hydroquebec.com/publications/en/comparison_prices/pdf/comp_2012_en.pdf>.

Index Mundi. 2013. "Crude Oil (Petroleum); Dated Brent Daily Price," 11 Mar. At: <www.indexmundi.com/commodities/?commodity=crude-oil-brent&months=120>.

Llewellyn, Stephen. 2008. "Nackawic Mill May Shut Down for 4–6 Weeks," *Daily Gleaner* (Fredericton), 11 Nov. At: <dailygleaner.canadaeast.com/rss/article/477340>.

McCarthy, Shawn. 2007. "Nfld. Grabs a Seat at Oil Table," *Globe and Mail*, 23 Aug. At: <www.reportonbusiness.com/servlet/story/RTGAM.20070823.whebron0823/BNStory/robNews/home>.

———. 2011. "Lower Churchill Power Plan Pushes Ahead," *Globe and Mail*, 11 Nov., B3.

MacDonald, Michael. 2012a. "Irving Shipbuilding Sets Deadline of January for Arctic Patrol Ship Work," *Vancouver Sun*, 19 Oct. At: <www.vancouversun.com/news/national/Irving+Shipbuilding+sets+deadline+January+Arctic+patrol+ship+work/7417496/story.html>.

———. 2012b. "Cod Making a Comeback in Newfoundland, Research Shows," Globe and Mail, 2 July. At: <www.theglobeandmail.com/news/national/cod-making-a-comeback-in-newfoundland-research-shows/article4385506/>.

MacKenzie, A.A. 1979. *The Irish in Cape Breton.* Antigonish, NS: Formac.

Macpherson, Alan G., ed. 1972. *The Atlantic Provinces: Studies in Canadian Geography.* Toronto: University of Toronto Press.

Macpherson, Joyce. 1997. "Cold Ocean," Newfoundland and Labrador Heritage, Memorial University of Newfoundland. At: <www.heritage.nf.ca/environment/ocean.html#amherst>.

Matthews, Ralph. 1983. *The Creation of Regional Dependency.* Toronto: University of Toronto Press.

Moore, Oliver. 2009. "Innu Reach Deal on Lower Churchill Project," *Globe and Mail*, 26 Sept., B1.

Natural Resources Canada. 2009a. "Key Facts in the Forest Industry, New Brunswick." At: <canadaforests.nrcan.gc.ca/keyfacts/nb>.

———. 2009b. "Canada's Forests: Statistical Data," *Forest Management.* At: <canadaforests.nrcan.gc.ca/statsprofile/forest/pe>.

———. 2012. "Mineral Production," 23 Aug. At: <mmsd.mms.nrcan.gc.ca/stat-stat/prod-prod/ann-ann-eng.aspx>.

North of 56. 2012. "Voisey's Bay Nickel Mine in Labrador Hums Along," 11 June. At: <northof56.com/minerals/article/voiseys-bay-nickel-mine-in-labrador-hums-along>.

Phillips, David. 1993. *The Day Niagara Falls Ran Dry!* Toronto: Canadian Geographic and Key Porter Books.

Physorg.com. 2006. "Ocean Study Predicts the Collapse of All Seafood Fisheries by 2050." At: <www.physorg.com/news81778444.html>.

Power, Thomas P., ed. 1991. *The Irish in Atlantic Canada, 1780–1900.* Fredericton: New Ireland Press.

Prince Edward Island. Department of Agriculture and Forestry. 2012. "Farm Cash Receipts," Statistics. At: <www.gov.pe.ca/photos/original/af_stat_tab2.pdf>.

Quinn, Greg, 2012. "Long-Distance Commutes the Normal Life for Many Canadians," Financial Post, 1 Aug. At: <business.financialpost.com/2012/08/01/long-distance-commutes-the-new-normal-for-many-canadians/>.

Roberts, Terry. 2009. "Construction Begins on $22B Nickel Plant in Long Harbour," 17 Apr. At: <www.dailybusinessbuzz.ca/2009/04/17/construction-begins-on-22b-nickel-plant-in-long-harbour/>.

Samson, Colin. 2003. *A Way of Life That Does Not Exist: Canada and the Extinguishment of the Innu.* St John's: ISER Books.

Statistics Canada. 2002a. *Census of Canada 2001—Census Geography. Highlights and Analysis: Canada's 2001 Population.* At: <www12.statcan.ca/English/census01>.

———. 2002b. "Population and Dwelling Counts, 2001 Census," 16 July. Catalogue no. 93F0050XCB2001013.

———. 2007a. "Population and Dwelling Counts, for Census Metropolitan Areas and Census Agglomerations, 2006 and 2001 Censuses—100% Data." At: <www12.statcan.ca/english/census06/data/popdwell/Table.cfm?T=201&S=3&O=D&RPP=150>.

———. 2007b. "Labour Force, Employed and Unemployed, Numbers and Rates." At: <www40.statcan.ca/l01/cst01/labor07c.htm>.

———. 2008. "Farm Population and Total Population by Rural and Urban Population, by Province," *2001 and 2006 Census of Agriculture and Census of Population.* At: http://www.statcan.gc.ca/tables-tableaux/sum-som/l01/cst01/agrc42a-eng.htm.

———. 2009a. "Chart 21.1 (data) Unemployment Rate," 16 Jan. At: <www41.statcan.gc.ca/2008/2621/grafx/htm/ceb2621_000_1-eng.htm#table>.

———. 2009b. "Distribution of Employed People, by Industry, by Province." At: http://www.statcan.gc.ca/tables-tableaux/sum-som/l01/cst01/labor21c-eng.htm.

———. 2009c. "Urban Area (UA)," *2006 Census Dictionary.* At: <www12.statcan.ca/census-recensement/2006/ref/dict/geo049-eng.cfm>.

———. 2009d. "Population, Age Distribution and Median Age by Province and Territory, as of July 1, 2008," Table 1. At: <www.statcan.gc.ca/daily-quotidien/090115/t090115c1-eng.htm>.

———. 2010. "Labour Force, Employed and Unemployed, Numbers and Rates, by Province," 29 Jan. At: http://www.statcan.gc.ca/tables-tableaux/sum-som/l01/cst01/labor07b-eng.htm.

———. 2012a. "Population and Dwelling Counts, for Canada, Provinces and Territories, 2011 and 2006 Censuses," 24 Jan. At: <www12.statcan.gc.ca/census-recensement/2011/dp-pd/hlt-fst/pd-pl/index-eng.cfm>.

———. 2012b. "Gross Domestic Product, Expenditure-based, by Province and Territory," 19 Nov. At: <www.statcan.gc.ca/tables-tableaux/sum-som/l01/cst01/econ15-eng.htm>.

———. 2012c. "Unemployment Rate, Canada, Provinces, Health Regions and Peer Groups,: CANSIM Table 109–5324,

24 May. At: <www5.statcan.gc.ca/cansim/pick-choisir?lang=eng&p2=33&id=1095324>.

———. 2012d. "Great Harbour Deep," Census Profile. 17 Oct. At: <www12.statcan.gc.ca/census-recensement/2011/dp-pd/prof/search-recherche/frm_res.cfm?Lang=E&TABID=1&G=1&Geo1=PR&Code1=10&Geo2=0&Code2=0&SearchType=Begins&SearchText=great+harbour+deep&PR=10>.

Stavely, Michael. 1987. "Newfoundland Economy and Society at the Margin," in L.D. McCann, ed., *Heartland and Hinterland: A Geography of Canada*, 2nd edn. Scarborough, Ont.: Prentice-Hall, 247–85.

Storey, Keith. 2009. "Help Wanted: Demographics, Labor Supply and Economic Change in Newfoundland and Labrador," *Challenged by Demography: A NORA Conference on the Demographic Challenges of the North Atlantic Region*, Alta, Norway, 20 Oct.

Tait, Carrie. 2013. "East Coast's Hebron Offshore Play Gets the Nod from Exxon," *Globe and Mail*, 4 Jan. At: <www.theglobeandmail.com/globe-investor/east-coasts-hebron-offshore-play-gets-the-nod-from-exxon/article6942298/>.

Taylor, Roger. 2012. "NDP Wise to Downplay Offshore Royalties," 3 Apr. *Chronicle Herald* (Halifax). At: <thechronicleherald.ca/business/80666-ndp-wise-downplay-offshore-royalties>.

Vale. 2012. *Long Harbour Overview*, Mar. At: <www.vbnc.com/Reports/LH%20Project%20Media%20Update%20March%2028%202012.pdf>.

Ware, Beverley. 2012. "Time Runs Out for Bowater Mill," *Chronicle Herald* (Halifax), 14 June. At: <thechronicleherald.ca/novascotia/107146-time-runs-out-for-bowater-mill>.

CHAPTER 10

Abele, Frances, Thomas J. Courchene, F. Leslie Seidle, and France St-Hilaire, eds. 2009. *Northern Exposure: Peoples, Powers and Prospects in Canada's North*. Montréal: Institute for Research on Public Policy.

Aboriginal Affairs and Northern Development Canada (AANDC). 2010. "The History of Land Claims and Self-government in the Yukon," 15 Oct. At: <www.aadnc-aandc.gc.ca/eng/1100100028417/1100100028418>.

———, Northern Oil and Gas Branch. 2012. *Northern Oil and Gas Annual Report 2011*, 2 May. At: <www.aadnc-aandc.gc.ca/eng/1335971994893/1335972853094>.

Agnew, John. 2005. "Sovereignty Regimes: Territoriality and Authority in Contemporary World Politics," *Annals, Association of American Geographers* 95, 2: 437–61.

BBC News. 2007. "Canada to Strengthen Arctic Claim," 10 Aug. At: <news.bbc.co.uk/2/hi/Americas/6941426.stm>.

Berger, Thomas R. 1977. *Northern Frontier, Northern Homeland: The Report of the Mackenzie Valley Pipeline Inquiry*, 2 vols. Ottawa: Minister of Supply and Services.

Bone, Robert M. 2012. *The Canadian North: Issues and Challenges*. Toronto: Oxford University Press.

Boswell, Randy, and Juliet O'Neill. 2010. "Clinton Blasts Canada for Exclusivity of Arctic Talks," *Montreal Gazette*, 29 Mar. At: <www.montrealgazette.com/news/Clinton+blasts+Canada+exclusive+Arctic+talks/2740399/story.html>.

Bumsted, J.M. 2010. *The Peoples of Canada: A Pre-Confederation History*, 3rd edn. Toronto: Oxford University Press.

Byers, Michael. 2009. *Who Owns the Arctic? Understanding Sovereignty Disputes in Canada's North*. Vancouver: Douglas & McIntyre.

Canada. 1985. *The Western Arctic Claim: The Inuvialuit Final Agreement*. Ottawa: Department of Indian Affairs and Northern Development.

———. 1991. "Comprehensive Land Claim Agreement Initialled with Gwich'in of the Mackenzie Delta in the Northwest Territories," Communiqué 1–9171. Ottawa: Department of Indian Affairs and Northern Development.

———. 1993a. "Formal Signing of Tungavik Federation of Nunavut Final Agreement," Communiqué 1–9324. Ottawa: Department of Indian Affairs and Northern Development.

———. 1993b. *Umbrella Final Agreement between the Government of Canada, Council for Yukon Indians and the Government of the Yukon*. Ottawa: Department of Indian Affairs and Northern Development.

———. 2004. "Agreements." Ottawa: Department of Indian and Northern Affairs. At: <www.ainc-inac.gc.ca/pr/agr/index_e.html#Comprehensive%20Claims%20Agreements>.

CBC News. 2007. "Mackenzie Gas Line Still 'Leading Case' Despite Bloating $16.2B Cost Outlook," 12 Mar. At: <www.cbc.ca/cp/business/070312/b031292A.html#skip300x250>.

———. 2009. "Yellowknife Diamond-Cutting Plant in Limbo," 11 June. At: <www.cbc.ca/canada/north/story/2009/06/11/nwt-diamonds.html?ref=rss>.

Chan, Laurie H.M. 2006. "Food Safety and Food Security in the Canadian Arctic," *Meridian* (Publication of the Canadian Polar Commission): 1–3.

Coates, Ken, P. Whitney Lackenbauer, William Morrison, and Greg Poelzer. 2008. *Arctic Front: Defending Canada in the Far North*. Toronto: Thomas Allen.

Crowe, Keith J. 1991. *A History of the Original Peoples of Northern Canada*, 2nd edn. Montréal and Kingston: McGill-Queen's University Press.

Danylchuk, Jack. 2007 "Giant, Glittering and Tarnished," *Up Here*, 18 Dec. At: <www.uphere.ca/node/175>.

Dickason, Olive Patricia. 2002. *Canada's First Nations: A History of Founding Peoples from Earliest Times*, 3rd edn. Toronto: Oxford University Press.

Drummond, K.J. 2009. *Northern Canada Distribution of Ultimate Oil and Gas Resources*. At: <drummondconsulting.com/NCAN09Report.pdf>.

Ebner, David. 2009. "Exxon Backs Alaska Gas Pipeline," *Globe and Mail*, 12 June, B14.

Elias, Peter Douglas. 1995. *Northern Aboriginal Communities: Economies and Development*. North York, Ont.: Captus Press.

Environment Canada. 2012. "Canadian Arctic Sea Ice Reached Record Low in Summer 2012," 5 Nov. At: < http://www.ec.gc.ca/glaces-ice/default.asp?lang=En&n=765F63E4-1>.

Globe and Mail. 1993. "Mackenzie Land Claim Settled," 7 Sept., A1.

Griffiths, Franklyn. 2009. "Canadian Arctic Sovereignty: Time to Take Yes for an Answer on the Northwest Passage," in Abele et al. (2009: 107–36).

Hamelin, Louis-Edmond. 1978. *Canadian Nordicity: It's Your North, Too*, trans. William Barr. Montréal: Harvest House.

Indian and Northern Affairs Canada (INAC). 2007. *Northern Oil and Gas Annual Report 2006*. At: <www.ainc-inac.gc.ca/oil/ann/ann2006/dev_e.html>.

———. 2009a. *Northern Oil and Gas Annual Report 2008*. At: <www.ainc-inac.gc.ca/nth/og/pubs/ann/ann2008/ann2008-eng.pdf>.

———. 2009b. *Framework Agreement*, 14 Jan. At: <www.ainc-inac.gc.ca/al/ldc/ccl/agr/dcf/dcf-eng.asp>.

Jordan, Pav. 2012. "Nunavut Mining Rush Attracts China's MMG," *Globe and Mail*, 5 Sept., B3.

———. 2013. "Baffinland Iron Mines Sharply Scales Back Mary River Project," *Globe and Mail*, 11 Jan. At: <www.theglobeandmail.com/globe-investor/baffinland-iron-mines-sharply-scales-back-mary-river-project/article7227358/>.

Koring, Paul. 2012. "In the Arctic, Drones Could Close the Gap," *Globe and Mail*, 9 July, A13.

Légaré, André. 2008a. "Canada's Experiment with Aboriginal Self-determination in Nunavut: From Vision to Illusion," *International Journal on Minority and Group Rights* 15: 335–67.

———. 2008b. "Inuit Identity and Regionalization in the Canadian Central and Eastern Arctic: A Survey of Writings about Nunavut," *Polar Geography* 31, 3 and 4: 99–118.

McGhee, Robert. 1996. *Ancient People of the Arctic*. Vancouver: University of British Columbia Press.

MacLachlan, Letha. 1996. *NWT Diamonds Project: Report of the Environmental Assessment Panel*. Ottawa: Canadian Environmental Assessment Agency.

McRae, Donald. 2008. "An Arctic Agenda for Canada and the United States," in *From Correct to Inspired: A Blueprint for Canada–US Engagement*. At: <www.carleton.ca/ctpl/conferences/documents/BackgroundPapers-Final.pdf>.

Marcus, Alan R. 1995. *Relocating Eden: The Image and Politics of Inuit Exile in the Canadian Arctic*. Hanover, NH: University Press of New England.

Natural Resources Canada. 2007. "High Metal Prices Spur Mineral Production to a Record $34 billion in 2006," *Information Bulletin*, Mar. At: <www.nrcan.gc.ca/mms/pdf/minprod-07_e.pdf>.

———. 2010. "The Global Recession Reduced Canada's Mineral Production in 2009." *Information Bulletin*, Mar. At: <mmsd.mms.nrcan.gc.ca/stat-stat/prod-prod/PDF/ib2010_e.pdf>.

Northern Gas Pipelines. 2009. *Northern Gas Pipelines: Mackenzie Valley Pipeline Project*. At: <www.arcticgaspipeline.com/Delta%20Route.htm>.

Northwest Territories Bureau of Statistics. 2011. *By the Numbers*, Nov. At: <www.statsnwt.ca/publications/bythenos/BytheNumbers2011.pdf>.

———. 2012. *Statistical Quarterly* 34, 2. At: <www.statsnwt.ca/publications/statistics-quarterly/sqjun2012.pdf>.

Nunavut. 1999. *The Bathurst Mandate Pinasuaqtavut: What We've Set Out to Do*. Iqaluit: Legislative Assembly.

Nunavut Tunngavik Incorporated. 2007. *Saqqiqpuq: Kindergarten to Grade 12 Education in Nunavut. Annual Report on the State of Inuit Culture and Society*. At: <www.tunngavik.com/documents/publications/2005-2007-Annual-Report-on-the-State-of-Inuit-Culture-and-Society-Eng.pdf>.

O'Neil, Peter. 2009. "Jean's Seal Meal Draws Praise, Criticism," *StarPhoenix* (Saskatoon), 27 May, C12.

Page, Robert. 1986. *Northern Development: The Canadian Dilemma*. Toronto: McClelland & Stewart.

Rowley, Graham W. 1996. *Cold Comfort: My Love Affair with the Arctic*. Montréal and Kingston: McGill-Queen's University Press.

Shadian, Jessica. 2007. "In Search of an Identity Canada Looks North," *American Review of Canadian Studies* 37, 3: 323–53.

Statistics Canada. 2004. "Study: Diamonds Are Adding Lustre to the Canadian Economy," *The Daily*, 13 Jan. At: <www.statcan.gc.ca/Daily/English/040113/d040113a.htm>.

———. 2008. "Aboriginal Identity Population by Age Groups, Median Age and Sex, 2006 Counts, for Canada, Provinces and Territories." At: <www12.statcan.gc.ca/english/census06/data/highlights/Aboriginal/pages/Page.cfm?Lang=E&Geo=PR&Code=01&Table=1&Data=Count&Sex=1&Age=1&StartRec=1&Sort=2&Display=Page>.

———. 2009a. "Population and Demography: Population Estimates and Projections." At: <cansim2.statcan.gc.ca/cgi-win/cnsmcgi.pgm?Lang=E&SP_Action=Result&SP_ID=3433&SP_TYP=5&SP_Sort=1&SP_Mode=2>.

———. 2009b. "Main Demographic Indicators for Canada, Provinces and Territories, 1981 to 2007," *Report on the Demographic Situation in Canada: 2005 and 2006*. At: <www.statcan.gc.ca/pub/91-209-x/2004000/tabl0-eng.htm>.

———. 2012a. "Population and Dwelling Counts, for Canada, Provinces and Territories and Population Centres," 11 Apr. At: <www12.statcan.gc.ca/census-recensement/2011/dp-pd/hlt-fst/pd-pl/Tables-Tableaux.cfm?LANG=Eng&T=800>.

———. 2012b. "Components of Population Growth, Canada, Provinces and Territories," CANSIM: Table 051–0004, 27 Oct. At: <www.statcan.gc.ca/tables-tableaux/sum-som/l01/cst01/demo33c-eng.htm>.

———. Statistics Canada. 2013. "Number and distribution of population reporting an Aboriginal identity and percentage of Aboriginal people in the population, Canada, provinces and territories, 2011", *Aboriginal Peoples in Canada: First Nations People, Métis and Inuit*. National Household Survey document 99-011-x. Table 2. 7 May. At:<http://www12.statcan.gc.ca/nhs-enm/2011/as-sa/99-011-x/2011001/tbl/tbl02-eng.cfm>.

Usher, Peter J., and George Wenzel. 1989. *A Strategy for Supporting the Domestic Economy of the Northwest Territories*. Report for the Legislative Assembly's Special Committee on the Northern Economy. Ottawa: P.J. Usher Consulting Services.

Weber, Bob. 2012. "Ottawa Set to Eye China's Nunavut Mine Plan," *Globe and Mail*, 28 Dec., B1.

Wente, Margaret. 2012. "Who's in Trouble, Polar Bears or People," *Globe and Mail*, 10 Apr., F9.

Widdowson, Frances, and Albert Howard. 2008. *Disrobing the Aboriginal Industry: The Deception behind Indigenous Cultural Preservation*. Montréal and Kingston: McGill-Queen's University Press.

Williams, Glyn. 2009. *Arctic Labyrinth*. Toronto: Viking Canada.

Williamson, Robert G. 1974. *Eskimo Underground: Socio-Cultural Change in the Canadian Central Arctic*. Occasional Papers II. Uppsala, Sweden: Almqvist & Wiksell.

Young, Oran. 2009. "Whither the Arctic: Conflict or Cooperation in the Circumpolar North," *Polar Record* 45, 232: 73–82.

CHAPTER 11

Adams, Michael. 2007. *Unlikely Utopia: The Surprising Triumph of Canadian Pluralism*. Toronto: Viking Press.

Agriculture and Agri-Food Canada. 2009. *Overview of the Canadian Agriculture and Agri-Food System 2008*. At: <www4.agr.gc.ca/AAFC-AAC/display-afficher.do?id=1228246364385&lang=eng#a2>.

Baker, Liana. 2009. "New Frontier in International Affairs," *CG Compass* (blog), 27 Aug. At: <www.canadiangeographic.ca/blog/posting.asp?ID=30>.

Cattaneo, Claudia. 2012. "Canada Embracing Its Inner Oil," *National Post*, 25 Jan. At: <business.financialpost.com/2012/01/25/obama-adopts-gas-for-future/?__lsa=38675950>.

Conway, Sir Gordon. 2009. "Geographical Crises of the Twenty-First Century," *Geographical Journal* 175, 3: 221–8.

Courchene, Thomas J., with Colin R. Telmer. 1998. *From Heartland to North American Region State: The Social, Fiscal and Federal Evolution of Ontario*. Monograph Series on Public Policy, Centre for Public Management. Toronto: University of Toronto Press.

De Blij, H.J., and Alexander B. Murphy. 2006. *Human Geography: Culture, Society, and Space*, 8th edn. Toronto: John Wiley.

Dobson, Wendy. 2009. *Gravity Shift: How Asia's New Economic Powerhouses Will Shape the Twenty-First Century*. Toronto: University of Toronto Press.

Economist, The. 2011. "Ten Years of China in the WTO: Shades of Grey," Dec. At: <www.economist.com/node/21541408>.

Florida, Richard. 2002. *The Rise of the Creative Class: And How It's Transforming Work, Leisure, Community and Everyday Life*. New York: Basic Books.

———. 2005. *The Flight of the Creative Class*. New York: HarperCollins.

———. 2008. *Who's Your City?: How the Creative Economy Is Making Where to Live the Most Important Decision of Your Life*. New York: Basic Books.

———. 2010. *The Great Reset*. Toronto: Random House.

———. 2012. *The Rise of the Creative Class—Revisited*. New York: Basic Books.

Foreign Affairs and International Trade Canada. 2012. *Canada's State of Trade: Trade and Investment Update—2011*, "V. Key Developments in Canadian Merchandise Trade," 12 Mar. At: <www.international.gc.ca/economist-economiste/performance/state-point/state_2011_point/2011_5.aspx?lang=eng&view=d>.

Frye, Northrop. 1971. *The Bush Garden: Essays on the Canadian Imagination*. Toronto: Anansi Press.

Garreau, Joel. 1981. *The Nine Nations of North America*. Boston: Houghton Mifflin.

Greenwood, John. 2007. "Surging Prices Separate Wheat from the Chaff," *National Post*, 15 June, FP1.

———. 2009. "Toronto Has Potential To Be Top Global Financial Centre, Report Says," *National Post*, 18 Nov. At: <network.nationalpost.com/np/blogs/toronto/archive/2009/11/18/toronto-has-potential-to-be-top-global-financial-centre-report-says.aspx>.

Hamilton, Graeme. 2007. "Welcome! Leave Your Customs at the Door," *National Post*, 30 Jan., A1, A7.

Hare, F. Kenneth. 1968. "Canada," in John Warkentin, ed., *Canada: A Geographical Interpretation*. Toronto: Methuen.

Ibbitson, John. 2010. "House Reform Boosts Fastest-Growing Provinces," *Globe and Mail*, 2 Apr., A1, A6.

Innis, Harold. 1930. *The Fur Trade in Canada: An Introduction to Canadian Economic History*. New Haven: Yale University Press.

International Energy Agency. 2012. "North America Leads Shift in Global Energy Balance, IEA says in Latest World Energy Outlook," 12 Nov. At: <www.iea.org/newsroomandevents/pressreleases/2012/november/name,33015,en.html>.

Li, Wei. 2009. *Ethnoburb: The New Ethnic Community in Urban America*. Honolulu: University of Hawaii Press.

McCarthy, Shawn. 2009. "Canada's Race for a High-Tech Strategy," *Globe and Mail*, 1 Aug., B1, B3.

Marotte, Bertrand. 2012. "Luxury Business Jet Order Buoys Bombardier," *Globe and Mail*, 27 Nov. At: <www.theglobeandmail.com/globe-investor/bombardier-signs-historic-78-billion-jet-deal-biggest-on-record/article5718399>.

Mohr, Patricia. 2009. *Scotiabank Commodity Price Index*, 23 Dec. At: <www.scotiacapital.com/English/bns_econ/bnscomod.pdf>.

Parkinson, David. 2009. "Despite Some Price Woes, Commodities Bull Market Is Thriving," *Globe and Mail*, 17 July, B8.

Perreaux, Les. 2010. "Asked to Remove Niqab, Quebec Woman Lodges Human-Rights Complaint," *Globe and Mail*, 3 Mar. At: <www.theglobeandmail.com/news/national/quebec/asked-to-remove-niqab-quebec-woman-lodges-human-rights-complaint/article1487526/>.

Saul, John Ralston. 1997. *Reflections of a Siamese Twin: Canada at the End of the Twentieth Century*. Toronto: Viking.

———. 2009. *The Collapse of Globalism and the Reinvention of the World*, 2nd edn. Toronto: Penguin Canada.

Scoffield, Heather. 2009. "Recession Toll Harsh, but Easing," *Globe and Mail*, 1 June. At: <www.globeinvestor.com/servlet/story/RTGAM.20090601.wgdp0601/GIStory/>.

Statistics Canada. 2009. *Annual Demographic Estimates: Canada, Provinces and Territories 2009*. At: <www.statcan.gc.ca/pub/91-215-x/91-215-x2009000-eng.htm>.

Wallerstein, Immanuel. 1979. *The Capitalist World Economy*. Cambridge: Cambridge University Press.

———. 1998. "Contemporary Capitalist Dilemmas, the Social Sciences, and the Geopolitics of the Twenty-First Century," *Canadian Journal of Sociology* 23, 2 and 3: 141–58.

INDEX

Note: Page numbers in *italic type* indicate figures or captions.

Aboriginal Affairs and Northern Development Canada (AANDC), 95
Aboriginal Fisheries Strategy (AFS), 290
Aboriginal/non-Aboriginal faultline, xvi, 14–15, 67, 89, 92–103, 162, 353, 419; in Atlantic Canada, 95, 369; in BC, 98–99, 268, 278, 307; bridging, 101–3; Indian Act and, 93–94; residential schools and, 94–95; in Québec, 99, 248; in Western Canada, 96, 353
Aboriginal peoples, 448; activists, 152; in Atlantic Canada, 368, 369, 370, 392, 394; band names and, 96; in BC, 278, 279, 280, 281, 296; birth rates and, 152, 402, 420; census and, *13*, 72, 95, 96, 152, 154; children and, 92, 94, 96, 155, 411; culture regions of, 72–73, *72*; defining, 95–96; demographic changes and, 152–55; diversity and, 15, 95; economy in Territorial North, 424; education and, 151, 155; European contact and, 72, 154, 410–11; families, 155; as first people, 68–74; homeland and, 151; housing shortages, 155; hunting and, 408; identity and, 13, 448; land disputes, 185; language families of, *73*, 95; mobility of, 421; "modernization" of, 412; in Ontario, 183–85, 212; population and, 72, 95–96, 122, 128, *151*, 150–55, *154*; in Québec, 217, 225, 227, 253, 353; relocation of, 411–13; rights and, 102, 83, 96, 97–98, 99, 290, 393–94, 448; self-government and, 95, 101, 151, 404, 411, 424, 425; self-identification, 95; socio-economic challenges and opportunities, 14, 21–22, 102–3, 152–3, 412–13; in Territorial North, 401, 402–3, 424; treaty rights, 96–97, 458; urban, 96, 151; voting rights, 14, 92, 152; in Western Canada, 326–27, 351, 352, 353; *see also* Inuit; land claims; Métis; treaties; specific groups
Acadians, 103, 366, 368–69
acculturation, 139, 149, 448
Act of Union, 113, 178, 232
aerospace industry: in Québec, 11, 159, 224, 238–39, 251, 252–53, 258
Afghanistan, 21
age dependency ratio, 122, 141, 448
Aglukkaq, Leona, 14, 417
agricultural fringe, 332–33, 335, 448
agriculture, 70–71; Aboriginal peoples and, 70, 71; in agricultural fringe, 335–36; in Atlantic Canada, 394–95, *394*; biofuel and, 336; continuous cropping, 335, 337, 450; dairy industry and, 226, 227, 251, 394; Doukhobor, 109–10; in Dry Belt, 334–35; fallowing, 330, 335, 337, *338*, 452; farm size and, 330, 331; feedlots and, 333, 340; in Fertile Belt, 333; global demand and higher prices, 336–37; grain economy and, 78; innovation in, 330–31; irrigation and, 335; loss of land, 336–37; in Manitoba, 108; mechanization and, 132, 331; in New France, 230–31; Ontario, 171–74, *183*; in Peace River country, 335–36; in Québec, 226, 251, 253–54; rural land-use changes, 173, 337; stubble fields, 330, 458; summer fallow, 325–26, 458; in Western Canada, 173, 316, 325–26, 330–40, 336; zero-tillage, 330, 459; *see also* livestock industry; specialty crops; specific regions
Ahmadiyya Muslim community mosque, *145*
air commuting, 431, 434, 448
air drainage, 174, 448
air masses, 51–52; continental, 48, 450; marine, 48, 454
air pollution, 60, 64, 175–76, 448
Aklavik, NWT, 42
Alaska, 80, 83; cruise ship industry, 296, *296*
Alaska Panhandle, 267, 277, 448
albedo effect, 407, 448
Alberta, 311, 312, 326; agriculture in, 335; boundaries of, 84; Confederation and, 83; economy of, xvii, 11, 327; GDP and, 223; as "have" province, 89; hydrocarbon resources, *341*; knowledge-based economy in, 329; mining industry in, 348; petroleum industry and, 38, 314, 327, 328, 341, 342–43, 346; population, 108, 109, 222; as province, 82, 84, 325; urban centers in, 351–53; unemployment in, 314; wealth of, *90*; *see also* oil sands; Western Canada
Alberta Badlands, 38, *39*
Alberta clippers, 315, 448
Alberta Energy, 348
Alberta Plateau, 39
Alcan, 253, 294–95, 468n7
Alcoa, 243
Algome steel mill (Essar Algoma), 211
Algonkian language family, *73*
Algonquins, 69, 71, 72
allophones, 143, 216, 217, 262, 448
aluminum, 58, 197, 224, 242–43, 294–95
Ambassador Bridge, 25
American Indian Movement, 102
American Revolution, 77, 369
American War of Independence, 74, 75, 112
ancestry, Aboriginal, 13, 448
anglophones, 216, 217, 448; *see also* French/English faultline
Annapolis Valley, 366, *366*, 394–95
anti-globalization movement, 123
Appalachian Mountains, 42, 93, 226, 230, 361
Appalachian Uplands, 30, 32, 42; in Atlantic Canada, 361, 391; in Québec, 226–27, 250–51; urban development in, 226, 251, 260

Aqualegend, 274
arable land: in Atlantic Canada, 42, 394; in BC, 271–72; population and, 123; in Ontario, 173, 179; in Québec, 226, 227, 254; in Western Canada, 78, 321–22, 323, 325, 335, 336
ArcelorMittal Dofasco, 207
ArcelorMittal Mines, 255, 257
Arctic Archipelago, 33, 41, 60, 79, *80*, *81*, 82, 85, 405, 418
Arctic Basin, 56, 57, 58, 406, 416, *416*
Arctic Circle, 41, 448
Arctic climate zone, 50, 60, 363, 405
Arctic Coastal Plain, 41
Arctic Co-operation Agreement, 418
Arctic Council, 417
Arctic culture region, 72, *72*
Arctic Lands, 32, 34, 41–42, 413; in Territorial North, 405, 406
Arctic Ocean, 405–6, 408, 421; global warming and, 407, 414; sovereignty and, 414–16, 418; toxic spills and, 418; transportation and, 418, 430
Arctic Platform, 41
Arctic seabed, 6, 413, 414, 416, *416*
Arctic Waters Pollution Prevention Act, 413, 418
arêtes, 37, 448
Arslanian Cutting Works, 435
Asbestos, QC, 251, 260
Asia: growing power of, 18; pork exports to, 336; *see also* Pacific Rim countries
Asia-Pacific Gateway and Corridor, 286, 287, 448
Assembly of First Nations (AFN), 152
assimilation, 67; Aboriginal peoples and, 92, 93, 94, 102, 106; Québec and, 113, 232, 325
Assiniboine Indians, 72
Athabasca Basin, 349
Athabasca Glacier, 37, *62*
Athabasca oil sands, 314, 343, 448; *see also* oil sands
Athabasca River, 39, 319
Athabaskan Council, 417
Athapaskans, 69, 71, 72, *73*, 408
Atlantic Basin, 56, 57–58
Atlantic Canada, xvi, 6, 7, 357–99, 442, 445; agriculture in, 389, 394–95; basic statistics, *361*, 361; within Canada, 358–60; climate in, 363; climatic zones in, 363; CMAs in, 397; coastal communities in, 363, 372, 379, 42; Confederation and, 370–73; cost of government in, 375; deindustrialization in, 372; distance to market, 372; as downward transitional region, 360; economic structure of, 18, 379–81; economy of, 357, 358, 360,

373–75, 375–76, 399; employment in, 379, 380; energy corridor in, 358; environment and, 366–67; equalization payments, 377; exports, 369; fishing industry in, 7, 381–87, *382*, 389; forest industry in, 389, 391–94; future of, 399; GDP of, 360; general characteristics of, 7; geographical fragmentation of, 362; as "have-not" region, 360; high-tech industries in, 379; as hinterland, 372; historical geography, 367–73; hydroelectric developments in, 395; industrial structure, 379–81; map, *359*, *368*; mining industry in, 390–91; megaprojects in, 389–90; out-migration and, 375, 376–77, 380–81; overfishing and, 387; petroleum industry, 387–90; physical geography of, 361–65, 377; physiographic regions in, 361; population, 132, 156, 375, 395–96; precipitation, 363; resources in, 360, 369, 379, 387–95, *388*; rural, 375; sense of place, 9, 148, 358; settlement of, 367–69; shipbuilding industry in, 377; social characteristics of, 13; sub-regions of, 362; today, 373–81; trade with New England, 369; unemployment rate, 360; urban centres in, 132, 376, 380, 398; urban geography of, 396–98; weather in, 361–63; *see also* New Brunswick; Newfoundland and Labrador; Nova Scotia; Prince Edward Island
Atlantic climate zone, 50, 363
Atlantic Energy Accord, 91
Atlantic Gateway, 358, 360, 397, 398, 399
Atlantic Ocean, 358, 363, 407
Atlantic Provinces Economic Council, 375
Atleo, Shawn, 152
Attawapiskat First Nation, 15, 22, 41, 212; De Beers and, 102–3
aurora borealis (northern lights), 405, 455
Australia, population density, 123
automobile industry, 192–99; assembly plants, 196–97, *197*; automobile parts firms, 197–99; Chinese-based firms, 193, 199; collapse of, 11, 166; economic crisis and, 193–94; electric, 196; employment in, *193*, 195, 198; federal bailout of, 11, 88, 90, 195; Japanese-based firms, 196, 197; just-in-time production, 198; mining industry and, 196; in Ontario, xvii, 182, 192–99, 210, 212 13; outsourcing, 198; pollution and, 60, 64, 175–78; production in Canada, 194; in Québec, 252; restructuring of, 188–89; transportation routes and, 198–99; US border and, 189, 198; *see also* Auto Pact
Automotive Innovation Fund, 195
Auto Pact (Canada–US Automotive Products Agreement), 19, 147, 181, 182, 189, 192–93, 440, 466n4

AV Nackawic mill, *393*
Ayed, Nahlah, 148

baby boom, 134, 135, 140
Baffin Island, 82, 408, 409
Baffinland Iron Mines Corporation, 406–7, 431
Bagnall, James, 209
Bakken formation, 329, 341, 448–49
Ballard Power Systems, 283
Banff National Park, 37, 314
Bangladesh, population density, 123
Barkerville, BC, 278
Barren Grounds, 412, 449
basalt columns, *41*
Basilica of Notre Dame, *144*
basins, 38, 449; drainage, 56–58, *56*, 451; sedimentary, 38, 387, 406
Bathurst Mandate, 422, 423
Batoche, Battle of, 323
Bayer, 327
BC Hydro, 294, 295
BC Rail, *280*, 281
BC Treaty Commission, 278
Beardy, Stan, 152
Beatty, Stephen, 188
Beaufort Sea, 418
Bell, Daniel, 160
Beothuk, 368, 449
Berger, Justice Thomas, 432
Berger Inquiry, 99, 432
Beringia, 68, 70
Bernier, Jacques, 156
Beyond Borders Agreement, 19, 20
BHP Billiton, 435
biculturalism, 115, 146
Big Bear, 106, 449
"Big Becky," 177
Big Commute, 395–96, 399, 449
Big Three, 182, 189, 195, 196, 198
bilingualism, 115
Bill 14, 217, 237, 262
Bill 101 (Québec), 218, 262, 465n9, 467n5
Bill C-45, 152
biomass, 386, 449; spawning, 386
biotechnology industry, 327, 330–31
birth rate, 134–35, 136, 139; Aboriginal, 152, 402, 420; in Atlantic Canada, 396; immigrants and, 137; in Québec, 219, 222, 231–32; in Territorial North, 402, 420
bitumen, 286, 319, 326, 328, 343, 346, 449; pipelines, 273; water required to process, 319–20, 347–48; *see also* oil sands
Black, David, 304
BlackBerry, 188
Blackfoot, 72, 323
Blair, Danny, 46
Blind Spot: What Happened to Canada's Aboriginal Fathers? 155

Blood people, 95
Blusson, Stewart, 434
Boeing, 238, 239, 327
Bombardier, 159, 224–25, 238–39, 252–53
Bone, Robert, 436; *The Canadian North*, 418, 437
boom-and-bust cycles, 402, 403, 449
borders: Canada–US, 189, 198
boreal forest, *201*, 350, 405
Bouchard, Lucien, 222, 262
Bouchard-Taylor Commission, 14, 16
boundaries: internal, 83–85, 87; national, 82–83
Boundaries Extension Act, 254
Bourassa, Henri, 115
Bourassa, Robert, 14, 245
Bourne, Larry S. and Damaris Rose, 150
Bow Valley, 30–31, 37
BP (before present), 70, 449
Bradsher, Keith, 336
brain drain, 188
Brandon, MB, 353
Brant, Joseph, 93, 112, 186
Britain: Aboriginal peoples and, 83, 93, 178; Atlantic Canada and, 367, 369, 370, 373; BC and, 278; colonization and, 74, 110; Conquest of New France, 9, 74, 110, 111, 231; dominance of, 110; empire and, 79–81, 83, 93, 103; as heartland, 17; immigrants from, 13, 76, 77–78, 103; immigration policy and, 78, 178, 233; in Territorial North, 413; Western Canada and, 322
British Colony, 231–34
British Columbia, xvi, 6, 7, 265–308, 441; Aboriginal population in, 278; American settlers in, 278; arable land in, 271–72; basic statistics, *268*; boundaries of, 277–78; within Canada, 266–68; climate of, 51–52, 268–69; climatic zones in, 272; CMAs in, 304; Confederation and, 79, 82, 279–80; demographic change in, 281; economic structure of, 18, 283–85; economy of, 267, 281–82, 283, 285, 308; employment in, 285; entertainment industry in, 283; environment and, 273–76; extreme weather events in, 273; fault-lines in, 307; First Nations ownership of land in, 278; fishing industry in, 288–90, *289*; foreign trade and, 285–86; forest regions in, 300–302, *301*; forestry industry in, 7, 297–303; GDP and, 267, *268*; general characteristics of, 7; gold rush, 278, 280; high-tech industry in, 285, 286; historical geography of, 277–82; hydroelectricity, 294–95; immigration and, 267–68; innovative industries in, 283; manufacturing sector in, 285; map, *266*; mining industry in, 290–92, *291*; National Policy and, 279; natural vegetation zones, 273;

oil and gas pipelines and, 284, 348; Pacific Rim trade, 265, 286–88; physical geography of, 268–73, 283; physiographic regions in, 269–72; population and, 156, 222, 267–68, 283; post-Confederation growth, 281; precipitation in, 52, 269, 272; railways in, *280*; resources in, 267, 282, 283; settlement of, 277; social characteristics of, 13; strength of, 265; today, 283–85; tourism in, 267, 283, 288, 295–97; trade and exports of, 282, 286; transportation in, 282, 286–87; Treaty Process in, 278; urban centres in, 305; urban geography, 303–7; urban population in, 132; wealth of, 285–97

British Columbia–Alberta–Saskatchewan Trade, Investment, and Labour Mobility Agreement (TILMA), 267, 449

British North America Act, 67, 79, 93, 115, 322, 464n1

British North American Exploring Expedition, 322

Brock, General, 75

Broecker, Wallace, 59

Bruce Trail, 171

buffalo, 69, 97, 106, 323

Bunting, Trudi, Pierre Fillion, and Ryan Walker, *Canadian Cities in Transition*, 163

Burney, Derek, 24

burqa, 15

Burrard Inlet, 268, 297, 305

Busch, Marc, 192

business cycle, 192, 257, 449

Cabot, John, 72, 367

Cabot Strait, 358, 377

CAE Inc, 239

Calder case, 98–99, 102, 244, 279, 424, 449

Caledonia dispute, 185, 186–87

Calgary, 130, *132*, 316, 352, 353

Calgary Declaration, 118

Calgary–Edmonton Corridor, 353

Cameron Hills oil production, *433*

Campbell River First Nation, 296, *296*

Canada: as Christian, 144; colonization and, 67, 74–78; as country of regions, 3; Dominion of, 79, 103, 235, 464n1; economic structure of, 18, 158–60; economy of, 157–61; as equal provinces, 77–78, 115–16; future of, 444–47; GDP and, 7; general characteristics of, 7; global economy and, 443–44; in global world, 19–23; historical geography of, 67–118; maps, *80*, *81*, *85*, *86*, *87*; physical geography of, 29–65; population, 8; population density, 30; Province of, 81, 113, 115, 232; Québec and, 237; regional character and structure, xvi, 3, 8, 26, 440–42; regional selection and, 5, 6–7; social characteristics of, 13; as "soft" country, 122, 137, 146, 149, 439, 443; territorial evolution of, 79–85; timeline of territorial expansion of, 82; trade with US, 23–24, 158, 162; as two founding peoples, 77, 115, 116; US geography and, 30

Canada East, 233

Canada Health Transfer (CHT), 91

Canada Pension Plan, 141

Canada Social Transfer, 91

Canada West, 178

Canadian Arctic Resources Committee, 433

Canadian Association of Petroleum Producers, 342–43

Canadian Atlantic Fisheries Scientific Advisory Committee, 386

Canadian Multiculturalism Act, 145

Canadian Pacific Railway (CPR), 82, 86, 106, 199, 200, 204, 279–80, 352; BC and, *280*, 281; land and, 78, 105; non-white labour, 103; route of, 325; Western Canada and, 323–24

Canadian Potash Exporters (Canpotex), 349

Canadian Shield, 30, 32, 35–36; in Atlantic Canada, 361, 390; Clay Belt, 226, 227; hydroelectricity and, 244; mining and, 202, 204, 256; in Ontario, 168, 169, 200; in Québec, 226, 227; in Territorial North, 405, 406; in Western Canada, 311, 314

Canadian Wheat Board, 331–32

Canaport LNG facility, *397*

canola, 311, 312, 330, 331, 337–39, *339*, 355

Canol Project, 414

cap-and-trade system, 89

Cape Breton Island, 42, 369, 397–98

Cape Breton Highlands National Park, *43*

capitalist world-systems theory, 17, 449

carbon capture and storage, 329

carbon sequestration, 329, 449

carbon tax, 89

Cardinal, Harold, *The Unjust Society*, 102

caribou, 407, 408

Caribou Inuit, 412

Carmack, George, 427

Carrick, Rob, 159

Carrier Indians, 72

Carrier-Sekani Tribal Council, 295

carrying capacity, 30

Carter, Frederic, 373

Cartier, Georg-Étienne, 234

Cartier, Jacques, 72, 228, 229, 427

"cartographic lying," 275

Cascade Mountains, 271

Cascadia, 268, 449

Casséus, Luc, "Canola: A Canadian Success Story," 355

Cassidy, Frank, 279

Caterpillar plant, 190, 213

CCGS *John G. Diefenbaker*, 430

census: Aboriginal peoples and, *13*, 72, 95, 96, 152, 154; ethnic origins in, 141, 280; French-speaking Canadians and, 14, 268

census metropolitan areas (CMAs), 129–32, 449; in Atlantic Canada, 397; in BC, 304; in Ontario, 204, 205; in Québec, 258

Central Canada, 86; dominance of, 86–88; erosion of position of power, 11, 12; hydroelectric power in, *243*; major urban centres in, *206*; map, *170*; see also Ontario; Québec

centralist/decentralist faultline, xvi, 11–13, 67, 89–92; federal role in, 89–91; major programs, 91–92

Centre for the Study of Living Standards, 22

Champlain, Samuel de, 72, 74, 103, 229

Champlain Sea, 43, *68*

Charest, Jean, 228, 235

Charter of Rights and Freedoms, 15; "notwithstanding clause," 14

Chartrand, David, 97

Château Clique, 112, 232, 449

chernozoic soils, 317, *318*; *see also* soil orders

Chilliwack, BC, 269, 304

China, 18, 19, 166; automobile industry in, 193; BC and, 277, 282, 286, 292; commodities super cycle and, 27; foreign trade, *444*; immigration from, 267, 268, *307*; oil sands and, 345–46; WTO and, 27, 237, 238, 443; textile manufacturing in, 238; see also Pacific Rim countries

China syndrome, 167, 449

chinooks, 315, 449

Chipewyan, 69, 323, 408

Christie Digital Systems Inc., 192

Chrysler, 166, *198*; bailout of, 11, 88, 90, 195; see also Big Three

churches, residential schools and, 94

Churchill, MB, 40–41

Churchill Falls, 57, 84, 244

Churchill Falls Agreement, 11–12, 379, 395

circumpolar countries, 416, 450

Circumpolar World, 417

cirques, 37, 450

cities, 8, 119, 130–32; Aboriginal peoples and, 96, 151; agriculture and, 336–37; in Atlantic Canada, 132, 376, 380, 397, 398; in BC, 132, 304, 305; capital, *128*; immigration and, 79, 138, 146; in Gaspé Peninsula, 261; in Québec, 260–61; in Ontario, 132, 173, 204–5, 210–12; population and, 126–27; in Western Canada, 331, 351–53; see also census metropolitan areas (CMAs)

citizenship, dual, 143

Citizenship and Immigration Canada, 145

Citizens Plus, 102

Clapp, Robert, 386

Clark, Christy, 307, 312–13

Clavet plant, 339

Clay Belt, *234*, 253

Clean Energy Fund, 329, 450

Cliff Natural Resources, 203

climate, 46–56, *47*, 65, 450; in Atlantic Canada, 363; in BC, 268–69; classification of, 49–50; definition, 46; factors,

46–49; Ontario, 171–74; in Québec, 225, 227–28; soils and vegetation and, 52; in Territorial North, 42, 405; types, 46, 48, 49–51; variation in, 30, 48–49; of Western Canada, 311, 312, 315, 326, 330; Western Canada agriculture and, 333; zones, 49–51, 65
climate change, 58–59, 60, 61, 276, 329; *see also* global warming
Climate Trend Mapper, 46
Clovis culture, 68, 69, 70
clusters/clustering, 160, 192
CN, 199, 200, 281; in North America, 21; in Territorial North, 429
CNOOC (China National Offshore Oil Corporation), 345
coal: global prices, *293*; in Atlantic Canada, 360, 367, 372, 390; in BC, 281, 291, 292, *293*; in Ontario, 176, 177, 191; pollution and, 176; in Western Canada, 312, 314, 329, 348
Coalition Avenir Québec (CAQ), 148
Coalition for Fair Lumber Imports, 302
Coast Mountains, 268, 270, 272
Coast Rainforest, 300, 301
Coates, Ken, 370, 416; *Arctic Front*, 416; *The Marshall Decision and Native Rights*, 399
Cold Lake oil sands field, 343
Cole, Trevor, 207
Columbia Forest, 300, 302
Columbia Mountains, 271, 273
Columbia River, 57, 294
Cominco, 429
Commission of Government, 373
Communitech, 188
Confederation, 76, 79, 110, 322; Atlantic Canada and, 81, 181, 370–73; BC and, 267, 279–80; map of Canada at, *80*; Ontario and, 84, 167, 179, 181; Québec and, 88, 216, 234–35; Western Canada and, 82, 114, 322
Confederation Bridge, 379, *380*
Conference Board of Canada, 237; *Metropolitan Outlook*, 207
Conrad, Cathy, T., 53
Conrad Business Entrepreneurship and Technology Centre, 188
Con-Rycon gold mine, 429
consciousness, regional, 4, 7, 362, 404, 456
Constitution Act (1791), 112–13
Constitution Act (1982), 91, 465n10; Aboriginal peoples in, 95; Aboriginal rights in, 102
construction industry, 148; in Atlantic Canada, 389; in Western Canada, 328
container shipments, 450; Atlantic Canada and, 376, 397; BC and, 287–88; St Lawrence River and, 220, 223
Continental Divide, 37, 56, 57
continental effect, 49
continentalism, 23, 450
continental shelves, 33
Cook, James, 277

Cordillera, 32, 34, 36–38; in BC, 266, 269, 270; mining in, 291; mountain ranges in, 270–71; physiography of, *270*; in Territorial North, 405, 406; in Western Canada, 314
Cordillera climate zone, 50, 272, 470n1
Cordillera Ice Sheet, 34, 37, *68*, 70
core, 6, 18, 450; regional, 258, 456
core/periphery model, xvi, 6, 7, 17, 26, 86, 162, 463n1; Atlantic Canada and, 360; BC and, 445; national version of, 18; Ontario and, 445; physical geography and, 29; Québec and, 242, 250, 445; Territorial North and, 401; Western Canada and, 342
Corn Belt, US, 336, 340
countries: "hard" and "soft," 67, 122, 453
country food, 426–27, 450
coureurs du bois, 230
creative class, 187, 188, 213, 225, 284–85, 445, 450
Cree, 72, 95, 225; James Bay Project and, 245, 246–48; in Québec, 227, *248*
Crispino, Len, 189
Crosbie, John, 148
Crow Benefit, 211, 327, 340, 450
Crow Rate, 337
crude birth/death rate, 133, 135, 136, 450
cultural adjustment, 16
cultural commuting, 431
cultural dualism, 115
culture, 141, 450; clash of 149; ethnicity and, 142; in Québec, 215, 216–18; religion and, 143–44
culture areas, 69, 450
Curry, Bill and Shawn McCarthy, 60
Cushing, Oklahoma, 348
Cuthand, Doug, 93
cyclic steam stimulation, 343
Cypress Hills, 40, 316, *316*

Dakota Territory, 321
Darlington nuclear plant, 176
David Suzuki Foundation, 300
Dawson City, YT, 277, 427, 428
Dearden, Philip and Bruce Mitchell, 59
De Beers, 15, 103, 204, 212
De Blij, H.J. and Alexander B. Murphy, 4
De Courcy, Diane, 263
Deep Panuke Project, 375
Deh Cho Bridge, 429, *429*
Delgamuukw case, 102, 244, 450
de Loë, R., 53
demographic transition theory, 138–40, 219, 450
demography, 121, 451; demographic change, 121, 162; new look for, 123; *see also* population
Dene, 323, 407, 414; land claims and, 424, 471n6
Denevan, William M., 154
density: physiological, 123; population, 123, 456

denudation, 32, 451
Department of Fisheries and Oceans (Fisheries and Oceans Canada), 289, 385
Department of Indian Affairs, 102
deposition, 32, 451; glacial, 35, 36
Detroit, 178
Detroit River International Crossing (DRIC), 199
Deutsch Commission on Maritime Union, 362
Development Road Program, 428
Diamond Cove, NFLD, 358, *358*
diamond mining, in Territorial North, 418, 431, 433–35, 436
Diavik mine, 431, 433, 434
Dickason, Olive Patricia, 410
Dickason, Olive Patricia and David T. McNab, 154
Dicken, Peter, 194
Diefenbaker, John, 428
Digital Media Zone, Ryerson University, 187, 188
dilbit, 343, 451
diluent, *348*, 451
Discovery Air, 327
diseases: "Dutch," xvii, 5, 88, 89, 162, 167, 347, 451; European, 72, 103, 154, 408, 410; SARS, 206–7
dispute settlement provisions, 302, 451
Distant Early Warning line, 414
diversification, 20, *158*, 162, 444; auto industry and; in BC, 283, 285; in Ontario, 188, 211; in Territorial North, 436; in Western Canada, 326, 328, 341, 345, 348
Dixon, Paul, 248
doctrine of Manifest Destiny, 451
Dogrib First Nation, 404, 408, 424, 425
dollar, Canadian, *158*, 167, 198, 302
Dominion Lands Act, 78, 110, 324
Dominion Land Survey, 322, 451
Donnacona, 72
Dorset people, 70, 71
"double contract," 115
Douglas, Sir James, 279
Douglas Creek Estates, 187
Doukhobors, 78, 104, 109–10
downward transitional region, 210, 360
draggers, 387
Drake, Francis, 277
drought, 318
Drucker, Peter, 160
drumlins, 36, *40*, 451
Drummond, Don and Derek Burleton, 166
Dry Belt, 318, 325, 326, 332, 333, 334–35, 451
dual loyalties, 143
Dumont, Gabriel, 106
Duplessis, Maurice, 116
Durham, Lord, 112–13, 232
Dust Bowl, 9
Dutch disease, xvii, 5, 88, 89, 162, 167, 347, 451

earthquakes, 36–37, 269
Eastern Deeps deposit, 391
Eastern Subarctic culture region, 72, *72*
Eastern Townships, 260, 451; *see also* Estrie
Eastern Woodlands culture region, 72, *72*, 73
Eastmain Diversion Project, 58
Eastmain River, *245*
Eaton Centre, *208*
economic leakage, 436, 451
economic structure, 451; in Atlantic Canada, 18, 379–80; in BC, 18, 283–85; Canada's, 18, 158–60; in Ontario, 18, 191; primary sector, 159, 456; quaternary sector, 159, 458; secondary sector, 159, 457; tertiary sector, 159, 458; traditional view of, 159; in Québec, 18, 240; in Territorial North, 18, 421–22; in Western Canada, 328–29
economies of scale, 23, 89, 451
economy, 162; Aboriginal, 424; in Atlantic Canada, 357, 358, 360, 373–75, 375–76, 399; in BC, 267, 283, 285, 281–82, 308; cyclical nature of, *158*; global, 17–18; knowledge-based, 160–61; in Ontario, xvii, 166, 189–91, 199–200; in Québec, xvii, 219, 223–24, 237–40; in Territorial North, 404–5; in Western Canada, 314, 326, 327–28, 354; westward shift, 185; *see also* knowledge-based economy
Ecotopia, 441, 451
ecumene, 29, 126, 451
Edmonton, 130, *134*, 352, 353
education, 115; Aboriginal peoples and, 94, 151, 155, 323, 392, 405; in Atlantic Canada, 381; in Québec, 116, 217, 235
Eeyou Istchee, 215, 451
Ekati mine, *428*, 431, 433, 434, 435
electricity: cost of, 201; in Ontario, 177, 191–92
Emerson, David, 288
employment: in Atlantic Canada, 379–81; auto parts industry and, 198; in BC, 285; in Ontario, 166, 191, *191*; in Québec, 240; in Territorial North, 421–22; in Western Canada, 327, 328–29; *see also* labour force
Enbridge, 275
England. *See* Britain
English/French faultline. *See* French/English faultline
entertainment industry, 283
environment, 58–59, 62, 63; air pollution and, 59, 60, 64, 175–76; in Atlantic Canada, 366–67; in BC, 273–76; fish farms and, 288; forest industry and, 201–2; human activities and, 30–31, 58, 273; Hydro-Québec and, 245; impact of physical, 24; megaprojects and, 432; mining and, 408; oil sands and, 346–47; in Ontario, 175–76, 177–78; in Québec, 228; in Territorial North, 407–8; tourism and, 297; water pollution and, 31, 175,

176; in Western Canada, 318–21; *see also* climate change; global warming
environmental determinism, 447, 451
equalization payments, 91, 92, 442; Atlantic Canada and, 357, 360, 377; Ontario and, 11, 166, *169*, 170; Territorial North and, 421, 422
Equalization Program, 91
equator, 45
Ericsson, 243
erosion, 34, 42, 451; glacial, 36, 452
Erskine, David, 372–73
eskers, 36, 451
Esquimalt and Nanaimo Railway, 281
Essex–Kent vegetable area, 173
Esso Resources Canada (Imperial Oil), 433
Estrie, *226*, 228, 250–51, 260, 451; dairy operations in, *252*
ethanol, 336
ethnic enclaves, 146, 150
ethnic groups, 141, 451
ethnicity, 141–42; language and, 142–43; religion and, 143–44
ethnic origins, 141, 142, 451
ethnoburbs, *446*, 451–52
ethnocentricity, 145, 452
Euclid-Hitachi Heavy Equipment, 196, 210
Europe: Atlantic fisheries and, 367–68, 387; immigrants from, 108–9
Europeans: initial contact with, 71–72; Territorial North and, 408–9; Western Canada and, 324–25
European Union, 157, 251, 336, 426–27
evapotranspiration, 316, 452
exports, 3; Atlantic Canada and, 369; BC and, 282, 286, 297–98; hydroelectricity, 236, 246, 294; lumber, 200; Ontario and, 166–67, 188–89, 200, 213; Québec and, 238; US and, 20, 23–24; water and, 57; Western Canada and, 328, 354; *see also* specific industries
Exxon, 389

F-35 fighter jets, 21
False Creek, 306, 307
Family Compact, 112, 232, 452
Fathers of Confederation, 80, 114
faulting, 36, 452
fault line (geological), 36, 452
faultlines, 10–16, 24, 67, 442–43, 452; in BC, 307; in Canada's early years, 85–89; concept of, 7; first appearance of, 85–86; Québec's place in Confederation, 88–89; *see also* Aboriginal/non-Aboriginal faultline; centralist/decentralist faultline; French/English faultline; immigrant faultline
federal gas tax transfers, 258
federal government, 90–91, 115; Aboriginal peoples and, 14, 92, 94, 95, 97, 106, 129, 155, 422; Atlantic Canada and, 5, 91, 377; automobile industry and, 11,

88, 90, 195; BC and, 268; as employer, 208; faultlines and, 85, 86, 88, 89, 307; interventions in marketplace, 89–90; mine cleanup and, 58, 408; multiculturalism and, 145; northern land claims and, 100, 392, 424, 425; powers of, 90, 115; railways and, 286, 324; regions and; scientific research and, 161; Sydney Tar Ponds and, 367; Territorial North and, 414, 418, 422; western alienation and, 89; Western Canada and, 11, 342, 350–51; Western settlement and, 78, 109, 325
Fenians, 80, 81, 464n2
Fertile Belt, 317, 326, 332, 333, 452
fertility: rate, 134, 140, 452; replacement, 136
filmmaking, 283
Finnish Forest Research Institute, 288
Fipke, Charles, 434
First Nations people, 95, 452; *see also* Aboriginal peoples; specific groups
first people, 68–74; *see also* Aboriginal peoples
Fischer, David, 229
Fisheries and Oceans Canada, 386
fishing industry, *289*, 288; aquaculture and, 288; in Atlantic Canada, 372, 381–87, *382*; Atlantic species and, 383; in BC, 288–90; cod, 384–87; crab, 383, 387; factory ships and, 387; fish habitat and, 468n6; groundfish, 367, 383, 384, 386, 452; inshore, 381, 383, 384, 386; international, 290; licences and, 290, 370, 387; lobster, 370, 383–84; management and, 290, 385–87; modernization of, 384; offshore, 384; overexploitation, 288–89, 385–87; overfishing, 387, 468n6; quotas, 289, 290, 385, 468n6; resource cycle and, 386; salmon, 288, 289–90; scallops, 383, 384; warming oceans and, 290, 468n6
flag, Canadian, 115
Flanagan, Thomas, 96
Fleming, Sir Sanford, 45
Flin Flon, Man., 350
floating production storage and offloading (FPSO) vessels, 389
floods, 53–54, 320
Florida, Richard, 160, 186, 187, 218, 225, 284, 445; *The Rise of the Creative Class*, 213
Foam Lake Terminal, 333
folding, 36, 452
Folsom culture, 69
Ford Motor Company, 182; *see also* Big Three
forest fires, 299; in BC, 275–76
forest industry: Aboriginal peoples and, 393–94; in Atlantic Canada, 391–94; in BC, 297–303, *299*, *301*; BC species and, 273, 298, 300, 302; challenges in, 201–2; clear-cutting and, 273, 300, *300*; commercial zone, 227; forest regions and, 300–302, *301*; NAFTA and,

INDEX

302–3; old-growth stands and, 300, 301; in Ontario, 200–202; pine beetle and, 273, 276, 299; private timberlands and, 392, 393; in Québec, 254–55; softwood forest and, 298, 457; softwood trade dispute, 302–3; sustainable, 201; transportation and, 302; US economy and, 202, 302; in Western Canada, 350–51; *see also* pulp and paper
forests, 450; by province, 350
Forks, the, *104*
Fort Bourbon, 230
Fort Chipewyan, 320
Fort Garry, 321
Fort McMurray, Alta., 129, 341, 344, 377
Fort Victoria, 278
Fort Victoria, BC, 279
forty-ninth parallel, 44, 76, 83
fossil fuels, 38, 59
Four Seasons Centre, *168*
fox, Arctic, 410
fracking, 228, 328, 452
France: Atlantic Canada and, 367; colonization and, 72, 74, 110, 229–30
francophones, 13, 116, 216, 452; *see also* French/English faultline
Franklin, John, 82, 409
Fraser Canyon, 271
Fraser Plateau Forest, 300, 301–2
Fraser River, 271, 290, 294
Fredericton, 397
Free Trade Agreement (FTA), 19, 24, 181, 182–83, 189, 258, 339, 440, 452
free trade agreements, 19, 20
French, H.M. and O. Slaymaker, 46
French Canadians, 13–14, 74, 76, 77, 111, 142, 143, 232, 452
French/English faultline, xvi, 13–14, 67, 110–18, 146, 156, 162, 216, 225, 228, 232; demographic shifts and, 113; language imbalance, 156; Manitoba Schools Question, 114; Northwest Rebellion and, 114; origin of, 111; in Québec, 262; Red River Rebellion and, 113–14
French language, 216, 217–18, 233, 262, 465n6, 465n9; laws, 237, 263; as mother tongue, 156
French Shore, 367, 368
friction of distance, 452
Friedmann, John, 17
Frobisher, Martin, 72, 82, 96, 408–9, 427
Frontenac Axis, 42
frontiers: northern, 403–4, 455; resource, 18, 26, 401, 403, 418, 457; strategic, 414
frosts, 322, 333
fur trade, 9, 57, 83, 103, 321, 323, 424, 427; in BC, 277; in New France, 103, 230; in Québec, 231; in Territorial North, 410–11, 414; in Western Canada, 321, 323

G20 meeting (2010), 20–21
Gable, Brian, "The Regions of Canada," 5

"Gaddafi disaster", 224, 239
garbage-to-energy programs, 175
Garreau, Joel, 441
Gaspé Peninsula, 42, 72, 227, 250, 260, 261
Gatineau, QC, 14, 16, 251, 257
Gaulle, Charles de, 117, *117*
General Motors, 166, 182; bailout of, 11, 88, 90, 195; Chevrolet Volt, 196; *see also* Big Three
Gentilcore, R. Louis, *Historical Atlas of Canada: Volume II*, 119
Gentilly-2 nuclear power plant, 228, 240–41
geographic location, 44–45
geography, 3; ancient Greeks and, 4; curiosity and, 4; as discipline, 4; "grooves of," 308, 452; regional, 4–5, 456–57; urban, 257–61, 303–7, 351–53, 396–98; *see also* historical geography; physical geography
Geological Survey of Canada, 256
geological time chart, *33*
George, Dudley, 185
Georges Bank, 381, 383, 384, 385
Gérin-Lajoie, Antoine, *Jean Rivard*, 233–34
Giant gold mine, 408
Gilbert, Anne, 156
glaciation, 35; alpine, 34, 37, 42; Arctic lands and, 42; in Atlantic Canada, 361; global warming and, 37; spillways, 39, 452; in Rocky Mountains, 37, 320; striations, 36, 452; in Territorial North, 405; troughs, 37, 452; worldwide change in volume, *63*
global circulation system, 47–48, 452
global economic crisis (2008–9), 3, 11, 121, 123, 156, 157–58, *161*, 162, 166, 260, 342; automobile industry and, 193–94; Ontario's manufacturing sector and, 183, 190
globalization, 18, 88, 167, 213, 255, 452; automobile industry and, 193; Ontario's economy and, 190; sense of place and, 8
global warming, 58–59, 60, 65, 273, 406, 407–8; albedo effect and, 407; Arctic Ocean and, 414; Canada and, 60–65; natural factors, 60; in Territorial North, 59, 407–8; whales and, 406; *see also* climate change
gold: early explorers and, 278, 280; placer, 278, 405, 427; in Territorial North, 427–28, 429
Golden Horseshoe, 205
Gordon M. Shrum generating facility, 58
government: public, 253, 404, 423; responsible, 113, 373; territorial, 91, 153, 404, 421, 422, 431; *see also* federal government; self-government; specific jurisdictions
governor, colonial, 231, 232
grain, 173, 337; Canadian Wheat Board and, 331–32; climate and, 38, 59; Dry Belt and, 326, 334; in Fertile Belt, 333; prices of, 108, 336; shipping of, 108, 211, 220, 282, 285, 312, 327, 339, 340; *see also* specific crops

Grand Banks, 375, 381, 383, 385
Grande Riviére Project, 58, 245
Grand Trunk Pacific Railway, 281
Granville Island, 306, 307
Great American Desert, 317, 319, 321, 322, 452
Great Depression, 281
Greater Toronto Area (GTA), 206, 207
Great Harbour Deep, 380, *381*
Great Lakes, 35, 171, 172; agriculture and, 72; exotic species in, 178; pollution and, 176–77, 178; reliance on, 176; transportation and, 57, 220, *223*, 230; water levels and, 178
Great Lakes Basin, *179*
Great Lakes–St Lawrence climate zone, 50
Great Lakes–St Lawrence Lowlands, 30, 42–44; agriculture in, 71, 72; in Québec, 226
Great Lakes Water Quality Agreement, 178
Great Melt, 68
Great Recycling and Northern Development Canal, 57
Great Sand Hills, 334, *334*
Great Whale River Project, 58, 245, 247
Greenbelt Plan, 175
Green Energy Act, 191–92, 452
greenhouse effect, 59, 452
greenhouse gases, 59, 61, 65, 161, 452; forestry and, 302; mining and, 62; reduction in emissions, 60, 192; Western Canada and, 62, 319, 329, 336
Green Lane Landfill, 175
green policies, 192
Green Transportation Plan, 350–51
gross domestic product (GDP), 7, 182, 452; Alberta and, 223; Atlantic Canada and, 7, 360, *361*; BC and, 7, 267, *268*, 296, *299*; Canada and, *237*; Manitoba and, 328; Ontario and, 7, *166*, *167*, 182, 223; Québec and, 7, 219, *222*, 223, 224, *237*; Territorial North, 7, *403*, 435; Western Canada and, 7, *313*, 314, 327, 328
Gulf of Mexico, 56, 57
Growth Plan for the Greater Golden Horseshoe Area, 175
Gulf Stream, 363, *364*, 452
Gupta, Amit, 161
Gwich'in agreement, 424, 425
Gwich'in Council, 417

habitants, 111, 231, 452
Haida Gwaii, 37, 96, 269, 467n1
hail, 326
Haldimand Grant, 93, 112, 178, 185, 186
Haldimand Proclamation, 183
Haldimand Tract, *184*
Halifax, 359, *360*, 363, 376, 397, 398
Hamilton, ON, 205, 207
Hans Island, 416
Hare, Kenneth, 147, 439
Harper, Stephen, 22, 118, 307, 416, *446*
Harris, Mike, 185

Harris, R. Cole, *Historical Atlas of Canada: Volume I*, 119
Harry Winston Diamond Corporation, 435
Harvey, David, 160
health: Aboriginal peoples and, 106, 129; aging population and, 135, 140, 141; pollution and, 64, 175, 177, 245; tar ponds and, 367; *see also* diseases
Health Canada, 367
health-care system, 11, 136, 141, 198; costs of, 195
Heap, Alan, "China—The Engine of a Commodities Super Cycle," 27
Hearne, Samuel, 427
heartland, 17, 453
Hell's Gate, *271*
Henco Industries Ltd, 187
Hernandez, Juan, 277
Héroux-Devtek, 238
Hérouxville, QC, 14, 16
Herschel Island, 410
"hewers of wood and drawers of water," 346, 453
Hibernia project, 389, *390*
High Lake, 407, 431
high-technology industries: in Atlantic Canada, 379, 398; in BC, 267, 285, 286; in Ontario, 209; in Québec, 238, 251–52, 259, 262; in Western Canada, 353
highways: Alaska, 414, 428; in BC, 282; Dempster, 428; Fort Smith, 428; Laurentian; Liard Trail, 428; Mackenzie, 428; population distribution and, 126; population zones and, *124*; Sea-to-Sky, 296–97, 305, 457; in Territorial North, 428–30; Trans-Canada, 199, 200, 212; Yellowknife, 428; *see also* roads
Hind, Henry, 322
hinterland, 6, 17, 288, 298, 453; resource, 6, 168, 169, 171, 199, 210, 250, 253, 257, 431, 432
historical geography, 67–118, 119; of Atlantic Canada, 367–73; of BC, 277–82; of Ontario, 178–85; of Québec, 228–35; of Territorial North, 408–14; of Western Canada, 321–29
Hobson's choice, 8, 453
Hochelaga, 72
"hollowing-out," 453
"Hollywood North," 283
Holocene epoch, 63, 453
homeland, 453; Territorial North as, 10, 93, 401, 403, 404
homesteaders, 322, 324–25, 453; geographic challenges facing, 325–26
Honda, 193, 197
Honoré Mercier Bridge, 258
horizontal (directional) drilling, 284, 341, 453
Horn River shale, 291
House of Commons, 86–88, 167
Hudson Bay, 82

Hudson Bay Basin, 56, 57, 58
Hudson Bay Lowlands, 32, 34, 39–41; in Ontario, 168, 169; in Québec, 226; in Western Canada, 314
Hudson Bay Railway, 326
Hudson's Bay Company (HBC), 78, 83, 105, 278, 321, 322, 323, 410, 413
hunting: Aboriginal peoples and, 63, 71, 97, 408; in Territorial North, 408, 410, 411, 424, 426
Huron people, 71, 72, 103
Huronia, 72
Hurricane Igor, 365
Hurricane Juan, 53, 362
Hurricane Leslie, 365
hurricanes, 52–53
Hutterite Brethren, 144
Hybrid Air Vehicles (HAV), 327
hydraulic fracturing, 284, 341, 453
hydroelectricity: in Arctic Basin, 58; in Atlantic Canada, 57–58, 395; in BC, 294–95; in Central Canada, *243*; in Ontario, 58, 200; in Québec, 200, 215, 227, 236, 244; in Western Canada, 348
hydrometallurgy, 376, 453
Hydro-Québec, 84, 116, 222, 236, 240–48, 262; Atlantic Canada and, 241; Churchill Falls Agreement and, 12–13, 395; exports to New England, 246; industrial strategy of, 241–44; James Bay Project, 241, 244–46; Romaine Hydro Complex, 241, *242*

ice: pack, 405–6, 430, 448; sea, 59, 60, *64*, 407, 437, 457
ice age, 36, 453
icebergs, 42, 363, *365*
ice storm, 228
identity: Aboriginal, 13, 448; Canadian, 146, 443; regional, 4, 457
Idle No More movement, 22, 93, 96, 123, 152
immigrants: challenges facing, 15; cultural adaptation of, 150; cultural differences, 149–50; destinations of, 138; country of origin and, 141–42; integration, 149; upward mobility of, 146
immigration, *133*, 137–40; Atlantic Canada and, 368–69; annual numbers, *138*; BC and, 267–68; children and, 149, 150; colonial-style, 103; demographic transition theory, 138–40; multiculturalism and, 147; population and, 122–23, 137; Québec and, 142, 237; reasons for encouraging, 137; recent, xvi, 143, 150; Sifton and, 78–79; sources of, *139*; Toronto and, 207, 209; Western Canada and, 78–79, 325
immigration faultline, xvi, 15–16, 67, 103–10, 148–50, 162, 446; making treaty, 105–6; Northwest Rebellion, 106; Red River Rebellion and, 104–5

impact benefit agreements, 15
Imperial Oil, 432
Inco, *391*, 392, 470n4
Indian, Inuit and Métis peoples of Canada, 14
Indian Act, 14, 89, 93–94, 95, 102
Indian Affairs department, 95
Indian agents, 93, 94
Indian Posse, 351
Indians, 69–71; agriculture, 70–71; in BC, 277; non-status, 95, 454; population at first contact, 72; population of New France, 74; status (registered), 95, 458; treaty, 95, 458; urban, 96, 151; in Western Canada, 323; *see also* Aboriginal peoples
Industrial Revolution, 17
information society, 160–61
Inland Passage, 271, 296, 453
Innis, Harold, 17
innovation, 186
Innovation Place, University of Saskatchewan, 161
Innu (Naskapi and Montagnais), 72, 392
Innuitian Ice Sheet, *68*
Innuitian Mountain Complex, 41
Insular Mountain Range, 270
Insular Mountains, 272
Intercontinental Railway, 372
Intergovernmental Panel on Climate Change, 52, 54
Interior Boreal Forest, 300–301
Interior Plains, 32, 34, 38–39; in BC, 266, 269–70, 271; mining and, 291; in Territorial North, 405, 406; in Western Canada, 311, 314, 316–17
Interior Plateau, 268, 272, 273
International Convention for the Regulation of Whaling, 407
International Court at The Hague, 383, 384
Inuit, 14, 71, 72, 93, 95, 215, 407, 414, 453; birth rate and, 402, 420; early James Bay Project and, 247; European contact and, 409, 410; Hydro-Québec and, 244, 245, 247; northern land claims and, 392, 424; Nunangat, *423*; Nunavut and, 422–24; population of, 72, 150, 154; Québec and, 225, 227, 253, 254; resources and, 255; sense of place and, 10; in Territorial North, 408; voting rights, 152; whaling and fur trade and, 410
Inupiat Inuit, 410
Inuvialuit, 410, 471n7
Inuvialuit Final Agreement (IFA), 15, 100, 425, 426, 427, 471n8
Inuvialuit Game Council, 425, 427
Inuvialuit Regional Corporation, 425
Inuvik, NT, 42, 414
Ipperwash dispute, 185
Iqaluit, 96, 128, 129, 422
Ireland: immigration from, 76, 369, 453;

iron ore: in Atlantic Canada, 372, 387, 390, 391; in Québec, 255, 256–57; shipping of, 220
Iron Ore Company, 255, 256–57
Iroquois, 71, 72
Irving Forest Corporation, 393
Irving Shipbuilding Inc., 377, 381
Islam, 15, 123, 143, 145; *see also* Muslim Canadians

James Bay and Northern Québec Agreement (JBNQA), 15, 97, 100, 225, 245, 246–48, 253, 254, 425, 453
James Bay Hydroelectric Project, 58, 99, 215–16, 241, 254; Aboriginal peoples and, 245, 248; environmental impacts of, 245
Japan: auto industry and, 196, 197; trade with, 286, 292, 441; *see also* Pacific Rim countries
Japan Current, 47
Jarislowsky, Stephen, 218
Jasper National Park, 314
Jean, Michaëlle, 426, *426*
Jeanne d'Arc Basin, 387
Jean Rivard (Gérin-Lajoie), 233–34
Joint Review Panel, 57
Jones, David, *Empire of Dust*, 317
Jordan Free Trade Agreement, 19
just-in-time principle, 453

Kahnawake, 95
Kananaskis Country Provincial Park, 314
Kangiqsujjuaq, 255
Kashechewan First Nation, 41, 212
Kativik Regional Government, 247, 254, 453
kayak, 408, 409
Kelowna, BC, 130, 275–76, 304
Kemano Completion Project, 54, 294–95, 468n7
Kennedy, John F., 23
Kenora, 211
Kerr, Donald and Deryck W. Holdsworth, *Historical Atlas of Canada: Volume III*, 119
Kettle First Nation, 185
Keystone XL Pipeline, 22, 62, 328, 348, 443
kimberlite pipes, 434, 453
Kinder Morgan, 273
Kipling, Rudyard, 103
Kitimat, BC, 58, 267, 273, 284, 294, 304
Kitimat LNG project, 275
Kitimat Works, 295, 304
Klein, Ralph, *90*
Klondike gold rush, 414, 427–28, 429
Kluane National Park Reserve, *38*
knowledge-based economy, 185–87, 453; in Atlantic Canada, 376; in BC, 283, 284; in Ontario, 185, 186–87, 207; in Québec, 225, 236, 239–40, 262; in Western Canada, 329, 353, 354
Köppen climate classification, 48, 49–50, 52
Kroeger, Arthur, *Hard Passage*, 317

Kutchin people, 96, 408
Kyoto Protocol, 19, 23, 60, 61

labour force: Aboriginal peoples and, 122; in Atlantic Canada, 372, 375, 379, 381; in knowledge-based economy, 160, 186–87; manufacturing, 156, 190; in Ontario, 190, 211–12; in Québec, 237, 240, 252, 257; sectoral changes in, 160; in Territorial North, 404–5; in Western Canada, 328, 350; *see also* employment
Labrador, 362, 363; mining and, 390; weather in, 361; Québec and, 377–78; *see also* Newfoundland and Labrador
Labrador City, NL, 36, 128, 129, 397
Labrador Current, 363, *364*, 453
Labrador Iron Mines, 390
Labrador Trough, 256
Lac la Ronge Cree, 96
La Grande Project, 245, 247
Lake Agassiz, 35, 39, *68*, 453
Lake Athabasca, 321, 347
Lake Barlow-Ojibway, 254
Lake Diefenbaker, 335
Lake Louise, *315*
land claims, 15; in BC, 100–101; comprehensive, 99–101, 424–27; megaprojects and, 432; modern, 15, 100, 247; rights and, 83, 98, 254, 424, 426; specific, 185, 458; in Territorial North, 225, 424–27, 450; *see also* James Bay and Northern Québec Agreement (JBNQA); treaties
Land Commission Act (BC), 272
landforms, 30, 31–32
land survey system, 78, 325
land use, 52; in BC, 273; in Ontario, 173, 175; in Western Canada, 337; *see also* arable land
language: Aboriginal, 73, *74*, 95; Chinese, 123; demographic shifts in, 76–77; English, 142; ethnicity and, 142–43; French/English imbalance, 13, 156; immigration and, 110, 137; in Manitoba, 114; official, 142–43; in Québec, 13, 14, 237; *see also* French/English faultline; French language
late Wisconsin ice advance, 34–35, *34*, 35–36, 37
latitude, 44, 45, 453
Laurentide Ice Sheet, 34, 35, 38–39, 40, 41, 43, *68*, 69, 71
Laurentides, 36
Laurier, Sir Wilfrid, 114
law: language, 237, 262; in Québec, 111; shariah, 15, 145, 457
Laxer, James, 75
Légaré, André, 423
Legault, François, 148
Lemay, Martin, 218
Leo, Geoff, 155
Lesage, Jean, 116, 235
Leslie, Megan, 61

Levesque, René, 117, 254
Li, Peter, 149
Liberal Party, 89
lieutenant-governors, colonial, 105, 112
life expectancy, 140
Lillooet, 72
linkages: backward, 17, 448; and final, 17, 452; forward, 17, 452
Lions Gate Bridge, 305
liquefied natural gas (LNG), 291, 304, 308, 376, 453–54
Little Ice Age, 70, 408
Little Pine, 323
Liverpool, NS, 393
livestock industry: environment and, 318; hog-farming and, 340; in Ontario, 173, 176; in Québec, 227, 251; slaughterhouses and, 340; in Western Canada, 318, 327, 333, 339–40
Lomonosov Ridge, *416*, 454
London, ON, 209–10
Long Harbour nickel processing facility, 376, *376*, 391
longitude, 44, 45, 454
Louisiana Purchase, 75, 80
Louisiana Territory, 80, 83, 103
Louis S. St Laurent, 430, *440*
Low, A.P., 256
Lower Canada, 75, 76, 112, 232
Lower Churchill Project, 378–79, 395, 399; map of, *378*
Lower Mainland, 278, 303, *303*, 307, 454
Loyalists, 74–75, 77, 112, 454; in Atlantic Canada, 178, 369; in Ontario, 178, 75; in Québec, 75, 250, 260
Lucky Man, 323

Maa-nulth First Nation, 101
McCain, Michael, 340
McCain Foods, 394, 395
McClure, Robert, 409
MacDonald, Dettweiler and Associates, 283
MacDonald, John, 283
Macdonald, Sir John A., 81, 82, 105, 181, 324
McDougall, William, 105
McGuinty, Dalton, *16*, 20, 62, *90*, 196
Mackenzie, Alexander, 277
Mackenzie, William Lyon, 112
Mackenzie Basin, 406
Mackenzie Gas Project, 403, 432
Mackenzie River, 39, 42, 58
Mackenzie Valley Pipeline Inquiry (Berger Inquiry), 99, 432
Mackenzie Valley Pipeline Project, 99, 432
McMaster University, 207
Macpherson, Alan G., 362
magma, 31
Magna International, 198
magnesium, 244
Maisonneuve, Paul de Chomedey de, 229
Maisons du Québec, 235

INDEX

"*maîtres chez nous*" (masters in our own house), 116, 217, 235
Makivik Corporation, 253, 254, 255, 454
Malartic, QC, 257
Malavoy, Marie, 217
Maliseet, 73, 368
malls, Asian theme, 208
Manic 5, 116
Manifest Destiny, doctrine of, 75, 79, 80
Manifest—pour un Québec lucide, 222, 262
Manitoba, 311, 312; agriculture in, 108, 340; boundaries and, 83–84; Confederation and, 82, 114; creation of, 105, 323; demographic change in, 114; economy of, 327, 328; electricity and, 342, 348; floods in, 320; GDP and, 314, 328; legislative assembly, 113–14; making treaty in, 105–6; mining in, 349–50; population of, 107–8; settlement of, 107–8; unemployment in, 314; *see also* Western Canada
Manitoba Act of 1870, 82, 96, 104, 113, 454
Manitoba Lowland, 39
Manitoba Métis Federation, 94, 96, 97
Manitoba Schools Question, 114
Mantario, 210
manufacturing, xvii, 18; in Atlantic Canada, 372; in BC, 285; branch plants and, 19, 182, 183, 189; changes in, 159–60; employment and, 157, *157*, 190, 191, 213; global economic crisis and, 17, 183, 190; global prices and, 17–18; hollowing-out, 157–58, *160*, 199; just-in-time, 189, 453; National Policy and, 86, 181, 182, 371; in Ontario, 180–83, *190*; in Québec, 224–25, 237–38, 251–53, 258; restructuring of, 188–89; in Territorial North, 421; in Western Canada, 328; *see also* specific industries
Maple Leaf Foods, 340
Maritimes, 358; Confederation and, 370–73; economy of, 76, 81, 372; fishing industry in, 383–84; forest industry in, 392; map of, *371*; population, 395, 396; settlement of, 367, 368; *see also* Atlantic Canada
Marois, Pauline, 217, 218, 228, 237, 243
Marquis wheat, 326, 331, 469n5
Marshall, Donald, Jr, 370, 399
Marshall case, 290, 370, 399
Martel, Yann, 143
Martin, Paul, *90*
Mary River Project, 406–7, 431, 436, 470–71n2
Massachusetts Institute of Technology, 188
Matonabbee, 427
Medieval Warming Period, 63
megaprojects: in Territorial North, 403, 431–36, 454; *see also* specific projects
Melancthon, ON, 173
Métis, 76, 95, 321, 323, 454; Aboriginal/non-Aboriginal faultline and, 14, 101; dispersal from Red River Valley, 96; French/English faultline and, 113–14; northern land claims and, 424, 425; origins of, 101; population of, 95, 107, 150; rebellions and, 104, 106, 113–14; rights of, 93, 94, 96, 152; Supreme Court case and, 97
Métis Provisional Government, 82
Microsoft, 284
Middleton, Frederick D., 106
migration: Atlantic Canada and, 375, 376–77, 380–81; French Canadians and, 253, 257, 260–61; net, 133; Old World Hunters, 68, 69; out-, 375, 262, 380–81; push-pull model, 133, 456; rural-to-urban, 129, 380; Territorial North and, 71, 420–21
Mi'kmaq, 73, 368, 369, 370, 393–94, 399
Miller, J.R., 94
Milne, 1995, 96
minerals: in Atlantic Canada, 389; in BC, 288, 291; Canada's leading, 294; of Canadian Shield, 36, 202, 203; explorers and, 427; land claims and, 422; in Ontario, 200, 202; in Québec, 255; in Territorial North, 406–7, 429; in Western Canada, 314; *see also* mining industry; specific minerals
mining industry, 36; asbestos, 228, 251; in Atlantic Canada, 390–91; in BC, *291*; cleanup and, 321, 408; closures, 257; environment and, 58, 408; equipment for, 196; in situ, 343; in Manitoba, 349–50; mineral production by province and territory, 202; in NFLD, 376; in Ontario, 202–4; open-pit, 62, 319, 343, 346–47, 348, 431; in Québec, 255–57; in Territorial North, 406–7, 428, 435; in Western Canada, 314, 348–50; *see also* petroleum industry; specific minerals
Minister of State for Multiculturalism, 145
missionaries, 72, 94, 103, 410, 411
MMG Ltd, 431
Moffat, Ben, 105
Mohawk people, 75, 95
Molloy, Tom, *The World Is Our Witness*, 308
Moncton, NB, 376, 397
Mongolia, population density, 123
Monsanto, 327, 331, 332, 339
Monsanto Canada Inc. v. Schmeiser, 331, 332
Montréal, 79, *125*, 130, 250, 251, 257, 258–59; cultural industry in, 225; infrastructure in, 258; knowledge-based economy and, 218; manufacturing in, 252, 238, 258; origins of, 229; population of, 251, 259; as port, 220; as regional core, 258; Toronto and, 258–59; tourism and, 249; Victoria Square, *236*
Mont-Wright Mining Complex, *256*
Mooney, James, 154
Moose Cree First Nation, 212
Moosonee, ON, 40–41
moratorium, cod, 386, 387
mortality rate, 136, 454
Mount Barbeau, 41
Mount Jacques Cartier, 42
Mount Logan, 38
Mulcair, Thomas, 62, 88, 143, 347, *347*
Muller, Richard, 59, 60; "The Conversion of a Climate-Change Skeptic," 65
Mulroney, Brian, 342, 418
multiculturalism, 79, 115, 139, 144–46, 150, 162; criticism of, 146; immigration and, 147
muskeg, 39–40, 454
Muslim Canadians, 137, 145

Nalcor Energy Corporation, 378, 379, 454
Nanaimo, BC, 304
Nanisivik mine, 406, 414, 429
Nanticoke Station, 60, 176
Nass River, 279
nation, 10, 454
National Energy Program, 89, 342, 454
National Household Survey, 95, 137, 141
nationalism: English-Canadian, 115–16; in Québec, 116, 235
National Milk Marketing Plan, 251
National Policy, 86, 181, 182, 371, 454
National Research Council of Canada, 161, 454
Native settlements, 128, 129, 454; *see also* Aboriginal peoples
natural gas, 286, 291, 376; *see also* petroleum industry
natural increase, 123, 134–37, 456; rate of, 133, 136, 456; *see also* population
natural resources: in Atlantic Canada, 360, 379, 387–95, *388*, 396; in BC, xvi, 267, *267*, 282, 283, 288, 291–92; centralist/decentralist faultline and, 11; economic growth of, 88; non-renewable, 401, 403, 435; in Ontario, 200; in Québec, 253, 260–61; in Territorial North, 401, 403, 417, 418, 427–31, *430*, 435; in Western Canada, 312, 313, 326, 340–51; *see also* specific resources
Natural Resources Canada, 316
Nechako Forest, 300, 302
Nechako River, 294, 295
Neil's Harbour, *386*
Nelson River, 39, 58, 314
Nemetz, Peter, 287
net migration, 454
New Brunswick, 7, 76, 83, 358, 397; agriculture in, 394–95; basic statistics, 361; as colony, 79; Confederation and, 81, 370, *371*; economy of, 369, 399; employment in, 380; energy corridor and, 241, 379; equalization payments and, 377; forestry in, 392, 393; French culture in, 115; natural gas industry in, 376; official languages in, 13, 156, 465n9; Québec and, 378; *see also* Atlantic Canada
newcomers, 121–22

INDEX

newcomers/old-timers faultline. *See* immigration faultline
New Dawn Agreement, 379, 392
New England: Atlantic Canada and, 246, 369
Newfoundland (island of), 42
Newfoundland and Labrador, 88, 148, 358; basic statistics, 361; boundaries and, 84; coastal communities in, 372; Confederation and, 373; economy of, 12, 399; equalization payments and, 5; farmland in, 395; fishing industry in, 372, 383, 385, 387; forest industry in, 392–93; as "have" province, 389, 396, 399; map of, *374*; mining industry in, 376; oil production in, 343, *375*, 389; population in, 395, 396; Québec and, 11–13, 84, 234–35; settlement of, 367, 368; unemployment rate in, 396; *see also* Atlantic Canada
New France, 228–31; British Conquest of, 9, 74, 110, 111, 231; Champlain's map of, *230*; population of, 74; historical milestones, 229
New Glasgow, 397–98
New International Trade Crossing, 25
New Orleans, Battle of, 75–76
newsprint, 298
"Next Year Country," 322
Niagara Escarpment, 43, 171, *172*
Niagara Fruit Belt, 173, 174, *174*, 175
Niagara Tunnel Project, 176, *177*
nickel, 15, 32, 36: in Atlantic Canada, 376, 380, 387, 390, 391, 399; in Ontario, 200, 202, 203, 210; in Western Canada, 348, 349, 350
niqab, 15
Nisga'a, Supreme Court case and, 98–99
Nisga'a Agreement, 101, 278, 308–9
Nootka (Nuu-chah-nulth), 72, 277
Nootka Convention, 277
Nootka Sound, 277, *277*
NORAD, 454
Norcliffe, Glen, 167
nor'easters, 363, 454
Norfolk Tobacco Belt, 173
Norman Wells oil field, 433, 414
Norman Wells, NT, *433*
Norman Wells Project, *434*
Norsk Hydro, 244
Nortel, 186, 209
North American Boundary Commission, 83, 84
North American Free Trade Agreement (NAFTA), 19, 189, 454–55; agriculture and, 332, 333, 395; BC forestry and, 302–3; water and, 57
North American Security Perimeter, 455
North American Water and Power Alliance, 57
Northeast Coal Project, 292
Northeast Energy Corridor, 376, 455
Northeastern Québec Agreement, 100

Northern Alberta Railway, *280*
Northern Divide, 57
Northern Forest, 300, 302
Northern Frontier, Northern Homeland, 432
Northern Gateway Pipeline, 20, 22, 32, *275*, 273, 286, 313, 328, 348, 444; proposed route, *444*
northern lights (aurora borealis), 405, 455
Northern Ontario Railway (Ontario Northland Railway), 200
Northern Pacific Railway, 324
"Northern Vision," 428
Northrup Grumman, *415*
Northwest Coast culture region, 72, *72*
North West Company, 83, 277, 321
North-Western Territory, 79, *81*, 321
North West Mounted Police (NWMP), 411
Northwest Passage, 60, *64*, 406, 407, 421, 437, 455; search for, 408, 409, 413; sovereignty and, 414–18
Northwest Rebellion, 106, 107, 114
Northwest Staging Route, 414
Northwest Territories, 83, *85*, 91, 353, 404, 422; highways and, 428–29; mining industry in, 428, 429, *430*, 431, 433, 434–35; out-migration from, 403, 421; population of, 79, 420; *see also* Territorial North
North-west Territories, 79, *81*, *85*, 107
North West Transportation Corridor, 304, 455
Norton, Richard, 427
Nova Scotia, 7, 75, 76, 91, 358, 363, 369; agriculture in, 394, 395; basic statistics, 361; coal mining in, 390; as colony, 76, 79; Confederation and, 81, 370, 371; economy of, 399; employment in, 380; fishing industry in, 383, 384; forestry industry in, 392, 393; Hydro-Québec and, 84, 379, 395; industrial base in, 372; oil and gas industry in, 375–76; origins of, 368; population of, 79, 113; steel industry in, 372, 390; *see also* Atlantic Canada
nuclear power, 191, 176
nunataks, 40
Nunavik, 215, 254, 455
Nunavik Inuit Offshore Land Claim, 254
Nunavik Regional Government, 254
Nunavut, 15, 22, 84–85, 146, 404, 422–24; Aboriginal/non-Aboriginal faultline and, 101; Inuit and, 422–24; meaning of, 96, 422; population of, 412, 420, 421; schooling system, 423–24; self-government and, 151–52, 153, 404; *see also* Territorial North
Nunavut Act, 422
Nunavut Assembly, 422
Nunavut Political Accord, 404
Nunavut Tunngavik, 423
Nuu-chah-nulth First Nations, 278
NWT Diamonds Project, 431, 433–35

Obama, Barack, 62, 328, 348
"O Canada" (national anthem), 115
Occupy movement, 123, 221
Ocean Ranger, 389
oceans, 47; climate and, 49, 52
Office Québécoise de la langue française, 217
Official Languages Act, 13, 466n2
oil and gas industry. *See* petroleum industry
oil sands, 23, 38, 62, *62*, 65, 157, 319–20, 326, 340–41, 342, 343, 354, 443; Atlantic Canada and, 376–77, 381, 395–96; challenges of, 346; deposits and proposed pipelines, *344*; dispute over, 20, 347; distance to market, 348; economic impact of, 344–45; environment and, 346–47; investment in, 345–46; spin-off industries, 345; tailing ponds and, *62*, 319–20, 347; *see also* petroleum industry
Okanagan Indians, 72
Okanagan Valley, 272, 275
Old Age Security, 141
Old World hunters, 68–69
Oleson, Brian, 336
Oliver, Frank, 110
Olymel, 340
Olympics, 2010 Vancouver, 20, 296
Ontario, xvi, 6, 6, 7, 165–213, 440–41; Aboriginal territory within, 183–85; agriculture and, 171–74; arts and, 167; automobile industry in, xvii, 7, 182, 192–99; basic statistics, 167; boundaries of, 84; within Canada, 166–68; cities in, 210–12; climate of, 171–74; CMAs in, 204, *205*; Confederation and, 179; as core region, 18; economic structure of, 18, 191, *191*; economy of, xvii, 166, 189–91, 212–13; employment, 166, 191, *191*; energy and, 176, 177, 191–92; environment and, 175–78; exports of, 200; forest industry in, 200–202; GDP and, 7, 166, *167*, 223; general characteristics of, 7; globalization and, 167; government debt, 166; as "have-not" province, 11, 166, 167, 170; historical geography of, 178–85; industrial base of, 180–83; land disputes in, 185; manufacturing in, 180–83, *190*; map, *169*; mining industry in, 202–4; music and film industry in, 187; northern, 6, 168–69, 171, 199–204; physical geography of, 168–71; political and cultural role of, 167–68; pollution and, 175–76; population density in, 123; population distribution, 126, 211; population of, 126, 132, 156, 211; settlement of northern, 200; shift in industrial sectors in, 240; snowbelts, 172; social characteristics of, 13; southern, 6, 168, 169–71, 171–72; southwestern, 209–10; territorial expansion, 179–80; today, 185–92; trade with US, 166–67, 188–89, 213; urban geography, 204–12; urban population in, 132; urban sprawl in, 173; water pollution in, 176–78

INDEX

Ontario Chamber of Commerce, 189
Ontario Medical Association, 175
Ontario Plain, 174
Ontario Power Authority, 192
Open Text, 192
Orange Lodge, 373
Orange Order, 105, 107
Oregon Boundary Treaty of 1846, 83
Oregon Territory, 80, 83, 277, 278, 321, 455
Organization of Petroleum Exporting Companies, 341
orogeny, 361, 455
orographic uplift, 52, 315, 455
Oshawa, ON, 205
Ostry, Bernard, 146
Ottawa-Gatineau, 45, 126, 129, 204, 208
Ottawa Valley, 208–9
outports, 9, 358, 395
outsourcing, 198, 455

Pacific Basin, 56, 57, 58
Pacific climatic zone, 50, 272
Pacific Great Eastern Railway, 280, 281, 282, 299, 336
Pacific Ocean, 51, 268–69
Pacific Rim countries, 20, 282, 441; BC and, 265, 267, 283, 286–88, 305; Western Canada and, 328
Pacific Ring of Fire, 37
Pacific Salmon Commission, 289
Pacific Salmon Treaty, 289, 468n5
Paix des Braves, 245, 248
Paleo-Eskimos, 70, 71
Paleo-Indians, 68, 69, 455
Palliser, John, 317, 319, 322
Palliser's Triangle, 316, 317, 318, 322, 325, 455
Panama Canal, 281, 305, 352
Pangnirtung, 413
Papineau, Louis-Joseph, 112
Parizeau, Jacques, 221
parkland, 108, 208, 316, 318, 322, 332
Parks Canada, 409
Parry, William, 409
Parti Québécois, 102, 117, 118, 217, 218, 222, 225, 228, 237, 262
Pastagate, 217
patriation, Constitution and, 118
patterned ground, 41–42, 455
Peace River, 39, 294, 343
Peace River country, 333, 335–36
Peachland wildfire, 276
Peggy's Cove, NS, 9
Pembina Institute, 62, 347
peneplain, 42, 455
Penney, Kirk, 396
Penticton, BC, 306
Percé, QC, 250
periphery, 17, 455; see also core/periphery model
Peritz, Ingrid, 260
permafrost, 39, 41, 55–56, 428, 455; alpine, 55, 448; continuous, 56, 405, 450;

discontinuous, 56, 451; sporadic, 56, 458; zones, 55
Peters, Russell, 150
PetroChina, 345–46
petroleum industry, 342; in Atlantic Canada, 387–90; in BC, 291–93, 348; conventional oil and gas, 341, 450; environment and, 319–20; light sweet crude and, 387, 453; oil prices and, 89, 328, 343, 345, 348, 455; refining and, 225; in Territorial North, 417, 432–33, 435; western alienation and, 89, 342; in Western Canada, 327–28, 329, 341–48; see also oil sands
Petronas, 345
Phillips, David, 365
physical geography, 29–65; in Atlantic Canada 361–65, 377; in BC, 268–73, 283; climate and, 46–56; drainage basins, 56–58; environmental challenges and, 58–59, 60–65; geographic location, 44–45; human activity and; landforms and, 30, 31–32; in Ontario, 168–71; physiographic regions and, 32–44, 405; in Québec, 225–28; in Territorial North, 405–7, 436; variations within Canada, 30–31; in Western Canada, 314–18
physiographic regions, 32–44, 64–65, 455; in Atlantic Canada, 361; in BC, 269–72; in Canada, 33; definition, 32; geological structure of, 32–34; topography, 34; in Western Canada, 314
physiography, 30, 455
Pincher Creek, 320
Pineapple Express, 272, 455
pine beetle, 273, 276, 299
Pine Point deposit, 429–30
pingos, 42, 455
pipelines, gas and oil, 11, 20, 22, 273, 284, 441, 442; in BC, 267, 269, 273, 283, 286; in Territorial North; in Western Canada, 311, 312, 342–48, 344; see also specific projects
Pitt River Bridge, 267
place, 4; identity and, 142; sense of, 4, 8–10, 147–57
placelessness, 8, 455
Places to Grow Act, 175
Plains Cree, 72
Plains culture region, 72, 72
Plains Indians, 104, 106, 323
Plains of Abraham, 111, 229
Plan Nord, 228
Plano culture, 69
Plasco Energy Group, 175
Plateau culture region, 72, 72
plate tectonics, 36–37, 38, 455
Pleistocene epoch, 34, 455
pluralism, 79, 453, 455
pluralistic society, 24, 89, 101, 443, 446, 456
polar drones, 414–15, 415
Polar Hawk drones, 415
Polaris mine, 406, 429

Polar Sea, 418
Pontiac, 93
Poon, Joyce, 188
population, 122–29; Aboriginal, 72, 95–96, 122, 128, 150–55; aging, 122, 140–42, 140; in British North America, 113; change, 132–37, 157; in CMAs, 129; components of change, 132–33; core zone and, 126, 127; density, 123, 124, 456; distribution, 123–29, 456; foreign-born, 137, 148, 149; French/English language imbalance, 156; geography of, 156–57; growth, 122, 123, 133–34, 162, 456; immigration and, 123, 137–38; increase, 133, 137; natural increase, 123, 134–37, 456; net migration, 133; rate of natural increase, 133, 136; regions and, 8, 124; secondary core zone and, 126–27; size, 123, 124; sparsely populated, 127, 128; strength, 456; urban, 129–32, 148; of Western Canada, 351; zones and, 124, 126–29; see also specific places
Port Metro Vancouver, 287
Port of Montréal, 220, 223
ports: in BC, 281, 282, 286–88, 291
postglacial uplift, 41, 456
post-industrial phase, 139, 159, 160, 161
Post-Secondary Student Support Program (PSSSP), 155
potash, 294, 441, 456, 470n10; Western Canada and, 148, 267, 285, 311, 313, 314, 328, 340, 341, 342, 348–49
Poundmaker, 106, 456
Power, Don, 386
Prairies, 9, 30; climate change and, 59; climate zone, 50
precipitation: annual, 51; in Ontario, 171; in Atlantic Canada, 363; in BC, 269, 272; convectional, 52, 450; frontal, 52, 452; orographic, 455; types of, 52; in Western Canada, 313, 315, 316, 318–19, 326
pre-Clovis people, 68
Prentice, Jim, 185
prices: commodity, 19, 27, 88, 354; energy, 19, 191, 379, 395; primary, 328, 456; see also specific products
Prince Albert, MB, 353
Prince Edward Island, 7, 42, 358; basic statistics, 361; agriculture and, 394, 394; Confederation and, 79, 82, 370, 371; forestry industry in, 393; hydroelectricity and, 442; and; population of, 79, 375; Québec and; unemployment, 166, 380; see also Atlantic Canada
Prince George, BC, 282, 304
Prince Rupert, BC, 267, 287, 304
Proceedings of the National Academy of Sciences, 346
producer services, 456; in BC, 285
products, primary, 26, 456
Project Crimson, 414
protests, 123; Québec students, 221, 221

provinces: equal, 77–78, 115–16, 118; fault-lines, 11–12; "have"/"have-not," xvi, 12, 88, 91, 157, 377; powers of, 90, 92; support for scientific research, 161; *see also* specific provinces
Provincial Agricultural Land Commission (BC), 272, 456
Provisional Government, 104, 105, 114, 456
Ptolemy, 4
public services, 14, 91, 129; outports and; Territorial North, 254, 421–22
pulp and paper industry: in Atlantic Canada, 391–93; in BC, 298–99; declining demand and, *299*; in Québec, 255; in Ontario, 201; *see also* forest industry

Québec, xvi, 6, 7, 215–62, 441, 445; Aboriginal peoples in, 225, 227, *248*; agriculture, 226, 253–54; Appalachian Uplands in, 250–51; basic statistics, 222; beyond urban cores, 260–61; Bill 14, 263; Bill 22, 14; Bill 101, 14; birth rate in, 222, 231–32; boundaries of, 84; under British rule, 231–34; within Canada, 219–25; Canada and, 237; cities in, 261; climate, 225, 227–28; Confederation and, 88–89, 216, 234–35; core/periphery model in, 250; as core region, 18; cultural industry in, 225; cultural identity and, 216–18; culture of, 115, 215; demographics of northern, 253; dispute with Newfoundland and Labrador, 11–13, 84, 234–35; as distinct society, 216; divided into Upper and Lower Canada, 112; economic structure of, 18, 240; economy of, xvii, 219, 223–24, 237–40; employment in, 240; environment and, 228; federal government and, 234, 235, 237; fertility rate, 231–32; forest industry, 224, 254–55; French/English faultline in, 14, 262; under French rule, 228–31; GDP and, *224*; general characteristics of, 7; as "have-not" province, 236; high-tech industry in, 238, 262; historical geography, 228–35; historical milestones, 235; hydroelectricity in, 7, 200, 244, identity and, 216–18, 443; immigration and, 142, 237; industrial strategy, 241–44; industrial structure, 240; knowledge-based economy in, 225, 236, 239–40, 262; labour force, 240; language and, 14; language laws, 237; literature of, 233–34; manufacturing sector in, 224–25; map, *219*; migration into urban, 257; mining industry in, 251, 255–57; nationalism, 116, 235; as nation within Canada, 118, 216; northern, 215, 217, 246, 250, 253; Official Language Act, 13; petroleum refining in, 225; physical geography, 225–28; population of, 156, 219–22, 262; population density in, 124; population distribution, 126; Quiet Revolution, 116, 117, 221, 235, 262, 456; reasonable accommodation in, 14, 16; reliance on foreign and US trade, 238; religion in, 110, 229, 232, 234; resources in, 253, 260–61; sense of place, 9; settlement of, 216, 228–31, 250; social characteristics of, 13; southern, 250–53, 260; sovereignty, 218; St Lawrence Lowland in, 251–53; student protests in, 123, 221, *221*; territorial expansion, 234–35; today, 236–37; tourism in, 224, 248–50, 253; urban geography of, 257–61
Québec Act (1774), 111, 232
Québec Cartier Mining Company, 255, 256, 257
Québec City, 72, 74, 225, 229, 250, 251, 257, 259–60, *261*
Québecers/Québécois: as terms, 143, 235, 456
Québécois, 115–16, 216, 217, 443
Québec Pension Plan, 235
Quiet Revolution, 116, 117, 221, 235, 262, 456

racism, 92, 146, 150
Radarsat Constellation Mission, *284*, 414
Rae, Bob, 370
Raglan Agreement, 255
Raglan nickel mine, 36
Rahnama, Hossein, 188
railways, *21*; BC and, *280*, 281, 286, 467n; goals for, 324; Manitoba and, 108; in Ontario, 200; Territorial North and, 41, 429; US, 321; Western Canada and, 324, 332, *333*; *see also* specific railways
rain shadow effect, 52, 456
Rainy River, 211
Reagan, Ronald, 418
reasonable accommodation, 14, 16
rebellions of 1837 and 1838, 112–13, 232, 233
rebound: isostatic, 40, 41, 453; restrained, 41, 457
Reciprocity Treaty, 81, 180, 181, 370
Redford, Alison, 20, 307, 313
"Red Paper," 102
Red River, 54, 57
Red River Colony, 105
Red River Floodway, 320
Red River Migration, 83, 456
Red River Rebellion, 82, 104–5, 323
Red River Settlement, 76, 77, 82, 83, 321, 322, 352
referendums (Québec), 10, 235; *1980*, 117, 258; *1995*, 117–18, 262
Regan, Gerald, 362
Regina, 314, 351, 352, 353
regionalism, 85, 457; in BC, 268
regions, xvi, 4, 8, 26, 456; *see also* Atlantic Canada; British Columbia; Ontario; Quebec; Territorial North; Western Canada

relief, 32, 457
religion, 143–44; freedom of, 144; immigration and, 137, 138, 146, 150; Québec and, 229, 232, 234; residential schools and, 94, 411; tensions and, 113
research, scientific, 161
Research in Motion (RIM), 186
reserves, 96–97, 457
residential schools, 14–15, 94–95, 411; federal apology for, 95
Residential Schools Settlement Agreement, 92, 94–95
residual uplift, 41, 457
resource base, of Western Canada, 340–51
resource cycle, 386, 442
"resource curse," 162
restructuring, 188–89, 457
"revenge of the cradles" (*revanche des berceaux*), 231
Reynolds Aluminum, 242–43
Richelieu River, 57
Richmond, BC, 267
Rideau Canal, *127*
Riel, Louis, 104, 105, 106, 113, 114, 323, *324*, 465n7
Rim of Fire, 199
Rimouski, QC, 250, 260
Ring of Fire, 36, *203*
Rio Tinto, 255
Rio Tinto Alcan, 295
Rio Tinto Iron Ore Company of Canada, 257
rivers, 57, 58; hydroelectricity and, 200, 244, 245, 294, 342; in Interior Plains, 39; salmon and, 290; transportation and, 209; *see also* specific rivers
Rivière-du-Loup, Que., 250, 260
roads: ice, 327; Territorial North and, 428, 429; winter, 327, 429, 459; *see also* highways
Roberts Bank terminal, 292
rocks, 31; igneous, 31, 453; metamorphic, 31, 454; sedimentary, 31, 38, 43, 457; physiographic region and, 35, 36, 38, 41, 43, 314
Rocky Mountains, 30, 270, 314
Rogers, Stan, "Northwest Passage," 46
Romaine Hydro Complex, 58, 241, *242*
Roman Catholic Church, 111, 143, *144*, 233, 233
Ross, John, 409
Royal Bank of Canada, 223, 224
Royal Canadian Mounted Police (RCMP), 411
Royal Commission on Aboriginal Peoples, 101
Royal Commission on Bilingualism and Biculturalism, 115, 144–45
Royal Proclamation of 1763, 83, 89, 93, 464n3
Rupert's Land, 79, *81*, 83, 234, 254, 321, 322, 413
Russel Metals, 191

INDEX

Russia: in BC, 277; fur traders, 83; immigrants from, 78
Ryerson University, 188

Saddle Lake Cree, 95–96
Sadlermiut people, 70, 410
Saguenay, QC, 257, 260
Sahtu/Métis land claim agreement, 424, 425
St Jean Baptiste, QC, 54
Saint John, NB, 369, 376, 397
Saint John River, 57
Saint John River Valley, 394
St John's, 363, 365, 376, 397
St. Lawrence Action Plan: 2011–2026, The, 228
St Lawrence Lowland, 260; climate of, 251; major cities in, 258; in Québec, 226, 251–53; *see also* Great Lakes–St Lawrence Lowland
St Lawrence River, 9, 43, 57, 220, *220*, 228
St Lawrence Seaway, 180, 220
St Lawrence Valley, agriculture in, 231
St Mary River Irrigation District, 335
Saint-Pierre and Miquelon, 369
Salish, 72, 73
Salluit, 255
Sanders, Doug, *The Myth of the Muslim Tide*, 137
sandwich generation, 141
Sarcee Indians, 72
Saskatchewan, 7, 11, 148, 311, 312, 326; agriculture in, 327, 331, 333, 334, 335, 337, *338*; boundaries and, 83, 84; creation of, 82, 84; economy of, xvii, 327–28; farm size in, 330; GDP and, 314, 328; government and uranium mines and, 349, 350; knowledge-based industry in, 329, 331; mining in, 314, 321, 348–49; oil production in, 343; population of, 78, 222, 328; potash industry in, 340, 348, 349; as province, 325; unemployment in, 314; *see also* Western Canada
Saskatchewan Plain, 39
Saskatchewan Research Council (SRC), 321
Saskatoon, 130, 316, *351*, 353
Saul, John Ralston, 10, 67, 122, 439, 443; *The Collapse of Globalism and the Reinvention of the World*, 447; *Reflections of a Siamese Twin*, 146, 216
Sault Ste Marie, ON, 210–11
Scandinavia, 78, 108
Scientific Research and Experimental Development (SRED), 161
Scott, Thomas, 105
Scottish Highland clearances, 369, 457
scrip, 96, 113, 457
seal hunt, 424, 426–27
Seaspan Shipyard, 283
Second Narrows railway bridge, 268, *269*
second people, 74–78
Sector Principle, 418
seigneurial system, 111, 231
seigneurs (lords), 231, 457

self-government, Aboriginal, 95, 101, 425; Nunavut and, 151; Territorial North and, 404, 411, 424, 425
self-interest, regional, 4–5, 457
Selkirk, Lord, 321
sense of place, 4, 8–10, 148, 457
separatism, 116–17, 118, 216, 235
September 11 terrorist attacks, 198
service centres, regional, 127, 457
settlement areas, 425, 457
sex ratio, 136, 457
Sgro, Judy, 143
Shea, Ambrose, 373
Shell's Quest Project, 329
Sherbrooke, QC, 228, 251, 257, 260
Shiell, Leslie and Robin Somerville, 195
shipbuilding industry, 81, 369, 377
Shuswap, 72
Sierra Club, 62, 245
Sifton, Clifford, 78, 79, 108, 325
Simcoe, John Graves, 186
Simon, Mary, 426
Simpson, Jeffrey, 10, 110–11
Sir Adam Beck Generating Station, 176
Six Nations of the Grand River, 95, 112, 185
Skeena River, 269, 290
Skookum Jim, 427
SkyHook International, 327
SkyTrain (BC), 224
Smallwood, Joey, 362
Smith, Ryan, 46
smog, 64, 175
Snap Lake mine, 431, 433
SNC-Lavalin, 224, 238, 239
snowbelts, 172
Softwood Lumber Agreement (SLA), 200
soil orders, 52, 457
soils: in Atlantic Canada, 366, *394*; in BC, 272, 273; climate and, 52; Ontario, 171, 173, 174; in St Lawrence Lowlands, 43, 44; in Western Canada, *317*, 317, *318*, 332; zones, 54
Solomon, Jonathan, 212
South Nahanni River, *404*
sovereignty: environmental legislation and, 418; Territorial North and, 20, 414–18, 436
sovereignty-association, 117, 457
Soviet Union, 414, 418; *see also* Russia
Spain, 277
Sparrow case, 290
specialty crops: in Atlantic Canada, 394; in Québec, 226; in Western Canada, 333, 335
Spence, Theresa, 22
sports teams, professional, 168
Sprague, D.N., 96
SS Manhattan, 418
Stadacona, 72
staples: thesis, 17, 458; "trap," 17, 435, 458
"Statement of the Government of Canada on Indian Policy." *See* White Paper on Indian Policy

Statistics Canada: Aboriginal peoples and, 95, 152; agriculture and, 330; CMAs and, 129, 205; ethnic origin and, 141, 142; population and, *133*; religion and, 144; regional characteristics and, 13; Territorial North and, 418–19; urban places and, 396
steam-assisted gravity drainage, 343, *344*
steel industry, 207; in Atlantic Canada, 372; in Québec, 256–57
Steyn, Mark, *America Alone*, 137
Stikine Plateau, 271
Stony Point First Nation, 185
strata, 31, 458
Stryker (film), 351, 458
Subarctic climate zone, 50–51, 405
Subarctic Mountain climate zone, 272
subsidence, 60, 458
subsidies, 11; agriculture, 336; in Atlantic Canada, 373; automobile, 195; "feed-in tariff," 192; freight, 372; megaprojects and, 432; in Québec, 195
suburbs, 130–32
Sudbury, ON, 36, 204, 205, 210, 211–12
Sum of Us, 275
Suncor Energy, 62, 343–44, 396
super cycle theory, 18, 19, 458
Supreme Court of Canada, 12, 14, 96, 102, 331; Aboriginal rights and, 393–94; comprehensive land claim agreements and, 424–25; Métis and, 96, 97; Québec and, 10–11
Sustainable Development Technology Canada, 175, 458
sustainable resource use, 273, 458
Sydney, NS, 372
Sydney Steel Co. (Sysco), 367
Sydney Tar Ponds, 366–67, 372
Syncrude Aurora, 344

Ta'an Kwach'an Council, 425, 426
Tagish Charley, 427
tailing ponds, 62, 319, 320, 347
Talon, Jean, 230, 231
Tamil demonstration, *147*
tar ponds. *See* oil sands
tariffs: "feed-in," 192; imported goods and, 86, 181, 252
tax: carbon, 89; export, 202, 303; gasoline, 258; scientific research and, 161, 238; softwood lumber and, 202, 303; Territorial North and, 421, 422, 431
Taykwa Tagamou Nation, 212
Taylor, Charles, 116, 145, 150; *Multiculturalism: Examining the Politics of Recognition*, 145
technology: agriculture and, 161, 330, 335, 336; firms, 186, 192, 243; fishing and, 387; forest industry and, 201, 303; green, 192; information, 238, 251; ocean, 379; oil sands and, 329, 343; petroleum industry and, 304; research and, 186, 188, 210; resources and, 283; transmission, 241, 458; *see also* high technology

INDEX

Technology Triangle, 161, 210
Temiskaming Shores, ON, 211
temperature: fluctuations in global, 63; global warming and, 59; seasonal, *49, 50*
terraces, *40*, 458
Terra Nova project, 389
terra nullis, 71, 458
Territorial Formula Financing, 91, 421
Territorial North, xvi, 6, 7, 401–36, 442; Aboriginal economy, 424; Aboriginal peoples in, 401, 402–3; Aboriginal vision of, 403–4; air commuting and, 431; basic statistics, *403*; within Canada, 402–5; cities in, 418–19, *419*; climate of, 405; climate zones of, 405; Cold War and, 414; Aboriginal land claims in, 424–27, 450; diamond mines in, 418, 431, 433–35, 436; economic structure of, 18, 421–22; economy in, 404–5; employment in, 421–22; environment and, 407–8; equalization payments, 422; Europeans in, 408–9, 427; federal government and, 414, 421, 422; general characteristics of, 7; geology of, 406–7; gold mining in, 429; government sector in, 421–22; government structure of, 422; highways in, 428–30; historical geography of, 408–14; as homeland, 403–4; map of, *402*; megaprojects in, 7, 403, 431–36; mining industry in, 406–7, 408, 428, 435; non-renewable resources in, 401, 403, 417, 435; as northern frontier, 403–4; physical geography of, 405–7, 436; population of, 402, 420–21; population density in, 123; precipitation in, 52; resource development in, 427–31, *430*; as resource frontier, 18, 401, 403, 418; settlement areas in, 425; settlements in, 419–20; social characteristics of, 13; as strategic frontier, 414; territorial expansion, 413; territorial government in, 404; today, 418–27; transportation and, 428–30; two visions of, 403–5; urban population in, 132; underemployment in, *403*, 412, 422, 436; vegetation of, 405; whaling and fur trade in, 409–11; World War II and, 414; *see also* Northwest Territories; Nunavut; Yukon
textile industry, 194, 238, 252
Thales, 4
Thetford Mines, QC, 250, 260
third people, 78–79
Thomas, David Hurst, 69
Thompson, David, 83
Thompson, MB, 350, 353
Thompson Valley, 272
Thule people, 63, 70, 71, *71*
Thunder Bay, ON, 176, 205, 211
till, 36, 458
time zones, 44–45
Timmins, ON, 204, 210
title, Aboriginal, 97, 244, 278, 448

Tlicho Final Agreement, 404
Tofino, BC, 269
topography, 23, 32, 34, 458
Torngat Mountains, 35, 361, *362*
Toronto, *125*, 130, 148, 171, *182*, 204, 205–8, *446*, 466–67n9; air pollution in, 175; cultural and entertainment centre, 206–7; immigration and, 207, 209; Montréal and, 258–59; neighbourhoods in, 207–8; population change in, 259; Pride Parade, 187; waste disposal in, 175
Toronto International Film Festival, 187
Toronto Island, 208
Toronto Stock Exchange, 145
tourism: in BC, 267, 283, 288, 295–97; in Ontario, 206–7; in Québec, 248–50, 253, 259
towns: resource, 127, 457; *see also* cities
township survey, 322
Toyota, 192, 197
Tracie, Carl, 110
trade, 26; Atlantic Canada and, 369; disputes, 24, 463n5; diversification, 20; global, 3, 19; Indians and, 71; liberalization of, 182, 237; policies and, 440; *see also* Free Trade Agreement (FTA); North American Free Trade Agreement (NAFTA); specific industries
traditional ecological knowledge, 427, 458
"tragedy of the commons," 289–90, 366, 385, 458
Trans-Alaska Pipeline System (TAPS), 471n9
TransAlta Plant, 60
Trans-Canada Highway, 199, 200, 212
transfer payments, 11; *see also* equalization payments
Trans Mountain pipeline, 273, 328, 348
Trans-Pacific Partnership, 20, 458
transportation: agriculture and, 327, 333, 337, 340; Arctic Ocean and, 430; in Atlantic Canada, 372; in BC, 268, *270*, 281, 282, 286–87, 292; blimp-like, 327; mining industry and, 36, 429; in Ontario, 180, 181, 196, 199, 211; in Québec, 220, 227, 236, 239, 253; in Territorial North, 414, 428–30; water, 30, 57, 407; in Western Canada, 311, 312, 326, 327; *see also* railways
treaties, Aboriginal, 14, 78, 83, 89, 96–97, 107; in Atlantic Canada, 369; in BC, 278; comprehensive, 97; Douglas, 278; modern, 97–101, *99*; numbered, 82, 97, *98*, 106, 183, 323; in Ontario, 183; Robinson, 106, 183, 457; unnumbered, 183; in Western Canada, 323; *see also* land claims
Treaty of Ghent, 76, 178
Treaty of Paris, 9, 83, 110, 111, 229, 232, 369
Treaty of Utrecht, 368
Trois-Riviéres, QC, 257
Trudeau, Pierre Elliott, 102

Truth and Reconciliation Commission, 92–93
Tsawwassen First Nation, 101, 278
tundra, 405
Tundra-Boreal Transition, 255
Turner, John, 143
Tyendinaga Reserve, 152
Tyrrell Sea, 39, 40, 458

Ukraine, 78
Umbrella Final Agreement, 425
underemployment, *403*, 412, 422, 436, 458–59
unemployment, 157, 459; Aboriginal peoples and, 227, 248; in Alberta, 380; in Atlantic Canada, 360, *361*, 379–80, 396, 399; in BC, 267, *268*, 281; by province, 166; in Ontario, 166, *167*, 199, 224; in Québec, *222*, 223, 252; in Territorial North, *403*; in Western Canada, *313*, 314, 327, 380
Union Nationale, 116
United Auto Workers (UAW), 196
United Nations: Canada's position on Kyoto and, 61; Durban climate change conference, 61
United Nations Convention on the Law of the Sea (UNCLOS), 20, 416, 418
United States: in Arctic waters, 417, 418; Atlantic Canada and, 369, 383, 384; auto industry and, 24, 189, 193, 194, 195, 196, 198; BC and, 265, 278, 283, 286, 297–98; biofuel industry in, 336; boundaries and, 83; economic crisis and, 123, 157, 158, 167; economy of, 20, *24*; exports to, 3, 20, 354; energy and, 336; expansion of, 75, 80; forestry industry and, 201, 299, 302, 303, 308, 350; geography of, 30, 44; housing industry, 302, 303; immigration from, 78, 178, 325; Ontario and, 166, 167, 188–89, 440; origins of, 75; population density, 123; Québec and, 224, 236, 238; in Territorial North, 414, 418; trade with, 3, 20, 23–24, 147, 188–89, 328, 442, 445; Western Canada and, 311, 328, 443
unit trains, 312, 459
Université Laval, 259
universities, 239, 288; auto industry and, 192, 210; knowledge-based economy and, 148, 159, 161, 187–88
University of Toronto, 188
University of Waterloo, 161, 187, 188
upgraders, 346, 347, 459
Upper Canada, 75, 76, 112, 113, 178, 232
upward mobility, 146
uranium, 32, 191, 314, 321, 326, 328, 341, 349, 350
urban areas, 129, 459; in Territorial North, *133*, 419; *see also* cities
urban geography: of Atlantic Canada, 396–98; BC's, 303–7; of Ontario, 204–12; of Québec, 257–61; of Western Canada, 351–53

urban place, 396; *see also* cities
urban sprawl, 31, 130–32, 175; in Ontario, 31, 173, 175
US Steel Canada Inc., 207

value-added production, 339, 459
Vancouver, George, 277
Vancouver, *126*, 281, 304–6; economy of, 281, 305–7; growth of, 304; physical geography of, 304–5; as port, *126*, 282, 287, 305; precipitation and, 49, 272
Vancouver Chinatown Night Market, *307*
Vancouver Island, 269, 303
vegetation: climate and, 52; in BC, 273; in Territorial North, 405; in Western Canada, 317
Victoria, BC, 272, 303, 304
Victor Mine, 203, 204, 212
Vigneault, Gilles, "Mon pays," 46
Vikings, 63, 70
visible minorities, 138, 146, 148, 150, 207
Voisey's Bay nickel mine, 376, 390, 391, *391*, 392, *392*, 399
volcanic activity, Cordilleran, 36–37

Walkerton, ON, 176
Wall, Brad, 88
War of 1812, 75–76, 178
water: as commodity, 57; contaminated, 176; forest industry and, 201–2; livestock industry and, 176; oil sands and, 319–20, 347–48; in Ontario, 176–78; polluted, 31, 175, 176, 212, 228; from Sydney Tar Ponds, 367; Western Canada and, 313
Watkins, M.H., 17
weather, 46; Atlantic Canada, 361–63, 365; in BC, 273; in Ontario, 172; in Québec, 227–28; in Territorial North, 405; in Western Canada, 316, 318
weather events, extreme, 52–54, 227–28
weathering, 32, 41, 42, 459
websites, 460–62
Webster-Ashburton Treaty, 83
Wei Wai Kum Terminal, 296
Welland Canal, 180, *180*

western alienation, 89, 181, 342, 459
Western Appalachian Uplands, 260
Western Canada, xvi, 6, 7, 311–54, 441–42; Aboriginal/non-Aboriginal faultline, 353; Aboriginal peoples and, 351, 353; agricultural fringe in, *319*, 332, 333, 335; agricultural regions in, *319*, 332–36; agriculture in, 7, 173, 311, 312, 316, 325–26, 330–40; basic statistics, 313, 328; within Canada, 312–14; cities in, 331; climate of, 311, 312, 315, 326, 330, 333; depopulation in rural areas, 331; Doukhobors in, 109–10; droughts in, 318; economic structure of, 18, 328–29; economy of, xvii, 326, 327–28, 354; employment in, 328–29; environment and, 318–21; ethnic groups and; exports of, 326, 354; farm size in, 331; forest industry in, 350–51; GDP and, *313*, 314; general characteristics of, 7; historical geography, 321–29; homesteaders, 325–26; immigration and, 107, 108–9; livestock industry in, 339–40; map, *312*; migration and, 156–57; mining in, 314, 348–50; oil and gas industry, 329, 343; original inhabitants of, 323; physical geography of, 311, 312, 314–18; physiographic regions in, 314; population of, 108, 109, 156, 222, 312, 314, 326; precipitation in, 313, 315, 316, 318–19, 326; regional core, 314; resources in, 313, 326, 340–51; rural land-use changes, 337; sense of place in, 9; settlement of, 78–79, 103–4, 321, 322, 324–25; social characteristics of, 13; today, 326–28; trade, 311, 328; transportation and, 312, 326, 327; unemployment rate in, 327; urban centers in, 132, 351–53; urban geography of, 351–53; urbanization in, 132; *see also* Alberta; Manitoba; Saskatchewan
Western Sedimentary Basin, 38, 270, 314, 406, 459
Western Subarctic culture region, 72, *72*
Westridge oil terminal expansion, *274*
Westshore Terminal, 292

whales: beluga, 228, 229; global warming and, 406; in Territorial North, 407–8
whaling, 409–10
wheat, 108, 330, 336, 339; climate and, 332–33; durum, 332; Marquis, 108, 326, 331; Red Fife, 108, 331, 469n5; spring, 312, 326, 330, 331, 332, 337, *339*; summer fallowing and, 335
Whistler Mountain, 297
White, Adam, 177
Whitecap First Nation, 102
Whitehorse, 127, 422
"White Man's Burden," 103
White Paper on Indian Policy, 98
White Rose project, 389
Whitworth, Jonathan, 283
wildlife: Aboriginal peoples and, 426, 427; megaprojects and, 245, 403, 432, 435; in Territorial North, 407, 411; *see also* specific species
Williams, Danny, 5, 91
Williams gold mine, *204*
Williston Lake, 294
Windsor, ON, 45, 169, 182
Winnipeg, 351, 352, 353
"wintering over," 409–10
Wolseley Expedition, 105
Woodland Cree, 323
workforce, immigrants in, 150
World Energy Outlook, 443
World Food Price Index, 336
World Trade Organization (WTO), 193, 237, 251
World War I, 322
World War II, 414
world systems theory, capitalist, 17, 449

Yellowknife, NWT, 8, 9, 127, 422, 434–35
York, ON, *181*
Yukon: First Nations Agreement, 424, 425; *see also* Territorial North
Yukon/Alaska offshore boundary, 416
Yukon River Valley, *406*

zebra mussels, 228, 459